Bone Histology

Bone Histology: A Biological Anthropological Perspective, Second Edition builds on the success of the first edition, recognizing the significant advances that have occurred in bone biology, histology, and histological techniques and methods in subsequent years.

Bones and teeth are of considerable importance for anthropological and related research, due to their nature as hard tissues. The physical remains of humans available to biological anthropologists, bioarchaeologists, paleopathologists, and paleontologists are, with exception to forensic anthropology, limited to skeletal material; fortunately, the same characteristics of hard tissues that lead to their persistence after death make them a storehouse of information about biological processes experienced during the life of the individual. This book covers important aspects of bone biology which underlie the microstructure of hard tissues that are crucial for histological analysis. This includes an overview of two major metabolic processes, bone remodeling and modeling, and their importance for understanding and interpreting bone histomorphology. Subsequent chapters apply histological methods to the biological profile, such as estimation of age and evaluation of pathological conditions that affect the skeleton, or to determine whether remains are human or nonhuman. Finally, there is a discussion of current research trends in bone histology, with a focus on technological advances in imaging and methods. Reviews of four well-documented skeletal collections – developed specifically for bone histological and imaging research – are discussed, as well as the importance of such collections for future research.

Bone Histology, Second Edition has assembled a collection of contributing authors, with extensive experience and expertise in various aspects of hard tissue biology, to provide readers with an overview of the current state of research and potential applications of histological analysis in biological anthropology, forensic anthropology, and skeletal biology. It serves as a valuable resource for students, researchers, and practitioners in these and related disciplines.

Bone Histology
A Biological Anthropological Perspective

Second Edition

Edited by
Sam D. Stout
Christian Crowder

CRC Press
Taylor & Francis Group
Boca Raton London New York

CRC Press is an imprint of the
Taylor & Francis Group, an **informa** business

Designed cover image: Corey Maggiano

Second edition published 2025
by CRC Press
2385 NW Executive Center Drive, Suite 320, Boca Raton FL 33431

and by CRC Press
4 Park Square, Milton Park, Abingdon, Oxon, OX14 4RN

CRC Press is an imprint of Taylor & Francis Group, LLC

First edition published by CRC Press 2012

Library of Congress Cataloging-in-Publication Data
Names: Crowder, Christian, editor. | Stout, Samuel D., editor.
Title: Bone histology : a biological anthropological perspective / edited by Christian Crowder and Sam D. Stout.
Description: Second edition. | Boca Raton, FL : CRC Press, 2025. | Includes bibliographical references and index.
Identifiers: LCCN 2024025541 (print) | LCCN 2024025542 (ebook) | ISBN 9781032473277 (hardback) | ISBN 9781032473239 (paperback) | ISBN 9781003385608 (ebook)
Subjects: LCSH: Bones--Histology. | Human remains (Archaeology) | Dental anthropology. | Forensic anthropology. | Paleopathology. | Paleoanthropology.
Classification: LCC GN70 .B656 2025 (print) | LCC GN70 (ebook) | DDC 573.7/635--dc23/eng/20240715
LC record available at https://lccn.loc.gov/2024025541
LC ebook record available at https://lccn.loc.gov/2024025542

ISBN: 978-1-032-47327-7 (hbk)
ISBN: 978-1-032-47323-9 (pbk)
ISBN: 978-1-003-38560-8 (ebk)

DOI: 10.4324/9781003385608

Typeset in Minion
by SPi Technologies India Pvt Ltd (Straive)

Dr. Sam D. Stout:
I dedicate this book to two mentors whose sage advice has guided me throughout my academic career.
John Eaton profoundly influenced me early in my career, revealing the excitement of doing research, and "thinking outside the box."
Steven Teitelbaum introduced me to the field of bone biology and histomorphological analysis, and inculcated within me an appreciation for the importance of rigor in research.
My appreciation goes, too, to Clark Larsen who brought me into the Department of Anthropology at The Ohio State University, supported my research, and continues to be a valuable colleague and friend.
Finally, I am thankful to the many outstanding graduate students with whom I have had the privilege to work with over the years, and who have gone on to become colleagues and collaborators. Indeed, several are authors of chapters in this book.

Dr. Christian Crowder:
I would also like to dedicate this book to two mentors who helped me find balance between my career paths as a forensic practitioner and a researcher / academic.
Dr. Dana Austin who took me under her wing as a graduate student and eventually as an intern at the Tarrant County Medical Examiner's Office. She showed me the exciting world of applied anthropology as a forensic anthropologist.
Dr. Susan Pfeiffer accepted me as her PhD student at the University of Toronto and influenced my passion for bone histology and skeletal biology. She created the most positive atmosphere within our cohort, and I fondly think back at our discussions during "Pfeiffer club" and the wisdom she passed on to us.
Finally, I would like to thank Dr. Sam Stout. Although I was never his student, he always had the time to discuss bone histology with me which led to us becoming colleagues and friends. I am also thankful for the graduate students and researchers who have worked with me on bone histology projects over the years, especially Janna Andronowski, Victoria Dominguez, Sophia Mavroudas, Deborrah Pinto, and Zoe Morris.

Contents

Preface

Because of their nature as hard tissues, bones and teeth are of considerable importance for anthropological and related research. Except for forensic anthropology and the relatively rare cases of mummification, the physical remains of humans available to bioarchaeologists, paleopathologists, and paleontologists are limited to skeletal material. Fortunately, the same characteristics of hard tissues that lead to their persistence after death make them a storehouse of information about biological processes experienced during the life of the individual. Unlike for the soft tissues of the body, the activity of cells involved in growth and development, tissue maintenance, and adaptation of hard tissues are encoded in their microstructure. Recognition of our ability to extract biologically important information from hard tissues emerged through the work of researchers such as Amprino, Currey, Enlow, Frost, Johnson, and Marotti, and for teeth the work of Dean, Reid, Boyde, Beynon, Bromage, and Shellis. We now understand that while bone serves an important metabolic function relating to mineral homeostasis, its primary function is biomechanical. The first edition of *Bone Histology: An Anthropological Perspective* addressed the need for a comprehensive volume covering theoretical and applied aspects in histological analysis of skeletal tissue. The stimulus for this updated and expanded second edition arose from the success of the first edition, and recognition that significant advances have occurred in bone biology, histology, and histological methodology since the first edition was published. While the focus of the book is the application of histological analysis in bioarchaeology, forensic anthropology, paleontology, and paleopathology, relevant aspects of basic bone biology that are essential for interpretation of histomorphometric data are also included.

Organization and Content

The first three chapters of the book cover important aspects of bone biology that underlie the microstructure of hard tissues and are crucial for histological analysis. They provide an overview of two major metabolic processes, bone remodeling and modeling, and their importance for understanding and interpreting bone histomorphology. Furthermore they discuss pediatric and adult bone histomorphology and its response to various intrinsic and extrinsic factors.

Chapter 1, *Bone Remodeling and Its Histomorphological Products*, provides a broad overview of the theoretical foundations, cellular activities, microstructural products, and histomorphometric interpretations of bone remodeling, including principles and models of bone functional adaptation, mechanotransduction of mechanical stimuli into bone remodeling, and a brief overview of bone cell activity during each phase of bone remodeling. It also presents considerations for interpreting remodeling activity from histomorphometry, including methodological guidelines, identifying osteon types, sources of intra- and inter-skeletal variation, techniques for static and dynamic histomorphometry, and method reliability.

Related, but distinct in important ways from bone remodeling, is bone modeling. In Chapter 2, *Making the Mold: A Microstructural Perspective on Bone Modeling during Growth and Mechanical Adaptation*, Cory and Isabel Maggiano define bone modeling, and discuss how it acts in concert with remodeling during growth and mechanical adaptation. This well-illustrated and innovative chapter concludes with a discussion of current methods employed to investigate how bones change in morphology and envisions possible future directions in research to be explored.

In Chapter 3, *Pediatric Bone Histomorphology and Environmental Stresses*, Goldman and Schug put forward a novel and timely methodology by which bone microstructure can be used to assess the effects of biocultural and environmental stresses on pediatric health in past populations and provide insights into the effects of modern climate change and other environmental issues.

The next two chapters cover important aspects of bone biomechanics and how bone adapts under mechanical loading. These chapters delve into structure and function, with insight into research models and applications of load history interpretation in forensic anthropology and bioarchaeology.

In Chapter 4, *Skeletal Fracture: Biomechanics and Forensic Perspectives*, Harden, Hunter, and Agnew discuss how the application of histological methods and analyses in skeletal trauma research can enhance our ability to relate bony responses to traumatic loading and fracture outcomes, and effectively substantiate the advantages of histological analysis as component of human osteological examinations, specifically in forensic anthropology and bioarchaeology. The inclusion of results stemming from research at the Injury Biomechanics Research Center at The Ohio State University is particularly compelling for the use of histological analysis in biomechanics and trauma analysis, and reveals a complex relationship between functional adaptation, bone biomechanics, and skeletal trauma for human pediatric bone.

In Chapter 5, *Biomechanical Foundations of Histological Analysis in Limb Bones*, John Skedros expands upon biomechanical and mechanical adaptation discussion covered in Chapters 2–4, by identifying correlations between structure, material, function, and load history. In this chapter, Skedros proposes that bone's adaptations for its load history can best be interpreted using a systematic approach that assesses both structural and material characteristics that considers load-complexity category, and shear resistance-priority hypotheses, modeling and remodeling working in concert, and the mechanical relationships of osteons, osteon morphotypes, and predominant collagen fiber orientation (CFO).

The next five chapters of the book focus on the application of histological methods to explore aspects of the biological profile, such as estimation of age and evaluation of pathological conditions that affect the skeleton, or to determine if remains are human or non-human. These chapters discuss how analyses are used to answer questions in forensic and bioarchaeological contexts.

Teeth also provide a record of their growth in their histological structures and are a valuable resource for skeletal analyses. In Chapter 6, *Histological Features of Dental Hard Tissues: Methodologies and Utilities in Forensic Anthropology and Bioarchaeology*, Chris Aris discusses the application of histological methodology to dental hard tissues. More specifically, he focuses on how histological analysis can be applied in the fields of bioarchaeology (and osteoarchaeology) and forensic anthropology. Included is a review of methods of analysis applicable to the three main hard dental tissues (enamel, dentine, and cementum) typically used in anthropological and forensic research and practice. Finally, the chapter highlights potential developments, and how they could impact the application of dental hard tissue histology for skeletal remains.

For those having to deal with extremely fragmented remains, the question of whether the bone is of human origin is often critical. In many cases, when diagnostic landmarks are not present, the fragmentation precludes the use of macroscopic methods to differentiate human bone from non-human bone. In such cases, histological analysis is a potential means to differentiate human from non-human bones. Chapter 7, *Human versus Non-Human: Bone Microstructure in Species Differentiation and Experimental Research*, by Victoria Dominguez, looks at the differentiation of human from non-human skeletal remains based upon bone microstructure from a broad perspective. The chapter begins with a short summary of the evolutionary history and origins of skeletal tissues as currently understood, followed by an introduction to bone microstructure in the context of human and non-human growth and development, including an overview of existing methods and related considerations for distinguishing between human and non-human bone at the microstructural level. The chapter concludes with a brief discussion of how bone microstructure can enhance the selection of appropriate non-human models for experimental research.

In Chapter 8, *Age-at-Death Estimation from Bone Microstructure*, Gocha, Mavrouda, and Goldstein describe and evaluate the utility of bone microstructure to estimate age-at-death from human skeletal remains and underscore the merits of histological age-at-death estimating methods. A chronological survey of some of the major developments in methods for age estimation through histological investigation of cortical bone microstructure is utilized to offer a temporal context to advances in the field as they arose, e.g., population-specific histological age-at-death estimation equations, offering less complicated and invasive methods, methods applicable to sub adult bone, application of three-dimensional data for age estimation, validation of histomorphometric definitions, use of novel technological, and more sophisticated statistical methodologies.

In Chapter 9, *The Histopathological Analysis of Human Skeletal Remains*, Assis, Gómez García-Donas, and de Boer provides a comprehensive overview of the benefits, limitations, and challenges for the histopathological analysis of dry bone, and offers a realistic assessment of the contribution of histology in the diagnostic process. Following a discussion of how histology should be incorporated in "the broader diagnostic work-up" is an overview of relevant bone diseases, and how histology may aid in their diagnosis, including various types of metabolic bone disease, infections, tumors, and trauma.

Histological methods can be effectively applied to bone derived from a broad range of contexts. In Chapter 10, *Histological Analyses of Human Bone from Archaeological Contexts*, Pfeiffer and Pinto discuss the application and benefits of histological analysis of human bone in archaeological contexts. The chapter focuses on research and methods based upon transmitted light microscopy. Whether determining if fragmentary remains are human, providing an age at death estimate, evaluating the extent of diagenesis during time since death, or evaluating skeletal health and disease, the authors demonstrate that the histological evaluation of bone provides useful information for bioarchaeological and paleopathological studies, as well as modern forensic casework. They also note the role that histological analysis can play in other types of analyses that require adequate preservation of bone tissue, such as assessing bone sample quality preliminary to ancient DNA or isotopic analyses. The authors make the important point that histological study of archaeologically derived material can contribute to research questions arising in forensic and biomedical fields. Given the potential for histological analysis in the broader scope of skeletal analysis, the authors propose that histology of normal cortical and cancellous human bone should be included in osteological training.

The final three chapters of the book discuss current research trends in bone histology, with a focus on technological advances in imaging and methods. Reviews of four well-documented skeletal collections developed specifically for bone histological and imaging research are discussed, as well as the importance of such collections for future research.

In Chapter 11, *Skeletal Collections for Histological and Imaging Research*, Andronowski and colleagues discuss skeletal collections that are amenable to histological and imaging research, including profiles of specific bone collections including backgrounds, ethical considerations, bone procurement processes, demographic compositions, and existing imaging and histological data available. In addition, they illustrate the emergence of digital repositories that allow sharing of data retrieved from skeletal collection reference banks and encourage national and international collaboration among researchers interested in human skeletal biology, biological anthropology, and related fields.

Chapter 12, *Three-Dimensional Microstructural Imaging of Bone: Technological Developments and Anthropological Applications*, Cooper and colleagues provide an overview of the development of 3D analytical techniques for the study of bone microstructure and their applications in biological anthropology, including an in-depth review of anthropological applications of 3D micro-CT and synchrotron radiation (SR) micro-CT, discussion of associated methodological issues, and suggested future uses of this technology in the field.

Chapter 13, *Visualization and Interpretation of Cortical Porosity*, by Mary E. Cole, provides an overview of the morphometry of vascular pores of cortical bone. It describes cellular processes that form the cortical pore network, summarizes microscopic and radiographic imaging techniques for visualizing pore morphometry in bone tissue, and concludes with a broader discussion of specific aspects of pore morphometry and interpretations of their possible mechanical origins and physiological modifications with age.

This volume has assembled authors with extensive experience and expertise in various aspects of hard tissue biology. Its intended goal is to provide readers with an overview of the current state of research and potential applications of histological analysis in biological anthropology, forensic anthropology, and skeletal biology. It is our hope that the contents provide a useful resource for students, researchers, and practitioners in these and related disciplines.

Editors

Sam D. Stout received his PhD in physical anthropology from Washington University in St. Louis in 1976 and is currently Professor Emeritus of anthropology at The Ohio State University. He is a skeletal biologist specializing in microstructural (histological) analysis of bone. His research has application in several disciplines, including bone biology, bioarchaeology, paleopathology, and forensic anthropology. During his career, he has been called upon to provide forensic analyses of human remains for local, state, and federal agencies, including histological age estimations of historically important skeletal remains, such as Francisco Pizarro, Janaab Pakal (Mayan Ruler of Palenque), and the "Sundance Kid." He is a Fellow of the American Academy of Forensic Sciences, and American Association for the Advancement of Science.

Christian Crowder, PhD, D-ABFA, received his BA from Texas A&M University, MA from the University of Texas at Arlington, and PhD in Biological Anthropology from the University of Toronto. Dr. Crowder is board certified in Forensic Anthropology by the American Board of Forensic Anthropology and has published numerous scientific articles and book chapters covering topics such as bone histology, skeletal trauma, and mass fatality response. He is the Chief of the Human Identification and Anthropology Laboratories for the Tarrant County Medical Examiner's Office in Fort Worth, TX. He has previously worked as a forensic anthropologist for international, federal, county, and city agencies. Over his career he has worked for the United Nations International Criminal Tribunal for the former Yugoslavia, the Department of Defense POW/MIA Accounting Agency (previously known as the Joint POW/MIA Accounting Command Central Identification Laboratory), the Office of Chief Medical Examiner – New York City, the Office of the Armed Forces Medical Examiner, the Harris County Institute of Forensic Sciences, and the Southwestern Institute of Forensic Sciences. In addition to his practitioner duties, Dr. Crowder is adjunct faculty at Pace University in New York City and at the University of Texas at Arlington.

Contributors

Janna M. Andronowski
Faculty of Medicine
Division of BioMedical Sciences
Memorial University of Newfoundland
St. John's, Newfoundland, Canada

Amanda M. Agnew
Injury Biomechanics Research Center
The Ohio State University
Columbus, Ohio, USA

Christopher Aris
Keele University
Keele, UK

Sandra Assis
NOVA University of Lisbon
Lisbon, Portugal
And
Center for Research in Anthropology
Faculty of Social and Human Sciences
NOVA University of Lisbon
Lisbon, Portugal

Hans H. de Boer
Victorian Institute of Forensic Medicine
Dept. of Forensic Medicine Monash University
Southbank, Victoria, Australia

Mary E. Cole
Division of Radiologic Sciences and Therapy
School of Health and Rehabilitation Sciences
The Ohio State University
Columbus, Ohio, USA

David M. L. Cooper
Department of Anatomy Physiology and Pharmacology
College of Medicine
University of Saskatchewan
Saskatoon, Saskatchewan, Canada

Christian Crowder
Office of Chief Medical Examiner and Forensic Laboratories and Department of Sociology and Anthropology
The University of Texas at Arlington
Fort Worth, Texas, USA

Victoria M. Dominguez
Department of Anthropology
Lehman College
City University of New York (CUNY)
New York, USA

Julieta Gómez García-Donas
Centre for Anatomy and Human Identification (CAHID)
School of Science and Engineering
University of Dundee
Dundee, UK

Haviva M. Goldman
Department of Neurobiology and Anatomy
College of Medicine
Drexel University
Philadelphia, Pennsylvania, USA

Timothy P. Gocha
Forensic Anthropology Center
Texas State University
San Marcos, Texas, USA

Justin Z. Goldstein
Forensic Anthropology Center
Texas State University
San Marcos, Texas, USA
And
Office of Chief Medical Examiner
New York City, New York, USA

Angela L. Harden
Injury Biomechanics Research Center
The Ohio State University
Columbus, Ohio, USA

Rita Hardiman
Melbourne Dental School
The University of Melbourne
Melbourne, Victoria, Australia

Randee L. Hunter
Injury Biomechanics Research Center
The Ohio State University
Columbus, Ohio, USA

Corey Maggiano
University of West Georgia
Carrollton, Georgia, USA

Isabel Maggiano
University of West Georgia
Carrollton, Georgia, USA

Sophia R. Mavroudas
Forensic Anthropology Center
Texas State University
San Marcos, Texas, USA

Susan Pfeiffer
Professor Emerita
University of Toronto
Toronto, Canada

And

Research Professor of Anthropology
Center for the Advanced Study of Paleobiology (CASHP)
George Washington University
Washington, DC, USA

Deborrah Pinto
Forensic Anthropologist
Southwestern Institute of Forensic Sciences
Dallas, Texas, USA

Natalie A. Sims
St. Vincent's Institute of Medical Research
Melbourne, Victoria, Australia
And
Melbourne Medical School
The University of Melbourne
Melbourne, Victoria, Australia

John G. Skedros
Adjunct Associate Professor of Orthopaedic Surgery
University of Utah
Salt Lake City, Utah, USA

Brian F. Spatola
National Museum of Health and Medicine
Armed Forces Institute of Pathology
Washington, DC, USA

Sam D. Stout
The Ohio State University Columbus
Ohio, USA

Gwen Robbins Schug
Department of Biology
The University of North Carolina, Greensboro
Greensboro, North Carolina, USA

Joshua T. Taylor
Faculty of Medicine
Division of BioMedical Sciences
Memorial University of Newfoundland
St. John's, Newfoundland, Canada

Xuan Wei
Department of Anatomy Physiology and Pharmacology
College of Medicine
University of Saskatchewan
Saskatoon, Saskatchewan, Canada

Bone Remodeling and Its Histomorphological Products

1

MARY E. COLE, CHRISTIAN CROWDER, AND SAM D. STOUT

1.1 Introduction

Skeletal analyses can be approached at several hierarchical levels. These range from organ-level macroscopic analyses of whole bones, to microscopic analyses of bone tissue structure, to nanoscale analyses of the tissue's collagen and mineral components, to molecular analyses of bone cell signaling pathways. Intermediate at the micro-scale is the histological analysis of bone tissue, which Harold Frost (Frost, 1986) described as the level of "skeletal intermediary organization," representing the collaborative activity of cells. The mineralized structures of this microscopic anatomy, or histomorphology,[1] are the observable and measurable products of bone cell activity over the lifespan. An anthropological perspective interprets the morphometry of these tissue structures as representing life history, including the cellular activity related to growth, aging, trauma, and pathology (Stout et al., 2019). Bone tissue microstructure is also an important component of bone quality, or the structural and material properties that contribute to bone strength (Paschalis et al., 2011). In order to interpret the biological information encrypted in the histomorphology of bone, it is essential to understand the basic biology underlying the creation of histomorphological structures.

Bone metabolism, whether for growth, adaptation, or homeostasis, involves two basic kinds of cells: bone-forming osteoblasts and their derivatives, and bone-resorbing osteoclasts. The normal activity of these same cells is associated with two distinct physiological processes referred to as modeling and remodeling. Bone modeling describes the independent activity of bone cells to either form bone (osteoblasts) or resorb bone (osteoclasts) on a surface (Frost, 2003; Hughes et al., 2020). Bone modeling can occur on both periosteal (external) and endosteal (internal) surfaces (Szulc & Seeman, 2009). Bone modeling is essentially restricted to the growing skeleton, where it works in concert with growth to adapt bones to their changing biomechanical environment by adjusting the amount and spatial distribution of bone tissue. Exceptions where modeling occurs in the adult skeleton are fracture repair and certain cases of pathological mechanical loadings (Peck & Stout, 2009). Testosterone also promotes bone formation responsible for periosteal expansion with age, particularly in males, in response to the loss in bone strength that follows age-associated endosteal resorption (Feik et al., 2000; Martin, 1993). Chapter 2 provides an in-depth discussion of the bone modeling process. Bone remodeling refers to the coordinated or "coupled" action of bone-resorbing osteoclasts and bone-forming osteoblasts at the same tissue location, focally replacing old or damaged bone (Frost, 2003; Hughes et al., 2020). A second type of remodeling that is stochastic and non-site-dependent is also proposed (Burr, 2002). This bone turnover occurs throughout the lifespan and is the predominant process in the adult skeleton responsible for its characteristic histomorphological features.

DOI: 10.4324/9781003385608-1

This chapter will focus on bone remodeling in adult humans with an anthropological perspective, suitable for application to bioarchaeology, forensic anthropology, paleopathology, and paleontology. Anthropologically related analyses typically focus on the histomorphometric products of cortical bone remodeling, which are well defined in bone cross-sections. Cortical bone is compact bone, forming the shafts of long bones and the thin shells surrounding flat, short, and irregular bones (Eriksen, 2010). Although cortical bone appears solid macroscopically, it is microscopically perforated with pore systems, which house vascular structures (cortical pores) and osteocytes (lacunar-canalicular networks) (Andronowski & Cole, 2021). Cortical bone surrounds cancellous bone, which is located inside the marrow cavity and composed of trabecular struts and plates, giving it a "spongy" appearance (Eriksen, 2010).

The histological analysis of skeletal remains is made possible because remodeling occurs throughout life and produces discrete, definable, and quantifiable microscopic features such as osteons (Andronowski & Cole, 2021; Stout et al., 2019). The application of some methods to other animals is possible, but problematic because of the highly variable expression of cortical remodeling due to biomechanical and physiological differences (Schaffler & Burr, 1984). See Chapter 7 for the use of histomorphology to distinguish between human and nonhuman skeletal remains. Most notably, in addition to age at death estimation, histomorphometry can provide insights into bone remodeling history that can be used to estimate remodeling rates, creating what can be called "paleophysiology."

The purpose of this chapter is to provide a broad overview of the theoretical foundations, cellular activities, microstructural products, and histomorphometric interpretations of bone remodeling. The first section of this chapter will discuss principles and models of bone functional adaptation. The second section will describe the mechanotransduction of mechanical stimuli into bone remodeling. The third section will provide a brief overview of bone cell activity during each phase of bone remodeling. The fourth section presents considerations for interpreting remodeling activity from histomorphometry, including methodological guidelines, identifying osteon types, sources of intra- and inter-skeletal variation, techniques for static and dynamic histomorphometry, and method reliability.

1.2 The Role of Remodeling in Bone Functional Adaptation

Mechanical loading alters the composition and function of cells in structural tissues of the body, ultimately shaping tissue development, maintenance, and repair. Mechanobiology studies the cellular mechanisms that transduce these mechanical forces into biological signaling cascades (Wang & Thampatty, 2006). More broadly, bone functional adaptation describes how the mechanical environment modifies bone shape and structure over the lifespan (Ruff et al., 2006).

1.2.1 Early Concepts of Bone Functional Adaptation

Skeletal adaptation to mechanical loading was recognized in the early 16th century by Galileo, who noted that bone width scales with load-bearing and not just with size (Skerry, 2006). The idea was not popularized until the 19th century, when orthopedic surgeon Julius Wolff introduced "the law of bone transformation" (Wolff, 1892). Wolff was influenced by anatomist Georg Hermann von Meyer (von Meyer, 1867) and engineer Karl Culmann

(Culmann, 1866) who both noted the alignment of trabecular struts in cancellous bone with what they calculated as principal stress trajectories. Wolff proposed that the internal architecture and external form of bones are altered by changes in the primary stressing of the bone. However, Wolff believed that stress trajectories modified bone structure according to strict mathematical rules, although he did not attempt to formulate these equations. His diagrams of trabecular struts were oversimplified, appearing more geometrically consistent than cancellous bone actually is (Skedros & Brand, 2011). Culmann's calculations of the corresponding stress trajectories also had some engineering inaccuracies. Moreover, Wolff believed that bone permanently adapts to static loads, rejecting the concept of dynamic remodeling as chaotic "structural upheaval." Prior to Wolff, but within the same century, it was Wilhelm Roux (Roux, 1885) who correctly identified bone structure as self-regulating in adaptation to functional mechanical stimuli (Lee & Taylor, 1999; Roesler, 1981). Still, Wolff's general observation that bone adapts to its mechanical environment has persisted in cultural memory as "Wolff's Law."

1.2.2 Principles of Bone Functional Adaptation

Starting in the 1970s, in vivo strain gage experiments of varied species, skeletal elements, and loading conditions established several important principles of bone functional adaptation. This framework quantifies mechanical loading using the dimensionless quantity of strain, or the fractional change in the length of a bone during its deformation (Pivonka et al., 2018).

Bone adapts to dynamic strains. Strain is the relative change in length when an external load causes the bone to deform, or change in size or shape (Pivonka et al., 2018). A static, continuously applied load will result in bone loss (Hert et al., 1971). Bone adapts to intermittent dynamic strains. Increasing mechanical strains stimulate bone apposition, while decreasing mechanical strains stimulate bone resorption. There is minimal net change in bone mass or architecture while mechanical strains remain at equilibrium, or the range of the bone tissue's "optimum customary strain level" (Lanyon et al., 1982). This optimum range is likely determined by a combination of skeletal location and genetics, and modified by physiological factors, including hormonal influences, aging, and pathology (Skerry, 2006). More recent animal models suggest that bone appears nonresponsive within this so-called "lazy zone" not because it is inherently unstimulated by the strain range, but because it has already structurally adapted to resist that strain. When limbs are loaded in novel configurations outside of their natural positioning, bone adapts linearly to increasing strain, without a nonresponsive strain range (Meakin et al., 2014; Sugiyama et al., 2012).

Bone adapts to local strains, not total strains. Most longitudinal forces on a bone (>70%) result from bending, which places varying regions of the same cross-section into tension or compression (Biewener & Bertram, 1993). Compressed regions of bone tissue experience higher strain (x1.9) than tensed regions (Lanyon & Baggott, 1976). Bone is modeled preferentially on the periosteal surface of the region under higher strain, rather than expanding around the periosteal surface uniformly (Lanyon et al., 1982). Compressive and tensile stresses decrease toward the neutral axis, where they meet with a net stress of zero (Robling, 2014). Strains are lowest at the endosteum closest to this neutral axis (Martin, 1993; Thomas et al., 2005). Skeletal biologists have long recognized that bone resorption increases toward the endosteum of long bone cross-sections in the form of elevated intracortical porosity (Andreasen et al., 2020; Atkinson, 1965; Bousson et al., 2001; Jowsey, 1960; Martin et al., 1980; Thomas et al., 2005; Zebaze et al., 2010; Zebaze & Seeman, 2015). Elevated intracortical porosity has also been observed in some tissue regions thought to be under lower-strain tension, compared to compressed regions of the same cross-section (Agnew et al., 2013; Agnew & Stout, 2012; Cole, Stout, et al., 2022; Dominguez & Agnew, 2016; Frost, 1990; Skedros et al., 1994, 2005). Tensed and compressed regions of the same cross-section have also been observed to vary in the morphometry and population density of secondary osteons, osteocyte lacunae, and microdamage, as previously reviewed (Andronowski & Cole, 2021; Stout et al., 2019).

Bone adaptation is stimulated both by high strains at low frequency and low strains at high frequency. Strain stimulus is proportional to strain rate, which is the product of strain magnitude, measured in microstrains (με) and strain frequency, measured in cycles/second (Hertz or Hz) (Turner, 1998). Peak functional strains in weight-bearing long bones are similar across many species, measuring approximately 2,000–3,000 με (Meakin et al., 2014). Bone experiences relatively few daily high-magnitude (2,000–3,000 με), low frequency (1–3 Hz) loading events. However, maintenance of body posture imposes continual low-magnitude (<5 με), high frequency (10–50 Hz) loads. This continual cyclic stimulation of bone declines with sarcopenia (muscle wasting), contributing to age-associated bone loss (Ozcivici et al., 2010). Loading creates pressure gradients, driving fluid flow through the lacunar-canalicular spaces that surround mechanosensory osteocytes embedded in bone tissue. Fluid flow shear stress is sensed by the osteocyte plasma membrane, and fluid pressure, which is amplified at higher strain rates, may be sensed by the osteocyte cytoskeleton (Liu et al., 2015).

Bone adaptation also becomes "saturated," or desensitized to strain, after a short number of cycles (Turner, 1998). Minimal gains in bone mass occur above about 40 cycles per day (Rubin & Lanyon, 1984) and 10- to 15-second breaks between loading cycles substantially improves adaptive response (Meakin et al., 2014). Osteocytes may not effectively sense repeated loading before their lacunar-canalicular fluid has redistributed (Scheiner et al., 2016).

1.2.3 The Utah Paradigm and the Mechanostat

Prior to the 1960s, skeletal physiology was thought to be determined primarily by bone effector cells responding non-mechanical influences, such as hormones, vitamins, minerals, genetics, sex, and age. This view was challenged by orthopedic surgeon Harold M. Frost and developed through Hard Tissue Workshops organized by W.S.S. Jee at the University of Utah, starting in 1965. These workshops discussed the unrecognized connections between skeletal biology research in numerous basic science and clinical disciplines, hampered by a lack of interdisciplinary communication. By 1995, the "Utah paradigm" in skeletal biology began to coalesce (Frost, 2001a, 2001b). The new paradigm explored the load-bearing skeletal elements at the organ, tissue, and cellular level focusing on the mechanical competence of bone (Frost, 1999a, 1999b, 2000). The Utah paradigm indicated that the driving force of load-bearing skeletal architecture and strength are mechanical factors, and that non-mechanical factors either hinder or help the process but do not replace it (Frost, 1999b, 2000; Frost & Schönau, 2000). Growth hormones have the greatest effect because they are directly associated with increasing muscle strength, body weight, and longitudinal bone growth (Frost, 1998). This paradigm has endured to the present day as a core concept of modern musculoskeletal research (Pivonka et al., 2018).

The leading hypothesis in the Utah paradigm is Frost's mechanostat, a mechanical feedback system that explains how load-bearing skeletal organs attain mechanical competence. Bone modifies its mass, geometry, and morphometry to keep its strain (deformation) within an equilibrium range. The mechanostat describes how bone coordinates modeling and remodeling to maintain this adapted range. Using a house thermostat as an analogy, the mechanism controlling bone mass turns on in response to stimulation outside the set range, and off when stimulation returns to this set range. Strain thresholds or set points called minimum effective strains (MES) activate or depress bone modeling and remodeling. The threshold strain ranges determine if, when, where, and how long biologic activity turns on or off. Bone tissue is added or removed through modeling or remodeling until the bone again deforms within the set point strange range (Frost, 1983, 1987b, 1987c).

In 2003, Frost published an update of the mechanostat hypothesis with approximate strange ranges derived from strain gage experiments. Strains below the threshold of the minimum effective strain for remodeling (MESr = 50–100 με) fall into the **disuse window**, and resorptive modeling removes bone from the endosteal envelope, or disuse-mode remodeling resorbs more bone than is formed, increasing intracortical porosity. This can lead to osteopenia and, in severe cases, disuse osteoporosis. Strains above the MESr fall into the **adapted window**, where conservation-mode remodeling retains bone. Increasing strain past the modeling threshold (MESm = 1,000–1,500 με) enters the **mild overload window**, where formative modeling begins, and conservation-mode remodeling is suppressed. Further strain increases enter the **pathological overload window**, where microdamage accumulates faster than it can be remodeled (MESp = ~3,000 με) until bone failure (~25,000 με in young adults) (Frost, 2003).

1.2.4 Explaining Intraskeletal Variation in Tissue Response

The mechanostat is an equilibrium model, meaning that a changing mechanical load triggers bone formation or resorption to maintain the bone's response to loading. A challenge for equilibrium models is that the mechanical environment and the bone's response substantially vary between skeletal sites and over the lifespan (Pearson & Lieberman, 2004). Skerry (2006) notes that the interspecies similarity in peak strains has led to the misconception that all bones have identical thresholds for strain magnitude which serve as the sole stimulus for the mechanostat. Skerry proposes the term "customary bone strain stimulus" (CSS) to include strain magnitude, frequency, duration, and the number and timing of loading cycles and rest periods. Set points for this CSS are genetically regulated and modified by the physiological environment created by age, sex, pathology, and local, systemic, and external biochemical influences. Importantly, the CSS is site-specific to the skeletal location and optimized for its typical dynamic loading. Skerry suggests bone curvature as one potential mechanism for this optimization (Skerry, 2006). As straighter bones are stronger in axial compression, bone curvature may be related to regulating customary strains or inducing predictable load distribution. Removal of an artificial load does not always revert changes in bone curvature, mass, and shape (Javaheri et al., 2020).

Following Frost's mechanostat, additional equilibrium models were proposed to directly address this question of site-specific adaptation. In Turner's principle of **cellular accommodation**, cells retain a memory of their customary strain stimulus. Therefore, the set points vary from site to site depending on the local strain. The set point in weight-bearing bones will be high, while the set point in non-weight-bearing bones will be much lower. This explains why non-weight-bearing bones, such as the skull, do not resorb away completely due to disuse remodeling (Turner, 1999). Cellular accommodation proposes that minimum effective strain thresholds for cellular response are not set values, but products of a cell's local strain history. Short-term cell memory may derive from the enhancement of cell signaling pathways and release of their biochemical products (Schriefer et al., 2005; Turner et al., 2002). Turner claims that Frost places biomechanics at the center of bone biology ignoring the fact that bone is "insensitive to mechanical loading in the absence of either parathyroid or growth hormones" (Turner, 2000). This indicates that endocrine and paracrine factors do not have a secondary role, but rather a primary role in bone formation. Long-term cell memory may be linked to structural changes, such as cytoskeleton reorganization, the expression of membrane receptors and the extracellular microenvironment, or alteration of cellular connectivity (Robling et al., 2006; Schriefer et al., 2005; Turner et al., 2002). Turner has also suggested the bone curvature created by previous modeling and remodeling events as a mechanism for predictable strain patterns (Turner, 1998). Carter and Beaupré's (2001) **mechanobiology** hypothesis is another equilibrium model, wherein bones initially model to experience optimum strain levels. Bone size and shape is the combination of genetic and hormonal factors that regulate skeletal patterning, and mechanical loading from bone length and body mass during growth (Carter & Beaupré, 2001). Muscle arrangements during initial skeletal patterning may induce site-specific loading, explaining consistency in the cross-sectional shape of certain bones across varying species size, gait, and locomotor posture (Carpenter & Carter, 2008).

Site-specific variation in tissue response has also been explained through **optimization** models. In equilibrium models, bone is added or removed to maintain the tissue's deformation response (strain) within a set range. Optimization models instead consider bone modeling as a trade-off, maximizing the benefit of bone strength while minimizing the energetic cost of adding and moving bone mass. Bone remodeling must also balance the benefit of repairing damaged bone with the cost of creating mechanically weaker porosity and under-mineralized new bone (Giorgio et al., 2021; Pearson & Lieberman, 2004). For example, limbs are generally observed to taper distally. In an optimization model, this distal repression of modeling could be explained by the higher energetic cost of swinging the same amount of bone mass at a further distance from the joint. Concurrently, remodeling would be predicted to increase distally to repair the microdamage incurred under this higher strain (Pearson & Lieberman, 2004). Cancellous bone might optimize a stiff, yet lightweight, structure by remodeling trabecular architecture for a constant strain energy density over time (Lekszycki, 2002, 2005) and in multiple loading conditions (Nowak et al., 2018, 2020).

1.3 Stimuli for Bone Remodeling

1.3.1 Types of Bone Remodeling

Burr (2002) states that bone remodeling fulfills three goals: (1) maintenance of mineral homeostasis, (2) adaptation to bone's mechanical environment, and (3) repair of microdamage to ensure mechanical integrity (Burr, 2002). **Stochastic remodeling** fulfills the first goal and is regulated by systemic (endocrine) factors in response to physiological influences. Such physiological stimuli include calcium–phosphate balance (parathyroid hormone, vitamin D), metabolism (insulin, leptin), the immune system (interleukins), stress (glucocorticoids), and fluctuations in sex steroids associated with growth, aging, or pathology (Bellido & Gallant, 2019; Stout et al., 2019). These endocrine factors modify the sensitivity and functioning of bone cells as they respond to mechanical stimuli (Pivonka et al., 2018). Stochastic remodeling is considered non-targeted remodeling because it is not directed at the mechanical needs of a specific bone tissue site (Eriksen, 2010).

Bone remodeling fulfills its mechanically adaptive goals by redistributing bone mass to suit the tissue's mechanical environment, and by repairing microdamage that compromises the tissue's mechanical strength. In Frost's mechanostat model, strains within the adapted window maintain remodeling at a baseline level, which Frost called **conservation-mode remodeling**. Remodeling forms as much bone as it resorbs, except the space left for the new Haversian canal, resulting in minimal net change in bone mass (Frost, 2003; Hughes et al., 2020; Hughes & Petit, 2010). Remodeling is increased above this baseline level when customary strain stimuli fall below or above this adapted window, although with diverging outcomes for net bone mass.

Below the adapted window, strains fall into the mechanostat's disuse window, and bone formation is repressed. Strains are lowest at the endosteum, increasing both resorptive modeling for endosteal resorption, and **disuse-mediated remodeling** for endocortical resorption. Remodeling forms less bone than it resorbs, resulting in increased porosity and a net loss in bone mass (Frost, 2003; Hughes et al., 2020). For example, in the human fibula, endosteal and periosteal pores may initiate the formation of new pores at similar rates, reflected in their similar pore densities. However, endosteal pores are significantly larger than periosteal pores, either due to incomplete infilling or expansion and convergence with adjacent pores. Large-diameter (>100 μm) endosteal pores result almost exclusively (97.5%) from this remodeling of existing pore systems, a process known as trabecularization (Andreasen et al., 2020).

Above the adapted window, strains fall into the mechanostat's mild overload window. Formation modeling results in bone apposition on periosteal and/or endosteal surfaces. Remodeling again increases above baseline in the form of **targeted remodeling** (Frost, 2003; Hughes et al., 2020). Bone tissue resists failure by dissipating energy through the creation of microcracks within the cortex (Ritchie, 2011). Resorption spaces are not randomly associated with microdamage, but specifically target its location within a 100–300 μm radius (Kennedy et al., 2012). Targeted remodeling is estimated to account for 10–30% of all remodeling activity (Burr & Martin, 1993; Li et al., 2001), although theoretically all remodeling could be "steering" toward microdamage (Martin, 2002).

1.3.2 Osteocyte Mechanotransduction

Mechanical signals are transduced into modeling and remodeling activity by osteocytes, the mechanosensory osteoblast derivatives embedded in bone tissue. Osteocytes compose approximately 95% of all bone cells, as they accumulate over many modeling and remodeling cycles (Franz-Odendaal et al., 2006). Osteocyte cell bodies are entombed in lacunae and extend their numerous dendritic processes (on average 89 ± 25 per cell) through narrow channels called canaliculi (Varga et al., 2015). These cellular spaces contribute an estimated 2% porosity to the bone matrix. Osteocyte communication is facilitated by their dense interconnection of within this lacunar-canalicular network (LCN), with an estimated average of 12.7 termini per dendrite, 1,128 cell–cell connections per osteocyte, and 23.4 trillion osteocytic connections throughout the human skeleton (Buenzli & Sims, 2015).

One mechanism of osteocyte mechanotransduction is the sensation of interstitial fluid flow within the LCN, driven by the pressure gradients of localized strain (Cowin et al., 1991; Weinbaum et al., 1994). Hughes and colleagues have proposed that fluid flow mechanotransduction primarily activates formation modeling (Hughes et al., 2020). Osteocytes are tethered to the bony enclosure of the LCN by pericellular transverse fibrils, and by direct focal adhesions of osteocyte transmembrane integrins. Fluid flow deflects the osteocytes against these tethers, amplifying cellular strain up to 100-fold (Liu et al., 2015; Verbruggen & McNamara, 2018). The stretched osteocyte plasma membrane results in an intracellular calcium influx, triggering osteocyte cell signaling pathways. The osteocyte upregulates its Cx43 hemi-channels for improved communication with other osteocytes, and transcribes and releases factors that inhibit bone resorption and promote bone formation (Pivonka et al., 2018; Uda et al., 2017).

A second mechanism of osteocyte mechanotransduction is osteocyte apoptosis. The diffusion of nutrients through interstitial fluid is not sufficient for osteocyte survival. When local strains are abnormally low, osteocyte apoptosis may result from nutrient deficiency and waste accumulation (Hughes & Petit, 2010). Linear microcracks can also trigger osteocyte apoptosis by disrupting fluid flow and destroying cell–cell connections, resulting in impaired nutrient transport and hypoxia followed by oxidative stress (Ru & Wang, 2020). In murine models, microdamage can result in osteocyte apoptosis within 24 hours, followed by resorption within 10 to 14 days (Bentolila et al., 1998; Cardoso et al., 2009; Verborgt et al., 2000). Hughes and colleagues accordingly link osteocyte apoptosis to resorption modeling and disuse-mediated remodeling, triggered by low strain, and to targeted remodeling of microcracks (Hughes et al., 2020). Frost identified this remodeling stimulus as early as his original proposal, suggesting that microcracks disrupt canalicular connections between osteocytes (Frost, 1960). Martin further proposed that osteocytes inhibit, rather than stimulate, remodeling, and that osteocyte apoptosis releases this containing mechanism so that resorption can proceed (Martin, 2002, 2003a, 2003b). Subsequent research has associated this remodeling stimulus not with the apoptotic osteocytes, but with osteocytes in a 100–300 μm "penumbra." Apoptotic release of extracellular ATP may trigger these "bystander" osteocytes to shift the balance of their own released factors toward osteoclastogenesis and subsequent bone resorption (Kennedy et al., 2012; McCutcheon et al., 2020).

1.4 Phases of Bone Remodeling

1.4.1 Basic Multicellular Unit Structure

Remodeling requires the tethered action of bone-resorbing osteoclasts and bone-forming osteoblasts. This cellular activity is coordinated within a basic multicellular unit (BMU), also called a bone remodeling unit (BRU). The microstructural product of this remodeling is a bone structural unit (BSU) (Parfitt, 1979). In cortical bone, the BMU tunnels into the cortex, producing a BSU called a secondary osteon. Osteoclasts resorb old bone in a "cutting cone," and osteoblasts follow behind forming new bone in a "closing cone." A bone remodeling compartment (BRC) encloses the BMU with a canopy of cells of mesenchymal origin, which is penetrated with a central blood vessel. Osteoblasts form concentric lamellae starting from the periphery of the resorption space, leaving a pore or "Haversian canal" surrounding the blood vessel (Eriksen, 2010). This extension of the vascular network transports nutrients, cell signaling factors, and bone cell precursors, sustaining active remodeling and facilitating future remodeling at this tissue location (Chen et al., 2021; Delaisse et al., 2020). In cancellous bone, the BMU remodels bone on the exposed surface of a trabecula, producing a BSU called a hemi-osteon. A trench is resorbed by osteoclasts and then filled with new bone by osteoblasts (Parfitt, 1994). The canopy enclosing the BMU is created when bone lining cells and overlying bone marrow envelope cells lift up from the bone surface, potentially in response to osteoclast precursors migrating underneath (Kristensen et al., 2014). Complete osteons can form in cancellous bone by tunneling through very thick trabeculae (Burr & Akkus, 2014) (Figure 1.1).

Histomorphometry requires accurate identification of the histomorphological features that are traces of earlier remodeling events. This requires an understanding of the remodeling process itself. Cortical bone remodels quicker (~120 days) than cancellous bone (~200 days) (Eriksen, 2010). It is useful to describe the remodeling in terms of three distinct phases: activation, resorption, and formation. This sequence is sometimes denoted by the acronym ARF (Martin et al., 2015). These phases can be further broken down into (1) **activation** by osteocytes, (2) **resorption** by osteoclasts (2–3 weeks), (3) **reversal** from bone resorption to bone formation (4–5 weeks), (4) **formation** by osteoblasts (3–4 months), (5) **mineralization** of the formed osteoid (initiated with a 1- to 4-week lag behind formation), and (6) **resting**, or quiescence until future activation (Agerbaek et al., 1991; Eriksen et al., 1984; Kenkre & Bassett, 2018; Sims & Martin, 2020).

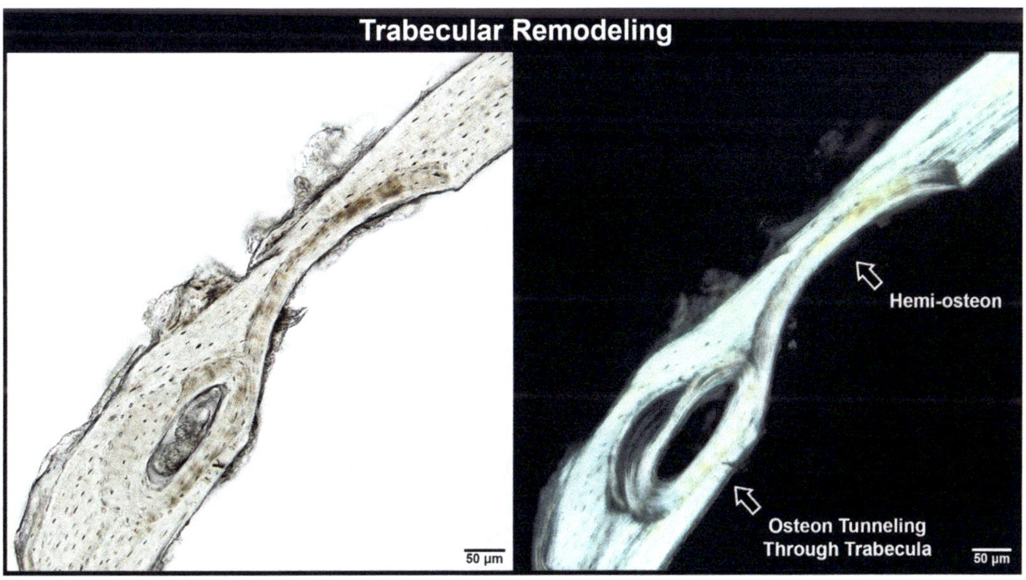

Figure 1.1 Microscopic images taken with brightfield (left) and circularly polarized light (right) show a trabecula with typical, trench-like hemi-osteon and a rare complete osteon.

1.4.2 Activation

Activation begins with the appearance of osteoclast progenitors at the site of BMU initiation. Osteoclast progenitors arise from hematopoietic stem cells in the bone marrow and arrive at the BMU through its vasculature. Osteoclasts are polykaryons, having multiple nuclei, which involves fusion of osteoclast progenitors during differentiation (Veis & O'Brien, 2023). Cell signaling pathways for osteoclastogenesis are activated when receptors on the osteoclast precursor cell membrane bind to external factors. Precursors proliferate in response to membrane integrin αvβ3 binding the extracellular matrix. Precursor differentiation is promoted by membrane receptors binding extracellular ligands M-CSF (Macrophage colony-stimulating factor), RANKL (receptor activator of nuclear factor kappa beta), TNF-α (Tumor necrosis factor alpha), and IL-1 (Interleukin-1). Binding of extracellular Wnt5a to membrane receptors Frizzled and ROR 2 (Receptor tyrosine kinase-like orphan receptor 2) promotes RANK expression in osteoclast precursors and actin ring formation in mature osteoclasts (Bellido et al., 2019; Kobayashi et al., 2000, 2018; Takayanagi, 2021).

Cytokines M-CSF and RANKL, secreted by osteoblasts and osteocytes, are required for osteoclastogenesis. M-CSF binding to osteoclast precursor receptor c-FMS (Colony-stimulating factor-1 receptor) activates multiple cell signaling pathways for proliferation, survival, differentiation, and cytoskeletal reorganization. M-CSF binding also induces osteoclast precursor cell membranes to express the receptor RANK. Subsequent binding of the circulating RANK ligand, RANKL, activates multiple cell signaling pathways for differentiation. However, RANKL can instead be bound by a circulating decoy receptor, osteoprotegerin (OPG). Osteoblast lineage cells release both RANKL and OPG, modulating bone resorption by adjusting the ratio of these factors in the local environment. When OPG is in excess, it binds up circulating RANKL, preventing interaction with osteoclast precursors. When RANKL is in excess, enough remains unbound by OPG to instead bind RANK on osteoclast precursors and promote their differentiation (Maeda et al., 2019; Plotkin & Bruzzaniti, 2019; Trichilo & Pivonka, 2018).

Osteoblast lineage cells, particularly osteocytes, regulate bone resorption by modulating the RANKL:OPG ratio (Veis & O'Brien, 2023). Low strain sensed by fluid flow stimulates osteocytes to increase the RANKL:OPG ratio, promoting bone resorption, while high strain decreases the RANKL:OPG ratio, inhibiting bone resorption (Kulkarni et al., 2010; You et al., 2008). Microdamage also stimulates osteocytes in the penumbra of the damage site to increase the RANKL:OPG ratio and activate bone resorption, with a response proportional to crack size (Mulcahy et al., 2011). These bystander osteocytes additionally release vascular endothelial growth factor (VEGF) to promote capillary growth toward the forming BMU, and protect themselves by expressing antiapoptotic protein B-cell lymphoma 2 (Bcl-2) (Kennedy et al., 2012). Osteoblast modulation of the RANKL:OPG ratio also reflects progression of the remodeling cycle. Osteoblasts produce more RANKL, promoting bone resorption, when calcium concentrations are low suggests the matrix was not recently resorbed (Takeyama et al., 2000). Osteoblasts that express Wnt16, which promotes osteoblast differentiation for bone formation, decrease their RANKL:OPG ratio to inhibit concurrent bone resorption (Alam et al., 2016; Movérare-Skrtic et al., 2014). Non-mechanical physiological influences additionally act on osteoblast lineage cells. The RANKL:OPG is increased in the presence of pro-resorptive hormones (PTH, calcitriol) and inflammation (TNF-α, inflammatory cytokines), and decreased in response to bone protective sex steroids (testosterone, estradiol) (Horwood et al., 1998; Mbalaviele et al., 2017; Michael et al., 2005).

1.4.3 Resorption

Following differentiation, the mature osteoclast seals to the bone surface with a podosome belt of actin rings (Figure 1.2). Within this sealing zone, vesicles fuse to the plasma membrane for the release of bone-resorbing products, giving the plasma membrane a "ruffled border" appearance (Figure 1.2). The osteoclast uses a carbonic anhydrase II enzyme to convert water and carbon dioxide into protons (H^+), which enter the sealing zone through a vacuolar H+/ATPase (ATP6I). The osteoclast also takes in extracellular chloride ions (Cl^-) through an HCO_3^-/CL^- exchanger, then pumps them into the sealing zone through a chloride channel (CIC-7). The resulting HCl in the sealing zone dissolves the hydroxyapatite mineral of existing bone tissue. After demineralization, the exposed type I collagen is digested through enzymes that osteoclast lysosomes release into the sealing zone, including cathepsin K (CatK), matrix metalloproteases (MMPs), and tartrate-resistant acid phosphatase (TRAP). Degradation products are removed from the sealing zone through transcytosis, where debris is absorbed into the osteoclast in vesicles and released extracellularly (Bellido et al., 2019; Ross, 2013). The sealing zone extends as the osteoclast migrates across the bone surface. However, podosomes will not cluster into a sealing zone on unmineralized collagen left over from previous resorption (Veis & O'Brien, 2023). Bone lining cells and osteal macrophages ("osteomacs") may release matrix metalloproteases to clean collagen from bone surfaces adjacent to remodeling, facilitating migration of the BMU (Batoon et al., 2021; Weivoda & Bradley, 2023).

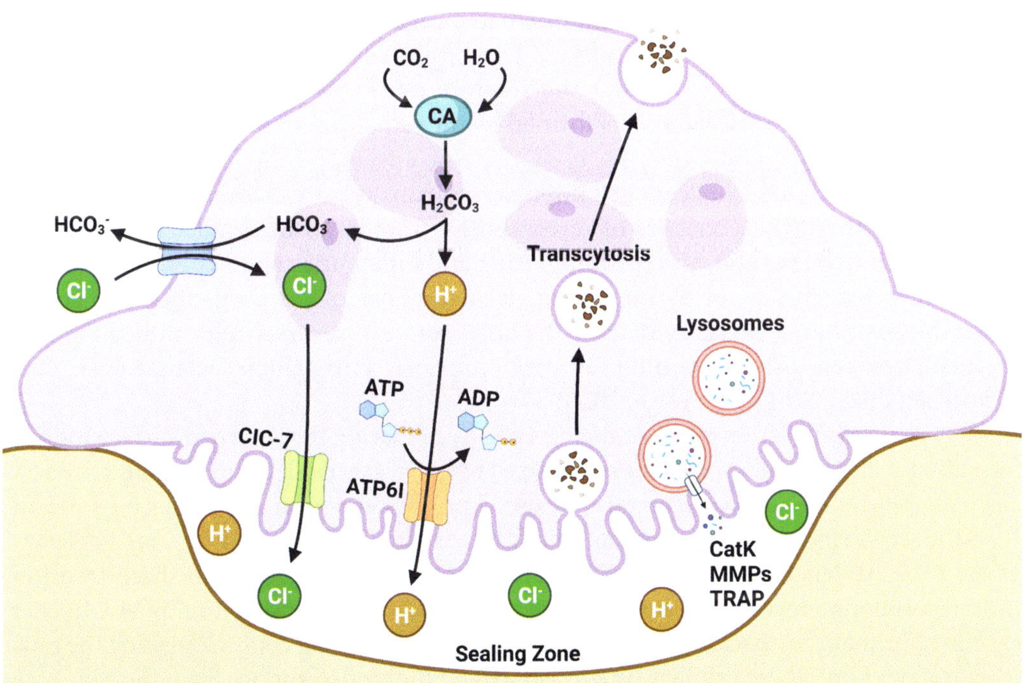

Figure 1.2 The mature osteoclast seals to the bone surface and releases bone-resorbing products into this sealing zone. Hydroxyapatite mineral is dissolved with HCl, and lysosome-released enzymes digest the exposed collagen. Figure created with BioRender.com and based on Ross (2013) and Bellido and Gallant (2019).

Excessive resorption is prevented by osteoclast detachment and apoptosis in response to indicators of remodeling cycle progression. During active osteoclast differentiation, osteoclast survival is promoted by M-CSF and RANKL (Veis & O'Brien, 2023). As resorption progresses, calcium released from the demineralized matrix activates the osteoclast calcium sensing receptor (CaSR). Rising calcium concentrations rearrange the osteoclast cytoskeleton, decreasing its resorbing activity, and eventually trigger osteocyte apoptosis (Bennett et al., 2001; Lorget et al., 2000). As osteoblasts differentiate for bone formation, they also inhibit further bone resorption by secreting FasL, which binds the osteoclast Fas death receptor (Kovačić et al., 2007; Wang et al., 2015).

1.4.4 Reversal

Reversal is the transition from bone resorption to bone formation. This phase is characterized by the presence of mononuclear cells between osteoclasts and osteoblasts, covering as much as 80% of eroded bone surfaces (Andersen et al., 2013). Although these reversal cells were discovered in the late 1970s (Baron, 1977), their exact origin and function as post-osteoclasts or pre-osteoblasts remained unclear for decades. Reversal cells can show immunoreactivity for osteoclast markers, but this is now understood to derive from their uptake of osteoclast products (Delaisse et al., 2020). Recent work has identified reversal cells as belonging to the osteoblastic lineage and expressing early osteoblast lineage markers (Andersen et al., 2009, 2013; Lassen et al., 2017). How reversal cells and osteoblasts actually arrive at the BMU remains a question of interest. Osteoblasts arise from mesenchymal stem cells, but their intermediate progenitors may be chondrocytes, bone marrow stromal cells, bone lining cells, and other fibroblasts, varying across the lifespan and between modeling, remodeling, and fracture repair (Mizoguchi & Ono, 2021). In cancellous bone, the BMU canopy is formed by the retraction of bone lining cells, which cover the trabeculae, and overlying bone marrow envelope cells, which surround the red bone marrow. These adjacent cell populations express early osteoblast differentiation markers and are likely reservoirs of reversal cells. Capillaries in contact with the BMU are another possible origin, either by migrating and differentiating the pericytes that surround their vessel walls, or by importing circulating osteoblast progenitors. In cortical bone, the penetrating capillary may be the only source of reversal cells, unless the canal is already covered with bone lining cells from previous remodeling (Delaisse et al., 2020; Kristensen et al., 2014; Root et al., 2020).

Reversal cells progress the remodeling cycle by preparing the resorbed surface for bone formation. Matrix metalloproteases synthesized by reversal cells clean up demineralized collagen on the eroded surface. Reversal cells also deposit osteopontin and collagen types I and III as the secondary osteon's cement line, marking the extent of bone resorption (Abdelgawad et al., 2016; Andersen et al., 2009, 2013; Lassen et al., 2017). There is no sharp transition from resorption to reversal phases, but rather a mixed "reversal-resorption" phase. In cortical bone, "primary" osteoclasts are located in the tip of the cutting cone, elongating the canal. In cancellous bone, these primary osteoclasts extend the eroded surface from the edge of the BMU. "Secondary" osteoclasts are scattered among reversal cells, widening the canal in cortical bone, or excavating the resorption bay deeper in cancellous bone. This secondary radial resorption contributes on average 83% of overall resorption in the BMU. When osteoid deposition begins, osteoclastic resorption is halted, and reversal cells begin to express factors (Runx2, OSX) characteristic of mature osteoblasts (Lassen et al., 2017). Reversal cells

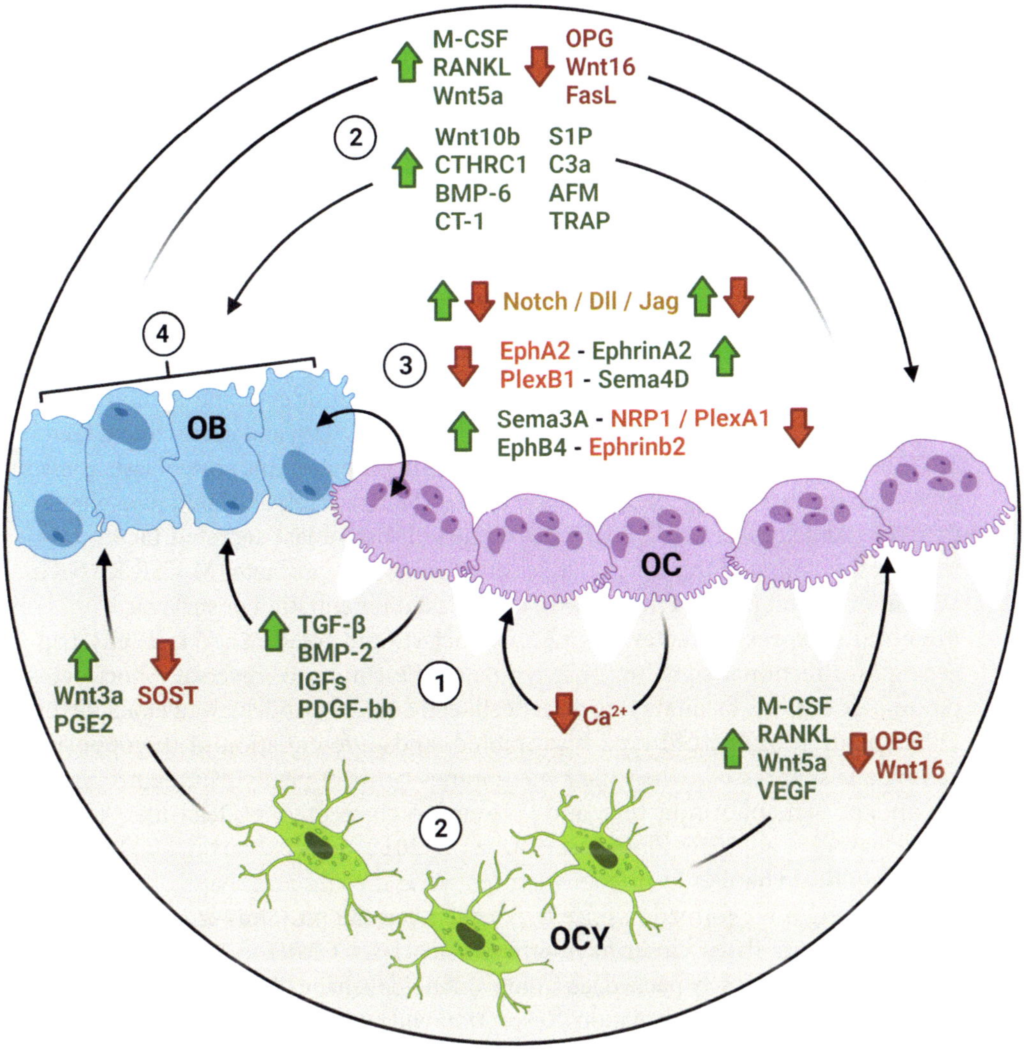

Figure 1.3 Four classes of factors that couple bone resorption to bone formation are illustrated in this figure: (1) Matrix-derived signals released from resorbed bone slow osteoclast function and promote osteoblast differentiation. (2) Factors secreted by osteoclasts and osteoblast-lineage cells promote or inhibit differentiation of the opposing cell type. (3) Membrane-expressed factors allow bi-directional signaling between osteoclasts and reversal cells, promoting differentiation of one cell type while repressing the other. (4) Topographical changes in reversal cell size and shape signal the critical cell density required to initiate bone formation. Figure created with BioRender.com.

and mature osteoblasts also progress the remodeling cycle through cell signaling pathways that "couple" osteoclast activity to osteoblast activity. Sims and Martin (2020) identify four main classes of coupling factors (Figure 1.3):

1) **Matrix-derived signals** released from resorbed bone slow osteoclast function and trigger osteoblastogenesis. Growth factors released from the bone matrix promote osteoblast progenitor replication, migration, and differentiation, including transforming growth factor β (TGF-β), bone morphogenetic protein 2 (BMP-2),

insulin-like growth factors (IGFs), and platelet-derived growth factor (PDGF-bb) (Sims & Martin, 2020). As discussed, osteoclasts are also self-regulating, decreasing resorption and then triggering apoptosis, as they sense rising calcium concentrations from the demineralized matrix (Bennett et al., 2001; Lorget et al., 2000). Components of the exposed matrix are also chemoattractants for osteoblast lineage cells, including TGF-β, fibronectin, and demineralized collagen remnants (Delaisse et al., 2020).

2) **Osteoclast-secreted factors** promote osteoblastogenesis. Non-resorbing osteoclasts secrete afamin (AFM), which stimulates the migration of osteoblasts progenitors (Kim et al., 2012). Differentiation is promoted by factors secreted by active and inactive osteoclasts, including Wnt10b, the Wnt signaling modifier Collagen Triple Helix Repeat Containing 1 (CTHRC1), BMP-6, the signaling cytokine cardiotrophin-1 (CT-1), the lipid mediator sphingosine-1-phosphate (S1P), and complement factor 3a (C3a) (Sims & Martin, 2020). Reversal cell differentiation into osteoblasts is also promoted by their update of osteoclast-expressed tartrate-resistant acid phosphatase (TRAP) (Abdelgawad et al., 2016). As discussed, **osteoblast-secreted factors** also regulate the timing of this reversal, releasing factors that promote (M-CSF, RANKL, Wnt5a) or inhibit (OPG, Wnt16, FasL) osteoclast differentiation or survival.
3) **Membrane-expressed factors** connect osteoclasts and adjacent reversal cells, triggering bi-directional signaling that regulates the timing of reversal. Short cytoplasmic extensions facilitate this direct cell–cell contact (Abdelgawad et al., 2016). Differentiation of one cell type is promoted, and differentiation of the opposing cell type is suppressed, when their membranes connect via ephrin, neutrophilin, plexin, and semaphorin factors, and various Notch receptors (Delta-like, Jagged) (Abdelgawad et al., 2016; Plotkin & Bruzzaniti, 2019; Sims & Martin, 2020).
4) **Topographical changes** in osteoprogenitors sense the size and shape of resorption pits, initiating bone formation once they have filled the pit (Sims & Martin, 2020). Reversal cells are flatter closer to resorbing osteoclasts. Closer to osteoblasts, reversal cells are more closely packed, attaining a cuboidal shape. A critical cell density is required to initiate bone formation. Resorption will continue as long as reversal cells and osteoblasts are still packing to this critical cell density (Abdelgawad et al., 2016; Andersen et al., 2013; Lassen et al., 2017). In osteoporotic bone, reversal cells are more frequently arrested, failing to reach a critical packing density or express markers of osteoblast differentiation. This interruption of the remodeling cycle may contribute to osteoporotic bone loss (Andersen et al., 2013, 2014; Jensen et al., 2015). Age-associated increases in porosity are also driven not by increased pore density, but by the enlargement of pores through delayed or arrested coupling (Andreasen et al., 2018).

1.4.5 Formation

Osteoblastogenesis is promoted by coupling factors exchanged during reversal. Mesenchymal stem cells express Runt-related transcription factor 2 (RUNX2, also called Cbfa1) as they commit to become pre-osteoblasts. The timing of this RUNX2 expression is regulated in part by coupling factors, such as those released from the resorbed matrix (TGF-β, BMPs, IGFs) or expressed by osteoclasts or osteoblast lineage cells during reversal (Wnt, Notch). As pre-osteoblasts mature, they express both RUNX2 and Osterix (OSX, also called Sp7). RUNX2 is a transcription factor, with products including fibroblast growth factors (FGFs), hedgehog signaling, parathyroid hormone related peptide (PTHrP), distal less homeobox 5 (Dlx5), and Wnts. These pre-osteoblast ligands and receptors promote proliferation and maturation into cells expressing RUNX2-OSX (Komori, 2019; Long, 2012).

Wnt signaling is key in both RUNX2 and OSX expression. **Canonical Wnt signaling** allows the protein β-catenin to escape its typical degradation within the pre-osteoblast cytoplasm and translocate to the nucleus. There, β-catenin complexes with TCF/LEF (T-cell factor/Lymphocyte enhancer factor) and CREB (Cyclic AMP-responsive element-binding protein) to target gene promoters that maintain transcription factors for osteoblast differentiation (e.g., RUNX2, OSX, Dlx5) while suppressing other differentiation pathways for the mesenchymal stem cells (PPARϒ, C/EBPα). Canonical Wnt signaling also self-regulates by transcribing genes for factors that antagonize or degrade its own receptors, including sclerostin (SOST) and ubiquitin E3 ligases (ZNRF3, RNF43) (Houschyar et al., 2019; Maeda et al., 2019). Topographical changes in the resorption pit during coupling may help regulate the timing of canonical Wnt signaling. As the pit is resorbed, localized strain increases (McNamara et al., 2006; Sims & Martin, 2020). Fluid flow mechanotransduction stimulates local osteocytes to transcribe and release OPG to halt bone resorption, and Wnt3a to initiate bone formation, while downregulating the Wnt antagonist SOST (Pivonka et al., 2018; Verbruggen & McNamara, 2018). As the resorption pit fills with proliferating osteoblasts, this strain is relieved and the stimulus for bone formation declines (Sims & Martin, 2020).

Noncanonical Wnt signaling pathways are also initiated by Wnt ligands, but do not involve β-catenin in the cell signaling cascade. Pathways such as the Wnt/Ca^{2+} pathway and planar cell polarity pathway promote structural changes in the differentiating osteoblast, including rearrangement of the cytoskeleton, cell polarity, cell attachment and migration, and energy production (Houschyar et al., 2019; Maeda et al., 2019; Pivonka et al., 2018; Plotkin & Bruzzaniti, 2019). Osteoblasts secrete a large amount of type I collagen and non-collagenous proteins including osteocalcin and alkaline phosphatase. Before mineralization, this deposited organic matrix is referred to as osteoid. Osteoblast cell polarity means that the cell membrane in contact with the bone surface extends numerous cytoplasmic processes into this osteoid (Long, 2012).

1.4.6 Mineralization

About five to ten days after osteoid deposition (Parfitt, 1987), the osteoid matrix begins to mineralize through deposition of hydroxyapatite, which is a crystalline phase of calcium phosphate (Murshed, 2018). These nanoscale crystals nucleate in the gap zones of collagen fibrils (Bala et al., 2013). While the precise mechanism is unknown, it is possible that the densely packed collagen induces its own hydroxyapatite nucleation by admitting calcium and phosphate ions, but not larger proteins that inhibit crystal growth (Murshed, 2018; Price et al., 2009; Toroian et al., 2007). The crystals first elongate into their typical long, thin platelet shape, and then begin to thicken (Fratzl et al., 1993). This primary mineralization establishes 50–70% of maximum matrix mineralization within a few days (Amprino & Engstrom, 1952; Marotti et al., 1972). Due to this lag, an osteoid seam forms between the matrix depositing osteoblasts and mineralized bone. In vivo labels of bone formation, such as tetracycline, are incorporated at the mineralization front where mineralization begins. Osteoid seam thickness is typically about 7µm (Pirok et al., 1966), but is increased in mineralization disorders such as rickets or osteomalacia (McCarthy, 2016), X-linked hypophosphatemia (Teitelbaum, 1980), and osteogenesis imperfecta (Rauch et al., 2000).

Secondary mineralization, during which the mineral crystal matures, has been measured as complete within one to three years (Bala et al., 2010; Fuchs et al., 2008; Huang et al., 2002; Marotti et al., 1972) or alternatively continuing up to thirty years (Akkus et al., 2003; Handschin & Stern, 1994), although this may also reflect hyper-mineralization of vascular and cellular voids within bone tissue (Fuchs et al., 2008). During secondary mineralization, the mineral increases in crystallinity, which refers to the size and perfection (order) of the crystal lattice. The apatitic core of the mineral is surrounded by a hydrated layer, which contains hydrogen phosphate, phosphate, and carbonate ions exchanged during mineral maturation. As the crystal matures, its apatitic core grows, and its hydrated layer shrinks. To fit within the matrix, crystal growth stops at about 95% of the theoretical maximum, at which point the matrix is said to be "fully mineralized" (Bala et al., 2013). Undersized crystals are less able to bear mechanical loads, as observed in diabetes (Einhorn et al., 1988) and osteogenesis imperfecta (Fratzl-Zelman et al., 2015), while oversized crystals are brittle (Gao, 2006), as has sometimes been observed with osteoporosis (Burnell et al., 1982).

Systemic levels of the contributing calcium and phosphate ions are one determinant of mineralization, and are regulated in part by vitamin D (1α,25-dihydroxyvitamin D or 1,25(OH)2D), which increases the efficiency of calcium and phosphate absorption in the small intestine (Holick, 2007). While calcium serum levels are tightly regulated due to their involvement in many physiological processes, mineralization defects have been linked to decreased phosphate levels (hypophosphatemia), marked by elevations in FGF23 (fibroblast growth factor 23), osteopontin, and parathyroid hormone (Murshed, 2018).

1.4.7 Osteoblast Fates

Whereas osteoclasts undergo apoptosis after completing their resorptive task, osteoblasts have three potential fates. Following bone formation, approximately 65% of osteoblasts undergo apoptosis, 10–30% are entombed in the forming bone matrix and differentiate into osteocytes, and the rest remain on the newly formed bone surface as bone lining cells (Franz-Odendaal et al., 2006). Bone lining cells cover quiescent bone surfaces and can convert to osteoblasts for subsequent remodeling as previously described. Bone lining cells are also connected to the extended dendrites of underlying osteocytes embedded in the matrix. Gap junctions facilitate communication and metabolite exchange between these osteoblast lineage cells above and below the bone surface (Guo & Bonewald, 2009; Robling & Bonewald, 2020).

The specific mechanism of osteocyte differentiation is unclear but may be linked to the condensation of osteoblasts at the remodeling site. This cell density halts osteoblast proliferation and produces hypoxic conditions known to induce osteocyte differentiation (Kim & Adachi, 2019). Differentiating osteocytes express podoplanin (E11/gp38), which rearranges the actin cytoskeleton of the osteoblast for dendrite formation and branching (Nagai et al., 2021). These dendrites extend and contract until they connect with a previously embedded osteocyte, while retaining connections with osteoblasts and bone lining cells on the remodeling surface (Robling & Bonewald, 2020). Early embedded "osteoid osteocytes" promote mineralization through expression of PHEX (phosphate regulating neutral endopeptidase on the chromosome X) and DMP1 (dentin matrix protein 1), which regulate circulating FGF23, with PHEX additionally binding mineralization-inhibiting osteopontin as a substrate. Mature osteocytes subsequently slow mineralization by expressing FGF23, which downregulates phosphate resorption in the kidney, and MEPE, a mineralization inhibitor (Murshed, 2018; Robling & Bonewald, 2020).

1.5 Interpreting Remodeling Activity from Histomorphometry

1.5.1 General Guidelines for Applying Histomorphometric Methods

Skeletal biologists employ histomorphometry in the analysis of both modern and ancient bone, particularly for discerning aging and pathology (reviewed in Stout et al., 2019). In short, histomorphometry is the quantification of microstructures in the skeletal tissue or their characteristics. Histomorphometric analysis of bone provides quantitative information, such as bone turnover (remodeling) and microarchitecture, which cannot be obtained from other approaches. Anthropologists have widely applied this technique for histological estimation of age at death (reviewed in Andronowski & Cole, 2021) and differentiation of human from non-human bone (Dominguez & Crowder, 2012; Hillier & Bell, 2007; Mulhern & Ubelaker, 2001, 2003, 2012), in both bioarchaeological and forensic contexts.

The quantification of bone turnover, microarchitecture, and static and dynamic cell activity is performed through measuring and counting of structures to characterize changes in bone histomorphology. Anthropology, as a discipline, is a relative newcomer in the quantification of bone structure and organization. As a result, analyses are often performed without full consideration of the underlying processes governing bone biology. To reliably evaluate bone at the histological level the analyst must interpret histomorphometric results in the context of bone biology. Anthropologists should also follow the standardization of nomenclature, symbols, and unit released by the ASBMR to assist in a universal understanding of bone histomorphometry and eliminate ambiguity in the reporting of data (Dempster et al., 2013).

To ensure the proper application of any method utilizing histomorphometric techniques it is important to consult the methods from the original publication. The proper magnification, sampling area, and other specified procedures must be followed as prescribed for each method. Ambiguous or subjective definitions of microstructures, such as intact versus fragmentary osteons, result in higher inter-observer error (Morris & Crowder et al. 2005). Crowder and colleagues (2022) provide definitions and descriptions for long used histological structures to reduce the amount of misclassification and observer error in analysis. Equations for histological age-at-death estimation are also specific to certain skeletal elements and sampling techniques. Intraskeletal variation in mechanical loading alters the local remodeling rate and microstructural size, which are components of most equations. Additionally, these equations reflect the skew of age and sex distributions in the original sample population, and accuracy may vary according to geographic location or population affinity (Andronowski & Cole, 2021).

Evaluating histological structures is a slow process and requires constant manipulation of the microscope's fine focus and light. Both polarized and non-polarized light should be used to provide full resolution of structures during assessment. The application of automated techniques may be beneficial and reduce evaluation time. However, current technology may not allow for the complete dismissal of direct viewing through the microscope. In practice, because of the effects of depth of field in light microscopy, reversal lines and lamellar patterns are more easily and reliably identified if the microscopist uses the fine focus to focus up and down when viewing sections (Figure 1.4). This procedure is well known by microscopists and provides some measure of three-dimensionality to microscopic structures. The lack of depth of field for digitally captured images, therefore, can make the

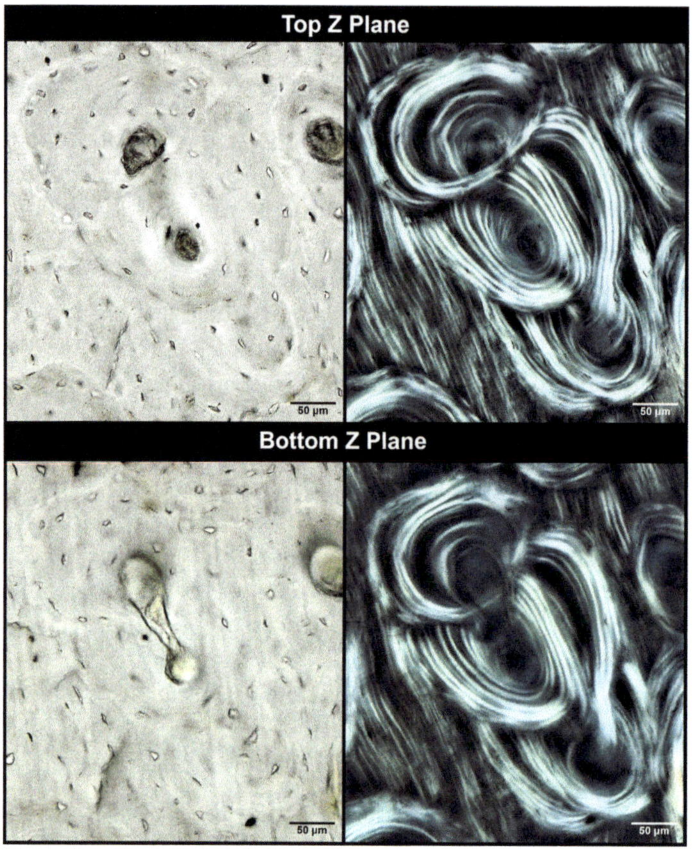

Figure 1.4 Microscopic images taken with brightfield (left) and circularly polarized light (right) show a transverse connection between adjacent canals that appears while focusing down through the field of view. A static digital image of the top Z plane would make it difficult to discern whether these osteons were connected.

accurate differentiation of histomorphological features such as reversal lines difficult. This is why we recommend that analyses be performed directly through the microscope as well as from digital images.

1.5.2 Osteon Types

In cortical bone, BMUs are observable histomorphologically as osteons (Figures 1.5 and 1.6), which are easily distinguished in the unstained section.[2] Osteons formed by remodeling are known as secondary osteons, also called Haversian systems. Primary osteons (Figure 1.5) are similar in appearance, but result from modeling, where lamellae surround a blood vessel lying on the bone surface. The key to accurately distinguishing secondary osteons from other similar structures, such as primary osteons and vascular canals, is the presence of a well-defined reversal line. This hypercalcified cement line is deposited by reversal cells and demarcates the BMU's reversal from resorption to formation. Fragmentary osteons that lack an intact Haversian canal because of their partial resorption by subsequent remodeling activity can also be distinguished from interstitial primary lamellar bone by the presence of a reversal line, and the concentric nature of their lamellae and osteocyte lacunae.

Secondary osteon types lend insight to local remodeling history. **Type I** osteons are common secondary osteons, and the most prevalent type in adult human bone (Figure 1.5). **Type II** or embedded osteons are inset with separate smaller osteons, with their own reversal lines (Figure 1.6). This suggests intra-osteonal remodeling, meaning remodeling within the bounds of an existing secondary osteon (Ericksen, 1991). Type II osteons have been suggested as indicators of remodeling to maintain mineral homeostasis, as with aging or dietary variation (Raguin & Drapeau, 2020). **Double-zonal** osteons (Figure 1.6) contain a hypercalcified ring within their concentric lamellae, representing an arrest line during the remodeling process, potentially due to the physiological stress of disease or nutrition. Double-zonal osteons were originally identified with microradiographs, but can also be

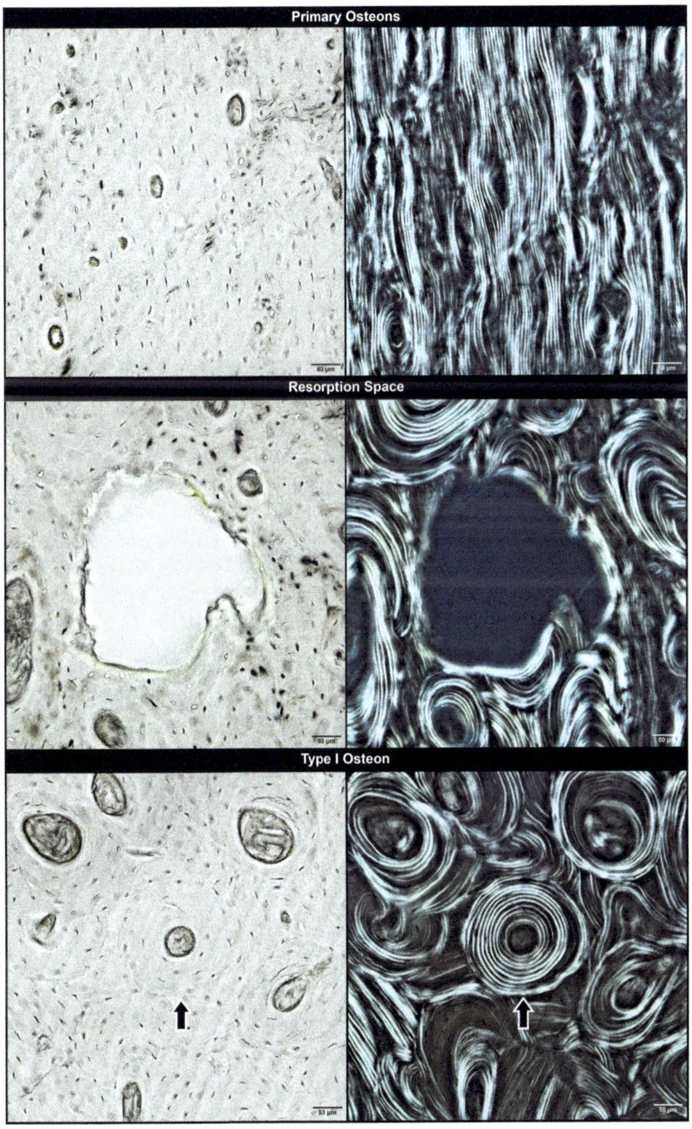

Figure 1.5 Microscopic images taken with brightfield (left) and circularly polarized light (right) show primary osteons with no cement line (top), a resorption space with a scalloped border (middle), and a Type I (secondary) osteon with an arrow indicating the cement line (bottom).

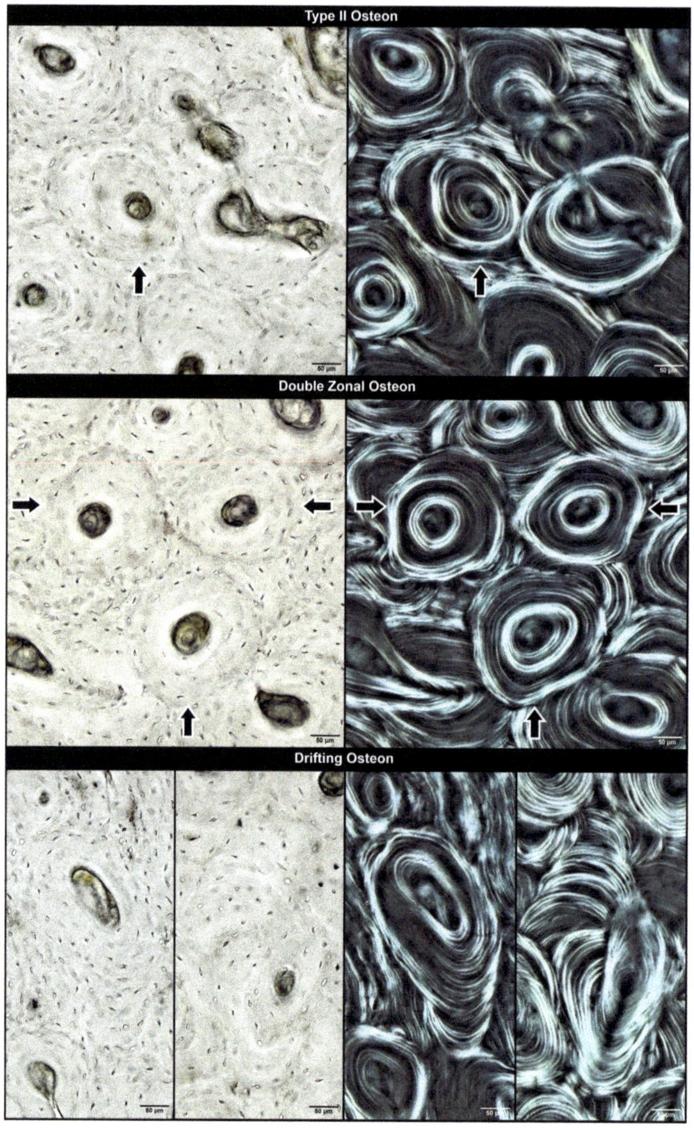

Figure 1.6 Microscopic images taken with brightfield (left) and circularly polarized light (right) show additional types of secondary osteons, including a Type II osteon with an offset, internal cement line (top), a double-zonal osteon with a hyper-mineralized ring within its concentric lamellae (middle), and two examples of drifting osteons with tails of semicircular lamellae (bottom).

observed with backscattered electrons in scanning electron microscopy, or even with linearly polarized light microscopy (Raguin & Streeter, 2018). Importantly for distinguishing these types with polarized light microscopy, the lamellae of double-zonal osteons are aligned, as they represent an arrest in the same remodeling cycle. In contrast, the lamellae and cement lines of type II osteons should be slightly offset between the internal and surrounding osteons, as they represent separate remodeling cycles.

Drifting osteons (Figure 1.6) display a tail of semicircular lamellae in cross-section. This results from lateral resorption on one side, and lateral formation on the other, angling the osteon as the BMU tunnels longitudinally (Robling & Stout, 1999). Models suggest that this results from a steep lateral gradient of strain, inducing resorption on the side with less strain, and formation on the side with more strain (van Oers et al., 2008). Osteons have long been observed to generally drift toward the lowest strains at the endosteum (Burton et al., 1989; Cohen & Harris, 1958; Epker & Frost, 1965; Koltze, 1951; Maggiano, 2011). However, strain gradients would need to be highly localized and frequently changing to precisely "steer" the drift. Most drifting osteons do not continually drift in a single direction, but change the angle of direction slightly or even significantly (Robling & Stout, 1999). Microcracks might provide this transient localized stimulus, causing the BMU to drift toward sites targeted for remodeling (Cooper et al., 2011). Drifting osteons are frequently observed in subadults (Epker & Frost, 1965; Streeter, 2010) but their frequency of formation significantly declines with age (Coutelier, 1976; Sedlin et al., 1963; Streeter, 2011).

1.5.3 Intraskeletal Variation in Remodeling Activity

Equations for histological age at death typically involve measuring the areal fraction, counts, or population density of secondary osteons that accumulate with remodeling over time (Andronowski & Cole, 2021; Gocha et al., 2018). These microstructural products of remodeling depend on remodeling rate, which is stimulated by the bone tissue's mechanical loading environment. Variation in this localized mechanical strain produces intraskeletal variation between and within skeletal elements, and even within the same cross-sectional plane. Consequently, it is essential to choose the correct skeletal element and carefully replicate the cross-sectional sampling technique associated with an equation for histological estimation of age at death.

Prediction of local remodeling rate is complex because remodeling can be activated under low strain (disuse-mode), moderate strain (conservation-mode), and high strain (targeted). For example, both human and quadruped ribs have been reported to have higher remodeling rates than some weight-bearing limb bones, commonly the femur (Cho & Stout, 2011; Fahy et al., 2017; Mulhern, 2000; Mulhern & Van Gerven, 1997; Skedros et al., 2003; Stout, 1984; Vajda et al., 1999; Wilson et al., 1998). The higher remodeling rate in the rib has been attributed to its low-strain permissiveness to stochastic remodeling (Frost, 1969) or resorption (Skedros et al., 2013), to elevated targeted remodeling of microdamage occurring more frequently under continual loading from respiration (Cho & Stout, 2011; Skedros et al., 2013), or to a potentially earlier effective age of adult compacta (Mulhern, 2000; Mulhern & Van Gerven, 1997).

Variation between skeletal elements in remodeling rate contributes to intraskeletal variation in mean tissue age (MTA), effective age of adult compacta, and the osteon population density (OPD) asymptote. Cortical osteons and osteon fragments observed in a bone cross-section do not represent BMUs that have accumulated since birth, but rather a later age characteristic of each bone called the effective age of adult compacta. Cortical drifts during growth remove some of the visible evidence of earlier BMU creations, osteons and their fragments, and the chronological age of an individual is usually greater than actual age of a given area of bone tissue. Based upon data from the use of in vivo tissue-time labeling, the MTA and effective age for the birth of adult compacta for the middle third of the rib is estimated to be approximately 12.5 years (Wu et al., 1970). The years over which the BSUs observed in the cortex of the rib have accumulated can be estimated by chronological age minus 12.5 years. It is not appropriate, therefore, to apply most currently available histological age estimation formulas derived primarily from adult skeletons to subadults.

Histological age-at-death estimation through secondary osteon accumulation is also limited in older individuals. Some existing BSUs are partially removed by the creation of new BMUs to produce what are referred to as osteon fragments, which are incorporated in some age estimation methods (Andronowski & Cole, 2021). However, the creation of a BMU can also completely remove an existing BSU, eliminating that evidence of past remodeling activity. After an individual's primary lamellar bone tissue has been completely remodeled, new secondary osteons necessarily remove existing secondary osteons. Consequently, OPD reaches an asymptote, no longer increasing with age in the cross-sectioned plane. Skeletal elements vary in the OPD required to reach an asymptote, depending on factors such as the remodeling activation rate, cortical area, and osteon size (Cho et al., 2002; Frost, 1987a; Jackes, 2000; Pfeiffer, 1992; Stout et al., 1996; Walker et al., 1994). The rib is reported to reach an asymptote at an OPD of about 30 osteons/mm^2 (Stout & Paine, 1994). The femur was reported by Frost (1987a) to accommodate an asymptote of about 50 osteons/mm^2. Crowder (2005) and Crowder and Pfeiffer (2010) estimated histological age at death for individuals of known age and sex using equations developed for the femur and rib. While the rib reached its OPD asymptote around 50 years, and increased twofold in inaccurate age estimation after 60 years, the paired femurs of individuals became inaccurate around 80 years (Crowder & Pfeiffer, 2010). It is unclear whether the OPD asymptote is reached earlier in the rib because it accumulates osteons more quickly due to its high remodeling rate, or because its smaller cortical area limits the potential density of osteons (Crowder, 2005; Crowder et al., 2016; Stout et al., 1996).

Elements with similar mechanical loading environments are more consistent in OPD and estimated age at death. Bones under minimal mechanical strain, such as cranial elements, have a low correlation between OPD or osteon areal fraction and age at death (Clarke, 1987; Cool et al., 1995; Curtis, 2004; Curtis & Nawrocki, 2010; Nawrocki, 2010). Midthoracic ribs do not significantly vary in OPD and estimated age at death (Crowder & Rosella, 2007). Weight-bearing bones such as the femur and tibia also develop similar OPD and areal fractions of remodeled bone (Drapeau & Streeter, 2006). Estimating age at death from both high-strain and low-strain skeletal elements of an individual may help compensate for this intraskeletal variation in remodeling (Stout et al., 1996).

Accuracy of histological age-at-death estimation also depends on replicating the method's placement of sampling fields (Gocha & Agnew, 2016; Stout & Gehlert, 1980). As previously discussed, cross-sections vary in mechanical strain. Strains are lowest near the endosteum due to proximity to the neutral axis, and regions tensed by bending experience lower strain than regions compressed by bending. In compressed regions, the microdamage incurred under higher strain elevates targeted remodeling, producing the higher OPD often observed in quadrupeds and humans (Lanyon et al., 1979; Mason et al., 1995; Mayya et al., 2013; Riggs et al., 1993; Skedros et al., 1994, 1997, 2009; Zedda et al., 2015). Conversely, tensed regions exhibit higher OPD in some species and skeletal elements, potentially due to combination with torsion or shear (Gocha & Agnew, 2016; Mason et al., 1995; Portigliatti Barbos et al., 1984; Skedros et al., 2001, 2009; Skedros, Kiser, Keenan, et al., 2011; Skedros, Kiser, & Mendenhall, 2011).

Previous observations of a higher remodeling rate in the tibia, compared to the femur, may have been confounded by such methodological differences in sampling field placement

(Evans, 1976; Evans & Bang, 1967). Significant variation has been observed in the femur between anterior, posterior, medial, and lateral regions in OPD (Gocha & Agnew, 2016) and percent osteonal bone (Drusini, 1987; Pfeiffer et al., 1995). OPD has also been observed to increase toward the endosteum in the femur (Gocha & Agnew, 2016; Skedros et al., 1994; Stout & Gehlert, 1982). Chan and colleagues (Chan et al., 2007) found that age at death was inconsistently estimated across five diaphyseal levels of the femur if the sampling field was placed outside of its anterior cortex. Similarly, Tibial OPD is more concentrated along its elongated anterior-posterior axis, as opposed to its medial-lateral axis (Drapeau & Streeter, 2006). Estimated age at death has also been observed to vary significantly along one centimeter intervals in the femur, tibia, and radius, but not humerus or fibula (Tersigni, 2005).

The sampling technique may also influence the OPD asymptote. Frost based the estimate for the femur on data from Kerley (1965), which sampled single microscope fields in the anterior, posterior, medial, and lateral cross-sections. Gocha (2014) conversely found no asymptote when counting every remodeling event in the femoral cross-section, accurately predicting age at death into the tenth decade. Age estimation for elderly individuals may also be strengthened by incorporating additional histomorphological features known to vary with age. For example, Stout and Streeter (2004, 2006) used osteon size and relative cortical area of the rib from a Mayan ruler to support a ninth decade of life age based upon inscriptions. New techniques for estimating histological age at death should consider combinations of histomorphometric variables, and innovative sampling techniques, which may extend beyond the OPD asymptote.

Some methods for histological age-at-death estimation additionally incorporate aspects of mean osteon size or shape (Andronowski & Cole, 2021; Gocha et al., 2018). Osteons tend to become smaller and more circular with age, even in elements that do not bear substantial body weight (Black et al., 1974; Britz et al., 2009; Burr et al., 1990; Currey, 1964; Ericksen, 1991; Evans, 1976; Goliath et al., 2016; Hennig et al., 2015; Iwamoto et al., 1979; Jowsey, 1964; Martin et al., 1980; Mulhern & Van Gerven, 1997; Ortner, 1975; Pfeiffer, 1998; Pfeiffer et al., 2006; Singh & Gunberg, 1970; Thompson, 1980; Thompson & Galvin, 1983; Tong et al., 2015; Watanabe et al., 1998; Yoshino et al., 1994). Smaller, more circular osteons in older age might reflect the higher strain on a thinning cortex (Hennig et al., 2015). Higher strain should repress the extent of resorption during remodeling (Abbott et al., 1996). These compact osteons could also be advantageous under high strain because they can pack together closely, creating a high density of cement lines to trap or deflect propagating microcracks (Dominguez & Agnew, 2019; van Oers et al., 2008). Conversely, these smaller osteons might merely be the remnant of a thinning cortex, where larger osteons are more likely to removed (Seeman, 2013). Intraskeletal patterning in osteon size and shape does not consistently reflect variation in strain environments. The weight-bearing human femur, which might be expected to form smaller osteons under high strain, actually forms larger osteons than the human rib (Goliath et al., 2016; Pfeiffer, 1998; Pfeiffer et al., 2006). Similarly, osteon circularity does not significantly vary between the femur, humerus, and rib (Britz et al., 2009; Crescimanno & Stout, 2012; Goliath et al., 2016). Cross-sectional regions under higher-strain compression would also be expected to form more compact osteons, which is observed in some species and skeletal elements, but not others (Keenan et al., 2017; Mason et al., 1995; Skedros et al., 1997, 2001, 2004; Yeni et al., 1997). Smaller cross-sectional areas may limit osteon size and elongation, even where the strain environment is more permissive to resorption (Dominguez & Agnew, 2016; Goliath et al., 2016).

Methods for histological age-at-death estimation occasionally quantify bone loss through cortical width, relative cortical area, and cortical porosity, pore density, or pore morphometry (Andronowski & Cole, 2021; Gocha et al., 2018). These metrics reflect age-associated endosteal and intracortical resorption. As early as the third decade of life, bone resorption begins to outpace bone formation due to a combination of reduced physical activity, sarcopenia (muscle wasting), declining sex steroids, and the senescence of bone cell sensitivity and function (Demontiero et al., 2012; Infante & Rodríguez, 2018). Cortical pores remain unfilled, and can converge with adjacent pores or the endosteum as they are further eroded, a process known as "trabecularization" (Andreasen et al., 2020; Zebaze & Seeman, 2015). Pathological alteration of bone remodeling often presents as an acceleration of bone resorption accompanied by the compensatory formation of disorganized woven bone, as we have previously reviewed in detail (Stout et al., 2019). When assessing histological age at death, methods should combine cortical porosity with osteonal evidence of remodeling, as the rate and extent of porous bone loss varies widely between individuals (Thomas et al., 2000, 2005). Declining cortical area due to porosity or endosteal erosion can impact the accuracy of OPD and percent osteonal bone, which use cortical area as a divisor (Pfeiffer et al., 1995, 2016). As previously discussed, cortical porosity tends to be elevated in cross-sectional regions under lower strain, increasing toward the endosteum and in regions under tension, as opposed to higher-strain compression. However, like osteon morphometry, cortical porosity does not clearly reflect intraskeletal strain patterns. The human midshaft femur, tibia, rib, and distal radius have been reported as histologically consistent in percent porosity, despite their varying strain environments (Cole, Stout, et al., 2022; Hunter & Agnew, 2016). For a detailed discussion of histological age estimation, refer to Chapter 8.

1.5.4 Inter-Skeletal Variation in Remodeling Activity

Histological age-at-death estimation should also consider whether the equations are differentiated by sex. Bone formation is promoted at the endosteum by estrogen and at the periosteum by testosterone (Turner et al., 1990). Age-associated bone loss is exacerbated in females due to declining estrogen in perimenopause, which increases resorption activation at the endosteum by 33% (Han et al., 1997). Studies of archaeological populations also suggest earlier bone loss in adult females, potentially associated with nutritional stress and nutrient deficiency during pregnancy and lactation (Cho & Stout, 2011; Ericksen, 1980; Martin & Armelagos, 1979, 1985). For a discussion of archaeological applications of histological analysis, see Chapter 11. Males exceed females in the cross-sectional area of long bones both during and after menopause (Feik et al., 2000; Schlecht et al., 2015). Males also retain or increase cortical area with age, due to periosteal bone apposition, in the femur, tibia, radius, and lumbar vertebral body (Burr & Martin, 1983; Martin & Atkinson, 1977; Mosekilde & Mosekilde, 1990; Ruff & Hayes, 1988). Smaller osteons observed in males, compared to females, have been hypothesized to reflect the higher strain of male body weight (Britz et al., 2009; Schlecht et al., 2012). In archaeological populations, smaller osteons in males have also been hypothesized to reflect cultural behaviors of more physically intensive subsistence (Burr et al., 1990; Mulhern, 2000; Mulhern & Van Gerven, 1997). However, some study populations show no significant sex difference in osteon size (Dominguez & Agnew, 2016; Dupras & Pfeiffer, 1996; Pfeiffer, 1998; Pfeiffer et al., 2006; Stout & Lueck, 1995).

Osteon size differences can impact osteon packing density and the extent of osteon overlap, which differentiates intact and fragmentary osteons. Gocha et al. (2018) noted that sex differences tend to be insignificant when considering total OPD, but variation does emerge when separating fragmentary and intact OPD.

Remodeling activity can also vary with population affinity. Mechanical stimulation of remodeling may vary with habitual physical behaviors, such as intensive foraging or farming compared to a more sedentary subsistence (Robling & Stout, 2003). Yet the comparative histomorphometry of archaeological and/or modern populations does not consistently reflect their variable levels of physical activity (Burr et al., 1990; Ericksen, 1976; Miszkiewicz & Mahoney, 2016; Pfeiffer, 1998; Pfeiffer et al., 2006; Streeter, 2010). Dietary differences and genetic factors may introduce confounding physiological effects. For example, a diet high in protein has been associated with increased remodeling and OPD (Richman et al., 1979; Thompson & Gunness-Hey, 1981). Remodeling activity can also appear artificially low in populations with later skeletal maturity, which increases the effective age of adult compacta and delays the recorded accumulation of osteons (Stout & Lueck, 1995). Application of a method outside of its reference population can increase inaccuracy in histological age-at-death estimation (Aiello & Molleson, 1993; Cho et al., 2002, 2006; Cho & Stout, 2011; Garcia-Donas et al., 2016; Karydi et al., 2022; Kranioti et al., 2020; Maggio & Franklin, 2019; Mnich et al., 2017; Pfeiffer et al., 2016; Thompson & Gunness-Hey, 1981; Watanabe et al., 1998).

1.5.5 Static Histomorphometry

Static histomorphometry allows for the quantification of histological structures being evaluated in a bone at a particular point in time. In most anthropological research and practical applications of bone histology methods (i.e., age estimation, human vs. nonhuman determination), static indices of bone microarchitecture are evaluated and calculated. Histological structures are manually or automatically measured through a variety of measurement or counting techniques. Typical variables (measurements) recorded include, but are not limited to, volume, length, perimeter, diameter, area, circularity, and osteon population density (osteon counts). As mentioned previously, the proper nomenclature should be followed when recording and reporting histological data. For example, the following is a list of commonly measured histological features for age estimation.

Total area of bone sampled (Sa.Ar.): Amount of cortical bone. Sa.Ar. is expressed in mm2, and can be determined either directly under the microscope using an appropriate eyepiece counting reticule, or from digital images using image analysis software.

Number of Intact Osteons (N.On.): A secondary osteon with an intact Haversian canal and the presence of a scalloped/irregular reversal line. Refer to Crowder et al. (2022) for an in-depth description of intact secondary osteons.

Number of Fragmentary Osteons (N.On.Fg.): A secondary osteon with a partially visible Haversian canal that has been breached either by a neighboring osteon or by a resorptive bay. This variable also includes remnants of preexisting secondary osteons that no longer contain a Haversian canal. Refer to Crowder et al. (2022) for a more in-depth description of intact secondary osteons.

Osteon Population Density in #/mm2 (OPD): This variable is the sum of N.On. and N.On.Fg. divided by Sa.Ar.

Relative Cortical Area (Ct.Ar./Tt.Ar.): The relative amount of cortical bone in cross-sectional area of bone, or the ratio of cortical bone area (Ct.Ar.) to total area (Tt.Ar.) of a rib cross-section.

Mean Osteonal Cross-sectional Area in mm2 (On.Ar.): the average area of bone contained within the cement lines of structurally complete osteons for each rib specimen. Osteons are considered structurally complete if their reversal lines were intact. Complete osteons with Haversian canals that deviated significantly from circular should be excluded. Mean area is calculated as the average cross-sectional area of 25–50 complete osteons per cross-section.

Several authors have used these variables, and others, to create algorithms in order to derive static indices of bone remodeling to be used for various purposes. Research by Stout and colleagues (1986, 1992, 1996) and Cho et al. (2002) have focused on developing equations to evaluate and predict age-related bone turnover in the rib. Frost (1987) developed an algorithm to estimate the number of missing osteons that correspond to observed osteon population densities (Stout and Paine 1994). The algorithm appeared promising to allow for more accurate age estimations when evaluating remodeling rates through static histomorphometry. Because new osteon creations eventually begin to remove all evidence of older turnover events, the algorithm allows for the estimation of the total number of osteon creations from the observed osteon population density of a cross-section. Frost provided a caveat, in that the lack of systematic studies to accurately determine the values for the variables within the algorithm, such as mean tissue age (MTA) and the fragment packing factor (k), limits its overall utility. Advances in dynamic histomorphometry will likely lead to a better understanding of how factors, such as biomechanics, affect the variables proposed by Frost. Dynamic histomorphometry, however, does not allow for the measure of bone remodeling rates averaged over the lifetime of an individual (Stout and Paine 1994) suggesting that a combination of static and dynamic research methods are needed to make advances in the field.

Basic static histomorphometry is typically performed through the preparation of undecalcified bone. Undecalcified[3] or decalcified bone samples may require special stains depending on the structures or processes that are being quantified. Traditionally used trichrome stains for paraffin-embedded tissue are not compatible with plastic-embedded bone (Villanueva and Mehr 1977), however, a number of stains are available for histochemistry of plastic embedded undecalcified bone (Anderson 1982). Various stains and staining techniques may be used to differentiate calcified bone from osteoid allowing the analyst to appreciate and measure the osteoid seam in order to quantify the amount of bone formation and resorption. Mineralized and unmineralized components of bone (i.e., osteoblasts, osteoclasts, and other cells) can easily be differentiated with the appropriate staining technique.[4] While evaluation of bone for age estimation or the determination of human vs. nonhuman may not require any staining protocol, the analysis of, for example, a fracture callus using specific stains to estimate the timing of the injury will assist in quantifying the area of the mineralized callus, area of cartilaginous tissue within the callus, and the area or proportion of fibrotic tissue within the callus.

1.5.6 Dynamic Histomorphometry

Dynamic histomorphometry allows for the quantification of more precise changes in cellular activity over time. In order to observe specific cellular responses, studies are performed in vitro or in vivo and the subject is exposed to various labeling techniques. One such procedure is to double label bone with a fluorochrome, such as tetracycline, and measure histomorphometric parameters involved in bone turnover and bone mineralization over a specific period of time (Bassett et al., 1980; Frost, 1969; Maggiano et al., 2006). While static histomorphometry averages bone remodeling activity over the tissue's age, dynamic histomorphometry records bone remodeling activity between the labeled timepoints. However, the skeletal remains typically studied by anthropologists, including bioarchaeologists, forensic anthropologists, and paleontologists, are unlikely to contain intravital tissue labels. Fortuitous labeling has been reported in archaeological samples through consumption of stored grains infected with tetracycline-producing streptomycetes (Bassett et al., 1980; Cook et al., 1989; Hummert & Van Gerven, 1982; Maggiano et al., 2006), and in modern samples using tetracycline as an antibiotic prior to death (Warner et al., 2022). In small animal models that form secondary osteons, such as rabbits, calcein injections can fluorescently label remodeling activity at multiple timepoints. This facilitates a more precise calculation of metrics such as activation frequency, osteon formation time, osteonal mineral apposition rate, and active remodeling centers. This approach can be combined with in-vivo synchrotron radiation micro-CT imaging of cortical pore systems at multiple timepoints, creating a four-dimensional imaging technique (Andronowski et al., 2022; Cole, Taylor, et al., 2022; Cooper et al., 2023; Harrison et al., 2020, 2022).

1.5.7 Reliability of Histological Analyses

Method reliability and validity are strongly emphasized in forensic anthropological research primarily owing to the evidentiary nature of analytical results and the courts may challenge the admissibility of a practitioner's planned testimony (Crowder 2005, Christensen and Crowder 2009, Lesciotto 2015). In the United States, the legal requirements for the evaluation of scientific evidence are set forth in *Daubert v. Merrel Dow Pharmaceuticals*, 113 S.Ct. 2786 (1993). Thus, admissibility hearings in jurisdictions that adopt the *Daubert* standard (opposed to the *Frye* standard) are referred to as *Daubert* hearings or challenges. According to the *Daubert* "criteria" for determining admissibility, a trial judge considers if a method has been adequately tested, has been subjected to peer review in a published journal, has known error rates, has standards for carrying out the provided with potential error rates, and enjoys general acceptance within the relevant scientific community (Daubert v. Merrel Dow Pharmaceuticals, 113 S.Ct. 2786, 1993). Following the 1993 *Daubert* ruling and subsequent court rulings expanding the scope of the *Daubert* criteria (General Electric Co v. Joiner 1997; Kumho Tire Co v. Carmichael 1998), many forensic disciplines have determined that there was a need to critically reevaluate some of the techniques and methods used in their examinations, as well as the validity of the underlying scientific theories (Christensen and Crowder, 2009). Methods with vague descriptions of samples, procedures, variables, or potential error rates should not be considered for use in the forensic setting (or any setting).

In the wake of the National Academy of Sciences (NAS) report on the current state of the Forensic Sciences, practitioners became concerned that some methods of examination may not meet these standards. Considering histological methods, at the time of the NAS report very few large validation studies had been performed; thus, error rates were unclear. Thus, new methods or clarification of current methods are likely needed to improve scientific standards within the field. As mentioned previously, the ASBMR standardized the nomenclature for hard-tissue histologists (Parfitt et al. 1987) to relieve confusion and to reduce the semantic barriers that they must cross within the field; however, few researchers in anthropology have adhered to the protocol.

Crowder (2005) identified specific methodological issues in histological analyses and suggested protocols to improve histological methods (for the estimation of age at death) so that they may become a conventional tool for anthropological analysis and address any evidentiary admissibility concerns. As mentioned previously, ambiguous and/or subjective variable definitions are a primary issue within histological methods that often result in inflated observer error values (Crowder et al. 2022). Recognizing and quantifying potential biases in histological analyses will support the development of best practices for histological analysis. Although various histological methods have been revised in the literature, the fundamental issues concerning the reliability and repeatability of these methods must be continually addressed. The application of qualitative and quantitative bone histology as a preferred method in anthropological analyses has been hindered by researchers' uncertainty regarding the accuracy and reliability of the methods, as well as, an incomplete understanding in the literature of the biological processes behind intracortical bone remodeling. For example, while microscopic age changes are considered to be universal the inconsistencies in reported accuracy of the methods when they are applied to individuals outside of the reference samples suggest that intrinsic and extrinsic biological factors, such as genetics and a wide range of suggested behaviors, have varying effects on bone microstructure. These issues can only be resolved through research designs that consider the dynamic nature of bone and evaluate results in the scope of skeletal biology.

1.6 Conclusions

Anthropological analysis of skeletal tissue should ideally employ both macroscopic and microscopic analyses, and multimodal microscopic and radiographic imaging techniques. Radiographic techniques (e.g., HR-pQCT, micro-CT, synchrotron radiation micro-CT) have been well established for three-dimensional imaging of voids within bone tissue, including trabecular architecture, cortical pores, and osteocyte lacunae. Yet microscopy remains the primary technique and the gold standard for quantifying the rate and frequency of bone remodeling, through the histomorphometry of secondary osteons. Future methodological development should consider how to expedite and automate histomorphometric analyses and translate their implementation to three- and four-dimensional imaging techniques.

Notes

1 The following three terms will be used throughout the chapter: **Histology** is the study of tissues, also called microscopal anatomy. **Histomorphology** is the morphology of tissues. **Histomorphometry** refers to quantitative histology, where histomorphological features are described in terms of mean dimensions of size or shape, spatial distribution, or number per unit area, analogous to osteometry.

2 The material nature of bone makes the use of histochemistry (staining) usually unnecessary for histomorphological and histomorphometric analyses of the bone samples available for anthropological age estimation.

3 Because the standard in most histology laboratories is to decalcify bone, the term “undecalcified” is often used rather than calcified bone.

4 Histochemical staining is essentially limited to relatively fresh specimens, such as those acquired through biopsy or at autopsy. Bone cells and osteoid tissue are not preserved in archaeological or unfixed skeletal remains, making the use of histological staining unnecessary except for enhancing certain histomorphological features such as cement lines.

References

Abbott, S., Trinkaus, E., & Burr, D. B. (1996). Dynamic bone remodeling in later Pleistocene fossil hominids. *American Journal of Physical Anthropology, 99*(4), 585–601. https://doi.org/10.1002/(SICI)1096-8644(199604)99:4<585::AID-AJPA5>3.0.CO;2-T

Abdelgawad, M. E., Delaisse, J.-M., Hinge, M., Jensen, P. R., Alnaimi, R. W., Rolighed, L., Engelholm, L. H., Marcussen, N., & Andersen, T. L. (2016). Early reversal cells in adult human bone remodeling: Osteoblastic nature, catabolic functions and interactions with osteoclasts. *Histochemistry and Cell Biology, 145*(6), 603–615. https://doi.org/10.1007/s00418-016-1414-y

Agerbaek, M. O., Eriksen, E. F., Kragstrup, J., Mosekilde, L., & Melsen, F. (1991). A reconstruction of the remodelling cycle in normal human cortical iliac bone. *Bone and Mineral, 12*(2), 101–112. https://doi.org/10.1016/0169-6009(91)90039-3

Agnew, A. M., Moorhouse, K., Kang, Y.-S., Donnelly, B. R., Pfefferle, K., Manning, A. X., Litsky, A. S., Herriott, R., Abdel-Rasoul, M., & Bolte, J. H. (2013). The response of pediatric ribs to quasi-static loading: Mechanical properties and microstructure. *Annals of Biomedical Engineering, 41*(12), 2501–2514. https://doi.org/10.1007/s10439-013-0875-6

Agnew, A. M., & Stout, S. D. (2012). Brief communication: Reevaluating osteoporosis in human ribs: The role of intracortical porosity. *American Journal of Physical Anthropology, 148*(3), 462–466. https://doi.org/10.1002/ajpa.22048

Aiello, L. C., & Molleson, T. (1993). Are microscopic ageing techniques more accurate than macroscopic ageing techniques? *Journal of Archaeological Science, 20*(6), 689–704. https://doi.org/10.1006/jasc.1993.1043

Akkus, O., Polyakova-Akkus, A., Adar, F., & Schaffler, M. B. (2003). Aging of microstructural compartments in human compact bone. *Journal of Bone and Mineral Research, 18*(6), 1012–1019. https://doi.org/10.1359/jbmr.2003.18.6.1012

Alam, I., Alkhouli, M., Gerard-O’Riley, R. L., Wright, W. B., Acton, D., Gray, A. K., Patel, B., Reilly, A. M., Lim, K.-E., Robling, A. G., & Econs, M. J. (2016). Osteoblast-specific overexpression of human WNT16 increases both cortical and trabecular bone mass and structure in mice. *Endocrinology, 157*(2), 722–736. https://doi.org/10.1210/en.2015-1281

Amprino, R., & Engstrom, A. (1952). Studies on x ray absorption and diffraction of bone tissue. *Acta Anatomica, 15*(1–2), 1–22. https://doi.org/10.1159/000140734

Anderson, Colin (1982). *Manual for the Examination of Bone*. Boca Raton, FL:CRC Press.

Andersen, T. L., Abdelgawad, M. E., Kristensen, H. B., Hauge, E. M., Rolighed, L., Bollerslev, J., Kjærsgaard-Andersen, P., & Delaisse, J.-M. (2013). Understanding coupling between bone resorption and formation: Are reversal cells the missing link? *The American Journal of Pathology, 183*(1), 235–246. https://doi.org/10.1016/j.ajpath.2013.03.006

Andersen, T. L., Hauge, E. M., Rolighed, L., Bollerslev, J., Kjærsgaard-Andersen, P., & Delaisse, J.-M. (2014). Correlation between absence of bone remodeling compartment canopies, reversal phase arrest, and deficient bone formation in post-menopausal osteoporosis. *The American Journal of Pathology, 184*(4), 1142–1151. https://doi.org/10.1016/j.ajpath.2013.12.005

Andersen, T. L., Sondergaard, T. E., Skorzynska, K. E., Dagnaes-Hansen, F., Plesner, T. L., Hauge, E. M., Plesner, T., & Delaisse, J.-M. (2009). A Physical Mechanism for Coupling Bone Resorption and Formation in Adult Human Bone. *The American Journal of Pathology, 174*(1), 239–247. https://doi.org/10.2353/ajpath.2009.080627

Andreasen, C. M., Bakalova, L. P., Brüel, A., Hauge, E. M., Kiil, B. J., Delaisse, J.-M., Kersh, M. E., Thomsen, J. S., & Andersen, T. L. (2020). The generation of enlarged eroded pores upon existing intracortical canals is a major contributor to endocortical trabecularization. *Bone, 130*, 115127. https://doi.org/10.1016/j.bone.2019.115127

Andreasen, C. M., Delaisse, J.-M., van der Eerden, B. C., van Leeuwen, J. P., Ding, M., & Andersen, T. L. (2018). Understanding age-induced cortical porosity in women: The accumulation and coalescence of eroded cavities upon existing intracortical canals is the main contributor. *Journal of Bone and Mineral Research*, *33*(4), 606–620. https://doi.org/10.1002/jbmr.3354

Andronowski, J. M., & Cole, M. E. (2021). Current and emerging histomorphometric and imaging techniques for assessing age-at-death and cortical bone quality. *WIREs Forensic Science*, *3*(2), e1399. https://doi.org/10.1002/wfs2.1399

Andronowski, J. M., Cole, M. E., Davis, R. A., Schuller, A. J., Tubo, G. R., LaMarca, A. R., & Taylor, J. T. (2022). Longitudinal Effects of Prolonged Opioid Use on Cortical Bone Remodeling in a Rabbit Model: Part II—Intraskeletal Histological Variability and Regional Differences. *Proceedings of the American Academy of Forensic Sciences, 28*.

Atkinson, P. J. (1965). Changes in resorption spaces in femoral cortical bone with age. *The Journal of Pathology and Bacteriology*, *89*(1), 173–178. https://doi.org/10.1002/path.1700890118

Bala, Y., Farlay, D., & Boivin, G. (2013). Bone mineralization: From tissue to crystal in normal and pathological contexts. *Osteoporosis International*, *24*(8), 2153–2166. https://doi.org/10.1007/s00198-012-2228-y

Bala, Y., Farlay, D., Delmas, P. D., Meunier, P. J., & Boivin, G. (2010). Time sequence of secondary mineralization and microhardness in cortical and cancellous bone from ewes. *Bone*, *46*(4), 1204–1212. https://doi.org/10.1016/j.bone.2009.11.032

Baron, R. (1977). Importance of the intermediate phases between resorption and formation in the measurement and understanding of the bone remodeling sequence. In P. J. Meunier (Ed.), *Bone histomorphometry: Second international workshop Lyon* (pp. 179–183). Armour Montagu.

Bassett, E. J., Keith, M. S., Armelagos, G. J., Martin, D. L., & Villanueva, A. R. (1980). Tetracycline-labeled human bone from ancient Sudanese Nubia (A.D. 350). *Science (New York, N.Y.)*, *209*(4464), 1532–1534. https://doi.org/10.1126/science.7001623

Batoon, L., Millard, S. M., Raggatt, L. J., Wu, A. C., Kaur, S., Sun, L. W. H., Williams, K., Sandrock, C., Ng, P. Y., Irvine, K. M., Bartnikowski, M., Glatt, V., Pavlos, N. J., & Pettit, A. R. (2021). Osteal macrophages support osteoclast-mediated resorption and contribute to bone pathology in a postmenopausal osteoporosis mouse model. *Journal of Bone and Mineral Research*, *36*(11), 2214–2228. https://doi.org/10.1002/jbmr.4413

Bellido, T. M., & Gallant, K. M. H. (2019). Hormonal effects on bone cells. In D. B. Burr & M. R. Allen (Eds.), *Basic and Applied Bone Biology* (pp. 299–313). Academic Press.

Bellido, T. M., Plotkin, L. I., & Bruzzaniti, A. (2019). Bone cells. In D. B. Burr & M. R. Allen (Eds.), *Basic and Applied Bone Biology* (2nd ed., pp. 37–55). Academic Press.

Bennett, B. D., Alvarez, U., & Hruska, K. A. (2001). Receptor-operated osteoclast calcium sensing. *Endocrinology*, *142*(5), 1968–1974. https://doi.org/10.1210/endo.142.5.8125

Bentolila, V., Boyce, T. M., Fyhrie, D. P., Drumb, R., Skerry, T. M., & Schaffler, M. B. (1998). Intracortical remodeling in adult rat long bones after fatigue loading. *Bone*, *23*(3), 275–281. https://doi.org/10.1016/s8756-3282(98)00104-5

Biewener, A. A., & Bertram, J. E. (1993). Mechanical loading and bone growth in vivo. In B. K. Hall (Ed.), *Bone, Volume VII* (pp. 1–36). Boca Raton, FL: CRC Press.

Black, J., Mattson, R., & Korostoff, E. (1974). Haversian osteons: Size, distribution, internal structure, and orientation. *Journal of Biomedical Materials Research*, *8*(5), 299–319. https://doi.org/10.1002/jbm.820080512

Bousson, V., Meunier, A., Bergot, C., Vicaut, E., Rocha, M. A., Morais, M. H., Laval-Jeantet, A. M., & Laredo, J. D. (2001). Distribution of intracortical porosity in human midfemoral cortex by age and gender. *Journal of Bone and Mineral Research*, *16*(7), 1308–1317. https://doi.org/10.1359/jbmr.2001.16.7.1308

Breyer, S. G. & Supreme Court of The United States. (1998). *U.S. Reports: Kumho Tire Co. v. Carmichael, 526 U.S. 137*. [Periodical] Retrieved from the Library of Congress, https://www.loc.gov/item/usrep526137/

Britz, H. M., Thomas, C. D., Clement, J. G., & Cooper, D. M. (2009). The relation of femoral osteon geometry to age, sex, height and weight. *Bone*, *45*(1), 77–83. https://doi.org/10.1016/j.bone.2009.03.654

Buenzli, P. R., & Sims, N. A. (2015). Quantifying the osteocyte network in the human skeleton. *Bone*, *75*, 144–150. https://doi.org/10.1016/j.bone.2015.02.016

Burnell, J. M., Baylink, D. J., Chestnut, C. H., Mathews, M. W., & Teubner, E. J. (1982). Bone matrix and mineral abnormalities in postmenopausal osteoporosis. *Metabolism: Clinical and Experimental*, *31*(11), 1113–1120. https://doi.org/10.1016/0026-0495(82)90161-5

Burr, D. B. (2002). Targeted and nontargeted remodeling. *Bone*, *30*, 2–4.

Burr, D. B., & Akkus, O. (2014). Chapter 1—bone morphology and organization. In D. B. Burr & M. R. Allen (Eds.), *Basic and Applied Bone Biology* (pp. 3–25). Academic Press. https://doi.org/10.1016/B978-0-12-416015-6.00001-0

Burr, D. B., & Martin, R. B. (1983). The effects of composition, structure and age on the torsional properties of the human radius. *Journal of Biomechanics*, *16*(8), 603–608. https://doi.org/10.1016/0021-9290(83)90110-0

Burr, D. B., Ruff, C. B., & Thompson, D. D. (1990). Patterns of skeletal histologic change through time: Comparison of an archaic native American population with modern populations. *The Anatomical Record*, *226*(3), 307–313. https://doi.org/10.1002/ar.1092260306

Burton, P., Nyssen-Behets, C., & Dhem, A. (1989). Haversian bone remodelling in human fetus. *Acta Anatomica*, *135*(2), 171–175. https://doi.org/10.1159/000146748

Cardoso, L., Herman, B. C., Verborgt, O., Laudier, D., Majeska, R. J., & Schaffler, M. B. (2009). Osteocyte apoptosis controls activation of intracortical resorption in response to bone fatigue. *Journal of Bone and Mineral Research*, *24*(4), 597–605. https://doi.org/10.1359/jbmr.081210

Carpenter, R. D., & Carter, D. R. (2008). The mechanobiological effects of periosteal surface loads. *Biomechanics and Modeling in Mechanobiology*, *7*(3), 227–242. https://doi.org/10.1007/s10237-007-0087-9

Carter, D., & Beaupré, D. (2001). *Skeletal Function and Form: Mechanobiology of Skeletal Development, Aging, and Regeneration*. Cambridge University Press.

Chan, A. H. W., Crowder, C. M., & Rogers, T. L. (2007). Variation in cortical bone histology within the human femur and its impact on estimating age at death. *American Journal of Physical Anthropology*, *132*(1), 80–88. https://doi.org/10.1002/ajpa.20465

Chen, M., Li, Y., Huang, X., Gu, Y., Li, S., Yin, P., Zhang, L., & Tang, P. (2021). Skeleton-vasculature chain reaction: A novel insight into the mystery of homeostasis. *Bone Research*, *9*(1), 1–20. https://doi.org/10.1038/s41413-021-00138-0

Cho, H., & Stout, S. D. (2011). Age-associated bone loss and intraskeletal variability in the Imperial Romans. *Journal of Anthropological Sciences*, *89*, 109–125. https://doi.org/10.4436/jass.89007

Cho, H., Stout, S. D., & Bishop, T. A. (2006). Cortical bone remodeling rates in a sample of African American and European American descent groups from the American Midwest: Comparisons of age and sex in ribs. *American Journal of Physical Anthropology*, *130*(2), 214–226. https://doi.org/10.1002/ajpa.20312

Cho, H., Stout, S. D., Madsen, R. W., & Streeter, M. A. (2002). Population-specific histological age-estimating method: a model for known African-American and European-American skeletal remains. *Journal of Forensic Sciences*, *47*(1), 12–18. http://doi.org/10.1520/JFS15199J

Christensen, A. M. and Crowder, C. M. (2009). Evidentiary standards for forensic anthropology. *Journal of Forensic Sciences*, 54(6), 1211–1216.

Clarke, D. F. (1987). *Histological and Radiographic Variation in the Parietal Bone in a Cadaveric Population* [Ph.D. Dissertation]. The University of Queensland.

Cohen, J., & Harris, W. H. (1958). The three-dimensional anatomy of haversian systems. *The Journal of Bone and Joint Surgery. American Volume*, *40-A*(2), 419–434.

Cole, M. E., Stout, S. D., Dominguez, V. M., & Agnew, A. M. (2022). Pore Extractor 2D: An ImageJ toolkit for quantifying cortical pore morphometry on histological bone images, with application to intraskeletal and regional patterning. *American Journal of Biological Anthropology, 179*(3), 365–385. https://doi.org/10.1002/ajpa.24618

Cole, M. E., Taylor, J. T., Hassan, M., & Andronowski, J. M. (2022). OsteoFlo: A FIJI / ImageJ toolkit for semi-automated identification and characterization of fluorescently labeled secondary osteons. *Proceedings of the American Academy of Forensic Sciences, 28*, 205.

Cook, M., Molto, E., & Anderson, C. (1989). Fluorochrome labelling in Roman period skeletons from Dakhleh Oasis, Egypt. *American Journal of Physical Anthropology, 80*(2), 137–143. https://doi.org/10.1002/ajpa.1330800202

Cool, S. M., Hendrikz, J. K., & Wood, W. B. (1995). Microscopic age changes in the human occipital bone. *Journal of Forensic Sciences, 40*(5), 789–796.

Cooper, D. M. L., Erickson, B., Peele, A. G., Hannah, K., Thomas, C. D. L., & Clement, J. G. (2011). Visualization of 3D osteon morphology by synchrotron radiation micro-CT. *Journal of Anatomy, 219*(4), 481–489. https://doi.org/10.1111/j.1469-7580.2011.01398.x

Cooper, D. M. L., Harrison, K. D., Hiebert, B. D., King, G. A., Panahifar, A., Zhu, N., Swekla, K. J., Pivonka, P., Chapman, L. D., & Arnason, T. (2023). Daily administration of parathyroid hormone slows the progression of basic multicellular units in the cortical bone of the rabbit distal tibia. *Bone, 176*, 116864. https://doi.org/10.1016/j.bone.2023.116864

Coutelier, L. (1976). Le remaniement interne de l'os compact chez l'enfant. *Bulletin de l'Association Des Anatomistes, 60*, 95–110.

Cowin, S. C., Moss-Salentijn, L., & Moss, M. L. (1991). Candidates for the mechanosensory system in bone. *Journal of Biomechanical Engineering, 113*(2), 191–197. https://doi.org/10.1115/1.2891234

Crescimanno, A., & Stout, S. D. (2012). Differentiating fragmented human and nonhuman long bone using osteon circularity. *Journal of Forensic Sciences, 57*(2), 287–294. https://doi.org/10.1111/j.1556-4029.2011.01973.x

Crowder, C. (2005). *Evaluating the use of quantitative bone histology to estimate adult age at death* [Ph.D. Dissertation, University of Toronto]…

Crowder C, Dominguez VM, Heinrich J, Pinto D, Mavroudas S. Analysis of histomorphometric variables: Proposal and validation of osteon definitions. *J Forensic Sci.* 2022 Jan;67(1):80–91. doi: 10.1111/1556-4029.14949. Epub 2021 Nov 25. PMID: 34821385.

Crowder, C., & Pfeiffer, S. (2010). The application of cortical bone histomorphometry to estimate age-at-death. In K. E. Latham & M. Finnegan (Eds.), *Age Estimation of the Human Skeleton* (pp. 193–215). Charles C. Thomas.

Crowder, C., & Rosella, L. (2007). Assessment of intra- and intercostal variation in rib histomorphometry: its impact on evidentiary examination. *Journal of Forensic Sciences, 52*(2), 271–276. https://doi.org/10.1111/j.1556-4029.2007.00388.x

Crowder, C. M., Heinrich, J. T., & Dominguez, V. M. (2016). Histological age estimation. In S. Blau & D. H. Ubelaker (Eds.), *Handbook of Forensic Anthropology and Archaeology* (pp. 293–304). https://doi.org/10.4324/9781315427775-29

Culmann, K. (1866). *Die Graphische Statik*. Verlag von Meyer & Zeller.

Currey, J. D. (1964). Some effects of ageing in human Haversian systems. *Journal of Anatomy, 98*(Pt 1), 69–75.

Curtis, J. (2004). *Estimation of Age at Death from the Microscopic Appearance of the Frontal Bone* [Master's Thesis]. University of Indianapolis.

Curtis, J., & Nawrocki, S. P. (2010). Skeletal aging using frontal bone histomorphometrics. In K. E. Latham & J. M. Finnegan (Eds.), *Age Estimation of the Human Skeleton* (pp. 216–231). Charles C Thomas.

Delaisse, J.-M., Andersen, T. L., Kristensen, H. B., Jensen, P. R., Andreasen, C. M., & Søe, K. (2020). Re-thinking the bone remodeling cycle mechanism and the origin of bone loss. *Bone, 141*, 115628. https://doi.org/10.1016/j.bone.2020.115628

Demontiero, O., Vidal, C., & Duque, G. (2012). Aging and bone loss: New insights for the clinician. *Therapeutic Advances in Musculoskeletal Disease*, *4*(2), 61–76. https://doi.org/10.1177/1759720X11430858

Dempster, D. W., Compston, J. E., Drezner, M. K., Glorieux, F. H., Kanis, J. A., Malluche, H., Meunier, P. J., Ott, S. M., Recker, R. R., & Parfitt, A. M. (2013). Standardized nomenclature, symbols, and units for bone histomorphometry: A 2012 update of the report of the ASBMR Histomorphometry Nomenclature Committee. *Journal of Bone and Mineral Research*, *28*(1), 2–17. https://doi.org/10.1002/jbmr.1805

Dominguez, V. M., & Agnew, A. M. (2016). Examination of factors potentially influencing osteon size in the human rib. *Anatomical Record*, *299*(3), 313–324. https://doi.org/10.1002/ar.23305

Dominguez, V. M., & Agnew, A. M. (2019). Microdamage as a bone quality component: practical guidelines for the two-dimensional analysis of linear microcracks in human cortical bone. *Journal of Bone and Mineral Research Plus*, *3*(6), e10203. https://doi.org/10.1002/jbm4.10203

Dominguez, V. M., & Crowder, C. M. (2012). The utility of osteon shape and circularity for differentiating human and non-human Haversian bone. *American Journal of Physical Anthropology*, *149*(1), 84–91. https://doi.org/10.1002/ajpa.22097

Drapeau, M. S. M., & Streeter, M. A. (2006). Modeling and remodeling responses to normal loading in the human lower limb. *American Journal of Physical Anthropology*, *129*(3), 403–409. https://doi.org/10.1002/ajpa.20336

Drusini, A. (1987). Refinements of two methods for the histomorphometric determination of age in human bone. *Zeitschrift Für Morphologie Und Anthropologie*, *77*(2), 167–176. JSTOR.

Dupras, T. L., & Pfeiffer, S. K. (1996). Determination of sex from adult human ribs. *Canadian Society of Forensic Science Journal*, *29*(4), 221–231. https://doi.org/10.1080/00085030.1996.10757065

Einhorn, T. A., Boskey, A. L., Gundberg, C. M., Vigorita, V. J., Devlin, V. J., & Beyer, M. M. (1988). The mineral and mechanical properties of bone in chronic experimental diabetes. *Journal of Orthopaedic Research*, *6*(3), 317–323. https://doi.org/10.1002/jor.1100060303

Epker, B. N., & Frost, H. M. (1965). The direction of transverse drift of actively forming osteons in human rib cortex. *JBJS*, *47*(6), 1211.

Ericksen, M. F. (1976). Cortical bone loss with age in three native American populations. *American Journal of Physical Anthropology*, *45*(3), 443–452. https://doi.org/10.1002/ajpa.1330450306

Ericksen, M. F. (1980). "Patterns of microscopic bone remodeling in three aboriginal American populations," in: *Early Native Americans: Prehistoric Demography, Economy and Technology* (D. L. Browman, ed.), pp. 239–270, Mouton, The Hague.

Ericksen, M. F. (1991). Histologic estimation of age at death using the anterior cortex of the femur. *American Journal of Physical Anthropology*, *84*(2), 171–179. https://doi.org/10.1002/ajpa.1330840207

Eriksen, E. F. (2010). Cellular mechanisms of bone remodeling. *Reviews in Endocrine & Metabolic Disorders*, *11*(4), 219–227. https://doi.org/10.1007/s11154-010-9153-1

Eriksen, E. F., Melsen, F., & Mosekilde, L. (1984). Reconstruction of the resorptive site in iliac trabecular bone: A kinetic model for bone resorption in 20 normal individuals. *Metabolic Bone Disease and Related Research*, *5*(5), 235–242. https://doi.org/10.1016/0221-8747(84)90065-1

Evans, F. G. (1976). Mechanical properties and histology of cortical bone from younger and older men. *The Anatomical Record*, *185*(1), 1–11. https://doi.org/10.1002/ar.1091850102

Evans, F. G., & Bang, S. (1967). Differences and relationships between the physical properties and the microscopic structure of human femoral, tibial and fibular cortical bone. *American Journal of Anatomy*, *120*(1), 79–88. https://doi.org/10.1002/aja.1001200107

Fahy, G. E., Deter, C., Pitfield, R., Miszkiewicz, J. J., & Mahoney, P. (2017). Bone deep: Variation in stable isotope ratios and histomorphometric measurements of bone remodelling within adult humans. *Journal of Archaeological Science*, *87*, 10–16. https://doi.org/10.1016/j.jas.2017.09.009

Feik, S. A., Thomas, C. D., Bruns, R., & Clement, J. G. (2000). Regional variations in cortical modeling in the femoral mid-shaft: Sex and age differences. *American Journal of Physical Anthropology*, *112*(2), 191–205. https://doi.org/10.1002/(SICI)1096-8644(2000)112:2<191::AID-AJPA6>3.0.CO;2-3

Franz-Odendaal, T. A., Hall, B. K., & Witten, P. E. (2006). Buried alive: How osteoblasts become osteocytes. *Developmental Dynamics*, *235*(1), 176–190. https://doi.org/10.1002/dvdy.20603

Fratzl, P., Fratzl-Zelman, N., & Klaushofer, K. (1993). Collagen packing and mineralization. An x-ray scattering investigation of turkey leg tendon. *Biophysical Journal*, *64*(1), 260–266.

Fratzl-Zelman, N., Schmidt, I., Roschger, P., Roschger, A., Glorieux, F. H., Klaushofer, K., Wagermaier, W., Rauch, F., & Fratzl, P. (2015). Unique micro- and nano-scale mineralization pattern of human osteogenesis imperfecta type VI bone. *Bone*, *73*, 233–241. https://doi.org/10.1016/j.bone.2014.12.023

Frost, H. M. (1960). In vivo osteocyte death. *The Journal of Bone and Joint Surgery. American Volume*, *42-A*, 138–143.

Frost, H. M. (1969). Tetracycline-based histological analysis of bone remodeling. *Calcified Tissue Research*, *3*(1), 211–237. https://doi.org/10.1007/BF02058664

Frost, H. M. (1983). A determinant of bone architecture—the minimum effective strain. *Clinical Orthopaedics and Related Research*, *175*, 286–292.

Frost, H. M. (1986). *Intermediary organization of the skeleton*. CRC Press.

Frost, H. M. (1987a). Secondary osteon populations—an algorithm for determining mean bone tissue age. *Yearbook of Physical Anthropology*, *30*, 221–238.

Frost, H. M. (1987b). The mechanostat—a proposed pathogenic mechanism of osteoporoses and the bone mass effects of mechanical and nonmechanical agents. Journal of *Bone and Mineral*, *2*(2), 73–85.

Frost, H. M. (1987). Bone Mass and the Mechanostat—A Proposal. *Journal of Anatomical Record*, *219*(1), 1–9. https://doi.org/10.1002/ar.1092190104

Frost, H. M. (1990). Skeletal Structural Adaptations to Mechanical Usage (Satmu). 1. Redefining Wolff Law—The Bone Modeling Problem. *Anatomical Record*, *226*(4), 403–413. https://doi.org/10.1002/ar.1092260402

Frost, H. M. (1998). From Wolff's Law to the mechanostat: A new "face" of physiology*. *Journal of Orthopaedic Science*, *3*(5), 282–286. https://doi.org/10.1007/s007760050054

Frost, H. M. (1999a). An approach to estimating bone and joint loads and muscle strength in living subjects and skeletal remains. *American Journal of Human Biology*, *11*(4), 437–455. https://doi.org/10.1002/(SICI)1520-6300(1999)11:4<437::AID-AJHB4>3.0.CO;2-K

Frost, H. M. (1999b). Why do bone strength and "mass" in aging adults become unresponsive to vigorous exercise? Insights of the Utah paradigm. *Journal of Bone and Mineral Metabolism*, *17*(2), 90–97. https://doi.org/10.1007/s007740050070

Frost, H. M. (2000). The Utah paradigm of skeletal physiology: An overview of its insights for bone, cartilage and collagenous tissue organs. *Journal of Bone and Mineral Metabolism*, *18*(6), 305–316. https://doi.org/10.1007/s007740070001

Frost, H. M. (2001a). From Wolff's law to the Utah paradigm: Insights about bone physiology and its clinical applications. *The Anatomical Record*, *262*(4), 398–419. https://doi.org/10.1002/ar.1049

Frost, H. M. (2001b). Why should many skeletal scientists and clinicians learn the Utah paradigm of skeletal physiology? *Journal of Musculoskeletal & Neuronal Interactions*, *2*(2), 121–130.

Frost, H. M. (2003). Bone's mechanostat: A 2003 update. *Anatomical Record Part A-Discoveries in Molecular Cellular and Evolutionary Biology*, *275a*(2), 1081–1101. https://doi.org/10.1002/ar.a.10119

Frost, H. M., & Schönau, E. (2000). The "muscle-bone unit" in children and adolescents: A 2000 overview. *Journal of Pediatric Endocrinology & Metabolism: JPEM*, *13*(6), 571–590. https://doi.org/10.1515/jpem.2000.13.6.571

Fuchs, R. K., Allen, M. R., Ruppel, M. E., Diab, T., Phipps, R. J., Miller, L. M., & Burr, D. B. (2008). In situ examination of the time-course for secondary mineralization of Haversian bone using synchrotron Fourier transform infrared microspectroscopy. *Matrix Biology*, *27*(1), 34–41. https://doi.org/10.1016/j.matbio.2007.07.006

Gao, H. (2006). Application of fracture mechanics concepts to hierarchical biomechanics of bone and bone-like materials. *International Journal of Fracture*, *138*(1), 101–137. https://doi.org/10.1007/s10704-006-7156-4

Garcia-Donas, J. G., Dyke, J., Paine, R. R., Nathena, D., & Kranioti, E. F. (2016). Accuracy and sampling error of two age estimation techniques using rib histomorphometry on a modern sample. *Journal of Forensic and Legal Medicine, 38*, 28–35. https://doi.org/10.1016/j.jflm.2015.11.012

Giorgio, I., Spagnuolo, M., Andreaus, U., Scerrato, D., Bersani, A.M. (2021). In-depth gaze at the astonishing mechanical behavior of bone: A review for designing bio-inspired hierarchical metamaterials. *Mathematics and Mechanics of Solids* 26, 1074–1103. https://doi.org/10.1177/1081286520978516

Gocha, T. P. (2014). *Mapping Spatial Patterns in Cortical Remodeling from the Femoral Midshaft using Geographic Information Systems Software: Implications for Age Estimation from Adult Human Skeletal Remains* [The Ohio State University]. https://etd.ohiolink.edu/!etd.send_file?accession=osu1416584228

Gocha, T. P., & Agnew, A. M. (2016). Spatial variation in osteon population density at the human femoral midshaft: Histomorphometric adaptations to habitual load environment. *Journal of Anatomy, 228*(5), 733–745. https://doi.org/10.1111/joa.12433

Gocha, T. P., Robling, A. G., & Stout, S. D. (2018). Histomorphometry of human cortical bone. In *Biological Anthropology of the Human Skeleton* (3rd ed., pp. 145–187). John Wiley & Sons, Ltd. https://doi.org/10.1002/9781119151647.ch5

Goliath, J. R., Stewart, M. C., & Stout, S. D. (2016). Variation in osteon histomorphometrics and their impact on age-at-death estimation in older individuals. *Forensic Science International, 262*, 282 e1–e6. https://doi.org/10.1016/j.forsciint.2016.02.053

Guo, D., & Bonewald, L. F. (2009). Advancing Our Understanding of Osteocyte Cell Biology. *Therapeutic Advances in Musculoskeletal Disease, 1*(2), 87–96. https://doi.org/10.1177/1759720X09341484

Han, Z.-H., Palnitkar, S., Rao, D. S., Nelson, D., & Parfitt, A. M. (1997). Effects of ethnicity and age or menopause on the remodeling and turnover of iliac bone: implications for mechanisms of bone loss. *Journal of Bone and Mineral Research, 12*(4), 498–508. https://doi.org/10.1359/jbmr.1997.12.4.498

Handschin, R. G., & Stern, W. B. (1994). Crystallographic and chemical analysis of human bone apatite (Crista Iliaca). *Clinical Rheumatology, 13 Suppl 1*, 75–90.

Kim D Harrison, Kim D., Hiebert, Beverly D, Panahifar, Arash, & Andronowski, Janna M (2020). Cortical bone porosity in rabbit models of osteoporosis. *Journal of Bone and Mineral Research*, 2211–2228.

Harrison, K. D., Sales, E., Hiebert, B. D., Panahifar, A., Zhu, N., Arnason, T., Swekla, K. J., Pivonka, P., Chapman, L. D., & Cooper, D. M. (2022). Direct assessment of rabbit cortical bone basic multicellular unit longitudinal erosion rate: a 4d synchrotron-based approach. *Journal of Bone and Mineral Research, 37*(11), 2244–2258. https://doi.org/10.1002/jbmr.4700

Hennig, C., Thomas, C. D. L., Clement, J. G., & Cooper, D. M. L. (2015). Does 3D orientation account for variation in osteon morphology assessed by 2D histology? *Journal of Anatomy, 227*(4), 497–505. https://doi.org/10.1111/joa.12357

Hert, J., Lisková, M., & Landa, J. (1971). Reaction of bone to mechanical stimuli. 1. Continuous and intermittent loading of tibia in rabbit. *Folia Morphologica, 19*(3), 290–300.

Hillier, M. L., & Bell, L. S. (2007). Differentiating human bone from animal bone: A review of histological methods. *Journal of Forensic Sciences, 52*(2), 249–263. https://doi.org/10.1111/j.1556-4029.2006.00368.x

Holick, M. F. (2007). Vitamin D deficiency. *The New England Journal of Medicine, 357*(3), 266–281. https://doi.org/10.1056/NEJMra070553

Horwood, N. J., Elliott, J., Martin, T. J., & Gillespie, M. T. (1998). Osteotropic agents regulate the expression of osteoclast differentiation factor and osteoprotegerin in osteoblastic stromal cells. *Endocrinology, 139*(11), 4743–4746. https://doi.org/10.1210/endo.139.11.6433

Houschyar, K. S., Tapking, C., Borrelli, M. R., Popp, D., Duscher, D., Maan, Z. N., Chelliah, M. P., Li, J., Harati, K., Wallner, C., Rein, S., Pförringer, D., Reumuth, G., Grieb, G., Mouraret, S., Dadras, M., Wagner, J. M., Cha, J. Y., Siemers, F., … Behr, B. (2019). Wnt pathway in bone repair and regeneration – what do we know so far. *Frontiers in Cell and Developmental Biology, 6*, 170. https://doi.org/10.3389/fcell.2018.00170

Huang, R. Y., Miller, L. M., Carlson, C. S., & Chance, M. R. (2002). Characterization of bone mineral composition in the proximal tibia of cynomolgus monkeys: Effect of ovariectomy and nandrolone decanoate treatment. *Bone, 30*(3), 492–497. https://doi.org/10.1016/s8756-3282(01)00691-3

Hughes, J. M., Castellani, C. M., Popp, K. L., Guerriere, K. I., Matheny, R. W., Nindl, B. C., & Bouxsein, M. L. (2020). The central role of osteocytes in the four adaptive pathways of bone's mechanostat. *Exercise and Sport Sciences Reviews, 48*(3), 140–148. https://doi.org/10.1249/JES.0000000000000225

Hughes, J. M., & Petit, M. A. (2010). Biological underpinnings of Frost's mechanostat thresholds: The important role of osteocytes. *Journal of Musculoskeletal & Neuronal Interactions, 10*(2), 128–135.

Hummert, J. R., & Van Gerven, D. P. (1982). Tetracycline-labeled human bone from a medieval population in Nubia's Batn El Hajar (550-1450 A.D.). *Human Biology, 54*(2), 355–363.

Hunter, R. L., & Agnew, A. M. (2016). Intraskeletal variation in human cortical osteocyte lacunar density: Implications for bone quality assessment. *Bone Reports, 5*, 252–261. https://doi.org/10.1016/j.bonr.2016.09.002

Infante, A., & Rodríguez, C. I. (2018). Osteogenesis and aging: Lessons from mesenchymal stem cells. *Stem Cell Research & Therapy, 9*(1), 244. https://doi.org/10.1186/s13287-018-0995-x

Iwamoto, S., Onuki, E., & Konishi, M. (1979). Study on the age-related changes of the compact bone and the age estimation. 2 On the humerus. *Acta Medica Kinki University, 3*, 203–208.

Jackes, M. (2000). Building the bases for paleodemographic analysis: adult age determination. In Katzenberg S, Saunders S, editors. *Biological Anthropology of the Human Skeleton*. New York: Wiley-Liss, pp. 417–466.

Javaheri, B., Razi, H., Gohin, S., Wylie, S., Chang, Y.-M., Salmon, P., Lee, P. D., & Pitsillides, A. A. (2020). Lasting organ-level bone mechanoadaptation is unrelated to local strain. *Science Advances, 6*(10), eaax8301. https://doi.org/10.1126/sciadv.aax8301

Jensen, P. R., Andersen, T. L., Hauge, E.-M., Bollerslev, J., & Delaissé, J.-M. (2015). A joined role of canopy and reversal cells in bone remodeling—Lessons from glucocorticoid-induced osteoporosis. *Bone, 73*, 16–23. https://doi.org/10.1016/j.bone.2014.12.004

Jowsey, J. (1960). Age changes in human bone. *Clinical Orthopaedics and Related Research, 17*, 210.

Jowsey, J. (1964). Variation in bone mineralization with age and disease. *Bone Dynamics., 661*. https://ci.nii.ac.jp/naid/10007089639/

Karydi, C., García-Donas, J. G., Tsiminikaki, K., Bonicelli, A., Moraitis, K., & Kranioti, E. F. (2022). Estimation of age-at-death using cortical bone histomorphometry of the rib and femur: a validation study on a British population. *Biology, 11*(11), 1615. https://doi.org/10.3390/biology11111615

Keenan, K. E., Mears, C. S., & Skedros, J. G. (2017). Utility of osteon circularity for determining species and interpreting load history in primates and nonprimates. *American Journal of Physical Anthropology, 162*(4), 657–681. https://doi.org/10.1002/ajpa.23154

Kenkre, J. S., & Bassett, J. (2018). The bone remodelling cycle. *Annals of Clinical Biochemistry, 55*(3), 308–327. https://doi.org/10.1177/0004563218759371

Kennedy, O. D., Herman, B. C., Laudier, D. M., Majeska, R. J., Sun, H. B., & Schaffler, M. B. (2012). Activation of resorption in fatigue-loaded bone involves both apoptosis and active pro-osteoclastogenic signaling by distinct osteocyte populations. *Bone, 50*(5), 1115–1122. https://doi.org/10.1016/j.bone.2012.01.025

Kerley, E. R. (1965). The microscopic determination of age in human bone. *American Journal of Physical Anthropology, 23*(2), 149–163. https://doi.org/10.1002/ajpa.1330230215

Kim, B.-J., Lee, Y.-S., Lee, S.-Y., Park, S.-Y., Dieplinger, H., Ryu, S. H., Yea, K., Choi, S., Lee, S. H., Koh, J.-M., & Kim, G. S. (2012). Afamin secreted from nonresorbing osteoclasts acts as a chemokine for preosteoblasts via the Akt-signaling pathway. *Bone, 51*(3), 431–440. https://doi.org/10.1016/j.bone.2012.06.015

Kim, J., & Adachi, T. (2019). Cell condensation triggers the differentiation of osteoblast precursor cells to osteocyte-like cells. *Frontiers in Bioengineering and Biotechnology, 7*. https://www.frontiersin.org/articles/10.3389/fbioe.2019.00288

Kobayashi, K., Takahashi, N., Jimi, E., Udagawa, N., Takami, M., Kotake, S., Nakagawa, N., Kinosaki, M., Yamaguchi, K., Shima, N., Yasuda, H., Morinaga, T., Higashio, K., Martin, T. J., & Suda, T. (2000). Tumor necrosis factor alpha stimulates osteoclast differentiation by a mechanism independent of the ODF/RANKL-RANK interaction. *The Journal of Experimental Medicine, 191*(2), 275–286. https://doi.org/10.1084/jem.191.2.275

Kobayashi, Y., Uehara, S., & Udagawa, N. (2018). Roles of non-canonical Wnt signaling pathways in bone resorption. *Journal of Oral Biosciences, 60*(2), 31–35. https://doi.org/10.1016/j.job.2018.03.001

Koltze, H. (1951). Studie zur äußeren Form der Osteone. *Zeitschrift für Anatomie und Entwicklungsgeschichte, 115*(6), 584–596. https://doi.org/10.1007/BF00536346

Komori, T. (2019). Regulation of proliferation, differentiation and functions of osteoblasts by Runx2. *International Journal of Molecular Sciences, 20*(7), E1694. https://doi.org/10.3390/ijms20071694

Kovačić, N., Lukić, I. K., Grčević, D., Katavić, V., Croucher, P., & Marušić, A. (2007). Fas/Fas ligand system inhibits differentiation of murine osteoblasts but has a limited role in osteoblast and osteoclast apoptosis. *Journal of Immunology (Baltimore, Md.: 1950), 178*(6), 3379–3389.

Kranioti, E. F., Michopoulou, E., Tsiminikaki, K., Bonicelli, A., Kalochristianakis, M., Xhemali, B., Paine, R. R., & García-Donas, J. G. (2020). Bone histomorphometry of the clavicle in a forensic sample from Albania. *Forensic Science International, 313*, 110335. https://doi.org/10.1016/j.forsciint.2020.110335

Kristensen, H. B., Andersen, T. L., Marcussen, N., Rolighed, L., & Delaisse, J.-M. (2014). Osteoblast recruitment routes in human cancellous bone remodeling. *The American Journal of Pathology, 184*(3), 778–789. https://doi.org/10.1016/j.ajpath.2013.11.022

Kulkarni, R. N., Bakker, A. D., Everts, V., & Klein-Nulend, J. (2010). Inhibition of osteoclastogenesis by mechanically loaded osteocytes: Involvement of MEPE. *Calcified Tissue International, 87*(5), 461–468. https://doi.org/10.1007/s00223-010-9407-7

Lanyon, L. E., & Baggott, D. G. (1976). Mechanical function as an influence on the structure and form of bone. *The Journal of Bone and Joint Surgery. British, 58-B*(4), 436–443.

Lanyon, L. E., Goodship, A. E., Pye, C. J., & MacFie, J. H. (1982). Mechanically adaptive bone remodelling. *Journal of Biomechanics, 15*(3), 141–154. https://doi.org/10.1016/0021-9290(82)90246-9

Lanyon, L. E., Magee, P. T., & Baggott, D. G. (1979). The relationship of functional stress and strain to the processes of bone remodelling. An experimental study on the sheep radius. *Journal of Biomechanics, 12*(8), 593–600. https://doi.org/10.1016/0021-9290(79)90079-4

Lassen, N. E., Andersen, T. L., Pløen, G. G., Søe, K., Hauge, E. M., Harving, S., Eschen, G. E. T., & Delaisse, J.-M. (2017). Coupling of bone resorption and formation in real time: new knowledge gained from human haversian BMUs. *Journal of Bone and Mineral Research, 32*(7), 1395–1405. https://doi.org/10.1002/jbmr.3091

Lee, T. C., & Taylor, D. (1999). Bone remodelling: Should we cry Wolff? *Irish Journal of Medical Science, 168*(2), 102–105. https://doi.org/10.1007/BF02946474

Lekszycki, T. (2002). Modelling of bone adaptation based on an optimal response hypothesis*. *Meccanica, 37*(4), 343–354. https://doi.org/10.1023/A:1020831519496

Lekszycki, T. (2005). Functional adaptation of bone as an optimal control problem. *Journal of Theoretical and Applied Mechanics, 43*(3), 555–574.

Lesciotto KM. The impact of Daubert on the admissibility of forensic anthropology expert testimony. J Forensic Sci. 2015 May;60(3):549–55. doi: 10.1111/1556-4029.12740. Epub 2015 Feb 26. PMID: 25716577.

Li, J., Mashiba, T., Burr, D.B. (2001). Bisphosphonate treatment suppresses not only stochastic remodeling but also the targeted repair of microdamage. *Calcif. Tissue Int.* 69, 281–286. https://doi.org/10.1007/s002230010036

Liu, C., Middleton, K., & You, L. (2015). Bone cell mechanobiology using micro- and nano-techniques. In C. A. Simmons, D.-H. Kim, & Y. Sun (Eds.), *Integrative Mechanobiology: Micro- and Nano- Techniques in Cell Mechanobiology* (pp. 245–265). Cambridge University Press. https://doi.org/10.1017/CBO9781139939751.015

Long, F. (2012). Building strong bones: Molecular regulation of the osteoblast lineage. *Nature Reviews Molecular Cell Biology*, *13*(1), 27–38. https://doi.org/10.1038/nrm3254

Lorget, F., Kamel, S., Mentaverri, R., Wattel, A., Naassila, M., Maamer, M., & Brazier, M. (2000). High extracellular calcium concentrations directly stimulate osteoclast apoptosis. *Biochemical and Biophysical Research Communications*, *268*(3), 899–903. https://doi.org/10.1006/bbrc.2000.2229

Maeda, K., Kobayashi, Y., Koide, M., Uehara, S., Okamoto, M., Ishihara, A., Kayama, T., Saito, M., & Marumo, K. (2019). The regulation of bone metabolism and disorders by Wnt signaling. *International Journal of Molecular Sciences*, *20*(22), E5525. https://doi.org/10.3390/ijms20225525

Maggiano, C. (2011). Making the mold: A microstructural perspective on bone modeling during growth and mechanical adaptation. In: Crowder C, Stout S, editors. *Bone Histology: Anthropological Perspectives*. Boca Raton: CRC Press, 45–90.

Maggiano, C., Dupras, T., Schultz, M., & Biggerstaff, J. (2006). Spectral and photobleaching analysis using confocal laser scanning microscopy: A comparison of modern and archaeological bone fluorescence. *Molecular and Cellular Probes*, *20*(3), 154–162. https://doi.org/10.1016/j.mcp.2005.11.009

Maggio, A., & Franklin, D. (2019). Histomorphometric age estimation from the femoral cortex: A test of three methods in an Australian population. *Forensic Science International*, *303*. https://doi.org/10.1016/j.forsciint.2019.109950

Marotti, G., Favia, A., & Zallone, A. Z. (1972). Quantitative analysis on the rate of secondary bone mineralization. *Calcified Tissue Research*, *10*(1), 67–81. https://doi.org/10.1007/BF02012537

Martin, B. (1993). Aging and strength of bone as a structural material. *Calcified Tissue International*, *53 Suppl 1*, S34–S39; discussion S39-40. https://doi.org/10.1007/bf01673400

Martin, D. L., & Armelagos, G. J. (1979). Morphometrics of compact bone: An example from Sudanese Nubia. *American Journal of Physical Anthropology*, *51*(4), 571–577. https://doi.org/10.1002/ajpa.1330510409

Martin, D. L., & Armelagos, G. J. (1985). Skeletal remodeling and mineralization as indicators of health: An example from prehistoric Sudanese Nubia. *Journal of Human Evolution*, *14*(5), 527–537. https://doi.org/10.1016/S0047-2484(85)80031-2

Martin, R. B. (2002). Is all cortical bone remodeling initiated by microdamage? *Bone*, *30*(1), 8–13. https://doi.org/10.1016/S8756-3282(01)00620-2

Martin, R. B. (2003a). Functional adaptation and fragility of the skeleton. In S. C. Agarwal & S. D. Stout (Eds.), *Bone Loss and Osteoporosis: An Anthropological Perspective* (pp. 121–138). Springer US. https://doi.org/10.1007/978-1-4419-8891-1_8

Martin, R. B. (2003b). Fatigue microdamage as an essential element of bone mechanics and biology. *Calcified Tissue International*, *73*(2), 101–107. https://doi.org/10.1007/s00223-002-1059-9

Martin, R. B., & Atkinson, P. J. (1977). Age and sex-related changes in the structure and strength of the human femoral shaft. *Journal of Biomechanics*, *10*(4), 223–231. https://doi.org/10.1016/0021-9290(77)90045-8

Martin, R. B., Burr, D. B., Sharkey, N. A., & Fyhrie, D. P. (2015). *Skeletal Tissue Mechanics* (2nd Edition). Springer Press.

Martin, R. B., Pickett, J. C., & Zinaich, S. (1980). Studies of skeletal remodeling in aging men. *Clinical Orthopaedics and Related Research*, *149*, 268–282.

Mason, M. W., Skedros, J. G., & Bloebaum, R. D. (1995). Evidence of strain-mode-related cortical adaptation in the diaphysis of the horse radius. *Bone*, *17*(3), 229–237. https://doi.org/10.1016/8756-3282(95)00213-w

Mayya, A., Banerjee, A., & Rajesh, R. (2013). Mammalian cortical bone in tension is non-Haversian. *Scientific Reports*, *3*, 2533. https://doi.org/10.1038/srep02533

Mbalaviele, G., Novack, D. V., Schett, G., & Teitelbaum, S. L. (2017). Inflammatory osteolysis: A conspiracy against bone. *The Journal of Clinical Investigation*, *127*(6), 2030–2039. https://doi.org/10.1172/JCI93356

McCarthy, E. F. (2016). The histology of metabolic bone disease. *Diagnostic Histopathology*, *22*(10), 378–383. https://doi.org/10.1016/j.mpdhp.2016.09.004

McCutcheon, S., Majeska, R. J., Spray, D. C., Schaffler, M. B., & Vazquez, M. (2020). Apoptotic osteocytes induce RANKL production in bystanders via purinergic signaling and activation of pannexin channels. *Journal of Bone and Mineral Research, 35*(5), 966–977. https://doi.org/10.1002/jbmr.3954

McNamara, L. M., Van der Linden, J. C., Weinans, H., & Prendergast, P. J. (2006). Stress-concentrating effect of resorption lacunae in trabecular bone. *Journal of Biomechanics, 39*(4), 734–741. https://doi.org/10.1016/j.jbiomech.2004.12.027

Meakin, L. B., Price, J. S., & Lanyon, L. E. (2014). The contribution of experimental in vivo models to understanding the mechanisms of adaptation to mechanical loading in bone. *Frontiers in Endocrinology, 5*, 154. https://doi.org/10.3389/fendo.2014.00154

Michael, H., Härkönen, P. L., Väänänen, H. K., & Hentunen, T. A. (2005). Estrogen and testosterone use different cellular pathways to inhibit osteoclastogenesis and bone resorption. *Journal of Bone and Mineral Research, 20*(12), 2224–2232. https://doi.org/10.1359/JBMR.050803

Miszkiewicz, J. J., & Mahoney, P. (2016). Ancient human bone microstructure in medieval England: Comparisons between two socio-economic groups. *Anatomical Record (Hoboken, N.J.: 2007), 299*(1), 42–59. https://doi.org/10.1002/ar.23285

Mizoguchi, T., & Ono, N. (2021). The diverse origin of bone-forming osteoblasts. *Journal of Bone and Mineral Research, 36*(8), 1432–1447. https://doi.org/10.1002/jbmr.4410

Mnich, B., Skrzat, J., & Szostek, K. (2017). Estimating age at death from an archaeological bone sample – a preliminary study based on comparison of histomorphometric methods. *Anthropological Review, 80*(1), 37–55. https://doi.org/10.1515/anre-2017-0002

Morris, Z., & Crowder, C. (2005). *Histomorphometrics: Osteon population density*. Annual Meeting of the Southern Anthropological Society, Chattanooga, TN.

Mosekilde, L., & Mosekilde, L. (1990). Sex differences in age-related changes in vertebral body size, density and biomechanical competence in normal individuals. *Bone, 11*(2), 67–73. https://doi.org/10.1016/8756-3282(90)90052-z

Movérare-Skrtic, S., Henning, P., Liu, X., Nagano, K., Saito, H., Börjesson, A. E., Sjögren, K., Windahl, S. H., Farman, H., Kindlund, B., Engdahl, C., Koskela, A., Zhang, F.-P., Eriksson, E. E., Zaman, F., Hammarstedt, A., Isaksson, H., Bally, M., Kassem, A., … Ohlsson, C. (2014). Osteoblast-derived WNT16 represses osteoclastogenesis and prevents cortical bone fragility fractures. *Nature Medicine, 20*(11), 1279–1288. https://doi.org/10.1038/nm.3654

Mulcahy, L. E., Taylor, D., Lee, T. C., & Duffy, G. P. (2011). RANKL and OPG activity is regulated by injury size in networks of osteocyte-like cells. *Bone, 48*(2), 182–188. https://doi.org/10.1016/j.bone.2010.09.014

Mulhern, D., & Ubelaker, D. (2012). Differentiating human from nonhuman bone microstructure. In C. M. Crowder & S. D. Stout (Eds.), *Bone Histology: An Anthropological Perspective*. Boca Raton: CRC Press, pp. 109–134.

Mulhern, D. M. (2000). Rib remodeling dynamics in a skeletal population from Kulubnarti, Nubia. *American Journal of Physical Anthropology, 111*(4), 519–530. https://doi.org/10.1002/(SICI)1096-8644(200004)111:4<519::AID-AJPA7>3.0.CO;2-7

Mulhern, D. M., & Ubelaker, D. H. (2001). Differences in osteon banding between human and non-human bone. *Journal of Forensic Sciences, 46*(2), 220–222. https://doi.org/10.1520/JFS14952J

Mulhern, D. M., & Ubelaker, D. H. (2003). Histologic examination of bone development in juvenile chimpanzees. *American Journal of Physical Anthropology, 122*(2), 127–133. https://doi.org/10.1002/ajpa.10294

Mulhern, D. M., & Van Gerven, D. P. (1997). Patterns of femoral bone remodeling dynamics in a Medieval Nubian population. *American Journal of Physical Anthropology, 104*(1), 133–146. https://doi.org/10.1002/(SICI)1096-8644(199709)104:1<133::AID-AJPA9>3.0.CO;2-S

Murshed, M. (2018). Mechanism of bone mineralization. *Cold Spring Harbor Perspectives in Medicine, 8*(12), a031229. https://doi.org/10.1101/cshperspect.a031229

Nagai, T., Hasegawa, T., Yamamoto, T., Hongo, H., Abe, M., Yoshida, T., Yokoyama, A., de Freitas, P. H. L., Li, M., Yokoyama, A., & Amizuka, N. (2021). Immunocytochemical assessment of cell differentiation of podoplanin-positive osteoblasts into osteocytes in murine bone. *Histochemistry and Cell Biology*, *155*(3), 369–380. https://doi.org/10.1007/s00418-020-01937-y

Nawrocki, S. P. (2010). The nature and sources of error in the estimation of age at death from the skeleton. In K. E. Latham & J. M. Finnegan (Eds.), *Age Estimation of the Human Skeleton* (pp. 79–101). Charles C Thomas.

Nowak, M., Sokołowski, J., & Żochowski, A. (2018). Justification of a certain algorithm for shape optimization in 3D elasticity. *Structural and Multidisciplinary Optimization*, *57*(2), 721–734. https://doi.org/10.1007/s00158-017-1780-7

Nowak, M., Sokołowski, J., & Żochowski, A. (2020). Biomimetic approach to compliance optimization and multiple load cases. *Journal of Optimization Theory and Applications*, *184*(1), 210–225. https://doi.org/10.1007/s10957-019-01502-1

Ortner, D. J. (1975). Aging effects on osteon remodeling. *Calcified Tissue Research*, *18*(1), 27–36. https://doi.org/10.1007/bf02546224

Ozcivici, E., Luu, Y. K., Adler, B., Qin, Y.-X., Rubin, J., Judex, S., & Rubin, C. T. (2010). Mechanical signals as anabolic agents in bone. *Nature Reviews. Rheumatology*, *6*(1), 50–59. https://doi.org/10.1038/nrrheum.2009.239

Parfitt, A. M. (1979). Quantum concept of bone remodeling and turnover: Implications for the pathogenesis of osteoporosis. *Calcified Tissue International*, *28*(1), 1–5. https://doi.org/10.1007/BF02441211

Parfitt, A. M. (1994). Osteonal and hemi-osteonal remodeling: The spatial and temporal framework for signal traffic in adult human bone. *Journal of Cellular Biochemistry*, *55*(3), 273–286. https://doi.org/10.1002/jcb.240550303

Parfitt, A.M., Drezner, M.K., Glorieux, F.H., Kanis, J.A., Malluche, H., Meunier, P.J., Ott, S.M., Recker, R.R. (1987). Bone histomorphometry: standardization of nomenclature, symbols, and units. *Report of the ASBMR Histomorphometry Nomenclature Committee. J Bone Miner Res* 2, 595–610. https://doi.org/10.1002/jbmr.5650020617

Paschalis, E. P., Mendelsohn, R., & Boskey, A. L. (2011). Infrared assessment of bone quality: a review. *Clinical Orthopaedics and Related Research*, *469*(8), 2170–2178. https://doi.org/10.1007/s11999-010-1751-4

Pearson, O. M., & Lieberman, D. E. (2004). The aging of Wolff's "law": Ontogeny and responses to mechanical loading in cortical bone. *American Journal of Physical Anthropology*, *Suppl 39*, 63–99. https://doi.org/10.1002/ajpa.20155

Peck, J. J., & Stout, S. D. (2009). The effects of total hip arthroplasty on the structural and biomechanical properties of adult bone. *American Journal of Physical Anthropology*, *138*(2), 221–230. https://doi.org/10.1002/ajpa.20921

Pfeiffer, S. (1992). Cortical bone age estimates from historically known adults. *Zeitschrift Fur Morphologie Und Anthropologie*, *79*(1), 1–10.

Pfeiffer, S. (1998). Variability in osteon size in recent human populations. *American Journal of Physical Anthropology*, *106*(2), 219–227. https://doi.org/10.1002/(SICI)1096-8644(199806)106:2<219::AID-AJPA8>3.0.CO;2-K

Pfeiffer, S., Crowder, C., Harrington, L., & Brown, M. (2006). Secondary osteon and Haversian canal dimensions as behavioral indicators. *American Journal of Physical Anthropology*, *131*(4), 460–468. https://doi.org/10.1002/ajpa.20454

Pfeiffer, S., Heinrich, J., Beresheim, A., & Alblas, M. (2016). Cortical bone histomorphology of known-age skeletons from the kirsten collection, Stellenbosch University, South Africa. *American Journal of Physical Anthropology*, *160*(1), 137–147. https://doi.org/10.1002/ajpa.22951

Pfeiffer, S., Lazenby, R., & Chiang, J. (1995). Brief communication: Cortical remodeling data are affected by sampling location. *American Journal of Physical Anthropology*, *96*(1), 89–92. https://doi.org/10.1002/ajpa.1330960110

Pirok, D., Ramser, J., Takahashi, H., Villanueva, A., & Frost, H. (1966). Normal histological, tetracycline and dynamic parameters in human, mineralized bone sections. *Henry Ford Hospital Medical Journal*, *14*(2), 195–218.

Pivonka, P., Park, A., & Forwood, M. R. (2018). Functional adaptation of bone: The mechanostat and beyond. In P. Pivonka (Ed.), *Multiscale Mechanobiology of Bone Remodeling and Adaptation* (pp. 1–60). Springer International Publishing. https://doi.org/10.1007/978-3-319-58845-2_1

Plotkin, L. I., & Bruzzaniti, A. (2019). Molecular signaling in bone cells: Regulation of cell differentiation and survival. *Advances in Protein Chemistry and Structural Biology, 116*, 237–281. https://doi.org/10.1016/bs.apcsb.2019.01.002

Portigliatti Barbos, M., Bianco, P., Ascenzi, A., & Boyde, A. (1984). Collagen orientation in compact bone: II. Distribution of lamellae in the whole of the human femoral shaft with reference to its mechanical properties. *Metabolic Bone Disease and Related Research*, *5*(6), 309–315. https://doi.org/10.1016/0221-8747(84)90018-3

Price, P. A., Toroian, D., & Lim, J. E. (2009). Mineralization by inhibitor exclusion. *The Journal of Biological Chemistry*, *284*(25), 17092–17101. https://doi.org/10.1074/jbc.M109.007013

Raguin, E., & Drapeau, M. S. M. (2020). Relation between cross-sectional bone geometry and double zonal osteon frequency and morphology. *American Journal of Physical Anthropology*, *171*(4), 598–612. https://doi.org/10.1002/ajpa.23954

Raguin, E., & Streeter, M. A. (2018). Brief communication: Test of a method to identify double-zonal osteon in polarized light microscopy. *American Journal of Physical Anthropology*, *167*(2), 407–415. https://doi.org/10.1002/ajpa.23616

Rauch, F., Travers, R., Parfitt, A. M., & Glorieux, F. H. (2000). Static and dynamic bone histomorphometry in children with osteogenesis imperfecta. *Bone*, *26*(6), 581–589. https://doi.org/10.1016/s8756-3282(00)00269-6

Rehnquist, W. H. & Supreme Court of The United States. (1997). *U.S. Reports: General Electric Co. v. Joiner, 522 U.S. 136*. [Periodical] Retrieved from the Library of Congress, https://www.loc.gov/item/usrep522136/

Richman, E. A., Ortner, D. J., & Schulter-Ellis, F. P. (1979). Differences in intracortical bone remodeling in three aboriginal American populations: Possible dietary factors. *Calcified Tissue International*, *28*(1), 209–214. https://doi.org/10.1007/BF02441238

Riggs, C. M., Vaughan, L. C., Evans, G. P., Lanyon, L. E., & Boyde, A. (1993). Mechanical implications of collagen fibre orientation in cortical bone of the equine radius. *Anatomy and Embryology*, *187*(3), 239–248. https://doi.org/10.1007/bf00195761

Ritchie, R. O. (2011). The conflicts between strength and toughness. *Nature Materials*, *10*(11), 817–822. https://doi.org/10.1038/nmat3115

Robling, A. (2014). Mechanical adaptation. In D. Burr & M. Allen (Eds.), *Basic and Applied Bone Biology* (pp. 175–203). Academic Press.

Robling, A., & Stout, S. (2003). Histomorphology, geometry, and mechanical loading in past populations. In S. Agarwal & S. Stout (Eds.), *Bone Loss and Osteoporosis: An Anthropological Perspective* (pp. 189–203). Kluwer Academic/Plenum Publishers.

Robling, A. G., & Bonewald, L. F. (2020). The osteocyte: New insights. *Annual Review of Physiology*, *82*, 485–506. https://doi.org/10.1146/annurev-physiol-021119-034332

Robling, A. G., Castillo, A. B., & Turner, C. H. (2006). Biomechanical and molecular regulation of bone remodeling. *Annual Review of Biomedical Engineering*, *8*, 455–498. https://doi.org/10.1146/annurev.bioeng.8.061505.095721

Robling, A. G., & Stout, S. D. (1999). Morphology of the drifting osteon. *Cells, Tissues, Organs*, *164*(4), 192–204. https://doi.org/10.1159/000016659

Roesler, H. (1981). Some historical remarks on the theory of cancellous bone structure (Wolff's Law). In S. C. Cowin (Ed.), *Mechanical properties of bone* (pp. 27–42). American Society of Mechanical Engineers,.

Root, S. H., Wee, N. K., Novak, S., Rosen, C. J., Baron, R., Matthews, B. G., & Kalajzic, I. (2020). Perivascular osteoprogenitors are associated with trans-cortical channels of long bones. *Stem Cells (Dayton, Ohio)*, *38*(6), 769–781. https://doi.org/10.1002/stem.3159

Ross, F. P. (2013). Osteoclast biology and bone resorption. In C. J. Rosen (Ed.), *Primer on the Metabolic Bone Diseases and Disorders of Mineral Metabolism*. x8thv pp. 25–33). John Wiley and Sons, Inc.

Roux, W. (1885). Beitrage zur morphologie der funktionnellen anspassung. *Arch Anat. Physiol. Anat. Abt.*, *9*, 120–158.

Ru, J.-Y., & Wang, Y.-F. (2020). Osteocyte apoptosis: The roles and key molecular mechanisms in resorption-related bone diseases. *Cell Death & Disease*, *11*(10), 846. https://doi.org/10.1038/s41419-020-03059-8

Rubin, C. T., & Lanyon, L. E. (1984). Regulation of bone formation by applied dynamic loads. *The Journal of Bone and Joint Surgery. American Volume*, *66*(3), 397–402.

Ruff, C., Holt, B., & Trinkaus, E. (2006). Who's afraid of the big bad Wolff?: "Wolff's law" and bone functional adaptation. *American Journal of Physical Anthropology*, *129*(4), 484–498. https://doi.org/10.1002/ajpa.20371

Ruff, C. B., & Hayes, W. C. (1988). Sex differences in age-related remodeling of the femur and tibia. *Journal of Orthopaedic Research*, *6*(6), 886–896. https://doi.org/10.1002/jor.1100060613

Schaffler, M. B., & Burr, D. B. (1984). Primate cortical bone microstructure: Relationship to locomotion. *American Journal of Physical Anthropology*, *65*(2), 191–197. https://doi.org/10.1002/ajpa.1330650211

Scheiner, S., Pivonka, P., & Hellmich, C. (2016). Poromicromechanics reveals that physiological bone strains induce osteocyte-stimulating lacunar pressure. *Biomechanics and Modeling in Mechanobiology*, *15*(1), 9–28. https://doi.org/10.1007/s10237-015-0704-y

Schlecht, S. H., Bigelow, E. M. R., & Jepsen, K. J. (2015). How does bone strength compare across sex, site, and ethnicity? *Clinical Orthopaedics and Related Research*, *473*(8), 2540–2547. https://doi.org/10.1007/s11999-015-4229-6

Schlecht, S. H., Pinto, D. C., Agnew, A. M., & Stout, S. D. (2012). Brief communication: The effects of disuse on the mechanical properties of bone: What unloading tells us about the adaptive nature of skeletal tissue. *American Journal of Physical Anthropology*, *149*(4), 599–605. https://doi.org/10.1002/ajpa.22150

Schriefer, J. L., Warden, S. J., Saxon, L. K., Robling, A. G., & Turner, C. H. (2005). Cellular accommodation and the response of bone to mechanical loading. *Journal of Biomechanics*, *38*(9), 1838–1845. https://doi.org/10.1016/j.jbiomech.2004.08.017

Sedlin, E. D., Frost, H. M., & Villanueva, A. R. (1963). Age changes in resorption in human Rib Cortex. *Journal of Gerontology*, *18*(4), 345–349. https://doi.org/10.1093/geronj/18.4.345

Seeman, E. (2013). Age- and MENOPAUSE-RELATED BONE LOSS COMPROMISE CORTICAL AND TRABECULAR MICROSTructure. *The Journals of Gerontology: Series A*, *68*(10), 1218–1225. https://doi.org/10.1093/gerona/glt071

Sims, N. A., & Martin, T. J. (2020). Osteoclasts provide coupling signals to osteoblast lineage cells through multiple mechanisms. *Annual Review of Physiology*, *82*(1), 507–529. https://doi.org/10.1146/annurev-physiol-021119-034425

Singh, I. J., & Gunberg, D. L. (1970). Estimation of age at death in human males from quantitative histology of bone fragments. *American Journal of Physical Anthropology*, *33*(3), 373–381. https://doi.org/10.1002/ajpa.1330330311

Skedros, J. G., & Brand, R. A. (2011). Biographical Sketch: Georg Hermann von Meyer (1815–1892). *Clinical Orthopaedics and Related Research*, *469*(11), 3072–3076. https://doi.org/10.1007/s11999-011-2040-6

Skedros, J. G., Grunander, T., & Hamrick, M. (2005). Spatial distribution of osteocyte lacunae in equine radii and third metacarpals: considerations for cellular communication, microdamage detection and metabolism. *Cells, Tissues, Organs*, *180*, 215–236. https://doi.org/10.1159/000088938

Skedros, J. G., Hunt, K. J., & Bloebaum, R. D. (2004). Relationships of loading history and structural and material characteristics of bone: Development of the mule deer calcaneus. *Journal of Morphology*, *259*(3), 281–307. https://doi.org/10.1002/jmor.10167

Skedros, J. G., Kiser, C. J., Keenan, K. E., & Thomas, S. C. (2011). Analysis of osteon morphotype scoring schemes for interpreting load history: Evaluation in the chimpanzee femur. *Journal of Anatomy*, *218*(5), 480–499. https://doi.org/10.1111/j.1469-7580.2011.01348.x

Skedros, J. G., Kiser, C. J., & Mendenhall, S. D. (2011). A weighted osteon morphotype score outperforms regional osteon percent prevalence calculations for interpreting cortical bone adaptation. *American Journal of Physical Anthropology, 144*(1), 41–50. https://doi.org/10.1002/ajpa.21365

Skedros, J. G., Knight, A. N., Clark, G. C., Crowder, C. M., Dominguez, V. M., Qiu, S., Mulhern, D. M., Donahue, S. W., Busse, B., Hulsey, B. I., Zedda, M., & Sorenson, S. M. (2013). Scaling of Haversian canal surface area to secondary osteon bone volume in ribs and limb bones. *American Journal of Physical Anthropology, 151*(2), 230–244. https://doi.org/10.1002/ajpa.22270

Skedros, J. G., Mason, M. W., & Bloebaum, R. D. (1994). Differences in osteonal micromorphology between tensile and compressive cortices of a bending skeletal system: Indications of potential strain-specific differences in bone microstructure. *The Anatomical Record, 239*(4), 405–413. https://doi.org/10.1002/ar.1092390407

Skedros, J. G., Mason, M. W., & Bloebaum, R. D. (2001). Modeling and remodeling in a developing artiodactyl calcaneus: A model for evaluating Frost's Mechanostat hypothesis and its corollaries. *The Anatomical Record, 263*(2), 167–185. https://doi.org/10.1002/ar.1094

Skedros, J. G., Mendenhall, S. D., Kiser, C. J., & Winet, H. (2009). Interpreting cortical bone adaptation and load history by quantifying osteon morphotypes in circularly polarized light images. *Bone, 44*(3), 392–403. https://doi.org/10.1016/j.bone.2008.10.053

Skedros, J. G., Su, S. C., & Bloebaum, R. D. (1997). Biomechanical implications of mineral content and microstructural variations in cortical bone of horse, elk, and sheep calcanei. *The Anatomical Record, 249*(3), 297–316. https://doi.org/10.1002/(SICI)1097-0185(199711)249:3 <297::AID-AR1>3.0.CO;2-S

Skedros, J. G., Sybrowsky, C. L., Parry, T. R., & Bloebaum, R. D. (2003). Regional differences in cortical bone organization and microdamage prevalence in Rocky Mountain mule deer. *The Anatomical Record Part A: Discoveries in Molecular, Cellular, and Evolutionary Biology, 274*(1), 837–850. https://doi.org/10.1002/ar.a.10102

Skerry, T. M. (2006). One mechanostat or many? Modifications of the site-specific response of bone to mechanical loading by nature and nurture. *Journal of Musculoskeletal & Neuronal Interactions, 6*(2), 122–127.

Stout, S. D. (1986). The Use of Bone Histomorphometry in Skeletal Identification: The Case of Francisco Pizarro. *Journal of Forensic Sciences, 31*(1), pp. 296–300.

Stout, S.D. (1992). Methods of Determining Age at Death Using Bone Microstructure. In Saunders, S.R., Katzenberg, editors. *Skeletal Biology of Past Peoples: Research Methods.* New York, pp. 21–35.

Stout, S. D., Cole, M. E., & Agnew, A. M. (2019). Histomorphology: Deciphering the metabolic record. In J. E. Buikstra (Ed.), *Ortner's Identification of Pathological Conditions in Human Skeletal Remains (Third Edition)* (pp. 91–167). Academic Press. https://doi.org/10.1016/B978-0-12-809738-0.00006-5

Stout, S. D., & Gehlert, S. J. (1980). The relative accuracy and reliability of histological aging methods. *Forensic Science International, 15*(3), 181–190. https://doi.org/10.1016/0379-0738(80)90132-2

Stout, S. D., & Gehlert, S. J. (1982). Effects of field size when using Kerley's histological method for determination of age at death. *American Journal of Physical Anthropology, 58*(2), 123–125. https://doi.org/10.1002/ajpa.1330580203

Stout, S. D., & Lueck, R. (1995). Bone remodeling rates and skeletal maturation in three archaeoloqical skeletal populations. *American Journal of Physical Anthropology, 98*(2), 161–171. https://doi.org/10.1002/ajpa.1330980206

Stout, S. D., & Paine, R. R. (1994). Bone Remodeling Rates—A Test of an Algorithm for Estimating Missing Osteons. *American Journal of Physical Anthropology, 93*(1), 123–129. https://doi.org/10.1002/ajpa.1330930109

Stout, S. D., Porro, M. A., & Perotti, B. (1996). Brief communication: A test and correction of the clavicle method of Stout and Paine for histological age estimation of skeletal remains. *American Journal of Physical Anthropology, 100*(1), 139–142. https://doi.org/10.1002/(SICI)1096-8644 (199605)100:1<139::AID-AJPA12>3.0.CO;2-1

Stout, S. D., & Streeter, M. A. (2004). Un análisis histomórfo-logico del hueso cortical de la costilla de Janaab'Pakal. In V. Tiesler & A. Cucina (Eds.), *Janaab'Pakal De Palenque: Vida Y Muerte De Un Gobernante Maya* (pp. 123–131). D.R. Universidad National Autonoma De Mexico.

Stout, S. D., & Streeter, M. A. (2006). A histomorphometric analysis of cortical bone of Janaab'Pakal's Rib. In V. Tiesler & A. Cucina (Eds.), *Janaab' Pakal of Palenque: Reconstructing the Life and Death of a Maya Ruler* (pp. 60–67). The University of Arizona Press.

Streeter, M. (2010). A four-stage method of age at death estimation for use in the subadult rib cortex. *Journal of Forensic Sciences*, *55*(4), 1019–1024. https://doi.org/10.1111/j.1556-4029.2010.01396.x

Streeter, M. (2011). Histological Age-at-Death Estimation. *Bone Histology: An Anthropological Perspective.* https://scholarworks.boisestate.edu/anthro_facpubs/56

Streeter, M., Stout, S., Trinkaus, E., & Burr, D. (2010). Brief communication: Bone remodeling rates in Pleistocene humans are not slower than the rates observed in modern populations: A reexamination of Abbott et al. (1996). *American Journal of Physical Anthropology*, *141*(2), 315–318. https://doi.org/10.1002/ajpa.21192

Sugiyama, T., Meakin, L. B., Browne, W. J., Galea, G. L., Price, J. S., & Lanyon, L. E. (2012). Bones' adaptive response to mechanical loading is essentially linear between the low strains associated with disuse and the high strains associated with the lamellar/woven bone transition. *Journal of Bone and Mineral Research*, *27*(8), 1784–1793. https://doi.org/10.1002/jbmr.1599

Szulc, P., & Seeman, E. (2009). Thinking inside and outside the envelopes of bone: Dedicated to PDD. *Osteoporosis International*, *20*(8), 1281–1288. https://doi.org/10.1007/s00198-009-0994-y

Takayanagi, H. (2021). Osteoimmunology as an intrinsic part of immunology. *International Immunology*, *33*(12), 673–678. https://doi.org/10.1093/intimm/dxab057

Takeyama, S., Yoshimura, Y., Shirai, Y., Deyama, Y., Hasegawa, T., Yawaka, Y., Kikuiri, T., Matsumoto, A., & Fukuda, H. (2000). Low calcium environment effects osteoprotegerin ligand/osteoclast differentiation factor. *Biochemical and Biophysical Research Communications*, *276*(2), 524–529. https://doi.org/10.1006/bbrc.2000.3498

Teitelbaum, S. L. (1980). Pathological manifestations of osteomalacia and rickets. *Clinics in Endocrinology and Metabolism*, *9*(1), 43–62. https://doi.org/10.1016/s0300-595x(80)80020-x

Tersigni, M. A. (2005). *Serial long bone histology: Inter- and intra -bone age estimation* [Ph.D. Dissertation]. The University of Tennessee, Knoxville.

Thomas, C. D. L., Feik, S. A., & Clement, J. G. (2005). Regional variation of intracortical porosity in the midshaft of the human femur: Age and sex differences. *Journal of Anatomy*, *206*(2), 115–125. https://doi.org/10.1111/j.1469-7580.2005.00384.x

Thomas, C. D. L., Stein, M. S., Feik, S. A., Wark, J. D., & Clement, J. G. (2000). Determination of age at death using combined morphology and histology of the femur. *Journal of Anatomy*, *196*(Pt 3), 463–471. https://doi.org/10.1046/j.1469-7580.2000.19630463.x

Thompson, D. D. (1980). Age changes in bone mineralization, cortical thickness, and haversian canal area. *Calcified Tissue International*, *31*(1), 5–11. https://doi.org/10.1007/bf02407161

Thompson, D. D., & Galvin, C. A. (1983). Estimation of age at death by tibial osteon remodeling in an autopsy series. *Forensic Science International*, *22*(2–3), 203–211. https://doi.org/10.1016/0379-0738(83)90015-4

Thompson, D. D., & Gunness-Hey, M. (1981). Bone mineral-osteon analysis of Yupik-Inupiaq skeletons. *American Journal of Physical Anthropology*, *55*(1), 1–7. https://doi.org/10.1002/ajpa.1330550102

Tong, X., Burton, I. S., Isaksson, H., Jurvelin, J. S., & Kröger, H. (2015). Cortical Bone Histomorphometry in Male Femoral Neck: The Investigation of Age-Association and Regional Differences. *Calcified Tissue International*, *96*(4), 295–306. https://doi.org/10.1007/s00223-015-9957-9

Toroian, D., Lim, J. E., & Price, P. A. (2007). The size exclusion characteristics of type I collagen: Implications for the role of noncollagenous bone constituents in mineralization. *The Journal of Biological Chemistry*, *282*(31), 22437–22447. https://doi.org/10.1074/jbc.M700591200

Trichilo, S., & Pivonka, P. (2018). Application of disease system analysis to osteoporosis: From temporal to spatio-temporal assessment of disease progression and intervention. In P. Pivonka (Ed.), *Multiscale Mechanobiology of Bone Remodeling and Adaptation* (pp. 61–121). Springer International Publishing. https://doi.org/10.1007/978-3-319-58845-2_2

Turner, C. H. (1998). Three rules for bone adaptation to mechanical stimuli. *Bone, 23*(5), 399–407. https://doi.org/10.1016/S8756-3282(98)00118-5

Turner, C. H. (1999). Toward a mathematical description of bone biology: The principle of cellular accommodation. *Calcified Tissue International, 65*(6), 466–471. https://doi.org/10.1007/s002239900734

Turner, C. H. (2000). Reply to Letter to the Editor on: Toward a mathematical description of bone biology: The principle of cellular accommodation. *Calcified Tissue International, 67*(2), 184–187. https://doi.org/10.1007/s00223001114

Turner, C. H., Robling, A. G., Duncan, R. L., & Burr, D. B. (2002). Do bone cells behave like a neuronal network? *Calcified Tissue International, 70*(6), 435–442. https://doi.org/10.1007/s00223-001-1024-z

Turner, R. T., Wakley, G. K., & Hannon, K. S. (1990). Differential effects of androgens on cortical bone histomorphometry in gonadectomized male and female rats. *Journal of Orthopaedic Research, 8*(4), 612–617. https://doi.org/10.1002/jor.1100080418

Uda, Y., Azab, E., Sun, N., Shi, C., & Pajevic, P. D. (2017). Osteocyte Mechanobiology. *Current Osteoporosis Reports, 15*(4), 318–325. https://doi.org/10.1007/s11914-017-0373-0

Vajda, E. G., Kneissel, M., Muggenburg, B., & Miller, S. C. (1999). Increased intracortical bone remodeling during lactation in beagle dogs. *Biology of Reproduction, 61*(6), 1439–1444. https://doi.org/10.1095/biolreprod61.6.1439

van Oers, R. F. M., Ruimerman, R., van Rietbergen, B., Hilbers, P. A. J., & Huiskes, R. (2008). Relating osteon diameter to strain. *Bone, 43*(3), 476–482. https://doi.org/10.1016/j.bone.2008.05.015

Varga, P., Hesse, B., Langer, M., Schrof, S., Männicke, N., Suhonen, H., Pacureanu, A., Pahr, D., Peyrin, F., & Raum, K. (2015). Synchrotron X-ray phase nano-tomography-based analysis of the lacunar–canalicular network morphology and its relation to the strains experienced by osteocytes in situ as predicted by case-specific finite element analysis. *Biomechanics and Modeling in Mechanobiology, 14*(2), 267–282. https://doi.org/10.1007/s10237-014-0601-9

Veis, D. J., & O'Brien, C. A. (2023). Osteoclasts, Master Sculptors of Bone. *Annual Review of Pathology, 18*, 257–281. https://doi.org/10.1146/annurev-pathmechdis-031521-040919

Verborgt, O., Gibson, G. J., & Schaffler, M. B. (2000). Loss of osteocyte integrity in association with microdamage and bone remodeling after fatigue in vivo. *Journal of Bone and Mineral Research, 15*(1), 60–67. https://doi.org/10.1359/jbmr.2000.15.1.60

Verbruggen, S. W., & McNamara, L. M. (2018). Chapter 6—Bone mechanobiology in health and disease. In S. W. Verbruggen (Ed.), *Mechanobiology in Health and Disease* (pp. 157–214). Academic Press. https://doi.org/10.1016/B978-0-12-812952-4.00006-4

Villanueva AR, Mehr LA. (1977) Modifications of the Goldner and Gomori one-step trichrome stains for plastic-embedded thin sections of bone. *Am J Med Technol.* 1977 Jun;43(6):536–8. PMID: 69401.

von Meyer, G. H. (1867). Die Architectur der Spongiosa. *Reichert Und Du Bois-Reymond's Archiv, 8*, 615–628.

Walker, R. A., Lovejoy, C. O., & Meindl, R. S. (1994). Histomorphological and geometric properties of human femoral cortex in individuals over 50: Implications for histomorphological determination of age-at-death. *American Journal of Human Biology, 6*(5), 659–667. https://doi.org/10.1002/ajhb.1310060515

Wang, J. H.-C., & Thampatty, B. P. (2006). An Introductory Review of Cell Mechanobiology. *Biomechanics and Modeling in Mechanobiology, 5*(1), 1–16. https://doi.org/10.1007/s10237-005-0012-z

Wang, L., Liu, S., Zhao, Y., Liu, D., Liu, Y., Chen, C., Karray, S., Shi, S., & Jin, Y. (2015). Osteoblast-induced osteoclast apoptosis by fas ligand/FAS pathway is required for maintenance of bone mass. *Cell Death & Differentiation, 22*(10), 1654–1664. https://doi.org/10.1038/cdd.2015.14

Warner, A. J., Hathaway-Schrader, J. D., Lubker, R., Davies, C., & Novince, C. M. (2022). Tetracyclines and bone: Unclear actions with potentially lasting effects. *Bone, 159*, 116377. https://doi.org/10.1016/j.bone.2022.116377

Watanabe, Y., Konishi, M., Shimada, M., Ohara, H., & Iwamoto, S. (1998). Estimation of age from the femur of Japanese cadavers. *Forensic Science International, 98*(1–2), 55–65. https://doi.org/10.1016/s0379-0738(98)00136-4

Weinbaum, S., Cowin, S. C., & Zeng, Y. (1994). A model for the excitation of osteocytes by mechanical loading-induced bone fluid shear stresses. *Journal of Biomechanics, 27*(3), 339–360. https://doi.org/10.1016/0021-9290(94)90010-8

Weivoda, M. M., & Bradley, E. W. (2023). Macrophages and bone remodeling. *Journal of Bone and Mineral Research, 38*(3), 359–369. https://doi.org/10.1002/jbmr.4773

Wilson, A. K., Bhattacharyya, M. H., Miller, S., Mani, A., & Sacco-Gibson, N. (1998). Ovariectomy-induced changes in aged beagles: Histomorphometry of rib cortical bone. *Calcified Tissue International, 62*(3), 237–243. https://doi.org/10.1007/s002239900423

Wolff, J. (1892). *Das Gesetz der Transformation der Knochen*. Verlag von August Hirschwald.

Wu, K., Schubeck, K. E., Frost, H. M., & Villanueva, A. (1970). Haversian bone formation rates determined by a new method in a mastodon, and in human diabetes mellitus and osteoporosis. *Calcified Tissue Research, 6*(3), 204–219. https://doi.org/10.1007/bf02196201

Yeni, Y. N., Brown, C. U., Wang, Z., & Norman, T. L. (1997). The influence of bone morphology on fracture toughness of the human femur and tibia. *Bone, 21*(5), 453–459. https://doi.org/10.1016/S8756-3282(97)00173-7

Yoshino, M., Imaizumi, K., Miyasaka, S., & Seta, S. (1994). Histological estimation of age at death using microradiographs of humeral compact bone. *Forensic Science International, 64*(2–3), 191–198. https://doi.org/10.1016/0379-0738(94)90231-3

You, L., Temiyasathit, S., Lee, P., Kim, C. H., Tummala, P., Yao, W., Kingery, W., Malone, A. M., Kwon, R. Y., & Jacobs, C. R. (2008). Osteocytes as mechanosensors in the inhibition of bone resorption due to mechanical loading. *Bone, 42*(1), 172–179. https://doi.org/10.1016/j.bone.2007.09.047

Zebaze, R., & Seeman, E. (2015). Cortical bone: a challenging geography. *Journal of Bone and Mineral Research, 30*(1), 24–29. https://doi.org/10.1002/jbmr.2419

Zebaze, R. M. D., Ghasem-Zadeh, A., Bohte, A., Iuliano-Burns, S., Mirams, M., Price, R. I., Mackie, E. J., & Seeman, E. (2010). Intracortical remodelling and porosity in the distal radius and post-mortem femurs of women: A cross-sectional study. *Lancet, 375*(9727), 1729–1736. https://doi.org/10.1016/S0140-6736(10)60320-0

Zedda, M., Lepore, G., Biggio, G. P., Gadau, S., Mura, E., & Farina, V. (2015). Morphology, Morphometry and Spatial Distribution of Secondary Osteons in Equine Femur. *Anatomia, Histologia, Embryologia, 44*(5), 328–332. https://doi.org/10.1111/ahe.12141

Making the Mold
A Microstructural Perspective on Bone Growth and Mechanical Adaptation through Bone Modeling

2

COREY MAGGIANO AND
ISABEL MAGGIANO

List of Abbreviations

PEM - Periosteal and Endosteal Membrane
BMU - Bone Multicellular Unit
BSU - Bone Structural Unit
ELP - Endosteal Lamellar Pocket
LCS - Lacunar-Canalicular System
BLC - Bone Lining Cells

2.1 Introduction

Variation in relative bone size and shape informs much of skeletal inquiry, whether medical, forensic, or archaeological. Therefore, a detailed knowledge of bone growth, development, and mechanical adaptation is required to understand features of diagnostic, identificational, or inferential import. It is common knowledge that bones grow from growth plates. But understanding variation in bone size and shape requires a bigger picture and a number of integrated processes. Some, more familiar and more easily envisioned than others.

In long bones, formation and growth begins in a cartilaginous bone precursor and elongates through growth plates, in a process called endochondral ossification. But what then accounts for growth and mechanical adaptation of diaphyses and intramembranous bones; the reduction of metaphyses and compaction of trabecular voids; and the achievement of mature bone curvature, orientation, and robusticity?

A deceptively simple process, called bone modeling, accounts for all these changes. Typically defined in juxtaposition with intracortical remodeling, modeling is responsible for all formation and resorption initiated by the periosteal and endosteal membranes, whereas remodeling ensures the bone's material integrity and vascularization, replacing old bone through resorption and subsequent formation at collections of cells called bone multicellular units (Frost, 1987). These processes ensure that, during skeletal maturation and (for as much as is possible) thereafter, bone achieves a delicate balance of adaptive architectural strength and efficiency of mass.

DOI: 10.4324/9781003385608-2

Unfortunately, modeling activity is not easily conceptualized, in part because it is not easily measurable or even directly observable. Compact bone is an optically dense, dynamically adapting, mineral and organic tissue of remarkable structural and chemical complexity. We cannot watch bone histology as it changes and so are left to infer processes from resulting fragments of stratigraphic microarchitecture. In this regard, observing modeling is more challenging than remodeling, with its relatively easily defined unit of depositions, the secondary osteon. Conceptualizing modeling accurately may be elusive, but it is necessary to truly understand bony changes at the microscopic scale.

Envision bone growth sped up; the lifetime of a femur could be viewed within a minute or two. The solid, static state of bone becomes relatively unimportant from our high-speed perspective. A mold of cartilage bubbles up and a sheath of bone wraps it, the ends of the bony collar hardening under elongating caps. Looking closer, the new layers are proliferating, seeming to ebb and flow in all directions; added at first to the entire surface, other times only to a specific region. If distinguishable, sites of activity would appear patch- or puddle-like, amorphous, forming in layers, like spreading clay over inner and outer surfaces, each new smear widening out from epicenters of activation. Original or slower formations would be overlain by those newer or faster, each one older at their epicenters than at their extremities. Other flows would appear to reverse, resorbed, seemingly evaporating from the external clay puddle margins, revealing deeper, older regions.

From this visualization, some useful questions should leap to the foreground. What cells mold these bony structures? What determines the number or speed of their formative or resorptive phases? How many of these "epicenters" of activity does it take to mold large bone surfaces? What impedes or alters the process? Unfortunately, we cannot view bone growth and modeling from start to finish in minutes, like a potter's work. Nor can we easily discern between separate osteogenic membrane activations, nor count them, nor measure their actual rate of formation or resorption rather than merely the rate at a particular transsection. In fact, for now, these activations are nearly heuristic in comparison with observable units formed during remodeling.

Thankfully, redoubled efforts of bone researchers have permitted significant gains in our understanding of bone modeling over the past 80 years, revealing some secrets of bone's mold maker (see for example Becker et al., 2020; Cambra-Moo et al., 2015; Enlow, 1962; Frost, 1963a, 1963b; Goldman et al., 2003, 2009; Maggiano, Maggiano & Tiesler et al., 2015; Maggiano et al., 2015; Maggiano, Maggiano, Tiesler, Kierdorf et al., 2011; McFarlin et al., 2016; Raguin et al., 2014; Robling et al., 2001; Taylor and Frost, 1966). Of primary importance, they revealed what modeling can accomplish during formation and mechanical adaptation: (1) diametric growth, curvature, and orientation, (2) medullary extension and centralization, (3) metaphyseal reduction, (4) lamellar compaction, and (5) mechanical adaptation. All this is performed by two membranes: the periosteum and the endosteum. Despite their sometimes assumed similarity, these membranes have unique developmental origins, dissimilar histologically, disparate environments, and different sensitivities to mechanical and chemical stimuli. Yet, they somehow maintain a concerted effort, cooperating even synergistically, in order to effect bony change. This chapter defines, describes, and explains bone modeling microstructure against the context of growth and development. It encourages perspectives that advance hypothesis driven analyses in skeletal biology for forensic and bioarchaeological applications alike.

2.2 Defining Modeling

Despite (or perhaps because of) the obvious importance of bone modeling, there has been significant contention and confusion regarding the term's usage, and even its basic definition. To develop a functional definition of "modeling", one needs to (1) differentiate modeling as a sub-process of general growth and development, (2) recognize that the process is similar in all bones, whether they are formed intramembranously or endochondrally, and (3) distinguish clearly between modeling and remodeling.

2.2.1 Growth and Modeling

An important defining characteristic of modeling is its conceptual distinction from other aspects of long bone development, growth, and fracture repair. Bone "growth" is a more general concept referring to overall increase in the size and/or mass of a developing skeleton and transpires via two main processes: intramembranous and endochondral bone formation. Modeling can be differentiated from growth as a subprocess, transpiring only at the periosteal and endosteal membranes (Frost, 1963b) (here referred to collectively as PEM) rather than within the physis or callus. A functional understanding of bone growth, however, recognizes that as a hard tissue, bone cannot grow via developmental formation alone. Unlike soft tissue, bone cannot be moved or reorganized in relative tissue-space after its formation. Even if it were possible, though, bone grown in such a manner would, be (1) less resistant to strains, owing to its highly disorganized matrix (Zimmermann et al., 2019), and (2) too heavy to satisfy its structural and functional role in the mobile musculoskeletal system. Therefore, growth necessitates significant resorption in addition to net formation. Modeling satisfies this requirement, too. In general, it removes bone from the medullary surfaces and deposits bone diametrically, permitting an overall increase in bone size and maintains or adjusts bone shape during growth. Detailed consideration demonstrates the underlying mechanisms behind modeling are similar regardless of whether the bone's origins are intramembranous or endochondral.

2.2.2 Intramembranous and Endochondral Bone Modeling

During intramembranous growth, primary bone formation takes place directly within or under mesenchymal condensations differentiating into osteoblasts (Allen & Burr, 2019; Percival & Richtsmeier, 2013). These cells secrete an unmineralized matrix called osteoid, hardened by binding calcium into quickly forming bone, eventually merging and entrapping the osteoblasts that made them. This first bone material is called woven bone and is characterized by a highly disorganized mineral and collagen matrix. Thus, is formed the primary ossification center, providing surfaces for new osteoblasts to create organized lamellar bone through continued deposition (Mizoguchi & Ono, 2021, Scheuer & Black, 2000). Several of these islands may be combined in the eventual formation of a complete intramembranous bone, flat skull bone being an example for this. In larger, or thicker bones, where blood and nutrient supply is insufficient, vascularization and subsequent osteoclastic activity may result in a medullary cavity or spongy interior (Allen & Burr, 2019). The bone's maturation includes both the formalization of the original mesenchymal surface into periosteum, and the resorption of original woven bone by osteoclasts and its replacement with lamellar bone. Though a complete discussion of the various differentiation pathways of all cell types would be beyond our goals here, recent work in non-humans has shown there are separate periosteal and endosteal stem cell lineages. Periosteal stem cells not only maintain their baseline intramembranous formation, but also acquire endochondral formation capabilities during fracture repair (Debnath et al., 2018).

In contrast, endochondral ossification begins with the development of a non-vascular cartilaginous precursor or "anlage" (Allen & Burr, 2019). An example for this are the diaphyses of long bones. Endochondral ossification is both evolutionarily more recent and more prevalent in larger organisms due to comparably faster bone formation (Mizoguchi & Ono, 2021). In this process, mesenchymal cells differentiate into chondroblasts that produce type II collagen and alkaline phosphatase, generating the anlage which is enveloped by the highly vascularized perichondrium. Some of its mesenchymal cells differentiate into osteoblasts, depositing osteoid in a "bone collar" around the anlage intramembranously (Berendsen & Olsen, 2015). This collar limits nutrient diffusion to the anlage which provokes vascular invasion and the formation of the nutrient vessel (Allen & Burr, 2019). Although many aspects are still unknown, the conversion of avascular cartilage into bone is intimately coupled with the development of blood supplying vessels (Maes, 2013). In vivo studies show chondrocytes there have two possible fates: they either die through apoptosis or further differentiate into osteoblasts, offering two sources for bone-making cells during growth (Mizoguchi & Ono, 2021, Yang et al., 2014). At some point, osteoclasts, or bone-eating cells, start resorbing newly formed woven bone within the bone collar to shape a hollow diaphysis and provide the bone with a medullary cavity. Subsequent lamellar apposition transpires simultaneously to the maturation of the perichondrium into the periosteum.

Still in the utero, one or two ends (depending on the bone) of the compact bone collar differentiate into chondroblasts comprising epiphyseal growth plates, responsible for the bone's longitudinal growth (Allen & Burr, 2019, Martin et al., 1998, Voller et al., 2020). Elongation involves coordinated chondrocyte proliferation, hypertrophy, apoptosis, calcifying cartilage, vascularization, and new bone from osteogenic cells. The growth plates that accomplish it are highly ordered structures of the metaphysis that can be separated into distinct zones, each with its own function (Figure 2.1). In the distal metaphysis, the first zone (furthest from the bone formation) is referred to as the reserve, germinal, or resting zone. Its main aspect is extracellular matrix, with a comparatively lower number of embedded chondroblasts. Both the perichondrium and the chondroblasts in this zone contribute to the extracellular matrix (Hallett et al., 2021). Next is the proliferative or columnar zone. Here, chondroblasts multiply, flatten, and are packed into columnar stacks that produce type II collagen. The hypertrophic zone is populated by enlarged hypertrophic chondrocytes, up to ten times bigger in comparison to their predecessors. They secrete type X collagen fibers unique to these cells, also occurring in bone calluses after fracture. Cartilage stiffens here as a result. It is at this hypertrophic zone that longitudinal growth occurs. Chondrocytes in this oxygen poor environment release calcium, providing provisional mineralization, and create a zone of calcified cartilage. This early calcification causes blood vessels to be introduced, thereby supplying nutrients and cells of various types to the new bone. Also in this zone, chondroclasts and osteoclasts remove cartilage, creating scaffolds in which spongy bone is formed, known as secondary spongiosa, and leave behind a zone of ossification (Voller et al., 2020). The latter zone is also where osteoclasts remove calcified cartilage and newly formed woven bone to make room for more intricate secondary lamellar bone, through the remodeling process.

Figure 2.1 Proximal epiphyseal plate of a rabbit embryo (4 wks.) The metaphysis is below (blue), cartilaginous growth plate is centered (purple), and epiphyseal bone is above. Columns of cells span resting, proliferative, and hypertrophic zones (top to bottom, toluidine blue, ×60 objective). (From Rivas R and Shapiro F, 2002, J Bone Joint Surg AM 84:85–100. Reproduced by permission, JBJS, Inc.)

Two important structures exist at the periphery of the growth plate, the ossification groove of Ranvier and the perichondral ring of Lacroix. The groove of Ranvier is a rim of cells surrounding the growth plate that supports its peripheral growth and adjusts its size to the ongoing apposition determining diaphyseal diameter (Shapiro, 2002, Voller et al., 2020). The perichondral ring of Lacroix consists of fibrous tissue overlying the groove of Ranvier to add stability. It also builds the connection between the metaphyseal and the epiphyseal periosteum (Lacroix, 1947).

In addition to the described diaphyseal development, secondary growth centers develop at both terminal ends of a bone, still within the cartilage anlage. These separate bone centers solicit their own vessels and form bone in the core of the epiphysis, eventually forming the articular surface of the bone. Later in adolescence and young adulthood, epiphyses will fuse with the diaphyseal section after extension ceases. There is a growing recent body of work suggesting a hypothetical "tertiary" type of growth center needs to be considered, which includes at least some bony protrusions, both articular and non-articular in non-human bones (Blitz et al., 2009; Hamrick et al., 2010; Killian, 2022). Whether this process looks similar in humans and what its relationship is with the periosteum, modeling, and mechanical adaptation awaits further research.

Longitudinal growth ends due to cessation of cell division and chondrocyte senescence in the growth plate's resting zone, which is followed by its ossification. Hormones play a key role in this process and studies show elevated concentrations of estrogen can lead to an irreversible acceleration thereof (Nilsson et al., 2014). Despite the cessation of longitudinal growth, both the periosteum and the endosteum continue to influence a bone's diaphyseal shape and size throughout life through bone modeling. Hence, though the recruitment, differentiation, and activation of cells continuing modeling processes can mimic these developmental processes in mechanism, however, typically the term "modeling" itself is reserved for events transpiring after initial bone formation.

2.2.3 Modeling and Remodeling

Once formed, bone tissue must be maintained. In some organisms (including humans), modeling's sister process, remodeling, accomplishes this function – one shaping, the other reconstituting. The two sister processes would seem easily separable. As Parfitt (1983) and Jee and colleagues (2001) summarize, modeling and remodeling have several different defining traits, including speed of formation (modeling is faster) and microstrain activation (modeling requires ~2x the strain to stimulate formation). Unfortunately, despite general differences, easy distinction on a case by case basis is often elusive. Primarily this is due to three main confounding issues: (1) Unclear distinction between periosteal/endosteal osteogenic envelopes and those formed during remodeling processes. (2) The assumption that evidence of prior resorption indicates *only* remodeling (equating resorption and remodeling on certain surfaces despite little or no evidence for BMU formation). (3) The historic (and even modern) clinical use of the word "remodeling" as a general term to describe nearly all bony changes during growth, healing, or in response to various stimuli. Often this leads to *modeling* processes being labeled "periosteal or endosteal *remodeling*" (e.g. Frost, 1963b; 1964), necessitating caution in the interpretation of some early literature (see also Barak, 2019). Starting as early as the 1970s, publications like Frost's (1973), attempted to clarify the distinction between modeling and remodeling processes.

Distinction can be drawn between "modeling" and "remodeling" both functionally and structurally. Modeling accounts for changes in bone size, shape, and position (relative to other tissues) via primary bone formation and resorption at the PEM. It can result in dramatically larger volumes of deposited or resorbed bone tissue, sometimes encasing entire bone surfaces in sheet upon sheet of bone, or removing them altogether. In contrast, remodeling is the process of bone maintenance and repair, accomplished by bone multicellular units (BMUs), collections of osteoclasts and osteoblasts arranged in a particular structure. This multicellular unit penetrates existing primary bone via resorption at its leading end "or cutting cone" and concentrically and appositionally replacing it, from the void's periphery, inward, with new lamellar bone. Often these secondary lamellae close around an accompanying vessel and nerve system at their center, forming a vascular canal (Bolamperti et al., 2022; Martin et al., 1998; Parfitt, 2003). BMUs result in smaller bone structural units (BSUs), referred to as Haversian systems, or "osteons" (transverse diameter ~200–350 ųm) in intracortical tissue (Cooper et al., 2011; Maggiano, Maggiano & Clement et al., 2016), and "hemi-osteons" (~60 ųm) in cancellous tissue (Parfitt 1994, 2003; van Oers et al., 2008). BMUs form in three sequential steps: activation (A), resorption (R), formation (F) at a single two-dimensional location. It is more difficult to remember, however, that each BMU is simultaneously accomplishing both "R" and "F" in three dimensions (and potentially "A" as well, during branching events) (Maggiano, Maggiano & Clement et al., 2016).

PEM and BMU activity are also distinguished by the degree of "coupling" between formative and resorptive processes. PEM activity requires no connection (in process or mechanism) between formation or resorption (Frost, 1973; Martin et al., 1998; Robling et al., 2006). Therefore, according to convention, modeling could progress from "A" to "F" and subsequently reverse to "R" when necessary (or vice versa). This stands in stark contrast to BMU activity which appears "coupled" with formation closely following resorption in both time and space (Erben, 2015; Lassen et al., 2017; Parfitt et al., 2000). Unfortunately, most observation of bone histologically is not in real time which can hide certain circumstances during development or disease that challenge the closeness of BMU coupling, ensuring that formation lags behind resorption significantly, leading to large resorptive bays (Andronowski & Cole, 2021; Cooper et al., 2016; Maggiano, Maggiano & Tiesler et al., 2016; Montoya-Sanhueza & Chinsamy, 2018). This phenomenon is sometimes called "trabecularization" of compact tissue, an occurrence blurring the lines between compact, trabecular, and endocortical tissues; and therefore, between osteonal and hemi-osteonal BSUs, and those resulting from endosteal modeling events, necessitating caution in histological interpretation (see Section 2.3.2.3).

In short, when referring to any process accomplished by the PEM rather than the BMU, the term "remodeling" can be abandoned in favor of "modeling", even when formation follows previous resorption. In addition, great care should accompany the use of processual terminology, "periosteal" or "endosteal", and positional terminology, "pericortical" or "endocortical". Conflating these terms creates error and oversimplifies bony responses, leaving important questions unasked. Confusion can also be averted here by using the acronym "PEM" to address the periosteal and endosteal membranes collectively; and by referring to bone tissue formed by these membranes by a cortical descriptor (i.e. pericortex or endocortex) when context permits (rather than only using these terms to refer to the current surfaces of the cortex, or as arbitrary "zones").

2.3 Modeling Form

2.3.1 Membrane Histology and Vascularization

The periosteal membrane (Figures 2.2 and 2.3) comprises an outer fibrous layer and an inner layer of osteogenic cells called the cambium, firmly adhered to the bone surface via Sharpey's fibers that extend deep into bone tissue (Allen et al., 2004; Bisseret et al., 2015; Dwek, 2010; Tang, 1986). A three zone model has also been suggested by Squier et al. (1990), based on the functional separation of three zones: an external region marked by mature fibroblasts and dense collagen fibers; a region with increased vasculature, pericytes, and a more amorphous extracellular matrix, and a thin layer of the precursor and osteogenic cells themselves. It is important to note that the degree of distinction between the cambium and any osteogenic layer would be dependent on the nature of deposition there, including the vascular morphology and rate of deposition.

Fibroblasts and their excreted matrix (densest in the external layer) lend the membrane its particular strength and exert surprising force on the bone itself via membrane insertion on the physis (Forriol & Shapiro, 2005), though it does not entirely enclose it (Jee et al., 2001). Recent studies underline the cellular heterogeneity of periosteal layers which include cells that function as tissue-specific stem cell reservoirs (Bisseret et al., 2015) and several types of macrophages, all playing important roles in tissue repair during fracture (Alexander et al., 2017; Wang et al., 2017; Yu et al., 2010). A higher concentration of osteoprogenitor cells in the periosteum compared to the endosteum was reported, although the histomorphology of the osteogenic layer changes regionally based on the activity of the constituent cells (Brighton et al., 1992). In dormancy, osteogenic cells are elongated; while in actively forming bone they adopt a more cuboidal shape, arranged as a simple stratified epithelium while secreting osteoid (Ellender et al., 1988; Chow et al., 1998; Wang et al., 2017).

Reports are varied in their interpretation of similarities between the periosteum and endosteum. The periosteum and endosteum are histologically similar according to Shapiro

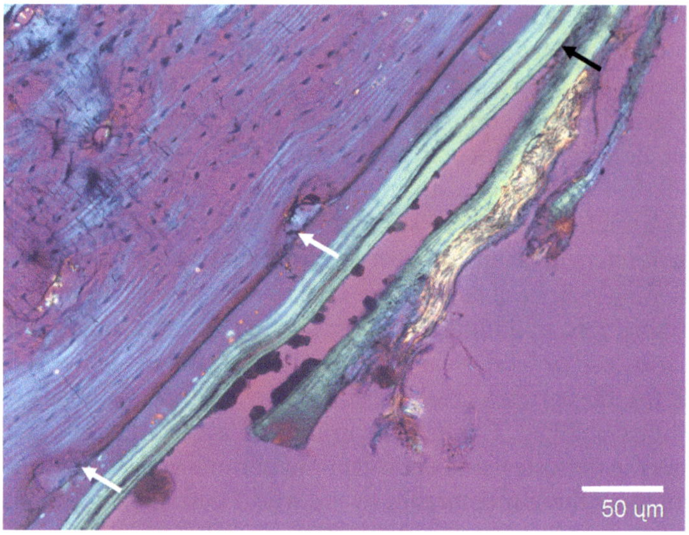

Figure 2.2 Juvenile periosteal tissue preserved by natural mummification (black solid arrow). The external bone surface was currently forming primary lamella at the time of death as evidenced by several partially enclosed primary vessel canals (white solid arrows). Thin-ground, undecalcified transection of the naturally mummified, archaeological, femoral mid-diaphysis, Dakhleh Oasis, Egypt. Micrograph taken with red quartz cross-polarized filter (530 nm compensated). Scale bar indicates 50 ųm.

(2002), but Hohmann and colleagues point out that the periosteum is densely innervated and contains lymphatic vessels (Hohmann et al., 1986). Still others emphasize their differences more strongly, labeling periosteal osteogenic tissue as fibroblast-like and in deep periosteal layers, whereas BLCs comprise the endosteum (Allen & Burr, 2019; Martin et al., 1998). It would seem that the periosteum is a more histologically complex membrane than the endosteum, whether currently depositing osteoid or not (Figure 2.4). Not only are the membranes distinct histologically, but their environments are completely different. The periosteum is constrained and interrupted by tendons, ligaments, and fibrocartilage, whereas the endosteum is awash with hematopoitic bone marrow (Jones et al., 1991). The periosteum accounts for the majority of blood flow to the bone and simultaneously achieves nearly all its venous drainage (Brookes & Revell, 2012; Maes, 2013; Simpson, 1985). Periosteal vessels create a complex network with fewer anastomoses in the diaphysis compared to the metaphysis (Crocker et al., 1970). On occasion, periosteal vessels can even affect a full cortical penetration to the medullary cavity (Rhinelander, 1972). In contrast, at least in the adult, the endosteum seems to penetrate bone tissue radially via Volkmann's canals which connect Haversian systems laterally within the intracortex (Enlow, 1962; Maggiano et al., 2011).

The periosteum and endosteum are also differentially affected by mechanical and chemical stimuli (Robinson et al., 2021). For example, the periosteum is well known for its capacity to induce formation in response to skeletal loading, particularly when higher loads are experienced dynamically, separated by resting or recovery periods (Mosley & Lanyon, 2002; Robling et al., 2001). Carpenter and Carter (2008, 2010) proposed a model in which pressures on periosteal surfaces, by adjacent muscles for example, can impede bone formation or induce bone resorption and affect bone cross-sectional shapes. Conversely, the endosteum has shown far reduced or non-existent formative adaptation in response to

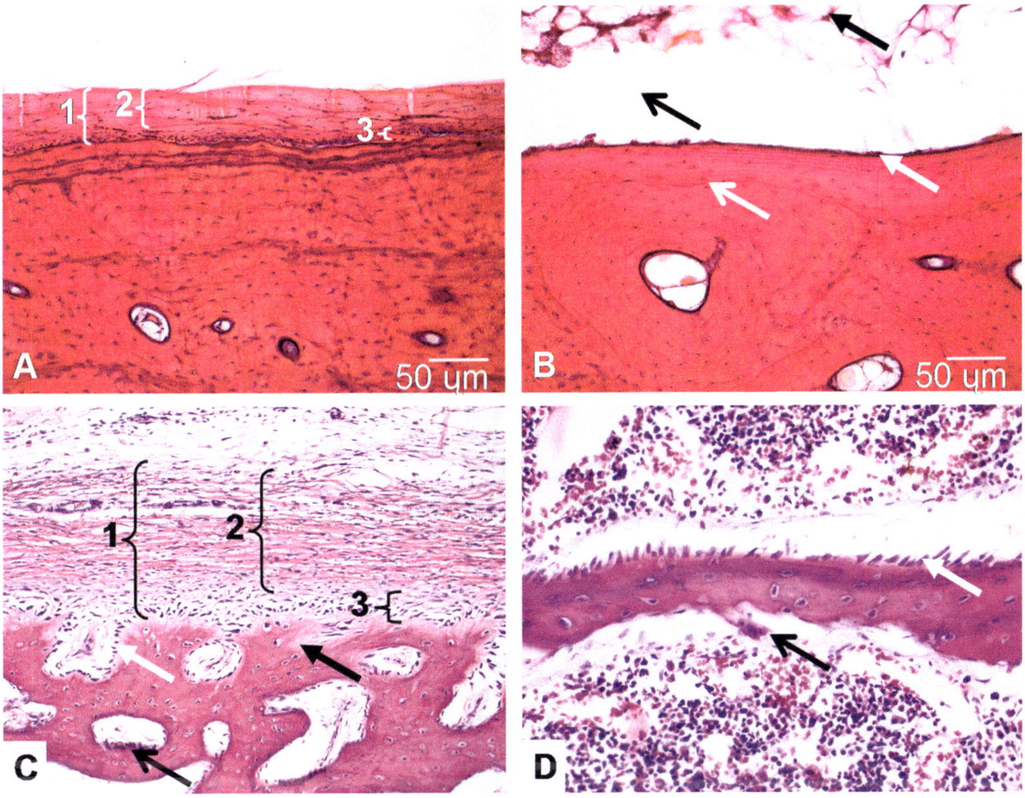

Figure 2.3 A) Mature periosteal membrane of the humerus (bracket 1), demonstrating clear distinction between the outer fibrous layer (bracket 2) and the smaller, darker osteogenic cellular layer (bracket 3). Note the most recent primary bone deposition has occurred in two thin phases (white arrow). B) Mature endosteal membrane of the humerus (white arrow). Top margin shows marrow lipid tissue (black arrow) and marrow space (black lined arrow). Note a cement line between Haversian tissue (below white lined arrow) and the most recent endosteal primary formation phase (above white lined arrow). "A" and "B" demonstrate a quiescent bone surface, as evidenced by the lack of unmineralized matrix. C) Fetal (6 mo. gestation) metatarsal bone collar demonstrating cellular and tissue level changes associated with osteogenesis at the periosteum, including columnar osteoblasts (white arrow). Periosteal deposition of woven bone transpires via significant secretion of unmineralized matrix (osteoid) (black arrow). Note the targeted intracortical resorption at internal void surfaces facing the endosteum (bottom of image) as demonstrated by the position of osteoclasts (black lined arrow). D) Endosteal formation in the same bone as "C" via columnar osteoblasts on one side of a trabecular spicule, while the other side appears newly resorbed with one remaining osteoclast (black lined arrow) and Howships lacunae left behind as evidence. All images are generated from paraffin embedded, decalcified, thin transections, stained with hematoxylin and eosin. Scale bar indicates 50 ųm for "A" and "B", taken at 100x magnification; "C" was taken at 200x magnification; "D" at 400x. Images courtesy of Lisa Lee.

mechanical loading (Jones et al., 1991; Riffault et al., 2020; Robinson et al., 2021; Srinivasan et al., 2002), though see Birkhold et al. (2016, 2017) for alternate perspectives. Endosteal estrogen stimulation generates bone formation during puberty (Martin, 2003), which reverses with decreasing estrogen levels (Nilsson et al., 2014). Venken et al. (2006) however, call attention to the complexity of these pubescent bony changes, specifically in that bony response is not only hormone specific, but also dose dependent. In later life, basic sex

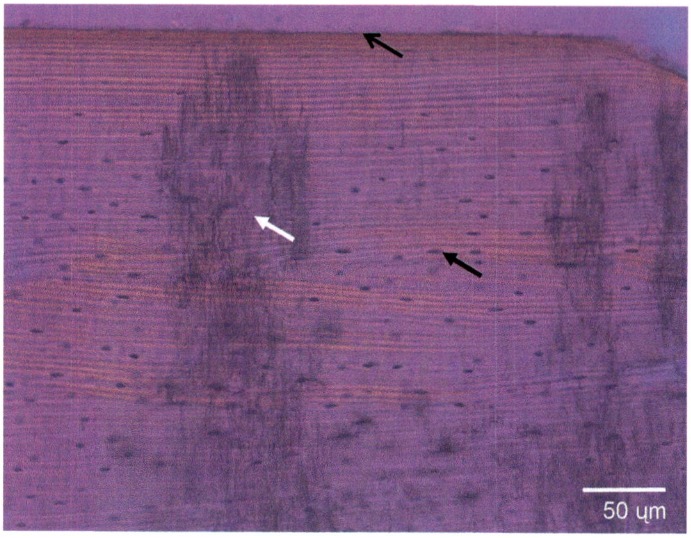

Figure 2.4 Adult, primary endosteal lamellae with interspersed lacunae, once housing osteocytes entombed during active apposition. Endocortical surface is marked by the black lined arrow. Structure of the lacunar-canalicular system is emphasized by a bacterial diagenic effect which has slightly enlarged and darkened canaliculi radiating from some lacunae (white solid arrow). Outside these regions canaliculi are less visible. Thin-ground, undecalcified transection of archaeological, ulnar mid-diaphysis, Dakhleh Oasis, Egypt. Micrograph taken with red quartz cross-polarized filter (530 nm compensated). Scale bar indicates 50 ɥm.

hormones affect each envelope differently and play a significant role in determining bone loss patterns (Gamsjaeger et al., 2021; Paschalis et al., 2017). Despite these differences and despite their clear physical separation within bone tissues, the periosteum and endosteum are capable of extremely fine-tuned cooperation during growth and adaptation; without which bone's compositional resilience and capacity for remodeling alone would insufficiently counter the loss of strength suffered by inappropriate morphology.

In the healthy, active individual, modeling is initiated and directed by a unique genetic and mechanical developmental history, accounting for most of what we recognize about a specific bone's morphology and function. This unique history is an important source of histomorphological variation and records itself stratigraphically within bone microarchitecture.

2.3.2 Modeling Formation and Resorption

2.3.2.1 Primary Bone Histology

Modeling formation at the PEM can occur in rapid deposition of woven bone with less well-organized collagen content; or as more slowly forming, lamellar bone with more regularly organized collagen. In lamellar bone, individual sheets are fairly uniform in thickness (Ascenzi et al., 1982, Mitchell & Van Heteren, 2016), ranging from 2 to 7 ɥm in cross-section transversely (Frost, 1963a; Reid, 1986; Reznikow et al., 2014) with lamellar dimensions unknown along other axes. A single lamella is a pair of layers each with differing collagen orientation (typically one bright and one dark under polarized light), resulting from calcification of osteoid (average osteoid seam, 15 ɥm), layered one upon the other

(Johnson, 1964; Marotti, 2010; Mitchell & van Heteren, 2016). Newer applications of FIB-SEM dual beam microscopy permit more detailed structural 3D analyses of their composition and have uncovered two distinct organizational levels (Reznikow et al., 2014). Specific to human bone, unidirectionally ordered fibrils are separate from disordered material, filling the spaces between ordered bundles. The latter create the space holding the canaliculi. Reznikow and colleagues' (2014) groundbreaking analysis found no structural difference between primary and secondary lamellar bone.

The biorhythm of Retzius lines in teeth closely aligns with appositional rates of lamellar bone (Bromage et al., 2009), suggesting both are regulated through hypothalamic signaling (Mahoney et al. 2016, Pitfield et al. 2019). The manifestation of such biorhythms in hard tissues enables the calculation of formation rates, with some species- and age-specific variation (Ricqles, 1991): In humans, Parfitt et al. (2000) report a rate of 1.04 μm/day in children 2–20 years old, while, as an example, Cerrito et al. (2022) measure 0.75–1.25 μm/day in rhesus macaques. Care should be taken in comparisons, though between similar species, ages, and primary bone type (particularly how many ranks of osteogenic cells there are – lamellar bone has one, but fibrolamellar or plexiform has more). Bromage et al. (2009, 2011) estimate that lamellar bone forms at 3–6 μm every 8–9 days (i.e., 0.33–0.75 μm/day; but note variability). Altogether, human lamellar bone develops relatively slow, with endosteal and secondary bone depositing faster than periosteal tissue (McFarlin et al. 2016). According to isotopic variation consistent with seasonality, human endosteal lamellar tissue in one individual had a relatively constant rate of 0.79 μm/day, less in areas affected by disruptions (Maggiano et al., 2019).

During their formation, lamellae encase some of the osteoblasts (and even vessels) from the osteogenic membrane that produces them. These cells then adopt a new morphology, identifying them as osteocytes, locked in small oblong cavities between lamellae called lacunae (Chen et al., 2015; Manolagas & Parfitt, 2010). Osteocytes intercommunicate through their processes within the so-called lacunar-canalicular system (LCS) (Figure 2.4). The LCS is a complex network, or syncytium, acting as a means for transmission of chemical and/or mechanical signals, which, as part of compact bone's mechano-transduction system, are vital for healthy bone growth, maintenance, and adaptation (Knothe Tate, 2003; Schaffler et al., 2014; Taylor & Lee, 2003). Part of this system are also bone lining cells (BLCs), covering most bone surfaces (Aubin, 2001, Bolamperti et al., 2022). Recently, their role as osteoblastic progenitors has also been emphasized, underlining their involvement as regulators of bone formation (Bolamperti et al. 2022). According to Martin (2000), another potential role for lining cells is to protect bone surfaces from a constant resorption signal, typically dampened in healthy, active tissue. To enact resorption, these cells must somehow pull away from the bone's surface, exposing it to resorption, though this is one of many theories. Osteoclastic destruction transpires via mineral acidification and organic enzymatic digestion of surface areas enclosed by osteoclasts' ruffled border (Vaananen et al., 2000; Weivoda & Bradley, 2023) which limits the scale of resorption to small regions per cell, leaving a "scalloped" surface due to Howships lacunae (each typically < 20 ųm). With efforts in identifying structural properties of lamellar bone still ongoing, fortunately, light microscopy easily reveals the transected stratigraphy of lamellae, detailing localized growth and adaptive processes at the histological scale.

Though this general description of modeling formation and resorption fits adult human lamellar bone, it is not accurately descriptive of other ages, and certainly is different across taxa (Reznikov 2014; Shahar & Weiner 2018). To organize terminology for discussions of primary bone in general, Locke (2004) encourages the use of the word "laminar" bone for several reasons, including that the descriptor "lamellar" fails to distinguish primary and secondary bone, and that the vertebrate norm is often appositional "laminae", sets of 4–20 lamellae (Locke 2004). These are layered on top of one another and contain porous horizons formed by the regular entombment of vascular networks, at one time associated with the formative membrane (Locke 2004). Laminae display dramatic variability in their spatial relationships with one another, and in the degree or orientation of vessel incorporation. From this generality we can approach various specific tissue type descriptors, including "plexiform" and "fibro-lamellar". Plexiform bone (Stein & Prondvai, 2014) is similar to laminar bone except that it is less concentrically organized, repeatedly folding back on itself (Figure 2.5). Fibro-lamellar tissue has more randomly distributed primary vascular voids, dominated by relatively interspersed primary osteons (Figure 2.6). Compare these to the even pace of appositional lamellar formation in a large reptile (Figure 2.7).

2.3.2.2 Primary Lamellae and Vessel Entrapment

Much of what distinguishes PEM primary tissue types depends on the organization of lamellar, vascular, and woven bone regions. In general, formative layers are separated from the surface of the bone by struts of woven bone spanning a highly vascularized space upon which the newest layer is formed. Within these vascular voids, lamellar bone is normally laid down more slowly. Sometimes woven scaffolding forms quickly, so that there are several tables of unfinished lamina upon one another and only deeper intracortical tissue displays more mature lamellar microstructure. To illustrate meaningful differences between these bone types, Currey (2006) cites a simple perspective outlined by Castanet and colleagues (1996), in which the relative speed of formation at the membrane determines structural variance in the mallard duck. In this example, the fastest formation is accomplished by circumferential laminar formation (especially in plexiform bone). Note, in Figure 2.5b, there are 19 potential deposition surfaces in just this view of a single cortex), each trapping whole generations of periosteal vascular sheaths. Slower forming fibro-lamellar or primary osteonal tissue captures vessels more rarely, particularly seeming to favor longitudinal vessels (Castanet et al., 1996). Human periosteal lamellar tissue typically traps few transverse vessels from the vascular sheath during osteogenesis and so cannot typically be referred to as laminar. However, rapid growth periods in humans can demonstrate laminar-like sets of pericortical woven or lamellar bone separated by longitudinal primary vessel horizons locked within "generations" of lamellar apposition (Figure 2.8). These can lead to osteon banding in some regional tissue in humans over certain developmental time periods and locations (Andronowski et al., 2017).

Entombed vessels within human primary bone are not trapped equally on both external and medullary surfaces. Periosteal vessels overlain by lamellae are typically longitudinal, and are often referred to as primary vascular canals, or primary "osteons" (Currey, 2006; Hage & Hamade 2014); despite that they are completely unrelated to secondary osteons (or Haversian systems) formed by BMUs (Figures 2.9 and 2.10). In fact, using this term conflates two completely different structures, both regarding their associated vessel, and their structure and function. Primary osteons do not require resorption prior to enclosure and infilling (Currey, 2006), and can range in the type of bone in-filling, from woven to organized lamellae. In transection, primary osteons appear "sandwiched" between periosteal lamellae that bend slightly over and under them, but typically contain few or even no lamellae of their own (sheathing the vessel) (Bright & Elmore, 1968; Chamay, 1970; Currey, 2006;

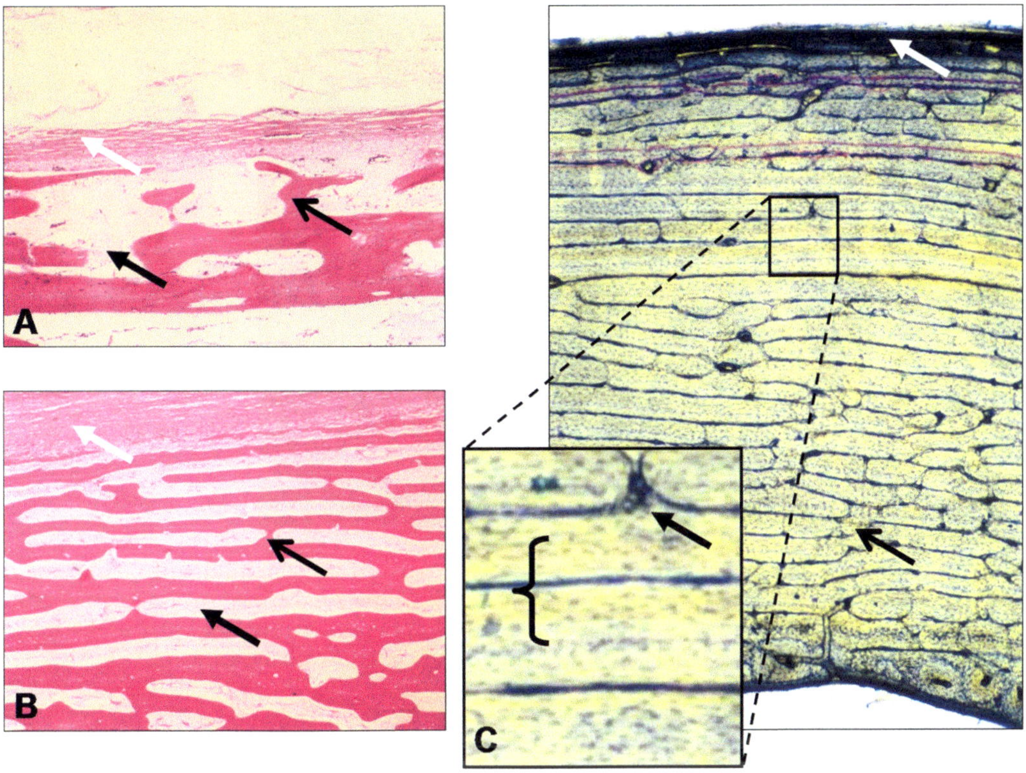

Figure 2.5 Periosteal woven bone deposition in the laminar fashion with plexiform vessel inclusion in the *perinatal* horse foal femur. White arrows in all images denote the periosteum; black arrows, the vascular void or canal; and black lined arrows mark bridging woven struts providing support for the newest of the forming external laminae. B) In this young individual (same as "A") vessels are still within many layers of large unfilled voids prior to bony compaction. C) Laminar bone exhibiting compacted plexiform structure in the 18 mo. old Sinclair Minipig tibia. The black arrow (expansion) marks a previously periosteal vascular network now deep within the cortex. Note that entombed vessels are dominantly circumferential but with short connections between generations of vascular sheath (black arrow in expansion). The black bracket indicates one full layer as described by Currey (2006), bordered by faint light lines of woven bone. Confocal laser scanning micrograph of archaeological bone fluorescence. Images "A" (100x magnification) and "B" (40x magnification) are microphotographs of paraffin embedded, decacified thin-sectioning treated with hematoxylin and eosin stain; "C" is a 40x magnification microphotograph of an undecalcified thin-ground section containing visible, though feint, fluorochrome labeling. Images courtesy of Steven Weisbrode.

Maggiano et al., 2015; Maggiano, Maggiano & Tiesler et al., 2016). When they contain only one or no layers of their own, the term "primary vascular canal" is often preferred. Both forms can be seen in completely woven tissue as well.

Owing to its dissimilar vascular environment, deposited endosteal lamellar bone rarely traps vasculature in the same way as the periosteum. Instead, it seems primary radial canals (similar to Volkmann's canals in secondary tissue), anchored in the intercortex, extend during growth to maintain the connection between the endosteal vascular network and deeper tissue (Figure 2.11). This has been observed in the rhesus tibia (Enlow 1962) as well as the human femur and humerus (Maggiano et al., 2015; Maggiano, Maggiano, Tiesler, Kierdorf

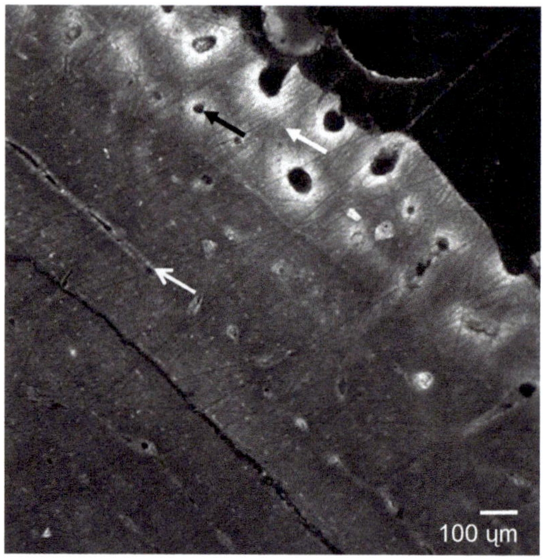

Figure 2.6 Fibrolamellar bone tissue from the goat pericortex. Longitudinal primary vessel canals (black solid arrow) are surrounded by woven bone (white solid arrow), while internal compact structure appears more laminar, enclosing vessel networks (white lined arrow). Confocal laser scanning micrograph of archaeological bone fluorescence. Scale indicates 100 μm.

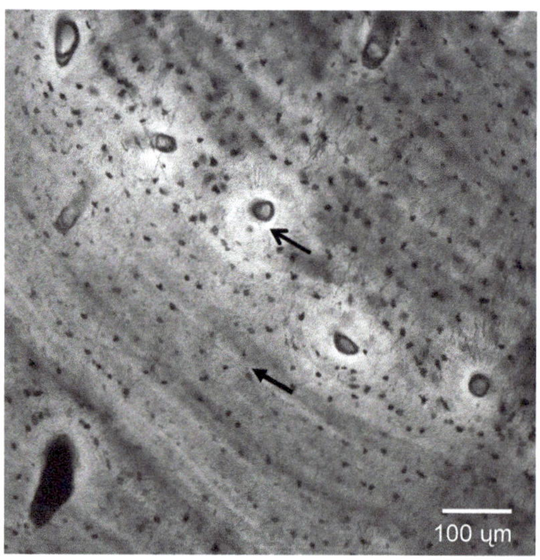

Figure 2.7 Alligator pericortex demonstrating regular concentric lamellar apposition (black solid arrow) as well as the occasional inclusion of longitudinally oriented primary vascular canals, surrounded by highly mineralized woven bone (black lined arrow). Micrograph taken in white light transmission. Scale indicates 100 μm. Image courtesy of Y. Castro, D. Vanmali, and T. Dupras.

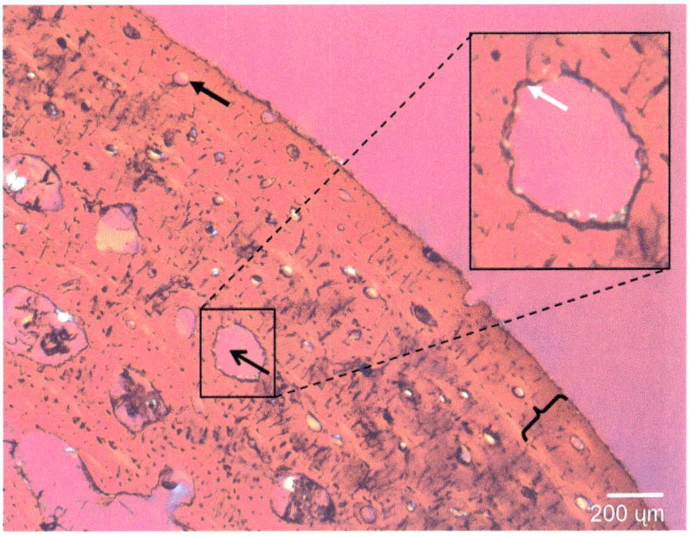

Figure 2.8 Human infant pericortex demonstrating occasional inclusion of longitudinal vessels, as primary osteons (black arrow) as in "fibrolamellar" or "primary osteonal" bone. In this case there exists a single "pseudo laminar" layer within the forming (mostly woven) pericortex (black bracket). The black lined arrow denotes a resorption bay destined for expansion during endocortical resorption or infilling via BMU-based remodeling. The white arrow marks one of many Howships lacuna roughening the bay's perimeter, evidence that resorption was active *antemortem*. Thin-ground, undecalcified transection of archaeological, femoral mid-diaphysis, Dakhleh Oasis, Egypt. Micrograph taken with red quartz cross-polarized filter (530 nm compensated). Scale bar indicates 200 ųm.

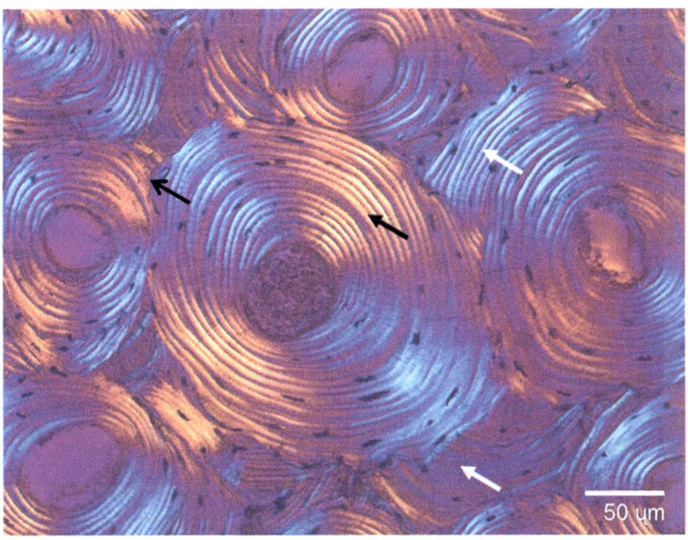

Figure 2.9 Primary "osteons", or primary vascular canals, from a juvenile (white solid arrows), have no lamellae or cement line of their own. Instead, they are entrapped by successive lamellar apposition at the periosteum. The black solid arrow denotes the bone surface most recently applied by the periosteum continuing to inhume several other longitudinal primary vessels. Thin-ground, undecalcified transection of archaeological, humeral mid-diaphysis, Dakhleh Oasis, Egypt. Mid-diaphyseal micrograph taken with red quartz cross-polarized filter (530 nm compensated). Scale bar indicates 50 ųm.

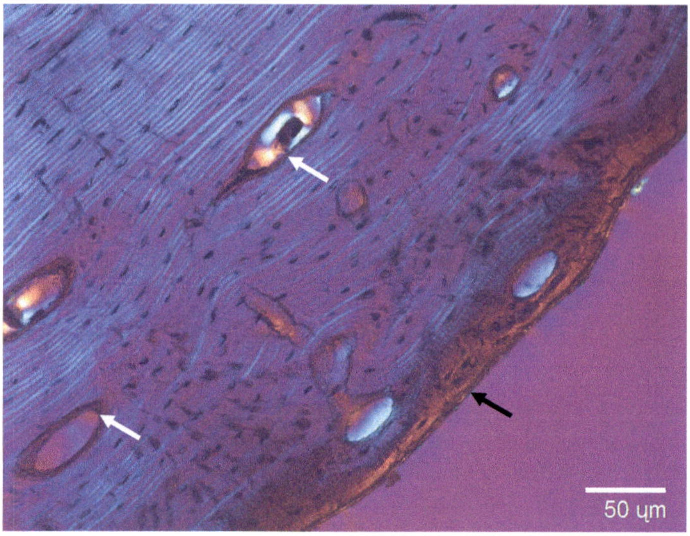

Figure 2.10 Adult secondary osteons, or Haversian systems, with concentric lamellae and a central canal, which in life would house vessels and nerves. Note the similarity of lamellar structure between BMU tissue and PEM tissue (Figures 4.3 and 4.5) in both lamellar thickness and lacunar inclusion (black solid arrow). The black striped arrow marks the central osteon's cement line (reversal line), the zone of maximum resorption during the resorptive phase of BMU activity. Subsequent formation progressed from this point toward the canal at the center. White solid arrows denote osteonal fragments from previous BMUs that passed through the region. No primary tissue is present in this field of view or in the larger region suggesting that the tissue can be classified as Haversian tissue and that the individual is likely mature or even of advancing age. Thin-ground, undecalcified transection of archaeological, humeral mid-diaphysis, Dakhleh Oasis, Egypt. Micrograph taken with red quartz cross-polarized filter (530 nm compensated). Scale bar indicates 50 ɥm.

et al., 2011; Maggiano, Maggiano & Tiesler et al., 2016), but could be expected as a normal trait of any sizable primary endosteal deposition.

Having considered human bone modeling microstructure in reference to that of some other animals we come to a conclusion of necessity: human bone biologists should maintain reference to human PEM bone tissue as "primary lamellar" qualified by the membrane of origin when possible (i.e. periosteal lamellae), rather than simply "lamellar" *or* "laminar". This is because (1) without delineating between primary and secondary lamellae we run the risk of confusing modeling and remodeling processes and resulting bone tissue, and (2) it is rare to find apparent, 4–20 lamellar sets by which we would otherwise define adult, PEM-origin human bone as "laminar" by Locke's (2004) definition (or fibro-lamellar, or plexiform, for that matter) (see e.g. Goldman et al., 2009; McFarlin et al., 2016). More detailed information on histological variance among species can be references from Hillier and Bell (2007) or Crowder et al. (2018).

2.3.2.3 *Lamellar Microstructural Variation*

Modeled tissue can be complex to interpret in part due to the above interactions between formative rate and local vasculature, and is further complicated by the fact that formation

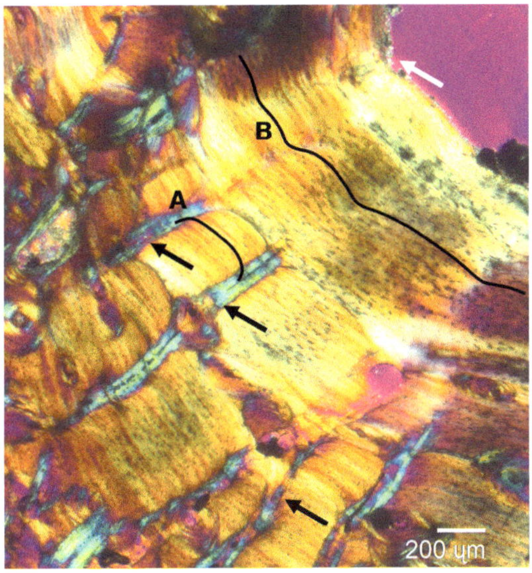

Figure 2.11 Dense primary endosteal lamellae, comprising nearly a fourth of the total cortical area, have failed to entrap a single longitudinal vessel from the endosteal surface. Instead, lamellae formed under endosteal vasculature, pushing it away, filling the spaces between transverse vessels (Volkmann's canals denoted by black solid arrows). These vessels extended to maintain connectivity to the intercortical Haversian network (out of view to the left). A) As bone is laid between transverse vessels its lamellae "pillow" slightly, perhaps due to slower rates of calcification adjacent to the pulsating vessels. B) This effect can be easily seen in lamellae merely adjacent to transverse vessels, even when these vessels are not within the plane of sectioning. Thin-ground, undecalcified transection of archaeological, humeral mid-diaphysis, Xcambó, Mexico. Micrograph taken with red quartz cross-polarized filter (530 nm compensated). Scale bar indicates 200 ɥm.

occurs independently from resorption. This means that reconstructing microstructural events requires considering the potential for significant time between events. Here, we simplify as much as possible, by focusing mostly on modeling of human primary lamellar bone of course, but questions still abound. What structures are generated by the entire membrane versus smaller regions? What structures are generated as the bone gains its curvature or alters position? What disruptions to growth can we see and what do they mean? Variation in lamellar microstructure holds valuable stratigraphic information regarding growth and adaptation processes or stimuli.

When the membrane is fully engaged in formation, lamellae can span nearly the entire periosteal or endosteal perimeter; generating "circumferential lamellae". This leads to the familiar histological landscape of a long bone transverse cross-section (especially in adolescents or young adults; Enlow, 1962): PEM lamellae "sandwiching" more highly remodeled intracortical bone (Figure 2.12). Other times the landscape of large lamellar depositions is less familiar.

For example, especially larger primary vessels included in modeled tissue are capable of introducing a "pillowing" of primary lamellae under them. On the surface of pathologically affected periosteum, this effect can be severe, forming series of structures named "polsters" (Schultz, 2001; Weston, 2009). In much less severe fashion a similar effect can be seen

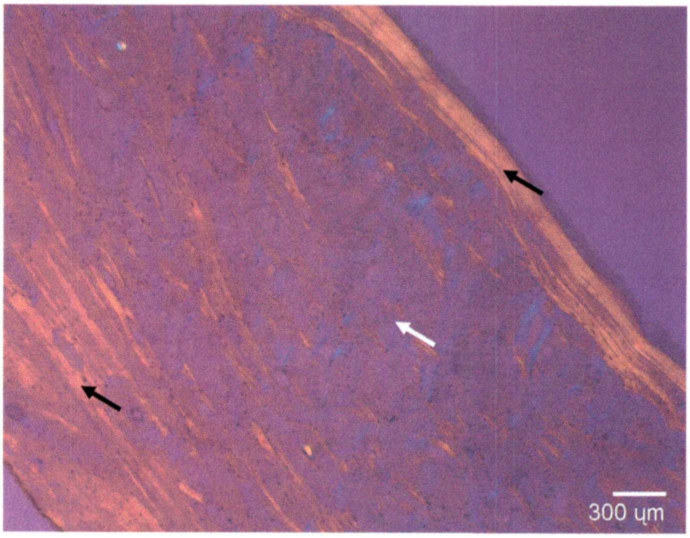

Figure 2.12 The traditionally described histological landscape of young compact bone. Primary lamellar bone (black solid arrows) formed by the periosteum (lower left) and endosteum (upper right). White solid arrow marks remodeled Haversian tissue formed by BMU activity. This remodeling transpired in the older intercortical primary tissue and has now turned over a large area, replacing primary osteons with secondary and increasing the general porosity of the region. Thin-ground, undecalcified transection of archaeological, humeral mid-diaphysis, Dakhleh Oasis, Egypt. Micrograph taken with red quartz cross-polarized filter (530 nm compensated). Scale bar indicates 300 ųm.

regularly on the endosteal surface, as in Figure 2.11. It is well known that static and pulsatile strains can inhibit bone formation, or even induce resorption of bone tissue (Du Boulay, 1966, Feik et al., 1987; Srinivasan et al., 2019). Therefore, one potential cause for transverse lamellar pillowing is the static or pulsatile pressures applied by the vessel, slowing calcification slightly relative to nearby tissue. What about when deposition occurs at a smaller scale?

Local or regional modeling domains sometimes form structural zones (not as well defined as bone structural units in remodeling, but identifiable none-the-less). Here, we use the term "phases", borrowed from geology and archaeology, meaning a specific frame of stratigraphic deposition. In 1973, Frost used "packets" as a general descriptor for the same phenomenon; and used the term, "foci", for sites of initial modeling formation or resorption, a term also used here. With these concepts in hand, we can address any variations in the amount, pace, and orientation of what otherwise might be treated generally as "circumferential lamellae".

For example, if the newest lamellae (adjacent to the bone surface) are concentric with a membrane of origin, that membrane was either (A) currently forming lamellar bone at the time of death or extraction, or (B) it had paused after formation with little or no subsequent membrane activity. Owing to anisotropy in the basic orientation of lamellar sheets themselves, lamellae or sets of lamellae can "disappear" along the field of view, even without other evidence of disruption (Bromage et al., 2009, 2011; Smolyar and Bromage, 2004). Alternatively, lamellar phases can be disrupted in their formation rate or direction; or they may be resorbed. Evidence of any of these occurrences can be collectively referred to as "cement lines", lines disrupting typical lamellar structure (Burr et al. 1988; Schaffler et al., 1987; Skedros et al., 2005). Some of these are slight and smooth, with no evidence of resorptive activity, and are called "arrest lines" (Figure 2.13). Under higher magnification, other lines display tell-tale evidence of prior osteoclastic activity (such as Howship's lacunae) cemented within the disruption itself. These are referred to as modeling "reversal lines" (Figure 2.14). In this situation lamellae often (but not always) appear cross-cut at some oblique angle. Presumably modeling reversal lines would demonstrate some of the same interesting attributes as their remodeling cousins, including increased mineral composition, unique mechanical properties, and biomechanical importance as described by Skedros et al. (2005). Arrest lines can denote annulations or seasonal periods of slowed growth in non-human animals (Becker et al. 2020, Calderón et al., 2021; Castanet et al., 2004; 2006, Köhler et al., 2012; Kolb et al., 2015; Sander and Andrássy, 2006). They have also been proposed as potential evidence of health insults (Schultz 2001), not unlike microstructural cousins of Harris lines formed by afflicted growth plates. Doing so, however, would necessitate means of distinguishing this type of arrest from more localized requirements of changing modeling rate, direction, or influences of local topography that could necessitate

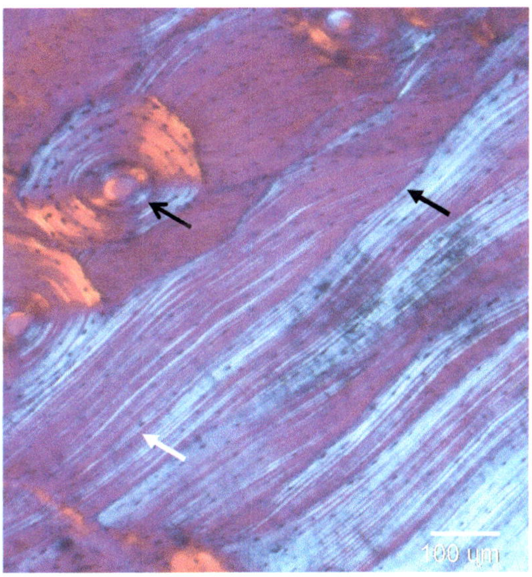

Figure 2.13 Endosteal primary lamella now occupying an intracortical position adjacent to several Haversian systems. These particular primary lamellae are laid in several formative phases. The white arrow marks consecutive, uninterrupted formation of the major phase, while the black arrow marks a cement line cutting across previous lamellae of the same major phase, splitting this local region into two separate minor formation phases due to a single, slight resorption focus. This presents stratigraphic evidence of a single phase of formation continuing in one region while at another reversal occurs followed by formation, thusly directing drift in a specific net direction. Subsequent phases to the bottom right of the image also bear evidence of targeted modeling resorption in the form of roughened cement lines that cross-cut previous lamellae. Also note the presence of an arrest line in a secondary osteon (black lined arrow) for comparison with cement lines with resorptive evidence. Thin-ground, undecalcified transection of archaeological, humeral mid-diaphysis, Dakhleh Oasis, Egypt. Micrograph taken with red quartz cross-polarized filter (530 nm compensated). Scale bar indicates 100 ụm.

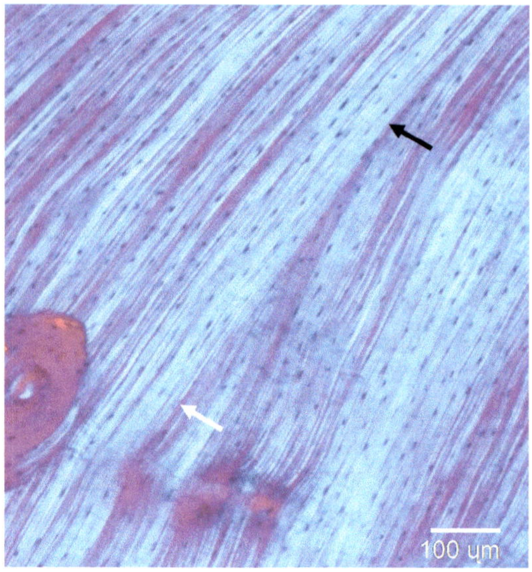

Figure 2.14 Endosteal lamellae demonstrating complex phase activity. The white solid arrow marks continued uninterrupted formation through the entire field of view. Conversely, the black solid arrow denotes a modeling arrest line. Though subtle in appearance, the effect is significant, marking a pause in phase formation in the upper right region while formation continues for hundreds of micrometers (past the lower field of view boundary). Thin-ground, undecalcified transection of archaeological, humeral mid-diaphysis, Dakhleh Oasis, Egypt. Micrograph taken with red quartz cross-polarized filter (530 nm compensated). Scale bar indicates 100 ųm.

both pauses and resorptions on parts of formation fronts as they deposit tissue elsewhere (Maggiano et al., 2019).

To illustrate the complexity of such local requirements, we can consider that a given phase focus consists of initially activated osteoblasts, but may later involve a greater or lesser number of adjacent cells, either changing or "migrating" the surface area of apposition to a new PEM region. Either occurrence should be visible in the bone's primary microstructure. For example, increasing surface areas of apposition should result in unique lamellar organization relative to those that are decreasing (Figure 2.15). Migrating surface areas of apposition would involve phases deposited in a stepwise fashion, potentially changing the foci between or even within a phase. In this way, the next team of osteogenic cells in or between a phase might not occupy the same surface area as their predecessors. Interestingly, however, the transverse section of a migrating formation surface could be confused for changing surface areas of apposition if the lamellar migration was strong and its direction was perpendicular to the plane of sectioning. Local microstructural context, such as surface topography would predictably alter expectation for the stratigraphic appearance of an increasing or decreasing surface area of formation. It is also important to remember that Howship's lacunae could provide evidence of changing surface areas of resorption as well, though without accompanying formation, this phenomenon is not viewable stratigraphically. Perhaps the best way to envision such an event in modeling is by reminding ourselves that migrating resorption (and formation) is also accomplished in remodeling by the drifting osteon (Robling and Stout, 1999). In fact, at small scales especially (< 300 ųm), the boundary between modeling and remodeling processes could be obscured. Consider

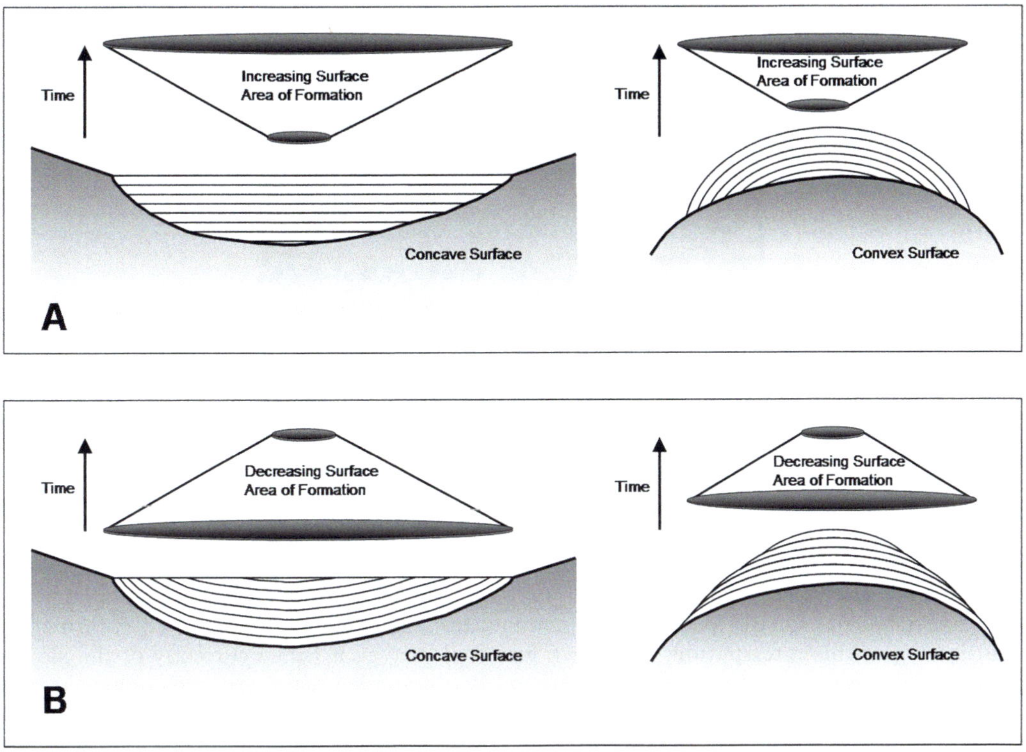

Figure 2.15 Two theoretically distinct lamellar phase microstructures depending on changing surface areas of apposition over time, represented in schematic transverse cross sections. A) An increasing surface area of formation on both a concave (left) and convex surface (right), requiring the activation of more and more previously dormant osteogenic cells from the membrane. B) A decreasing surface area of formation on both a concave (left) and convex surface (right), requiring fewer and fewer of the surface cells to be active after each deposited layer. These changes in surface area should accompany meaningful changes in some aspect of the formation signal that initiates or directs modeling activity. The size of the formation phase or its local micro-topography could also be a factor influencing surface areas of formation. The prevalence or meaning of this type of microstructural variation is unreported.

an endocortical region in which a remnant of trabecular structure separates two modeling formation foci (Figure 2.16). Later intracortical remodeling could eventually obscure the interposing feature entirely, the endosteal surface will have a new position, and the original formative context could be difficult to interpret. In other circumstances, phases are small and numerous, with evidence for prior resorption (Figure 2.17). But prior resorption does not necessarily indicate coupled processes enacted by a BMU. At these scales, distinguishing between modeling events and hemi-osteons could prove challenging (Smit, 2021). To understand the role of the hemi-osteon, researchers typically focus on trabecular tissue, the investigation of endosteal modeling phases during growth could contribute to our understanding of the complex circumstances in spongy tissue (see Section 2.2.3 for more detail).

How (or if) trabecular modeling and hemi-osteonal remodeling could be differentiated histologically is a question raised by Jee and fellows (2001). They answer in the affirmative by supporting Frost's (1987) usage of the term "minimodeling" to differentiate "packets" (here, "phases") of bone formation on trabecular surfaces with no evidence of prior

Figure 2.16 The black solid arrows mark two different sets of modeling foci separated by an irregular trabecular spicule (potentially a remnant from an earlier growth period). If formation continued past this interruption, these two foci could merge if timed correctly or the slower of the two would abut the faster. Thin-ground, undecalcified transection of archaeological, humeral mid-diaphysis, Dakhleh Oasis, Egypt. Merged cross-polarized micrograph. Comprising images taken when primary lamellae were at 45° to normal x, y stage orientation. Scale bar indicates 300 ųm.

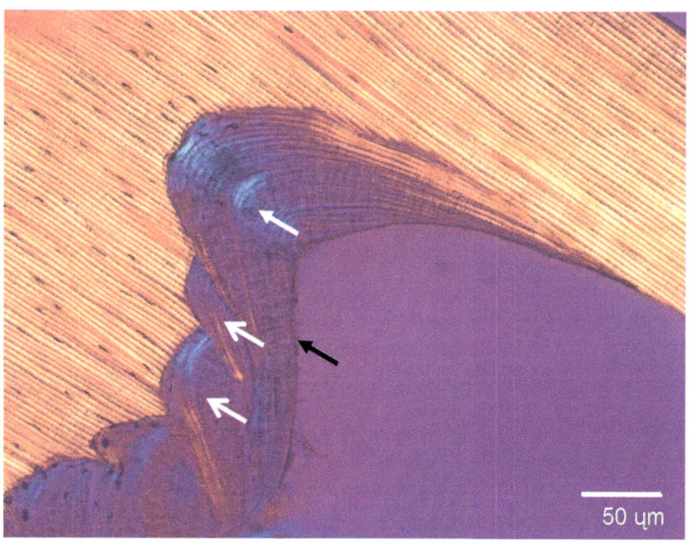

Figure 2.17 A series of endocortical modeling formation phases present as adjustments to a much larger phase (past the lower field of view boundary). The white solid arrow marks the most recent phase, whereas the white lined arrows mark overlain fragments, indicating prior resorption took place before the final phase was formed. These phases are small enough to permit confusion with hemi-osteons, although the latter are typically only reported in trabecular bone. Black solid arrow denotes medullary surface. Thin-ground, undecalcified transection of archaeological, humeral mid-diaphysis, Dakhleh Oasis, Egypt. Micrograph taken with red quartz cross-polarized filter (530 nm compensated). Scale bar indicates 50 ųm.

resorption due to the presence of "smooth cement lines", or arrest lines (Jee et al. 2001). Modeling is a wonderfully complex process. Jee and colleagues (2001) call attention to this fact, even in trabecular space, by recognizing some packets of trabecular minimodeling represent "mixed remodeling-minimodeling".

Altogether, modeling microstructural complexity explains why a typical volume of "circumferential" lamellae can be constructed of smaller non-circumferential phases. More than that though, its correct interpretation reveals the history of growth and adaptation in the region. More research is necessary to differentiate these processes more clearly.

2.4 Modeling Function

2.4.1 Growth

To envision how small changes in lamellar orientation, apposition, and resorption account for modeling during growth and mechanical adaptation, we need to expand our perspective, considering the concerted efforts of both surface membranes simultaneously. It is true that growth typically requires periosteal expansion and endosteal resorption (Stoker & Epker, 1971; Epker et al., 1965; Maggiano et al., 2015; Maggiano, Maggiano & Clement et al., 2016). However, modeling cannot be this simple if the goal is to affect complex morphological change. So growth also requires significant resorption of bone. Moreover, growth requires deposition of bone also at the endosteum. According to Garn (1970), more than 25% of total cortical thickness comes from endosteal bone deposition, continuing (especially in males) even into the fourth decade of life. Without concerted and targeted PEM formation *and* resorption, longitudinal extension would fail to form an effective compact diaphysis; instead, metaphyses would grow unaltered volumes of trabecular tissue. In addition, bones would be unable to modify their morphology, curvature, or relative orientation to other elements (Maggiano et al., 2015).

2.4.2 Metaphyseal Reduction and Medullary Extension

In addition to diaphyseal growth, modeling processes account for the shaping of bone during its elongation at the growth plate (Figure 2.18B, white checkering). This occurs, in part, via metaphyseal reduction, Frost's (1973) "necking down" or "metaphyseal drift system", where periosteal resorption reduces and smooths the diameter of metaphyseal tissue destined for eventual inclusion in the diaphysis (Enlow, 1962; Jee & Frost, 1992; Rauch et al., 2001; Rauch, 2012; Isojima and Sims, 2021). In this fashion, the growing metaphysis undergoes periosteal resorption almost immediately following the formation necessary to stabilize the new tissue forming at the widening physis (Enlow, 1962). Slowly, as the bone elongates, the flared diameter, now more distant from the growth plate, reduces to match the expanding diaphysis. Only the cessation of longitudinal extension "freezes" the metaphyseal morphology, precluding continued large scale resorption at the metaphyseal surface.

Simultaneously, both in the metaphysis (trabeculae) and within the expanding diaphysis (compacta), the medullary cavity is forming, widening, and elongating (Enlow, 1962, Garn, 1970). Diaphyses accomplish expansion or alteration of medullary space in two ways:

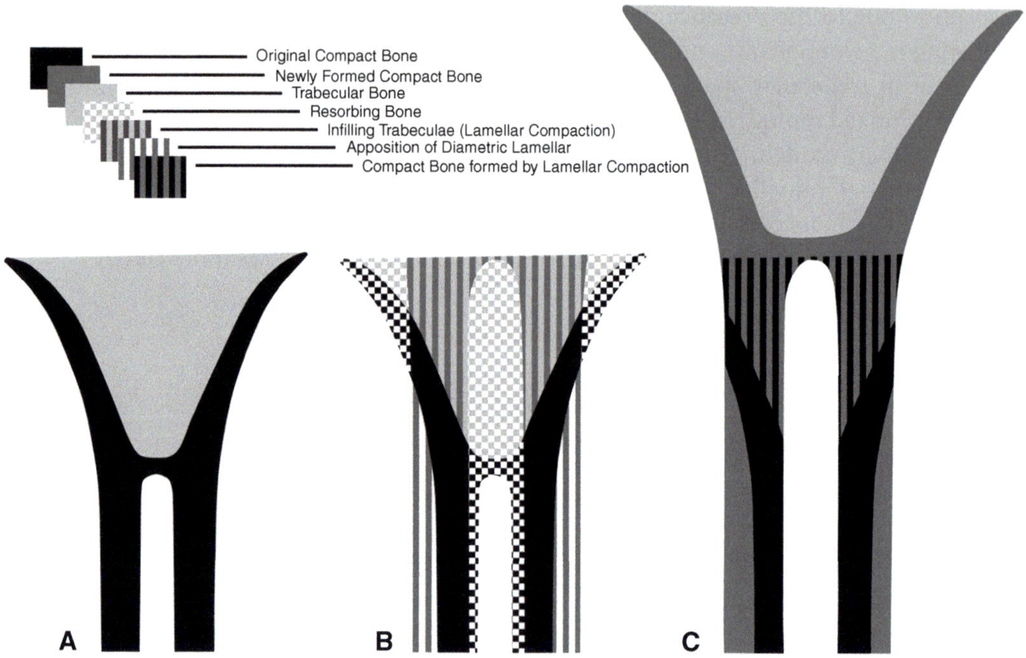

Figure 2.18 A) Schematic of immature cortical (black) and trabecular (gray) regions destined for modeling alteration during long bone extension. B) Bone is resorbed at the metaphyseal periphery and within the medullary cavity, affecting both compact and trabecular tissue alike (white checkering). New compact tissue must also be constructed, both within trabecular voids during lamellar compaction, and at the periosteum via diametric apposition (gray bars). C) Even in the adult there remains evidence of these changes as the bone extends in length through endochondral ossification. Adapted from Enlow D (1962) A study of the postnatal growth and remodeling of bone. Am J Anat 110:79–101.

(1) resorption directly on the endosteal surface, (2) resorption in ever expanding intracortical resorptive bays. Especially once the second process is initiated, medullary expansion in cortical bone could speed up exponentially due to the ever-increasing surface area created (a process reversed for trabecular bone) (Ensrud et al., 1995). Trabecularization of compact bone has been connected most commonly with age associated bone loss, but is also present in abundance during dramatic periods of growth, as in Figure 2.19.

Initially, metaphyseal reduction and medullary expansion seem paradoxical, opposite to the growth model of a long bone, with tissues being resorbed on the periosteum and inner surfaces. Recent work, however, has begun to untangle the complex systems of coordinating cell signaling pathways capable of guiding this process (Isojima and Sims 2021). Isojima and Sims document a body of work on genetically altered mouse models for defective cortical structure, by which an understanding of normal processes can be inferred. From an anthropological perspective, as this knowledge leads to better understanding of human bone, our interpretations of healthy and pathological bone morphology improves. Indeed the synergy between modeling and remodeling process is astounding (and confounding). But this is not the only example of well-orchestrated, complex manipulation of both sister processes simultaneously.

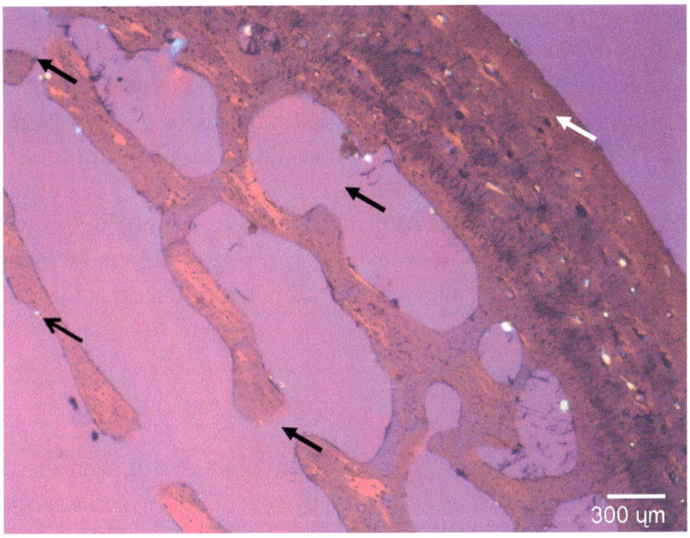

Figure 2.19 Infant cortical woven expansion at the periosteum (white solid arrow), is countered by dramatic endocortical resorption (marked by Howship's lacunae roughening bone surfaces) both at the endosteal margin (black lined arrow) and in expanding, large, resorption bays (black solid arrows). Black solid arrows also denote loss of several "trabecularized" cortical connections. Thin-ground, undecalcified transection of archaeological, femoral mid-diaphysis, Dakhleh Oasis, Egypt. Micrograph taken with red quartz cross-polarized filter (530 nm compensated). Scale bar indicates 300 ɥm.

2.4.3 Lamellar Compaction

The process Enlow (1962) termed "lamellar compaction" (Figure 2.18B, light and dark gray stripes) is also transpiring intracortically, simultaneously with metaphyseal reduction and medullary expansion. The specific process of lamellar compaction has been very rarely researched from a modern perspective, in general (an important exception is Parfitt et al., 2000). It has been even less addressed in long bones.

According to Enlow (1962), lamellar compaction is the infilling of trabecular void spaces to form compact tissue at several different sites at different periods of long bone development. The term is used to discuss resulting compact tissues whether trabeculae were thin or more coarse. The infilling tissue must not be strictly lamellar, despite the term. The process is often overlooked in long bones, in part because lamellar compaction is much easier to view in the "flat bones" and the rib. These elements maintain some number of coarse medullary trabeculae, most of which are formed by prior external periosteal growth and subsequent expansion of resorptive bays that later need infilled (Parfitt et al., 2000). Compaction can take on several appearances from disorganized "whorls" or "convolutions", to well organized, inward (endosteal) lamellations entombing trabecular spicules (Enlow 1962).

The mechanism behind this process is suggested through experimental and computerized modeling data examined in an interesting work contributed by Tanck et al. (2006), who argues lamellar compaction near the spongy cortex transpires via mechanical adaptation phenomena. Therefore, it is possible that lamellar compaction is actually a secondary effect of longitudinal growth. Perhaps, as Salmon (2015) suggests, understanding the complex, emergent interactions between PEM activities during elongation might be aided by considering the complex patterns that can emerge from relatively simple signaling optimization between cell types. In actuality, it's unlikely these are "different" processes at all. Of course, after reduction and compaction, this tissue exists at a relatively low volume for a short period of time during growth and so could be challenging to study or utilize in histomorphometric methodology. It could be quickly remodeled or even lost to endosteal resorption during subsequent phases of growth (which is probably more likely). Its brief existence does not imply non-importance, however, since it serves as yet another way that adjacent tissues can be of differing ages at formation, and microstructural complexity could be misunderstood.

2.4.4 Drift

One of the most important modeling processes involves the uneven distribution of bone on different sides of the diaphysis, as described by several early works (Foote, 1916; Tomlin et al., 1953). For example, Enlow (1962) notes that in the rhesus tibia, lateral periosteal formation is accompanied by medial resorption, ensuring that its adult position is far removed from its immature position (Figure 2.20). Conceivably, modeling drift alone (with no turnover from remodeling) could account for a bone that microscopically contains almost entirely new bone tissue, the immature tissue having been resorbed by one or several envelopes. This phenomenon, in part, accounts for the fact that mean bone tissue age is significantly younger than chronological age (Wu et al., 1970). Enlow (1962) called the general

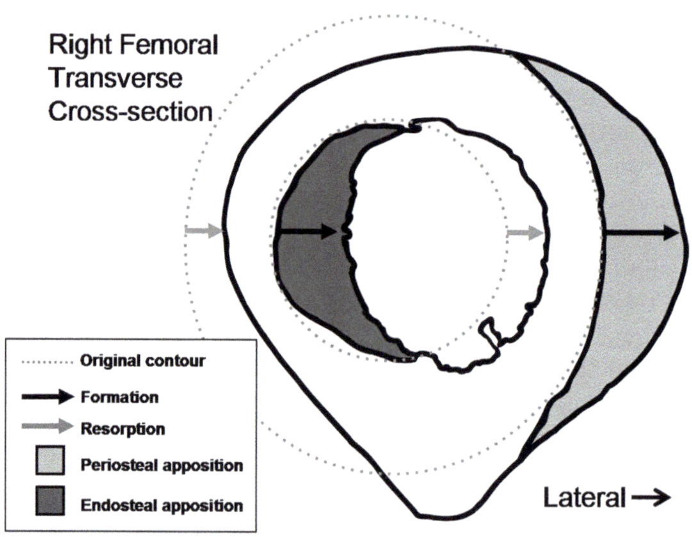

Figure 2.20 Schematic of a right femoral transverse cross-section demonstrating significant net linear drift. Lateral periosteal apposition is accompanied by medial endosteal apposition, shifting the bone laterally relative to its original position. Resorption transpires on the medial periosteal and lateral endosteal surfaces, maintaining the centralized position of the medullary cavity. Endosteal lamellae (dark gray) have relatively few secondary osteons in comparison to periosteal lamellae (light gray) which are more mechanically strained and could remodel more quickly (see text for details). Adapted from Maggiano et al. (2011).

process "osseous drift" whereas Frost's term, "modeling drift" has been preferred in the literature (Epker & Frost, 1965; Frost, 2001).

Modeling drift at the periosteum alone cannot account for adult long bone morphology, because the described pattern would yield an eccentric medullary cavity. According to Garn (1970), more than 25% of total cortical thickness comes from endosteal bone deposition, continuing (especially in males) even into the fourth decade of life. Therefore, periosteal formation and resorption are often mirrored in process on the endosteum. That is, during lateral diaphyseal drift, the endosteum must apply new bone lamellae to the medial surface while resorbing those on the lateral surface. Only in this way can the medullary cavity maintain its central position within the drifting diaphyses. In fact, depending on the element and the location of sectioning, a strong drift may also involve the nutrient foramen itself, which, like the medullary cavity, must mirror periosteal drift to stay in relative cross-sectional position (Maggiano, Maggiano, Tiesler, Stout, 2011). Close observation of the resorbing endocortex indicates not only Howship's lacunae on the endosteal surface itself, but also dramatic resorption occurring in large intracortical resorption bays. Exactly how much tissue is removed by the endosteal membrane itself versus the proliferation (Parfitt 1988) and expansion (Han et al., 1997; Parfitt, 2003) of these bays (Figure 2.21) is unknown.

Conceptually, there are two types of long bone drift: linear and curvilinear. In linear drift, a bone's position (or a portion thereof) is shifted in any net direction well summarized

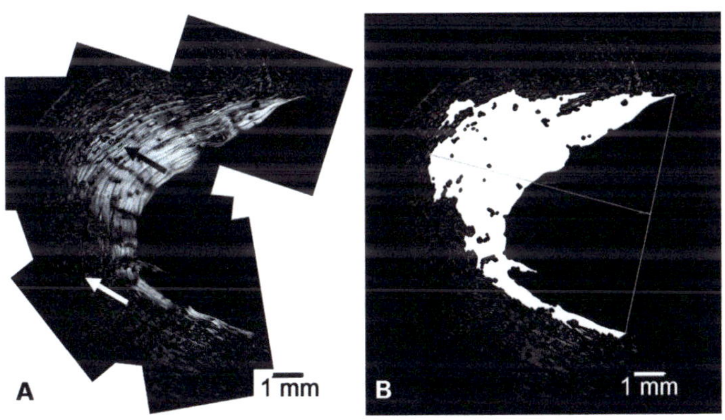

Figure 2.21 A) Potential merged micrograph technique using cross-polarization to emphasize endosteal lamellae. Each image was taken of endosteal lamellae when they were at 45° orientation and therefore brightest birefringence in polarized light. This method ensures primary lamellae remain bright by removing darkening artifacts from cross-polarization interference. Haversian tissue retains this artifact, keeping it relatively dark in comparison (white solid arrow). Nearly the entire cortex on the left side is *endosteal* in origin, indicating strong linear drift. Several features common to this endocortical region include hemi-circumferential lamellar orientation, reduced osteonal presence, and radially oriented Vokmann's canals. Together, they can be considered a "meta-feature", referred to as the endosteal lamellar pocket (ELP) (black solid arrow), currently under evaluation as a quantitative indicator of modeling processes (Maggiano et al., 2011). B) The ELP is automatically recognized via computerized image analysis using pixel intensity and region of interest size thresholding to create a quantifiable indicator of endosteal drift (white mask) (Maggiano, Maggiano, Tiesler, Kierdorf et al., 2011). Thin-ground, undecalcified transection of archaeological, humeral mid-diaphysis, Dakhleh Oasis, Egypt. Merged cross-polarized micrograph. Comprising images taken when primary lamellae were at 45° to normal x, y stage orientation. Scale bar indicates 1 mm.

by a straight line (Enlow, 1962; Epker & Frost, 1965; Streeter & Stout, 2003; Frost, 1973). The typical assumption would be that in a transverse long bone section, linear drift would be transverse, but a superiorly or inferiorly migrating focus of formation would be indistinguishable from transverse drift in sampled diaphyseal cross-sections. Curvilinear drift can be visualized in the transverse cross-section as stepwise alteration in the direction of drift at one or both surface envelopes, through the complex manipulation of primary PEM phases (Maggiano, Maggiano & Clement et al., 2016). This can be accomplished via changes in the size, shape, or relative location of their surfaces of apposition, for example.

In some cases, typically in youth, drift could be so significant that most if not all of a cortex is formed by only one of the envelopes. Some care should be taken when referring to primary lamellar tissue by region or nearest membrane to include consideration of microstructural characteristics (i.e., lamellar form, vasculature, circumferential agreement with current osteogenic surfaces, etc.; Maggiano, Maggiano & Tiesler et al., 2016). As part of this caution, it may decrease accuracy histomorphometrically compared two endocortical regions where primary bone may not be entirely endosteal in origin. Of course, more complex drifts than those described here could be envisioned in intramembranous bones. In long bones, drift complexity moves beyond diaphyseal cross-sectional morphology and can account for mature curvature or relative orientation.

To start, we might expect that because the mid-diaphysis is the site for maximum load induced deformation (Biewener, 1993), and because bone is mechanically responsive, that the diaphysis should demonstrate the most dramatic evidence of drift. However, the same shape or orientation for a long bone can be formed by several different means (Figure 2.22). A curved diaphysis could be formed by drift occurring at each end of the diaphysis, only at the mid-diaphysis, or in all these regions simultaneously (Goldman et al. 2009; Maggiano et al., 2015). Without histological examination of cortical tissues, this remains hidden variation that could be useful for anthropological applications.

By now several other studies have put forward methods for quantifying drift magnitude and direction, and assessed it in both femora and humeri (Cambra-Moo et al., 2015; Goldman et al., 2009; Maggiano et al., 2015; Maggiano, Maggiano, Tiesler, Kierdorf et al., 2011; Maggiano, Maggiano & Tiesler et al., 2016; McFarlin et al. 2016; Raguin et al. 2014). Rubin (1964) described resorption on some ribs' pleural periosteum, opposed by formation on the cutaneous surface, an observation commented and elaborated upon in other works (Epker & Frost, 1965; Streeter & Stout, 2003, Agnew et al., 2013), with the effect diminished in older age (Epker & Frost, 1965). Likewise, Parfitt et al. (2000) and later Schnitzler and Mesquita (2013) note that mirrored periosteal and endosteal modeling is necessary to expand the pelvis during growth, whereas Balena et al. (1992) found no evidence for drift in their investigation of the adult transilium. Preliminary results on the second metacarpal confirmed a significant lateral drift in two population samples (Raguin et al., 2014). The adult human histomorphological patterns that indicate drift have also been observed at the femoral (Maggiano et al., 2015; Maggiano, Maggiano, Tiesler, Kierdorf et al., 2011), humeral (Maggiano, Maggiano & Tiesler et al., 2016), and ulnar mid-diaphysis (Maggiano unpublished results). This work attempted to utilize predictable phenomena of significant diaphyseal drift, namely asymmetry of endosteal depositions maintaining medullary centricity, as a cross-sectional meta-feature (or BSU) summarizing drift. We then characterized this feature's presence and microstructure (Maggiano, Maggiano, Tiesler, Kierdorf et al., 2011), size, shape, position (Maggiano et al., 2016), and longitudinal variation (Maggiano et al., 2015). We used several techniques to aid our efforts, including 530 nm compensated,

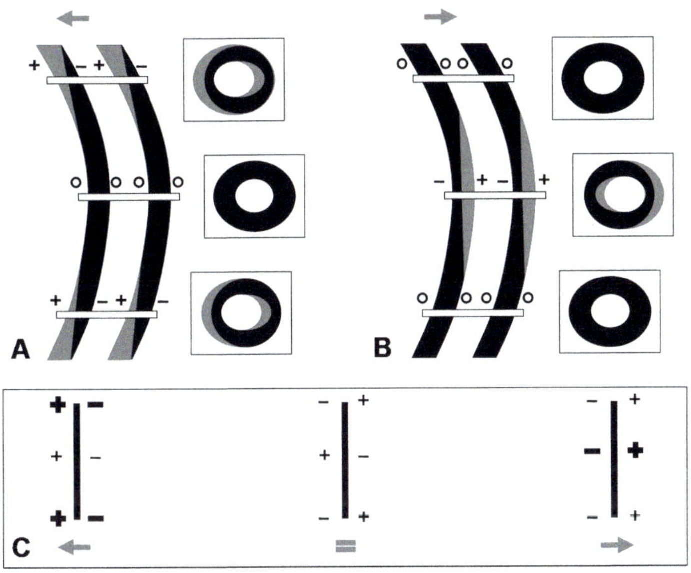

Figure 2.22 Schematic representing curvature achievement during long bone modeling. Large and small plusses and minuses indicate varying magnitudes of modeling formation and resorption, respectively. Arrows indicate the direction of net drift. A) This schematic, curved diaphysis results from a modeling drift to the left at regions distant from the mid-shaft. B) A net right drift can result in the exact same curvature when only the mid-shaft is undergoing modeling activity (upper right). Transverse cross-sections at each region demonstrate how the positioning and orientation of primary lamellae can reveal which occurrence accounts for the curve. C) At least three other meaningfully different circumstances could model a previously straight long bone into the same curved shape. Curvature could be gained during net drift of the entire diaphysis (bottom left and right), or could be balanced with no net drift whatsoever (bottom center).

cross-polarization stitched microphotography to emphasize lamellar concentricity or variation (Maggiano, Maggiano, Tiesler et al., 2016); and "starburst" point-count or hand-drawn analyses to assess ELP size and relative position (Maggiano et al.; 2015). In brief summary, the ELP was easily identified by its unique lamellar appearance and orientation, present in the majority of considered individuals of various ages (74–84%), relatively large (with mean percentages of up to 50% in some regions of the cortical bone, variable along the shaft and largest distally), has reduced disruptions due to primary canals/osteons and secondary osteons relative to periosteal tissues at similar age-at-formation; and was typically vascularized by primary Volkmann's canals in larger deposits. In the humerus, the ELP (positioned anterolaterally) indicated a general posteromedial drift with significant posterior rotational elements; whereas in the femur, the ELP (positioned medially; with a posterior trend distally) indicated a net lateral drift with anterior tendency. In the humerus, periosteally indicated drift largely agreed with that indicated by the ELP, whereas the femur was more complex indicating that the net drift suggested by the ELP could potentially summarize several drift demands placed on the periosteum (lest, otherwise the medullary cavity become acentric).

Net drift directions have yet to be identified for most human skeletal elements and only a few data are available on drift variability per element, across or between populations. Without such data, we discuss bone curvature or shape but have relatively no idea how it came to be. The theory of modeling drift also implies adjacent tissues could be far more separated by age-at-formation than previously realized. For example, endosteal tissue formed during the adolescent growth spurt can on occasion abut Haversian tissue laid down within the sixth decade of life. It is in this microstructural environment that we attempt to estimate age or measure stable isotopic ratios and some studies have recognized this limitation (Agnew et al., 2013; Gocha & Agnew, 2016; Gocha et al., 2018; Maggiano et al., 2019). The concept of tissue age-at-formation is a subtle extension of the modeling drift concept and accounts for the fact that adjacent bony tissues could be years apart in formation (Wu et al., 1970). This occurrence must be factored into equations used to estimate age (Dominguez & Mavroudas, 2019; Gocha et al. 2018).

2.4.5 Functional Adaptation

That bone morphology is responsive to mechanical loading has been well documented. Sufficient evidence has even been collected to argue how vital fetal movements are for the achievement of functional morphology (Hall and Herring, 1990; Rodríguez et al., 1988; Verbruggen et al., 2016). That said, Wong et al. (1993) showed that even without fetal movements required for normal development, a primary bone collar forms (see also Nowlan et al. 2010). If bone modeling occurs originally as an aspect of growth and development, exactly *how* it happens is owed primarily to mechanical adaptation. For example, modeling drift could account for the presence of a significant deposit of endosteal tissue on one side of the medullary cavity – but variation in exactly the amount of tissue and its position is likely affected by how the element is being used.

Attempting to directly measure the differential effects of growth and mechanical adaptation on modeling is challenging. One important contribution to the effort is to uncover the genetic sources of modeling variation. Hansen et al. (2009), for example, measured the heritability of femoral cross-sectional geometry in baboons. They found that genetic effects accounted for only about 15–23% of cross-sectional geometric phenotypical variance. Likewise, Volkman's (2003) research group determined that genetics accounted for 2.9–15.4% of the variance observed for mechanical traits they measured in the mouse femur. Robling and Turner (2002) point out that mechanosensation is also not without its own genetic component. For comparison, however, approximately 75% of human height is estimated to be heritable (Liu et al., 2006; Silventoinen, 2003), providing yet another reason to investigate modeling processes on their own merit, distinct from the process of long bone elongation.

Since mechanical demands influence modeling, we can expect modeling activity to continue after longitudinal growth cessation at around 20 years of age, despite well-known decreases in mechano-sensitivity of bone after maturation (Ruff et al., 2006, Szulc & Delmas 2007). This is precisely the observation of Garn's (1972) radiological study, showing endocortical formation continuing into the fourth decade of life. Corroboration was provided by Sumner and Andriacchi (1996) who reported diametric growth continued into the third decade during their assessment of cross-sectional properties of the humerus and femur, and by Szulc (2010) in an even older human sample. The potential cause for such a phenomenon is that muscular strength increases well into the third decade (Parker et al., 1990), and the skeletal system must adjust accordingly. This means modeling is mechanically adaptive during time periods critically sensitive to the environment and nutrition (during growth), and to the body's work and habit socially and culturally (in early adulthood). Because achievement of peak bone mass and its retention in late adulthood is also related to physical activity (Razi et al., 2015; Weaver et al., 2016), even the sunset years of our life are registered in some way by modeling activity. This is true for primarily resorptive activations, but periosteal formation can also be initiated specifically due to mechanical strain caused by age associated endosteal bone loss (Lazenby, 1990; Beck et al., 2001; Carpenter & Carter, 2008). Exactly how and where our bones are sensitive to exertion is complex and important for understanding specific patterns of modeling activity, different per element and individual.

Some non-human animal studies have investigated the mechanical effects of muscle activity on bone directly by removing a necessary muscular or skeletal element surgically, and/or labeling bone with fluorescent markers for response observation. Epker and Frost (1965) related the work of Avis (1961) and Washburn (1947), studies in which the removal of certain muscles inhibits the formation of the mandibular angle and coronoid process, respectively. Mosley and fellow researchers (1998) have demonstrated that the ulnae of growing rats responds dramatically to physical loading by increasing the overall diameter of apposition, whereas active controls demonstrated only normal lateral modeling drift. More specifically, they found that at lower strains, overall bone formation decreased, rather than increased and an adaptive response was, instead, accomplished by a change in long bone shape via straightening of the diaphysis. At higher strains, total bone modeling increased, both by increased lateral drift apposition and reversal to formation on opposing envelope surfaces as well, affecting a general diametric increase in total cortical area. Similar studies on rat ulnae by Robling et al. (2001) showed that even brief static loads could stunt modeling apposition as well as longitudinal growth, and that this effect is proportional to the magnitude of loading. They also found a positive bone growth effect from applied dynamic loading (Robling et al., 2001). To summarize the net findings of similar research, it seems that bone formation is stimulated by specific activity circumstances. Dynamic, supernormal strains, with higher strain rates interspersed by periods of recovery are more osteogenic than static, lower strain rates during sustained activity (Mosely et al., 1997, Burr, 2002, Fluckey et al., 2002). Further work should expand on the possibility that modeling response (formation or resorption) may be even more sensitive to tensile strains than compressive (Birkhold et al. 2017), and on the variable set points for modeling activity, region-to-region within flat (Pivonka et al., 2018) and long bones (Miller et al., 2021).

It is important to remember, however, that the "primary" influence of physical activity is not the only mechanical influence on the modeling process. Body mass and height, or the length of the bone in question can have their own effects (Robling & Stout, 2003). Also, the "secondary" mechanical environment can account for unique morphology, due to static and pulsatile loading of vessels, arteries, or even space constraints for muscle bellies. As an excellent example, Carpenter and Carter (2008) showed that static loading from the muscle bellies surrounding the tibia could account, in large part, for its triangular cross-section. At a smaller scale, there are local microstructural/topographical circumstances that could heavily influence structure and position of modeled tissue, at times at odds with expectation for grander phenomena of the bone, appendage, or skeleton itself. Birkhold et al. (2017) examined regional mechanical adaptation in the mouse tibia and argue that heterogeneity in response can be due to non-mechanical determinants at specific sites, and propose local nutrition supply as a potential example of causality. For all these reasons the mechanical influence on modeling is both significant and complex and requires a detailed understanding of the primary and secondary effects of the local mechanical environment.

2.5 Current and Future Directives

Bone modeling accounts for a considerable portion of achieved bone mass and morphology over the lifetime of an individual. Compared to remodeling, its microstructure is not as well documented. However, a better understanding could aid anthropological efforts of uncovering new information about lives lived long ago, leading to methodological improvement and innovation. Along the way, anthropological knowledge gained from bone modeling theory and methodology will certainly yield transdisciplinary returns (Maggiano et al., 2019; Maggiano & Maggiano, 2022).

Several researchers focused on histological age assessment have already noted both the complicating effect drift has on results interpretation, and the benefits of techniques that could take advantage of distributed tissues' varying in age-at-formation (Andronowski & Cole, 2021; Chan et al., 2007; Dominguez & Mavroudas, 2019; Gocha & Agnew, 2016; Gocha, 2018). A similar philosophy, encouraging more light on modeling processes, is being encouraged in the investigation of comparative microstructural variability from anatomical and evolutionary perspectives both within primates (Goldman et al. 2014; McFarlin et al. 2016; Raguin and Drapeau 2020), and elsewhere (Baag & Lee, 2022; Becker et al., 2020; Montoya-Sanhueza et al. 2021). In general, an understanding of modeling drift has the potential to improve histomorphometric sampling site selection and interpretation for a variety of other efforts as well, including assessment of vascularity/porosity, osteon typology, osteocytic lacunae, and microfracture. Together with the body of research bringing forward a more holistic understanding of bone through three-dimensional analyses (Andronowski et al., 2023; Carter et al., 2013; Maggiano, Maggiano & Clement et al., 2016; Repp et al. 2017), and structural levels only more recently accessible (see e.g. Reznikow et al. 2014; Roschger et al. 2020), exciting future applications and improved methodology await.

As a distinct feature indicating modeling activity, the endosteal lamellar pocket (ELP) serves as a potential summary for drift that could inform on growth, physical activity, and life history (Maggiano, Maggiano, Tiesler, Kierdorf et al., 2011; Maggiano, Maggiano, Tiesler, et al., 2016, Raguin et al. 2014). Employing a quantitative standard for drift direction in periadolescent bones yields important information, especially combined with cross-sectional geometric information or studies using diaphyseal shape analyses to reconstruct activity levels of individuals or populations (see e.g. Ruff & Hayes, 1983; Maggiano et al., 2008; Niinimäki et al., 2017). Future investigations thereof could correlate aspects of activity with ELP position or magnitude. Not only could the ELP be methodologically informative about variation in human growth patterns and direction of drift, but also differences in the impact of developmental insults and growth arrest due to disease and/or nutrition or other environmental factors. The complexity of a bone's shape is more and more recognized (Macintosh et al., 2013; OMahoney et al., 2023; Turcotte et al., 2022); adding intracortical variation will be another step along the way. Employing modeling theory and an understanding of drift, we may also find improved or new ways to assess other lifestyle characteristics. A better understanding of systemic and local drivers of bone growth and modeling mechanisms could help develop regimes for the prevention of public health risks such as osteopenia or osteoporosis (Agnew et al., 2013; Langdahl et al., 2016; Walker et al. 2023). Modeling research has even led to surprising applications in isotopic analysis. In a recent interdisciplinary collaboration, Maggiano and colleagues (2019) demonstrated that from even a tiny bone fragment, it was theoretically and methodologically possible to measure changes in weather or landscape mobility (at monthly intervals) within a single individual's lifetime, despite their death long ago. They employed secondary ion mass spectrometry along with traditional bulk, and micro-drilled, isotopic analyses to assess oxygen isotopic variation across the human humeral ELP. Though currently simply a proof-of-concept study, the work encourages future efforts that could improve our understanding of lines of arrest and other aspects of human growth, health, and physical adaptation important for reconstructing life histories in bioarchaology and even identifying forensic or war dead unknown remains. With this much potential, we anticipate continued efforts in histological analyses of bone modeling and eagerly await the results of future research.

Acknowledgements

We thank S. Stout and C. Crowder and other authors contributing to this edited volume. Research contributing directly to the current work results from cooperative efforts between the Department of Anthropology, The Ohio State University (OSU), Columbus, OH; and the following institutions, co-authors, and contributors: Department of Biology, University of Hildesheim, Germany (H. Kierdorf, S. Flohr), the Anatomy Center, University of Göttingen, Göttingen, Germany (M. Schultz), the Facultad de Ciencias Antropológicas of the Universidad Autonóma de Yucatán, Mérida, Mexico (V. Tiesler), the Anthropology and Biology Departments, University of Central Florida (UCF), Orlando, FL (T. Dupras), and the Center for Environmental Biotechnology, University of Tennessee, Knoxville, TN (J. Biggerstaff). Many of the pristine archaeological samples showcased here are the results of tireless archaeological efforts of the Dakhleh Oasis Project (A.J. Mills, C.A. Hope, and E. Molto, T. Dupras) with the permissions of the Egyptian Supreme Council of Antiquities, but in particular, T. Dupras' support has been unwavering, invaluable, and inspiring. Thin-ground sectioning and imaging for many of these samples was performed at the University of Göttingen thanks to the collaborative efforts of M. Schultz, his research group, and M. Brandt. Direct contributors of images used in this work are L. Lee (College of Medicine, Division of Anatomy, OSU); Y. Castro, D. Vanmali, and T. Dupras (Department of Anthropology, UCF); S. Weisbrode (Department of Veterinary Biosciences, OSU). Thank you all for your time and contributions.

References

Agnew, A. M., Moorhouse, K., Kang, Y. S., Donnelly, B. R., Pfefferle, K., Manning, A. X., ... & Bolte, J. H. (2013). The response of pediatric ribs to quasi-static loading: mechanical properties and microstructure. *Annals of Biomedical Engineering, 41*, 2501–2514.

Alexander, K. A., Raggatt, L. J., Millard, S., Batoon, L., Chiu-Ku Wu, A., Chang, M. K., ... & Pettit, A. R. (2017). Resting and injury-induced inflamed periosteum contain multiple macrophage subsets that are located at sites of bone growth and regeneration. *Immunology and Cell Biology, 95(1)*, 7–16.

Allen, M. R., & Burr, D. B. (2019). Bone growth, modeling, and remodeling. *In Basic and applied Bone biology* (pp. 85–100). Academic Press.

Allen, M. R., Hock, J. M., & Burr, D. B. (2004). Periosteum: biology, regulation, and response to osteoporosis therapies. *Bone, 35(5)*, 1003–1012.

Andronowski, J. M., & Cole, M. E. (2021). Current and emerging histomorphometric and imaging techniques for assessing age-at-death and cortical bone quality. *Wiley Interdisciplinary Reviews: Forensic Science, 3(2)*, e 1399.

Andronowski, J. M., Cole, M. E., Davis, R. A., Tubo, G. R., Taylor, J. T., & Cooper, D. M. (2023). A multimodal 3D imaging approach of pore networks in the human femur to assess age-associated vascular expansion and Lacuno-Canalicular reduction. *The Anatomical Record, 306(3)*, 75–493.

Andronowski, J. M., Pratt, I. V., & Cooper, D. M. (2017). Occurrence of osteon banding in adult human cortical bone. *American Journal of Physical Anthropology, 164(3)*, 635–642.

Ascenzi, A., Benvenuti, A., & Bonucci, E. (1982). The tensile properties of single osteonic lamellae: technical problems and preliminary results. *Journal of Biomechanics, 15(1)*, 29–37.

Aubin, J. E. (2001). Regulation of osteoblast formation and function. *Reviews in Endocrine and Metabolic Disorders, 2*, 81–94.

Avis, V. (1961). The significance of the angle of the mandible: an experimental and comparative study. *American Journal of Physical Anthropology, 19(1)*, 55–61.

Baag, S. J., & Lee, Y. N. (2022). Bone histology on Koreaceratops hwaseongensis (Dinosauria: Ceratopsia) from the Lower Cretaceous of South Korea. *Cretaceous Research, 134*, 105150.

Balena, R., Shih, M. S., & Parfitt, A. M. (1992). Bone resorption and formation on the periosteal envelope of the ilium: a histomorphometric study in healthy women. *Journal of Bone and Mineral Research, 7(12)*, 1475–1482.

Barak, M. M. (2019). Bone modeling or bone remodeling: That is the question. American *Journal of Physical Anthropology, 172(2)*, 153–155.

Beck, T. J., Stone, K. L., Oreskovic, T. L., Hochberg, M. C., Nevitt, M. C., Genant, H. K., & Cummings, S. R. (2001). Effects of current and discontinued estrogen replacement therapy on hip structural geometry: the study of osteoporotic fractures. *Journal of Bone and Mineral Research, 16(11)*, 2103–2110.

Becker, M., Witzel, C., Kierdorf, U., Frölich, K., & Kierdorf, H. (2020). Ontogenetic changes of tissue compartmentalization and bone type distribution in the humerus of Soay sheep. *Journal of Anatomy, 237(2)*, 334–354.

Berendsen, A. D. & Olsen, B. R. (2015). Bone development. *Bone, 80*, 14–18.

Biewener, A. A. (1993). Safety factors in bone strength. *Calcified tissue international*, 53, S68–S74.

Birkhold, A. I., Razi, H., Duda, G. N., Checa, S., & Willie, B. M. (2017). Tomography-based quantification of regional differences in cortical bone surface remodeling and mechano-response. *Calcified Tissue International, 100*, 255–270.

Birkhold, A. I., Razi, H., Duda, G. N., Weinkamer, R., Checa, S., & Willie, B. M. (2016). The periosteal bone surface is less mechano-responsive than the endocortical. *Scientific Reports, 6(1)*, 1–11.

Bisseret, D., Kaci, R., Lafage-Proust, M. H., Alison, M., Parlier-Cuau, C., Laredo, J. D., & Bousson, V. (2015). Periosteum: characteristic imaging findings with emphasis on radiologic-pathologic comparisons. *Skeletal Radiology, 44*, 321–338.

Blitz, E., Viukov, S., Sharir, A., Shwartz, Y., Galloway, J. L., Pryce, B. A., ... & Zelzer, E. (2009). Bone ridge patterning during musculoskeletal assembly is mediated through SCX regulation of Bmp 4 at the tendon-skeleton junction. *Developmental Cell, 17(6)*, 861–873.

Bolamperti, S., Villa, I., & Rubinacci, A. (2022). Bone remodeling: An operational process ensuring survival and bone mechanical competence. *Bone Research, 10(1)*, 48.

Bright, R. W., & Elmore, S. M. (1968, January). Physical properties of epiphyseal plate cartilage. In *Surgical forum* (Vol. 19, pp. 463–464).

Brighton, C. T., Lorich, D. G., Kupcha, R., Reilly, T. M., Jones, A. R., & Woodbury, I. I. (1992). The pericyte as a possible osteoblast progenitor cell. *Clinical Orthopaedics and Related Research®, 275*, 287–299.

Bromage, T. G., Juwayeyi, Y. M., Smolyar, I., Hu, B., Gomez, S., & Chisi, J. (2011). Enamel-calibrated lamellar bone reveals long period growth rate variability in humans. *Cells Tissues Organs, 194(2–4)*, 124–130.

Bromage, T. G., Lacruz, R. S., Hogg, R., Goldman, H. M., McFarlin, S. C., Warshaw, J., … & Boyde, A. (2009). Lamellar bone is an incremental tissue reconciling enamel rhythms, body size, and organismal life history. *Calcified Tissue International, 84*, 388–404.

Brookes, M., & Revell, W. J. (2012). *Blood supply of bone: scientific aspects.* Springer Science & Business Media.

Burr, D. B. (2002). Targeted and nontargeted remodeling. *Bone, 30(1)*, 2–4.

Burr, D. B., Schaffler, M. B., & Frederickson, R. G. (1988). Composition of the cement line and its possible mechanical role as a local interface in human compact bone. *Journal of Biomechanics, 21(11)*, 939–945.

Calderón, T., Arnold, W., Stalder, G., Painer, J., & Köhler, M. (2021). Labelling experiments in red deer provide a general model for early bone growth dynamics in ruminants. *Scientific Reports, 11(1)*, 14074.

Cambra-Moo, O., Nacarino-Meneses, C., Díaz-Güemes, I., Enciso, S., García Gil, O., Rodríguez, L., Rodríguez Barbero, M. Á., de Aza, A. H. & González Martín, A. (2015). Multidisciplinary characterization of the long-bone cortex growth patterns through sheep's ontogeny. *Journal of Structural Biology, 191(1)*, 1–9.

Carpenter, R. D., & Carter, D. R. (2008). The mechanobiological effects of periosteal surface loads. *Biomechanics and Modeling in Mechanobiology, 7*, 227–242.

Carpenter, R. D., & Carter, D. R. (2010). Computational simulation of spontaneous bone straightening in growing children. *Biomechanics and Modeling in Mechanobiology, 9*, 317–328.

Carter, Y., Thomas, C. D. L., Clement, J. G., & Cooper, D. M. (2013). Femoral osteocyte lacunar density, volume and morphology in women across the lifespan. *Journal of Structural Biology, 183(3)*, 519–526.

Castanet, J. (2006). Time recording in bone microstructures of endothermic animals; functional relationships. *Comptes Rendus Palevol, 5(3–4)*, 629–636.

Castanet, J., Croci, S., Aujard, F., Perret, M., Cubo, J., & De Margerie, E. (2004). Lines of arrested growth in bone and age estimation in a small primate: Microcebus murinus. *Journal of Zoology, 263(1)*, 31–39.

Castanet, J., Francillon-Vieillot, H., & Bruce, R. C. (1996). Age estimation in desmognathine salamanders assessed by skeletochronology. *Herpetologica*, 160–171.

Cerrito, P., Hu, B., Goldstein, J. Z., Kalisher, R., Bailey, S. E., & Bromage, T. G. (2022). Elemental composition of primary lamellar bone differs between parous and nulliparous rhesus macaque females. *Plos One, 17(11)*, e0276866.

Chamay, A. (1970). Mechanical and morphological aspects of experimental overload and fatigue in bone. *Journal of Biomechanics, 3(3)*, 263–270.

Chan, A. H., Crowder, C. M., & Rogers, T. L. (2007). Variation in cortical bone histology within the human femur and its impact on estimating age at death. *American Journal of Physical Anthropology, 132(1)*, 80–88.

Chen, H., Senda, T., & Kubo, K. Y. (2015). The osteocyte plays multiple roles in bone remodeling and mineral homeostasis. *Medical Molecular Morphology, 48*, 61–68.

Chow, J. W. M., Wilson, A. J., Chambers, T. J., & Fox, S. W. (1998). Mechanical loading stimulates bone formation by reactivation of bone lining cells in 13-week-old rats. *Journal of Bone and Mineral Research, 13(11)*, 1760–1767.

Cooper, D. M., Erickson, B., Peele, A. G., Hannah, K., Thomas, C. D. L., & Clement, J. G. (2011). Visualization of 3D osteon morphology by synchrotron radiation micro-CT. *Journal of Anatomy, 219(4)*, 481–489.

Cooper, D. M. L., Kawalilak, C. E., Harrison, K., Johnston, B. D., & Johnston, J. D. (2016). Cortical bone porosity: what is it, why is it important, and how can we detect it?. *Current Osteoporosis Reports, 14*, 187–198.

Crocker, D. J., Murad, T. M., & Geer, J. C. (1970). Role of the pericyte in wound healing: an ultrastructural study. *Experimental and Molecular Pathology, 13(1)*, 51–65.

Crowder, C. M., Andronowski, J. M., & Dominguez, V. M. (2018). Bone histology as an integrated tool in the process of human identification. In *New perspectives in forensic human skeletal identification* (pp. 201–213). Academic Press.

Currey, J. D. (2006). *Bones: structure and mechanics*. Princeton university press.

Debnath, S., Yallowitz, A. R., McCormick, J., Lalani, S., Zhang, T., Xu, R., Li, N., Liu, Y., Yang, Y. S., Eiseman, M., Shim, J. H., Hameed, M., Healey, J. H., Bostrom, M. P., Landau, D. A., Greenblatt, M. B. (2018). Discovery of a periosteal stem cell mediating intramembranous bone formation. *Nature, 562(7725)*, 133–139.

Dominguez, V. M., & Mavroudas, S. (2019). Bone histology for skeletal age-at-death estimation. In *Age Estimation* (pp. 145–159). Academic Press.

Du Boulay, G. H. (1966). Pulsatile movements in the CSF pathways. *The British journal of radiology, 39(460)*, 255–262.

Dwek, J. R. (2010). The periosteum: what is it, where is it, and what mimics it in its absence?. *Skeletal Radiology, 39(4)*, 319–323.

Ellender, G. R. A. H. A. M., Feik, S. A., & Carach, B. J. (1988). Periosteal structure and development in a rat caudal vertebra. *Journal of Anatomy, 158*, 173.

Enlow, D. (1962). A study of the postnatal growth and remodeling of bone. *The American Journal of Anatomy, 110*, 79–101.

Ensrud, K. E., Palermo, L., Black, D. M., Cauley, J., Jergas, M., Orwoll, E. S., … & Cummings, S. R. (1995). Hip and calcaneal bone loss increase with advancing age: longitudinal results from the study of osteoporotic fractures. *Journal of Bone and Mineral Research, 10(11)*, 1778–1787.

Epker, B. N., & Frost, H. M. (1965). A histological study of remodeling at the periosteal, haversian canal, cortical endosteal, and trabecular endosteal surfaces in human rib. *The Anatomical Record, 152(2)*, 129–135.

Erben, R. G. (2015). Hypothesis: coupling between resorption and formation in cancellous bone remodeling is a mechanically controlled event. *Frontiers in Endocrinology, 6*, 82.

Feik, S. A., Storey, E., & Ellender, G. (1987). Stress induced periosteal changes. *British journal of experimental pathology, 68(6)*, 803.

Fluckey, J. D., Dupont-Versteegden, E. E., Montague, D. C., Knox, M., Tesch, P., Peterson, C. A., & Gaddy-Kurten, D. (2002). A rat resistance exercise regimen attenuates losses of musculoskeletal mass during hindlimb suspension. *Acta Physiologica Scandinavica, 176(4)*, 293–300.

Foote, J. S. (1916). *A contribution to the comparative histology of the femur*. Smithsonian institution.

Forriol, F., & Shapiro, F. (2005). Bone development: interaction of molecular components and biophysical forces. *Clinical Orthopaedics and Related Research®, 432*, 14–33.

Frost, H. (1963a). Measurement of human bone formation by means of tetracycline labeling. *Canadian Journal of Biochemistry and Physiology, 41*, 31–42.

Frost, H. M. (1963b). Dynamics of Bone Remodeling. In H. M. Frost (Ed.), *Bone Biodynamics.* its clinical applications. *The Anatomical Record: An Official Publication of the American* Little, Brown, and Co.

Frost, H. M. (1964). *The laws of bone structure*. Charles C. Thomas.

Frost, H. M. (1973). *Bone modeling and skeletal modeling errors*. Charles C. Thomas.

Frost, H. M. (1987). Bone "mass" and the "mechanostat": a proposal. *Anatomical Record, 219*, 1–9.

Frost, H. M. (2001). From Wolff's law to the Utah paradigm: insights about bone physiology. *Association of Anatomists, 262(4)*, 398–419.

Gamsjaeger, S., Eriksen, E. F., & Paschalis, E. P. (2021). Effect of hormone replacement therapy on bone formation quality and mineralization regulation mechanisms in early postmenopausal women. *Bone Reports, 14*, 101055.

Garn, S. M. (1970). The Earlier Gain and the Later Loss of Cortical Bone *in Nutritonal Perspectives.*

Garn, S. M. (1972). The course of bone gain and the phases of bone loss. *Orthopedic Clinics of North America, 3(3)*, 503–520.

Gocha, T. P., & Agnew, A. M. (2016). Spatial variation in osteon population density at the human femoral midshaft: histomorphometric adaptations to habitual load environment. *Journal of Anatomy, 228(5)*, 733–745.

Gocha, T. P., Robling, A. G., & Stout, S. D. (2018). Histomorphometry of human cortical bone: applications to age estimation. *Biological Anthropology of the Human Skeleton*, 145–187.
Goldman, H. M., Bromage, T., Thomas, C. D., Clement, J. G. (2003). Preferred collagen fiber orientation in the human mid-shaft femur. *Anatomical Record, 272A*, 434–445.
Goldman, H. M., Hampson, N. A., Guth, J. J., Lin, D., & Jepsen, K. J. (2014). Intracortical remodeling parameters are associated with measures of bone robustness. *The Anatomical Record, 297(10)*, 1817–1828.
Goldman, H. M., McFarlin, S. C., Cooper, D. M., Thomas, C. D., Clement, J. G. (2009). Ontogenetic patterning of cortical bone microstructure and geometry at the human mid-shaft femur. *Anatomical Record, 292(1)*, 48–64.
Hage, I. S., & Hamade, R. F. (2014, February). Toward quantifying geometric microstructural differences between primary and secondary osteons via segmentation. In *2nd Middle East Conference on Biomedical Engineering* (pp. 371–374). IEEE.
Hall, B. K., & Herring, S. W. (1990). Paralysis and growth of the musculoskeletal system in the embryonic chick. *Journal of Morphology, 206(1)*, 45–56.
Hallett, S. A., Matsushita, Y., Ono, W., Sakagami, N., Mizuhashi, K., Tokavanich, N., Nagata, M., Zhou, A., Hirai, T., Kronenberg, H. M., Ono, N. (2021). Chondrocytes in the resting zone of the growth plate are maintained in a Wnt-inhibitory environment. *Elife, 26(10)*, e64513.
Hamrick, M. W., McNeil, P. L., & Patterson, S. L. (2010). Role of muscle-derived growth factors in bone formation. *Journal of Musculoskeletal & Neuronal Interactions, 10(1)*, 64.
Han, Z. H., Palnitkar, S., Rao, D. S., Nelson, D., & Parfitt, A. M. (1997). Effects of ethnicity and age or menopause on the remodeling and turnover of iliac bone: implications for mechanisms of bone loss. *Journal of Bone and Mineral Research, 12(4)*, 498–508.
Hansen, H. L., Bredbenner, T. L., Nicolella, D. P., Mahaney, M. C., & Havill, L. M. (2009). Cross-sectional geometry of the femoral midshaft in baboons is heritable. *Bone, 45(5)*, 892–897.
Hillier, M. L., & Bell, L. S. (2007). Differentiating human bone from animal bone: a review of histological methods. *Journal of Forensic Sciences, 52(2)*, 249–263.
Hohmann, E. L., Elde, R. P., Rysavy, J. A., Einzig, S., & Gebhard, R. L. (1986). Innervation of periosteum and bone by sympathetic vasoactive intestinal peptide-containing nerve fibers. *Science, 232(4752)*, 868–871.
Isojima, T., & Sims, N. A. (2021). Cortical bone development, maintenance and porosity: genetic alterations in humans and mice influencing chondrocytes, osteoclasts, osteoblasts and osteocytes. *Cellular and Molecular Life Sciences, 78(15)*, 5755–5773.
Jee, W. S. (2001). Integrated bone tissue physiology: anatomy and physiology. In S. C. Cowin (Ed.) *Bone mechanics handbook*. CRC Press.
Jee, W. S., & Frost, H. M. (1992). Skeletal adaptations during growth. *Triangle; The Sandoz Journal of Medical Science, 31(2/3)*, 77–88.
Johnson, L. C. (1964). Morphologic analysis in pathology: the kinetics of disease and general biology of bone. In H. M. Frost (Ed.) *Bone Biodynamics*. Little, Brown, and Co.
Jones, D. B., Nolte, H., Scholübbers, J. G., Turner, E., & Veltel, D. (1991). Biochemical signal transduction of mechanical strain in osteoblast-like cells. *Biomaterials, 12(2)*, 101–110.
Killian, M. L. (2022, March). Growth and mechanobiology of the tendon-bone enthesis. In *Seminars in Cell & Developmental Biology* (Vol. 123, pp. 64–73). Academic Press.
Köhler, M., Marín-Moratalla, N., Jordana, X., & Aanes, R. (2012). Seasonal bone growth and physiology in endotherms shed light on dinosaur physiology. *Nature, 487(7407)*, 358–361.
Kolb, C., Scheyer, T. M., Veitschegger, K., Forasiepi, A. M., Amson, E., Van der Geer, A. A., … & Sánchez-Villagra, M. R. (2015). Mammalian bone palaeohistology: a survey and new data with emphasis on island forms. *PeerJ, 3*, e1358.
Lacroix, P. (1947). Organizers and the growth of bone. *The Journal of Bone & Joint Surgery, 29(2)*, 292–296.
Langdahl, B., Ferrari, S., & Dempster, D. W. (2016). Bone modeling and remodeling: potential as therapeutic targets for the treatment of osteoporosis. *Therapeutic advances in musculoskeletal disease, 8(6)*, 225–235.

Lassen, N. E., Andersen, T. L., Pløen, G. G., Søe, K., Hauge, E. M., Harving, S., ... & Delaisse, J. M. (2017). Coupling of bone resorption and formation in real time: new knowledge gained from human Haversian BMUs. *Journal of Bone and Mineral Research, 32(7)*, 1395–1405.

Lazenby, R. A. (1990). Continuing periosteal apposition I: Documentation, hypotheses, and interpretation. *American Journal of Physical Anthropology, 82(4)*, 451–472.

Liu, Y. Z., Xiao, P., Guo, Y. F., Xiong, D. H., Zhao, L. J., Shen, H., ... & Deng, H. W. (2006). Genetic linkage of human height is confirmed to 9q22 and Xq24. *Human Genetics, 119*, 295–304.

Locke, M. (2004). Structure of long bones in mammals. *Journal of Morphology, 262(2)*, 546–565.

Macintosh, A. A., Davies, T. G., Ryan, T. M., Shaw, C. N., & Stock, J. T. (2013). Periosteal versus true cross-sectional geometry: A comparison along humeral, femoral, and tibial diaphyses. *American Journal of Physical Anthropology, 150(3)*, 442–452.

Maes, C. (2013). Role and regulation of vascularization processes in endochondral bones. *Calcified Tissue International, 92(4)*, 307–323.

Maggiano, C., & Maggiano, I. (2022). Histomorphology of Un-Decalcified Bone in Mesoamerican Bioarchaeology. *The Routledge Handbook of Mesoamerican Bioarchaeology*, 601–616.

Maggiano, C., Maggiano, I., Tiesler, V., & Stout, S. (2011). The endosteal lamellar pocket as an indicator of childhood and adolescent modeling drift direction and magnitude during long bone growth. In *American Journal of Physical Anthropology* (Vol. 144, pp. 203–203), USA: Wiley-Blackwell.

Maggiano, C. M., Maggiano, I. S., Tiesler, V. G., Chi-Keb, J. R., Stout, S. D. (2016). Methods and theory in bone modeling drift: comparing spatial analyses of primary bone distributions in the human humerus. *Journal of Anatomy, 228(1)*, 190–202.

Maggiano, C. M., White, C. D., Stern, R. A., Peralta, J. S., & Longstaffe, F. J. (2019). Focus: Oxygen isotope microanalysis across incremental layers of human bone: Exploring archaeological reconstruction of short term mobility and seasonal climate change. *Journal of Archaeological Science, 111*, 105028.

Maggiano, I. S., Maggiano, C. M., Clement, J. G., Thomas, C. D. L., Carter, Y., & Cooper, D. M. (2016). Three-dimensional reconstruction of Haversian systems in human cortical bone using synchrotron radiation-based micro-CT: morphology and quantification of branching and transverse connections across age. *Journal of Anatomy, 228(5)*, 719–732.

Maggiano, I. S., Maggiano, C. M., Tiesler, V., Kierdorf, H., Stout, S. D., Schultz, M. (2011). A distinct region of microarchitectural variation in femoral compact bone: Histomorphology of the endosteal lamellar pocket. *International Journal of Osteoarchaeology, 21(6)*, 743–750.

Maggiano, I. S., Maggiano, C. M., Tiesler, V. G., Chi-Keb, J. R., Stout, S. D. (2015). Drifting diaphyses: Asymmetry in diametric growth and adaptation along the humeral and femoral length. *Anatomical Record, 298(10)*, 1689–1699.

Maggiano, I. S., Schultz, M., Kierdorf, H., Sosa, T. S., Maggiano, C. M., & Tiesler Blos, V. (2008). Cross-sectional analysis of long bones, occupational activities and long-distance trade of the Classic Maya from Xcambó—Archaeological and osteological evidence. *American Journal of Physical Anthropology, 136(4)*, 470–477.

Mahoney, P., Miszkiewicz, J. J., Pitfield, R., Schlecht, S. H., Deter, C., & Guatelli-Steinberg, D. (2016). Biorhythms, deciduous enamel thickness, and primary bone growth: a test of the Havers-Halberg Oscillation hypothesis. *Journal of Anatomy, 228(6)*, 919–928.

Manolagas, S. C., & Parfitt, A. M. (2010). What old means to bone. *Trends in Endocrinology & Metabolism, 21(6)*, 369–374.

Marotti, G., Zaffe, D., Ferretti, M., & Palumbo, C. (2010). Static osteogenesis and dynamic osteogenesis: Their relevance in dental bone implants and biomaterial osseointegration. *Journal of Osteology and Biomaterials, 1*, 133–139.

Martin, R. B. (2000). Does osteocyte formation cause the nonlinear refilling of osteons? *Bone, 26(1)*, 71–78.

Martin, R. B. (2003). Functional adaptation and fragility of the skeleton. *Bone loss and Osteoporosis: An Anthropological Perspective*, 121–138.

Martin, R. B., Burr, D. B., and Sharkey, N. A. (1998). *Skeletal tissue mechanics*. Springer.

McFarlin, S. C., Terranova, C. J., Zihlman, A. L., & Bromage, T. G. (2016). Primary bone microanatomy records developmental aspects of life history in catarrhine primates. *Journal of Human Evolution, 92*, 60–79.

Miller, C. J., Trichilo, S., Pickering, E., Martelli, S., Delisser, P., Meakin, L. B., & Pivonka, P. (2021). Cortical thickness adaptive response to mechanical loading depends on periosteal position and varies linearly with loading magnitude. *Frontiers in Bioengineering and Biotechnology, 9*, 671606.

Mitchell, J., & van Heteren, A. H. (2016). A literature review of the spatial organization of lamellar bone. *Comptes Rendus Palevol, 15(1-2)*, 23–31.

Mizoguchi, T. & Ono, N. (2021). The diverse origin of bone-forming osteoblasts. *Journal of Bone and Mineral Research, 36(8)*, 1432–1447.

Montoya-Sanhueza, G., Bennett, N. C., Oosthuizen, M. K., Dengler-Crish, C. M., & Chinsamy, A. (2021). Long bone histomorphogenesis of the naked mole-rat: histodiversity and intraspecific variation. *Journal of Anatomy, 238(6)*, 1259–1283.

Montoya-Sanhueza, G., & Chinsamy, A. (2018). Cortical bone adaptation and mineral mobilization in the subterranean mammal Bathyergus suillus (Rodentia: Bathyergidae): effects of age and sex. *PeerJ, 6*, e4944.

Mosley, J. R., & Lanyon, L. E. (1998). Strain rate as a controlling influence on adaptive modeling in response to dynamic loading of the ulna in growing male rats. *Bone, 23(4)*, 313–318.

Mosley, J. R., & Lanyon, L. E. (2002). Growth rate rather than gender determines the size of the adaptive response of the growing skeleton to mechanical strain. *Bone, 30(1)*, 314–319.

Mosley, J. R., March, B. M., Lynch, J., & Lanyon, L. E. (1997). Strain magnitude related changes in whole bone architecture in growing rats. *Bone, 20(3)*, 191–198.

Niinimäki, S., Narra, N., Härkönen, L., Abe, S., Nikander, R., Hyttinen, J., … & Sievänen, H. (2017). The relationship between loading history and proximal femoral diaphysis cross-sectional geometry. *American Journal of Human Biology, 29(4)*, e22965.

Nilsson, O., Weise, M., Landman, E. B. M., Meyers, J. L., Barnes, K. M., Baron, J. (2014). Evidence That Estrogen Hastens Epiphyseal Fusion and Cessation of Longitudinal Bone Growth by Irreversibly Depleting the Number of Resting Zone Progenitor Cells in Female Rabbits. *Endocrinology, 155(8)*, 2892–2899.

Nowlan, N. C., Bourdon, C., Dumas, G., Tajbakhsh, S., Prendergast, P. J., & Murphy, P. (2010). Developing bones are differentially affected by compromised skeletal muscle formation. *Bone, 46(5)*, 1275–1285.

O'Mahoney, T. G., Lowe, T., Chamberlain, A. T., & Sellers, W. I. (2023). Endostructural and periosteal growth of the human humerus. *The Anatomical Record, 306(1)*, 60–78.

Parfitt, A. M. (1983). The physiologic and clinical significance of bone histomorphometric data. In R. R. Recker (ed.) *Bone Histomorphometry: Techniques and Interpretation*. CRC Press.

Parfitt, A. M. (1988). Bone remodeling: relationship to the amount and structure. *Osteoporosis. Etiology, Diagnosis and Management*, 45–93.

Parfitt, A. M. (1994). Osteonal and hemi-osteonal remodeling: the spatial and temporal framework for signal traffic in adult human bone. *Journal of Cellular Biochemistry, 55(3)*, 273–286.

Parfitt, A. M. (2003). New concepts of bone remodeling: A unified spatial and temporal model with physiologic and pathophysiologic implications. In C. Crowder and S. Stout (eds.) *Bone Loss and Osteoporosis: An Anthropological Perspective*, 3–17.

Parfitt, A. M., Travers, R., Rauch, F., & Glorieux, F. H. (2000). Structural and cellular changes during bone growth in healthy children. *Bone, 27(4)*, 487–494.

Parker, D. F., Round, J. M., Sacco, P., & Jones, D. A. (1990). A cross-sectional survey of upper and lower limb strength in boys and girls during childhood and adolescence. *Annals of Human Biology, 17(3)*, 199–211.

Paschalis, E. P., Gamsjaeger, S., Hassler, N., Klaushofer, K., & Burr, D. (2017). Ovarian hormone depletion affects cortical bone quality differently on different skeletal envelopes. *Bone, 95*, 55–64.

Percival, C. J., Richtsmeier, J. T. (2013). Angiogenesis and intramembranous osteogenesis. Developmental Dynamics, *242(8)*, 909–922.

Pitfield, R., Miszkiewicz, J. J., & Mahoney, P. (2019). Microscopic markers of an infradian biorhythm in human juvenile ribs. *Bone, 120*, 403–410.

Pivonka, P., Park, A., & Forwood, M. R. (2018). Functional adaptation of bone: the mechanostat and beyond. *Multiscale Mechanobiology of Bone Remodeling and Adaptation*, 1–60.

Raguin, E., & Drapeau, M. S. (2020). Relation between cross-sectional bone geometry and double zonal osteon frequency and morphology. *American Journal of Physical Anthropology, 171(4)*, 598–612.

Raguin, E., Streeter, M. A., Lazenby, R., & Drapeau, M. S. (2014, March). Histomorphology of endosteal lamellar pocket: comparison of the second metacarpal in two genetically distinct populations. In *American Journal of Physical Anthropology, 153*, 215–215.

Rauch, F. (2012). The dynamics of bone structure development during pubertal growth. *Journal of Musculoskeletal and Neuronal Interactions, 12(1)*, 1–6.

Rauch, F., Neu, C., Manz, F., & Schoenau, E. (2001). The development of metaphyseal cortex—implications for distal radius fractures during growth. *Journal of Bone and Mineral Research, 16(8)*, 1547–1555.

Razi, H., Birkhold, A. I., Weinkamer, R., Duda, G. N., Willie, B. M., & Checa, S. (2015). Aging leads to a dysregulation in mechanically driven bone formation and resorption. *Journal of Bone and Mineral Research, 30(10)*, 1864–1873.

Reid, S. A. (1986). A study of lamellar organisation in juvenile and adult human bone. *Anatomy and Embryology, 174*, 329–338.

Repp, F., Kollmannsberger, P., Roschger, A., Berzlanovich, A., Gruber, G. M., Roschger, P., … & Weinkamer, R. (2017). Coalignment of osteocyte canaliculi and collagen fibers in human osteonal bone. *Journal of Structural Biology, 199(3)*, 177–186.

Reznikov, N., Shahar, R., & Weiner, S. (2014). Three-dimensional structure of human lamellar bone: the presence of two different materials and new insights into the hierarchical organization. *Bone, 59*, 93–104.

Rhinelander, F. W. (1972). *Circulation in bone. The Biochemistry and Physiology of Bone*, 2, 1–77.

Ricqles, D. (1991). Comparative microstructure of bone. *Bone matrix and bone specific products*, 1–78.

Riffault, M., Johnson, G. P., Owen, M. M., Javaheri, B., Pitsillides, A. A., & Hoey, D. A. (2020). Loss of Adenylyl Cyclase 6 in Leptin Receptor Expressing Stromal Cells Attenuates Loading Induced Endosteal Bone Formation. *Journal of Bone and Mineral Research Plus, 4(11)*, e10408.

Rivas, R., & Shapiro, F. (2002). Structural stages in the development of the long bones and epiphyses: a study in the New Zealand white rabbit. *JBJS, 84(1)*, 85–100.

Robinson, S. T., Shyu, P. T., & Guo, X. E. (2021). Mechanical loading and parathyroid hormone effects and synergism in bone vary by site and modeling/remodeling regime. *Bone, 153*, 116171.

Robling, A. D., Duijvelaar, K. M., Geevers, J. V., Ohashi, N., Turner, C. H. (2001). Modulation of appositional and longitudinal bone growth in the rat ulna by applied static and dynamic force. *Bone, 29(2)*, 105–113.

Robling, A. G., Castillo, A. B., & Turner, C. H. (2006). Biomechanical and molecular regulation of bone remodeling. *Annual Review of Biomedical Engineering, 8*, 455–498.

Robling, A. G., & Stout, S. D. (1999). Morphology of the drifting osteon. *Cells Tissues Organs, 164(4)*, 192–204.

Robling, A. G., & Stout, S. D. (2003). Histomorphology, geometry, and mechanical loading in past populations. In *Bone loss and osteoporosis: An anthropological perspective* (pp. 189–205). Boston, MA: Springer US.

Robling, A. G., & Turner, C. H. (2002). Mechanotransduction in bone: genetic effects on mechanosensitivity in mice. *Bone, 31(5)*, 562–569.

Rodríguez, J. I., Garcia-Alix, A., Palacios, J., & Paniagua, R. (1988). Changes in the long bones due to fetal immobility caused by neuromuscular disease. *A radiographic and histological study. JBJS, 70(7)*, 1052–1060.

Roschger, A., Wagermaier, W., Gamsjaeger, S., Hassler, N., Schmidt, I., Blouin, S., … & Fratzl, P. (2020). Newly formed and remodeled human bone exhibits differences in the mineralization process. *Acta Biomaterialia, 104*, 221–230.

Rubin, P. (1964). Dynamic classification of bone dysplasias. *Academic Medicine, 39(11)*, 1059.

Ruff, C., Holt, B., & Trinkaus, E. (2006). Who's afraid of the big bad Wolff?:"Wolff's law" and bone functional adaptation. *American Journal of Physical Anthropology, 129(4)*, 484–498.

Ruff, C. B., & Hayes, W. C. (1983). Cross-sectional geometry of Pecos Pueblo femora and tibiae—A biomechanical investigation: I. Method and general patterns of variation. *American Journal of Physical Anthropology, 60(3)*, 359–381.

Salmon, P. (2015). Non-linear pattern formation in bone growth and architecture. *Frontiers in Endocrinology, 5*, 239.

Sander, P. M., & Andrássy, P. (2006). Lines of arrested growth and long bone histology in Pleistocene large mammals from Germany: what do they tell us about dinosaur physiology?. *Palaeontographica A, 277(1)*, 143–159.

Schaffler, M. B., Burr, D. B., & Frederickson, R. G. (1987). Morphology of the osteonal cement line in human bone. *The Anatomical Record, 217(3)*, 223–228.

Schaffler, M. B., Cheung, W. Y., Majeska, R., & Kennedy, O. (2014). Osteocytes: master orchestrators of bone. *Calcified Tissue International, 94*, 5–24.

Scheuer, L. & Black, S. (2000). *Developmental juvenile osteology*. Academic Press.

Schnitzler, C. M., & Mesquita, J. M. (2013). Cortical porosity in children is determined by age-dependent osteonal morphology. *Bone, 55(2)*, 476–486.

Schultz, M. (2001). Paleohistopathology of bone: a new approach to the study of ancient diseases. *American Journal of Physical Anthropology: The Official Publication of the American Association of Physical Anthropologists*, 116(S33), 106–147.

Shahar, R., & Weiner, S. (2018). Open questions on the 3D structures of collagen containing vertebrate mineralized tissues: A perspective. *Journal of Structural Biology, 201(3)*, 187–198.

Shapiro, F. (2002). *Pediatric orthopedic deformities. Basic science, diagnosis, and treatment.* Elsevier.

Silventoinen, K. (2003). Determinants of variation in adult body height. *Journal of Biosocial Science, 35(2)*, 263–285.

Simpson, A. H. (1985). The blood supply of the periosteum. *Journal of Anatomy, 140(Pt 4)*, 697.

Skedros, J. G., Holmes, J. L., Vajda, E. G., & Bloebaum, R. D. (2005). Cement lines of secondary osteons in human bone are not mineral-deficient: New data in a historical perspective. *The Anatomical Record Part A: Discoveries in Molecular, Cellular, and Evolutionary Biology: An Official Publication of the American Association of Anatomists, 286(1)*, 781–803.

Smit, T. H. (2021). Closing the osteon: Do osteocytes sense strain rate rather than fluid flow?. *BioEssays, 43(8)*, 2000327.

Smolyar, I. V., & Bromage, T. G. (2004). Discrete model of fish scale incremental pattern: A formalization of the 2D anisotropic structure. *ICES journal of Marine Science, 61(6)*, 992–1003.

Squier, C. A., Ghoneim, S., & Kremenak, C. R. (1990). Ultrastructure of the periosteum from membrane bone. *Journal of Anatomy, 171*, 233.

Srinivasan, S., Balsiger, D., Huber, P., Ausk, B. J., Bain, S. D., Gardiner, E. M., & Gross, T. S. (2019). Static preload inhibits loading-induced bone formation. *JBMR Plus, 3(5)*, e10087.

Srinivasan, S., Weimer, D. A., Agans, S. C., Bain, S. D., & Gross, T. S. (2002). Low-magnitude mechanical loading becomes osteogenic when rest is inserted between each load cycle. *Journal of Bone and Mineral Research, 17(9)*, 1613–1620.

Stein, K., & Prondvai, E. (2014). Rethinking the nature of fibrolamellar bone: An integrative biological revision of sauropod plexiform bone formation. *Biological Reviews, 89(1)*, 24–47.

Stoker, N. G., & Epker, B. N. (1971). Age changes in endosteal bone remodeling and balance in the rabbit. *Journal of Dental Research, 50*(6), 1570–1574.

Streeter, M., & Stout, S. D. (2003). The histomorphometry of the subadult rib: Age-associated changes in bone mass and the creation of peak bone mass. *Bone Loss and Osteoporosis: An Anthropological Perspective*, 91–101.

Sumner, D. R., & Andriacchi, T. P. (1996). Adaptation to differential loading: Comparison of growth-related changes in cross-sectional properties of the human femur and humerus. *Bone, 19(2)*, 121–126.

Szulc, P. (2010). Changes in bone size and geometry with aging. In *Osteoporosis in men* (pp. 193–206). Academic Press.
Szulc, P., & Delmas, P. D. (2007). Bone loss in elderly men: increased endosteal bone loss and stable periosteal apposition. The prospective MINOS study. *Osteoporosis International, 18*, 495–503.
Tanck, E., Hannink, G., Ruimerman, R., Buma, P., Burger, E. H., & Huiskes, R. (2006). Cortical bone development under the growth plate is regulated by mechanical load transfer. *Journal of Anatomy, 208(1)*, 73–79.
Tang, X. M., & Chai, B. P. (1986). Effect of CO2 laser irradiation on experimental fracture healing: a transmission electron microscopic study. *Lasers in Surgery and Medicine, 6(3)*, 346–352.
Tate, M. L. K. (2003). "Whither flows the fluid in bone?" An osteocyte's perspective. *Journal of Biomechanics, 36(10)*, 1409–1424.
Taylor, D., & Lee, T. C. (2003). Microdamage and mechanical behaviour: predicting failure and remodelling in compact bone. *Journal of Anatomy, 203(2)*, 203–211.
Taylor, T., Frost, H. M. (1966). The existence of a zone of finite thickness during tetracycline labeling of bone. *Henry Ford Hospital Medical bulletin, 14*, 397–403.
Tomlin, D. H., Henry, K. M., & Kon, S. K. (1953). Autoradiographic study of growth and calcium metabolism in the long bones of the rat. *British Journal of Nutrition, 7(3)*, 235–252.
Turcotte, C. M., Rabey, K. N., Green, D. J., & McFarlin, S. C. (2022). Muscle attachment sites and behavioral reconstruction: An experimental test of muscle-bone structural response to habitual activity. *American Journal of Biological Anthropology, 177(1)*, 63–82.
Vaananen, H. K., Zhao, H., Mulari, M., & Halleen, J. M. (2000). The cell biology of osteoclast function. *Journal of Cell Science, 113(3)*, 377–381.
Van Oers, R. F., Ruimerman, R., Tanck, E., Hilbers, P. A., & Huiskes, R. (2008). A unified theory for osteonal and hemi-osteonal remodeling. *Bone, 42(2)*, 250–259.
Venken, K., De Gendt, K., Boonen, S., Ophoff, J., Bouillon, R., Swinnen, J. V., ... & Vanderschueren, D. (2006). Relative impact of androgen and estrogen receptor activation in the effects of androgens on trabecular and cortical bone in growing male mice: A study in the androgen receptor knockout mouse model. *Journal of Bone and Mineral Research, 21(4)*, 576–585.
Verbruggen, S. W., Vaughan, T. J., & McNamara, L. M. (2016). Mechanisms of osteocyte stimulation in osteoporosis. *Journal of the mechanical behavior of biomedical materials*, 62, 158–168.
Volkman, S. K., Galecki, A. T., Burke, D. T., Paczas, M. R., Moalli, M. R., Miller, R. A., & Goldstein, S. A. (2003). Quantitative trait loci for femoral size and shape in a genetically heterogeneous mouse population. *Journal of Bone and Mineral Research, 18(8)*, 1497–1505.
Voller, T., Cameron, P., Watson, J., Phadnis, J. (2020). The growth plate: anatomy and disorders. *Orthopaedics and Trauma, 34(3)*, 135–140.
Walker, M. M., Oxenham, M. F., Nguyen, T. M. H., Trinh, H. H., Minh, T. T., Nguyen, L. C., ... & Miszkiewicz, J. J. (2023). Primary bone retention in a young adult male with limb disuse: A bioarchaeological case study. *Historical Biology, 35(2)*, 235–241.
Wang, T., Zhang, X., & Bikle, D. D. (2017). Osteogenic differentiation of periosteal cells during fracture healing. *Journal of Cellular Physiology, 232(5)*, 913–921.
Washburn, S. L. (1947). The relation of the temporal muscle to the form of the skull. *The Anatomical Record, 99(3)*, 239–248.
Weaver, C. M., Gordon, C. M., Janz, K. F., Kalkwarf, H. J., Lappe, J. M., Lewis, R., ... & Zemel, B. (2016). *The National Osteoporosis Foundation's position statement on peak bone mass development and lifestyle factors: a systematic review and implementation recommendations. Osteoporosis International, 27*, 1281–1386.
Weivoda, M. M., & Bradley, E. W. (2023). Macrophages and bone remodeling. *Journal of Bone and Mineral Research, 38(3)*, 359–369.
Weston, D. A. (2009). Paleohistopathological analysis of pathology museum specimens: can periosteal reaction microstructure explain lesion etiology. *American Journal of Physical Anthropology, 140*, 186–193.

Wong, M., Germiller, J., Bonadio, J., & Goldstein, S. A. (1993). Neuromuscular atrophy alters collagen gene expression, pattern formation, and mechanical integrity of the chick embryo long bone. *Progress in Clinical and Biological Research*, *383*, 587–597.

Wu, K., Schubeck, K. E., Frost, H. M., & Villanueva, A. (1970). Haversian bone formation rates determined by a new method in a mastodon, and in human diabetes mellitus and osteoporosis. *Calcified Tissue Research*, *6*, 204–219.

Yang, L., Tsang, K. Y., Tang, H. C., Chan, D., Cheah, K. S. (2014). Hypertrophic chondrocytes can become osteoblasts and osteocytes in endochondral bone formation. *Proceedings of the National Academy of Sciences*, *111(33)*, 12097–12102.

Yu, Y. Y., Lieu, S., Lu, C., & Colnot, C. (2010). Bone morphogenetic protein 2 stimulates endochondral ossification by regulating periosteal cell fate during bone repair. *Bone*, *47(1)*, 65–73.

Zimmermann, E. A., Riedel, C., Schmidt, F. N., Stockhausen, K. E., Chushkin, Y., Schaible, E., Gludovatz, B., Vettorazzi, E., Zontone, F., Püschel, K., Amling, M., Ritchie, R. O., Busse, B. (2019). Mechanical Competence and Bone Quality Develop During Skeletal Growth. *Journal of Bone and Mineral Research*, *34(8)*, 1461–1472.

Pediatric Bone Histomorphology and Environmental Stresses

3

HAVIVA M. GOLDMAN AND
GWEN ROBBINS SCHUG

3.1 Introduction

Throughout childhood, adolescence, and into early adulthood, our bones undergo dramatic changes in size, shape, and organization through modeling and remodeling. These processes work in concert to allow the individual to attain and maintain a skeleton that can withstand normal physiological loads. The quantity and quality of the bone we acquire as children have long-lasting consequences for adult bone structure and function, potentially either protecting or predisposing an individual to bone fragility and increased fracture risk later in life (Proia et al., 2021). Research has shown how adequate bone accrual during growth is dependent on functional loading (Currey, 1984; Lanyon, 1984; Biewener et al., 1996; Lanyon, 1996, Rizzoli et al., 2010), with genetics also playing a key role in the determination of bone architecture (Bachrach, 2001, Pocock et al., 1987, Heaney et al., 2001). Environmental, nutritional, and metabolic influences are significant factors affecting modeling and remodeling processes as well, resulting in a complex interplay of factors influencing bone accrual that reflects an individual's life history, and affects bone health through development and as we age.

The obvious clinical implications of the complex interplay between mechanics, genetics, and environment have spurred extensive research using clinical imaging techniques (Chevalley et al., 2017, Digby et al., 2016, Levine, 2012, Gabel et al., 2018, Mata-Mbemba et al., 2019). Unfortunately, these *in vivo* techniques do not allow for the visualization of underlying cellular and histomorphology that directly reflects modeling and remodeling processes. The examination of histomorphology of juvenile skeletal remains from archaeological contexts provides a unique opportunity to shed light on functional adaptation from a life history perspective, to better understand the impact of nutrition, environment, and biocultural factors on bone architecture. From a bioarchaeological perspective this is important as researchers seek to shed light on the health and lifeways of past populations (Agarwal, 2016). From a modern bone health perspective such studies can provide clues not only about the factors affecting fracture risk but also about issues that could face populations today due to effects of modern climate change and environmental issues.

In this chapter, we will begin with a brief review of the processes of bone modeling and remodeling as well as the genetic and functional determinants of bone architecture that are more extensively reviewed in previous chapters of this volume. We then explore the literature focused on the potential effects that biocultural and environmental stresses can have on these bone growth processes, how such effects can be linked to climate and environmental changes in the past, and why it may be important for our understanding of bone health and adaptation in current and future populations.

 DOI: 10.4324/9781003385608-3

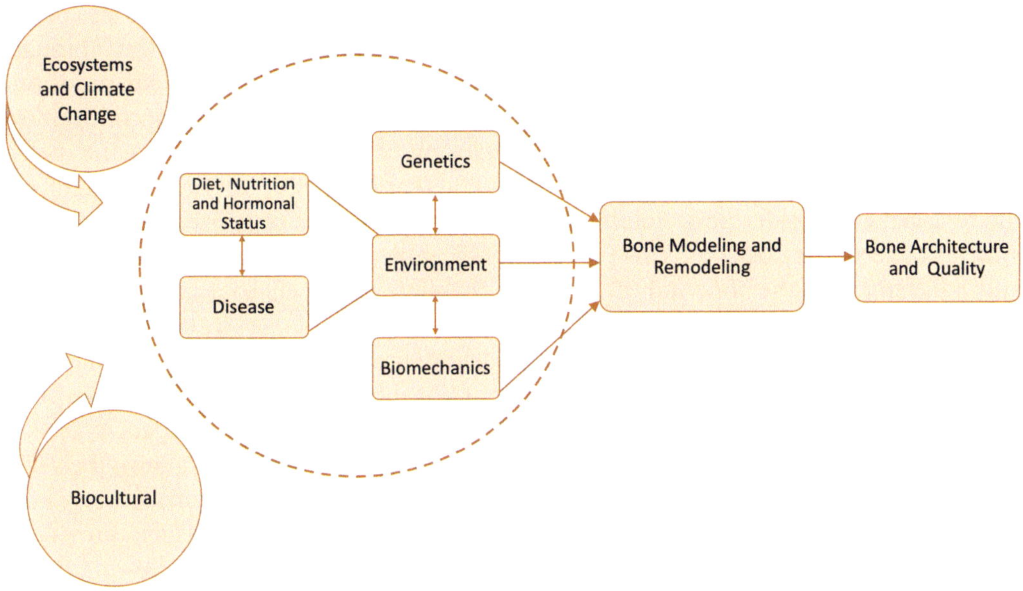

Figure 3.1 The complex interplay between factors that influence adult skeletal architecture and quality. H.M. Goldman.

Figure 3.1 illustrates the complex interplay between the factors that influence adult skeletal architecture and quality. Our *background section* will explore each of these in some detail, followed by a deeper dive in the *Biocultural Perspectives* section into examples from the archaeological record that can shed light on how environmental change, climate, and social and biocultural events may shape human skeletal health.

3.2 Background

3.2.1 Bone Growth Processes and the Attainment of Adult Bone Architecture

In previous chapters, the basic processes of bone growth and modeling, as well as bone remodeling have been presented in detail in Chapter 2 and Chapter 1, respectively. As discussed, the processes of endochondral and intramembranous ossification set the stage for how different regions of the skeleton will be transformed from mesenchymal tissues to mineralized bone tissue, while the process of modeling, classically called 'growth remodeling' (Enlow, 1963, 1966) determines the size, shape, and curvature of long bone diaphyses during ontogeny (Frost, 1973). This process involves coordinated patterns of deposition and resorption on complementary bone surfaces during growth (Enlow, 1962; Enlow, 1963; Enlow et al., 1982; Enlow and Hans, 1996), and to a lesser extent throughout life (Seeman, 2003; Smith and Walker, 1964). Modeling can result in periosteal expansion and endosteal resorption and often results in 'drift' of the whole bone cortex. This is significant during the lifespan of an individual because it causes sequential changes in the position of bone tissue laid down during earlier growth stages (Enlow, 1963; Biewener and Bertram, 1994) reflecting the local adjustments made to bone shape as size increases (McFarlin et al., 2016). Studying these histological 'signatures' of modeling makes it possible to reconstruct aspects of the growth history of a given bone that may relate to aspects of life history (McFarlin et al., 2016).

Bone remodeling, classically called 'secondary remodeling' (Enlow, 1976), is also occurring through development, particularly from birth through adolescence and then throughout adult life. It is a process of bone turnover in which bone resorption and bone deposition occur at the same site, involving two stages. First, osteoclasts lining a cutting cone remove bone tissue, and second, osteoblasts lining a closing cone deposit the new bone matrix along the walls of the resorption bay (Frost, 1986; Parfitt, 1988). This process results in the replacement of older bone volumes with new bone material organized into Haversian systems or secondary osteons. This process significantly affects the distribution of mineral and collagen within the bone and the porosity. Therefore, it too will have lasting effects into adulthood.

The processes of bone modeling and remodeling through ontogeny are influenced by a mix of genetic, cultural, environmental, and functional factors (Duren et al., 2013). These factors are intertwined, and not entirely understood, in part because of the paucity of skeletal samples available to study (Goldman et al., 2009). Despite such limitations, the variation in processes associated with bone modeling and remodeling, and consequently bone organization through the lifespan can be used by biological anthropologists to provide an important window into the past (Gosman et al., 2011; Ruff, 2013). The following review summarizes the influence of each of these factors with an emphasis on these processes during growth and development.

3.2.2 Functional Determinants of Bone Architecture

Bone architecture is strongly influenced by the functional demands of the skeleton, which include its roles in posture, locomotion, protection, and homeostasis. The development of the normal architecture of bone, including the size and shape of whole bones, their processes, and articular surfaces, is dependent on functional loading (Currey, 1984; Lanyon, 1984; Biewener et al., 1996; Lanyon, 1996). Specifically, mechanically induced strain, in its varying magnitude, rate and mode, is thought to be the primary stimulus for the modeling or remodeling response (Bouvier & Hylander, 1981; Frost, 2003, Rubin & Lanyon, 1984), with osteocytes playing a role in sensing this strain (Lanyon, 1993) via fluid flow through canaliculi (Han et al., 2004; Nicolella et al., 2006). The relationship between bone morphology and orientation of load, often cited as Wolff's law (Wolff, 1892), is far more complex than simply that load orientation influences bone shape (Pearson, 2004). The response of bone cells to load and their resultant activity does not lead to precise predictability in terms of magnitude of osteogenic response. Rather, bone's response to mechanical forces is complicated by genetic factors (see discussion in this chapter), and varies depending on the developmental stage and health of the individual (Pearson et al., 2004, Ruff et al., 2006) with bone response to mechanical stimuli shown to be much more pronounced in juveniles relative to adults (Forwood and Burr, 1993), and significantly less responsive with aging (Riggs and Melton, 1986) and a general pattern of loss of bone mass (Khosla and Riggs, 2005), bone strength (Vilayphiou et al., 2016), and bone quality (Boskey et al., 2017).

The relationship between bone loading and adult bone shape has been used extensively in the field of biological anthropology as the basis of making interpretations about the activity types and activity levels of past populations. Experimental data demonstrating that the quantity and preferred distribution of bone around a long bone cortex can be related to mechanical loading and response to stress (e.g., Lanyon et al., 1982, Robling et al., 2006), and studies of athletes demonstrating distinct side differences in bone robusticity between loaded vs. non-loaded limbs (Jones et al., 1977; Shaw and Stock, 2009a and 2009b), led to numerous archaeological studies using long bone geometry data to shed light on patterns of activity and behavior in past populations. These studies focused on differentiating more sedentary agricultural populations from those with more mobile, hunter gatherer subsistence patterns, with the agricultural populations demonstrating more circular long bone geometries, versus more mobile populations more anteroposterior expansion (Ruff and Hayes 1983a and 1983b, Ruff, Larsen, and Hayes, 1984). As more studies were carried out, more variability in morphological patterns was identified that researchers attempted to link to specific behaviors involved in maize agriculture (Bridges, 1989), use of terrestrial- vs. marine-based subsistence foods (Stock and Pfeiffer, 2001), the importance of taking into account sexually dimorphic roles in hunter-gatherer subsistence (Wescott, 2006) that may differ depending on environmental resources (Stock and Pfeiffer, 2004). What these studies began to highlight is that while mechanical loading influenced bone morphology in different populations – with broad recognizable trends in decreasing cross-sectional areas and increased circularity with increasing sedentism, once details of environmental factors were considered, the story became much more complex (Weiss, 2005, Lieberman et al., 2004). Many other factors also need to be considered – from genetic to environmental and nutritional, to cultural – in the interpretation of changes in geometry between populations and through time.

The study of the relationship between bone geometry and architecture and functional adaptation in children is less well studied owing to limitations in obtaining juvenile bone samples. Most longitudinal studies of childhood bone growth are based on X-ray data (Garn et al., 1967, 1970, Duren et al., 2011, 2013) and helped establish what we know about the timing and patterning of growth. A few studies of iliac crest biopsies have focused on tissue-level bone growth dynamics (Glorieux et al., 2000, Parfitt et al., 2000, Rauch et al., 2007) but these do not reflect bones involved in locomotor functions. CT (computed tomography)-based studies are also available and increasing in prevalence (Macdonald et al., 2006; Petit et al., 2005, Vilayphiou et al., 2016, Stagi et al., 2013, Zemel, 2011), but resolution is limited to what is clinically feasible. Thus, much of what we know about the histomorphology of juvenile bone is from samples derived from archaeological contexts (Beresheim et al., 2020, Ruff et al., 1994, Sumner and Andriacchi, 1996) or autopsy (Goldman et al., 2009) as these allow for thin section–based and/or high-resolution micro-CT-based study of geometric properties, trabecular organization, and histomorphology, though most samples from archaeological collections suffer from potential diagenetic changes and lack of known age at death.

From a mechanical standpoint change in bone architecture in children have been correlated to the timing of locomotor development, beginning at birth with the exposure to gravity, progressing through the first year to sitting without support, standing with assistance, crawling on hands and knees, standing without assistance, and walking alone (Swan et al., 2020). Even after the onset of walking, gait patterns continue to change through the toddler phase. Several previous studies have demonstrated changes in architecture and organization of cancellous bone as it relates to the onset of walking and maturation of gait pattern (Gosman & Ketcham, 2009; Milovanovic et al., 2017; Raichlen et al., 2015; Ryan & Krovitz, 2006; Saers, Ryan, & Stock, 2020). Other studies have focused on geometric changes in the distribution of cortical bone in early vs. mature walkers (Cowgill et al., 2010, Goldman et al., 2009, Gosman et al., 2013, Swan et al., 2020). Cowgill et al., 2010 found that ground reaction forces in children with immature gaits were much higher mediolaterally relative to relative to adults with a mature gait, explaining the more mediolaterally widened femoral shaft seen in cross-sectional geometric analysis of archaeological skeletal samples and reflective of a minimal bicondylar angle and broad body of young children. Both Cowgill et al. (2010) and Swan et al. (2020) noted an increase in bone placed further from the centroid of the diaphyseal cross-section, along with increased endocortical resorption in the first months after birth prior to substantial locomotor-based load bearing. Swan et al. (2020) explained these findings as likely related to decreasing exposure to estrogen postnatally during rapid growth. This was shown to be followed by a mediolateral widening of the midshaft as the child progressed to walking, finally becoming more circular as gait matures.

3.2.3 Genetic Determinants of Bone Architecture

The general form of a bone, including its attachments and anatomical relationships, is genetically determined and will develop even without functional influences. The genetic basis of bone architecture is complex, as human development is governed by the interaction between thousands of different genes, the proteins they encode, and the environment they are subjected to. And, as mentioned earlier, heritability is only one of many influences on bone architecture.

Longitudinal studies of peak adult bone mass (Gueguen et al., 1995, Jouanny et al., 1995, Picard et al., 2001, Seeman et al., 1989, Slemenda et al., 1991) and fracture incidence (Fox et al., 1998) in related cohorts suggest that it is a heritable trait explaining more than 50–80% of variation. Early longitudinal studies by Garn and colleagues (Garn et al., 1967) focused on familial correlations in geometric properties (bone width and cortical thickness) using X-ray data, which have since been refined via continued studies of the same sample (Duren et al., 2013) showing bone length had high heritability from a young age and throughout life, whereas bone diameter is most influenced by genetics in early childhood and diminishes later in life. Understanding the timing and influence of genetics on different bone traits is important for identifying the effects of environmental variables on the skeleton depending on underlying genetics.

Another measure of heritability stems from serum markers of osteoblast and osteoclast activities. Identification of genetic factors in these measures of bone turnover from infancy onwards (Ferrari et al., 1998) suggests that genetic susceptibility to diseases like osteoporosis can be determined early in life (Duren et al., 2011). Studies suggest a high (up to 90%) heritability of osteoporosis (Tse et al., 2009) with studies identifying several candidate genes influencing susceptibility (Richards et al., 2008, Cheung et al., 2009, Ermakov et al., 2012).

Genetic QTL (Quantitative Trait Loci) mapping has been used to identify genes that affect bone mass and size. Over 350 QTLs have been mapped for bone traits, some of which have been shown to be regulators of bone mass and bone morphology (see Jepsen 2009 for summary). Studying functional interaction among multiple traits in recombinant inbred mouse strains, Jepsen et al. (2007) found co-segregation of traits depending on which set of genes from the cross the offspring inherited. Specifically, genetic variation in transverse bone growth relative to longitudinal growth was functionally related to cortical thickness and matrix mineralization. Therefore, bones that were slenderer had increased cortical thickness and matrix mineralization to compensate, resulting in more brittle cortices, whereas bones that were more robust had thinner cortices and lower mineralization to avoid excessive addition of bone weight. Studies of morphology in human tibiae supported the inverse association of robustness and tissue mineralization (Tommasini et al., 2008) and porosity distribution (Goldman et al., 2014).

3.2.4 Influence of Diet, Nutrition, and Hormonal Status on Bone Processes

As seen in the previous sections, it is difficult to tease apart aspects of the bone structure that result from genetically controlled growth and development, versus those that represent the effects of the adaptive response to functional load bearing. The complexity increases significantly once one begins to consider the influence that hormonal, dietary, and health-related variability can add.

3.2.4.1 Micronutrients

Calcium, vitamin D, and other micronutrients influence bone modeling and remodeling and play a key role in skeletal development and maintenance (Proia et al., 2021). Vitamin D is necessary for homeostatic maintenance of serum calcium and phosphorus, mediated through parathyroid hormones and calcitonin (Bell et al., 2010). It is synthesized from a prohormone, which is successively hydroxylated in the liver and kidneys to produce a physiologically active form that we call vitamin D. It promotes normal bone mineralization by moderating the deposition of mineral apatite on the Type I collagen matrix of the skeleton (Pitt, 2002). Vitamin D also plays a role in stimulating osteolysis as part of the mobilization of stored calcium and phosphorus in bone tissue.

In children, vitamin D deficiency can lead to rickets, characterized by decreased mineralization of bone tissue and the formation of bioapatite crystals, making them larger and irregularly shaped. In rickets, bones may end up curving leading to bow-leggedness and knock-knees (Chanchiani et al., 2020). Rickets can often be avoided by adequate sunlight exposure. However, this is subject to variation in skin color due to microevolutionary processes, selective pressures posed by latitudinal variation in UV B intensity (Jablonski and Chaplin, 2013). Evolutionarily speaking, the condition was rare for our ancestors until they migrated to regions 35° North latitude or higher and a lack of UV radiation drove the evolution of lighter skin tones (Elder et al., 2014). Unfortunately, vitamin D deficiency is currently reaching pandemic proportions worldwide; one billion people are currently suffering from insufficiency (Elder et al., 2014). Climate and environmental changes can lead to morbidity from even low-frequency climate oscillations in sea level, sunshine duration, and cloud cover (Majeed and Moore, 2017).

Vitamin D deficiency can also manifest as osteomalacia in children and adults, affecting mature compact and trabecular bone through an imbalance between osteoid production and subsequent mineralization (Pitt, 2002).

Adequate ingestion of dietary calcium, with vitamin D and even without, has been shown to increase bone mineral density in children (Johnston et al., 1992, Ward et al., 2007), but calcium supplementation beyond daily recommended requirements and its effects on fracture risk, particularly in adults, have been questioned (Iuliano et al., 2019). But evidence does suggest that reducing dairy intake below recommended doses can have negative impacts on bone – not just due to low calcium levels, but also due to a reduction in other micronutrients like phosphorus, potassium, magnesium, or vitamins B2, B12, A, and D (Dror et al., 2014).

3.2.4.2 Macronutrients

Macronutrients (e.g., protein, fat and carbohydrates) are also essential factors in promoting bone health particularly in sub-adults (Proia et al., 2021) with adequate protein essential for normal bone growth (Chevalley et al., 2014), excessive fat and sugar in the diet linked to poor calcium absorption, and decreased bone strength (Tian et al., 2017), while a diet high in polyunsaturated fatty acids (PUFAs, e.g., omega-3s) can reduce bone resorption and increase bone formation (Watkins et al., 2003). Increased understanding of bone's role in the regulation of energy metabolism via peptides like osteocalcin has led to studies examining the importance of gastrointestinal tract derived peptides and insulin in bone health (Proia et al., 2021). Nutrition influences the release of gut peptides, which in turn can modulate the secretion of parathyroid hormone, thus affecting bone remodeling processes.

Given the variability in the metabolic needs of the body through life history, nutritional requirements vary from one time point to another, making nutritional deficiencies more or less detrimental to skeletal health depending on when they occur. Notable time points of rapid skeletal growth where nutrition would play a particularly significant role in attainment of bone strength include in utero, where inadequate material nutrition has been shown to lead to poor skeletal shape and mineralization (Lanham et al., 2011), and the onset of puberty where nutritional requirements monumentally increase to accommodate rapid growth (Clark et al., 2006; Elhakeem et al., 2019). Growth at this point of time is also crucially important for obtaining peak bone mass that ultimately could be protective against bone loss that may occur with age (Oliver et al., 2007, Gille, 2010, Peters and Martini, 2010).

3.2.4.3 Hormones

Normal bone growth is dependent on the actions of numerous hormones with both systemic and local effects. Growth hormone (GH) produced by the pituitary gland stimulates insulin-like growth factor which is crucial for the regulated cell division, differentiation, and activity of cells involved in bone growth, particularly those involved in longitudinal growth and the growth plates. Imbalances of GH during growth can significantly affect adult stature and peak bone mass (Bouillon and Prodonova, 2000). Parathyroid hormone (PTH) regulates calcium levels in the blood through the regulation of osteoblast stimulation of osteoclast differentiation and activity. Thus, PTH can stimulate both bone resorption and formation and plays a key role in the regulation of bone remodeling (Silva et al., 2011). Sex hormones (estrogen and testosterone) play a crucial role during puberty in regulating the pubertal growth spurt and eventual closure of growth plates and are protective against bone loss. Thyroid hormone (Bassett and Williams, 2016) plays a vital role in regulating skeletal development and bone maintenance. Cortisol produced by the adrenal gland is also important in regulating bone maintenance with either insufficient or excess production having potential skeletal effects (Zhou et al., 2013).

With such a complex interplay of hormones involved in the regulation of bone growth and maintenance it is not surprising that hormonal disturbances due to disease - some associated with underlying genetic disorders - are likely to have significant effects on the skeletal health of both growing and adult individuals.

3.2.4.4 Influence of Infectious Disease on Bone Growth and Maintenance

As we saw in the previous section, the processes of bone modeling and remodeling, and their action on bone growth and maintenance, are tightly regulated by numerous hormones. Infections can affect this balance leading to dysregulation of osteoblast and osteoclast activities by a variety of processes. Response to infection via inflammatory cytokines (e.g., interleukin-6, IL-6) and tumor necrosis factor is thought to play a role in the bone loss that is seen in patients with viral infections such as hepatitis B and C viruses

(Chen et al., 2019). Some pathogens additionally impact bone regulation by colonizing bone, using receptors to invade into osteoblasts, and leading to the disruption of their function and cell death. At the same time, pro-inflammatory factors can lead to increased bone resorption leading to the dysregulation of bone homeostasis (Oliveira et al., 2020). A common example of this is seen with *Staphylococcus aureus* in osteomyelitis infections. *Mycobacterium tuberculosis* acts in a similar fashion.

3.3 Biocultural Perspectives on Histomorphology

Having reviewed the myriad of factors – biomechanical, genetic, environmental (nutritional/metabolic), and hormonal – we can see there is substantial evidence that (1) genetic influence is strongest in the early decades of life with environmental factors increasing variation as the decades go on, (2) the effect of mechanical loading on bone strength is highest during periods of growth and development because that is when bone is being laid down at the highest rate, and (3) there are specific time points in development where nutritional needs are highest and nutritional disturbances can potentially have the greatest effect. Since all these factors are hard to untangle, interpretations of population level variation would improve if we had a better understanding of the variation within populations most broadly and then look at situations where individuals fall outside of the range of variation to look at effects of biocultural factors and environment.

3.3.1 Biocultural Factors

Anthropologists are interested in bone histomorphology, geometry, and pathology as they apply questions about human biology and evolution. Evolutionary processes, climate, and environmental changes have shaped human diversity at every level over hundreds of thousands of years (Larsen, 2023; Robbins Schug et al., 2023; Stock et al., 2023). To understand these processes, we rely on a foundational understanding of genetics and gene regulation, cell biology, physiology, functional morphology, ontogeny, paleoecology, and evolution. However, biological anthropologists recognize that human biology must always be understood within our historical, political, economic, ideological, and sociocultural contexts (Leatherman and Goodman, 2020; Zeder, 2018; Zuckerman and Martin, 2016).

This section describes recent research on bone histomorphology as a tool for understanding past environmental change, biocultural adaptations, and health challenges. In 2001, Michael Schultz famously claimed that rickets, hematogenous osteomyelitis, and treponemal and mycobacterial infections should be investigated histologically as they affect the microstructure of the compact bone (Schultz, 2001). In fact, the author went as far as to claim that these diseases will only be differentiated in bioarchaeology using histological techniques. This sweeping publication described the available techniques for microscopic investigation in paleopathology, the fundamentals of bone microarchitecture, taphonomic processes that could present as pseudo-pathologies, and how to use evidence for bone turnover to diagnose archaeological skeletons. It is not possible to say that this promise of diagnostic certainty has come to fruition in paleopathology but with the increasing availability of new technologies in paleosciences, we felt it was time to revisit what is known about pathohistology of pediatric bone and future directions for research, particularly in the context of understanding the health impacts of climate and environmental change in the past.

It is beyond the scope of this chapter to cover all the potential applications for pediatric bone histology as a tool for investigating climate and environmental change as there are far too many ways which evidence from the past suggests that our health will be affected by global warming (Buikstra et al., 2022; Robbins Schug, 2020; Robbins Schug et al., 2023). Provided here will be one example from each category of health impact on pediatric bone histology, using one example for each:

1. Direct impact: Micronutrients: Vitamin D deficiency
2. Ecosystem mediated impact – Macronutrients: malnutrition and food insecurity
3. Social institution mediated impact – Infectious disease: Tuberculosis and leprosy

In the following sections, we will give a brief overview of the literature and our perspective on which topics require more research.

3.4 Bone Microarchitecture in Ontogeny: Evaluating Climate and Environmental Change

3.4.1 Micronutrients: Vitamin D deficiency in Changing Environments

Vitamin D deficiency is a topic of concern for bioarchaeologists who study health in the more recent past when humans began living in industrial urban environments, polluting the air, wearing clothing that covers the majority of the skin, or in places where people prefer indoor lifestyles (Brickley et al., 2014; Colombo et al., 2021; Lamer et al., 2023; Lockau et al., 2019; Mays et al., 2018; Peacock et al., 2019; Veselka et al., 2021). This section describes bioarchaeological research on rickets in immature bone histology and potential areas of inquiry for anthropologists working on this topic.

Archaeologically, researchers have identified specific dry-bone features of vitamin D deficiency that relate to poor mineralization and mechanical deformation (Brickley et al., 2007, 2010, 2018, 2020; Mays and Brickley, 2018; Mays et al., 2006; Ortner and Ericksen, 1997; Ortner and Mays, 1998). These features are all based on anthroposcopic observations of skeletal gross anatomy; the changes include increased porosity, thickening of cortices, and deformities resulting from poor mineralization. Specifically, rickets in archaeological bone is diagnosed by cranial vault and orbital roof porosity, costochondral rib porosity, metaphyseal and growth plate porosity on the long bones; cranial vault and compact bone thickening in the long bones; and deformities in mandibular, rib, iliac, and metaphyseal morphology.

While there is research on interglobular dentin in human teeth as an indication of rickets in the past (Brickley et al., 2017; Colombo et al., 2019; D'Ortenzio et al., 2016; Hillson, 1996), much less work has been done on histological indicators of the disease in pediatric archaeological bone. Clinical research on pediatric bone in cases of rickets can be conducted at autopsy, but it is rare to find published cases where histomorphological features are assessed in the compact bone. Histological features of vitamin D deficiency rickets occur primarily as disorganization at the zone of maturation in the epiphyseal growth plates (Pitt, 2002). Typically, radiographs are taken to demonstrate severe osteopenia throughout the skeletal system and histological examination is limited to marrow hyperplasia in the cranium, hypertrophy in the ribs, and changes to the epiphyseal growth plates of long bones (e.g., Kepron and Pollanen, 2015).

Rickets is associated with osteopenia, fields of Howship's lacunae, and enlarged osteocyte lacunae in children (Schamall et al., 2003; Schamall and Teschler-Nicola, 2010). Defective cement lines and hypomineralization are visible in archaeological bone as voids left by a postmortem loss of osteoid (Brickley et al., 2007). The latter may be a pathognomonic sign, but additional research is required. Osteopenia, enlargement of the Haversian canals, and increased numbers of Howship's lacunae are also visible histologically in hyperparathyroidism, which may be distinctive from vitamin D deficiency due to the presence of tunneling resorption in subperiosteal and cancellous tissue (Cotran et al., 1999; McCarthy et al., 2010). Anemia must also be considered in the differential diagnosis of rickets, osteomalacia, and scurvy, and typically this is accomplished through the presence of bone marrow hypertrophy (Resnick and Kransdorf, 2004; Wapler et al., 2004).

Osteomalacia is well defined in archaeological contexts (Brickley et al., 2005) and includes accumulation of unmineralized bone leading to wide bands (> 12.5 μm in adults) of osteoid inside the Haversian canals, subperiosteal areas, and between concentric lamellae, such that the osteoid may exceed 10% of bone volume in the cross section (Beighton et al., 1993). Histologically and radiographically, osteomalacia is observed as osteopenia, a decline in bone mass and microstructural integrity (Welsh et al., 2020). However, these signs are pathognomonic for any condition that promotes high bone turnover, or which affects vitamin D metabolism, including hyperparathyroidism, metastatic disease, Paget's disease, and renal osteodystrophy (Mays et al., 2007; Resnick, 2002).

Additional research is necessary in using histomorphology to differentiate rickets and osteomalacia from related conditions and in teasing apart the manifestations of common comorbidities. Osteomalacia is often seen in infants and children with low birth weight, combined calcium, and phosphorus deficiency, and X-linked hypophosphatemic rickets (De Boer et al., 2013; Mays and Brickley, 2022; Pitt, 2002). Skeletal growth and development are occurring very rapidly in the last trimester of pregnancy, and 80% of bone mineralization occurs in this time (Cunningham et al., 2016). In cases of preterm birth, breast milk is inadequate to provide vitamin D, calcium, and phosphorus sufficient to maintain homeostasis and preterm babies may also suffer malabsorption disorders. Rickets and osteomalacia in low-birth-weight infants cause increased bone turnover, mineralization lag times that cause thick osteoid seams in the trabecular and cortical bone. Focal deposits of osteoid may occur in the cross-section, or thick regions of unmineralized bone may also surround osteocytes. Fractures and skeletal deformity often result.

3.4.2 Macronutrients: Evaluating Malnutrition in the Context of Environmental Change

As adequate dietary intake of macronutrients including protein, carbohydrates, and fats is needed for adequate skeletal growth (see review earlier), populations experiencing malnutrition, hunger, and food insecurity would be expected to show skeletal manifestations of these stresses, particularly in juvenile bone samples. There are not many published studies documenting famine using histological observations in archaeological pediatric bone, but their study is important as they can elucidate the impacts of climate and environmental change, which are heavily modulated through social institutions in cases where communities are founded in social inequality, structural violence, and a lack of basic resources. In small-scale societies that are not hierarchical, the risk of famine in the context of environmental change is greatest when a substantial proportion of the subsistence base is agricultural (Slavin, 2016). In fact, extreme weather is the principal driver of world hunger as rising temperature, sea level, heat waves, drought, flooding, fires, and storms make farming difficult to sustain (US Embassy, 2022). Archaeological and historical perspectives on famine can elucidate connections between ecological, biological, social, cultural, and historical factors involved in past famines, a topic which bioarchaeologists have long been interested in exploring (Horocholyn and Brickley, 2017).

3.5 Biocultural Perspectives: Examples from the Archaeological Record of Effects of Climate and Culture Change in Human Populations Occupying Semiarid Regions

3.5.1 The Late Holocene in West and South Asia

This example represents an excellent natural laboratory for observing the effects of climate and culture change in human populations occupying semiarid regions. At the beginning of the second millennium BCE, circa the disintegration of the Indus Valley Civilization, hundreds of agrarian villages sprang up at the margins of Indus territory (what is today India and Pakistan) and down into west-central peninsular India. Despite an increasingly inhospitable and unpredictable arid monsoon climate (Bryson, 1981; Dhavalikar, 1988; Enzel, 1999; Jain and Tandon, 2003; Phadtare, 2000; Prasad and Enzel, 2006; see also Robbins Schug, 2011), human populations thrived in the region known today as Maharashtra from 2200 to 1000 BCE, a time known as the Deccan Chalcolithic (DC).

3.5.1.1 Early Jorwe Phase of the Deccan Chalcolithic (1400–1000 BCE)

During the phase of the DC, settlements expanded rapidly, and people established far-reaching exchange relationships with communities from the west coast of India to the post-urban Indus communities to the northwest (Dhavalikar, 1984; Dhavalikar et al., 1988; Shirvalkar and Prasad, 2016). DC people constructed villages of stone, mudbrick, wattle, and daub. Copper and bronze technology was used only sparingly, as the raw materials were not available in this new landscape. DC people did use copper fishhooks, spearheads, chisels, beads, bangles, and pins (Shinde, 1989). Beads were also made of imported shell, steatite, carnelian, agate, and other stones. Finely made pottery was fired in kilns and painted with slip designs.

They pursued a mixed economic strategy, blending agriculture and stock-raising traditions from the Northwest with cultivation of some South Indian crops, hunting local deer and other animals in nearby forested tracts, and exploiting lacustrine resources (Korisettar, 2020; Nayak et al., 2022). They used double-cropping to harvest drought-resistant barley (*Hordeum vulgare*) and less tolerant wheat (*Triticum spp.*), rice (*Oryza sativa*), three species of African millet, peas, beans, lentils, and seeds including *Lens esculanta, Pisum arvensa, Macrotyloma uniflorum, Dolichos lablab, Vigna mungo, Vigna radiata, Linum usitatissimum, and Carthamus tictorius.* They kept buffalo (*Bos bubalus*), cattle (*Bos indicus*), sheep and goat (*Capra/Ovis*), and dogs (*Canis familiaris*). Wild species found in food remains included many fish and lacustrine molluscs, Indian jujube (*zizyphus spp.*), Blackbuck (*Antelope cervicapra*), chital (*Axis axis*), four-horned antelope (*Tetracerus quadricornus*), and sambar (*Cervus unicolor*).

After 1000 BCE, most of the villages in this region were abandoned. Larger climatic trends in South Asia during this time confirm peak arid conditions affected South Asia during 2000–1000 BCE (Burdanowitz et al., 2021; Dash et al., 2022; Jain and Tandon, 2003; Joshi et al., 2021; Kulkarni et al., 2020; Mishra et al., 2020; Phadtare, 2000; Sarkar et al., 2016; Sharma et al., 2004; Sinha et al., 2006; Srivastava et al., 2021). Monsoon runoff from the Indus River was at a marked low by 1100 BCE (Korisettar, 2020). Perhaps more importantly, these shifts in monsoon rainfall abundance were met with rapid population growth in the Early Jorwe phase, combined with unsustainable agricultural practices that led to soil salinization, which in turn led to local environmental degradation and a shift away from drought resistant barley agriculture and cattle keeping throughout the region (Kajale, 1988; Nayak et al., 2022; Robbins Schug, 2011). Much of this region was depopulated at this time, but a few villages remained occupied with a smaller population.

3.5.1.2 Late Jorwe (1000–700 BCE)

The final phase of the DC was known as the Late Jorwe (1000–700 BCE) and of all the sites with human skeletal remains – Chandoli, Daimabad, Nevasa, Songaon, and Walki – Late Jorwe burials only occur at Inamgaon (Naik and Mishra, 1997). The people of Inamgaon adjusted their economic activities after 1000 BCE and adaptive diversity allowed them to occupy their village for 300 years longer despite the abandonment of most of this region. The archaeological record of the Late Jorwe is characterized by an increasingly 'degenerate' ceramic style, greater numbers of hastily constructed 'flimsy' round houses, declines in agricultural production, shift in species preference from drought-resistant barley to saline-tolerant lentils, and a shift from cattle to sheep/goat bones in the faunal remains (Dhavalikar, 1988; Dhavalikar et al., 1988; Dutta, 2006; Kajale, 1988; Shinde, 1989; Pawankar and Thomas, 1997). Changes in artifact assemblages indicate a shift in emphasis away from implements associated with farming and a new prominence of lacustrine food resources, tools, and ornamentation as trade with coastal people became increasingly important during this time (Robbins Schug, 2011).

Paleoclimate and archaeological evidence support the hypothesis that the Late Jorwe represents a time of significant culture change and subsistence transition – away from reliance on agriculture toward emphasis on mobile pastoralism, hunting, foraging, lacustrine resources, and starvation foods (Dhavalikar, 1988; Robbins Schug, 2011). Mortuary analysis of 266 burials at Inamgaon was conducted to evaluate chronological variation and sociocultural change across the Early to Late Jorwe transition (Raczek, 2003). In Early Jorwe, infants were buried at many DC villages, including Inamgaon, typically in single or twin urns, capped, and placed under the house floors, with the feet toward the South. Late Jorwe burials demonstrate much continuity with the earlier phase, but there are a few key differences. In this phase, new kinds of funerary ornaments were included, and some of the infants were buried with their heads to the South instead of their feet. It is possible that the community was practicing a greater degree of mobility or bands of mobile hunters and gatherers joined the community (Panja, 1999). After 700 BCE, Inamgaon was also abandoned.

Bioarchaeologists have struggled to understand the impact of these changes in the human skeletal remains. The *Climate and Culture Change* model, developed by archaeologists, predicted dire health consequences for the Late Jorwe people of Inamgaon as plow agriculture was abandoned (Dhavalikar, 1984; Dhavalikar, 1988; Dhavalikar et al., 1988; Dhavalikar, 1994). Bioarchaeological investigators initially proposed an antithetical *Subsistence Transition Model* – that declines in agricultural production during the Late Jorwe phase would lead to an improvement in health status due to declining population density, improved sanitation, and increasingly diverse diet (Kennedy, 1984; Lukacs and Walimbe, 2007; Walimbe, 2016).

Comparisons of the dental anthropology profiles of Early versus Late Jorwe Inamgaon have been inconclusive in their support for either model (Lukacs, 1985a,b, 2001; Lukacs and Walimbe, 1998, 2000, 2005, 2007; Lukacs et al., 2001; Robbins Schug, 2011; Walimbe and Kulkarni, 1993). A significant difference between the Early and Late Jorwe was demonstrated in the frequency of dental caries and an enamel defect, known as localized hypoplasia of the primary canine (LHPC). However, these pathophysiological markers do not necessarily indicate poor health, and there was no systematic trend in biocultural stress markers that indicate growth disruption (tooth size and linear enamel hypoplasia, or LEH).

Osseous remains serve as an excellent source of evidence to address hypotheses about human health during the Deccan Chalcolithic, but this evidence has been neglected in the bioarchaeology of Inamgaon. For 40 years, cross-sectional geometric properties in adult skeletons have been used to infer locomotor behavior (Frayer, 1980, 1981, 1984; Larsen, 1981; Larsen et al., 1996; Lukacs and Pal, 2003; Ruff, 1981, 1982), subsistence practices (Boyd and Boyd, 1989; Collier, 1993; Goodman, 1984; Jacobs, 1993; Ruff, 1990; Sokal et al., 1991), and functional adaptations (Ruff et al., 2006; Wescott, 2006; Wescott and Cunningham, 2006).

Femoral torsional rigidity (*J*) is influenced by both mechanical and biological forces beginning early in life (Biewener, 1994; Carter and Beaupre, 2001; Carter et al., 1996; Moro et al., 1996; Pearson and Lieberman, 2004; Ruff, 2003a,b, 2005, 2007; Stock and Pfeiffer, 2004; Van Der Meulen et al., 1993). Furthermore, there is a strong scaling relationship among *J* and body mass in the subadult femur and body mass accounts for 98% of the variation in torsional rigidity for children less than eight years of age (Robbins et al., 2010; Robbins Schug et al., 2012) and accounts for 80% of the variation in cross-sectional geometry in adults (Davies, 2013). Based on this strong correlation between body mass and cross-sectional properties of the midshaft femur, Robbins Schug et al. (2010) developed a set of age-structured generalized, least squares regression equations to predict body mass for children 2 months to 12.5 years of age from the Denver Longitudinal Growth Study data (provided by Chris Ruff, 2008). These formulas were tested on the Franklin, Ohio, cadaver sample, and results indicate that they predict body mass with a high degree of accuracy and precision (Robbins et al., 2010). Later work provided a single panel regression formula for immature body mass that does not require age estimates (Robbins Schug et al., 2022), providing a robust new statistical approach to age structured data (Yapuncich et al., 2018).

When body mass is estimated from the midshaft femur and then scaled to estimates for stature squared, population-level differences in body size and shape (BMI) can be detected (Cowgill, 2018). Globally, prehistoric populations differ based on expectations for climate, activity level, and biocultural stress levels (Cowgill, 2010, 2018; Cowgill and Rob'yn, 2018). Robbins Schug applied this method to Deccan Chalcolithic samples, long bone length for dental age, and body mass for stature, and the distribution and volume of pores in infants from Inamgaon demonstrated significant differences between Early and Late Jorwe phases (Robbins Schug, 2011; Robbins Schug and Goldman, 2014; Robbins Schug and Blevins, 2016). Bone length was used to estimate stature and body mass was calculated using femoral midshaft cross-section geometry (torsional rigidity, or *J*) (Robbins et al., 2010; Robbins Schug et al., 2012). Body mass index (BMI=body mass divided by stature squared) was compared through time. Growth profiles for infants and children from the Early Jorwe were not significantly different from contemporary reference standards for stature and body mass for developmental age, and for BMI (Walimbe and Gambhir, 1994). In contrast, 22% of the children (0–5 years of age) who died during the Late Jorwe phase at Inamgaon had evidence for wasting – BMI was greater than 2 standard deviations below the median for developmental stage (using ages estimated from dental development).

The combined work on the DC population of Inamgaon has demonstrated (1) dental caries rates declined in the Late Jorwe, when cereal agriculture was largely abandoned; (2) deciduous tooth size increased over time, which Lukacs and Walimbe (2005) interpreted as biologically significant and related to a shift to a coarser diet; (3) the frequency of an enamel defect in the primary teeth (LHPC) declined in the Late Jorwe; (4) the Late Jorwe saw an increase in fertility rates and infant mortality that led to a declining population growth rate; (5) stature and body mass index were reduced in the Late Jorwe, particularly for infants who died in the first year of life; (6) infants who died in the first year of life were acquiring bipedal locomotion as expected, between 9 and 13 months of age; but (7) infants with small body size in the Late Jorwe also suffered from osteopenia, providing support for the hypothesis that nutritional deficiency, diarrhea, disease, and other biocultural stressors affected the Late Jorwe population from Inamgaon more heavily.

If small body size and osteopenia in the Late Jorwe infants are due to skeletal emaciation, this result supports the *Climate and Culture Change* model – that children in the Late Jorwe phase would suffer from acute stressors and growth disruption related to the failure of the subsistence system and socio-sanitation issues. The *Subsistence Transition* model – improvements in health follow when agriculture was abandoned – does not appear to be supported by the data if small body size is a result of disruption in homeostasis and growth faltering. More research is needed on micromorphological variation in typically developing infants and children and the process of growth disruption on a microstructural scale to evaluate the claim that the infants and children from Inamgaon suffered from osteopenia (Welsh and Brickley, 2023). Additional research is also needed to understand the functional impacts of the observed microstructural variation.

3.6 Human Variation in Bone Microarchitecture for Infants And Children

Despite ample research interest in bone quality for adult humans, for a variety of practical reasons, there have been few studies in biological anthropology, anatomy, or clinical fields on the ontogeny of compact bone from a histological perspective. This lack of basic research on human variation means that little is known about bone microarchitecture for infants and children who are meeting growth and development benchmarks or about the process of growth derangement. Important questions remain about human variation, functional morphology, and how these relate to geometric and histological properties during early development, when many different tissue types with different mechanical properties are present (Currey, 2003; Frost, 2001; Zioupos, 2001). For example, it is logical that locomotor skill acquisition will be visible in the cross section of lower limb bones in predictable ways as bone geometry shifts in response to a range of behaviors, from incipient waddling behavior to a mature bipedal gait (Cowgill, 2010). Bone geometry and microstructure will have a predictable relationship to body mass and behavior according to Wolff's law, and growth derangement due to immobility, malnutrition, and disease will also be visible (Cowgill, 2018; Robbins et al., 2010; Robbins Schug et al., 2012; Sparacello et al., 2016; Yim et al., 2021). It is also expected that bone strength and stiffness will be altered by the remodeling process, as osteoblast and osteoclast activity becomes decoupled (Bonivtch et al., 2007; Cooper et al., 2007; Ebacher et al., 2007; Jacobs et al., 2010; Robling et al., 2006). However, none of these relationships are straightforward (Agnew et al., 2013; Hunter and Agnew, 2016; Cardoso et al., 2021).

The UNCG (University of North Carolina Greensboro) Human Variation Lab (PI Gwen Robbins Schug, UNCG Biology) is working to address hypotheses related to these issues in collaboration with Haviva Goldman, Megan Perry, Megan Brickley, and Haley Welsh. We are using a Nikon XT H 225 Micro CT at the Joint School of Nanoscience and Nanotechnology (UNCG/NC A&T) to conduct scans of 48 immature bone blocks from Bordeaux, France (10th–13th centuries BCE) and 36 midshaft femoral blocks from Umm-al-Jamal, Jordan (1000 BCE) with a goal of evaluating the relationship between biological age/bone length/maturational stage, remodeling parameters, cross-sectional geometric properties, and microstructural characteristics. This project is characterizing the range of variation in Bone Volume/Total Volume, Total Cortical Porosity, Pore Density, Volume, and Distribution in the cross section (see Figure 3.2). We are also using slice analysis to get basic geometric parameters, including polar second moments, total area, cortical area, cortical thickness, major and minor diameters, and average thickness. These data will address predictions about locomotor skill acquisition (Cowgill, 2010), ontogenetic patterning of BV/TV (Robbins Schug and Goldman, 2014; Welsh and Brickley, 2023), pore distribution and osteon population density (Beresheim et al., 2020), patterns of deposition and resorption on complementary surfaces (Foote, 1916; Enlow, 1962, 1963; Enlow and Bang, 1965), and pore volume (as it relates to bone strength) across subadult age categories (Robbins Schug and Goldman, 2014; Welsh and Brickley, 2023). We will also be using the comparative method to evaluate 48 femoral thin sections of infants and children from India (1400–1000 BCE), 22 of whom demonstrate evidence of long bone growth disruption.

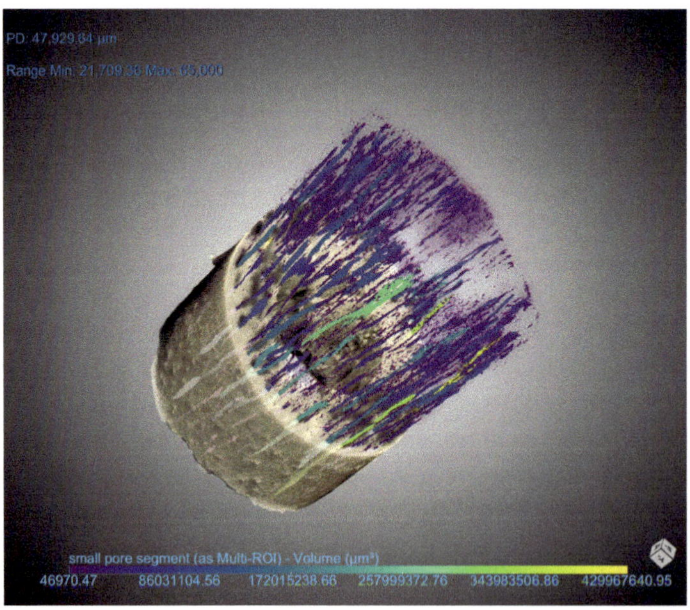

Figure 3.2 Microstructural analysis of skeletal architecture and quality includes an examination of pore volume and distribution. © G. Robbins Schug.

We expect that ontogenetic variation in the femoral midshaft cross section will demonstrate predictable patterns of variation in ontogeny, which include those related to locomotor skill acquisition (Cowgill, 2010), nutritional and disease states (Robbins Schug and Goldman, 2014; Sparacello et al., 2016; Welsh and Brickley, 2023), microarchitectural remodeling (Beresheim et al., 2020), and age (Robbins Schug, 2011; Welsh and Brickley, 2023). Osteon population density and maximum pore density will be found in coordinated patterns of deposition and resorption on complementary surfaces across the cross section (Foote, 1916; Enlow, 1962, 1963), but pore volume (being related to bone strength) will remain below specific thresholds across age intervals: < 5% in 0–6 months, < 15% from 6 to 12 months, and < 10% in older age categories (Robbins Schug and Goldman, 2014; Welsh and Brickley, 2023). This research will be a major step in the path to understanding the range of variation in bone growth, the process of derangement, and effects of health and disease on bone quality.

3.7 Infectious Disease and Socially Mediated Impacts of Environmental Change: TB and Leprosy

Climate and environmental change will have wide-ranging consequences, but the most disruptive is the potential for disease spillover. Contemporary global warming is occurring in the context of the Third Epidemiological Transition (Roberts, 2020), meaning that emerging and re-emerging disease and antibiotic resistance are likely to cause epidemic- and pandemic-level outbreaks that may not be readily solved by biological knowledge and medical interventions; a biocultural approach will be required (Robbins Schug and Halcrow, 2022). Tuberculosis and leprosy become human pathogens and spread into Asia from Africa during a period of changing climate conditions but strongly shaped by socially mediated responses to these environmental challenges (Robbins, 2009; Robbins Schug et al., 2012). That moment in the Bronze Age and the different evolutionary trajectories of these closely related mycobacteria may provide important lessons for understanding disease spillover in the future, and an excellent source of evidence on this topic could be bone histology, including in immature bone.

3.7.1 Tuberculosis

Tuberculosis is a growing threat worldwide in the context of environmental changes due to resistant strains combined with increased global economic inequality, more frequent and more devastating natural disasters, and potentially an increase in environmental refugees. It is a chronic, relapsing infection caused by *M. tuberculosis*, *M. Africanum*, *M. bovis*, and other members of the MTBC (Mycobacterium tuberculosis complex). Often considered to be a respiratory infection, skeletal tuberculosis can affect the skeleton in people of all ages, and this type of TB may be increasing in frequency, accounting for 10–20% of extrapulmonary TB (Kritsaneepaiboon et al., 2017; Martini, 1988; Teo and Peh, 2004).

Tuberculosis is rare in children less than one year of age (Resnick, 2002: 2524); however, the pathogen may be congenitally transmitted to newborns through the placenta, umbilical cord, or the amniotic fluid (Lewis, 2017:156). In older infants and children, whose immune systems are still immature, hematogenous osteomyelitis resulting from infection with TB (miliary TB) can spread much more rapidly throughout the skeleton than it typically does in adults (Schultz and Schultz, 2015). Hematogenous spread of mycobacteria in children leads to spondylitis, arthritis, and osteomyelitis in the appendicular skeleton and osteopenia in regions rich in red bone marrow (Lewis 2017:155–164). Chronic osteomyelitis, recurring inflammation, and periostitis (osteitis) affect the neighboring compact bone, which becomes sclerotic and dominated by Haversian systems and remodeling (e.g., Schultz, 1997, 2001, 2003; Roberts and Schultz, 2002; Schultz and Teschler-Nicola, 1985; Schultz and Schmidt-Schultz, 2015; Vekszler et al., 2022).

Histological changes in the femur associated with miliary tuberculosis result from a slowing of the bone growth process (Schultz and Teschler-Nicola, 1985). Specifically, these include osteopenia, diffuse isolated resorption spaces, and small pores (up to 1 mm in diameter) with and without lamellar borders, situated in the periosteal and the compact bone regions of the cortex, and thin marginal layers of woven bone (Schultz and Schultz, 2015). In infants, the lower limb bones are more likely to be affected than the bones of the axial skeleton (Kritsaneepaiboon et al., 2017). Long bones become infected through the metaphysis or nutrient canals and inflamed periosteal tissue causes bone thickening in the tibia and femur. Tuberculosis also often involves the synovial joints, particularly of the hips, and knees, with periarticular osteopenia, cortical irregularity, lytic and prolific lesions occurring at the metaphysis and in the epiphyseal bone in clinical histological analysis (Teo and Peh, 2004). Perfuse nonspecific subperiosteal new bone formation, clearly distinct from the underlying cortex, should also be visible histologically (Santos and Roberts, 2001; Schultz, 2001, 2003; Weston, 2009).

Vestiges of inflammation in the meninges and the dura have also been described in histological analysis of pediatric miliary TB. These occur as demarcated erosive lesions on the endocranial surface, visible macroscopically (Lewis, 2007:157; Schultz, 2001), and isolated or confluent round impressions on the endocranial surface, visible in the scanning electron microscope (Schultz, 2001; Schultz and Schultz, 2015). LM investigation of these smooth features demonstrated they were surrounded by mature lamellar bone and collagen fibers that followed their contours, suggesting they were not a taphonomic artifact but formed in life.

The effect of tuberculosis on bone histomorphology has not been investigated much in bioarchaeology for adult or children's skeletal remains (De Boer et al., 2013). However, some general statements can be made that infectious lesions due to osteomyelitis will present primarily as a destructive, lytic process with irregular microarchitecture, a relative increase in Howship's lacunae, and irregular bone proliferation (Wapler et al., 2004). Coqueugniot et al. (2015) used micro CT to investigate a case of tuberculosis in one five-year-old child and two adults from an early Neolithic site in Syria (dated to approximately 8500 BCE). They describe trabecular bone reorganization in the region of a focal infection in the lumbar vertebrae, but there is no overall loss of bone volume. Unfortunately, this study did not investigate any changes to compact bone histomorphometry.

Childhood tuberculosis is common in living populations, and it has been documented in historical collections of children's skeletons (Lewis, 2017:163-164; Santos and Roberts, 2001); however, the majority of documented cases of pediatric tuberculosis have only mild traces of the disease, likely due to children dying of the disease prior to skeletal manifestations (Lewis, 2017: 162-163). Histological analyses and investigations of both compact bone microarchitecture and intracranial hemorrhage in documented cases of pediatric tuberculosis will be instructive as to the potential for identifying pathognomonic lesions in the past. More progress in this area has been made for histopathology of infection with

Mycobacterium leprae, an organism which causes leprosy in humans and which has a pattern of diagnostic features in compact bone that complement its very recognizable macroscopic effects on the skeleton (Blondiaux et al., 1994).

3.7.2 Leprosy

Leprosy is also a disease of poverty and poor living conditions, but it is much less of an imminent threat today; 95% of people are immune to the pathogen that causes this disease. Globally, approximately 250,000 people have leprosy, and due to the highly stigmatized nature of the disease, it is still an important, though neglected, tropical disease. The most significant research question surrounding leprosy and tuberculosis, though, is how these two very closely related pathogens co-evolved with humans to have such different epidemiological impacts. Tuberculosis is a highly infectious, deadly, and re-emerging disease affecting millions of people today and predicted to be a substantial threat in the context of global warming; leprosy has gone through an extreme level of reductive evolution that makes it a very unlikely threat in a warming world. These questions about the origin, evolution, and early migrations of these two diseases can be investigated using macroscopic skeletal lesions, histology, and ancient DNA.

Leprosy is caused by infection with *Mycobacterium leprae* or *M. lepramatosis*. Like tuberculosis, an initial infection does not immediately or necessarily lead to skeletal lesions; only 10% of people infected with leprosy today will express any symptoms at all, and fewer will have signs of the disease in their skeleton. If leprosy is contracted congenitally, through the placenta, it can result in spontaneous abortion, low birth weight, and greater risk of infant mortality (Lewis, 2017:164–165). Macroscopic changes in immature skeletons of people with leprosy include atrophy of the anterior nasal spine, erosion of the nasal aperture, and resorption of the dental alveoli preceding loss of the anterior dentition. Abnormal porosity in the nasal aperture and on the hard palate and active bone formation (lamellar bone) in the maxillary and frontal sinuses are also common. Lepromatous leprosy also causes peripheral neuropathy, which can cause inflammation in the lower leg bones, claw-hand and foot changes, palmar and volar grooves on the phalanges, concentric remodeling of hand and foot bones, septic arthritis, ankylosis, and osteomyelitis in the hands, feet, and lower limb bones.

Histological changes in adult skeletons with leprosy include changes to the compact bone inside of the hypertrophied cortex (Blondiaux et al., 1994). The endosteal and periosteal surfaces of metacarpals and tarsals affected by the peripheral neuropathy of leprosy demonstrate disseminated osteoclast activity, endosteal and periosteal erosions, and little evidence of healing. Large lacunae and resorption cavities are visible in the compact bone, and there are occasional newly built, poorly mineralized osteons. Schultz (2001) and Schultz and Roberts (2002) documented periosteal new bone formation; osteoporotic changes to the lamellar bone probably related to evidence for periosteal inflammation and possibly also from disuse; fields of tangential lamellae in the midst of the cortex, which likely resulted from disuse; and up to a millimeter of circumferential lamellar bone external to the cortical surface of the tibiae, which appeared to have built up into a 'bark-like layer' slowly over time and which was undergoing active remodeling. This periosteal new bone formation was separated from the original compact bone surface by a thin border of newly built bone and a layer of tangentially oriented blood vessels. In a few cases, polsters like those seen in treponemal infection can also be seen in leprosy.

The macroscopic skeletal changes for people with leprosy are similar for adults and children, but there has been no systematic investigation of the histological changes we see in adult skeletons for immature bones. Fortunately, there are other sources of evidence to reconstruct the evolution and natural history of MTBC and to understand how human–environmental relations through time shape pathogen evolution in ways that make it impactful in human history. The most recent common ancestor of the MTBC existed in Africa between 100 and 70,000 years ago (Gagneux, 2018) and *M. leprae* diverged from its closest relative, *M. lepramatosis*, around 13 million years ago. Most of its genetic diversity began to accumulate 4,000–5,000 years ago, related to its migration to Asia (Cole et al., 2001; Robbins et al., 2009; Robbins Schug et al., 2013). Since becoming a human pathogen, the bacterium has seen reduction in the size of its genome compared to its close relative *M. tuberculosis*, and it contains a high percentage of pseudogenes, regions of the genome, which were once genes, but have since lost their function. This pattern is called reductive evolution and is seen in many human parasites, including Rickettsia and Chlamydia. It is indicative of the pathogen adapting to a highly specific niche where it no longer needs many of the genes it once required to survive (Cole et al., 2001). In fact, *M. leprae* is such an obligate parasite that it cannot be cultured in a lab except for on the footpads of mice.

While this makes it difficult to study in the lab, a small, stable genome such as this one makes piecing together its evolution much easier. In 2009, a large study was done by Marc Monot to find parts of the *M. leprae* genome that could be used to identify strains of the bacteria (Monot et al., 2009). These single nucleotide polymorphisms (SNPs) mutate at a predictable rate that allows evolutionary biologists to determine when one strain diverged from another. In this study, they identified almost 100 SNPs that were useful in genotyping these strains. They grouped the strains into 4 main types based on their SNP profiles, numbered 1 through 4, and 16 subtypes, denoted by a letter after the type number. After comparing hundreds of modern samples and several ancient samples, they formed several hypotheses about the spread of leprosy based on the evolutionary tree they computed.

Their hypotheses relied on one central idea: that *M. leprae* began from a single clone that had passed through a population bottleneck, as evidenced by the relatively little genomic diversity it exhibits compared to other pathogens. Type 2 began in East Africa, spread west into Asia, and east into the Middle East and Europe as type 3. Leprosy in Asia came from two sources, they believed. Type 1 traveled south through India and SE Asia, while a subtype from the Mediterranean, type 3K, spread to China via the Silk Road, and later to Japan and Korea. Europeans brought leprosy to North America with settler colonialism and the Atlantic slave trade. Type 4 originated in West Africa after branching off from a European strain and was brought to the Caribbean and Brazil. Leprosy in the rest of the Americas is related to a subtype from Europe. While these hypotheses have been revised over the last decade, this model and the genotypes that Monot categorized have served as a framework for evolutionary biologists and paleopathologists studying this pathogen.

Access to more diverse samples combined with improvements in DNA sequencing and bioinformatics has led to the discovery of more SNPs and for detection of the pathogen in older samples, including one from Egypt around 2,200 years ago from which we have learned that there is more diversity in certain regions than previously thought (Neukamm et al. 2020). In Europe, for example, it was thought that leprosy was introduced once. With the addition of dozens of medieval European samples, strains from most of the branches were detected, indicating that either leprosy was introduced there earlier than previously thought, or that it was introduced several times (Pfrengle et al., 2021; Schuenemann et al., 2013).

We also have possible new skeletal evidence of leprosy emerging around 3600 BCE in Hungary (Köhler et al., 2017) and a recent paper finding evidence for a new strain '0' in Hungarians from the first millennium, which is typically found in East Asia and the Pacific (Köhler et al., 2017; Mendum et al., 2018). While the abundance of skeletal and genetic samples added to the tree for leprosy in Europe has been helpful, more questions have come up than have been answered.

Less work has been done in Asia. Recent evidence from Europe has suggested a Eurasian origin for this pathogen, contradicting the African origin model (Schuenemann et al., 2013). To test this hypothesis, we need more ancient samples from outside of Europe, as most of the current ancient samples we currently have are from this region. One obstacle in paleogenomics has been the unreliable preservation of DNA in bone and dentine, especially in Asia. Dental calculus, the mineralized biofilm that forms from plaque on teeth, has been shown to preserve DNA incredibly well due to its lack of porosity compared to dentine and bone, shielding it from environmental contamination and degradation (Warinner et al., 2015). It contains mostly bacterial DNA, but occasionally human genomic DNA and DNA from plant and animal material eaten by the individual. In situations where leprosy is suspected based on skeletal evidence, calculus may prove to be an excellent resource even when *M. leprae* is not detected in the bone.

In 2017, researchers detected *M. leprae* DNA and proteins in dental calculus from an individual in 16th-century Norway (Fotakis et al., 2020). It is believed that the bacteria were trapped in the calculus due to their presence in the mucosa or soft palate. Since then, another group found *M. leprae* in dental calculus from an Egyptian mummy around the late first millennium BCE (Neukamm et al., 2020). This individual did not exhibit bone deformations indicative of leprosy infection, though the DNA evidence is strong enough to indicate that the individual was infected with the pathogen. As we are trying to expand our knowledge of *M. leprae*'s genome further back in time and across a wide region, dental calculus may prove to be the key to uncovering this information. Currently, the Robbins Schug lab is working with dental calculus from a case of leprosy from India (r) that could shed light on some of these issues. Skeletal histology on immature remains from Harappa and other sites in South and West Asia with documented evidence for periosteal new bone formation (e.g., Robbins Schug and Blevins, 2016) will be an important correlate to that work.

3.8 Conclusions

In this chapter we focused on the factors that can impact the growth and development of bone and how study of the histomorphology of pediatric bone can be used to better understand the impacts of climate, environment, and biocultural change on life history. The bioarchaeology of climate and environmental change has gained attention relatively recently, and there are a large number of scholars working on this topic (Robbins Schug et al., 2023). There is a lot more work that needs to be done to document variability in bone ontogeny in human populations to better understand historic and prehistoric biological responses to climate and culture change. By describing some of our ongoing work in this area, we hope to inspire collaborations and new areas of inquiry, particularly among graduate students.

This area of research takes on increasing relevance as we struggle to develop responses to global warming and its potential impacts on the global community. Poverty, social and economic inequalities, and historical injustices structure the health impacts of climate change. This includes potential effects on bone growth from the accumulations of environmental pollutants such as endocrine-disrupting chemicals, changing disease ecology, lifestyle and dietary changes related to extreme weather. Understanding the complexities of human–environmental interactions, strategies adopted by past people, and their ultimate consequences will be important to identify preventative measures and sustainable lifestyle modifications to ensure bone health today and in the future.

Acknowledgements

The authors thank the editors for including us in this volume and for their helpful comments on the chapter. We would also like to thank members of the Human Diversity Lab at UNCG – Annisianna Egerton, Cameron Jackson, Irena Mustafaj, Lelise Weyessa, Noemi Salinas, and Ruby Arteaga – for their participation in the scanning and analysis of bone microarchitecture in spring 2023.

References

Agarwal, S. C. (2016). Bone morphologies and histories: Life course approaches in bioarchaeology. *American Journal of Physical Anthropology*, *159*(S61), 130–149. https://doi.org/10.1002/ajpa.22905

Agnew, A. M., Moorhouse, K., Kang, Y.-S., Donnelly, B. R., Pfefferle, K., Manning, A. X., Litsky, A. S., Herriott, R., Abdel-Rasoul, M., & Bolte, J. H. (2013). The response of pediatric ribs to quasi-static loading: Mechanical properties and microstructure. *Annals of Biomedical Engineering*, *41*(12), 2501–2514. https://doi.org/10.1007/s10439-013-0875-6

Bachrach, L. K. (2001). Acquisition of optimal bone mass in childhood and adolescence. *Trends in Endocrinology & Metabolism*, *12*(1), 22–28. https://doi.org/10.1016/S1043-2760(00)00336-2

Bassett, J. H. D., & Williams, G. R. (2016). Role of thyroid hormones in skeletal development and bone maintenance. *Endocrine Reviews*, *37*(2), 135–187. https://doi.org/10.1210/er.2015-1106

Beighton, P., Sujansky, E., Patzak, B., & Portele, K. A. (1993). Genetic skeletal dysplasias in the Museum of Pathological Anatomy, Vienna. *American Journal of Medical Genetics*, *47*(6), Article 6.

Bell, T. D., Demay, M. B., & Burnett-Bowie, S.-A. M. (2010). The biology and pathology of vitamin D control in bone. *Journal of Cellular Biochemistry*, *111*(1), Article 1.

Beresheim, A. C., Pfeiffer, S., & Grynpas, M. (2020). Ontogenetic changes to bone microstructure in an archaeologically derived sample of human ribs. *Journal of Anatomy*, *236*(3), 448–462. https://doi.org/10.1111/joa.13116

Biewener, A. A., & Bertram, J. E. (1994). Structural response of growing bone to exercise and disuse. *Journal of Applied Physiology*, *76*(2), 946–955. https://doi.org/10.1152/jappl.1994.76.2.946

Biewener, A. A., Fazzalari, N. L., Konieczynski, D. D., & Baudinette, R. V. (1996). Adaptive changes in trabecular architecture in relation to functional strain patterns and disuse. *Bone*, *19*(1), 1–8. https://doi.org/10.1016/8756-3282(96)00116-0

Blondiaux, J., Duvette, J.-F., Vatteoni, S., & Eisenberg, L. (1994). Microradiographs of leprosy from an osteoarchaeological context. *International Journal of Osteoarchaeology*, *4*(1), Article 1. https://doi.org/10.1002/oa.1390040104

Bonivtch AR, Bonewald LF, Nicolella DP. Tissue strain amplification at the osteocyte lacuna: a microstructural finite element analysis. *J Biomech*. 2007; *40(10)*, 2199–2206.

Boskey, A. L., & Imbert, L. (2017). Bone quality changes associated with aging and disease: A review. *Annals of the New York Academy of Sciences*, *1410*(1), 93–106. https://doi.org/10.1111/nyas.13572

Bouillon, R., & Prodonova, A. (2000). Growth hormone deficiency and peak bone mass: Laboratory for experimental medicine and endocrinology, Catholic University of Leuven, Gasthuisberg, Leuven, Belgium. *Journal of Pediatric Endocrinology and Metabolism, 13*(s2), 1327–1342. https://doi.org/10.1515/jpem-2000-s604

Bouvier, M., & Hylander, W. L. (1981). Effect of bone strain on cortical bone structure in macaques (Macaca mulatta). *Journal of Morphology, 167*(1), 1–12. https://doi.org/10.1002/jmor.1051670102

Boyd, D. C., & Boyd Jr, C. C. (1989). A comparison of Tennessee Archaic and Mississippian maximum femoral lengths and midshaft diameters: Subsistence change and postcranial variability. *Southeastern Archaeology, 8*(2) 107–116.

Brickley, M. B., D'Ortenzio, L., Kahlon, B., Schattmann, A., Ribot, I., Raguin, E., & Bertrand, B. (2017). Ancient vitamin D deficiency: Long-term trends. *Current Anthropology, 58*(3), Article 3.

Brickley, M. B., Ives, R., & Mays, S. (2020). *The bioarchaeology of metabolic bone disease*. Academic Press.

Brickley, M. B., Mays, S., George, M., & Prowse, T. L. (2018). Analysis of patterning in the occurrence of skeletal lesions used as indicators of vitamin D deficiency in subadult and adult skeletal remains. *International Journal of Paleopathology, 23*, 43–53.

Brickley, M. B., Moffat, T., & Watamaniuk, L. (2014). Biocultural perspectives of vitamin D deficiency in the past. *Journal of Anthropological Archaeology, 36*, 48–59.

Brickley, M., Mays, S., & Ives, R. (2005). Skeletal manifestations of vitamin D deficiency osteomalacia in documented historical collections. *International Journal of Osteoarchaeology, 15*(6), Article 6.

Brickley, M., Mays, S., & Ives, R. (2007). An investigation of skeletal indicators of vitamin D deficiency in adults: Effective markers for interpreting past living conditions and pollution levels in 18th and 19th century Birmingham, England. *American Journal of Physical Anthropology: The Official Publication of the American Association of Physical Anthropologists, 132*(1), Article 1.

Brickley, M., Mays, S., & Ives, R. (2010). Evaluation and interpretation of residual rickets deformities in adults. *International Journal of Osteoarchaeology, 20*(1), Article 1.

Bridges, P. S. (1989). Changes in Activities with the Shift to Agriculture in the Southeastern United States. *Current Anthropology, 30*(3), 385–394. https://doi.org/10.1086/203756

Bryson, R. A., & Swain, A. M. (1981). Holocene variations of monsoon rainfall in Rajasthan. *Quaternary Research, 16*(2), Article 2.

Buikstra, J. E., DeWitte, S. N., Agarwal, S. C., Baker, B. J., Bartelink, E. J., Berger, E., Blevins, K. E., Bolhofner, K., Boutin, A. T., & Brickley, M. B. (2022). Twenty-first century bioarchaeology: Taking stock and moving forward. *American Journal of Biological Anthropology, 178*, 54–114.

Burdanowitz, N., Rixen, T., Gaye, B., & Emeis, K.-C. (2021). Signals of Holocene climate transition amplified by anthropogenic land-use changes in the westerly–Indian monsoon realm. *Climate of the Past, 17*(4), Article 4.

Cardoso, H.F., Spake, L., Rios, L., & Albanese, J. (2021). Population variation in diaphyseal growth and age estimation of juvenile skeletal remains. In *Remodeling Forensic Skeletal Age* (pp. 99–138). Academic Press.

Carter, D. R., & Beaupre, G. S. (2001). Skeletal function and form: Mechanobiology of skeletal development, aging, and regeneration. *Growth Development and Aging, 65*(1), Article 1.

Carter, D. R., Van der Meulen, M. C. H., & Beaupre, G. S. (1996). Mechanical factors in bone growth and development. *Bone, 18*(1), Article 1.

Chanchlani, R., Nemer, P., Sinha, R., Nemer, L., Krishnappa, V., Sochett, E., Safadi, F., & Raina, R. (2020). An overview of rickets in children. *Kidney International Reports, 5*(7), 980–990. https://doi.org/10.1016/j.ekir.2020.03.025

Chen, Y.-Y., Fang, W.-H., Wang, C.-C., Kao, T.-W., Chang, Y.-W., Yang, H.-F., Wu, C.-J., Sun, Y.-S., & Chen, W.-L. (2019). Crosssectional assessment of bone mass density in adults with hepatitis B virus and hepatitis C virus infection. *Scientific Reports, 9*(1), 5069. https://doi.org/10.1038/s41598-019-41674-4

Cheung, C.-L., Livshits, G., Zhou, Y., Meigs, J. B., McAteer, J. B., Florez, J. C., Cupples, L. A., Demissie, S., Kiel, D. P., & Karasik, D. (2009). Hip geometry variation is associated with bone mineralization pathway gene variants: The framingham study. *Journal of Bone and Mineral Research*, 091125104452062–32. https://doi.org/10.1359/jbmr.091102

Chevalley T, Bonjour JP, van Rietbergen B, Ferrari S, Rizzoli R. Tracking of environmental determinants of bone structure and strength development in healthy boys: an eight-year follow up study on the positive interaction between physical activity and protein intake from prepuberty to mid-late adolescence. *J Bone Miner Res*. 2014 Oct; *29*(10), 2182–2192. doi: 10.1002/jbmr.2247. PMID: 24715534.

Chevalley, T., Bonjour, J.-P., Audet, M.-C., Merminod, F., van Rietbergen, B., Rizzoli, R., & Ferrari, S. (2017). Fracture prospectively recorded from prepuberty to young adulthood: Are they markers of peak bone mass and strength in males?: Fracture during growth. *Journal of Bone and Mineral Research*, *32*(9), 1963–1969. https://doi.org/10.1002/jbmr.3174

Clark, E. M., Tobias, J. H., & Ness, A. R. (2006). Association between bone density and fractures in children: A systematic review and meta-analysis. *Pediatrics*, *117*(2), e291–e297. https://doi.org/10.1542/peds.2005-1404

Cole, S. T., Eiglmeier, K., Parkhill, J., James, K. D., Thomson, N. R., Wheeler, P. R., Honoré, N., Garnier, T., Churcher, C., Harris, D., Mungall, K., Basham, D., Brown, D., Chillingworth, T., Connor, R., Davies, R. M., Devlin, K., Duthoy, S., Feltwell, T., … Barrell, B. G. (2001). Massive gene decay in the leprosy bacillus. *Nature*, *409*(6823), 1007–1011. https://doi.org/10.1038/35059006

Collier, S. (1993). Sexual dimorphism in relation to big-game hunting and economy in modern human populations. *American Journal of Physical Anthropology*, *91*(4), 485–504. https://doi.org/10.1002/ajpa.1330910406

Colombo, A., Coqueugniot, H., & Dutour, O. (2021). Can the palaeoepidemiology of rickets during the industrialisation period in France be studied through bioarchaeological grey literature and French medico-historical literature of the 18th-early 20th centuries? Preliminary examination of a complex topic. *International Journal of Paleopathology*, *34*, 76–81.

Colombo, A., d'Ortenzio, L., Bertrand, B., Coqueugniot, H., Knüsel, C. J., Kahlon, B., & Brickley, M. (2019). Micro-computed tomography of teeth as an alternative way to detect and analyse vitamin D deficiency. *Journal of Archaeological Science: Reports*, *23*, 390–395.

Cooper, D. M. L., Thomas, C. D. L., Clement, J. G., Turinsky, A. L., Sensen, C. W., & Hallgrímsson, B. (2007). Age-dependent change in the 3D structure of cortical porosity at the human femoral midshaft. *Bone*, *40*(4), 957–965. https://doi.org/10.1016/j.bone.2006.11.011

Coqueugniot, H., Dutailly, B., Desbarats, P., Boulestin, B., Pap, I., Szikossy, I., Baker, O., Montaudon, M., Panuel, M., Karlinger, K., Kovács, B., Kristóf, L. A., Pálfi, G., & Dutour, O. (2015). Three-dimensional imaging of past skeletal TB: From lesion to process. *Tuberculosis*, *95*, S73–S79. https://doi.org/10.1016/j.tube.2015.02.004

Cotran, R. S., Kumar, V., Collins, T., & Robbins, S. L. (1999). *Robbins pathologic basis of disease* (6th ed.). Saunders.

Cowgill, L. W. (2010). The ontogeny of Holocene and Late Pleistocene human postcranial strength. *American Journal of Physical Anthropology*, *141*(1), 16–37. https://doi.org/10.1002/ajpa.21107

Cowgill, L. W. (2018). Juvenile body mass estimation: A methodological evaluation. *Journal of Human Evolution*, *115*, 78–84. https://doi.org/10.1016/j.jhevol.2017.07.007

Cowgill, L. W., & Rob'yn, A. J. (2018). Biomechanical implications of the onset of walking. *Journal of Human Evolution*, *122*, 133–145.

Cunningham, C., Scheuer, L., & Black, S. (2016). *Developmental juvenile osteology*. Academic press.

Currey, J. D. (2003). Role of collagen and other organics in the mechanical properties of bone. *Osteoporosis International*, *14*(0), 29–36. https://doi.org/10.1007/s00198-003-1470-8

Currey, J. D. (1984). *The mechanical adaptation of bones*. Princeton University Press.

Dash, C., Shankar, R., Pati, P., Jose, J., Seong, Y. B., Dhal, S. P., Manjunatha, B. R., & Sandeep, K. (2022). An environmental magnetic record of Holocene climatic variability from the Chilika Lagoon, Southern Mahanadi Delta, east coast of India. *Journal of Asian Earth Sciences*, *230*, 105190.

Davies, T. (2013). *Cross-sectional variation in the human femur and tibia: The influence of physique and habitual mobility on diaphyseal morphology* [PhD Thesis]. University of Cambridge.

De Boer, H. H., Van Der Merwe, A. E., & Maat, G. J. R. (2013). The diagnostic value of microscopy in dry bone palaeopathology: A review. *International Journal of Paleopathology*, *3*(2), 113–121. https://doi.org/10.1016/j.ijpp.2013.03.004

Dhavalikar, M. K. (1984). Toward an ecological model for chalcolithic cultures of central and western India. *Journal of Anthropological Archaeology*, *3*(2), Article 2.

Dhavalikar, M. K. (1988). *The first farmers of the Deccan*. Ravish Publishers.

Dhavalikar, M. K. (1994). Chalcolithic architecture at Inamgaon and Walki: An ethnoarchaeological study. In *Living Traditions: Studies in the Ethnoarchaeology of South Asia* (pp. 31–52). New Delhi: Oxford and IBH Publishing Company.

Dhavalikar, M. K., Sankalia, H., & Ansari, Z. (1988). Excavations at Inamgaon. I (Parts i and ii). *Pune: Deccan College Post-Graduate and Research Institute.*

Digby, M. G., Bishop, N. J., Paggiosi, M. A., & Offiah, A. C. (2016). HR-pQCT: A non-invasive 'biopsy' to assess bone structure and strength. *Archives of Disease in Childhood - Education & Practice Edition*, *101*(5), 268–270. https://doi.org/10.1136/archdischild-2015-309455

D'Ortenzio, L., Ribot, I., Kahlon, B., Bertrand, B., Bocaege, E., Raguin, E., Schattmann, A., & Brickley, M. (2018). The rachitic tooth: The use of radiographs as a screening technique. *International Journal of Paleopathology*, *23*, 32–42.

D'Ortenzio, L., Ribot, I., Raguin, E., Schattmann, A., Bertrand, B., Kahlon, B., & Brickley, M. (2016). The rachitic tooth: A histological examination. *Journal of Archaeological Science*, *74*, 152–163.

Dror, D. K., & Allen, L. H. (2014). Dairy product intake in children and adolescents in developed countries: Trends, nutritional contribution, and a review of association with health outcomes. *Nutrition Reviews*, *72*(2), 68–81. https://doi.org/10.1111/nure.12078

Duren, D. L., Blangero, J., Sherwood, R. J., Seselj, M., Dyer, T., Cole, S. A., Lee, M., Choh, A. C., Chumlea, Wm C., Siervogel, R. M., Czerwinski, S. A., & Towne, B. (2011). Cortical bone health shows significant linkage to chromosomes 2p, 3p, and 17q in 10-year-old children. *Bone*, *49*(6), 1213–1218. https://doi.org/10.1016/j.bone.2011.08.024

Duren, D. L., Seselj, M., Froehle, A. W., Nahhas, R. W., & Sherwood, R. J. (2013). Skeletal growth and the changing genetic landscape during childhood and adulthood. *American Journal of Physical Anthropology*, *150*(1), 48–57. https://doi.org/10.1002/ajpa.22183

Dutta, A. (2006). A Critical Review of the Economy of the Chalcolithic People of Inamgaon. *Ancient Asia*, *1*.

Ebacher, V., Tang, C., McKay, H., Oxland, T. R., Guy, P., & Wang, R. (2007). Strain redistribution and cracking behavior of human bone during bending. *Bone*, *40*(5), 1265–1275. https://doi.org/10.1016/j.bone.2006.12.065

Elder, C. J., & Bishop, N. J. (2014). Rickets. *The Lancet*, *383*(9929), 1665–1676. https://doi.org/10.1016/S0140-6736(13)61650-5

Elhakeem, A., Frysz, M., Tilling, K., Tobias, J. H., & Lawlor, D. A. (2019). Association between age at puberty and bone accrual from 10 to 25 years of age. *JAMA Network Open*, *2*(8), e198918. https://doi.org/10.1001/jamanetworkopen.2019.8918

Enlow, D. H. (1962). A study of the post-natal growth and remodeling of bone. *American Journal of Anatomy*, *110*(2), 79–101. https://doi.org/10.1002/aja.1001100202

Enlow, D. H. (1963). *Principles of bone remodeling: An account of post-natal growth and remodeling processes in long bones and the mandible*. Charles C. Thomas.

Enlow, D. H. (1966). A morphogenetic analysis of facial growth. *American Journal of Orthodontics*, *52*(4), 283–299. https://doi.org/10.1016/0002-9416(66)90169-2

Enlow, D. H. (1976). The remodeling of bone. *Yearbook of Physical Anthropology*, *20*, 19–34.

Enlow, D. H., & Bang, S. (1965). Growth and remodeling of the human maxilla. *American Journal of Orthodontics*, *51*(6), 446–464. https://doi.org/10.1016/0002-9416(65)90242-3

Enlow, D. H., & Hans, M. G. (1996). *Essentials of facial growth*. Saunders.

Enlow, D. H. (1982). *Handbook of Facial Growth*. W.H. Saunders.

Enzel, Y., Ely, L. L., Mishra, S., Ramesh, R., Amit, R., Lazar, B., Rajaguru, S. N., Baker, V. R., & Sandler, A. (1999). High-resolution Holocene environmental changes in the Thar Desert, northwestern India. *Science, 284*(5411), Article 5411.

Ermakov, S., Trofimov, S., Malkin, I., & Livshits, G. (2012). A significant association exists between receptor tyrosine kinase-like orphan receptor 2 gene variants and the OPG/RANKL ratio in human plasma. *Osteoporosis International, 23*(7), 1899–1907. https://doi.org/10.1007/s00198-011-1820-x

Ferrari, S., Rizzoli, R., Slosman, D., & Bonjour, J.-P. (1998). Familial resemblance for bone mineral mass is expressed before puberty1. *The Journal of Clinical Endocrinology & Metabolism, 83*(2), 358–361. https://doi.org/10.1210/jcem.83.2.4583

Foote, J.S. (1916). *Contribution to the Comparative Histology of the Femur.* Smithsonian Institution.

Forwood, M. R., & Burr, D. B. (1993). Physical activity and bone mass: Exercises in futility? *Bone and Mineral, 21*(2), 89–112. https://doi.org/10.1016/S0169-6009(08)80012-8

Fotakis, A. K., Denham, S. D., Mackie, M., Orbegozo, M. I., Mylopotamitaki, D., Gopalakrishnan, S., Sicheritz-Pontén, T., Olsen, J. V., Cappellini, E., Zhang, G., Christophersen, A., Gilbert, M. T. P., & Vågene, Å. J. (2020). Multi-omic detection of *Mycobacterium leprae* in archaeological human dental calculus. *Philosophical Transactions of the Royal Society B: Biological Sciences, 375*(1812), 20190584. https://doi.org/10.1098/rstb.2019.0584

Fox, K. M., Cummings, S. R., Powell-Threets, K., & Stone, K. (1998). Family history and risk of osteoporotic fracture. *Osteoporosis International, 8*(6), 557–562. https://doi.org/10.1007/s001980050099

Frayer, D. W. (1980). Sexual dimorphism and cultural evolution in the Late Pleistocene and Holocene of Europe. *Journal of Human Evolution, 9*(1), Article 5.

Frayer, D. W. (1981). Body size, weapon use, and natural selection in the European Upper Paleolithic and Mesolithic. *American Anthropologist, 83*(1), Article 1.

Frayer, D. W. (1984). Biological and cultural change in the European late Pleistocene and early Holocene. *The Origins of Modern Humans: A World Survey of the Fossil Evidence, 21*, 1–50.

Frost, H. M., *Bone remodeling and its relation to metabolic bone diseases*, C. Thomas, Springfield, IL, 1973.

Frost, H. M. (1986). *Intermediary organization of the skeleton*. CRC Press.

Frost, H. M. (1987). Bone "mass" and the "mechanostat": A proposal. *The Anatomical Record, 219*(1), 1–9. https://doi.org/10.1002/ar.1092190104

Frost, H. M. (2001). Cybernetic aspects of bone modeling and remodeling, with special reference to osteoporosis and whole-bone strength. *American Journal of Human Biology: The Official Journal of the Human Biology Association, 13*(2), Article 2.

Frost, H. M. (2003). Bone's mechanostat: A 2003 update. *The Anatomical Record Part A: Discoveries in Molecular, Cellular, and Evolutionary Biology, 275A*(2), 1081–1101. https://doi.org/10.1002/ar.a.10119

Gabel, L., Macdonald, H. M., Nettlefold, L. A., & McKay, H. A. (2018). Sex-, ethnic-, and age-specific centile curves for pQCT- and HR-pQCT-derived measures of bone structure and strength in adolescents and young adults: PQCT AND HR-PQCT CENTILE CURVES FOR ADOLESCENTS. *Journal of Bone and Mineral Research, 33*(6), 987–1000. https://doi.org/10.1002/jbmr.3399

Gagneux, S. (2018). Ecology and evolution of Mycobacterium tuberculosis. *Nature Reviews Microbiology, 16*(4), 202–213. https://doi.org/10.1038/nrmicro.2018.8

Garn, S. M., Rohmann, C. G., Wagner, B., & Ascoli, W. (1967). Continuing bone growth throughout life: A general phenomenon. *American Journal of Physical Anthropology, 26*(3), 313–317. https://doi.org/10.1002/ajpa.1330260306

Garn, S.M. (1970). *The Earlier Gain and the Later Loss of Cortical Bone in Nutritional Perspective.* Charles C. Thomas.

Gille, D. (2010). Overview of the physiological changes and optimal diet in the golden age generation over 50. *European Review of Aging and Physical Activity, 7*(1), 27–36. https://doi.org/10.1007/s11556-010-0058-5

Glorieux, F. H., Travers, R., Taylor, A., Bowen, J. R., Rauch, F., Norman, M., & Parfitt, A. M. (2000). Normative data for iliac bone histomorphometry in growing children. *Bone*, *26*(2), 103–109. https://doi.org/10.1016/S8756-3282(99)00257-4

Goldman, H. M., Hampson, N. A., Guth, J. J., Lin, D., & Jepsen, K. J. (2014). Intracortical remodeling parameters are associated with measures of bone robustness: Intracortical remodeling and bone robustness. *The Anatomical Record*, *297*(10), 1817–1828. https://doi.org/10.1002/ar.22962

Goldman, H. M., Mcfarlin, S. C., Cooper, D. M. L., Thomas, C. D. L., & Clement, J. G. (2009). Ontogenetic patterning of cortical bone microstructure and geometry at the human Mid-Shaft Femur. *The Anatomical Record: Advances in Integrative Anatomy and Evolutionary Biology*, *292*(1), 48–64. https://doi.org/10.1002/ar.20778

Goodman, A. H., Lallo, J., Armelagos, G. J., & Rose, J. C. (1984). Health changes at Dickson Mounds, Illinois (AD 950-1300). In: *Paleopathology At the Origins of Agriculture*. New York: Academic Press, Inc., 271–305.

Gosman, J. H., Hubbell, Z. R., Shaw, C. N., & Ryan, T. M. (2013). Development of cortical bone geometry in the human femoral and tibial diaphysis. *Anatomical Record (Hoboken, N.J.: 2007)*, *296*(5), 774–787. https://doi.org/10.1002/ar.22688

Gosman, J. H., & Ketcham, R. A. (2009). Patterns in ontogeny of human trabecular bone from SunWatch Village in the Prehistoric Ohio Valley: General features of microarchitectural change. *American Journal of Physical Anthropology*, *138*(3), 318–332. https://doi.org/10.1002/ajpa.20931

Gosman, J. H., Stout, S. D., & Larsen, C. S. (2011). Skeletal biology over the life span: A view from the surfaces. *American Journal of Physical Anthropology*, *146*(S53), 86–98. https://doi.org/10.1002/ajpa.21612

Guéguen, R., Jouanny, P., Guillemin, F., Kuntz, C., Pourel, J., & Siest, G. (1995). Segregation analysis and variance components analysis of bone mineral density in healthy families. *Journal of Bone and Mineral Research: The Official Journal of the American Society for Bone and Mineral Research*, *10*(12), 2017–2022. https://doi.org/10.1002/jbmr.5650101223

Han, Y., Cowin, S. C., Schaffler, M. B., & Weinbaum, S. (2004). Mechanotransduction and strain amplification in osteocyte cell processes. *Proceedings of the National Academy of Sciences of the United States of America*, *101*(47), 16689–16694. https://doi.org/10.1073/pnas.0407429101

Heaney, R. P., Abrams, S., Dawson-Hughes, B., Looker, A., Looker, A., Marcus, R., Matkovic, V., & Weaver, C. (2001). Peak Bone Mass. *Osteoporosis International*, *11*(12), 985–1009. https://doi.org/10.1007/s001980070020

Hillson, S. (1996). *Dental anthropology*. Cambridge University Press.

Horocholyn, K., & Brickley, M. B. (2017). Pursuit of famine: Investigating famine in bioarchaeological literature. *Bioarchaeology International*, *1*(3/4), Article 3/4.

How climate change affects the food crisis. (2022, October 16). U.S. Embassy & Consulates in Italy. https://it.usembassy.gov/how-climate-change-affects-the-food-crisis/

Hunter, R. L., & Agnew, A. M. (2016). Intraskeletal variation in human cortical osteocyte lacunar density: Implications for bone quality assessment. *Bone Reports*, *5*, 252–261. https://doi.org/10.1016/j.bonr.2016.09.002

Iuliano, S., & Hill, T. R. (2019). Dairy foods and bone health throughout the lifespan: A critical appraisal of the evidence. *British Journal of Nutrition*, *121*(7), 763–772. https://doi.org/10.1017/S0007114518003859

Jablonski, N. G., & Chaplin, G. (2013). Epidermal pigmentation in the human lineage is an adaptation to ultraviolet radiation. *Journal of Human Evolution*, *65*(5), Article 5.

Jacobs, C. R., Temiyasathit, S., & Castillo, A. B. (2010). Osteocyte mechanobiology and pericellular mechanics. *Annual Review of Biomedical Engineering*, *12*(1), 369–400. https://doi.org/10.1146/annurev-bioeng-070909-105302

Jacobs, K. (1993). Human postcranial variation in the Ukrainian Mesolithic-Neolithic. *Current Anthropology*, *34*(3), Article 3.

Jain, M., & Tandon, S. K. (2003). Quaternary alluvial stratigraphic development in a desert margin river, western India. *Current Science*, *84*(8), Article 8.

Jepsen, K. J. (2009). Systems analysis of bone. *Wiley Interdisciplinary Reviews. Systems Biology and Medicine*, *1*(1), 73–88. https://doi.org/10.1002/wsbm.15

Jepsen, K. J., Hu, B., Tommasini, S. M., Courtland, H.-W., Price, C., Terranova, C. J., & Nadeau, J. H. (2007). Genetic randomization reveals functional relationships among morphologic and tissue-quality traits that contribute to bone strength and fragility. *Mammalian Genome: Official Journal of the International Mammalian Genome Society*, *18*(6–7), 492–507. https://doi.org/10.1007/s00335-007-9017-5

Johnston, C. C., Miller, J. Z., Slemenda, C. W., Reister, T. K., Hui, S., Christian, J. C., & Peacock, M. (1992). Calcium supplementation and increases in bone mineral density in children. *New England Journal of Medicine*, *327*(2), 82–87. https://doi.org/10.1056/NEJM199207093270204

Jones, H. H., Priest, J. D., Hayes, W. C., Tichenor, C. C., & Nagel, D. A. (1977). Humeral hypertrophy in response to exercise. *The Journal of Bone and Joint Surgery. American Volume*, *59*(2), 204–208.

Joshi, P., Phartiyal, B., & Joshi, M. (2021). Hydro-climatic variability during last five thousand years and its impact on human colonization and cultural transition in Ladakh sector, India. *Quaternary International*, *599*, 45–54.

Jouanny, P., Guillemin, F., Kuntz, C., Jeandel, C., & Pourel, J. (1995). Environmental and genetic factors affecting bone mass. Similarity of bone density among members of healthy families. *Arthritis and Rheumatism*, *38*(1), 61–67. https://doi.org/10.1002/art.1780380110

Jouanny, P., Jeandel, C., Pourel, J., Guillemin, F., & Kuntz, C. (1995). Environmental and genetic factors affecting bone mass similarity of bone density among members of healthy families: Factors affecting bone mass. *Arthritis & Rheumatism*, *38*(1), 61–67. https://doi.org/10.1002/art.1780380110

Kajale, M. D. (1988). Plant economy, Excavations at Inamgaon, 1 (2). *Deccan College Post Graduate and Research Institute.*

Kennedy, K. A. (1984). Growth, nutrition and pathology in changing paleodemographic settings in South Asia. *Paleopathology at the Origins of Agriculture*, 169–192.

Kepron, C., & Pollanen, M. S. (2015). Rickets or abuse? A histologic comparison of rickets and child abuse-related fractures. *Forensic Science, Medicine, and Pathology*, *11*, 78–87.

Khosla, S., & Riggs, B. L. (2005). Pathophysiology of age-related bone loss and osteoporosis. *Endocrinology and Metabolism Clinics of North America*, *34*(4), 1015–1030. https://doi.org/10.1016/j.ecl.2005.07.009

Köhler, K., Marcsik, A., Zádori, P., Biro, G., Szeniczey, T., Fábián, S., Serlegi, G., Marton, T., Donoghue, H. D., & Hajdu, T. (2017). Possible cases of leprosy from the Late Copper Age (3780-3650 cal BC) in Hungary. *Plos One*, *12*(10), e0185966. https://doi.org/10.1371/journal.pone.0185966

Korisettar, R. (2020). Ancient agriculture in the Indian subcontinent: The archaeobotanical evidence. *A Companion to Ancient Agriculture*, 575–610.

Kritsaneepaiboon, S., Andres, M. M., Tatco, V. R., Lim, C. C. Q., & Concepcion, N. D. P. (2017). Extrapulmonary involvement in pediatric tuberculosis. *Pediatric Radiology*, *47*(10), 1249–1259. https://doi.org/10.1007/s00247-017-3867-0

Kulkarni, A., Sabin, T. P., Chowdary, J. S., Rao, K. K., Priya, P., Gandhi, N., Bhaskar, P., Buri, V. K., Sabade, S. S., & Pai, D. S. (2020). *Precipitation changes in India. Assessment of climate change over the Indian region*. Springer, Singapore.

Lamer, M., Veselka, B., Schrader, S., Hoogland, M., & Brickley, M. B. (2023). Precarious adolescence: Adolescent rickets and anterior sacral angulation in two Dutch skeletal collections from the 18th–19th centuries. *International Journal of Paleopathology*, *40*, 63–69.

Lanham-New, S. A., Buttriss, J. L., Miles, L. M., Ashwell, M., Berry, J. L., Boucher, B. J., Cashman, K. D., Cooper, C., Darling, A. L., Francis, R. M., Fraser, W. D., De Groot, C. P. G. M., Hyppönen, E., Kiely, M., Lamberg-Allardt, C., Macdonald, H. M., Martineau, A. R., Masud, T., Mavroeidi, A., … Williams, C. M. (2011). Proceedings of the rank forum on Vitamin D. *British Journal of Nutrition*, *105*(1), 144–156. https://doi.org/10.1017/S0007114510002576

Lanyon LE, Goodship AE, Pye CJ, MacFie JH. Mechanically adaptive bone remodelling. *J Biomech.* 1982;15(3):141–154. doi: 10.1016/0021-9290(82)90246-9. PMID: 7096367.

Lanyon, L. E. (1984). Functional strain as a determinant for bone remodeling. *Calcified Tissue International, 36*(S1), S56–S61. https://doi.org/10.1007/BF02406134

Lanyon, L. E. (1993). Osteocytes, strain detection, bone modeling and remodeling. *Calcified Tissue International, 53*(S1), S102–S107. https://doi.org/10.1007/BF01673415

Lanyon, L. E. (1996). Using functional loading to influence bone mass and architecture: Objectives, mechanisms, and relationship with estrogen of the mechanically adaptive process in bone. *Bone, 18*, 37S–43S.

Larsen, C. S. (1981). Skeletal and dental adaptations to the shift to agriculture on the Georgia coast. *Current Anthropology, 22*(4), Article 4.

Larsen, C. S. (2023). The past 12,000 years of behavior, adaptation, population, and evolution shaped who we are today. *Proceedings of the National Academy of Sciences, 120*(4), e2209613120. https://doi.org/10.1073/pnas.2209613120

Larsen, C. S., Ruff, C. B., & Griffin, M. C. (1996). Implications of changing biomechanical and nutritional environments for activity and lifeway in the eastern Spanish borderlands. *Bioarchaeology of Native American Adaptation in the Spanish Borderlands.* Gainesville: University Press of Florida. p, 95–125.

Leatherman, T., & Goodman, A. (2020). Building on the biocultural syntheses: 20 years and still expanding. *American Journal of Human Biology, 32*(4), Article 4.

Levine, M. A. (2012). Assessing bone health in children and adolescents. *Indian Journal of Endocrinology and Metabolism, 16*(Suppl 2), S205–212. https://doi.org/10.4103/2230-8210.104040

Lewis, M. (2017). *Paleopathology of Children: Identification of Pathological Conditions in the Human Skeletal Remains of Non-Adults.* Academic Press.

Lewis M.E. (2007). *The bioarchaeology of children: Perspectives from biological and forensic anthropology* (Vol. 50). Cambridge University Press.

Lieberman, D. E., Polk, J. D., & Demes, B. (2004). Predicting long bone loading from cross-sectional geometry. *American Journal of Physical Anthropology, 123*(2), 156–171. https://doi.org/10.1002/ajpa.10316

Lockau, L., Atkinson, S., Mays, S., Prowse, T., George, M., Sperduti, A., Bondioli, L., Wood, C., Ledger, M., & Brickley, M. B. (2019). Vitamin D deficiency and the ancient city: Skeletal evidence across the life course from the Roman period site of Isola Sacra, Italy. *Journal of Anthropological Archaeology, 55*, 101069.

Lukacs, J. R. (1985a). Crown dimensions of permanent teeth from chalcolithic inamgaon: An anthropological study of tooth size variation in prehistoric India. In *Dental anthropology: Application and methods* (pp. 293–309). New Delhi: InterIndia Publications.

Lukacs, J. R. (1985b). Tooth size variation in prehistoric India. *American Anthropologist, 87*(4), Article 4.

Lukacs, J. R., Nelson, G. C., & Walimbe, S. R. (2001). Enamel hypoplasia and childhood stress in prehistory: New data from India and Southwest Asia. *Journal of Archaeological Science, 28*(11), Article 11.

Lukacs, J. R., & Pal, J. N. (2003). Skeletal variation among Mesolithic people of the Ganga Plains: New evidence of habitual activity and adaptation to climate. *Asian Perspectives*, 329–351.

Lukacs, J. R., & Walimbe, S. R. (1998). Physiological stress in prehistoric India: New data on localized hypoplasia of primary canines linked to climate and subsistence change. *Journal of Archaeological Science, 25*(6), Article 6.

Lukacs, J. R., & Walimbe, S. R. (2000). Health, climate, and culture in prehistoric India: Conflicting conclusions from archaeology and anthropology. *South Asian Archaeology, 1*, 363–381.

Lukacs, J. R. & Walimbe, S. R. (2007). Climate, subsistence and health in prehistoric India: The biological impact of a short-term reversal. In *Ancient health: Skeletal indicators of agricultural and economic intensification.* (pp. 237–249). Gainesville: University Press of Florida.

Lukacs, J. R., & Walimbe, S. R. (2005). Biological responses to subsistence transitions in prehistory: Diachronic dental changes at Chalcolithic Inamgaon. *Man and Environment*, *30*(2), Article 2.

Macdonald, H., Kontulainen, S., Petit, M., Janssen, P., & McKay, H. (2006). Bone strength and its determinants in pre- and early pubertal boys and girls. *Bone*, *39*(3), 598–608. https://doi.org/10.1016/j.bone.2006.02.057

Majeed, H., & Moore, G. W. K. (2017). Impact of multidecadal climate variability on United Kingdom rickets rates. *Scientific Reports*, *7*(1), 15764. https://doi.org/10.1038/s41598-017-16058-1

Martini, M., & Ouahes, M. (1988). Bone and joint tuberculosis: A review of 652 cases. *Orthopedics*, *11*(6), 861–866. https://doi.org/10.3928/0147-7447-19880601-04

Mata-Mbemba, D., Rohringer, T., Ibrahim, A., Adams-Webberc, T., Moineddin, R., Doria, A. S., & Vali, R. (2019). HR-pQCT imaging in children, adolescents and young adults: Systematic review and subgroup meta-analysis of normative data. *PloS One*, *14*(12), e0225663. https://doi.org/10.1371/journal.pone.0225663

Mays, S., & Brickley, M. B. (2018). Vitamin D deficiency in bioarchaeology and beyond: The study of rickets and osteomalacia in the past. *International Journal of Paleopathology*, *23*, 1–5.

Mays, S., & Brickley, M. B. (2022). Is dietary deficiency of calcium a factor in rickets? Use of current evidence for our understanding of the disease in the past. *International Journal of Paleopathology*, *36*, 36–44.

Mays, S., Brickley, M., & Ives, R. (2006). Skeletal manifestations of rickets in infants and young children in a historic population from England. *American Journal of Physical Anthropology: The Official Publication of the American Association of Physical Anthropologists*, *129*(3), Article 3.

Mays, S., Brickley, M., & Ives, R. (2007). Skeletal evidence for hyperparathyroidism in a 19th century child with rickets. *International Journal of Osteoarchaeology*, *17*(1), Article 1.

Mays, S., Prowse, T., George, M., & Brickley, M. (2018). Latitude, urbanization, age, and sex as risk factors for vitamin D deficiency disease in the Roman Empire. *American Journal of Physical Anthropology*, *167*(3), Article 3.

McCarthy, E. (2010). *Metabolic bone disorders*. In *Essentials in Bone and Soft-Tissue Pathology* (10th ed.). Springer. Springer.

McFarlin, S. C., Terranova, C. J., Zihlman, A. L., & Bromage, T. G. (2016). Primary bone microanatomy records developmental aspects of life history in catarrhine primates. *Journal of Human Evolution*, *92*, 60–79. https://doi.org/10.1016/j.jhevol.2015.12.004

Mendum, T. A., Taylor, G. M., Donoghue, H. D., Wu, H., Szalontai, C., Marcsik, A., Molnár, E., Pálfi, G., & Stewart, G. R. (2018). The genome sequence of a SNP type 3K strain of *Mycobacterium leprae* isolated from a seventh-century Hungarian case of lepromatous leprosy. *International Journal of Osteoarchaeology*, *28*(4), 439–447. https://doi.org/10.1002/oa.2673

Milovanovic, P., Djonic, D., Hahn, M., Amling, M., Busse, B., & Djuric, M. (2017). Region-dependent patterns of trabecular bone growth in the human proximal femur: A study of 3D bone microarchitecture from early postnatal to late childhood period: MILOVANOVIC et al. *American Journal of Physical Anthropology*, *164*(2), 281–291. https://doi.org/10.1002/ajpa.23268

Mishra, P. K., Chauhan, P. R., Diwate, P., Parth, S., & Anoop, A. (2020). Holocene climate variability and cultural dynamics in the Indian subcontinent. *Episodes Journal of International Geoscience*, *43*(1), Article 1.

Monot, M., Honoré, N., Garnier, T., Zidane, N., Sherafi, D., Paniz-Mondolfi, A., Matsuoka, M., Taylor, G. M., Donoghue, H. D., Bouwman, A., Mays, S., Watson, C., Lockwood, D., Khamesipour, A., Dowlati, Y., Jianping, S., Rea, T. H., Vera-Cabrera, L., Stefani, M. M., … Cole, S. T. (2009). Comparative genomic and phylogeographic analysis of Mycobacterium leprae. *Nature Genetics*, *41*(12), Article 12. https://doi.org/10.1038/ng.477

Moro, M., Van der Meulen, M. C. H., Kiratli, B. J., Marcus, R., Bachrach, L. K., & Carter, D. R. (1996). Body mass is the primary determinant of midfemoral bone acquisition during adolescent growth. *Bone*, *19*(5), Article 5.

Naik, S., & Mishra, S. (1997). The Chalcolithic Phase in the Bhirna Basin, Maharashtra: A Review. *Man and Environment*, *22*, 1.

Nayak, A., Boivin, N., & Roberts, P. (2022). Multidisciplinary Perspectives on the Origins of Past Foodways and Farming Practice in South Asia. *Archaeology of Food and Foodways*, *1*(1), Article 1.

Neukamm, J., Pfrengle, S., Molak, M., Seitz, A., Francken, M., Eppenberger, P., Avanzi, C., Reiter, E., Urban, C., Welte, B., Stockhammer, P. W., Teßmann, B., Herbig, A., Harvati, K., Nieselt, K., Krause, J., & Schuenemann, V. J. (2020). 2000-year-old pathogen genomes reconstructed from metagenomic analysis of Egyptian mummified individuals. *BMC Biology*, *18*(1), 108. https://doi.org/10.1186/s12915-020-00839-8

Nicolella, D. P., Moravits, D. E., Gale, A. M., Bonewald, L. F., & Lankford, J. (2006). Osteocyte lacunae tissue strain in cortical bone. *Journal of Biomechanics*, *39*(9), 1735–1743. https://doi.org/10.1016/j.jbiomech.2005.04.032

Oliveira, T. C., Gomes, M. S., & Gomes, A. C. (2020). The crossroads between infection and bone loss. *Microorganisms*, *8*(11), 1765. https://doi.org/10.3390/microorganisms8111765

Oliver, H., Jameson, K. A., Sayer, A. A., Cooper, C., & Dennison, E. M. (2007). Growth in early life predicts bone strength in late adulthood: The Hertfordshire Cohort Study. *Bone*, *41*(3), 400–405. https://doi.org/10.1016/j.bone.2007.05.007

Ortner, D. J., & Ericksen, M. F. (1997). Bone changes in the human skull probably resulting from scurvy in infancy and childhood. *International Journal of Osteoarchaeology*, *7*(3), Article 3.

Ortner, D. J., & Mays, S. (1998). Dry-bone manifestations of rickets in infancy and early childhood. *International Journal of Osteoarchaeology*, *8*(1), Article 1.

Panja, S. (1999). Mobility and subsistence strategies: A case study of inamgaon, a chalcolithic site in Western India. *Asian Perspectives*, 154–185.

Parfitt, A. M. (1988). Bone remodeling. *Henry Ford Hospital Medical Journal*, *36*(3), 143–144.

Parfitt, A. M., Travers, R., Rauch, F., & Glorieux, F. H. (2000). Structural and cellular changes during bone growth in healthy children. *Bone*, *27*(4), 487–494. https://doi.org/10.1016/S8756-3282(00)00353-7

Pawankar, S. J., & Thomas, P. K. (1997). Fauna and subsistence pattern in the Chalcolithic culture of Western India, with special reference to Inamgaon. *Anthropozoologica*, *25*, *26*, 737–746.

Peacock, T., Bourbou, C., D'Ortenzio, L., Kahlon, B., Prowse, T., & Brickley, M. B. (2019). Mobility and rickets: Investigating Vitamin D deficiency and regional mobility in Aventicum, Roman Switzerland (first to third century CE). *International Journal of Osteoarchaeology*, *29*(4), Article 4.

Pearson, O. M., & Lieberman, D. E. (2004). The aging of Wolff's law?: Ontogeny and responses to mechanical loading in cortical bone. *American Journal of Physical Anthropology*, *125*(S39), 63–99. https://doi.org/10.1002/ajpa.20155

Peters, B. S. E., & Martini, L. A. (2010). Nutritional aspects of the prevention and treatment of osteoporosis. *Arquivos Brasileiros de Endocrinologia & Metabologia*, *54*(2), 179–185. https://doi.org/10.1590/S0004-27302010000200014

Petit, M. A., Beck, T. J., & Kontulainen, S. A. (2005). Examining the developing bone: What do we measure and how do we do it? *Journal of Musculoskeletal & Neuronal Interactions*, *5*(3), 213–224.

Pfrengle, S., Neukamm, J., Guellil, M., Keller, M., Molak, M., Avanzi, C., Kushniarevich, A., Montes, N., Neumann, G. U., Reiter, E., Tukhbatova, R. I., Berezina, N. Y., Buzhilova, A. P., Korobov, D. S., Suppersberger Hamre, S., Matos, V. M. J., Ferreira, M. T., González-Garrido, L., Wasterlain, S. N., … Schuenemann, V. J. (2021). Mycobacterium leprae diversity and population dynamics in medieval Europe from novel ancient genomes. *BMC Biology*, *19*(1), Article 1. https://doi.org/10.1186/s12915-021-01120-2

Phadtare, N. R. (2000). Sharp decrease in summer monsoon strength 4000–3500 cal yr BP in the Central Higher Himalaya of India based on pollen evidence from alpine peat. *Quaternary Research*, *53*(1), Article 1.

Picard, D., Imbach, A., Couturier, M., Lepage, R., & Picard, M. (2001). Familial resemblance of bone mineral density between females 18 years and older and their mothers. *Canadian Journal of Public Health*, *92*(5), 353–358. https://doi.org/10.1007/BF03404978

Pitt, M.J. (2002). Rickets and osteomalacia. In *Diagnosis of bone and joint disorders.* (4th ed., pp. 1901–1945). WB Saunders Co.

Pocock, N. A., Eisman, J. A., Hopper, J. L., Yeates, M. G., Sambrook, P. N., & Eberl, S. (1987). Genetic determinants of bone mass in adults. A twin study. *Journal of Clinical Investigation, 80*(3), 706–710. https://doi.org/10.1172/JCI113125

Prasad, S., & Enzel, Y. (2006). Holocene paleoclimates of India. *Quaternary Research, 66*(3), Article 3.

Proia, P., Amato, A., Drid, P., Korovljev, D., Vasto, S., & Baldassano, S. (2021). The Impact of diet and physical activity on bone health in children and adolescents. *Frontiers in Endocrinology, 12,* 704647. https://doi.org/10.3389/fendo.2021.704647

Raczek, T. P. (2003). Subsistence strategies and burial rituals: Social practices in the late Deccan chalcolithic. *Asian Perspectives, 42*(2), 247–266.

Raichlen, D. A., Gordon, A. D., Foster, A. D., Webber, J. T., Sukhdeo, S. M., Scott, R. S., Gosman, J. H., & Ryan, T. M. (2015). An ontogenetic framework linking locomotion and trabecular bone architecture with applications for reconstructing hominin life history. *Journal of Human Evolution, 81,* 1–12. https://doi.org/10.1016/j.jhevol.2015.01.003

Rath Bonivtch, A., Bonewald, L. F., & Nicolella, D. P. (2007). Tissue strain amplification at the osteocyte lacuna: A microstructural finite element analysis. *Journal of Biomechanics, 40*(10), 2199–2206. https://doi.org/10.1016/j.jbiomech.2006.10.040

Rauch, F., Travers, R., & Glorieux, F. H. (2007). Intracortical remodeling during human bone development—A histomorphometric study. *Bone, 40*(2), 274–280. https://doi.org/10.1016/j.bone.2006.09.012

Resnick, D. L. (2002). *Diagnosis of bone and joint disorders* (4th ed.). WB Saunders Co.

Resnick, D. L., & Kransdorf, M. J. (2004). *Bone and joint imaging.* Elsevier Health Sciences.

Richards, J., Rivadeneira, F., Inouye, M., Pastinen, T., Soranzo, N., Wilson, S., Andrew, T., Falchi, M., Gwilliam, R., Ahmadi, K., Valdes, A., Arp, P., Whittaker, P., Verlaan, D., Jhamai, M., Kumanduri, V., Moorhouse, M., van Meurs, J., Hofman, A., Spector, T. (2008). Bone mineral density, osteoporosis, and osteoporotic fractures: A genome-wide association study. *The Lancet, 371*(9623), 1505–1512. https://doi.org/10.1016/S0140-6736(08)60599-1

Riggs, B. L., & Melton, L. J. (1986). Involutional osteoporosis. *New England Journal of Medicine, 314*(26), 1676–1686. https://doi.org/10.1056/NEJM198606263142605

Rizzoli, R., Bianchi, M. L., Garabédian, M., McKay, H. A., & Moreno, L. A. (2010). Maximizing bone mineral mass gain during growth for the prevention of fractures in the adolescents and the elderly. *Bone, 46*(2), 294–305. https://doi.org/10.1016/j.bone.2009.10.005

Robbins Schug, G. (2011). *Bioarchaeology and climate change: A view from South Asian prehistory,* Gainsville: University Press of Florida.

Robbins Schug, G. (2020). *The Routledge handbook of the bioarchaeology of climate and environmental change.* Routledge.

Robbins Schug, G., & Blevins, K. E. (2016). The center cannot hold: A bioarchaeology of environmental crisis in the second millennium BCE, South Asia. In G. Robbins Schug, & S.R. Walimbe (Eds.), *A Companion to South Asia in the Past* (pp 255–273). Boston: Wiley Blackwell.

Robbins Schug, G., Brandt, E. T., & Lukacs, J. R. (2012). Cementum annulations, age estimation, and demographic dynamics in Mid-Holocene foragers of North India. *HOMO, 63*(2), 94–109. https://doi.org/10.1016/j.jchb.2012.01.002

Robbins Schug, G., Buikstra, J. E., DeWitte, S. N., Baker, B. J., Berger, E., Buzon, M. R., Davies-Barrett, A. M., Goldstein, L., Grauer, A. L., & Gregoricka, L. A. (2023). Climate change, human health, and resilience in the Holocene. *Proceedings of the National Academy of Sciences, 120*(4), Article 4.

Robbins Schug, G., & Goldman, H. M. (2014). Birth is but our death begun: A bioarchaeological assessment of skeletal emaciation in immature human skeletons in the context of environmental, social, and subsistence transition. *American Journal of Physical Anthropology, 155*(2), 243–259. https://doi.org/10.1002/ajpa.22536

Robbins Schug, G., Gupta, S., Cowgill, L. W., Sciulli, P. W., & Blatt, S. H. (2013). Panel regression formulas for estimating stature and body mass from immature human skeletons: A statistical approach without reference to specific age estimates. *Journal of Archaeological Science, 40*(7), Article 7.

Robbins Schug, G., & Halcrow, S. (2022). Building a bioarchaeology of pandemic, epidemic, and syndemic diseases: Lessons for understanding COVID-19. *Bioarchaeology International, 6*(1–2). https://doi.org/10.5744/bi.2022.6000

Robbins Schug, G., Sciulli, P. W., & Blatt, S. H. (2010). Estimating body mass in subadult human skeletons. *American Journal of Physical Anthropology, 143*(1), Article 1.

Robbins Schug, G., Tripathy, V. M., Misra, V. N., Mohanty, R. K., Shinde, V. S., Gray, K. M., & Schug, M. D. (2009). Ancient skeletal evidence for leprosy in India (2000 B.C.). *PLoS One, 4*(5), Article 5. https://doi.org/10.1371/journal.pone.0005669

Roberts, C. A. (2020). Exploring the third "epidemiological transition": Paleopathology's contribution to understanding health and well-being today and for the future. In *The Routledge Handbook of the Bioarchaeology of Climate and Environmental Change* (pp. 19–42). Routledge.

Roberts, C. (2002). Health and disease in past British populations, prehistory to the later Medieval period: some perspectives on the evidence and its problems. In M. Schultz (Ed.), *Homo: unsere Herkunft und Zukunft* (pp. 414–422).

Robling, A. G., Castillo, A. B., & Turner, C. H. (2006). Biomechanical and molecular regulation of bone remodeling. *Annual Review of Biomedical Engineering, 8*, 455–498. https://doi.org/10.1146/annurev.bioeng.8.061505.095721

Rubin, C. T., & Lanyon, L. E. (1984). Regulation of bone formation by applied dynamic loads. *The Journal of Bone and Joint Surgery. American Volume, 66*(3), 397–402.

Ruff, C. B. (1981). *Structural changes in the lower limb bones with aging at Pecos Pueblo*. University of Pennsylvania.

Ruff, C. B. (1990). Postcranial biomechanical adaptation to subsistence strategy changes on the Georgia coast. *Papers of the American Museum of Natural History, 68*, 94–120.

Ruff, C. B. (2003a). Growth in bone strength, body size, and muscle size in a juvenile longitudinal sample. *Bone, 33*(3), Article 3.

Ruff, C. B. (2003b). Ontogenetic adaptation to bipedalism: Age changes in femoral to humeral length and strength proportions in humans, with a comparison to baboons. *Journal of Human Evolution, 45*(4), Article 4.

Ruff, C. B. (2005). Growth tracking of femoral and humeral strength from infancy through late adolescence. *Acta Paediatrica, 94*(8), Article 8.

Ruff, C. B. (2007). Body size prediction from juvenile skeletal remains. *American Journal of Physical Anthropology, 133*(1), Article 1.

Ruff, C. B., Garofalo, E., & Holmes, M. A. (2013). Interpreting skeletal growth in the past from a functional and physiological perspective. *American Journal of Physical Anthropology, 150*(1), 29–37. https://doi.org/10.1002/ajpa.22120

Ruff, C. B., & Hayes, W. C. (1982). Subperiosteal expansion and cortical remodeling of the human femur and tibia with aging. *Science, 217*(4563), Article 4563.

Ruff, C. B., & Hayes, W. C. (1983a). Cross-sectional geometry of Pecos Pueblo femora and tibiae—A biomechanical investigation: I. Method and general patterns of variation. *American Journal of Physical Anthropology, 60*(3), 359–381. https://doi.org/10.1002/ajpa.1330600308

Ruff, C. B., & Hayes, W. C. (1983b). Cross-sectional geometry of Pecos Pueblo femora and tibiae—A biomechanical investigation: II. Sex, age, and side differences. *American Journal of Physical Anthropology, 60*(3), 383–400. https://doi.org/10.1002/ajpa.1330600309

Ruff, C. B., Holt, B., & Trinkaus, E. (2006). Who's afraid of the big bad Wolff?:"Wolff's law" and bone functional adaptation. *American Journal of Physical Anthropology: The Official Publication of the American Association of Physical Anthropologists, 129*(4), Article 4.

Ruff, C. B., Larsen, C. S., & Hayes, W. C. (1984). Structural changes in the femur with the transition to agriculture on the Georgia coast. *American Journal of Physical Anthropology, 64*(2), 125–136. https://doi.org/10.1002/ajpa.1330640205

Ruff, C. B., Walker, A., & Trinkaus, E. (1994). Postcranial robusticity in *Homo*. III: Ontogeny. *American Journal of Physical Anthropology, 93*(1), 35–54. https://doi.org/10.1002/ajpa.1330930103

Ryan, T. M., & Krovitz, G. E. (2006). Trabecular bone ontogeny in the human proximal femur. *Journal of Human Evolution, 51*(6), 591–602. https://doi.org/10.1016/j.jhevol.2006.06.004

Saers, J. P. P., Ryan, T. M., & Stock, J. T. (2020). Baby steps towards linking calcaneal trabecular bone ontogeny and the development of bipedal human gait. *Journal of Anatomy, 236*(3), 474–492. https://doi.org/10.1111/joa.13120

Santos, A. L., & Roberts, C. A. (2001). A picture of tuberculosis in young Portuguese people in the early 20th century: A multidisciplinary study of the skeletal and historical evidence. *American Journal of Physical Anthropology, 115*(1), Article 1. https://doi.org/10.1002/ajpa.1054

Sarkar, A., Mukherjee, A. D., Bera, M. K., Das, B., Juyal, N., Morthekai, P., Deshpande, R. D., Shinde, V. S., & Rao, L. S. (2016). Oxygen isotope in archaeological bioapatites from India: Implications to climate change and decline of Bronze Age Harappan civilization. *Scientific Reports, 6*(1), Article 1.

Schamall, D., Kneissel, M., Wiltschke-Schrotta, K., & Teschler-Nicola, M. (2003). Bone structure and mineralization in a late antique skeleton with osteomalacia. *American Journal of Physical Anthropology, 120*(S36)184.

Schamall, D., & Teschler-Nicola, M. (2010). Osteomalacia microarchitecture exemplified by a late antiquity skeleton and Verified by bse-mode in a sem. *Anthropologie (1962-), 48*(2), Article 2.

Schuenemann, V. J., Singh, P., Mendum, T. A., Krause-Kyora, B., Jäger, G., Bos, K. I., Herbig, A., Economou, C., Benjak, A., Busso, P., Nebel, A., Boldsen, J. L., Kjellström, A., Wu, H., Stewart, G. R., Taylor, G. M., Bauer, P., Lee, O. Y.-C., Wu, H. H. T., Krause, J. (2013). Genome-Wide Comparison of Medieval and Modern *Mycobacterium leprae*. *Science, 341*(6142), 179–183. https://doi.org/10.1126/science.1238286

Schultz, M. (1997). Microscopic investigation of excavated skeletal remains: A contribution to paleopathology and forensic medicine. In *Forensic taphonomy: The postmortem fate of human remains* (pp. 201–222). CRC Press.

Schultz, M. (2001). Paleohistopathology of bone: A new approach to the study of ancient diseases. *American Journal of Physical Anthropology, 116*(S33), 106–147. https://doi.org/10.1002/ajpa.10024

Schultz, M. (2002). Diagnosis of leprosy in skeletons from an English later Medieval hospital using histological analysis. In *The past and present of leprosy: Archaeological, historical, palaeopathological and clinical approaches* (pp. 89–104). Archaeopress.

Schultz, M. (2003). Light microscopic analysis in skeletal paleopathology. In *Identification of pathological conditions in human skeletal remains* (pp. 73–111). Academic Press.

Schultz, M., & Schmidt-Schultz, T. H. (2015). Is it possible to diagnose TB in ancient bone using microscopy? *Tuberculosis, 95*, S80–S86. https://doi.org/10.1016/j.tube.2015.02.035

Schultz, M. & Teschler-Nicola, M. (1985). *Krankhafte Veränderungen an den Skeletten aus dem Karner der St. Martins-Kirche in Klosterneuburg, Niederösterreich Teil I-IV*. 225–311.

Seeman, E. (2003). Periosteal Bone Formation—A Neglected Determinant of Bone Strength. *New England Journal of Medicine, 349*(4), 320–323. https://doi.org/10.1056/NEJMp038101

Seeman, E., Hopper, J. L., Bach, L. A., Cooper, M. E., Parkinson, E., McKay, J., & Jerums, G. (1989). Reduced bone mass in daughters of women with osteoporosis. *New England Journal of Medicine, 320*(9), 554–558. https://doi.org/10.1056/NEJM198903023200903

Sharma, S., Joachimski, M., Sharma, M., Tobschall, H. J., Singh, I. B., Sharma, C., Chauhan, M. S., & Morgenroth, G. (2004). Lateglacial and Holocene environmental changes in Ganga plain, Northern India. *Quaternary Science Reviews, 23*(1–2), Article 1–2.

Shaw, C. N., & Stock, J. T. (2009a). Habitual throwing and swimming correspond with upper limb diaphyseal strength and shape in modern human athletes. *American Journal of Physical Anthropology, 140*(1), 160–172. https://doi.org/10.1002/ajpa.21063

Shaw, C. N., & Stock, J. T. (2009b). Intensity, repetitiveness, and directionality of habitual adolescent mobility patterns influence the tibial diaphysis morphology of athletes. *American Journal of Physical Anthropology, 140*(1), 149–159. https://doi.org/10.1002/ajpa.21064

Shinde, V. (1989). New light on the origin, settlement system and decline of the Jorwe culture of the Deccan, India. *South Asian Studies, 5*(1), Article 1.

Shirvalkar, P., & Prasad, E. (2016). The archaeology of the Late Holocene on the Deccan Plateau (The Deccan Chalcolithic). *A Companion to South Asia in the Past*, 240–254.

Silva, B.C., A.G. Costa, N.E. Cusano, S. Kousteni, & J.P. Bilezikian. (2011). Catabolic and anabolic actions of parathyroid hormone on the skeleton. *Journal of Endocrinological Investigation, 34*(10). https://doi.org/10.3275/7925

Sinha, R., Smykatz-Kloss, W., Stüben, D., Harrison, S. P., Berner, Z., & Kramar, U. (2006). Late Quaternary palaeoclimatic reconstruction from the lacustrine sediments of the Sambhar playa core, Thar Desert margin, India. *Palaeogeography, Palaeoclimatology, Palaeoecology, 233*(3–4), Article 3–4.

Slavin, P. (2016). Climate and famines: A historical reassessment. *Wiley Interdisciplinary Reviews: Climate Change, 7*(3), Article 3.

Slemenda CW, Miller JZ, Hui SL, Reister TK, Johnston CC Jr. Rolc of physical activity in the development of skeletal mass in children. *J Bone Miner Res.* 1991 Nov; 6(11):1227–1233. doi: 10.1002/jbmr.5650061113. PMID: 1805545.

Smith, R. W., & Walker, R. R. (1964). femoral expansion in aging women: Implications for osteoporosis and fractures. *Science, 145*(3628), 156–157. https://doi.org/10.1126/science.145.3628.156

Sokal, R. R., Oden, N. L., & Wilson, C. (1991). Genetic evidence for the spread of agriculture in Europe by demic diffusion. *Nature, 351*(6322), Article 6322.

Sparacello, V. S., Roberts, C. A., Canci, A., Moggi-Cecchi, J., & Marchi, D. (2016). Insights on the paleoepidemiology of ancient tuberculosis from the structural analysis of postcranial remains from the Ligurian Neolithic (northwestern Italy). *International Journal of Paleopathology, 15*, 50–64.

Srivastava, J., Manjunatha, B. R., Balakrishna, K., Prajith, A., Manjunatha, H. V., Jose, J., & Kumar, N. (2021). Quantitative pollen-based reconstruction of the vegetation diversity in response to the late-Holocene climate change near Karwar, south-west coast of India. *Quaternary International, 599*, 95–106.

Stagi, S., Cavalli, L., Iurato, C., Seminara, S., Brandi, M. L., & de Martino, M. (2013). Bone health in children and adolescents: The available imaging techniques. *Clinical Cases in Mineral and Bone Metabolism: The Official Journal of the Italian Society of Osteoporosis, Mineral Metabolism, and Skeletal Diseases, 10*(3), 166–171.

Stock, J., & Pfeiffer, S. (2001). Linking structural variability in long bone diaphyses to habitual behaviors: Foragers from the southern African Later Stone Age and the Andaman Islands. *American Journal of Physical Anthropology, 115*(4), 337–348. https://doi.org/10.1002/ajpa.1090

Stock, J. T., & Pfeiffer, S. K. (2004). Long bone robusticity and subsistence behaviour among Later Stone Age foragers of the forest and fynbos biomes of South Africa. *Journal of Archaeological Science, 31*(7), Article 7.

Stock, J. T., Pomeroy, E., Ruff, C. B., Brown, M., Gasperetti, M. A., Li, F.-J., Maher, L., Malone, C., Mushrif-Tripathy, V., Parkinson, E., Rivera, M., Siew, Y. Y., Stefanovic, S., Stoddart, S., Zariņa, G., & Wells, J. C. K. (2023). Long-term trends in human body size track regional variation in subsistence transitions and growth acceleration linked to dairying. *Proceedings of the National Academy of Sciences, 120*(4), e2209482119. https://doi.org/10.1073/pnas.2209482119

Sumner, D. R., & Andriacchi, T. P. (1996). Adaptation to differential loading: Comparison of growth-related changes in cross-sectional properties of the human femur and humerus. *Bone, 19*(2), 121–126. https://doi.org/10.1016/8756-3282(96)00166-4

Swan, K. R., Ives, R., Wilson, L. A. B., & Humphrey, L. T. (2020). Ontogenetic changes in femoral cross-sectional geometry during childhood locomotor development. *American Journal of Physical Anthropology, 173*(1), 80–95. https://doi.org/10.1002/ajpa.24080

Teo, H. E. L., & Peh, W. C. G. (2004). Skeletal tuberculosis in children. *Pediatric Radiology*, *34*(11), 853–860. https://doi.org/10.1007/s00247-004-1223-7

Tian, L., & Yu, X. (2017). Fat, sugar, and bone health: A complex relationship. *Nutrients*, *9*(5), 506. https://doi.org/10.3390/nu9050506

Tommasini, S. M., Nasser, P., Hu, B., & Jepsen, K. J. (2008). Biological co-adaptation of morphological and composition traits contributes to mechanical functionality and skeletal fragility. *Journal of Bone and Mineral Research: The Official Journal of the American Society for Bone and Mineral Research*, *23*(2), 236–246. https://doi.org/10.1359/jbmr.071014

Tse, K. Y., Macias, B. R., Meyer, R. S., & Hargens, A. R. (2009). Heritability of bone density: Regional and gender differences in monozygotic twins. *Journal of Orthopaedic Research*, *27*(2), 150–154. https://doi.org/10.1002/jor.20721

Van Der Meulen, M. C. H., Beaupré, G. S., & Carter, D. R. (1993). Mechanobiologic influences in long bone cross-sectional growth. *Bone*, *14*(4), 635–642. https://doi.org/10.1016/8756-3282(93)90085-O

Vekszler, G., Granner, M., Nebot Valenzuela, E., Winter, E., Dockner, M., Weber, G. W., … & Pietschmann, P. (2022). Microarchitecture of historic bone samples with tuberculosis. *Wiener Klinische Wochenschrift*, *134*(11), 449–457.

Veselka, B., Brickley, M. B., & Waters-Rist, A. L. (2021). A joint medico-historical and paleopathological perspective on vitamin D deficiency prevalence in post-Medieval Netherlands. *International Journal of Paleopathology*, *32*, 41–49. https://doi.org/10.1016/j.ijpp.2020.10.010

Vilayphiou, N., Boutroy, S., Sornay-Rendu, E., Van Rietbergen, B., & Chapurlat, R. (2016). Age-related changes in bone strength from HR-pQCT derived microarchitectural parameters with an emphasis on the role of cortical porosity. *Bone*, *83*, 233–240. https://doi.org/10.1016/j.bone.2015.10.012

Walimbe, S. R. (2016). Human skeletal studies: Changing trends in theoretical and methodological perspectives. *A Companion to South Asia in the Past*, 482–495.

Walimbe, S. R., & Gambhir, P. B. (1994). *Long bone growth in infants and children: Assessment of the nutritional status*. Mujumdar Publications.

Walimbe, S. R. & Kulkarni, S. S. (1993). *Biological Adaptations in Human Dentition: An Odontometric Study on Living and Archaeological Populations in India*. Deccan College Research Institute.

Wapler, U., Crubezy, E., & Schultz, M. (2004). Is cribra orbitalia synonymous with anemia? Analysis and interpretation of cranial pathology in Sudan. *American Journal of Physical Anthropology: The Official Publication of the American Association of Physical Anthropologists*, *123*(4), Article 4.

Wapler, U., Crubézy, E., & Schultz, M. (2004). Is cribra orbitalia synonymous with anemia? Analysis and interpretation of cranial pathology in Sudan. *American Journal of Physical Anthropology*, *123*(4), 333–339. https://doi.org/10.1002/ajpa.10321

Ward, K. A., Roberts, S. A., Adams, J. E., Lanham-New, S., & Mughal, M. Z. (2007). Calcium supplementation and weight bearing physical activity—Do they have a combined effect on the bone density of pre-pubertal children? *Bone*, *41*(4), 496–504. https://doi.org/10.1016/j.bone.2007.06.007

Warinner, C., Speller, C., & Collins, M. J. (2015). A new era in palaeomicrobiology: Prospects for ancient dental calculus as a long-term record of the human oral microbiome. *Philosophical Transactions of the Royal Society B: Biological Sciences*, *370*(1660), Article 1660. https://doi.org/10.1098/rstb.2013.0376

Watkins, B. A., Li, Y., Lippman, H. E., & Feng, S. (2003). Modulatory effect of omega-3 polyunsaturated fatty acids on osteoblast function and bone metabolism. *Prostaglandins, Leukotrienes and Essential Fatty Acids*, *68*(6), 387–398. https://doi.org/10.1016/S0952-3278(03)00063-2

Weiss, E. (2005). Humeral cross-sectional morphology from 18th century Quebec prisoners of war: Limits to activity reconstruction. *American Journal of Physical Anthropology*, *126*(3), 311–317. https://doi.org/10.1002/ajpa.20069

Welsh, H., & Brickley, M. B. (2023). Pathology or expected morphology? Investigating patterns of cortical porosity and trabecularization during infancy and early childhood. *The Anatomical Record, 306*(2), 354–365. https://doi.org/10.1002/ar.25081

Welsh, H., Nelson, A. J., van der Merwe, A. E., de Boer, H. H., & Brickley, M. B. (2020). An investigation of micro-CT analysis of bone as a new diagnostic method for paleopathological cases of osteomalacia. *International Journal of Paleopathology, 31*, 23–33.

Wescott, D. J. (2006). Effect of mobility on femur midshaft external shape and robusticity. *American Journal of Physical Anthropology: The Official Publication of the American Association of Physical Anthropologists, 130*(2), Article 2.

Wescott, D. J., & Cunningham, D. L. (2006). Temporal changes in Arikara humeral and femoral cross-sectional geometry associated with horticultural intensification. *Journal of Archaeological Science, 33*(7), Article 7.

Weston, D. A. (2009). Brief communication: Paleohistopathological analysis of pathology museum specimens: Can periosteal reaction microstructure explain lesion etiology? *American Journal of Physical Anthropology, 140*(1), 186–193. https://doi.org/10.1002/ajpa.21081

Wolff, J. (1892). *Das gresetz der transformation der knochen*. Hirschwald.

Yapuncich, G. S., Churchill, S. E., Cameron, N., & Walker, C. S. (2018). Morphometric panel regression equations for predicting body mass in immature humans. *American Journal of Physical Anthropology, 166*(1), Article 1.

Yim, A., Konigsberg, L. W., Hwa, H., Chang, C., Chen, J., & Liu, H. (2021). Allometric scaling and growth: Evaluation and applications in subadult body mass estimation. *American Journal of Physical Anthropology, 175*(3), 577–588. https://doi.org/10.1002/ajpa.24329

Zeder, M. A. (2018). Why evolutionary biology needs anthropology: Evaluating core assumptions of the extended evolutionary synthesis. *Evolutionary Anthropology: Issues, News, and Reviews, 27*(6), Article 6.

Zemel, B. S. (2011). Quantitative computed tomography and computed tomography in children. *Current Osteoporosis Reports, 9*(4), 284–290. https://doi.org/10.1007/s11914-011-0076-x

Zhou, H., Cooper, M. S., & Seibel, M. J. (2013). Endogenous glucocorticoids and bone. *Bone Research, 1*(2), 107–119. https://doi.org/10.4248/BR201302001

Zioupos, P. (2001). Accumulation of in-vivo fatigue microdamage and its relation to biomechanical properties in ageing human cortical bone. *Journal of Microscopy, 201*(2), Article 2.

Zuckerman, M. K., & Martin, D. L. (2016). *New directions in biocultural anthropology*. John Wiley & Sons.

Skeletal Fracture
Biomechanics and Forensic Perspectives

4

ANGELA L. HARDEN,
RANDEE L. HUNTER, AND
AMANDA M. AGNEW

4.1 Introduction

Understanding fracture mechanisms and the multi-scale biological contributions to skeletal response to loading is essential for studies in skeletal biology. The implications of this knowledge can be applied across numerous disciplines, including, but not limited to, forensic anthropology, bioarchaeology, paleopathology, paleoanthropology, injury biomechanics, and clinical orthopedics. Insight into how and why bone fractures occur can reveal valuable information about health and disease, trauma, and activity in both modern and past human populations. Additionally, by considering fracture etiology and behavior, researchers may be able to contextualize fractures (e.g., reconstructing the series of events in a forensic case) as well as apply this knowledge to better prevent future fractures (e.g., osteoporosis research). The goal of this chapter is to introduce foundational concepts of functional adaptation, bone biomechanics, and skeletal trauma analysis with an emphasis on employing a multi-scale approach to support a more holistic understanding of bone fracture.

4.2 Bone Functional Adaptation

The skeletons of large mammals epitomize an evolutionary compromise with a structural framework that is strong but with limited mass to optimize mechanical properties while facilitating movement (Currey et al., 2007). Modeling processes drive the development and alteration of gross architecture (e.g., size and shape) of bone in response to loads throughout life (Frost, 1982). However, remodeling allows the skeleton to maintain a minimal size while extending its fatigue life (Martin, 2003). The hierarchical nature of bone allows for these processes to be coordinated across scales (i.e., macrostructure to nanostructure) and achieve a balance of biomechanical efficiency and resistance to damage due to strain, a process known as bone functional adaptation. Evidence suggests that bone as a complex adaptive system can maintain function under physiological loads across a large range of morphological variation observed in modern humans (Jepsen et al., 2011; Schlecht et al., 2014).

Bones are adapted to withstand physiological loads (Parfitt, 2002) through bone functional adaptation processes occurring throughout the lifespan; however, when loads exceed

 DOI: 10.4324/9781003385608-4

this capacity, fracture can occur. Currey (2003) argues this does not necessarily mean that the purpose of bone is simply to resist fracture. In fact, bone's response to mechanical and systemic influences across the skeleton and throughout life history result in intra-skeletal variability in fracture risk, and some individuals are more vulnerable than others despite having similar assessment values of fracture risk (e.g., bone mineral density or bone mass from medical imaging modalities). Although increasing age has often been blamed for fractures (Zioupos & Currey, 1998), due to its effects on multi-scale traits resulting in metabolic dysregulation and inefficiencies in adaptive or compensatory processes, fragility, or high fracture risk, is not solely determined by age (Heaney, 2003). In fact, age alone has been shown to poorly predict mechanical properties in certain skeletal elements such as ribs (Agnew et al., 2018; Schafman et al., 2016). Age-related bone loss can lead to osteopenia (i.e., low bone mass and moderate fracture risk) or osteoporosis (i.e., severely low bone mass and high fracture risk); however, these conditions are defined by the World Health Organization (WHO) using bone mineral density values (WHO, 1994) which fail to capture geometric or microstructural traits. The complex relationships between scales that underpin functional adaptation and are not currently represented by measures of bone mass or quantity may explain scenarios when individuals with previous fractures (even those with higher bone mass) are more likely to sustain subsequent fractures than someone with a lower bone mass who does not have a history of fracture (Kanis et al., 2004; Sornay-Rendu et al., 2005). This suggests there are additional factors to consider as contributors to a propensity for fracture and alternative perspectives on risk that appreciate adaptive properties of bone as a system.

Recent work investigating multi-scale functional adaptation with a holistic approach to the coordination of whole-bone geometry and both cross-sectional and microstructural traits has revealed alternative explanations for sources of increasing fracture risk with age across skeletal phenotypes. Often phenotypes are classified into categories of robustness, a measure accounting for both longitudinal and transverse growth, within a skeletal element using the ratio of cross-sectional total area and total length (Figure 4.1). Although the mechanistic link between scales has yet to be elucidated, age-related changes in microstructural traits that are predicted by external size groups (i.e., slender/narrow or robust/wide) have been identified in the femoral neck (Jepsen et al., 2017), femoral diaphysis (Bolger et al., 2020), and midshaft radius (Bigelow et al., 2019; Hunter et al., 2020). For example, when examining a pooled dataset of wide and narrow midshaft femora, Bolger et al. (2020) found increased cortical porosity, total area, and marrow area with age. However, when analyzing

Figure 4.1 Example of differences in slender and robust human rib cross-sections. Histological images processed using ImageJ MomentMacro to identify cortical bone. Images to scale.

size-based subgroups, varying trends were observed for wide femora (increased cortical porosity only) compared to narrow femora (increased total area and marrow area but not cortical porosity) with age. These findings suggest differences in mechanisms of bone loss and adaptation to loss between slender and robust phenotypes that produce tangible effects on age-related strength-decline trajectories. However, it is important to note that not all bones appear to demonstrate the ability to employ functional compensatory mechanisms in the same way. Bigelow et al. (2019) found cortical porosity increased with age at similar rates between phenotypes in the radius diaphysis; however, preliminary data in human mid-level ribs suggest robust ribs demonstrate significantly larger increases in pore size and overall porosity with age than slender ribs. Concomitantly, adult robust ribs appear to show no decrease in cortical area with age which may be evidence of a compensatory mechanism to offset the larger increases in porosity compared to slender ribs. The resulting effects of these adaptive and compensatory relationships in ribs on mechanical properties and their similarities and differences to appendicular bones are currently being investigated. Bone functional adaptation perspectives provide crucial insight to the biological underpinnings of bone's biomechanical response to loading.

4.3 Bone Biomechanics

Bone biomechanical properties are influenced by multi-scale traits affecting both material and structural properties, which are defined and discussed in the next section.

4.3.1 Foundational Concepts

4.3.1.1 Structural and Material Property Definitions

Researchers perform numerous mechanical tests (representing loading in tension, compression, shear, and torsion) in an attempt to document and compare properties of bone. In most materials, when a force is applied that exceeds the strength of that material, failure results. Since bone is constantly altering its composition in response to physiological, environmental, and mechanical loading stimuli making it extremely variable both within and across individuals, it is necessary to assess the complexity of properties to fully characterize bone's resistance to fracture (Felsenberg & Boonen, 2005). In order to do so, mechanical properties, which can be divided into structural properties and material properties, are often quantified from experimental testing to represent overall bone strength.

Structural mechanical properties of bone are influenced by bone size and can depend upon both intrinsic material properties and bone geometry. Such properties include, but are not limited to, force, displacement/deflection, structural stiffness, and energy/work. These properties are demonstrated on a force versus displacement curve from a typical tensile test (Figure 4.2). Ultimate force is how much load bone material can support prior to failure. Although this load is important, it is also critical that bone be stiff enough to withstand various loading conditions (Currey, 2002); however, stiffness is not necessarily a direct measurement of propensity for fracture or fragility (Turner, 2002; Turner & Burr, 1993). Stiffness, k, is related to a bone's ability to resist plastic deformation and is calculated as the slope of the elastic portion of the force-displacement curve prior to reaching the yield point. The yield point, YP, is the point at which the bone exits its elastic phase and begins to deform plastically (i.e., permanently). Compared with many other materials bone is often

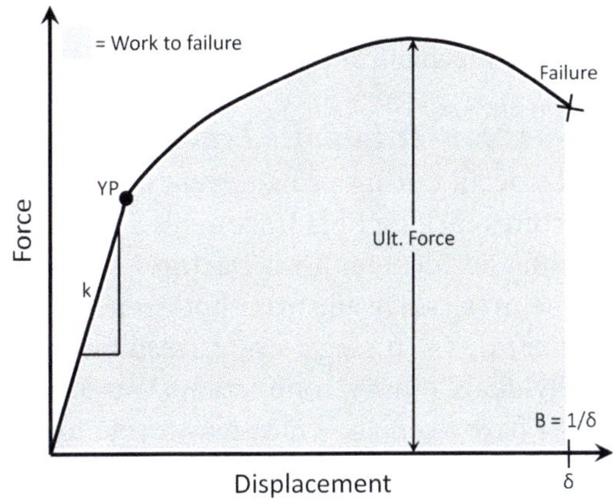

Figure 4.2 Force versus displacement curve for a typical tensile test.

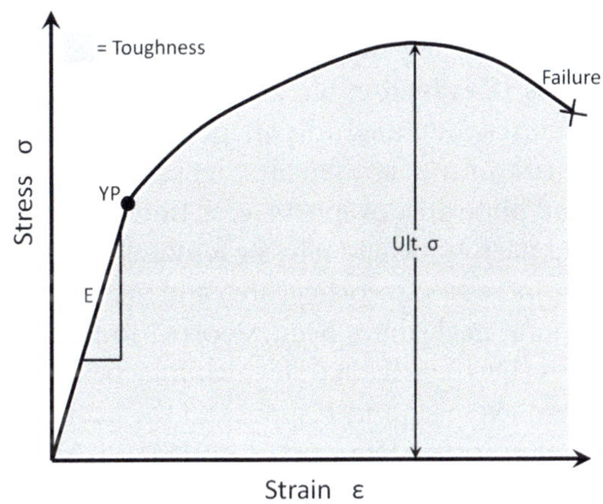

Figure 4.3 Stress versus strain curve for a typical tensile test.

considered brittle because it can fracture before plastic deformation occurs and may absorb little energy. As is typical for brittle materials, bone is strong in compression but weaker in tension (Currey, 2002). Finally, energy (work) to fracture is calculated as the area under the force-displacement curve until failure (shown in Figure 4.2).

Intrinsic properties of bone are the core material properties that characterize the composition of the organic and mineral matrix independent of size or distribution of bone. Material properties are based on a stress-strain curve (Figure 4.3) calculated from mechanically testing identically sized and shaped samples of material. Common material properties include Young's modulus, yield stress and strain, ultimate stress and strain, and toughness. In a tensile test, Young's modulus, E, is a measure of the elasticity of the bone material calculated from the slope of a stress-strain curve in the pre-yield phase. Yield stress is the amount of stress at which bone begins to deform plastically (the elastic limit), shown as YP in Figure 4.2. Ultimate strength is the point of maximum stress and is achieved as the material undergoes gross deformation before ultimate failure. Finally, toughness is a measure of bone's ability to absorb energy to resist fracture, determined by calculating the area under the stress-strain curve up to failure, shown in Figure 4.3. Both structural and material properties characterize bone's response in a given loading scenario. The following discussion will reference many of the mechanical properties outlined here.

4.3.1.2 Traumatic versus Non-Traumatic Fractures

Traumatic fractures typically occur during a single event when the bone is loaded to failure. Examples of traumatic fractures include tibia fracture in a pedestrian hit by a motor vehicle, radius fracture in a skiing accident, and rib fractures due to cardiopulmonary resuscitation. Non-traumatic fractures can result from both repetitive loading causing fatigue damage (e.g., stress fractures) or from single events resulting in a fragility fracture. For example, in osteopenic individuals' muscle contraction (Warden et al., 2002) and coughing (De Maeseneer et al., 2000) have been shown to play a role in rib fracture development. While not all fragility fractures occur in osteoporotic individuals, including a large proportion that occur in those deemed low risk for fracture (Bolotin, 2007; Mussawy et al., 2017; Samelson et al., 2018; Sornay-Rendu et al., 2007), a common site of fracture is the femoral neck. A thinning cortex and loss of trabecular bone and connectivity are primarily responsible for "hip" fracture vulnerability (Blain et al., 2008). Additionally, geometric changes of the femoral neck may be partly responsible for the reported increased fracture risk in modern populations (Sievänen et al., 2007), and may contribute to observations of fewer osteoporotic fractures identified in past populations. To further complicate the interpretation of fracture risk in modern females, Jepsen et al. (2017) found that changes in bone mineralization and bone area over a 14-year time period varied between external size phenotypes (wide and narrow) where narrow femoral necks showed smaller losses in mineral content but larger increases in cortical area and vice versa for wide femoral necks. Geometrically "wide" femoral necks have been reported to have a greater risk of fracture in postmenopausal females (Duan et al., 2003) which may be attributed to lack of adaptive processes such as periosteal expansion in response to the accumulation of large amounts of particularly endocortical porosity (Jepsen et al., 2017). It is important to note that these structural changes with age may result in individuals at high risk for non-traumatic fractures due to dysregulation or inefficiencies related to functional adaptation.

Bone stress injuries (BSI), or non-traumatic fatigue damage, occur due to excessive repetitive loading without sufficient recovery time as often experienced in military or elite athlete training. Recent work examining military recruits has identified significantly lower cortical area of the tibia at baseline (i.e., prior to initiating basic combat training) in males who sustained a BSI during basic training compared to those who did not fracture (Eastman et al., 2023). There were no differences in trabecular microstructure or cortical porosity, cortical thickness, or size of cortical pores between these groups suggesting the larger cortical area may provide a protective mechanism to axial loading of the tibial cortex. Similarly, other high-resolution imaging of the tibia in both males and females has identified geometrical vulnerabilities (e.g., lower cortical area and total area) that often characterize female tibiae prone to developing BSI during strenuous, prolonged periods of repetitive loading (Evans et al., 2008; Koltun et al., 2022). Additionally, this population provides an interesting longitudinal study in bone functional adaptation beyond quantifying traits that contribute to fracture incidence. Evidence for adaptive formation of bone during intensive physical activity performed as a part of basic combat training has been documented for both male and female recruits (Hughes et al., 2023). Females demonstrated larger percent increases in trabecular traits of the distal tibia after 8–14 weeks of training, while males experienced a larger percent increase in cortical traits (Hughes et al., 2023; O'Leary et al., 2022).

The magnitude of increase in female tibia geometrical and microstructural traits was larger over an extended period of training (44 weeks) resulting in a 2.5% increase in estimated failure load (O'Leary et al., 2021). Sex-specific differences in bone response are important to consider when investigating functional adaptation as well as its influence on fracture vulnerability.

4.3.2 Biological Contributors to Mechanical Properties

Bone's hierarchical structure influences its mechanical response (Currey, 2002). When defining material properties, the assumption of homogeneity is commonly applied; however, most biological tissues demonstrate heterogeneous material organization. The material properties of bone are determined by the quantity, quality, and spatial arrangement of its organic (collagen) and mineral components, which fluctuate throughout the lifespan in response to a multitude of local (e.g., mechanical loading) and systemic (e.g., disease states, aging processes) factors. The relative contributions to loading response of each component of the collagen-mineral composite matrix are difficult to assess independently (Burr, 2002). Generally, the mineral content of bone assists in defining strength and stiffness (Currey et al., 1996) but does not handle tensile loads well, while the organic content, organized in lamellae, contributes to its tensile strength (Augat & Schorlemmer, 2006) and toughness (Zioupos, 2001). The collagen-mineral composite matrix functions do allow bone to withstand varying loading types and directions associated with physiologic loading.

It is widely accepted that variations in mineralization affect the mechanical properties of bone (Martin et al., 1998). Increasing mineralization with age has been associated with an increase in Young's modulus (Currey et al., 1996), stiffness, and strength (Follet et al., 2004); however, excessive mineralization will cause bone to become brittle (Zioupos et al., 2000), resulting in less energy before failure (Augat & Schorlemmer, 2006). Remodeling activity regulates the arrangement and progression of mineralization, which in turn affects the directionally dependent mechanical properties of bone (Wagner & Weiner, 1992). Interstitial bone, due to its higher biological age, has experienced a larger number of loading cycles with higher levels of mineralization which can be an ideal location for microcracks to occur (Seref-Ferlengez et al., 2015). Newly remodeled bone is less mineralized than surrounding older tissue, but a decrease in bone turnover efficiency, as can occur with age, can result in relatively hypermineralized and brittle bone vulnerable to damage accumulation. (For a more detailed discussion of the bone remodeling process, see Chapter 1 by Cole, Crowder, and Stout in this volume.) Accumulating microdamage has been categorized as a component of bone quality and described as a "two-edged sword" (Seref-Ferlengez et al., 2015), given that although damage occurs, this process allows for energy dissipation and prevents larger failures by stimulating remodeling events. This phenomenon has been described as a critical adaptive process to repair damaged bone with reduced strength and stiffness to mitigate further failure (Burr, 2011). Thus, the complexity of degrees of mineralization and the efficiency of bone's natural repair mechanisms (i.e., remodeling) influence the micro-scale response to loading.

Individual collagen fibers composing lamellar bone are said to be anisotropic (respond differently according to the direction of applied force); however, the complex arrangement and orientation of collagen fibers in lamellar bone as a composite material further influence directional dependency. Therefore, collagen orientation plays an important role in how lamellar bone responds to applied loads (Ziv et al., 1996). For example, longitudinally oriented collagen fibers are shown to relate to increased bone strength, but not necessarily fatigue life (Martin et al., 1996). Collagen degradation can also adversely influence mechanical properties (Wynnyckyj et al., 2009) and increase bone fragility (Paschalis et al., 2004). Specifically, a decrease in bone toughness has been observed as collagen is degraded (Currey, 2003; Wang et al., 2002). It has even been suggested that a major role of remodeling is not only to prevent hypermineralization or microcrack accumulation (both of which can have negative mechanical effects), but also to maintain the integrity of the collagen network and thereby increase work to fracture (Burr, 2008). As bone turnover decreases with age, the quality of collagen in the older tissue declines, mineralization becomes more homogenous, and water content declines, resulting in brittleness (Burr, 2014). See Viguet-Carrin et al. (2006) and Garnero (2015) for a more thorough review of collagen's influence on bone mechanical properties.

At a slightly larger scale, bone strength is partially dependent on the types of bone tissue present throughout the structure (Frost, 2001). Woven, fibro-lamellar, primary and secondary lamellar, and plexiform are examples of cortical bone types. One study was able to successfully isolate bone types and found primary lamellar bone to be harder than secondary lamellar bone (Liu et al., 2000). The unique presence of primary lamellar bone in the growing skeleton makes it a rational target to assess the role which different bone types play in determining cortical bone strength. Mechanical properties vary between primary and secondary lamellar bone (Currey, 2002), and past studies on animal and adult human bones have suggested that secondary osteon accumulations are associated with reduced bone strength (Liu et al., 2000; Vincentelli & Grigorov, 1985), despite their positive role in extending the fatigue life of bone.

Studies by the authors provide evidence that human pediatric bone may reveal a more complex relationship. Pediatric rib sections were subjected to quasi-static (2.5 mm/min) 3-point bend tests and subsequent histomorphometric analysis near the location of failure (Agnew et al., 2013). Results suggest that strength parameters are more closely related to bone geometry/size than bone composition. However, a significant relationship ($p < 0.001$) was found between Young's modulus and the relative amount of secondary lamellar (osteonal) bone in cross-section, as distinct from primary bone (including woven, fibro-lamellar, and primary lamellar) (Figure 4.4). The lack of a significant relationship between percent area of porosity and Young's modulus may provide evidence that variation in mineralization and/or collagen arrangement, which may vary with bone type (in addition to geometry), could play a more critical role in determining fracture behavior of pediatric bone. In a subsequent study, Agnew et al. (2018) found that ribs from skeletally immature individuals (i.e., pediatric) were more likely to have a ductile response (measured by plastic energy) to dynamic loading than adults. This is likely explained by differences in material composition and distribution of bone types within growing ribs, as demonstrated by Agnew et al. (2013), i.e., the ratio of primary unremodeled bone to secondary remodeled bone; however, this has not yet been quantified in the dynamically tested sample. Similarly, Zioupos and colleagues (Zioupos et al., 2020) found that immature femoral specimens demonstrated more ductile behavior across a range of rates when compared to mature specimens. Future emphasis should be placed on mechanical testing of immature human bone under different loading conditions to better understand fracture mechanics and the influence of bone microstructure during development.

The contribution of cortical porosity to overall loading response has been investigated across skeletal elements particularly due to the known increase in age-related bone loss and concomitant increase in fracture risk. Evidence suggests substantial biomechanical effects

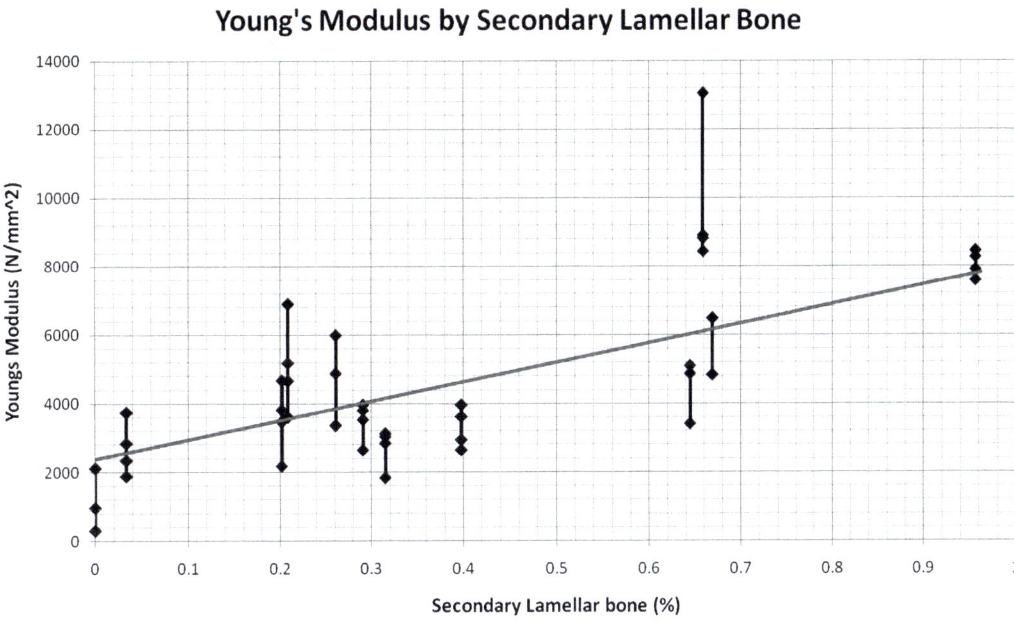

Figure 4.4 Plot showing significant positive relationship between Young's modulus and relative secondary lamellar bone area from mechanical testing of pediatric ribs. A univariate mixed model was used to create the mean estimate line.

of number, size, and location of resorption cavities (see Hernandez, 2008 for discussion) independent of bone mass. However, due to integrated response of traits, porosity effects on response to loading should not be characterized in isolation. Cole et al. (2022) demonstrated the importance of correcting cortical area for porosity in explaining structural properties of experimentally tested ribs. Similar perspectives were employed by Bigelow et al. (2019), who tested multivariate models in the male midshaft radius and found a combination of whole bone size, mineralization, age, height, and cortical porosity were the strongest predictors of bone strength (maximum bending moment). As porosity has been shown to vary within a skeletal element, the effects of variation in regional levels of porosity on mechanical properties have been investigated. Region-specific increases in porosity and declines in tissue-level mechanical properties with age were attributed to differential or asymmetrical loading in the cortex of male femora (Mandair et al., 2022). Consequently, cortical porosity was highest in the anterior and posterior regions of the midshaft and explained larger amounts of variance (46–51%) in tissue-level mechanical properties, including post-yield strain, compared to the lateral or medial quadrants (Mandair et al., 2022). This suggests the utility for region-specific porosity investigations to supplement cross-sectional analyses when quantifying whole bone response to loading.

4.4 Forensic Perspective of Skeletal Trauma

Skeletal trauma analysis is a key component of human osteological examinations, specifically in forensic anthropology and bioarchaeology. The precise and accurate interpretation of these analyses is critical, particularly in a forensic context, as the impact can be consequential and widespread considering interpretations of skeletal trauma are the majority of forensic anthropological scientific medicolegal testimony (Blau, 2016; Christensen & Passalacqua, 2018; Crowder et al., 2016; Dempsey & Blau, 2020; A. Harden et al., 2023; Hulse et al., 2019; Kroman & Symes, 2013; Lesciotto, 2015; Murray & Anderson, 2007; Simon et al., 2022; Symes et al., 2012; Ubelaker, 2018). A common approach to skeletal trauma analysis is to "reverse engineer," or rather to analyze fracture characteristics to reconstruct trauma events (A. Harden et al., 2023; Love & Symes, 2004). However, standardized training and methods, as well as current experimental research, are lacking and unable to link gross and microstructural fracture characteristics to loading mechanisms for improved trauma interpretation (A. Harden et al., 2023; Pechnikova et al., 2015).

While both forensic casework and trauma reconstructions from bioarchaeological contexts are valuable, the exact loading mechanisms are largely unknown and sample sizes are small; therefore, they are unable to provide any statistical significance to support future interpretations. Experimental skeletal trauma research can help to fill in these gaps of knowledge and statistical strength but is challenging to conduct due to high costs and difficulties in sample procurement. However, there has been an increase in skeletal trauma research on human specimens in recent years (Agnew et al., 2020; Bolte IV et al., 2020; Fenton & Isa, 2019; Goots et al., 2018; Harden, et al., 2021; Harden et al., 2020; Isa et al., 2018; Isa et al., 2020; Isa et al., 2019; M. I. Isa et al., 2018; Watson et al., 2018). The focus of these studies has been to evaluate fracture outcomes and their relationship with loading mechanisms. These studies provide critical data to support skeletal trauma interpretations, but a gap remains in understanding which intrinsic and extrinsic variables contribute to variation in fracture outcomes. Addressing this gap requires the integration of foundational knowledge and education in bone biology, bone functional adaptation, fracture biomechanics coupled with an exploration of relationships between individual characteristics (age, sex, stature, etc.) and bone characteristics, both gross and microstructural variables, and fracture characteristics.

Understanding the context in which fractures occur is critical to support accurate skeletal trauma interpretations. Establishing fracture context encapsulates identifying traumatic versus non-traumatic fractures, fracture timing, loading mechanism, and contributors to possible variation in fracture outcomes (adult versus pediatric, age, sex, bone quality, microstructural variation, etc.). While complicated, fracture context is an important component of skeletal trauma analyses across scientific fields, especially in forensic anthropology and bioarchaeology. The following sections will focus on fracture outcomes and fracture timing to provide an overview of various components of fracture context and the contributions of histological analyses.

4.4.1 Fracture Outcomes

General assumptions regarding fracture patterns (e.g., transverse, oblique, wedge, and spiral) and their association with specific loading mechanisms (e.g., tension, compression, bending, and torsion) prevail across scientific fields (Christensen et al., 2019; Galloway et al., 2014a; Harden, Kang, et al., 2022; Schmitt et al., 2004). As such, concepts such as transverse fractures are typically from impacts perpendicular to the bone, oblique fractures are indirect fractures that are the result of angulation and compressive loading, and wedge fractures are the result of bending persist in skeletal trauma analyses (Christensen et al., 2019; Galloway et al., 2014a; Harden, et al., 2021, Harden, Kang, et al., 2022). Along with these assumptions is the expectation that if the loading mechanism is the same, the fracture outcomes will be identical or similar (Berryman et al., 2018). Experimental research has demonstrated that fracture outcomes vary even when the skeletal element and loading event remain the same. In human ribs (n=68) subjected to identical dynamic anterior-posterior bending, fracture locations ranged from 20% to 96% of the total rib length, and six different fracture types were observed (Harden & Agnew, 2018) (Figure 4.5). Of the observed fracture outcomes, simple transverse fractures were observed in 40% of the sample, whereas, the frequency of wedge fractures (both intact and multifragmentary) was 32%. In this sample with identical loading conditions, not only did fracture characteristics vary, but the expected fracture pattern in a bending mechanism, a wedge fracture, was *not* the most frequently observed outcome. To test the specific assumption that a bending loading mechanism results in wedge fractures in a different element, Harden et al. (2022) experimentally tested 50 human tibiae in an identical loading direction (lateral-medial) and rate (6 m/s) (Figure 4.6). Variation was observed in both the number of fractures per tibia and fracture classifications, refuting the assumption that the same loading mechanism will result in the same fracture pattern in the same element across individuals (Harden, et al., 2021; Harden, Kang, et al., 2022). These findings suggest that intrinsic characteristics of the individual and/or the bone are contributing to the variance in fracture outcomes. It has been broadly stated that variation

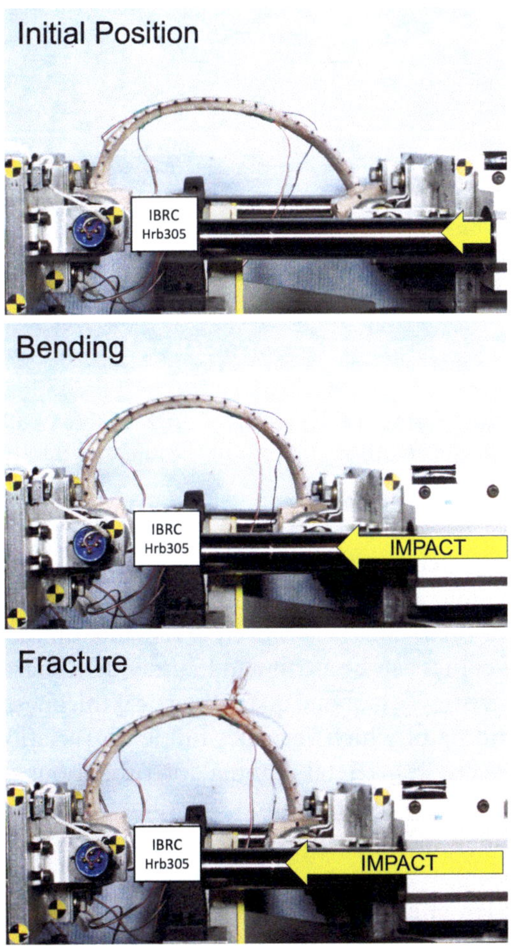

Figure 4.5 Experimental loading of human rib in dynamic anterior-posterior bending from initial position (left) to during the event (center) and finally at fracture (right).

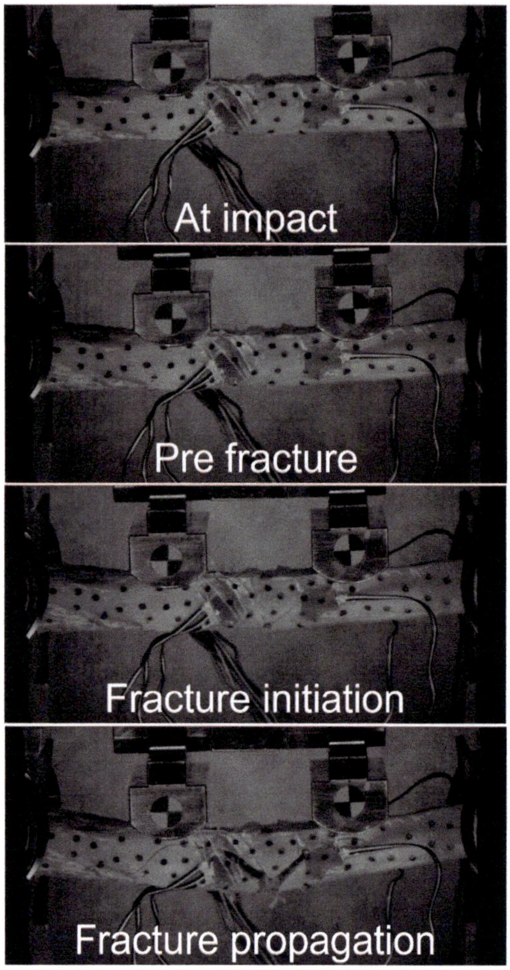

Figure 4.6 High-speed video frames of dynamic bending loading of a human tibia in a medial-lateral direction at impact, pre fracture, fracture initiation, and fracture propagation (from left to right).

in bone response and ultimately fracture outcomes is dependent upon age, sex, health, bone quality, impact direction and angle, mechanism, force, micromorphology, etc. (Berryman et al., 2018; Dempsey & Blau, 2020). However, it remains unknown what percentage of variation in fracture outcomes can be attributed to individual characteristics (e.g., age, sex, health) or bone characteristics (e.g., bone quality, cortical thickness, porosity).

A deeper understanding of which variables influence variation in fracture outcomes is critical to support accurate skeletal trauma interpretations. Experimental research conducted utilizing human tissue is required to explore this topic. Previous and ongoing research in the Injury Biomechanics Research Center, The Ohio State University, has focused on relationships between individual characteristics and fracture outcomes in human ribs (A. Harden et al., 2019), tibiae (Harden et al., 2021; Harden, Stull, et al., 2022), and radii to gain insight on why fracture characteristics vary between the same element across individuals, in duplicated loading mechanisms. All specimens were ethically acquired through The Ohio State Body Donation Program. Harden et al. (2019) examined the predictability of fracture characteristics (classification, location, number of fractures) from individual characteristics (age, sex, stature, weight, and BMI) in human ribs (n=346). Overall, no significant correlations or trends were found between fracture location, fracture classification, or number of fractures and any of the individual characteristics included in the study (A. L. Harden et al., 2019). The variation observed in the fracture characteristics within this study of human ribs suggests that loading mechanisms may not be determined based on fracture outcomes alone, and further exploration of these potential relationships should continue to be explored across skeletal elements.

More recently, the Injury Biomechanics Research Center team has conducted experimental research, funded by the National Institute of Justice, on tibiae in pedestrian impact scenarios (C. M. Goden et al., 2023; Harden, 2023; Harden, et al., 2021, Harden, Kang, et al., 2022; A. L. Harden, Y.-S. Kang, et al., 2023; Harden et al., 2021; Harden, Stull, et al., 2022; A. L. Harden, K. E. Stull, et al., 2023) to improve the accuracy of skeletal trauma analysis and expert testimony. One of the core aims of the overarching project is to analyze relationships between skeletal fracture characteristics and intrinsic variables of the individual (individual characteristics) (e.g., age, sex, stature) or the tibia (bone characteristics) (e.g., total length, cortical area, percent porosity). Harden et al. (2022) examined relationships between age, sex, and fracture type and number in human tibiae loaded in a lateral-medial direction in diaphyseal bending at 6 m/s (n=60). Variation was observed in both the number of fractures per tibia and fracture types (Simple, Wedge, Multifragmentary) (Harden, Stull, et al., 2022). No significant relationships were found between sex and either of the fracture outcome variables. Significant differences were observed in age and both fracture characteristic variables (Harden, Stull, et al., 2022). For more details about the observed trends between age, sex, number of fractures, and fracture type, see Harden, Stull, et al. (2022). These preliminary findings demonstrate that individual characteristics alone cannot explain the variation observed in fracture outcomes; therefore, histological methods and bone microstructural variables should be incorporated to improve skeletal trauma research and analyses. In general, research investigating relationships between bone microstructure and fracture characteristics is sparse. Since previous research has demonstrated that age, sex, weight, or BMI cannot explain the variance observed in fracture characteristics of the same element loaded in the same conditions, exploration of relationships between bone microstructure and fracture characteristics is a critical next step. Many studies have examined relationships between bone microstructure and bone response; however, few studies have investigated the relationship between microstructure and fracture outcomes. Harden et al. (2019) quantified relationships between structural properties and fracture characteristics in human ribs (n=347) loaded in an anterior-posterior direction in a simplified dynamic 2D bending scenario (Agnew et al., 2018). When examining differences in structural properties between types of fractures, simple fractures demonstrated less displacement, total energy, and plastic energy than wedge fractures. Additionally, this study found that ribs with one fracture demonstrated less displacement and energy than ribs that fractured in two distinct locations (A. Harden et al., 2019). In the same sample of human ribs, Agnew et al. (2020) explored sources of variability on rib fracture outcomes. These ribs were all loaded to failure in the exact same conditions, and despite common assumptions about bending of long bones resulting in wedge fractures, only 23% of all ribs resulted in wedge fractures (Christensen et al., 2019; Galloway et al., 2014a; Harden, et al., 2021; Harden, Kang, et al., 2022; Schmitt et al., 2004). They also found that the number of fractures and fracture complexity increased with age. In fact, pediatric ribs in the sample were more likely to result in zero fractures or incomplete fractures if they did fail, likely due to their complex material composition. Material properties are likely contributing to fracture types, which explains the relationships observed with age and lack of relationships with sex since rib material properties are known to be strongly associated with age but not sex (Albert et al., 2021; Katzenberger et al., 2020).

Bone type can also influence fracture behavior. Liu et al. (2000) found that post-failure, primary lamellar bone tends to break cleanly while lamellar bone containing osteon accumulations remains attached, providing an additional adaptive explanation for secondary lamellar bone function. Lynn and Fairgrieve (2009) found lamellar bone separation to be typical for fracture surfaces under dynamic loading conditions. They also found that osteon pullout is more likely to occur in fleshed than unfleshed bone, providing useful information for understanding fracture context. Osteon pullout, an interesting adaptive process that debonds the osteon at its interface, allows for the fatigue life and toughness of the bone to be extended. Osteon pullout has been shown to be affected by load type and vary by location, but how this phenomenon affects bone integrity requires further investigation (Hiller et al., 2003). Relationships between bone microstructure, specifically cortical porosity, and fracture type have been preliminarily investigated in human tibiae (Cole et al., 2023). This study utilized Pore Extractor 2D (Cole et al., 2021) to facilitate computer-assisted identification of cortical borders and pore spaces in cross-sections at the 66% site of tibiae that were impacted at the 50% site at 5 m/s in a lateral-medial direction. The quantification of pore morphometry included percent porosity, pore density, and mean pore size and shape descriptors. Exemplars of tibia cross-sections and cortical porosity data are provided in Figure 4.7. The results of this study demonstrated trends of variation in cortical area and

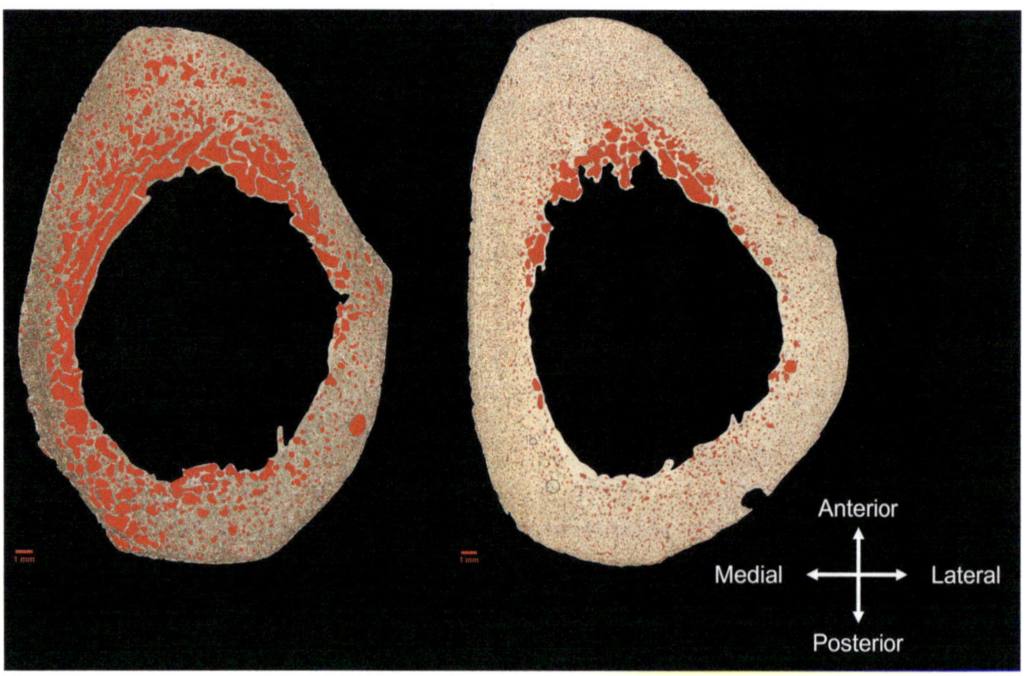

Figure 4.7 Exemplar images of variable cortical porosity in tibia cross-sections. Cortical pores shown in red. Left image demonstrates a tibia cross-section with a higher percent porosity, while the right image is a tibia cross-section with a lower percent porosity.

pore morphometry between fracture types (Cole et al., 2023). These preliminary findings suggest that bone microstructure, specifically porosity, influences fracture outcomes, but further research is needed to continue to explore relationships between bone microstructure and fracture outcomes in a larger sample, for various loading mechanisms, and in other skeletal elements.

4.4.2 Fracture Timing

Current standards and expectations of skeletal trauma analysis state that assessment of trauma timing and/or mechanism of the observed trauma should be attempted ((SWGANTH) 2011) (SWGANTH), 2011) since they can have a significant impact on further interpretations and ultimately the determination of cause and manner of death. The potential for pivotal impacts within the criminal justice system is considerable, yet this aspect of trauma analysis remains challenging for multiple reasons. One of the foundational challenges within assessment of trauma timing is that the definitions of the terminology used for classifying fracture timing vary across scientific fields. Specifically, perimortem and postmortem have distinctly different connotations in forensic pathology and forensic anthropology. While not yet published at the time this chapter was written, the draft of ASB Standard 147, Standard for Analyzing Skeletal Trauma in Forensic Anthropology (ASB Standard 147, First Edition 2023) provides the following definitions for antemortem, perimortem, and postmortem trauma:

Antemortem Trauma: A skeletal defect that occurred before an individual's death and shows evidence of osteological reaction

Perimortem Trauma: A skeletal defect that occurred when the bone is in a biomechanically fresh (visco-elastic) state with no evidence of an osteological reaction

Postmortem Damage: A skeletal defect that occurred after a bone has lost its biomechanically fresh (visco-elastic) properties

Beyond the ambiguous nature of the perimortem and postmortem definitions, standardized criteria for assessing timing of skeletal trauma are lacking (Galloway et al., 2014b; C. Goden et al., 2023; Winter-Buchwalder et al., 2022). More specifically, the lack of methods and/or presence of ambiguous methods to estimate trauma timing pose difficulties in differentiating perimortem versus postmortem trauma. Cappella et al. (2014) presented findings of the determination of trauma timing on 210 fractures by two forensic anthropologists and found a higher success rate for both observers in identification of postmortem fractures than perimortem fractures. Additionally, higher error rates were observed on flat bones versus long bones (Cappella et al., 2014). These findings are not surprising based on the current gross characteristics utilized in trauma timing assessment, such as color differentiation, evaluation of fracture morphology, presence of staining, and plastic deformation (Blau, 2016; Christensen et al., 2019; Galloway et al., 2014b; L'Abbe et al., 2022; Wieberg & Wescott, 2008), and many of the established methods are limited to long or cranial bones, restricting applications to skeletal elements with thinner cortices (e.g., ribs) (C. Goden et al., 2023). Current methods in skeletal trauma timing assessment focus on gross characteristics, highlighting the lack of histological approaches to trauma timing in forensic anthropology (Winter-Buchwalder et al., 2022).

Hard tissue histological methods are rarely applied in trauma timing research, yet the ASB Standard 147 states that, when possible, skeletal trauma examination should include microscopic observations (ASB Standard 147, First Edition 2023). While a clear statement regarding the value of these methods is not provided, specifically in the estimation of trauma timing, the general guidance is that all available methods should be utilized for a thorough analysis and holistic interpretations. A common theme in trauma timing research is the differentiation between perimortem and postmortem, which is also reflected in research incorporating histological methods (Pechníková et al., 2011; Winter-Buchwalder et al., 2022). Pechníková et al. (2011) examined differences in the pattern of osteon fracturing and found that the patterns observed in wet and dry bone were similar and did not allow for distinguishing between perimortem and postmortem fractures. In contrast, Winter-Buchwalder (2022) examined microcracking patterns for distinguishing between perimortem and postmortem fractures and found different patterns of microcracking between fractures in fresh/wet versus dry bone. Limitations of both studies include small sample sizes and the elements utilized in the fresh/wet versus dry cohorts were not antimeres, which may contribute to any observed variation between samples. Additional previous studies have investigated osteon fracture patterns and structure-microcracking to investigate the applications of bone histology to fracture analysis (Ebacher et al., 2012; Hanaue et al., 2007). While these studies are not specific to fracture timing, they provide histological insight on patterns of fracture propagation and relationships between microcracking and osteons' lamellar and sub-lamellar structures in fresh bone that may inform future studies on differentiating between perimortem and postmortem trauma (Ebacher et al., 2012; Hanaue et al., 2007).

More research utilizing histological methods should be completed to address current gaps in contextualizing fractures for forensic applications. The gold standard of assessing fracture timing should incorporate histological analyses to allow for a holistic approach resulting in a thorough analysis of all available features to assist in the accurate assessment of trauma timing. Histological research can present several obstacles, including obtaining a large sample size; however, there are several collections currently available for future research. The Skeletal Biology Research Laboratory (SBRL) at The Ohio State University houses one of the largest collections of undecalcified bone sections and slides from numerous elements throughout the human skeleton ethically obtained from willed anatomical tissue donors in Ohio. For more information about this collection and contact information, see *Research Resources* in this chapter.

4.5 Conclusions

As research in skeletal biology progresses, it is crucial to examine multiple contributors to fracture risk and embrace the role which bone quality, in addition to bone quantity, plays in determining bone strength. To improve understanding of bone biomechanics, material and structural properties, as well as histomorphometric traits of bone, should all be considered. As this chapter reveals, the analysis of bone behavior traverses many disciplines and is truly multidimensional. Individuals from many backgrounds (those studying past population's biology and behavior, injury prevention in traffic crashes or a clinical setting, forensic investigation, etc.) all benefit from increased knowledge on the interaction of bone biology with fracture. It is vitally important to study *how* and *why* skeletal trauma occurs as well.

By examining the biological mechanisms involved in bone and fracture biomechanics researchers can make sounder interpretations of fracture cause and context in past and present populations as well as in future prevention.

Skeletal trauma interpretations have become a fundamental component of forensic anthropological analyses due to the increase of both requests for skeletal trauma analyses and subsequent medicolegal testimony (Blau, 2016; Christensen & Passalacqua, 2018; Crowder et al., 2016; Dempsey & Blau, 2020; Hulse et al., 2019; Kroman & Symes, 2013; Lesciotto, 2015; Murray & Anderson, 2007; Simon et al., 2022; Symes et al., 2012; Ubelaker, 2018). As such, experimental research is becoming more prevalent within the field, whereas it has traditionally been conducted in physics, biomedical engineering, and injury biomechanics (Berryman et al., 2018; A. Harden et al., 2023). Multidisciplinary collaborations between forensic anthropologists and engineers are becoming more customary; however, this is still not the current standard for experimental skeletal trauma research. Within current skeletal trauma research, questions arise about the translation of results from non-human to human specimens (Christensen et al., 2012; DeLand et al., 2012; Dempsey & Blau, 2020; Dempsey et al., 2018; A. Harden et al., 2023; Hillier, 2007; Kulin et al., 2011; Passalacqua & Fenton, 2012; Reber & Simmons, 2015; Zephro et al., 2014). The incorporation of histological methods and analyses in the assessment of skeletal trauma is scarce across scientific fields and applications. To continue to move forward and close both knowledge and application gaps, a focus on increasing the understanding of relationships between individual and bone characteristics, bone response, fracture characteristics, and loading mechanisms is critical. The following section provides information on research resources with a focus on bone histology and skeletal trauma analyses to assist in supporting future research to advance the accuracy of skeletal trauma analyses and interpretations.

4.6 Research Resources for Histological and Skeletal Trauma Analyses

The histological collection housed in the Skeletal Biology Research Laboratory (SBRL) (www.ibrc.osu.edu) at The Ohio State University includes undecalcified bone sections and slides from numerous elements throughout the human skeleton ethically obtained from over 1,300 willed anatomical tissue donors in Ohio. The collection includes sections from adult and pediatric ribs, as well as adult femur, tibia, fibula, clavicle, humerus, radius, and ulna sections from various locations within each element. Ages for the sample range from 0 to 108 years with both sexes represented. Basic demographic information is available including sex, age, height, and weight at the time of death. Some records also include DXA scan, digital X-rays, high-resolution clinical CT, nano-CT, micro-CT, and/or 2D histological images up to 100x total magnification at specific regions of interest (ROIs). Many samples are from the same individuals, thus making the collection particularly valuable for exploring intra-skeletal variation.

4.6.1 Forensic Anthropology Skeletal Trauma Database

The Forensic Anthropology Skeletal Trauma (FAST) database (Harden, Stull, et al., 2023) is a novel resource, funded by the National Institute of Justice, which provides trauma analysis data for education, training, and case comparison applications. Students, academics, and practitioners can gain an interdisciplinary perspective of skeletal trauma through an examination of experimental research utilizing human specimens with known loading mechanisms. FAST features pre- and post-test imaging, data collected from advanced instrumentation during the impact event, and fracture analysis data. Additionally, undecalcified bone sections from all specimens included in FAST are part of the Skeletal Biology Research Laboratory's histological collection. For more information on FAST and to request access to the database, visit: https://ibrc.osu.edu/fast-database/.

For a list of additional research collections see Chapter 11.

References

Agnew, A., Moorhouse, K., Kang, Y., Donnelly, B., Pfefferle, K., Manning, A., Litsky, A., Herriott, R., Abdel-Rasoul, M., & Bolte IV, J. H. (2013). The response of pediatric ribs to quasi-static loading: Mechanical properties and microstructure. *Annals of Biomedical Engineering, 41*(12), 2501–2514.

Agnew, A., Murach, M., Dominguez, V., Sreedhar, A., Misicka, E., Harden, A., Bolte IV, J., Stammen, J., Moorhouse, K., & Kang, Y.-S. (2018). Sources of variability in structural bending response of pediatric and adult human ribs in dynamic frontal impacts. *Stapp Car Crash Journal, 62*, 119–192.

Agnew, A. M., Harden, A. L., Sreedhar, A., Bolte IV, J., & Kang, Y.-S. (2020). Rib Fractures: An Experimental Approach to Identifying Sources of Variability. American Academy of Forensic Sciences, Anaheim, California, USA.

Albert, D. L., Katzenberger, M. J., Agnew, A. M., & Kemper, A. R. (2021). A comparison of rib cortical bone compressive and tensile material properties: Trends with age, sex, and loading rate. *J Mech Behav Biomed Mater, 122*, 104668. https://doi.org/10.1016/j.jmbbm.2021.104668

Augat, P., & Schorlemmer, S. (2006). The role of cortical bone and its microstructure in bone strength. *Age Ageing, 35 Suppl 2*, ii27–ii31. https://doi.org/10.1093/ageing/afl081

Berryman, H., Berryman, J. B., & Saul, T. B. (2018). Bone trauma analysis in a forensic setting: Theoretical basis and a practical approach for evaluation. In C. C. Boyd Jr. & D. C. Boyd (Eds.), *Forensic Anthropology: Theoretical Framework and Scientific Basis* (pp. 213–232). John Wiley & Sons Ltd.

Bigelow, E. M. R., Patton, D. M., Ward, F. S., Ciarelli, A., Casden, M., Clark, A., Goulet, R. W., Morris, M. D., Schlecht, S. H., Mandair, G. S., Bredbenner, T. L., Kohn, D. H., & Jepsen, K. J. (2019). External Bone Size Is a Key Determinant of Strength-Decline Trajectories of Aging Male Radii. *Journal of Bone and Mineral Research*, e3661. https://doi.org/10.1002/jbmr.3661

Blain, H., Chavassieux, P., Portero-Muzy, N., Bonnel, F., Canovas, F., Chammas, M., Maury, P., & Delmas, P. D. (2008). Cortical and trabecular bone distribution in the femoral neck in osteoporosis and osteoarthritis. *Bone, 43*(5), 862–868. http://journals.ohiolink.edu/ejc/article.cgi?issn=87563282&issue=v43i0005&article=862_catbdifnioao

Blau, S. (2016). How traumatic: a review of the role of the forensic anthropologist in the examination and interpretation of skeletal trauma. *Australian Journal of Forensic Sciences, 49*(3), 261–280. https://doi.org/10.1080/00450618.2016.1153715

Bolger, M. W., Romanowicz, G. E., Bigelow, E. M. R., Ward, F. S., Ciarelli, A., Jepsen, K. J., & Kohn, D. H. (2020). External bone size identifies different strength-decline trajectories for the male human femora. *Journal of Structural Biology, 212*(3), 107650. https://doi.org/10.1016/j.jsb.2020.107650

Bolotin, H. H. (2007). DXA in vivo BMD methodology: an erroneous and misleading research and clinical gauge of bone mineral status, bone fragility, and bone remodelling. *Bone, 41*(1), 138–154. https://doi.org/10.1016/j.bone.2007.02.022

Bolte IV, J., Stark, D., Agnew, A. M., & Kang, Y.-S. (2020). Skull Fracture Risk: An Experimental Comparison Between Elastic (Drone) and Inelastic (Wood Block) Impacts Focused on Fracture Type and Tolerance Using Instrumented Postmortem Human Subjects (PMHS). *American Academy of Forensic Sciences*, Anaheim, California, USA.

Burr, D. B. (2002). The contribution of the organic matrix to bone's material properties. *Bone, 31*(1), 8–11. http://journals.ohiolink.edu/ejc/article.cgi?issn=87563282&issue=v31i0001&article=8_tcotomtbmp

Burr, D. B. (2008). Bone quality and fractures: Influence of remodeling on structural and material properties. *Bone, 43*, S20. http://journals.ohiolink.edu/ejc/article.cgi?issn=87563282&issue=v43inone_s1&article=s20_bqafiorosamp

Burr, D. B. (2011). Why bones bend but don't break. *J Musculoskelet Neuronal Interact, 11*(4), 270–285.

Burr, D. B. (2014). The complex relationship between bone remodeling and the physical and material properties of bone. *Osteoporosis International*. https://doi.org/10.1007/s00198-014-2970-4

Cappella, A., Amadasi, A., Castoldi, E., Mazzarelli, D., Gaudio, D., & Cattaneo, C. (2014). The difficult task of assessing perimortem and postmortem fractures on the skeleton: a blind text on 210 fractures of known origin. *J Forensic Sci, 59*(6), 1598–1601. https://doi.org/10.1111/1556-4029.12539

Christensen, A. M., & Passalacqua, N. V. (2018). *A Laboratory Manual for Forensic Anthropology* (1st ed.). Academic Press.

Christensen, A. M., Passalacqua, N. V., & Bartelink, E. J. (2019). *Forensic Anthropology: Current Methods and Practice* (2nd ed.). Academic Press.

Christensen, A. M., Smith, V. A., Ramos, V., Shegogue, C., & Whitworth, M. (2012). Primary and secondary skeletal blast trauma. *J Forensic Sci, 57*(1), 6–11. https://doi.org/10.1111/j.1556-4029.2011.01938.x

Cole, M., Agnew, A., McAdow, S., Goldsmith, C., & Harden, A. (2023). *Preliminary Investigation of Relationships between Cortical Porosity and Fracture Type in Human Tibiae. American Academy of Forensic Sciences*, Orlando, Florida.

Cole, M., Kang, Y.-S., Harden, A., Dominguez, V., & Agnew, A. (2022). Quantifying 2D cortical porosity improves prediction of human rib structural bending response. *Journal of Bone and Mineral Research*, 38, Suppl 1.

Cole, M., Stout, S., & Agnew, A. (2021). Pore Extractor 2D: An ImageJ plugin for identification, classification, and regional characterization of cortical pores on histological images. *American Journal of Physical Anthropology, 174*, 20–21.

Crowder, C. M., Wiersema, J. M., Adams, B. J., Austin, D. E., & Love, J. C. (2016). The Utility of Forensic Anthropology in the Medical Examiner's Office. *Acad Forensic Pathol, 6*(3), 349–360. https://doi.org/10.23907/2016.037

Currey, J. (2003). Role of collagen and other organics in the mechanical properties of bone. *Osteoporosis International, 14*(Suppl 5), S29–S36.

Currey, J. D. (2002). *Bones: Structure and Mechanics*. Princeton University Press.

Currey, J. D., Brear, K., & Zioupos, P. (1996). The effect of porosity and mineral content on the Young's modulus of elasticity of compact bone. *Journal of Biomechanics, 21*, 131–139.

Currey, J. D., Pitchford, J. W., & Baxter, P. D. (2007). Variability of the mechanical properties of bone, and its evolutionary consequences. *Journal of The Royal Society Interface, 4*(12), 127–135. https://doi.org/10.1098/rsif.2006.0166

De Maeseneer, M., De Mey, J., Debaere, C., Meysman, M., & Osteaux, M. (2000). Rib fractures induced by coughing: An unusual cause of acute chest pain. *American Journal of Emergency Medicine, 18*(2), 194–197. http://journals.ohiolink.edu/ejc/article.cgi?issn=07356757&issue=v18i0002&article=194_rfibcaucoacp

DeLand, T., Kendell, A., Fenton, T., & Haut, R. (2012, 20–23 June 2012). Fracture patterns from three-point bending of sheep femora ASME 2012 Summer Bioengineering Conference, Fajardo, Puerto Rico.

Dempsey, N., & Blau, S. (2020). Evaluating the evidentiary value of the analysis of skeletal trauma in forensic research: A review of research and practice. *Forensic Sci Int, 307*, 110140. https://doi.org/10.1016/j.forsciint.2020.110140

Dempsey, N., Gilbert, F. M., Miszkiewicz, J., & Oxenham, M. F. (2018). Biomechanical analysis of controlled tibial blunt force trauma. *Australian Journal of Forensic Sciences, 51*(5), 538–548. https://doi.org/10.1080/00450618.2018.1429017

Duan, Y., Beck, T. J., Wang, X.-F., & Seeman, E. (2003). Structural and biomechanical basis of sexual dimorphism in femoral neck fragility has its origins in growth and aging. *Journal of bone and mineral research: the official journal of the American Society for Bone and Mineral Research, 18*(10), 1766–1774. https://doi.org/10.1359/jbmr.2003.18.10.1766

Eastman, K., O'Leary, T. J., Carswell, A., Walsh, N., Izard, R., Fraser, W., & Greeves, J. (2023). Distal Tibial Bone Properties and Bone Stress Injury Risk in Young Men Undergoing Arduous Physical Training. *Calcified Tissue International.* https://doi.org/10.1007/s00223-023-01111-1

Ebacher, V., Guy, P., Oxland, T., & Wang, R. (2012). Sub-lamellar microcracking and roles of canaliculi in human cortical bone. *Acta Biomaterialia, 8*(3), 1093–1100.

Evans, R., Negus, C., Antczak, A., Yanovich, R., Istaeli, E., & Moran, D. (2008). Sex differences in parameters of bone strength in new recruits: beyond bone density. *Medicine and Science in Sports and Exercise, 40*(S11), S645–S653.

Felsenberg, D., & Boonen, S. (2005). The bone quality framework: determinants of bone strength and their interrelationships, and implications for osteoporosis management. *Clinical Therapeutics, 27*(1), 1–11. http://journals.ohiolink.edu/ejc/article.cgi?issn=01492918&issue=v27i0001&article=1_tbqfdoiaifom

Fenton, T. W., & Isa, M. I. (2019). A decade of experimental research on blunt force cranial fracture. Federation of American Societies for Experimental Biology.

Follet, H., Boivin, G., Rumelhart, C., & Meunier, P. J. (2004). The degree of mineralization is a determinant of bone strength: a study on human calcanei. *Bone, 34*(5), 783–789. http://journals.ohiolink.edu/ejc/article.cgi?issn=87563282&issue=v34i0005&article=783_tdomiasasohc

Frost, H. M. (1982). Mechanical determinants of bone modeling. *Metabolic Bone Disease & Related Research, 4*, 217–229.

Frost, H. M. (2001). From Wolff's law to the Utah paradigm: Insights about bone physiology and its clinical applications. *The Anatomical Record, 262*(4), 398–419.

Galloway, A., Zephro, L., & Wedel, V. L. (2014a). Classification of Fractures. In V. L. Wedel & A. Galloway (Eds.), *Broken Bones: Anthropological Analysis of Blunt Force Trauma* (2nd ed., pp. 59–71). Charles C Thomas.

Galloway, A., Zephro, L., & Wedel, V. L. (2014b). Diagnostic Criteria for the Determination of Timing and Fracture Mechanism. In V. L. Wedel & A. Galloway (Eds.), *Broken Bones: Anthropological Analysis of Blunt Force Trauma* (2nd ed., pp. 47–58). Charles C Thomas.

Garnero, P. (2015). The role of collagen organization on the properties of bone. *Calcified Tissue International, 97*(3), 229–240.

Goden, C., Brzezinski, E., Agnew, A., & Harden, A. (2023). Accuracy and Observer Agreement in the Determination of Trauma Timing in Human Ribs. *American Journal of Biological Anthropology, 180*, 64.

Goden, C. M., Agnew, A., Kang, Y.-S., Stull, K. E., & Harden, A. L. (2023). *A Preliminary Investigation of Fractographic Method Accuracy in Determining Fracture Propagation Direction in Human Tibiae. American Academy of Forensic Sciences*, Orlando, Florida, USA.

Goots, A. C., Isa, M. I., Fenton, T. W., Watson, E. O., Vaughan, P. E., Wei, F., & Haut, R. C. (2018). *Estimating Points of Impact in Multiple Blunt Force Cranial Trauma: Lessons from Experimental Impacts. American Academy of Forensic Sciences*, Seattle, Washington, USA.

Hanaue, K., Katakura, A., Kasahara, K., Kamiyama, I., Takaki, T., Shibahara, T., Abe, S., & Ide, Y. (2007). Course of fracture line in sagittal splitting of human mandible. *Bull Tokyo Dent Coll, 48*(4), 163–170. https://doi.org/10.2209/tdcpublication.48.163

Harden, A., Kang, Y.-S., Moorhouse, K., Stammen, J., & Agnew, A. (2019). Human rib fracture characteristics and relationships with structural properties International Research Council on Biomechanics of Injury, Florence, Italy.

Harden, A., Stull, K., Bolte, J., Kang, Y.-S., & Agnew, A. (2023). Blunt Force Skeletal Trauma Research Methods: A Multidisciplinary Perspective. *Forensic Anthropology*, 1–13–11–13.

Harden, A. L. (2023). Development of the Forensic Anthropology Skeletal Trauma (FAST) Database. National Institute of Justice Forensic Science Research & Development Symposium, Orlando, Florida, USA.

Harden, A. L., & Agnew, A. M. (2018). Classification of fractures in human ribs subjected to dynamic bending. *American Journal of Physical Anthropology, 168*(S68), 97–98.

Harden, A. L., Dominguez, V. M., & Agnew, A. (2019). Relationships between individual level variables and fracture characteristics in human ribs subjected to anterior-posterior loading. *American Journal of Biological Anthropology, 168*, 97–98.

Harden, A. L., Kang, Y.-S., Bendig, A., Bolte IV, J., Goldsmith, C., & Agnew, A. M. (2022). The Frequency of Wedge Fractures in Experimental Bending of the Tibia. American Academy of Forensic Sciences, Seattle, Washington.

Harden, A. L., Kang, Y.-S., Bolte IV, J., Stull, K. E., & Agnew, A. A. (2023). *An Examination of Strain Mode at Fracture in Experimentally Loaded Human Tibiae. American Academy of Forensic Sciences*, Orlando, Florida, USA.

Harden, A. L., Kang, Y.-S., Hunter, R. L., Bendig, A., Bolte IV, J., Eckstein, N. I., Smith, A. G. F., & Agnew, A. (2021). *Preliminary Sex-specific Relationships between Peak Force and Cortical Bone Morphometrics in Human Tibiae Subjected to Lateral Loading* International Research Council on the Biomechanics of Injury, virtual.

Harden, A. L., Kang, Y.-S., Sreedhar, A., & Agnew, A. (2020). *Variation in Human Rib Failure Mechanisms in Experimental Anterior-Posterior Loading. American Academy of Forensic Sciences*, Anaheim, California, USA.

Harden, A. L., Stull, K. E., Kang, Y.-S., Bendig, A., Bolte IV, J., & Agnew, A. M. (2022, September 9–11). *Relationships between Age, Sex, and Number and Type of Fractures in Human Tibiae International Research Council on Biomechanics of Injury*, Portugal, Spain.

Harden, A. L., Stull, K. E., Kang, Y.-S., Bolte IV, J., & Agnew, A. A. (2023). *Forensic Anthropology Skeletal Trauma Database* Figshare. https://doi.org/10.6084/m9.figshare.21948419

Heaney, R. P. (2003). Is the paradigm shifting? *Bone, 33*, 457–465.

Hernandez, C. J. (2008). How can bone turnover modify bone strength independent of bone mass? *Bone, 42*(6), 1014–1020. http://journals.ohiolink.edu/ejc/article.cgi?issn=87563282&issue=v42i0006&article=1014_hcbtmbsiobm

Hiller, L. P., Stover, S. M., Gibson, V. A., Gibeling, J. C., Prater, C. S., Hazelwood, S. J., Yeh, O. C., & Martin, R. B. (2003). Osteon pullout in the equine third metacarpal bone: effects of ex vivo fatigue. *Journal of Orthopaedic Research, 21*(3), 481–488. http://journals.ohiolink.edu/ejc/article.cgi?issn=07360266&issue=v21i0003&article=481_opitetbeoevf

Hillier, M. L., Bell LS. (2007). Differentiating human bone from animal bone: A review of histological methods. *Journal of Forensic Sciences, 52*(2), 249–263.

Hughes, J. M., Taylor, K. M., Guerriere, K. I., Smith, N. I., Staab, J. S., Walker, L. A., Staab, J. E., Bartlett, P. M., Spiering, B. A., Nguyen, V. T., Proctor, S. P., Foulis, S. A., Bouxsein, M. L., & Popp, K. L. (2023). Changes in Distal Tibial Microarchitecture During Eight Weeks of U.S. Army Basic Combat Training Differ by Sex and Race. *JBMR Plus, 7*(4), e10719. https://doi.org/10.1002/jbm4.10719

Hulse, C., Stull, K., & Weaver, A. (2019). Rib fracture frequency and location using vehicular crash data. *Forensic Anthropology, 2*(1), 1–8.

Hunter, R., Briley, K., & Agnew, A. (2020). Whole bone geometry effects on functional compensation in the radius. *Journal of Bone and Mineral Research*, 35, 58.

Isa, M., Fenton, T., Deland, T., & Haut, R. (2018a). Assessing impact direction in 3-point bending of human femora: Incomplete butterfly fractures and fracture surfaces. *Journal of Forensic Sciences, 63*(1), 38–46.

Isa, M. I., Fenton, T. W., Antonelli, L. S., Vaughan, P. E., & Wei, F. (2020). *The Application of Fractography in Trauma Analysis of Complex Long Bone Fractures. American Academy of Forensic Sciences*, Anaheim, California, USA.

Isa, M. I., Fenton, T. W., Goots, A. C., & Wei, F. (2019). *An Experimental Investigation of Blunt Force Fracture in the Human Mandible. American Academy of Forensic Sciences*, Baltimore, Maryland, USA.

Isa, M. I., Fenton, T. W., Watson, E. O., Vaughan, P. E., Wei, F., & Haut, R. C. (2018b). *Initiation and Propagation of Fractures in Blunt Impacts to Unconstrained Human Cadaver Heads. American Academy of Forensic Sciences*, Seattle, Washington, USA.

Jepsen, K. J., Centi, A., Duarte, G., Galloway, K., Goldman, H. M., Hampson, N., Lappe, J. M., Cullen, D., Greeves, J., Izard, R., Nindl, B., Kraemer, W., Negus, C., & Evans, R. (2011). Biological constraints that limit compensation of a common skeletal trait variant lead to inequivalence of tibial function among healthy young adults. *Journal of Bone and Mineral Research, 26*(12), 2872–2885.

Jepsen, K. J., Kozminski, A., Bigelow, E. M., Schlecht, S. H., Goulet, R. W., Harlow, S. D., Cauley, J. A., & Karvonen-Gutierrez, C. (2017). Femoral Neck External Size but not aBMD Predicts Structural and Mass Changes for Women Transitioning Through Menopause. *J Bone Miner Res, 32*(6), 1218–1228. https://doi.org/10.1002/jbmr.3082

Kanis, J. A., Johnell, O., De Laet, C., Johansson, H., Oden, A., Delmas, P., Eisman, J., Fujiwara, S., Garnero, P., Kroger, H., McCloskey, E. V., Mellstrom, D., Melton, L. J., Pols, H., Reeve, J., Silman, A., & Tenenhouse, A. (2004). A meta-analysis of previous fracture and subsequent fracture risk. *Bone, 35*(2), 375–382. http://journals.ohiolink.edu/ejc/article.cgi?issn=87563282&issue=v35i0002&article=375_amopfasfr

Katzenberger, M. J., Albert, D. L., Agnew, A. M., & Kemper, A. R. (2020). Effects of sex, age, and two loading rates on the tensile material properties of human rib cortical bone. *J Mech Behav Biomed Mater, 102*, 103410. https://doi.org/10.1016/j.jmbbm.2019.103410

Koltun, K. J., Sekel, N. M., Bird, M. B., Lovalekar, M., Mi, Q., Martin, B. J., & Nindl, B. C. (2022). Tibial Bone geometry is associated with bone stress injury during military training in men and women [Original Research]. *Frontiers in Physiology, 13*. https://doi.org/10.3389/fphys.2022.803219

Kroman, A. M., & Symes, S. (2013). Investigation of skeletal trauma. In E. A. DiGangi & M. K. Moore (Eds.), *Research Methods in Human Skeletal Biology* (pp. 219–240). Elsevier Inc.

Kulin, R. M., Jiang, F., & Vecchio, K. S. (2011). Effects of age and loading rate on equine cortical bone failure. *J Mech Behav Biomed Mater, 4*(1), 57–75. https://doi.org/10.1016/j.jmbbm.2010.09.006

L'Abbe, E. N., Symes, S., Stull, K. E., Lacroix, M., & Pokines, J. T. (2022). Taphonomy and the Timing of Bone Fractures in Trauma Analysis. In J. T. Pokines, E. N. L'Abbe, & S. Symes (Eds.), *Manual of Forensic Taphonomy* (2nd ed., pp. 473–501). CRC Press.

Lesciotto, K. M. (2015). The impact of Daubert on the admissibility of forensic anthropology expert testimony. *J Forensic Sci, 60*(3), 549–555. https://doi.org/10.1111/1556-4029.12740

Liu, D., Wagner, H. D., & Weiner, S. (2000). Bending and fracture of compact circumferential and osteonal lamellar bone of the baboon tibia. *Journal of Materials Science: Materials in Medicine, 11*(1), 49–60. http://journals.ohiolink.edu/ejc/article.cgi?issn=09574530&issue=v11i0001&article=49_bafocclbotbt

Love, J. C., & Symes, S. A. (2004). Understanding rib fracture patterns: incomplete and buckle fractures. *Journal of Forensic Sciences, 49*(6), 1153–1158.

Lynn, K. S., & Fairgrieve, S. I. (2009). Microscopic indicators of axe and hatchet trauma in fleshed and defleshed mammalian long bones. *Journal of Forensic Sciences, 54*(4), 793–797. http://journals.ohiolink.edu/ejc/article.cgi?issn=00221198&issue=v54i0004&article=793_mioaahfadmlb

Mandair, G. S., Bigelow, E. M. R., Viswanathan, G., Ward, F. S., Patton, D. M., Schlecht, S. H., Jepsen, K. J., & Kohn, D. H. (2022). Region-specific associations among tissue-level mechanical properties, porosity, and composition in human male femora. *Journal of Biomechanics, 139*, 111144. https://doi.org/10.1016/j.jbiomech.2022.111144

Martin, R. Bruce (2003). Functional Adaptation and Fragility of the Skeleton. In: Agarwal, Sabrina C. & Stout, Sam D. *Bone Loss and Osteoporosis: An Anthropological Perspective*. Boca Raton, FL:CRC Press.

Martin, R. B., Burr, D., & Sharkey, N. (1998). *Skeletal Tissue Mechanics*. Springer.

Martin, R. B., Lau, S. T., Mathews, P. V., Gibson, V. A., & Stover, S. M. (1996). Collagen fiber organization is related to mechanical properties and remodeling in equine bone. A comparison of two methods. *Journal of Biomechanics, 29*(12), 1515–1521. http://journals.ohiolink.edu/ejc/article.cgi?issn=00219290&issue=v29i0012&article=1515_cfoirtbacotm

Murray, E., & Anderson, B. E. (2007). *Forensic anthropology in the courtroom: trends in testimony* 59th Annual Meeting of the American Academy of Forensic Sciences, San Antonio, Texas.

Mussawy, H., Ferrari, G., Schmidt, F. N., Schmidt, T., Rolvien, T., Hischke, S., Rüther, W., & Amling, M. (2017). Changes in cortical microarchitecture are independent of areal bone mineral density in patients with fragility fractures. *Injury, 48*(11), 2461–2465. https://doi.org/10.1016/j.injury.2017.08.043

O'Leary, T. J., Izard, R. M., Tang, J. C. Y., Fraser, W. D., & Greeves, J. P. (2022). Sex differences in tibial adaptations to arduous training: An observational cohort study. *Bone, 160*, 116426. https://doi.org/10.1016/j.bone.2022.116426

O'Leary, T. J., Wardle, S. L., Gifford, R. M., Double, R. L., Reynolds, R. M., Woods, D. R., & Greeves, J. P. (2021). Tibial macrostructure and microarchitecture adaptations in women during 44 weeks of arduous military training. *Journal of Bone and Mineral Research, n/a*(n/a). https://doi.org/10.1002/jbmr.4290

Paschalis, E. P., Shane, E., Lyritis, G., Skarantavos, G., Mendelsohn, R., & Boskey, A. L. (2004). Bone fragility and collagen cross-links. *Journal of Bone and Mineral Research, 19*(12), 2000–2004. https://doi.org/10.1359/JBMR.040820

Passalacqua, N. V., & Fenton, T. W. (2012). Developments in Skeletal Trauma: Blunt-Force Trauma. In D. C. Dirkmaat (Ed.), *A Companion to Forensic Anthropology* (pp. 400–411). Blackwell Publishing.

Pechnikova, M., Mazzarelli, D., Poppa, P., Gibelli, D., Scossa Baggi, E., & Cattaneo, C. (2015). Microscopic Pattern of Bone Fractures as an Indicator of Blast Trauma: A Pilot Study. *J Forensic Sci, 60*(5), 1140–1145. https://doi.org/10.1111/1556-4029.12818

Pechníková, M., Porta, D., & Cattaneo, C. (2011). Distinguishing between perimortem and postmortem fractures: are osteons of any help? *Int J Legal Med, 125*(4), 591–595. https://doi.org/10.1007/s00414-011-0570-9

Reber, S. L., & Simmons, T. (2015). Interpreting Injury Mechanisms of Blunt Force Trauma from Butterfly Fracture Formation. *J Forensic Sci, 60*(6), 1401–1411. https://doi.org/10.1111/1556-4029.12797

Samelson, E. J., Broe, K. E., Xu, H., Yang, L., Boyd, S., Biver, E., Szulc, P., Adachi, J., Amin, S., Atkinson, E., Berger, C., Burt, L., Chapurlat, R., Chevalley, T., Ferrari, S., Goltzman, D., Hanley, D. A., Hannan, M. T., Khosla, S., … Bouxsein, M. L. (2018). Cortical and trabecular bone microarchitecture as an independent predictor of incident fracture risk in older women and men in the Bone Microarchitecture International Consortium (BoMIC): a prospective study. *The Lancet Diabetes & Endocrinology*. https://doi.org/10.1016/S2213-8587(18)30308-5

Schafman, M., Kang, Y., Moorhouse, K., White, S., Bolte IV, J., & Agnew, A. (2016). Age and sex alone are insufficient to predict human rib structural response to dynamic A-P loading. *Journal of Biomechanics, 49*, 3516–3522.

Schlecht, S., Bigelow, E., & Jepsen, K. J. (2014). Mapping the natural variation in whole bone stiffness and strength across skeletal sites. *Bone, 67*, 15–22.

Schmitt, K.-U., Niederer, P., & Walz, F. (2004). *Trauma Biomechanics: Introduction to Accidental Injury*. Springer.

Scientific Working Group for Forensic Anthropology (SWGANTH). Trauma Analysis –2011. https://www.nist.gov/system/files/documents/2018/03/13/swganth_trauma.pdf (accessed falseJanuary 5, 2020).

Seref-Ferlengez, Z., Kennedy, O., & Schaffler, M. (2015). Bone microdamage, remodeling and bone fragility: how much damage is too much damage? *BoneKEy Reports, 4*. https://doi.org/10.1038/bonekey.2015.11

Sievänen, H., Józsa, L., Pap, I., Järvinen, M., Järvinen, T. A., Kannus, P., & Järvinen, T. L. (2007). Fragile external phenotype of modern human proximal femur in comparison with medieval bone. *Journal of Bone and Mineral Research, 22*(4), 537–543. https://doi.org/10.1359/jbmr.070102

Simon, L. V., Lopez, R. A., & King, K. C. (2022). Blunt Force Trauma. [Updated 2023 Aug 7]. In: StatPearls [Internet]. Treasure Island (FL): StatPearls Publishing, 2024 Jan_. Available from https://www.ncbi.nlm.nih.gove/books/NBK470338.

Sornay-Rendu, E., Boutroy, S., Munoz, F., & Delmas, P. D. (2007). Alterations of cortical and trabecular architecture are associated with fractures in postmenopausal women, partially independent of decreased BMD measured by DXA: the OFELY study. *J Bone Miner Res, 22*(3), 425–433. https://doi.org/10.1359/jbmr.061206

Sornay-Rendu, E., Munoz, F., Garnero, P., Duboeuf, F., & Delmas, P. D. (2005). Identification of osteopenic women at high risk of fracture: The OFELY study. *Journal of Bone and Mineral Research, 20*(10), 1813–1819. https://doi.org/10.1359/JBMR.050609

Symes, S., L'Abbe, E. N., Chapman, E. N., Wolff, I., & Dirkmaat, D. C. (2012). Traumatic Injury to Bone in Medicolegal Investigations. In D. C. Dirkmaat (Ed.), *A Companion to Forensic Anthropology* (1st ed., pp. 340–389). John Wiley & Sons.

Turner, C. H. (2002). Biomechanics of bone: Determinants of skeletal fragility and bone quality. *Osteoporosis International, 13*(2), 97–104.

Turner, C. H., & Burr, D. B. (1993). Basic biomechanical measurements of bone: A tutorial. *Bone, 14*, 595–608.

Ubelaker, D. H. (2018). A history of forensic anthropology. *Am J Phys Anthropol, 165*(4), 915–923. https://doi.org/10.1002/ajpa.23306

Viguet-Carrin, S., Garnero, P., & Delmas, P. D. (2006). The role of collagen in bone strength. *Osteoporosis International, 17*(3), 319–336. http://journals.ohiolink.edu/ejc/article.cgi?issn=0937941x&issue=v17i0003&article=319_trocibs

Vincentelli, R., & Grigorov, M. (1985). The effect of haversian remodeling on the tensile properties of human cortical bone. *Journal of Biomechanics, 18*(3), 201–207.

Wagner, H. D., & Weiner, S. (1992). On the relationship between the microstructure of bone and its mechanical stiffness. *Journal of Biomechanics, 25*, 1311–1320.

Wang, X., Shen, X., Li, X., & Mauli Agrawal, C. (2002). Age-related changes in the collagen network and toughness of bone. *Bone, 31*(1), 1–7. http://journals.ohiolink.edu/ejc/article.cgi?issn=87563282&issue=v31i0001&article=1_acitcnatob

Warden, S. J., Gutschlag, F. R., Wajswelner, H., & Crossley, K. M. (2002). Aetiology of rib stress fractures in rowers. *Sports Medicine, 32*(13), 819–836. http://journals.ohiolink.edu/ejc/article.cgi?issn=01121642&issue=v32i0013&article=819_aorsfir

Watson, E. O., Fenton, T. W., Isa, M. I., Goots, A. C., Vaughan, P. E., Wei, F., & Haut, R. C. (2018). *The Influence of Implement Shape on Fracture Pattern and Defect Size in Experimental Blunt Cranial Impacts. American Academy of Forensic Sciences*, Seattle, Washington, USA.

Wieberg, D. A. M., & Wescott, D. J. (2008). Estimating the timing of long bone fractures: correlation between the postmortem interval, bone moisture content, and blunt force trauma fracture characteristics. *Journal of Forensic Sciences, 53*(5), 1028–1034. http://journals.ohiolink.edu/ejc/article.cgi?issn=00221198&issue=v53i0005&article=1028_ettolbabftfc

Winter-Buchwalder, M., Schwab, N., Galtés, I., Ortega-Sánchez, M., Scheirs, S., & Jordana, X. (2022). Microcracking pattern in fractured bones: new approach for distinguishing between peri- and postmortem fractures. *International Journal of Legal Medicine*. https://doi.org/10.1007/s00414-022-02875-1

World Health Organization (WHO). (1994). Assessment of fracture risk and its application to screening for postmenopausal osteoporosis: report of a WHO study group [meeting held in Rome from 22 to 25 June 1992].

Wynnyckyj, C., Omelon, S., Savage, K., Damani, M., Chachra, D., & Grynpas, M. D. (2009). A new tool to assess the mechanical properties of bone due to collagen degradation. *Bone, 44*(5), 840–848. http://journals.ohiolink.edu/ejc/article.cgi?issn=87563282&issue=v44i0005&article=840_anttatobdtcd

Zephro, L., Galloway, A., & Wedel, V. L. (2014). Theoretical Considerations in Designing Experimental Trauma Studies and Implementing Their Results. In V. L. Wedel & A. Galloway (Eds.), *Broken Bones: Anthropological Analysis of Blunt Force Trauma* (2nd ed., pp. 73–90). Charles C Thomas.

Zioupos, P. (2001). Ageing human bone: Factors affecting its biomechanical properties and the role of collagen. *Journal of Biomaterials Applications, 15*(3), 187–229. http://journals.ohiolink.edu/ejc/article.cgi?issn=08853282&issue=v15i0003&article=187_ahbfaipatroc

Zioupos, P., & Currey, J. (1998). Changes in the stiffness, strength, and toughness of human cortical bone with age. *Bone, 20*(1), 57–66.

Zioupos, P., Currey, J., & Casinos, A. (2000). Exploring the effects of hypermineralisation in bone tissue by using an extreme biological example. *Connective Tissue Research, 41*(3), 229–248.

Zioupos, P., Kirchner, H. O. K., & Peterlik, H. (2020). Ageing bone fractures: The case of a ductile to brittle transition that shifts with age. *Bone, 131*, 115176. https://doi.org/10.1016/j.bone.2019.115176

Ziv, V., Wagner, H. D., & Weiner, S. (1996). Microstructure-microhardness relations in parallel-fibered and lamellar bone. *Bone, 18*(5), 417–428. http://journals.ohiolink.edu/ejc/article.cgi?issn=87563282&issue=v18i0005&article=417_mripalb

Biomechanical Foundations of Histological Analysis in Limb Bones

The Crucial Role of Load-Complexity Categorization and Collagen Fiber Orientation Analysis When Interpreting Bone Adaptation

5

JOHN G. SKEDROS

5.1 Introduction

In vivo strain data, preferably obtained from controlled experimentation, is the strongest level of evidence for interpreting and more fully understanding how bones are adapted for their load history (Figures 5.1 and 5.2).[1] However, due to ethical and technical issues, obtaining *in vivo* strain data from human and non-human primate bones, commonly studied by anthropologists, is challenging. Consequently, in anthropological studies of appendicular long-bone adaptation (see Appendix for definition), variations and/or changes in material (e.g., histological) and structural characteristics are often used to interpret the load history and evaluate functional adaptations of a bone or bone region (Table 5.1).[2,3] This chapter aims to demonstrate the necessity of microscopic analysis for accurately interpreting limb-bone adaptation in many cases, particularly when strain data are lacking or insufficient.

This chapter is organized into three main sections:

1. Guiding principles of the chapter, a checklist to aid the design of future bone adaptation studies and comparisons between them, and a brief reflection on evaluating the quality of available evidence in the bone adaptation field.
2. Overview of structural *vs.* material adaptation in bone with emphasis on the stress-strain curve for understanding bone material adaptation especially in terms of energy absorption (where regional variations in osteon collagen/lamellar morphotypes and predominant collagen fiber orientations are most relevant).
3. Detailed foray into concepts that play fundamental roles in understanding bone adaptation with an emphasis on interpreting load history using histological analysis in the context of four load-complexity categories.

 DOI: 10.4324/9781003385608-5

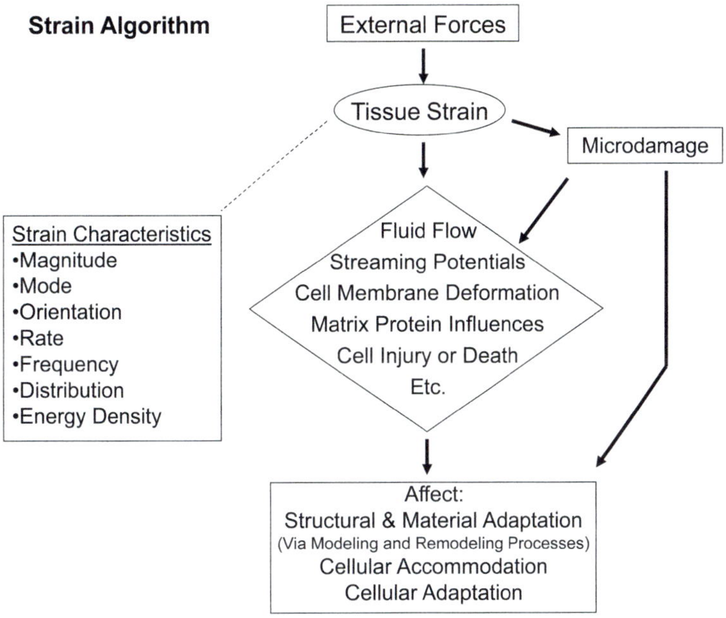

Figure 5.1 Proposed bone stimulus-response algorithm. Tissue strain may have less influence on bone adaptation compared to fluid-flow dynamics and other biophysical or biochemical stimuli. "Cellular accommodation" refers to bone cells adjusting to their physical and biochemical environment (e.g., C. Rubin, et al., 2002; C. H. Turner, 1999). Note that the various mechanisms listed in the central diamond have been proposed as candidate processes mediating bone remodeling (Currey, 2002; Ehrlich & Lanyon, 2002; Robling & Bonewald, 2020).

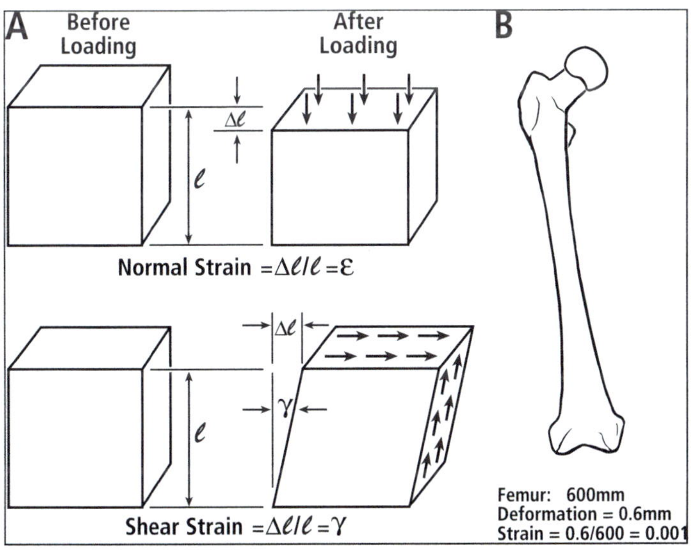

Figure 5.2 Proposed bone stimulus-response algorithm. Tissue strain may have less influence on bone adaptation compared to fluid-flow dynamics and other biophysical or biochemical stimuli. "Cellular accommodation" refers to bone cells adjusting to their physical and biochemical environment (e.g., C. Rubin, et al., 2002; C. H. Turner, 1999). Note that the various mechanisms listed in the central diamond have been proposed as candidate processes mediating bone remodeling (Currey, 2002; Ehrlich & Lanyon, 2002; Robling & Bonewald, 2020).

Table 5.1 Structural and material characteristics of limb-bone diaphyses. Further discussion and description of microstructural vs. nanostructural characteristics can be found in Barth et al. (2010) and Buss et al. (2022)

Biomechanically Important Structural and Material Characteristics in Diaphyseal Bone Hierarchical Organization*

1. Structural Characteristics
- Bone length
- Diaphyseal Curvature
- Cross-sectional shape and robusticity
  - (e.g., moments and axes of area (inertia), relative cortical area)
- Average and regional cortical thickness variations

2. Material Characteristics
- Mineral content (% ash)
- Microstructure
  - Secondary osteon population density and fractional area (On.N/T.Ar, On.Ar/T.Ar);
  - Secondary osteon cross-sectional area, shape, and orientation;
  - Secondary osteon morphotypes (e.g. bright, alternating, parallel-fibered, hooped)
  - Mineral heterogeneity (relatively highly mineralized interstitial bone, young osteons, etc.);
  - Collagen fiber heterogeneity;
  - Porosity (e.g., Haversian canals, primary vascular canals)
  - Lamellar organization of various osteon morphotypes;
  - Variations in primary histologic organization (e.g., laminar vs. reticular vascular patterns in fibrolamellar/plexiform bone);
  - Osteocyte population density, osteocyte lacuna-canalicular geometries;
- Nanostructure
  - Predominant collagen fiber orientation (CFO), collagen density;
  - Types and densities of collagen molecular cross-links;
  - Mineral crystallite and "tesselle" orientation, size, and heterogeneity;
  - Spatial distribution of non-collagenous proteins (e.g., osteopontin and osteocalcin)

* Cancellous (trabecular) bone is not considered here.
See Chapter 7 for further discussion of these terms: laminar, plexiform, fibrolamellar.

5.1.1 Seven Important Rules

1. The stiffness of bones must be managed. Limb bones (our focus here) must not be too stiff, or they will tend toward being brittle, and must not be too flexible because this results in too much elasticity (Currey, 2002; J. G. Skedros, Keenan, Cooper, & Bloebaum, 2014). Notably, in limb-bone adaptation, stiffness and strength requirements of a bone are primarily achieved by the bone modeling process, not the remodeling process (see Appendix and Section 5.4).[4]
2. Bone microdamage is not necessarily a "bad thing." This is because microdamage is an inevitable consequence of the natural loading of bones; but it must be kept in check. However, microdamage should occur infrequently to ensure optimal bone performance, as demonstrated by Reilly and co-worker's (1997) *in vivo* study (see sections 5.2.3 and 5.6.5). The remodeling process is responsible for microdamage repair in bones of many animals, which is especially true in primates and most animals that are generally larger than rats (Lad, 2023).[5] The remodeling process is also usually responsible for strain-related histological adaptation. The modeling process might also be involved in strain-mode-specific[6,7] (or other strain-related)

histological adaptation, especially earlier in ontogeny (even when secondary osteons are few; discussed in Section 5.9).

3. Be cautious of studies that conclude that bone adaptation did not occur when the methods employed only allowed for the analysis of structural characteristics (e.g., cross-sectional shape, cortical thickness, bone curvature) and microscopic examination was not done. The microscope is underused in many bone adaptation studies, especially in the field of anthropology. As emphasized below, the use of a microscope with circularly polarized light is crucial for discerning regional variations in predominant collagen fiber orientation (CFO) and/or secondary osteon (Haversian system)[8] collagen/lamellar "morphotypes." These two material characteristics are extremely helpful in interpreting load history when *in vivo* strain data are lacking (Figures 5.3–5.5). These characteristics are crucial in bone biomechanics because they mediate microdamage repair, restrict the length of microdamage entities (e.g., microcracks), and curb the rate of microdamage formation. This enhances the toughness (energy absorption) of a bone (discussed in Sections 5.2.3 and 5.6.4). This "toughening" cannot be similarly achieved by the modeling process—hence, a microscope is needed to detect this type of material adaptation.

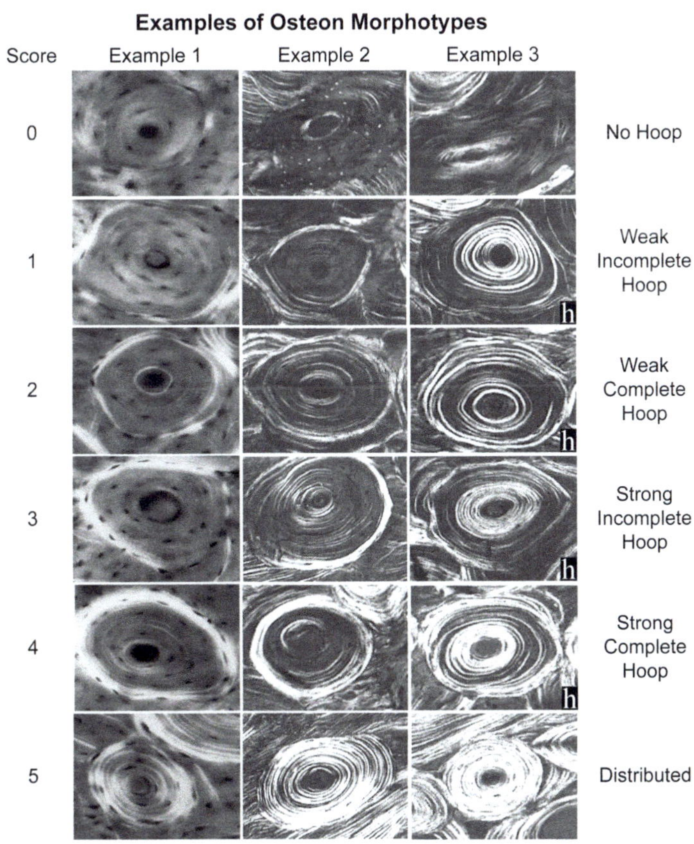

Figure 5.3 The six-point scoring scheme with examples of each osteon birefringence pattern (osteon collagen/lamellar "morphotype") in circularly polarized light. The numerical values of the six morphotypes are used to calculate the osteon morphotype score (MTS) of entire microscopic images that contain many osteons (J. G. Skedros, C. J. Kiser, et al., 2011; J. G. Skedros, et al., 2009).

(Continued)

Figure 5.3 (Continued) Four of the six numerical scores shown include consideration of the completeness and birefringence strength (brightness) of the peripheral ring "O" or hoop:

0 = category N, a dark osteon with no birefringent lamellae;
1 = category OWI, a combination of OI and OW;
2 = category OW, similar to O but the birefringent ring is weak (W);
3 = category OI, similar to O but the birefringent peripheral ring is incomplete (I);
4 = category O osteon with dark interior and strongly birefringent peripheral lamellae;
5 = "distributed" osteon group. This group includes "bright" osteons (less frequent) and "alternating" osteons (more frequent).

Note there can be osteons that are considered to be "hybrid" (h) morphotypes because they have variable gray level patterns within the osteon wall (see the middle four osteon images in the far-right column and see Figure 5.21). These osteons are not as easily scored with the six-number scheme. Hybrid osteon morphotypes might be adapted for predominant/prevalent shear stresses (Robert F. Bigley, et al., 2006; J. G. Skedros, C. J. Kiser, et al., 2011).

The images in the left column are reproduced from Martin et al. (1996) with permission of Elsevier Science, Inc.

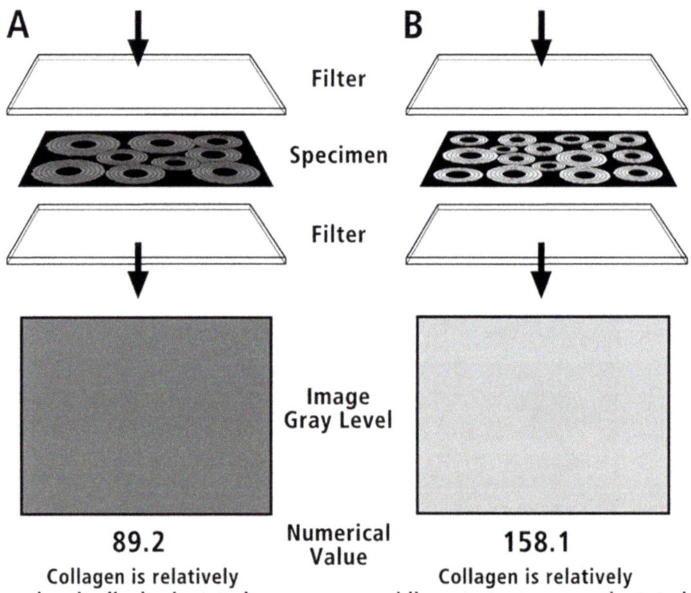

Figure 5.4 Illustration showing how polarizing filters are configured for detecting differences in predominant collagen fiber orientation (CFO) in terms of "weighted mean gray levels" (WMGLs) (Mason, et al., 1995; J. G. Skedros, et al., 2009). Darker WMGLs indicate more longitudinally oriented collagen (typically an adaptation for prevalent/predominant tension); brighter WMGLs indicate more transverse-to-oblique collagen (typically an adaptation for prevalent/predominant compression) (Bromage, et al., 2003; Riggs, et al., 1993). "Hybrid" osteon morphotypes (not shown here) have more heterogeneous variations in gray levels and might be adaptations for shear (Figures 5.3 and 5.22).

4. A bone or bone region (e.g., a diaphyseal location) can typically be placed into four load-complexity categories (Table 5.2). This concept is one of the most important for the design and comparison of studies of limb bone adaptations, as discussed in

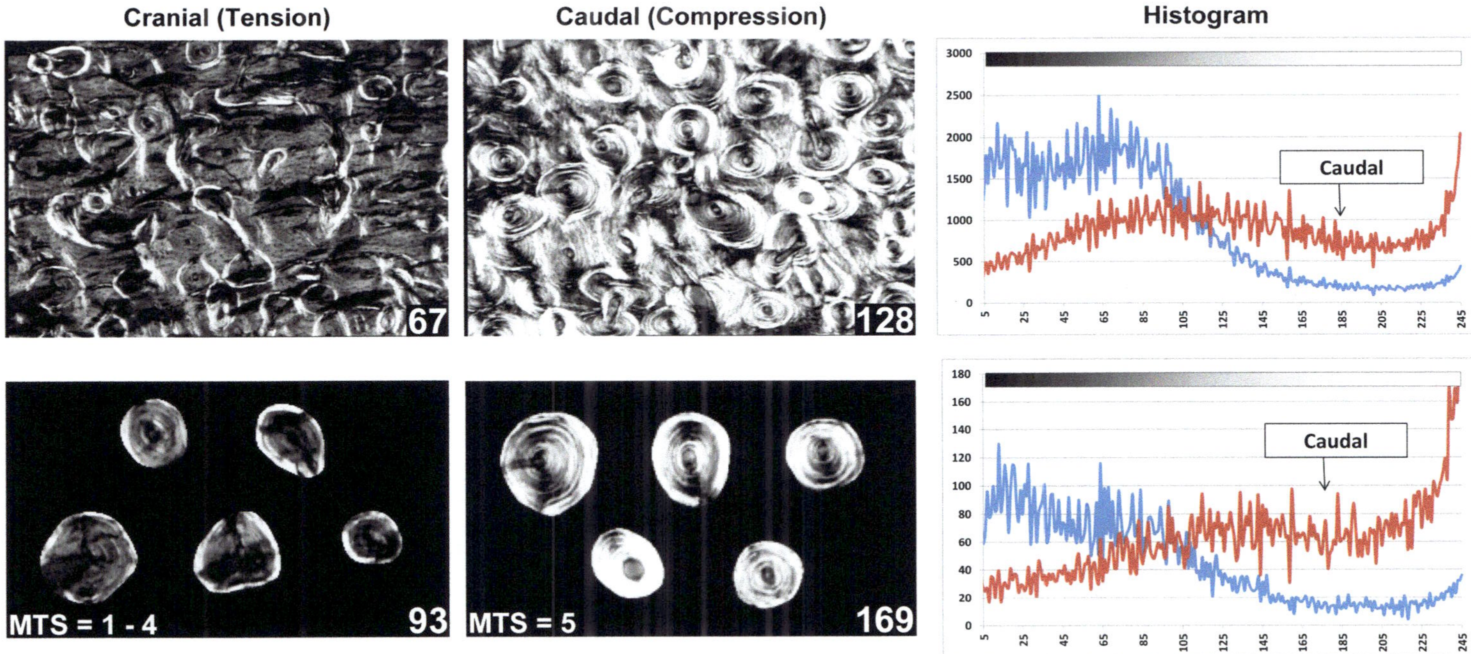

Figure 5.5 Circularly polarized light (CPL) images and gray level profiles of a horse radius (five-year-old animal, mid diaphysis) are shown at **top**. Five isolated osteons and their gray level profiles are shown at **bottom**. The dark-to-bright gradient over each of the gray level profiles (representing predominant collagen fiber orientation (CFO); see at **far right**) show how the gray levels change from left to right on the x axis. Darker gray levels, which are seen more so in the microscopic image at **top left**, represent more longitudinally oriented collagen (an adaptation for tension) (Mason, et al., 1995). Oblique-to-transverse collagen fibers that represent adaptation for habitual compression are seen as increased overall brightness/whiteness in the microscopic image at **top right** (caudal cortex). Abundant osteonal bone is seen in the caudal "compression cortex." The cranial "tension cortex" has a majority of non-osteonal bone (i.e., it is markedly <u>in</u>completely remodeled with secondary osteons). Notably, the secondary osteons and primary (non-osteonal) bone in the image of the cranial cortex have relatively longitudinal CFO (darker gray level) consistent with adaptation for tension. The mean gray level value (excluding the black background, values 0–5 of 256) for each image is shown as the number in the lower right corner. The osteon MTSs (morphotype scores) (Figure 5.3) assigned to the osteons in each of the images at **bottom** are at the lower left of those images. The specimens were embedded in polymethyl methacrylate in an unstained and fully calcified state, and ultramilled to 100 microns thickness (Mason, et al., 1995). All images were taken under the same illumination and 50x magnification.

Table 5.2 **The four load-complexity categories**

Complexity Category	Examples
(Based on N.A. rotation during middle portion of typical load phase.)	
Low	**1. artiodactyl and perissodactyl calcanei**
(N.A.: <10° rotation)	(Lanyon, 1974[‡]; Su et al., 1999; Skedros et al., 2019)
(Tension and compression minimally overlap; Shear is localized near N.A.)	**2. potoroo calcaneus** (Biewener et al., 1996[‡])
	3. chicken tarsometatarsus (TMT)[*] (Judex et al.,1997; Skedros et al., 2003a[‡])
	4. iguana tibia & femur (Blob and Biewener, 2001)
	5. alligator tibia (Blob and Biewener, 2001)
Moderate A	**6. dog, sheep and horse radii[§, **]**
(N.A.: 10°–20° rotation)	(Carter et al., 1980[‡]; Lanyon et al., 1982[‡], Coleman et al., 2002; Takano et al., 1999[‡])
	7. dog ulna (Carter et al., 1980[‡])
	8. macaque ulna (Demes et al., 1998; Skedros et al., 2003a[‡])
	9. macaque tibia (Demes et al., 2001)
	10. river cooter turtle femur (Butcher and Blob, 2008; Aiello et al., 2013)
Moderate B	**11. human fibula**
(N.A.: 20°–45° rotation)	(Lambert, 1971; Thomas et al., 1995; Weaver and Skedros, 2016)
	12. human tibia[§] (Lanyon et al., 1975[‡]; Burr et al., 1996[‡]; Milgrom et al., 2000[‡]; Peterman et al., 2001; Yang et al., 2014)
	13. human femur proximal diaphysis[§] (Cochran et al., 1980[‡]; Aamodt et al., 1997[‡]; Skedros et al., 1999[‡]; Skedros and Baucom, 2007[‡]; Skedros et al., 2012[‡])
	14. chimpanzee femoral neck[§] (Kalmey and Lovejoy, 2002[‡]; Skedros et al., 2008[‡])
	15. dog tibia (Bouvier and Hylander, 1984[‡])
	16. dog proximal femur (Page et al., 1993[‡]; Shahar et al., 2003[‡])
	17. horse third metacarpal[§] (Gross et al., 1992; Skedros et al., 1996)
	18. sheep metatarsal (Lieberman et al., 2004)
	19. opossum femur (Butcher et al. 2011; Gosnell et al., 2011)
	20. armadillo femur (Copploe et al., 2015)
	21. bat metacarpal (Swartz and Middleton, 2008[‡]; Skedros and Doutré, 2019)

Table 5.2 (*Continued*) The four load-complexity categories

	22. immature turkey ulna[§] (Skedros and Hunt, 2004)
	23. chicken tibiotarsus (Biewener et al., 1986[‡]; Biewener and Bertram 1993[‡]; Vitienes et al., 2023[‡])
	24. tegu lizard femur (Sheffield et al., 2011)
High (N.A.: >45° rotation)	**25. human tibia (in some athletes)**[§] (Lanyon et al., 1975[‡]; Burr et al., 1996[‡]; Milgrom et al., 2000[‡]; Peterman et al., 2001; Yang et al., 2014)
(Tension and compression overlap extensively; Shear is relatively more diffusely distributed across the cortex when compared to other categories.)	**26. human femur mid-diaphysis**[§] (Duda et al., 1998[‡]; Goldman et al., 2003[‡]; Drapeau and Streeter, 2006[‡]; Edwards et al., 2008[‡])
	27. human femoral neck (Pidaparti and Turner, 1997[‡]; Skedros et al., 1999[‡]; Skedros and Baucom, 2007[‡]; Edwards et al., 2008[‡])
	28. chimpanzee femoral neck[§] (Kalmey and Lovejoy 2002[‡]; Skedros et al., 2008[‡])
	29. dog femur mid-diaphysis (Schatzker 1980[‡]; Szivek et al., 1992; Shahar et al., 2003[‡])
	30. horse third metacarpal[§] (Skedros et al., 2006; Rubin et al., 2013[‡])
	31. goat radius[§, **] (Main and Biewener, 2004; Main, 2007; Moreno et al., 2008[‡])
	32. sheep tibia (Lanyon and Bourn, 1979[‡]; Lieberman et al., 2004; Gautier et al., 2000[‡])
	33. free-flying bat humerus (Swartz et al., 1992[‡]; Swartz and Middleton, 2008[‡]; Skedros and Doutré, 2019)
	34. mature turkey ulna[§] (Rubin and Lanyon, 1985; Skedros and Hunt, 2004)
	35. emu femur and tibiotarsus (Main and Biewener, 2007[‡])
	36. pigeon humerus (Biewener and Dial, 1995; Skedros and Doutré, 2019)
	37. chicken femur (Carrano and Biewener, 1999[‡]; Skedros, 2002[‡])
	38. alligator femur (Blob and Biewener, 1999; Blob and Biewener, 2001; Lee, 2004[‡])

‡ These studies, despite not reporting neutral axis rotation, contain strain gauge and other types of data that allow for reasonable inference as to what category they belong.

§ (with color highlight) These are bones or bone regions that are in more than one category because of, for example, differences caused by habitual activity changes during ontogeny (e.g., domesticated turkey ulna), habit (e.g., goat vs. sheep radius, **), or volitional activity (e.g., humans or racing horses).

* Generally high shear strains can confound these categorizations in some cases (e.g., chicken TMT). This might also help explain why the human femoral neck is in the high complexity category (Skedros et al. 2023).

Source: Studies that either support segregating these bones or bone regions into these categories using the hypothesized neutral axis rotation criterion, or support the comments in the footnote of this table, are listed here: (Aamodt, et al., 1997; Aiello, Blob, & Butcher, 2013; Biewener & Bertram, 1993b; Biewener & Dial, 1995; Biewener, Fazzalari, Konieczynski, & Baudinette, 1996; Biewener, et al., 1986; Blob & Biewener, 1999, 2001; Bouvier & Hylander, 1984; Burr et al., 1996; Butcher & Blob, 2008; Butcher et al., 2011;

(Continued)

Table 5.2 (*Continued*) The four load-complexity categories

Carrano & Biewener, 1999; Carter, Smith, Spengler, Daly, & Frankel, 1980; Cochran, Zickel, & Fielding, 1980; Coleman, Hart, Owan, Takano, & Burr, 2002; Copploe, Blob, Parrish, & Butcher, 2015; Demes, et al., 2001; Demes et al., 1998; Drapeau & Streeter, 2006; Duda, et al., 1998; Edwards, et al., 2008; Gautier, Perren, & Cordey, 2000; Goldman, et al., 2003; Gosnell, Butcher, Maie, & Blob, 2011; Gross, McLeod, & Rubin, 1992; Judex, Gross, & Zernicke, 1997b; Kalmey & Lovejoy, 2002; Lambert, 1971; Lanyon, 1974; Lanyon & Bourn, 1979; Lanyon, et al., 1982; Lanyon, Hampson, Goodship, & Shah, 1975; A. H. Lee, 2004; Lieberman, et al., 2004; R. P. Main, 2007; R. P. Main & Biewener, 2004, 2007; Milgrom et al., 2000; Moreno, et al., 2008; Page et al., 1993; Peterman, et al., 2001; Pidaparti & Turner, 1997; C. T. Rubin & Lanyon, 1985; C. T. Rubin, Seeherman, Qin, & Gross, 2013; Schatzker, Manley, & Sumner-Smith, 1980; Shahar, Banks-Sills, & Eliasy, 2003; Sheffield, Butcher, Shugart, Gander, & Blob, 2011; J. G. Skedros, 2002; J. G. Skedros & Baucom, 2007; J.G. Skedros, et al., 2008; J. G. Skedros, Dayton, et al., 2006c; J. G. Skedros, Demes, et al., 2003; J. G. Skedros & Doutré, 2019; J. G. Skedros, et al., 1999; J. G. Skedros & Hunt, 2004; J. G. Skedros, et al., 2012; J. G. Skedros, et al., 1996; J.G. Skedros, et al., 2019; Su, Skedros, Bachus, & Bloebaum, 1999; Swartz, Bennett, & Carrier, 1992; Swartz & Middleton, 2008; Szivek, Johnson, & Magee, 1992; Takano, et al., 1999; Thomas, Harris, Willis, Lu, & MacEwen, 1995; Vitienes, et al., 2023; Weaver & Skedros, 2016; Yang, et al., 2014).

more detail below. Load-complexity segregation of a bone or bone region helps one attain a better understanding of how and why adaptation occurred during ontogeny and how adaptation might occur after skeletal maturity. In some cases, actual or potential differences in the "adaptability" of bone in the pre-maturity *vs.* post-maturity phases of ontogeny can confuse one's expectations and interpretations, especially when differences in genetic endowment are at play (I.J. Wallace, et al., 2017). Clarity can be gained in such cases if one is able to: (1) place a bone or bone region in a load-complexity category, and (2) perform adequate histological analysis.

5. Bending is a very common loading condition in many limb bones of many species. As shown in Table 5.2, bending can be exclusively or often unidirectional (i.e., low- and intermediate-complexity load categories). Bending also produces the majority (>70%) of longitudinal strains occurring during peak loading of controlled *in vivo* activity (Biewener & Bertram, 1993a; Biewener, Swartz, & Bertram, 1986; Milne, 2016; Moreno, Main, & Biewener, 2008; Yang et al., 2014). But bending must be regulated so that strains do not become too high, or fracture will occur. However, it is important to emphasize that bending, though must be reduced in many cases, must not be eliminated by muscle "counter-bending" activity (e.g., in accordance with the now-archaic "flexion neutralization theory" proposed by H. Frost (Milne, 2016)) because this nullification can become deleterious in the biological and mechanical functioning of a long bone. This is because even though controlled unidirectional bending produces generally mutually exclusive regions with net tension (seemingly a "bad thing") and net compression, bending that is typically unidirectional is a "good thing" because it is linked to load predictability (J. E. A. Bertram & Biewener, 1988).

   In the adaptation of limb-bone diaphyses (our focus here), curvature of these regions helps ensure load predictability that concurrently optimizes strength (J. E. A. Bertram & Biewener, 1988). There are four additional consequences of load predictability that help to ensure the emergence of a well-adapted bone. First, load predictability allows for differential histological adaptation for the spatially and temporally predictably non-uniform strain mode/magnitude distribution (e.g., populating different regions of a bone cross-section with different secondary osteon collagen/lamellar morphotypes as a means to histologically adapt the bone is a slow process) (G. C. Reilly & Currey, 1999). Second, a predictably non-uniform strain distribution ensures predictable fluid-flow dynamics (e.g., via strain gradients) that are essential for nutrient delivery to bone cells (Ehrlich & Lanyon, 2002; Judex, Gross, & Zernicke, 1997b; R. P. Main, 2007; J. G. Skedros, Mason, Nelson, & Bloebaum, 1996). Third, load predictability ensures predictability of manageable microdamage formation. Finally, load predictability provides bioelectric/biophysical fluid-flow-mediated signals that bone cells likely track as the more proximate stimulus of the strain environment (Aiello et al., 2015; J. E. A. Bertram & Biewener, 1988; Burger, Klein-Nulend, & Smit, 2003; Carter, Caler, Spengler, & Frankel, 1981; deVet, Jhirad, Pravato, & Wohl, 2021; Francillon-Vieillot et al., 1990; Judex, Gross, Bray, & Zernicke, 1997a; T. C. Lee, Staines, & Taylor, 2002; C. T. Rubin, Fritton, Sun, & McLeod, 1996; J. G. Skedros, et al., 1996; J. G. Skedros, Sorenson, Hunt, & Holyoak, 2007a; I.J. Wallace, et al., 2017).

6. There is a "division of labor" in the ultimate mechanical goals of modeling and remodeling, where modeling is the process that governs the achievement of structural requirements, primarily stiffness and strength (Figure 5.6). By contrast, remodeling is the process that governs the achievement of material (i.e., tissue) requirements, primarily toughness and fatigue resistance. When interpreting load history using a bone's structural and material characteristics, it is important to think through the modeling/remodeling distinction (discussed in Section 5.4).
7. Assess the level(s) of the strength of evidence of your study, and, if possible, do this when designing your study. The field of anthropology-related bone adaptation often has studies that state conclusions or promote concepts regarding load-history-related bone adaptation that are based only, or primarily, on a bone's <u>structural</u> characteristics. Such studies often do <u>not</u> provide robust evidence, primarily because they typically lack stronger levels of evidence such as *in vivo* strain data or adequate histological analysis (discussed in Section 5.1.3).

5.1.2 Checklist and Guiding Principles for Bone Adaptation Studies

A good place to start any endeavor that seeks to interpret the load history of a bone or bone region is to use a "checklist of considerations." The checklist shown in Table 5.3 can be used as a worksheet for many studies of bone adaptation, including studies of ribs and short bones. Possibly the most important, and <u>least obvious</u>, consideration is that there are several "load-complexity categories" within which a bone or bone region can be placed (Table 5.2).[9] *In vivo* strain measurements from bones of various mammalian and avian species demonstrate that predominant bending (low- to intermediate-complexity loading) or torsion/multi-directional (high-complexity loading) are most prevalent during controlled ambulation and are therefore the most representative load histories of bone diaphyses, with most other natural/physiological load regimes being somewhere between these load scenarios (see references in Table 5.2). The load-complexity categories are differentiated by magnitudes of rotations/shifts of the neutral axis (Figures 5.7 and 5.8), which is applicable to all types of bone load conditions (Figure 5.9).

This chapter demonstrates that load-complexity categorization is a crucial and highly heuristic hypothesis/consideration for all bone adaptation studies, especially in the field of anthropology where the lack of *in vivo* strain data is the norm. Investigators of bone adaptation are urged to consider designating (in the Methods section of their studies) ssstudyingsis(are) most likely in. Following these guidelines will facilitate

Table 5.3 Checklist/worksheet of important considerations for interpreting adaptation in the appendicular skeleton and ribs

A CONSIDERATIONS		Human femur	Human tibia	Horse radius	Sheep radius	Goat radius	Deer calcaneus	Alligator femur	Human rib	Turkey Ulna	Rat femur	Rat ulna	Pigeon humerus
Examination of Bone Diaphyses of "Similar Volume" (left of rib) and Small Volumes (right of rib), and Different Species													
Animal Size/Mass Category (*mammalian* locomotor behavior; skeletal form/scaling issues)													
	Mammal: Large (1–300 kg)	X	X	X	X	X	X	X	X	X			
	Medium (0.1–1 kg)												X
	Small (0.001–0.1 kg)										X	X	
	Reptile							X					
	Bird												X
[See Biewener, 1990]	Other												
Bone Volume/Size (Stressed Volume Effect)													
	"Large" (larger than "small")	X	X	X	X	X	X	X					
	"Small"									?	X	X	X
Skeletal Division[‡]													
	Appendicular	X	X	X	X	X	X	X		X	X	X	X
	Axial								X				
Skeletal Location (proximal vs. distal) (regional functional morphology; e.g., humerus vs. metacarpal)													
	Proximal location	X						X					
	Intermediate or Distal location		I	I	I	I	D			I	D	I	D
	Other location or issue								X				

Muscle, Ligament, Tendon "Protection"/Influences												
High	X						X			X		X
Medium		X	X	X	X			X	X		X	
Low (mostly surrounding tendons)						X						
Within Bone Level (includes cortical vs. cancellous issues)												
Epiphyseal/metaphyseal						X						
Diaphyseal	X	X	X	X	X	X	X	NA	X	X	X	X
In Vivo Strain Environment												
Well characterized			X	X	X				X			X
Fair/poorly characterized	X	X				X	X				X	
No data								X		X		
Load Complexity Category												
High = reduced load predictability												
↕ High	X	X								?		X
Intermediate B	X	X	X	X	X		X		X	X		
Intermediate A			X	X	X			?	X		X	
Low						X		?				
Low = increased load predictability												
Volitional Changes in Habitual Load Complexity (e.g. human exercise)												
Yes, No, Possibly (P)	Y	Y	P	N	N	N	N	N	N	N	N	N
Maturation-related Changes in Habitual Load Complexity												
Yes, No, Possibly (P)	N	P	N	N	N	N	N	N	Y	N	N	N

Table 5.3 (*Continued*) **Checklist/worksheet of important considerations for interpreting adaptation in the appendicular skeleton and ribs**

Examination of Bone Diaphyses of "Similar Volume" (left of rib) and Small Volumes (right of rib), and Different Species												
B CONSIDERATIONS	Human femur	Human tibia	Horse radius	Sheep radius	Goat radius	Deer calcaneus	Alligator femur	Human rib	Turkey Ulna	Rat femur	Rat ulna	Pigeon humerus
Bone Selected for: (based primarily on K values from bird and mammal data (Currey and Alexander, 1985))[†]												
Yield Strength												
Fatigue Strength												
Ultimate Strength												
Impact Strength												
Stiffness												
Other												
Importance of Toughness or Energy Absorption												
High	X	X	X	X	X	X						
Intermediate												
Low							X	X	X	X	X	X
Cortical Envelope[a]												
Circumferential lamellar bone												
Periosteal derived; Pericortex												
Middle												
Endosteal derived; Endocortex												
All envelopes												
Anticipated Histology												
Fibrolamellar/Plexiform or Other Primary							X		X	X	X	X
< 25% Osteonal							X		X			
> 25% Osteonal	X	X	X	X	X	X		X				
Other												

Metabolism-mediated Remodeling												
High								X				
Intermediate												
Low	X	X	X	X	X	X	X		X	X	X	X
Sexual Dimorphism; Race/Ethnicity												
Yes, No, Possibly (P)	P	P	N	N	N	N	N	P	N	N	N	N
Growth Rate, Bone/Region												
High												
Medium			X	X	X	X		X	X	X	X	X
Low	X	X				X	X	X		X	X	
Phase of Ontogeny[b]												
Attainment												
Maintenance	X	X	X	X	X	X	X	X	X	X	X	X
Senescence												
Aging/Senescence Effects[c]												
Strong												
Medium	X	X						X				
Weak			X	X	X		X		X	X	X	X
Possibly						X	X					
[see Pearson & Lieberman, 2004] None												

‡ Cranial-facial, vertebral, and sternal adaptations are not considered here.
† K values (top of second page) are measures of robusticity; these are not marked above because they had not all been determined.
a = see Chapter 2 for discussion of the "PEM" (periosteal and endosteal membranes);
b = non-elderly adults are in the maintenance phase.
c = these are marked here as if some of the humans had advanced age (e.g., 65–75 years). (see Zebaze et al. 2010). **NA** = not applicable.
? = unclear or unknown.

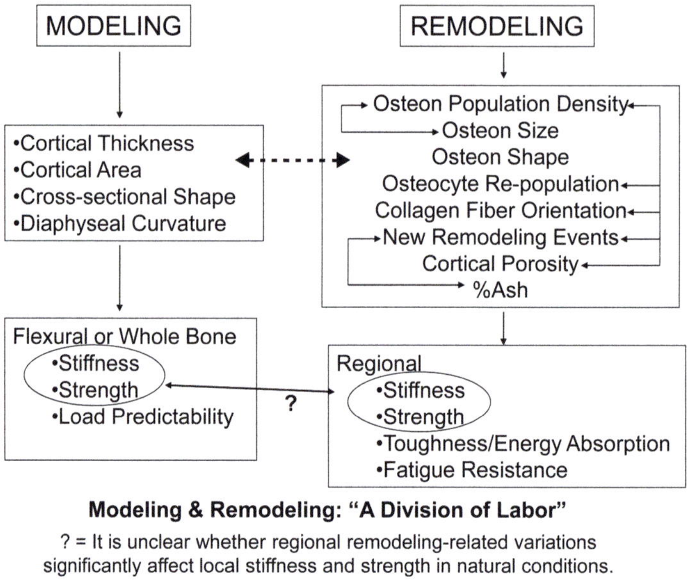

Figure 5.6 A structural *vs.* material "division of labor" is suggested between modeling and remodeling processes in limb bone development and adaptation. In this context, modeling processes seem sufficient for achieving requisite stiffness, strength, and load predictability in a whole bone and/or discrete region (e.g., in the middle portion of the diaphysis, or between anterior cortex *vs.* posterior cortex of this region) (J. E. A. Bertram & Biewener, 1988; J. G. Skedros, et al., 1996). The remodeling process primarily influences regional material properties related to toughness and fatigue resistance. Interrelationships among the material characteristics are shown with arrows and in the context of biomechanically relevant adaptation. The question mark (?) indicates that it is controversial whether or not necessary adjustments in regional stiffness and strength in a bone diaphysis are actually achieved by the remodeling process in natural conditions because these adjustments can be satisfied by the modeling process during growth. The horizontal dashed line indicates that there are studies suggesting that exceptions might occur that reflect some degree of structural/material compensation, such as described by Jepsen and co-workers (Goldman, et al., 2014; Jepsen, 2011).

making accurate assessments and appropriate comparisons in studies of bone adaptation, especially those dealing with load history.[10]

5.1.3 Levels of Evidence

Figure 5.10 is a "pyramid of evidence" for types of evidence typically utilized in biological anthropology and related fields and are ranked from strongest to weakest. The pyramid is based upon levels of evidence guidelines set-forth by the *Journal of Bone and Joint Surgery*. Note that because anthropological research regarding load-history-related bone adaptation is based primarily on a bone's structural characteristics and typically also lacks *in vivo* strain data and/or adequate histological analysis, such structure-based studies are shown as being in the lower levels of our (i.e., bone adaptation field) pyramid of evidence.

Note that in Figure 5.10 the four "load-complexity categories" are considered to be based on reasonably good evidence (Table 5.2). Where the current evidence might be deemed rather fair/poor in less-studied bones, the level of evidence will likely improve when they are given sufficient attention and funding to adequately examine them. For example, the recent *in vivo* strain study of Vitienes et al. (2023) on the chicken tibiotarsus lead to the

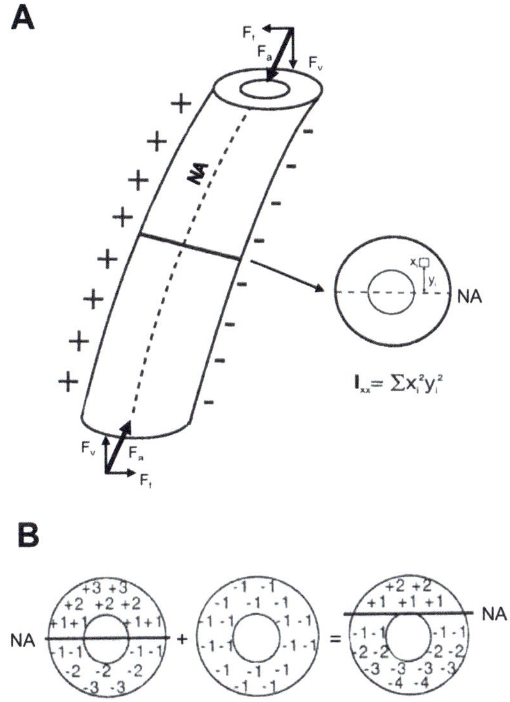

Figure 5.7 **A**. Idealized curved cylindrical beam-like bone subject to loading at proximal end from body mass and at distal end from ground reaction force. Axial forces (Fa), which generate compression, can be resolved into transverse (Ft) and vertical (Fv) components. **NA**, neutral axis around which compressive (–) and tensile (+) strains occur. Bending moment at mid-shaft cross section is counteracted by second moment of area, I. **B**. Effect of axial compression on NA position (the NA region is a "shear region"). At **left**, a section subjected to pure bending, with hypothetical values of compression (–) and tension (+). Superimposition of an evenly distributed compressive force of -1 causes NA to shift towards the cortex subject to tension (**far right**), thus making the strain magnitudes relatively lower there. (Figure reproduced from Lieberman et al. (2004) with permission from John Wiley & Sons, Inc.)

inclusion of this bone in the load-complexity table even though neutral axis rotation data were not explicitly provided. These investigators will report these data in a follow-up study (personal communication, I. Vitienes). Achieving the strongest level of evidence (i.e., *in vivo* strain studies) is very time- and technology-intensive, and expensive. Perhaps one day bone adaptation researchers will find it beneficial to explicitly state the strength of the level of evidence so that readers know the capacity of a study to establish a causal relationship (e.g., of a "typical" load history of a bone with its structural and material characteristics).

5.2 The Stress-Strain Curve and Other Basic Mechanical Concepts

There are online videos (e.g., on YouTube) that provide excellent tutorials for learning about load-deformation curves (from whole-bone tests) *vs.* stress-strain curves (from bone material tests) (Figure 5.11A). For readers who want to brush up on these concepts they can consider these resources or others (Cowin, 2001; Currey, 2002; Martin, Burr, & Sharkey, 1998; C.H. Turner & Burr, 1993).

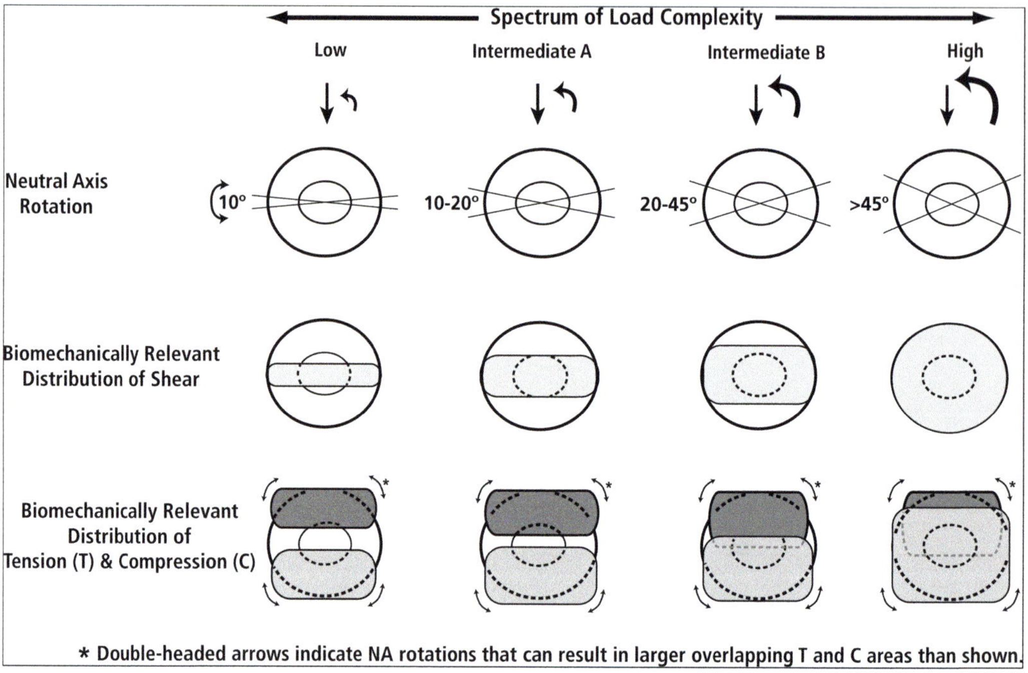

Figure 5.8 Diagrammatic cross-sections of mid-diaphyseal regions of long bones of a medium-size animal (e.g., sheep or chimpanzee) showing magnitudes of hypothesized neutral axis (**NA**) rotations that differentiate the four load-complexity categories (**Table 5.2**). These include: (1) simple (low-complexity) loading where the three strain modes (compression, tension, and shear) exhibit minimal overlap (drawings at **left**), (2) two intermediate load-complexity categories where the strain modes moderately overlap, and (3) torsion/multi-directional (high-complexity) loading (drawings at **right**) where the three strain modes exhibit substantial overlap; thus, shear strains are more diffusely distributed across the bone cross-section. In this hypothesized scheme, the diffuse distribution of shear strains is a hallmark of the high-complexity load category; regional variations in material adaptations (**Table 5.1**) would not be expected in a region with one diffusely distributed strain mode.

5.2.1 The Structural vs. Material Distinction (Table 5.1)

For most studies of bone adaptation, the following mechanical properties are the most important to thoroughly understand: stiffness, strength, toughness, and fatigue resistance. These properties are usually considered in terms of the behavior of bone material, but to some degree they can describe how well the whole bone can bear loads, which is measured via structural strength tests. As argued in Section 5.4 (The Modeling/Remodeling Distinction), the structural *vs.* material distinction is important for limb-bone adaptation studies; especially because a bone's structural requirements may have little to do with regional variations in its material/compositional requirements (Figure 5.6).

5.2.2 Solutions in Engineered Structures: I-beam Analogy

An engineering solution for a structure that must accommodate unidirectional bending is an I-beam (Figure 5.12A). Among the various bones that we have studied, those that tend to resemble I-beams are sheep and deer calcanei.[11] When bending is unpredictable, occurring in multiple directions, a shifting neutral axis must be accommodated in order to avoid catastrophic failure. In bones, such loading is typically the result of a relatively complex load

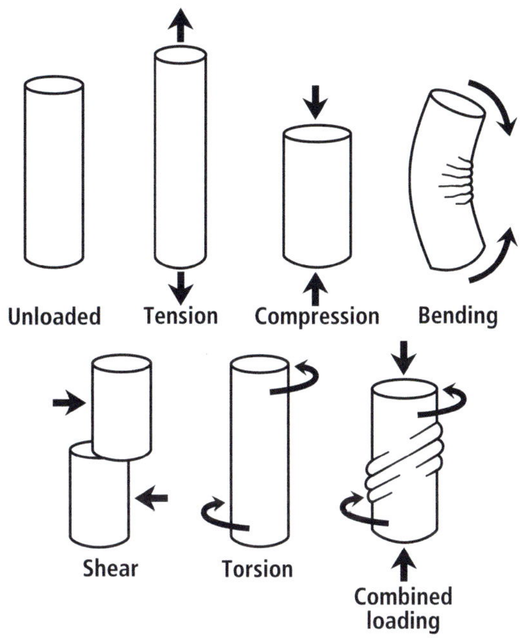

Figure 5.9 Various idealized load conditions that can be imparted to a bone or bone region. Although shear would also exist in a bone subject to bending (mostly toward the neutral axis, **Figures 5.7 and 5.8**), it is more prevalent/predominant and diffusely distributed in a torsion/multi-directional loading environment (Figure 5.8).

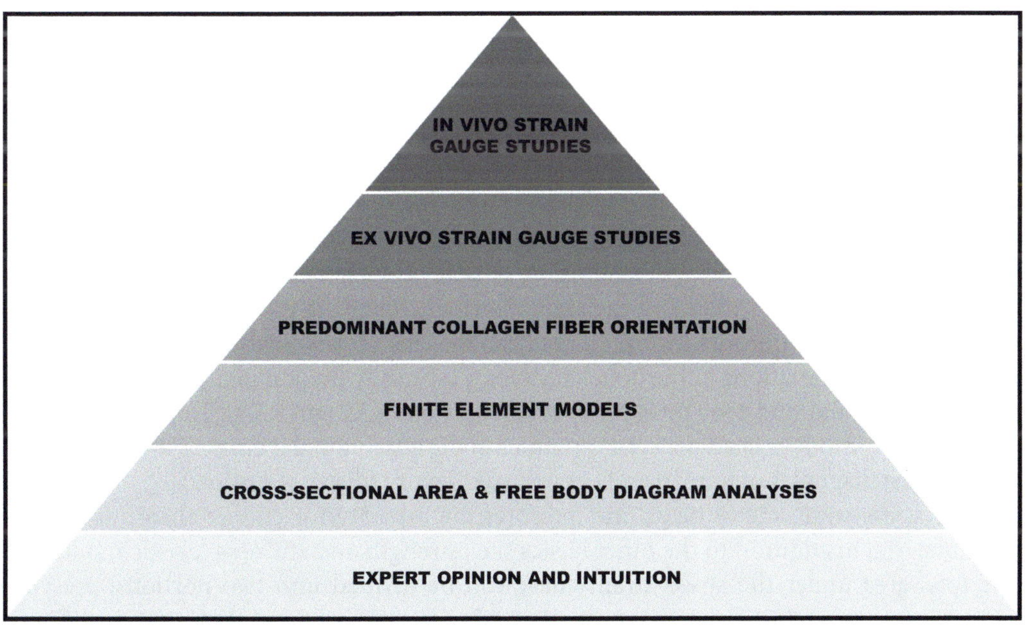

Figure 5.10 Proposed pyramid of levels of evidence for bone adaptation studies. Note that the quantification of predominant CFO can be more important than some finite element studies. If a finite element study is based on *in vivo* strain gauge data (as in (Gross, et al., 1992)), then that would be toward the top of the pyramid. However, it is common that finite element models used in adaptation studies of long bones do not incorporate *in vivo* strain data. This is primarily because these models are often made of the <u>human</u> femur where there is only <u>one</u> study reporting *in vivo* strain data (and only on the lateral aspect of the upper diaphysis/metaphysis region) (Aamodt, et al., 1997).

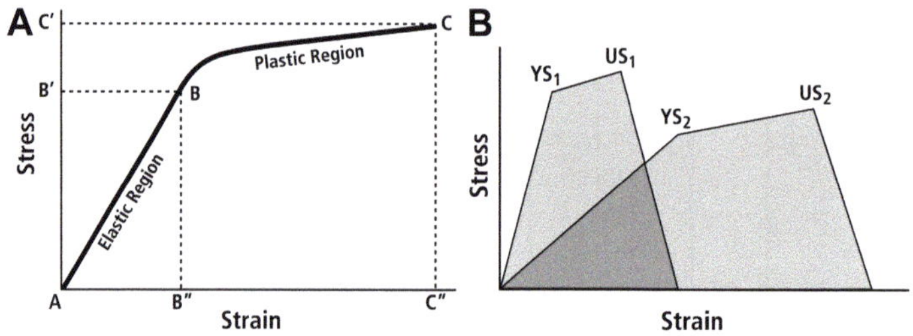

Figure 5.11 **(a)** Typical stress-strain curve obtained from a material test in tension (pulled). Beyond the yield point (B) is where permanent deformation of the specimen occurs. Yield stress (B′) is the load per unit area sustained by the specimen at the yield point, and yield strain (B″) is the amount of deformation withstood by the specimen before plastic deformation occurred. The strain at any point in the elastic region of the curve is proportional to the stress at that point. Therefore, the slope of the elastic region (A-B) is the material stiffness or elastic modulus; increased steepness of the A-B slope represents increased stiffness. The ultimate failure point (C) is the point beyond which failure of the specimen occurred. Ultimate stress (C′) and strain (C″) are shown. The area A-B-B″ is the elastic energy absorbed; the area B″-B-C-C″ is the plastic energy absorbed. Total energy absorbed is the sum of elastic and plastic energy. Brittle materials have shorter B-C regions than more ductile materials (longer B-C regions). **(b)** Two stress-strain curves obtained from specimens with significantly different histocomposition. The more compliant material (at **right**) requires more energy to break it. Energy absorption is proportional to the area under each curve; hence the less compliant material (at **left**) absorbs less energy than the more compliant material even though it has higher yield stress (YS) and ultimate stress (US).

history that is often characterized by prevalent/predominant torsion. A structural solution for this loading might be an array of radially arranged I-beams; however, the intersections of the I-beams that result (Figure 5.12B) are not efficient because the material toward the center is not needed. What then emerges is an annulus (Wainwright, Biggs, Currey, & Gosline, 1982). Bone biomechanics teaches that prevalent torsion (e.g., often more typical toward the mid-diaphysis of some bones) is associated with an annular cross-sectional shape—in limb diaphyses this morphology suggests the presence of a broadly shifting neutral axis (Biewener & Bertram, 1993a; Lieberman, et al., 2004).

But be aware that torsion and bending are typically concurrent (e.g., intermediate-B load-complexity category) and, therefore, the adaptation for bending of a quasi-annular cross-section might be "hidden" within its material organization. This means that there can be bending-adapted regional variations in the bone's histology, especially predominant collagen fiber orientation (CFO) and/or osteon morphotypes (see Sections 5.3.2 and 5.3.4). This can explain the regional strain-mode-related material variations in the quasi-annular cross-sectional shapes of the proximal diaphyses of human and chimpanzee femora (discussed below).

The stress-strain curve helps one interpret the capacity for energy absorption of the bone material in addition to the other basics (i.e., strength and stiffness (elastic modulus)). The total area under the stress-strain curve can be divided into two portions: pre-yield energy absorption (elastic energy) and post-yield energy absorption (plastic energy) (Figure 5.11B). As described below, regional CFO variations typically correlate more strongly with energy absorption than with stiffness or strength (J.G. Skedros, Dayton, Bachus, Bloebaum, & Cronin, 2023; J. G. Skedros, Dayton, Sybrowsky, Bloebaum, & Bachus, 2006).

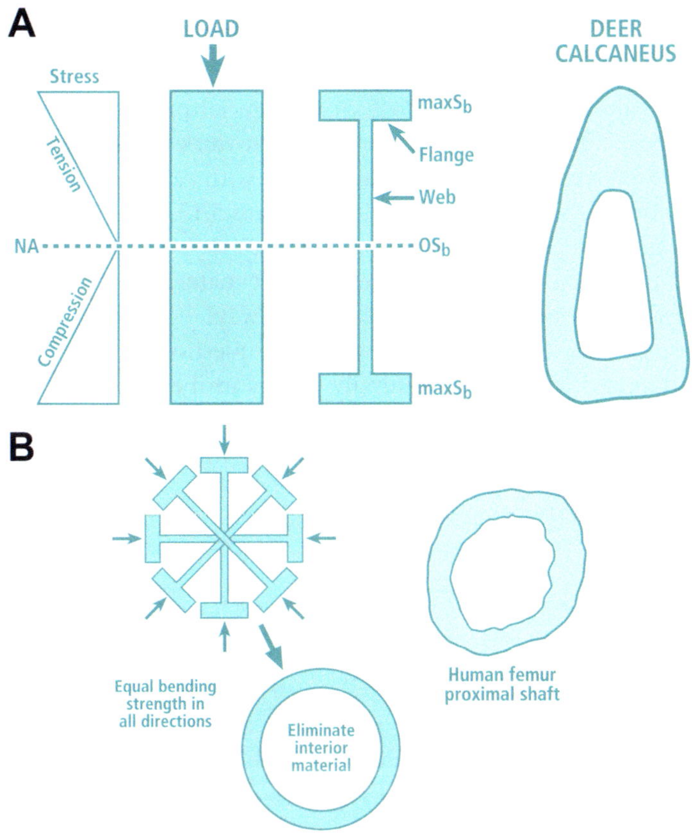

Figure 5.12 **(a)** A simple structural solution for unidirectional loading is an I-beam. The neutral axis (NA) is where the longitudinal strains fall to zero, which is in the central portion of a rectangular beam (middle) and the I-beam. The I-beam exhibits an economy of material where the maximum beam stresses (Sb) are in the flanges; these longitudinal stresses are nearly zero at the center of the web (0Sb). The section at the far right is from a mule deer calcaneus, which resembles a cantilevered beam and is in the low-complexity load category (**Table 5.2**) (J.G. Skedros, Cronin, et al., 2023a). **(b)** Analogy of torsion/multi-directional loading being accommodated by multiple intersecting I-beams, which is most efficiently accommodated by making the structure tubular (annular cross-section). The section at the far right is a transverse section from a human proximal femoral diaphysis. Even though this section is quasi-circular, statistically significant variations in predominant CFO that correspond to habitual medial-lateral bending have been reported. Consequently, a section with this shape does not necessarily rule out the coexisting presence of habitual bending. Note that the proximal human femoral diaphysis is in an intermediate-complexity load category.

In addition to stiffness, strength, and toughness, fatigue resistance is one of the four most important mechanical parameters that must be considered when interpreting the load history and adaptation of a bone or bone region. For detailed discussion of fatigue behavior, failure, and resistance in bone, readers are referred to Currey's book (2002) or to online video tutorials.

5.2.3 Use of Stress-Strain Curve for Considering Mechanisms of Adaptation in the Context of the Role Played by Osteon Morphotypes

Consider this question: How are the post-yield material properties of a bone enhanced when it would seem that natural selection would be at work in the pre-yield portion of the stress-strain curve (where bone tissue usually exists with regard to pre-fracture deformation throughout the course of an animal's life) (Figure 5.13A)? Can this be accomplished through quasi-independent natural selection for pre- *vs.* post-yield behavior, or is post-yield behavior simply a consequence of the inherent material properties of bone? Or are pre- and post-yield behavior linked in a less obvious way?

Three important facts that help to answer these questions are: (1) remodeling can be activated by bone microdamage that naturally occurs in the pre-yield phase (G. C. Reilly, 2000) (remember that tough materials allow some microdamage formation as long as these microdamage entities remain small by not propagating much; osteon interfaces help here);

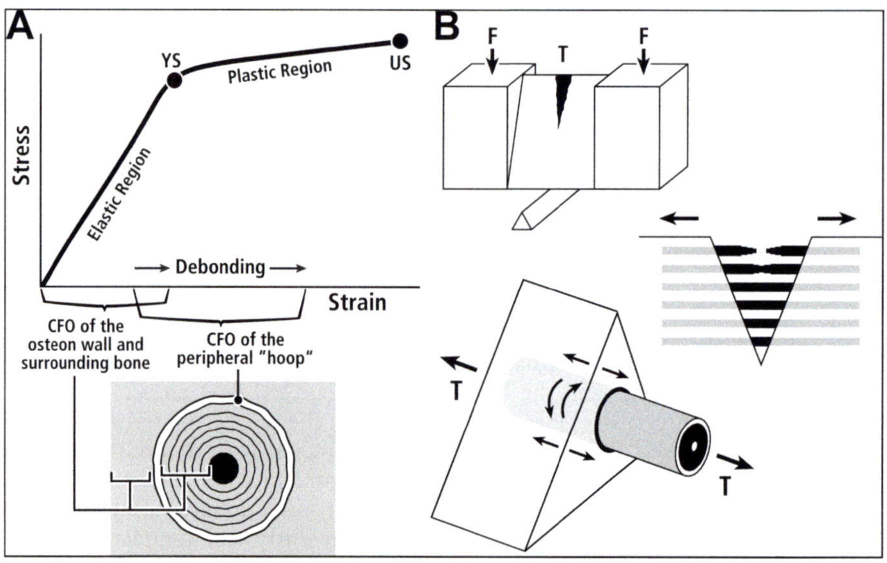

Figure 5.13 **(a)** Stress-strain curve of a tension test that shows the hypothesis that the longitudinal collagen in "tension-adapted osteons" (Figure 5.3) satisfies the material requirements of the bone-mineral composite for elastic behavior by curbing excessive strains in order to avoid irreversible damage. The oblique-to-transverse collagen fiber orientation (CFO) of the peripheral lamellae ("hoops") of these osteons (Figure 5.3) allows controlled debonding at the cement line as the yield point (YS) is approached, and this can continue beyond this point into the initial portion of the plastic region (between YS and US). This "beneficial" debonding dissipates energy that helps to avoid reaching the US point where fracture then occurs. This depiction is based on data and observations described in several studies (Robert F. Bigley, et al., 2006; Martin, et al., 1998; J.G. Skedros, Dayton, et al., 2023b; J. G. Skedros, Keenan, et al., 2013a). **(b)** Three drawings depicting an energy absorption (toughness) test in tension (T) of a machined specimen (at top) loaded at each end with a force (F). Energy is absorbed as the osteons debond, pullout, and bridge the forming crack (middle drawing). Osteon morphotypes that appear to be adapted to accommodate habitual tension and resist tension overload have a peripheral lamella with highly oblique/transverse collagen ("hoop") that probably also absorbs energy by resisting shear stresses (curved lines in lower drawing) during debonding and pull out. (Adapted from Martin et al. (1998), Figure 5.14, pg. 203.)

(2) remodeling can probably be activated by strains that are greatly amplified at the osteocyte level but may not have caused classical microdamage (e.g., microcracks) (see footnote no. 5); and (3) osteonal interfacial debonding (e.g., at the cement line) can be a "good thing" because it allows for the dissipation of energy, which helps to avoid ultimate failure (Allahyari et al., 2023; Robert F. Bigley, Griffin, Christensen, & Vandenbosch, 2006; Hiller et al., 2003; G. C. Reilly, et al., 1997; J.G. Skedros, Dayton, et al., 2023b; J. G. Skedros, Keenan, Williams, & Kiser, 2013) (Figure 5.13B). Adaptations that enhance the performance of bone tissue that are triggered by microdamage (an "extra-genetic" factor, thus are most readily subject to natural selection) include the formation of osteons and osteon morphotypes and/or remodeling-mediated adjustments in bone microstructure/nanostructure (e.g., interfaces and bonding characteristics of predominant CFO, and/or patterns of microvascular anatomy and their associated histological interfaces). These can be thought of as "primary adaptations"; namely, those that are primarily aimed at enhancing pre-yield behavior.

These primary adaptations can have beneficial effects for post-yield behavior. In turn, even though the post-yield (plastic) phase is not often experienced, the possibility that strains would become high enough to enter this phase must be anticipated (or "pre-adapted") so that failure is not rapidly catastrophic. From this perspective, it is plausible that the differences in tension-adapted osteon morphotypes (e.g., the degree of the completeness of the peripheral bright "hoop") could help enhance "beneficial" debonding when post-yield strains are experienced. By contrast, the majority of the wall of tension-adapted osteon morphotypes (i.e., the longitudinal collagen fibers) accommodates the requirements for pre-yield behavior, which is where usual/typical loading occurs (J.G. Skedros, Dayton, et al., 2023b).

It might seem that osteon size should also play a role in this context (Robert F. Bigley, et al., 2006; Hiller, et al., 2003), and could be linked to osteon morphotype (e.g., larger osteons having brighter and more complete peripheral "hoops"). However, we did not find evidence of this in bones where it should have been obvious (J. G. Skedros, Keenan, et al., 2013a). Nevertheless, this coupled pre-yield/post-yield adaptation idea warrants additional investigation in order to better understand how a bone's nanostructural and microstructural properties contribute to its toughness across the pre-yield/post-yield transition (J. G. Skedros, Keenan, et al., 2014a). Osteon morphotypes and predominant CFO are thought to be intimately involved in this process (J.G. Skedros, Dayton, et al., 2023b).

5.3 Crucial Role of Load-Complexity Categories

To understand the genesis of these load-complexity categories it is important to remember that bending of a bone, or any beam-like structure, creates a neutral axis, which is a region where longitudinal strains (i.e., those along the long axis of the bone or beam) drop to zero in materials that are ideal (isotropic, uniform, and continuous). In other words, the neutral axis is the location (or a "zone" because the neutral axis is never stationary in the natural loading of bones) where there is a transition from tension to compression (Figures 5.7 and 5.8). In neutral axis regions, shear strains become large because the maximum strains are obliquely oriented with respect to the long axis of the bone. Therefore, neutral axis rotations/variations can be used to provide criterion for differentiating the load-complexity categories.

As shown in **Table 5.2**, the load-complexity categories are differentiated by differences in the magnitude of neutral axis rotations/shifts during habitual/typical loading (Figure 5.8). Another less available criterion could be coefficients of variation in peak strains (R. P. Main & Biewener, 2004) or other measures of variation of the strain milieu. More research is needed to explore these or other criteria in order to advance the validity and utility of the load-complexity category as an important "working hypothesis."

As mentioned above, the low-complexity load category has bones that are habitually loaded in unidirectional bending; hence they have relatively stationary neutral axes. In contrast, bones in the high-complexity load category have broadly shifting/rotating neutral axes as the result of their habitual torsion/multi-directional loading. The remainder of this chapter will show that these categories contain bones or bone regions (e.g., mid diaphyses) that exhibit regional variations in bone material adaptation (typically see in terms of histomorphological characteristics) that can range from very obvious (e.g., low- to intermediate-complexity loading) to deceptively subtle or seemingly absent (e.g., high-complexity loading) (Figure 5.14).

Because of their relevance to the load-complexity category hypothesis, predominant collagen fiber orientation (CFO) and osteon morphotypes, which are strong correlates of load history, are important to consider. In nearly all bones that have been studied, osteon morphotypes produce regional variations in CFO (Beckstrom, Skedros, Kiser, & Keenan, 2010; Mason, Skedros, & Bloebaum, 1995; J. G. Skedros & Hunt, 2004; J. G. Skedros, Hunt, & Bloebaum, 2004; J. G. Skedros, Keenan, et al., 2013a; J.G. Skedros, Kiser, & Mendenhall, 2011c; J. G. Skedros & Kuo, 1999; J. G. Skedros, et al., 1996; J. G. Skedros, Mendenhall, Kiser, & Winet, 2009; J. G. Skedros, Sorenson, Hunt, et al., 2007). When *in vivo* strain data are lacking or insufficient, the distribution of osteon morphotypes and/or predominant CFO (Figures 5.3–5.5). may be the most reliable characteristic for differentiating a bending load history (i.e., mostly unidirectional) from a load history typified by torsion/shear (i.e., produced by multi-directional loading as is common in many long bones) (J. G. Skedros & Doutré, 2019; J.G. Skedros & Keenan, 2016; J. G. Skedros, Keenan, et al., 2013a; J.G. Skedros, et al., 2011; J. G. Skedros, et al., 2009; Warshaw, Bromage, Terranova, & Enlow, 2017). Traditionally, attempts to interpret the load history of a bone relied on population densities of secondary osteons and other osteon characteristics (e.g., cross-sectional size and shape). However, these approaches yielded limited success due to inconsistencies among bones and species, unrecognized differences in load complexity, and the absence of osteon morphotype recognition (see Section 5.7 for additional discussion).

These drawings are reproduced with permission of John Skedros from Keenan et al. (2017).

5.3.1 Shear Resistance-Priority Hypothesis vs. Tension Resistance-Priority Hypothesis

In order to optimize the use of the load-complexity category hypothesis, situations where the shear resistance-priority hypothesis (Figure 5.15) is at play must be detected (J. G. Skedros, Keenan, Mears, & Langston, 2015; J.G. Skedros, Su, Knight, Bachus, & Bloebaum, 2019). Are the stark disparities in the mechanical properties of bone in tension, compression, and shear reflected in the bone's histomorphological organization? Consider the likelihood that

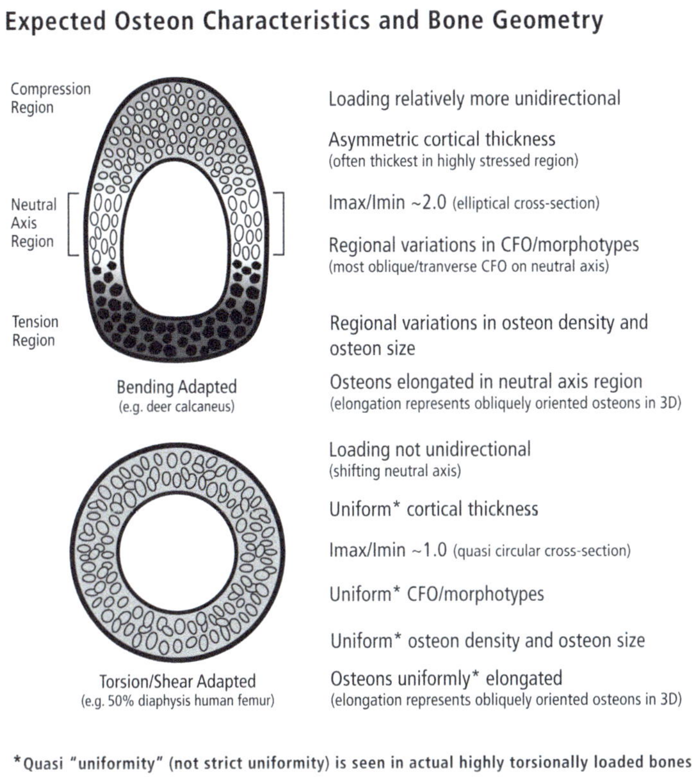

Figure 5.14 These simplified bone cross-section drawings show strain-related regional material variations in a bone subject to habitual unidirectional bending (low-complexity loading) (at **top**) and lack of regional material variations in a bone subject to habitual torsion/multi-directional loading (at **bottom**). The **top** section represents strain modes (compression at top, shear in the middle, and tension at bottom) observed in artiodactyl (sheep and deer) calcaneus models (J.G. Skedros, Cronin, et al., 2023a). The artiodactyl calcaneus model can serve as a "control bone" for studying bone adaptation under low-complexity loading (Sinclair et al., 2013; J.G. Skedros, Cronin, et al., 2023a). The **bottom** drawing resembles the mid-diaphyseal region of human or chimpanzee femora (Beckstrom, et al., 2010; Goldman, et al., 2003). The regional strain-related microstructural variations for the non-uniform strain distribution in the top drawing are not needed for the more uniform strain distribution experienced by the bone section in the bottom drawing (see also bottom images in Figure 5.19).

shear in many cases, and perhaps tension in others, have priority in mediating adaptations in the bone or bone region that is being examined (see additional discussion in (J.G. Skedros, et al., 2019; Warshaw, et al., 2017)).

The shear resistance-priority hypothesis is an important consideration when one designates the load-complexity category of the bone or bone region that is being studied. This is because, as shown in Figure 5.8, shear strains lack local/regional priority in effecting adaptation when a bone is in the high-complexity load category (Figure 5.14 at bottom). In contrast, shear-related adaptation is localized in the zone of the neutral axis in bones that are in the low-complexity load category (Figure 5.14 at top).

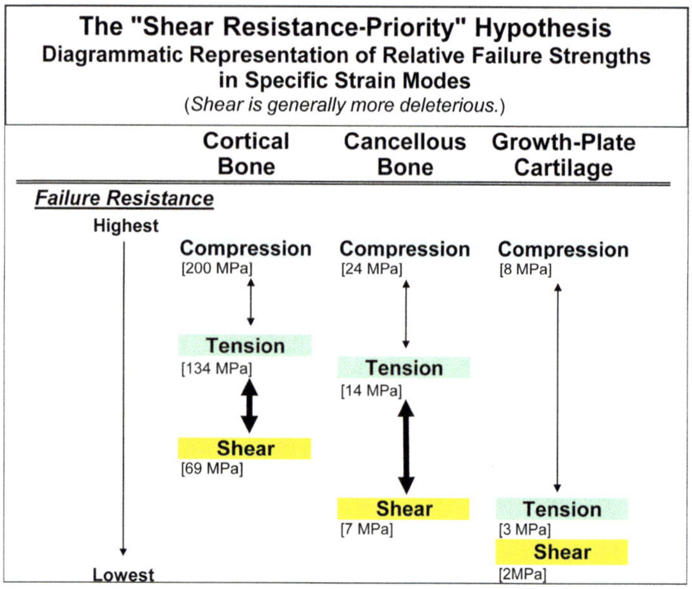

Figure 5.15 Diagrammatic representation of the "shear resistance-priority hypothesis." This shows that cortical and cancellous bone types are disproportionately weaker in shear than in tension or compression (increased vertical separation in the diagram). Although the disparity in cartilage is less marked, this tissue has poor tensile and shear strengths. This suggests that tension and shear are important in driving the ontogenetic adaptation of these tissue types. Values for this figure were obtained from these sources: **Cortical bone**: Cowin (1989) as shown for human bone, and values for bovine bone include: compression 197 MPa, tension 130 MPa, and shear 70 MPa; **Cancellous bone**: Estimated from Keaveny et al. (2001) for bovine bone using strength anisotropy ratios (longitudinal ÷ transverse strength) and bone volume fractions between 0.3 and 0.5; **Cartilage**: The compression value is estimated from human articular cartilage (Yamada, 1970). The values for tension and shear are from bovine tibia growth plates (Williams, Do, Eick, & Schmidt, 2001; Williams, Vani, Eick, Petersen, & Schmidt, 1999).

5.3.2 Between-Bone Examples of Using the Load-Complexity Categories

5.3.2.1 Example 1. Designating Load-Complexity Categories Helps Minimize Confusion

In a sample of 18 adult human femora and tibiae (age range: 18–50), Drapeau and Streeter (2006) examined relationships between cross-sectional geometric properties (modeling-related parameters) with osteon population densities (OPD) and percent areas of osteon bone (remodeling-related parameters). The histomorphometric analyses were performed in the anterior, posterior, medial, and lateral cortices at the mid-diaphysis. They sought to determine if there were statistically significant differences in these osteon parameters along the geometric anterior-posterior (AP) and medial-lateral (ML) axes (approximating the axes of the cross-sectional moments of area (inertia)). Statistical significance was found only in the tibiae: (1) in the more circularly shaped cross-section of the femur, OPD and percent of remodeled bone in the AP and ML axes were not significantly different, and (2) in the tibia, OPD was 25% greater ($p < 0.003$) in the AP axis than in the ML axis, and the percent of remodeled bone was 35% greater ($p < 0.001$). The authors seemed perplexed by the apparent inconsistencies of these findings; they could not be reconciled within the conventional paradigm of modeling/remodeling synergism/compensation (discussed in Section 5.4.1).

The "unexpected" lack of strain-mode-related regional histomorphological organization (i.e., regional CFO patterns) in adult mid-diaphyseal human femora was also noted in the study of Goldman et al. (2003). Manandhar et al. (2023) similarly reported what they considered to be the unexpected absence of strain-mode-related patterns in material density (obtained via micro-CT scans; no CFO or osteon data were obtained) in the curved femoral diaphysis. Although studying the sheep femoral diaphysis, they explicitly referred to the human femoral diaphysis to argue that regional tension *vs.* compression adaptation for habitual bending would be expected (i.e., expected in the curved diaphyses of both sheep and human femora (note that *in vivo* strain data are not available for these bone regions)).

The problem with the similar expectations in these three studies is that regional variations in these remodeling parameters would not be expected in these mid-diaphyseal femora because available evidence suggests they are in the high-complexity load category (**Table 5.2**). In turn, the human tibial mid-diaphysis is in the intermediate-B load-complexity category for most individuals; but the tibiae in individuals who regularly participate in twisting/turning sports would be in the high-complexity load category (the "§" superscripts in **Table 5.2** indicate cases where a bone region can be placed in more than one category, typically because of volitionally imposed differences in habitual activities).

5.3.2.2 *Example 2. Highly Complex Loading Often Produces Diffusely Distributed Shear*

In a microscopic study of human femoral mid-diaphyses (45 male, 43 female; 17–97 years; 12,690 osteons), Britz et al. (2009) found an inverse relationship between body weight and osteon size. Although the causal basis of this relationship could not be determined, the potential influence of the habitual loading of this region should be considered in addition to the effects of aging and other biomechanical/metabolic issues. Again, because the mid-diaphysis of the human femur is complexly loaded (combined compression, bending, and torsion) (Table 5.2, Figure 5.9), shear strains are diffusely distributed (Figure 5.8). Consequently, as argued above, there should be little to no regional differences in osteon characteristics (e.g., population density, fractional area, size, and cross-sectional shape) attributable to differences in strain modes/magnitudes between various cortical regions (e.g., anterior *vs.* posterior; medial *vs.* lateral) such as shown at bottom in Figure 5.14. Although these investigators did not conduct this regional analysis on their data, influences of regional strain modes or magnitudes would not be expected. Because of this (assuming aging effects have a minor influence in this context) the mid-diaphyseal human femur is a good choice for the analysis of potential relationships between osteonal characteristics and body weight because influences of habitual/local non-uniform strain variations are minimized.

In contrast, if the relatively more simply loaded tibial mid-diaphysis had been examined, then the potential influences of body weight could have been more difficult to detect because of the potent/superimposed influence of the locally prevalent/predominant tension and compression stresses in the anterior and posterior cortices, respectively. This is because, in contrast to the mid-diaphyseal femur, the mid-diaphyseal tibia is typically in the intermediate-B load-complexity category where bending produces a quasi-habitual "tension region" and "compression region" (Macdonald, Cooper, & McKay, 2009; Peterman, Hamel, Cavanagh, Piazza, & Sharkey, 2001; Yang, et al., 2014)—thus distinctive osteon morphotypes would be expected in these regions of the tibia (Vincentelli, 1978) (discussed further in section 5.4).

5.3.2.3 Additional Intra- and Inter-Species Examples

Figure 5.16 shows additional examples of regions of some bones that can be placed in one of the four load-complexity categories. Notably, the cross-sectional shapes of these bones and others (Figure 5.17) are so variable with respect to habitual loading directions/conditions that they do not help differentiate habitual bending from habitual torsion. This is why examination of their histological organization is so important.

At **bottom** are the hypothesized multi-“domains” for a human femur (**left**) (also see Figure 5.18) and chimpanzee femur (**right**); the letters within the femora correspond to the basic load conditions in the horse bones. The intertrochanteric regions of these bones (indicated by ellipses with oblique lines) are transition zones between the neck and proximal diaphyseal domains. In accordance with studies of regional variations in predominant CFO (Beckstrom, et al., 2010; Mears, et al., 2018; J. G. Skedros, et al., 1999), these drawings show the speculation that the chimpanzee femoral neck (also more robust and elliptical in cross-sectional shape) receives more prevalent/predominant bending than the human femoral neck.

5.3.3 Bending: Strain Mode and Magnitude

Habitual unidirectional bending creates a dilemma because the mechanical properties, including fracture and microdamage mechanics, of cortical bone markedly differ in compression, tension, and shear. As shown diagrammatically in Figure 5.14 (at top), differential histological modifications are necessary to better resist and/or accommodate fatigue microdamage in non-uniform strain environments that are common in habitual unidirectional bending (Akkus, Knott, Jepsen, Davy, & Rimnac, 2003; G. C. Reilly, et al., 1997). To reduce fracture susceptibility in regions subject to mechanical degradation from physiologic loading, regional variations in bone nanostructure (e.g., CFO and collagen molecular cross-links) and microstructure (e.g., osteon morphotypes, osteon population density, osteon percent area, and osteon diameter) are employed as the means for modifying matrix mechanical properties (e.g., stiffness, strength, fatigue resistance, and energy absorbed to failure (toughness)) (Allahyari, et al., 2023; Carter, Hayes, & Schurman, 1976; Corondan & Haworth, 1986; Diab & Vashishth, 2007; F.G. Evans, 1976; F. G. Evans & Bang, 1966, 1967; F. G. Evans & Vincentelli, 1974; Hiller, et al., 2003; Moyle, Welborn, & Cooke, 1978; Simkin & Robin, 1974; J. G. Skedros, Dayton, et al., 2006c; Stockhausen et al., 2021; Vincentelli & Grigorov, 1985; Yeni, Brown, Wang, & Norman, 1997).

5.3.3.1 Material Adaptation for Specific Aspects of the Strain Environment: Strain Magnitude vs. *Strain Mode*

Variations in ambient strain magnitudes may be more important than strain modes in evoking osteonal remodeling where their population density is more important than the introduction of specific osteon morphotypes. The coexistence of strain magnitude variations with strain mode variations (e.g., lower strains in “tension regions” and higher strains in “compression regions”) creates a potential dilemma when trying to decipher the relationships between bone structural and material organization and load history. What also may add to this confusion is a potentially significant role that load frequency has in affecting bone microstructural adaptation (more on this below in terms of OPD variations). But you should find solace in remembering that strain magnitudes (and the rate of strain application = strain rates) are relatively more important than strain mode in the attainment and

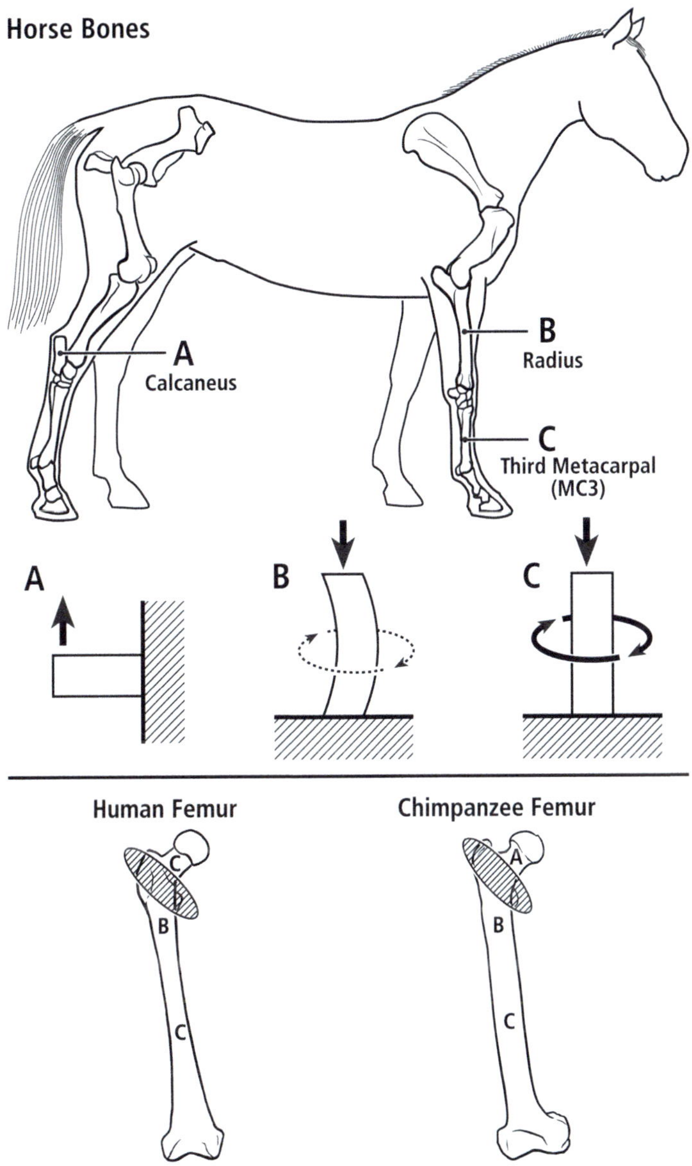

Figure 5.16 At **top** are lateral-to-medial views of the right forelimb and hindlimb skeletons of an adult horse showing a spectrum from simple loading to complex loading, respectively: calcaneus (**A**), radius (**B**), and third metacarpal (MC3) (**C**). The drawings below are simplified renditions of each bone type, showing: (A) the calcaneus as a cantilevered beam, (B) the radius as a curved beam with longitudinal loading; the curvature accentuates bending. Torsion (dotted line) is also present in the radius but is less than the torsion in the MC3 (solid circular line in C). Several studies reporting *in vivo* strain data were used to create these drawings (Biewener, Thomason, Goodship, et al., 1983a; Biewener, Thomason, & Lanyon, 1983b; Gross, et al., 1992; Lanyon, 1974; C. T. Rubin & Lanyon, 1982; R. K. Schneider, Milne, Gabel, Groom, & Bramlage, 1982; A. S. Turner, Mills, & Gabel, 1975).

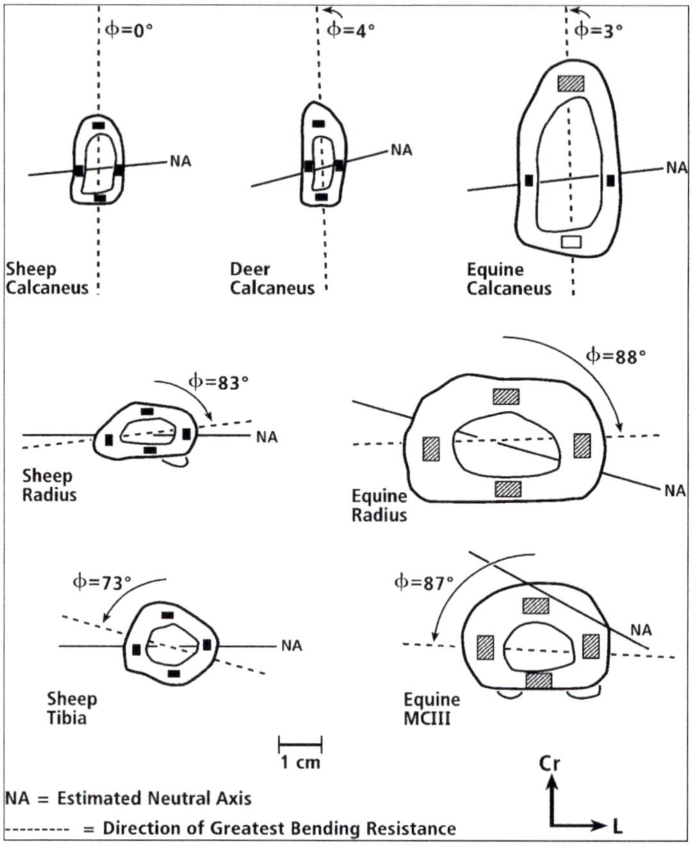

Figure 5.17 Representative drawings of cross sections of "tension/compression bones" (low-complexity load category) (at **top**), and more complexly loaded bones (**middle** and **bottom**). The phi (Φ) angles are the acute angles formed by the direction of greatest bending resistance (DGBR) (-----) and the sagittal plane (vertical orientation in each drawing) (DGBR is co-linear with axis of Imin, and 90° to axis of Imax). Small rectangles within the cortices are locations where nano-structural/microstructural analyses were conducted in our previous study (J. G. Skedros, et al., 2009) and are the locations of the images shown in Figure 5.24. "Anterior" and "posterior," respectively, are used as surrogates for the anatomical locations of the opposing sagittal-plane cortices in the calcanei (dorsal and plantar), radii (cranial and caudal), and equine third metacarpals (MCIII; dorsal and palmar). Cr = cranial = anterior = dorsal; L = lateral.

maintenance of skeletal mass and structural "build" of a bone (which are primarily mediated by the modeling process that is active primarily during growth) (additional discussion is in Section 5.4.3).

5.3.3.2 Strain-Mode-Specific Predominant CFO Not Linked to Strain Magnitude

A causal relationship of predominant CFO and strain mode (with magnitude having no apparent/clear effect) has been clearly shown in a prospective controlled *in vivo* study. In that study, Takano, Turner, and co-workers (1999) used the dog radius model with and without excision of a segment of the ulna (in order to change the neutral axis location, as done by Lanyon et al. (1982)). These investigators employed two methods for quantifying predominant CFO with respect to the bone's long axis: acoustic microscopy and CPL

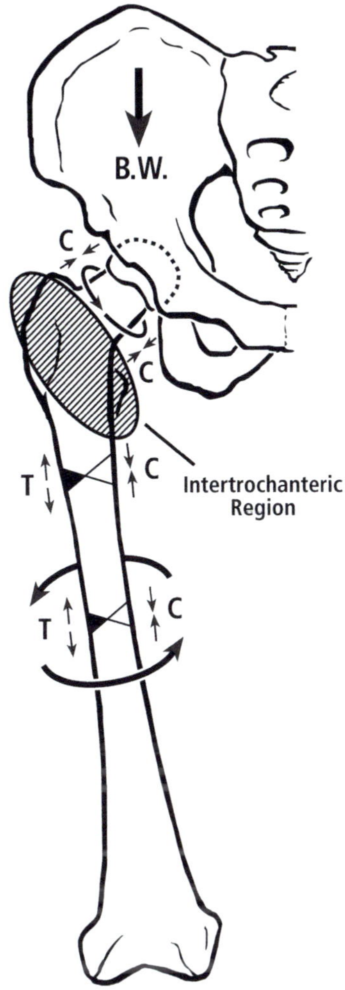

Figure 5.18 Habitual loading of a modern human femur showing the multi-domain load hypothesis. Prevalent/predominant torsion is depicted by the curved arrows. B.W. = body weight. In the lower portion of Figure 5.16 are drawings that contrast the hypothesized stark differences in the habitual loadings of the femoral neck domains between chimpanzee and human femora (see legend of Figure 5.16 for references).

measurements in thin transverse sections. They then analyzed the cranial "tension cortex" and caudal "compression cortex" before (i.e., sham surgery group) and after removal of a segment of the ulna. This allowed for comparisons between the original and the newly created "compression region" and "tension region." Their experiment induced a change in the strain-mode distribution (hence changed the location of the habitual neutral axis) and caused strain-mode-specific matrix formation; namely, more oblique-to-transverse CFO in the newly created "compression region" and more longitudinal CFO in the newly created "tension region." Perusal of their data shows that the experimentally induced strain-mode-specific matrix changes do not correspond to the trans-cortical variations in strain magnitude but appear to be responses to the strain mode that prevails across the entire cortex. There are additional notable examples in other bones where regional variations in strain magnitude are not correlated with regional variations in CFO (Mason, et al., 1995;

J. G. Skedros, et al., 1996). These findings also lead to the important conclusion that regional CFO variations are not confounded by strain magnitude variations.

5.3.3.3 Osteon Population Density (OPD) and Load History

There are cases where relative uniformity in CFO across a bone's mid-diaphysis has been observed even though there are locally increased concentrations of osteons in one region of the cross-section. For example, this has been observed in the cranial (anterior) cortex of the highly torsion-loaded sheep tibiae (J. G. Skedros, et al., 2009); we speculated that this locally increased osteon density reflects the relatively high tension strain magnitudes in this complexly loaded region. This finding shows that regional OPD can provide clues about prior adaptation for load history. However, again, available evidence suggests that variations in predominant CFO are generally the strongest correlate of load history differences.

Strain magnitude and bone adaptation is also discussed later in this chapter in sections 5.4.3 and 5.8 with respect to the important experimental adaptation study conducted using tibiae of sheep by Ian Wallace and co-workers (2014).

5.3.4 The Multi-Domain Load Hypothesis: Regional Load Complexity Differences within a Single Bone

Evidence of regional differences in load-complexity categories within the same bone led to the development of the multi-domain load hypothesis (Figure 5.18). Figure 5.16 (at bottom) shows the application of these categories to the hypothesized presence of different load domains within human and chimpanzee femora. In these examples, we argue that the region of the greater and lesser trochanters (the "intertrochanteric region") in each bone is a transition zone between one proximal and two distal "load domains" (Beckstrom, et al., 2010; J. G. Skedros, 2011; J. G. Skedros & Baucom, 2007; J.G. Skedros, Beckstrom, Kiser, & Bloebaum, 2008; J. G. Skedros, Hughes, Nelson, & Winet, 1999).

In the human femur the "proximal domain"—the femoral neck—is habitually loaded in net compression by the gluteus medius/minimus and in torsion/shear by the actions of other muscles, body weight, and variable joint reaction forces (Lovejoy, 1988; J. G. Skedros & Baucom, 2007; J.G. Skedros, Cronin, Dayton, Bloebaum, & Bachus, 2023; Tovar-López, Domínguez-Hernández, Diez-García, & Araujo-Monsalvo, 2015) (the femoral head is a separate domain but this is not considered here). In contrast, it has been argued that the upper portion of the "distal domain" of the human femur—the proximal metaphyseal-diaphyseal region—is customarily loaded in bending and torsion (but with relatively less torsion than seen in the mid-neck and mid-diaphysis), which is restrained to some extent by the iliotibial band (Edwards, Gillette, Thomas, & Derrick, 2008; J. G. Skedros & Baucom, 2007). *In vivo* strain data obtained during walking at peak loading of stance phase reveal that there is net tension on the lateral aspect of the proximal metaphysis-diaphysis of the human femur (Aamodt et al., 1997). This is the most important (and only!) *in vivo* support for considering the human proximal femoral diaphysis as a habitual lateral-medial "tension-compression environment" (J. G. Skedros & Baucom, 2007) (Figure 5.18); hence, it is in an intermediate-complexity load category (Table 5.2). Quantitative data showing the clear presence of tension *vs.* compression adaptation patterns of CFO in the sub-trochanteric region and proximal diaphysis of adult human femora (and also adult chimpanzee (Figure 5.19) and baboon femora) are consistent with this interpretation (Beckstrom, et al., 2010; Mears, Keenan, & Skedros, 2018; J.G. Skedros, et al., 2008; J. G. Skedros, et al., 1999; J. G. Skedros, Keenan, et al., 2013a; J.G. Skedros, et al., 2011).

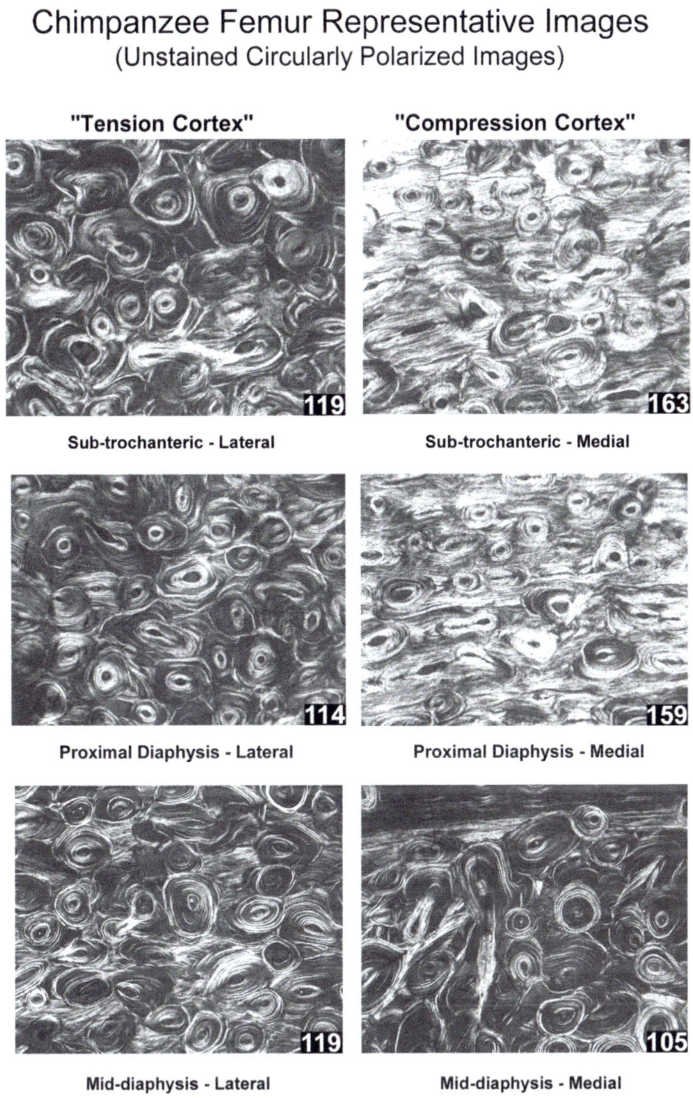

Figure 5.19 Representative circularly polarized light (CPL) images taken from the proximal diaphyseal cross-sections (**top four images**) and mid diaphysis (**bottom two images**) of an adult chimpanzee femur. In the proximal diaphysis, the images from the lateral "tension cortex" have generally darker osteons, which are quantified as lower osteon morphotype scores using our scoring scheme (e.g., values less than "5" in Figure 5.3). At right, the medial "compression cortex" has generally brighter osteons, which are quantified as higher osteon morphotype scores using our scoring scheme ("5" in Figure 5.3). The numerical mean gray level for each image is shown at the lower right. The difference in CFO between the medial and lateral cortices is less obvious at the mid diaphysis. The width of each image is 1.4mm. The specimens were embedded in polymethyl methacrylate in an unstained and fully calcified state, and ultramilled to 100 microns thickness. All images were obtained in the same imaging session using the same illumination and 50x magnification (J. G. Skedros, C. J. Kiser, et al., 2011).

5.3.4.1 *Mid-Diaphyseal "Load Domain" of the Human Femur*

The mid-diaphyseal "load domain" in the human femur extends from the proximal (subtrochanteric) domain to the distal third of the diaphysis where the strain distribution again changes (Duda et al., 1998; Edwards, et al., 2008). Several studies of CFO patterns in the mid-diaphysis of the human femur have attempted to detect the "expected" history of habitual medial-to-lateral bending (Goldman, et al., 2003; Portigliatti Barbos, Bianco, Ascenzi, & Boyde, 1984; Portigliatti Barbos, Carando, Ascenzi, & Boyde, 1987). This hypothesized load history is based on the idea that the habitual anterior-lateral (net tension) to posterior-medial (net compression) direction of bending seen in the proximal diaphysis would be transmitted distally and would have sufficient intensity/duration to evoke similar regional strain-mode-specific CFO patterns at the mid-diaphysis. However, these CFO-based studies failed to detect evidence of this load history (Figure 5.19 bottom two images). Why is the mid-diaphysis (no regional variations) so different from the proximal diaphysis (clear regional variations; Figure 5.19 top four images)?

Although the strain environment at the mid-diaphyseal femur has not been measured *in vivo*, there are *ex vivo* strain data and finite element models of femora loaded in simulated single-legged stance that reject the idea of a habitual typically unidirectional bending moment at the mid diaphysis. These indirect types of data suggest that there is both a reduction in the magnitude of the medial-to-lateral bending moment at mid-diaphysis (the bending moment is substantially greater in the sub-trochanteric area (Fetto, Leali, & Moroz, 2002; Koch, 1917)) and increased inter-specimen variability of the strain distribution at mid-diaphysis (see **Table 5.2** references and discussion in Goldman et al. (2003)). Consequently, the relatively complex load environment of the human and chimpanzee femoral mid-diaphysis (where torsion > bending) might be why "expected" tension/compression (anterior-lateral/posterior-medial) CFO differences are typically absent there but are present in the proximal diaphysis and subtrochanteric regions (where bending > torsion) (Beckstrom, et al., 2010; J.G. Skedros, et al., 2008; J. G. Skedros, et al., 1999; J.G. Skedros, et al., 2011).

Hence, there appears to be sufficient evidence in support of the idea that load complexity changes along the femoral diaphysis of these bones. The hypothesized difference in load history from the proximal femoral diaphysis (bending > torsion) to the mid-diaphysis (torsion > bending) in humans and chimpanzees is also consistent with the CFO data that we have reported in the baboon femur (Mears, et al., 2018). Additional support for proximal-distal changes in habitual load complexity within a bone's diaphysis, and how this could be mediated by strain-related site-specific differences in thresholds for modeling/remodeling activities, can be found in several studies (Espinoza Orias, Deuerling, Landrigan, Renaud, & Roeder, 2009; Goodship, Lanyon, & McFie, 1979; Hsieh, Robling, Ambrosius, Burr, & Turner, 2001; Skerry, 2006); and is also suggested by Vainionpää et al. (2007) and Gosman et al. (2013) in their ontogenetic studies of human femora.

5.3.5 Misleading Influence of Cross-Sectional Shape and Cortical Thickness When Interpreting Load History

The relatively circular shape of the human femoral mid-diaphysis reflects expected cross-sectional morphology in a torsion/multi-directional loading environment (Carter & Spengler, 1982; C. B. Ruff, 1981; Wainwright, et al., 1982). But a quasi-circular shape is also typical in the proximal diaphysis (Gosman, et al., 2013) where the bending moment (typically in the medial-lateral direction) is believed to be the greatest in the entire femur. Again, this reinforces the idea that cross-sectional shapes can be very misleading when interpreting load history (hence, cross-sectional shape is low-level evidence for interpreting load history; Figure 5.10).

Figure 5.17 shows additional examples of bones in several of the load-complexity categories where the cross-sectional shapes and/or cortical thickness asymmetries do not consistently reflect the direction of greatest bending resistance (DGBR). For example, the DGBR of sheep and deer calcanei is in the direction of bending whereas it is approximately 90° to the direction of bending in the radii of horses and sheep. Additionally, the thickest cortices in these calcanei are on the "compression side" of these bones (i.e., the upper cortices in Figure 5.17) whereas equine and sheep radii have similar thicknesses of the "compression cortex" and "tension cortex." Additional examples showing that principal axes of cross-sectional moments of inertia (which correspond to the DGBR) can be unreliable for interpreting bending load directions and neutral axis locations can be found in Lieberman et al. (2004), Demes (2007), and Ruff et al. (2006).

5.4 The Modeling/Remodeling Distinction

Remodeling can produce different histological/material modifications in discrete cortical regions of the same cross section (e.g., between the opposing "tension cortex" and "compression cortex" of a bone that is habitually loaded in bending). These remodeling-mediated modifications can include porosity (produced from the introduction of central [Haversian] canals), osteon population density (OPD) and size, predominant CFO, and osteon morphotypes (Figures 5.5, 5.6, and 5.15). Active remodeling adjusts bulk and tissue mineral content by introducing porosity and increasing the percentage of younger osteons, which are incompletely mineralized. For example, Figure 5.20 shows the consequences of a high remodeling rate (i.e., increased basal rate of osteon formation) that is known to occur in the plantar "tension cortex" of an adult deer calcaneus in normal conditions. This is seen has high percentages of younger osteons and resorption spaces, which results in the largest tissue mineral content difference that has been described between regions of the same bone in any vertebrate regardless of skeletal maturity (J. G. Skedros, Bloebaum, Mason, & Bramble, 1994; J. G. Skedros, et al., 2004). However, this sort or regional variation in tissue mineralization is uncommon and, therefore, is not helpful for interpreting load history and, furthermore, it is not an efficient way to help enhance load predictability—modeling-mediated changes in bone shape are much more effective.

5.4.1 Three Hypothetical Load Scenarios in the Adult Human Tibia Cortex

The various osteon morphotypes shown diagrammatically in Figure 5.21 (and used in Figure 5.22), provide examples of different remodeling-mediated (material) adaptations for different load histories that can occur within bones even though modeling-mediated (structural) adaptations might not occur (i.e., no adjustments in cortical thickness or cross-sectional shape). Figure 5.3 (far right column) and Figure 5.21 also show osteons that appear to be "hybrids" of those from habitual tension and compression cortical regions, and observations suggest that these hybrids tend to predominate in neutral axis and other bone regions where shear is prevalent/predominant (J. G. Skedros, Kiser, Keenan, & Thomas, 2011b).

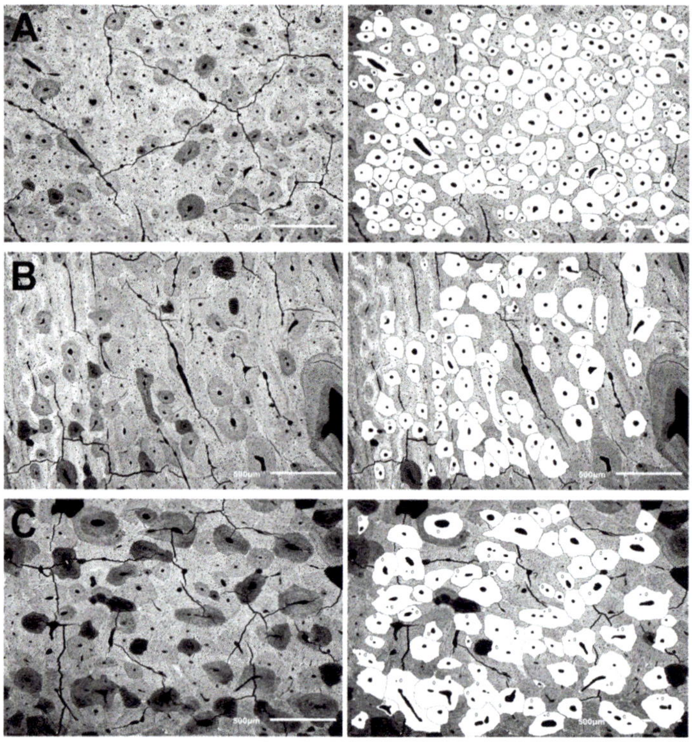

Figure 5.20 Backscattered electron (BSE) images from the: **A**. dorsal "compression cortex," **B**. lateral "neutral axis" cortex, and **C**. plantar "tension cortex" of a transversely sectioned skeletally mature mule deer calcaneus from Keenan et al. (2017). Their study examined osteon cross-sectional shape and population density of the deer calcaneus in order to detect potential relationships with cortical location (e.g., dorsal *vs.* plantar), perhaps reflecting strain-mode-specific adaptations. (BSE images show variation in bone mineral content, not CFO.) The artiodactyl (e.g., sheep and deer) calcaneus can be used as a control bone for studies of bone adaptation especially for the low- to intermediate-complexity load categories (Sinclair, et al., 2013; J.G. Skedros, Cronin, et al., 2023a). The meaning of the symbols within some of the osteons and other details of the use of the mule deer calcaneus for this analysis can be found in Keenan et al. (2017). These drawings are reproduced with permission of John Skedros from Keenan et al. (2017).

(See Section 5.6.4 for some discussion regarding the hypothesis that birefringent rings resembling peripheral "hoops" of dark/tension osteon morphotypes might form within the wall of hybrid osteons as an adaptation of prevalent/predominant shear.)

Figure 5.22 diagrammatically illustrates the use of circularly polarized light microscopy to specifically evaluate osteon morphotypes in the anterior cortex of mid-diaphyseal human tibiae of three young adult males: (A) an individual who habitually participates in distance running, (B) an individual who habitually participates in twisting/turning sports, and (C) an individual who is relatively sedentary activity (hence has fewer osteons than A and B because of a lower frequency of microdamage-mediated repair). Even though the conventional osteonal histomorphological characteristics are the same in the first two cases of Figure 5.22 (i.e., individuals A and B have equivalent secondary osteon fractional area and population density (OPD), and osteon cross-sectional sizes and shapes), the aggregate osteon morphotype scores (MTS) are markedly different for each group. Most notably, the percentage of osteon

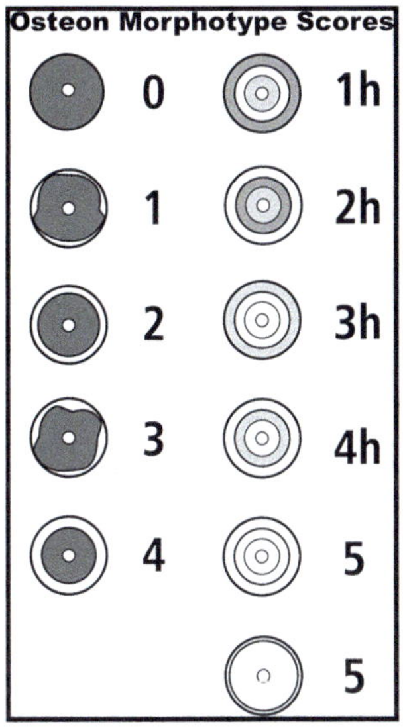

Figure 5.21 These are idealized drawings of secondary osteon morphotypes (Figure 5.3) that are used in the three loading scenarios in Figure 5.22. The numerical value assigned to each typical (non-hybrid) osteon morphotype is either 0, 1, 2, 3, 4, or 5, where lower numbers are darker osteons in circularly polarized light (Figures 5.3 and 5.5). MTS = the osteon morphotype score for the entire image. The darker osteons (0–4) are typically good at accommodating tension stresses whereas the brightest osteons (fully bright or alternating bright lamellae) are good at accommodating compression stresses. In this drawing there are also osteon morphotype "hybrids" ("h") that are believed to help accommodate shear stresses (Robert F. Bigley, et al., 2006; J. G. Skedros, C. J. Kiser, et al., 2011). These hybrid osteon morphotypes can also be scored using the six-number scoring system. A 12-number scoring scheme has been developed that specifically differentiates hybrids, although this was found to be unnecessary in our study of the chimpanzee femoral diaphysis (J. G. Skedros, C. J. Kiser, et al., 2011). More research on this issue is needed using various bones representing a broad range of shear stress magnitudes and distributions during physiological loading.

morphotypes that are best at accommodating shear (% hybrids ("h")) is substantially higher in the individual with habitual multi-directional loading of the tibia (Figure 5.22B).

5.4.2 Modeling/Remodeling Synergism/Compensation

Many investigators have suggested that strong correlations shown in some bones or bone regions between modeling outcomes (e.g., adjustments in cortical thickness, relative cortical area, or principal axes of cross-sectional moments of inertia (e.g., Imax and Imin)) and the remodeling process (e.g., increased porosity or osteon density; reduced mineral content) reflect their synergism/compensation in affecting local or whole-bone performance (Beck, Looker, Ruff, Sievanen, & Wahner, 2000; Burr, Ruff, & Thompson, 1990; Goldman, Hampson, Guth, Lin, & Jepsen, 2014; Lazenby, 1986; Tommasini, Nasser, Schaffler, & Jepsen, 2005; Ural & Vashishth, 2006; van der Meulen, Jepsen, & Mikić, 2001). (This does not refer to metaphyseal/epiphyseal regions where overall structural strength might be achieved by synergism/compensation in the relative amounts of cortical and cancellous bone mass (J.G. Skedros, Cronin, et al., 2023a)). However, synergism/compensation between modeling and remodeling in cortical bone as being generally applicable in bone adaptation is <u>not</u> well

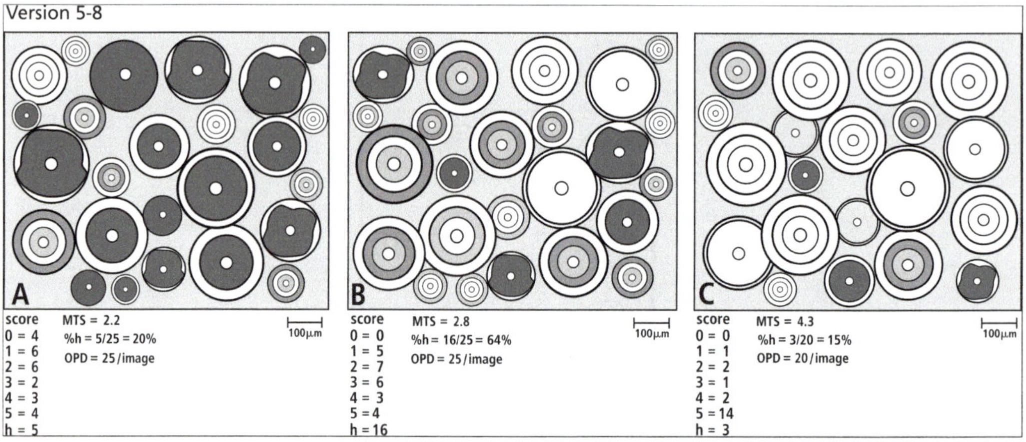

Figure 5.22 These diagrammatic depictions represent microscopic images of the anterior tibial cortices in three healthy 30-year-old males in a hypothetical scenario that likely mirrors reality (perhaps this could be tested in an animal model like the adult sheep tibia or with a futuristic non-invasive very high-resolution imaging technology in living humans). Load-history perspectives and strain data used as the basis for creating this graphic are based on Yang et al. (2014) and Macdonald et al. (2009) and studies discussed therein. All of these images are populated with osteon morphotypes shown in Figure 5.21. The individuals have different habitual activities: **A** runs 10 kilometers every other day, **B** participates in twisting/turning sports, and **C** is relatively sedentary. The images show osteon morphotypes associated with different activities. **A** shows tension-adapted morphotypes, **B** exhibits a mix of morphotypes with hybrids for multi-directional/shear loading, and **C** shows a relatively uniform mix of brighter morphotypes that might reflect a "default" organization (Zimmermann, et al., 2019). In all three individuals in this hypothesized scenario, significant structural adaptation of the tibia did not occur (hence modeling-mediated adaptation did not occur, which is generally consistent with findings of O'Leary et al. (2021) and Vainionpää (2007)). In contrast, microstructural adaptation would be expected to occur in the tibiae of the exercise groups (Lieberman, Pearson, Polk, Demes, & Crompton, 2003) and is shown to be dramatic in this hypothetical scenario.

supported. For example, as mentioned above, previous studies examining human tibiae and femora have reported significant correlations between modeling and remodeling parameters in some specific bone regions, but then similar relationships do not occur in different regions of the same bone or in neighboring bones (Drapeau & Streeter, 2006; Ural & Vashishth, 2006). Again, these conflicting results are likely confounded when the bones/bone regions examined are in different load-complexity categories.

5.4.3 Modeling/Remodeling "Division of Labor"

Considering these refutations and inconsistencies, it can be argued that structural/material associations in limb-bone diaphyses often do not reflect biomechanically important compensation/synergism. Existing knowledge of intercellular communication in bone suggests it cannot achieve the necessary coordination between modeling and remodeling to accomplish synergism/compensation across the breadth or length of a diaphysis. This would be especially true with advancing age as osteocytic functions become senescent. Even in youth, this idea is consistent with data reported by Tommasini et al. (2007), where sexual dimorphism was found to affect tibial size and shape but not tissue-level (remodeling parameters) mechanical properties in a sample of 14 young human tibiae (36.9±8.1 years). However, a rigid refutation of the possibility of cellular "crosstalk" in bone adaptation is unwise because structural/material synergism/compensation might occur in atypical cases (see dashed line in Figure 5.6); for example, where size–shape relationships occur in a broader size range of individuals of the same species (e.g., see Tommasini et al. (2005)), or in cases of precocial *vs.* altricial limb development and/or ambulation (e.g., artiodactyls, musk oxen, jackrabbits, and seagulls) (Carrier, 1983; Carrier & Leon, 1990; Currey & Pond, 1989; Heinrich, Ruff, & Adamczewski, 1999). See our recent study for more discussion of the idea that cellular crosstalk might be involved in development and maintenance of functional adaptation in the limb skeleton (J.G. Skedros, Cronin, et al., 2023a).

The modeling process is primarily responsible for the achievement of whole-bone structural performance. (Whole-bone structural performance is the overall strength and stiffness of the bone when end loads are imparted to it.) It is relatively easy to achieve this whole-bone goal by adjusting cortical thickness and/or bone shape (e.g., curvature or cross-sectional dimensions) as long as the material is suitable. Again, achieving this goal via regional material adjustments produced by the remodeling process is not as efficient or effective. For example, it would not be energetically efficient to modify local strains via osteon-mediated reductions in mineralization and increases in porosity (by increasing the density of osteon central canals). During the attainment of a bone's adapted morphology (when periosteal/endosteal bone apposition/resorption is still possible) (Figure 5.23), regional adjustments in local strains could be achieved more readily and efficiently by minor adjustments in cortical thickness by the modeling process (Goodship, et al., 1979; Lanyon, et al., 1982). This is an example of the modeling/remodeling "division of labor" depicted in Figure 5.6.

Failure to recognize the modeling/remodeling distinction explains why investigators seem perplexed when "expected" developmental/functional adaptation in limb diaphyses in adults that experience significant loading differences were not detected. For example, consider the results of studies by Carlson co-workers (2006; 2008) where attempts were made to find relationships between diaphyseal shape (e.g., principal moments of inertia) and locomotor behaviors along with other non-behavioral factors in femora and humeri of free-ranging chimpanzees and other primates. The failure to detect a relationship in this solely structural context caused the authors to conclude that "… diaphyseal shape may be unresponsive to mechanical demands of these [six] specific locomotor modes" (page 394) (Carlson, et al., 2006). Others have reported similarly (seemingly) perplexing results in studies of non-human primate long bones (Cosnefroy et al., 2022; Morimoto, Ponce de Leon, & Zollikofer, 2011). In an *in vivo* study of several groups of young adult female athletes who participated in seven different sports, Niinimäki et al. (2017) found that shape ratios of the proximal femoral diaphysis could not differentiate between the sports that loaded the proximal femur with "directional consistency" from those where the loading was "directionally inconsistent." This result and the conclusions of the other aforementioned studies should be viewed cautiously because none of these studies conducted microscopic analysis on the bones. If they had been able to perform microscopic analyses (e.g., using some futuristic non-invasive imaging technology), then it seems likely that regional variations (or lack thereof) in the distribution of secondary osteon morphotypes and/or predominant CFO would detected and would reveal adaptations of the bone material that correlate with specific locomotor activities / loading regimes (e.g., see differences in the microstructural consequences of the loading regimes in Figure 5.24). Without microscopic data or other types of data that provide a stronger level of evidence than an analysis that uses only structural/geometric characteristics (Figure 5.10),

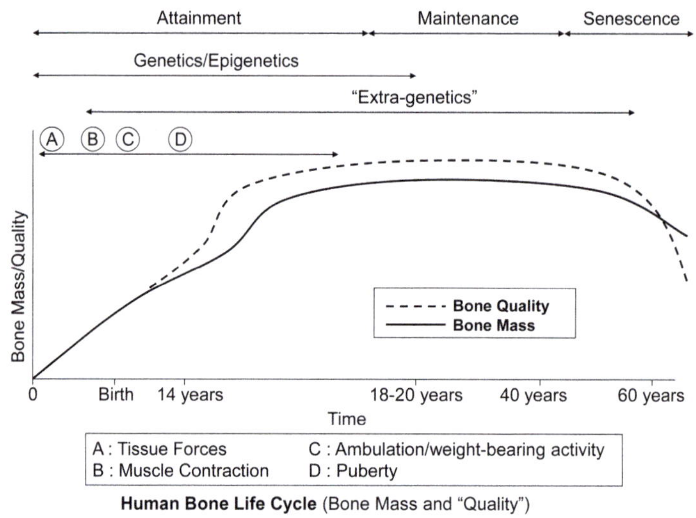

Figure 5.23 This is a stylized depiction of ontogenetic changes in bone mass and "quality" (e.g., bone tissue mechanical properties) in humans; this is from Skedros et al. (2007) and was adapted from several sources (Biewener & Bertram, 1993a; Carter & Beaupré, 2001; Kassem, Melton, & Riggs, 1996). This figure helps to conceptualize the putative shifts in the temporal importance of genetic, epigenetic, and extra-genetic influences, especially with respect to varying histocompositional characteristics within or between bones. In this context, for example, the regional material variations that emerge in the "tension cortices" and "compression cortices" during mid-to-late phases of growth of horse radii (Riggs, et al., 1993) and sheep radii (Figure 5.27) may serve to enhance the toughness and fatigue resistance for these habitual non-uniform strain environments. From the pre-natal phase and well into the attainment phase, the adaptive growth response of cartilage, or chondral modeling (not discussed in this chapter), has a profound influence on the growth and form of limb bones, especially at their epiphyses where articulations are formed with adjacent bones (Hammond, Ning, Ward, & Ravosa, 2010; Pantinople, et al., 2017). Shown is also the second bone-mass growth spurt that occurs in humans, which has not been demonstrated in any other amniote.

it is best to heed the recommendation of Demes et al. (2001) to "avoid broad behavioral conclusions derived from long bone cross-sectional shape."[12]

This view reflects the fact that a bone's shape appears to be highly constrained genetically/developmentally (Hansen, Bredbenner, Nicolella, Mahaney, & Havill, 2009; I.J. Wallace, et al., 2017; I. J. Wallace et al., 2010) unless unusual loading circumstances occur (e.g., increased robusticity of the humerus in the racket arm when an avid tennis player is skeletally immature) (J. E. Bertram & Swartz, 1991; Haapasalo et al., 2000). (See Ruff and Hayes (1988) for an example showing that seemingly minor, though clinically relevant, changes in relative cortical area have been reported in a sexually dimorphic context with aging in modern human femora.)

5.4.4 Effects of Genetics and Ontogeny on Modeling/Remodeling

Genetic, epigenetic, and extra-genetic influences also play a significant role in a bone's structure and material organization and are primarily at play prior to skeletal maturity (Figure 5.23) (I.J. Wallace, et al., 2017). Extra-genetic (environmental) influences that affect bone adaptation include microdamage events and strains that exceed hypothesized thresholds.

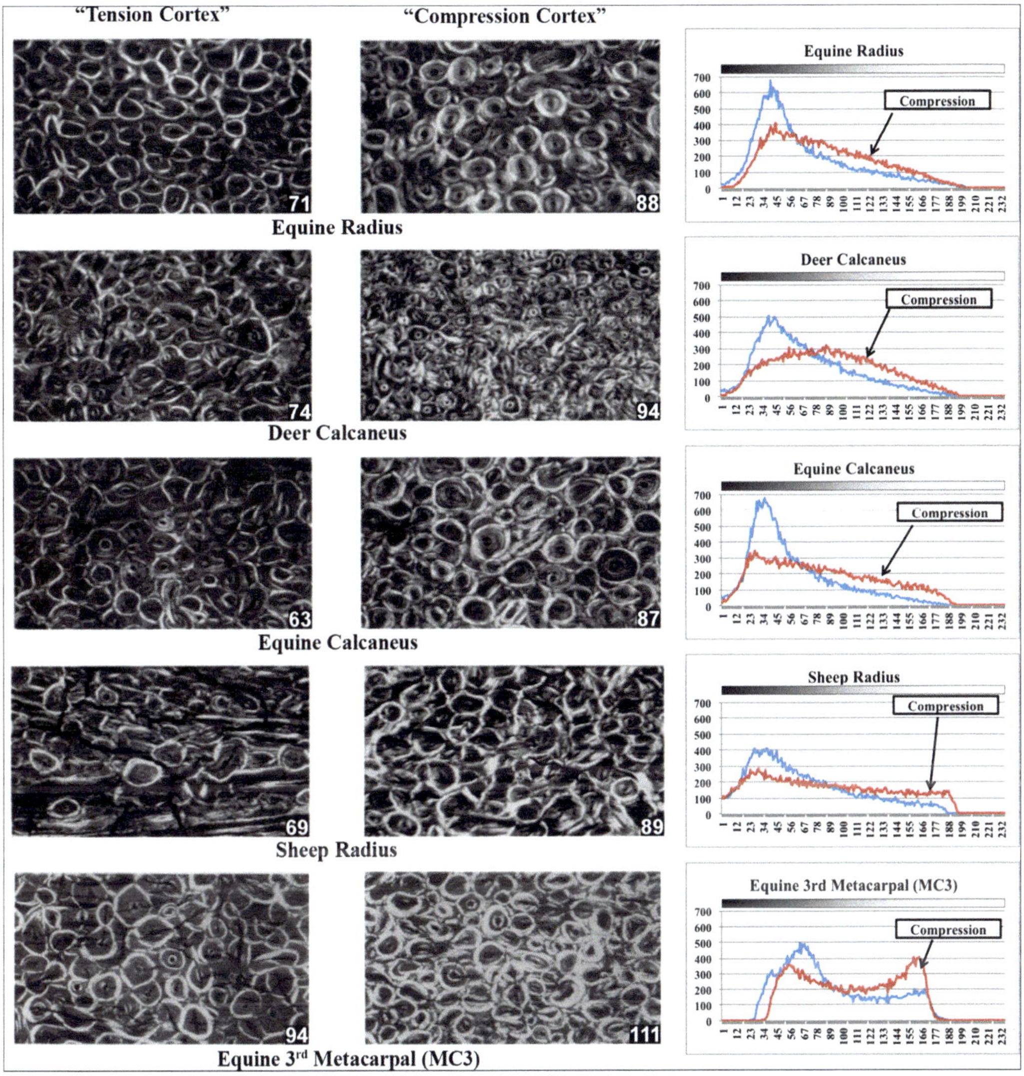

Figure 5.24 Representative circularly polarized light (CPL) images of bone from "tension cortices" (at **left**) and "compression cortices" (at **right**) of several non-human species (transverse sections) (from (J. G. Skedros, et al., 2009)) (see Figure 5.17 for image locations). The dark-to-bright gradient over each of the gray level profiles (at **far right**) show how the gray levels (predominant CFO) change from left to right on the x axes of these profiles. Although the equine third metacarpal (MC3), when compared to the other bones shown above, has relatively more prevalent/predominant torsion/multi-directional loading (i.e., shear strains), the regions selected for analysis are (like the other bones) from opposite sides of a relatively habitual neutral axis (hence are from a "tension region" *vs.* a "compression region") (J. G. Skedros, Dayton, et al., 2006c). This place the equine MC3 it in the intermediate-B load-complexity category. The numerical mean gray level for each image is shown at the **lower right**. The width of each image is 1.52mm. The specimens were embedded in polymethyl methacrylate in an unstained and fully calcified state, and ultramilled to 100 microns thickness. All images were obtained in the same imaging session using the same illumination and 50x magnification.

It is suggested that extra-genetic factors can have important influences on bone's structural and material organization by activating/modifying bone modeling and, more prominently, remodeling processes (J. G. Skedros, Mason, & Bloebaum, 2001a; J. G. Skedros, Sorenson, Hunt, et al., 2007). While extra-genetic stimuli may not be heritable (in contrast to genetic and epigenetic influences), they do act on genetic and cellular machinery that is heritable.

Wallace et al. (2017) have written a very astute review article on the ontogenetic and genetic influences on a bone's responsiveness to mechanical signals. Unfortunately, reading their article may leave you feeling hopeless as they mention numerous ways that researchers might be misled when attempting to interpret a bone's load history using seemingly straightforward functional adaptation approaches (e.g., the use of structural and material characteristics to decipher prior loading events). This hopelessness reflects numerable variables and unknown influences, especially genetic endowments that can be inexorably confounding. But there is a glimmer of hope in all of this. Rigorous histological analyses, such as espoused in this chapter, is a way forward especially in retrospective analysis of adaptations for load history. In this perspective, remain optimistic when reading this important summary from Wallace and co-worker's article, which conveys seemingly well-founded pessimism:

> Adult bones are much less responsive to loading than growing bones, which means that any functional signal present in adult bone morphology is primarily the result of childhood activity (Pearson & Lieberman, 2004). Thus, a researcher's ability to infer anything about adult activity from bone structure is evidently limited. Furthermore, because of genetic influences on bone mechanoresponsiveness, it is possible for the magnitude of the functional signal to vary in the bones of different individuals despite identical physical activity patterns during life (I. J. Wallace, Judex, & Demes, 2015). Two populations with similar bone structure may have exhibited dramatic differences in activity levels during life; and a population with gracile bones may have actually been more physically active than a population characterized by more robust bone structure. This complexity is inconvenient, but it is unavoidable and should be carefully considered when skeletal remains are used to reconstruct past human physical activity. Alas, skeletal morphology may indeed emit a signal related to functional loading history, but our ability to decipher this functional signal amidst the noise caused by other determinants of bone structure is limited.

5.5 The Stressed "Volume Effect"

The volume effect teaches that larger structures are more prone to failure than small structures (R. F. Bigley et al., 2007, 2008; Martin, 2003; Taylor, 2000). This assertion is based on the statistical probability that microdamage is more likely to form in a larger volume of tissue (all other variables being equal). Consequently, a larger bone is more prone to fatigue failure than a very small bone. This could help explain why an osteon-based repair mechanism evolved in some species and/or why marked histological differences might exist in some cases (perhaps as a means for managing microdamage more effectively). In studies that compare adaptations in bones of very small *vs.* larger animals, it is important to decide if the volume effect is applicable.

Luckily, in most anthropological studies of bone adaptation the volume effect does not need to be considered because it is most applicable in small animals like mice and rats, and the little-studied very small primates like mouse lemurs and tarsiers. Nevertheless, be aware of studies of bone adaptation in small-volume bones that the authors sometimes boldly extrapolate to anthropoid bones. For example, compare and contrast results of the studies of small-volume bones by Schneider et al. (2007), Tami et al. (2003), Carlson and Judex (2007), Kumar et al. (2010), Pinto et al. (2010), Wallace et al. (I. J. Wallace, et al., 2015, 2010), and Peacock et al. (2018a) in terms of how they (or you might) extrapolate their results to functional adaptation, or adaptability, of hominid limb bones. The debate regarding the validity and implications of Peacock et al.'s (2018b) study on the adaptation of femoral cross-sectional geometry in mice serves as a cautionary example for extrapolating findings from small-volume bones to larger bones in primates and other larger animals (Peacock, Garland, & Middleton, 2018; C. B. Ruff, Warden, & Carlson, 2018). See Lad (2023) for additional discussion of the myriad issues related to the presence or absence of osteonal remodeling in small-bodied animals.

5.6 Mechanical Relevance of Osteon Size and Population Density, Osteon Morphotypes, and Predominant Collagen Fiber Orientation (CFO)

5.6.1 The Old Idea That Osteons "Create an Inherently Weaker Structure"

In early mechanical testing studies, Currey (1975) and Carter et al. (1976) showed that primary bone is stronger and more fatigue-resistant than the more extensively remodeled bone. They attributed these results to the presence of secondary osteons, which reduce mineral content by increasing porosity and reducing mean tissue age, which in turn, creates an "inherently weaker structure" by increasing cement line and lamellar interfaces. For nearly two decades these results puzzled many bone investigators because they showed that prevalent osteonal bone, as seen in humans, is mechanically inferior to the fibrolamellar/plexiform bone that is prevalent in fast growing species, like artiodactyls. But it was eventually recognized that the interfaces, microscopic mineralization variations, CFO variations, and/or lamellar patterns/heterogeneity introduced by osteons enhanced the tissue's capacity to attenuate microcrack propagation (Ciarelli et al., 2009; Katsamenis, Jenkins, & Thurner, 2015; Martin & Burr, 1989; G. C. Reilly & Currey, 1999; G. C. Reilly, et al., 1997; Yeni & Norman, 2000). This appears to be most typically correlated with the higher density of cement lines provided by an increased population density of osteons (Gibson, Stover, Gibeling, Hazelwood, & Martin, 2006; Liu, Wagner, & Weiner, 2000; O'Brien, Taylor, & Lee, 2005). Additional beneficial consequences of osteons include the toughness provided by the formation of osteon morphotypes (see Section 5.6.5) and the re-population of tissue with viable osteocytes. However, increases in intracortical porosity associated with increased population densities of osteons can reduce fatigue life; this issue is not discussed further here and is more relevant in the context of comparing osteonal with non-osteonal bone (e.g., fibrolamellar bone) (Loundagin, Pohl, & Edwards, 2021).

5.6.2 Osteon Morphotypes and Matrix "Motifs"

Ascenzi and Bonucci (1967, 1968) were the first to report strong correlations between the mechanical properties of individual secondary osteons and the birefringence patterns that they identified: predominantly bright, predominantly dark, or alternating bright and dark lamellae. They observed that the dark osteons had greater strength in tension than those having bright or "alternating lamellae." Following the interpretation of lamellar structure postulated by von Ebner (1887) and Gebhardt (1906), osteons with alternating lamellae, as seen when viewed between crossed polarizers, were assumed to have plies in which the collagen fibers alternated between longitudinal and transverse directions (Figure 5.25). Osteons appearing entirely bright or dark were assumed to have mostly transversely or longitudinally oriented fibers, respectively (and with respect to the long axis of the bone) (Bromage et al., 2003). (See Marotti (1996), Burr and Turner (1998), Yamamoto et al. (2000), and Ascenzi and co-workers (2003; 2006) for discussions regarding alternative explanations that have been considered, but are not as well supported, for the physical bases of these differences in lamellar birefringence.) A brief review of observations of osteon morphotypes in polarized light reported from the 1930s through the 1980s can be found in Skedros et al. (1996).

5.6.3 Role of Osteoblast Cilia in Producing Strain-Mode-Specific Differences in Predominant CFO

We have proposed a mechanistic basis for the biosynthesis of the different osteon morphotypes based on osteoblast cilia deformation caused by physiological load-induced fluid flow (J.G. Skedros, et al., 2016). This proposed mechanism highlights the different matrix lamellar "motifs" that are created by osteoblasts for, presumably, different strain environments. For example, compression-adapted osteon morphotypes exhibit a "twisted plywood design," while tension-adapted osteon morphotypes feature an "orthogonal plywood design" (as diagrammatically generalized in Figure 5.25 at top middle and right, respectively). There are advanced imaging technologies with sufficient resolution to allow for the non-invasive/non-destructive (i.e., without bone sectioning) examination of these motifs and other three-dimensional orientation patterns of mineralized collagen fibrils (e.g., synchrotron X-ray phase nano-tomography) (Varga et al., 2013).[13] However, the vivid detail that conventional CPL imaging provides of secondary osteon morphotypes when examined in thin plane-parallel sections of bone has not yet been surpassed by any high-resolution non-destructive imaging technology.

5.6.4 Predominant CFO: "Toughening" Not "Strength Enhancement"

Older conventional wisdom (A. Ascenzi, 1988; F. G. Evans, 1958; F. G. Evans & Vincentelli, 1974; Martin & Boardman, 1993; Martin & Ishida, 1989; Martin, Mathews, Lau, Gibson, & Stover, 1996; D. T. Reilly & Burstein, 1974, 1975; Vincentelli & Evans, 1971), supported the idea that strength enhancement is the goal of regionally predominant CFO and, by extrapolation, to the more recently described specific osteon morphotypes (Figure 5.3). However, this conclusion has been disproven by more recent data showing that toughening is the aim of regional CFO-based variations in limb bones. The earlier conventional paradigm stemmed from studies of the relationship between CFO and mechanical properties of macroscopic cortical bone specimens without respect to regional habitual loads (i.e., strain-mode-specific loading experiments were not performed in these earlier studies (see footnote 7)) (J.G. Skedros, Dayton, et al., 2023b; J. G. Skedros, Dayton, et al., 2006c). For example, more advanced studies using strain-mode-specific loading have shown that predominant CFO more strongly influences regional material toughness (energy absorption)

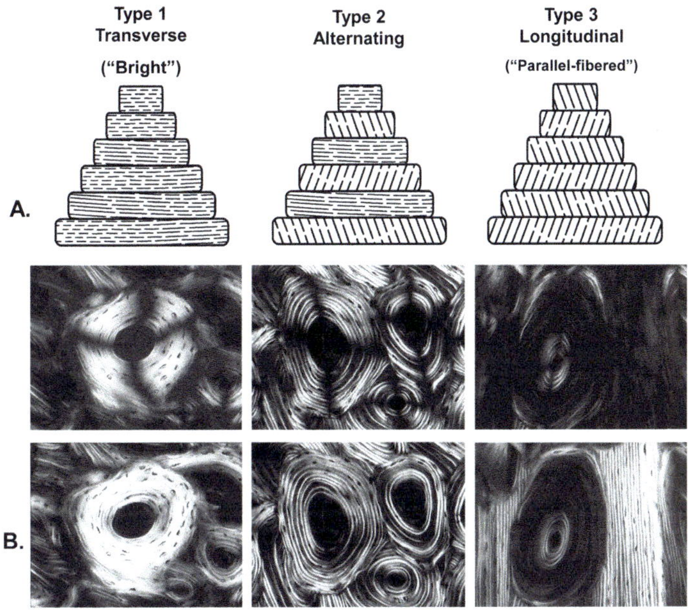

Figure 5.25 Examples of three osteon morphotypes in linearly polarized light (**top** microscopic images; dark cross-shaped extinction patterns are present) and circularly polarized light (**bottom** images; dark cross-shaped extinction patterns are absent). The illustrations at the **top** of this figure are diagrammatic depictions of the predominant collagen fiber orientation (CFO) patterns that result in the presence or absence of the birefringence patterns seen in the lamellae of the osteons in the microscopic images. The osteon images are from Bromage et al. (2003), and the layout of the figure is based both on Bromage et al. (2003) and Ascenzi and Bonucci (1968). These investigators ascribe to the view that the birefringence (gray level) variations, seen for example in the "alternating osteon," are primarily based on lamellar variations in predominant CFO. Other investigators have argued that variations in collagen fiber density account for the gray level variations in the alternating osteons (Marotti, 1996); but this view is <u>not</u> supported by experimental data and also appears to be influenced by artifacts of specimen preparation techniques (J.G. Skedros, et al., 2016; Yamamoto, et al., 2000). (Images are courtesy of Timothy Bromage.)

(J.G. Skedros, Dayton, et al., 2023b; J. G. Skedros, Dayton, et al., 2006c). These results add to a growing body of experimental data that reveal an important role for predominant CFO, collagen content, and/or intermolecular collagen cross-links in affecting energy absorption (a measure of toughness) (Banse, Sims, & Bailey, 2002; Burr, 2002; Nyman et al., 2007; J. G. Skedros, Sybrowsky, Bloebaum, Bachus, & Wang, 2006; Wang, Masilamani, Mabrey, Alder, & Agrawal, 1998).[14]

In a study of strain-mode-specific mechanical testing of an equine radius (12-year-old thoroughbred), Reilly and Currey (1999) showed that the patterns of microdamage formed, and the mode of failure of specimens from the cranial "tension cortex" and caudal "compression cortex," suggest that each cortex was adapted to its respective habitual loading mode (pg. 551):

> ... bone can be adapted to be better at resisting the growth of one type of microcrack (either tension or compression) but that this then causes it to be bad at resisting the other type. Therefore, it is better for bone to be loaded in one mode only, so that it can adapt histologically to that mode.

This observation has important implications for the pathogenesis of some stress fractures (see Section 5.6.5).

Predominant CFO and osteon pullout is beneficial for energy absorption in bone. In comparison to these studies addressing the mechanical importance of regional variations in averaged CFO values, there are only a few studies that specifically evaluate the mechanical consequences of differences in the distributions of specific osteon morphotypes. Hiller et al. (2003) showed that variations in the distributions of "distributed" (i.e., "bright" and "alternating") and "hooped" osteons (Figure 5.3) in dorsal (anterior), medial, and lateral cortices of equine third metacarpals can affect regional fatigue life and toughness by differentially dissipating energy (via osteonal pullout) (Figure 5.13). In mechanical testing studies of equine third metacarpals, we have also implied that regional variations in the prevalence of these various osteon morphotypes produce the regional differences in CFO that strongly influenced energy absorption in strain-mode-specific testing (J. G. Skedros, Dayton, et al., 2006c). However, we subsequently showed that CFO (as measured in entire microscopic regions; e.g., Figure 5.5 at top) and intermolecular collagen cross-link densities were more important than osteon morphotype scores (which are dependent only on osteons, not on entire microscopic regions) in explaining energy absorption properties in material tests of equine third metacarpals (J. G. Skedros et al., 2014; J. G. Skedros, Sybrowsky, et al., 2006a).

The reason for the subsidiary, though still statistically significant, importance of osteon morphotype scores in those studies was because of the rather low osteon densities in some of the tested regions (e.g., as shown in the "tension cortex" of a horse radius in Figure 5.5). The quantification of an osteon "morphotype score" (MTS) (e.g., Figures 5.3, 5.5 and 5.22) for a region of interest is a significant advance in interpreting bone adaptation because the regional prevalence of these osteon morphotypes appears to be common, having been observed in various bones of diverse species (Beraudi, Stea, Bordini, Baleani, & Viceconti, 2010; Currey, 2002; Martin, Gibson, Stover, Gibeling, & Griffin, 1996; McFarlin, Terranova, Zihlman, Enlow, & Bromage, 2008; Riggs, Lanyon, & Boyde, 1993; J. G. Skedros, Keenan, et al., 2013a; J.G. Skedros, et al., 2011; J. G. Skedros, et al., 1996; J. G. Skedros, et al., 2009; Vincentelli, 1978; Warshaw, et al., 2017).

5.6.4.1 *Mechanical Importance of Different Osteon Morphotypes and Hybrids*

In another study, Bigley et al. (2006) performed pushout tests on individual osteons isolation and with consideration of their surrounding matrix histomorphology. They used the diaphysis of the third metacarpal of an adult thoroughbred horse and made observations in both polarized light and scanning electron microscopy. They found clear differences in interfacial strength (at cement line and interlamellar "interfaces") associated with variations in CFO that typify the four osteon morphotypes that they evaluated (dark, hooped, alternating, and bright). In their pushout tests, "interfacial" failure was typically observed at the cement line "interface" or within the osteon wall (i.e., interlamellar interface). For alternating osteons, the cement line is weaker (12.4 MPa) than the interlamellar interface (33.3 MPa, p = 0.04). By contrast, the cement line is stronger for the bright osteon (47.1 MPa) than the interlamellar interface (33.4 MPa, p = 0.04). Overall, bright osteons had the highest interfacial debonding strength (40.3 MPa), and dark osteons had the lowest strength (22.8 MPa) ($p < 0.05$). In terms of maximum interfacial shear stress, the bright osteons also had highest stress (82.6 MPa), the dark osteons had the least (63.6 MPa) ($p < 0.05$), and alternating and hooped osteons had intermediate values (76.3 MPa and 71.0 MPa, respectively). Their linear regression analyses also suggest that there may be a relationship between the amount of oblique-to-transverse collagen fibers (compression-adapted collagen in the osteon wall) and these important mechanical properties, independent of the presence of alternating rings. (This supports the idea that "hybrid" osteon morphotypes (Figures 5.3 and 5.21) are beneficial for accommodating shear.) The histomorphology of the surrounding matrix also profoundly influenced the shear strength of the osteons. For example, as the fractions of bright, dark, and hooped osteons increased, the debonding shear strength increased, whereas when alternating osteons increased, the shear debond strength decreased. In view of these mechanical test results, it is clear that modified/advanced osteon MTSs that are based in mechanical contexts, as attempted in our study of the chimpanzee femur (J. G. Skedros, C. J. Kiser, et al., 2011), should consider differentiating bright from alternating osteons in addition to differentiating hooped from non-hooped osteons. This should prove to be important in anthropological studies because the variability of gait/ physical behaviors and activities often places the limb bones of these species in relatively high-complexity loading (Table 5.2).

5.6.5 Stress Fractures and Osteon Morphotypes

A change in a bone's load history that significantly changes the strain distribution could lead to catastrophic failure if there is insufficient time for microstructural toughening via osteonal remodeling. This likely explains why the majority of stress fractures in the equine third metacarpals occur in the dorsal-lateral cortex. Nunamaker (2001) suggests that the shift from prevalent tension in the dorsal cortex during race training at lower speeds to compression at higher racing speeds may provide an explanation for these fractures in thoroughbred horses. He suggests that tension-related adaptations produced by remodeling and modeling activities[15] would not be expected to accommodate compression strains produced at greater speeds: "It seems obvious that bone that models and remodels for tensile forces on the dorsal aspect of MCIII [third metacarpal] will be poorly adapted for the large compressive strains that are seen during racing" (p. 213 (Nunamaker, 2001)). In turn, when the tension-adapted dorsal cortex receives more prevalent, high magnitude compression strains microcrack formation is enhanced. This probability is supported by Reilly and Currey (2000), who showed that microdamage formed in compression is highly detrimental to tensile mechanical properties. In tension *vs.* compression, microdamage not only occurs at different thresholds (more readily forms in tension), but can exhibit different morphologic characteristics (e.g., length, shape, and orientation) (Boyce, Fyhrie, Glotkowski, Radin, & Schaffler, 1998; George & Vashishth, 2005; Joo, Jepsen, & Davy, 2004; Muir, Johnson, & Ruaux-Mason, 1999; G. C. Reilly & Currey, 1999; G. C. Reilly, et al., 1997; J. G. Skedros, Sybrowsky, Anderson, & Chow, 2011a).

An equine third metacarpal that is primarily/habitually exposed to sub-racing speeds would be expected to become adapted primarily to tension and shear in the dorsal/lateral cortices, reflecting the oblique neutral axis known to occur in that region (hence would be in the intermediate-A load-complexity category). However, if higher speeds are acutely and more consistently sustained, then the neutral axis rotates significantly (this is why this bone is in the intermediate-B load-complexity category). In turn, the dorsal/lateral cortex of the third metacarpal is then exposed to high-magnitude compression for which it has not microstructurally/nanostructurally adapted. Could the idea of an activity-induced shifting neutral axis also help to better explain the occurrence of stress fractures in human military recruits or in human femoral neck fragility fractures in the elderly (Mayhew et al., 2005; J.G. Skedros, Cronin, et al., 2023a; Tang, Cripton, Guy, McKay, & Wang, 2018)?

5.7 Poor Reliability of Many Often-Studied Osteonal/Histologic Characteristics for Interpreting Load History in Limb Bones

As discussed above, regional variations (or lack thereof) in predominant CFO and/or osteon morphotypes are the strongest characteristics for interpreting load history—much stronger than mineral content, osteon and osteocyte population densities, and osteon size (see below). Regional variations in the regional heterogeneity of CFO has also been shown to be weak in this respect (Dakoulas, Keenan, & Skedros, 2011; K. Keenan & Skedros, 2011). Osteon population density and other osteon morphologic characteristics (size, shape, etc.) have received great attention in anthropological studies that are aimed at differentiating species or deciphering the load history of limb bones and ribs (Abbott, Trinkaus, & Burr, 1996; Burr, et al., 1990; Martiniaková, Grosskopf, Omelka, Vondraková, & Bauerová, 2006; Mulhern & Ubelaker, 2009; Pfeiffer, Crowder, Harrington, & Brown, 2006; Robling & Stout, 2003; Schaffler, Burr, Jungers, & Ruff, 1985; Urbanová & Novotný, 2005) even though metabolic-related issues can confound osteon-based comparisons, especially between ribs and limb bones (J. G. Skedros et al., 2013). But it seems that OPD must be considered in most cases because secondary osteons tend to be concentrated (although not always (Gibson, et al., 2006; Mason, et al., 1995; J. G. Skedros, Keenan, et al., 2013a; J. G. Skedros, et al., 2009)) in the more highly strained regions ("compression cortices") of bones that are loaded in habitual bending. However, empirical studies show much stronger correlations of CFO with strain-related parameters of load history in limb bones (J.G. Skedros, 2012; J.G. Skedros, Cronin, et al., 2023a; J.G. Skedros, Dayton, et al., 2023b; J. G. Skedros, Keenan, et al., 2013a; J. G. Skedros, Sorenson, Takano, & Turner, 2006; Takano, et al., 1999).

Consequently, typically studied osteon parameters (i.e., OPD, and osteon size and shape) are a distant second to the distribution of osteon morphotypes and/or regional patterns of predominant CFO for interpreting load history (K. E. Keenan, Mears, & Skedros, 2017; J. G. Skedros, Keenan, et al., 2013a). This is likely related to non-mechanical factors that can affect OPD and/or related parameters (e.g., osteon area and the fractional area of secondary osteonal bone), including endocrine, nutrition, age, ethnicity, genetic, and some pathological influences (Cho, Stout, & Bishop, 2006; Feng & McDonald, 2011; Han, Palnitkar, Rao, Nelson, & Parfitt, 1997; Havill et al., 2013; Heaney et al., 1999; Kerley, 1965; Raisz, 1999; I.J. Wallace, et al., 2017). Also, there might be differential influences of load history on OPD that have not been fully appreciated until recently. For example, there are rather limited data (in limb bones) suggesting that, in some cases, osteonal remodeling is more strongly influenced by loading frequency than strain magnitude (Lad, Daegling, & McGraw, 2016; Lad, Kowalkowski, Liggio, Ding, & Ravosa, 2023; Lad, McGraw, & Daegling, 2019).

Another study worth mentioning is that of Britz et al. (2009), who found an inverse relationship between osteon size and body weight in modern human femoral mid-diaphyseal specimens. Notably, that relationship was superimposed on the more dominant pattern of decreasing osteon size with age. Although there was no evidence that this relationship was related to strain magnitude, a prior study reported smaller osteons in the femora and tibiae of late Pleistocene adult hominids that was interpreted as reflecting increased strains produced by increased strenuous activity (Abbott, et al., 1996). However, a mathematical error was detected in that study—reexamination of their data by Streeter et al. (2010) found that the osteon population densities are not higher and the osteons are not smaller for the presumably more strenuously loaded Pleistocene hominid limb bones. Additionally, the negative correlation of osteon size with strain magnitude has not been found in bone regions where this should have been obvious (K. E. Keenan, et al., 2017; Mason, et al., 1995; J. G. Skedros, et al., 2009). Therefore, osteon size is generally unreliable for interpreting load history. Although there is evidence that osteon size tends to increase toward the endosteum (Britz, et al., 2009), inconsistency in this relationship suggests that, when present, it might be an effect associated with the proximity to the marrow and not a consequence of low strain (Birkhold et al., 2016; Chang & Liu, 2022; J. G. Skedros, Mason, & Bloebaum, 2001b; J. G. Skedros, Su, & Bloebaum, 1997).

5.7.1 Population Densities of Osteocyte Lacunae and Various Non-Collagen/Lamellar Osteon Variants

Differences in the regional distribution of osteocyte lacuna population density also caught our attention because of the possibility that this characteristic might strongly correlate with load history especially in differentiating bending from torsion. We postulated this relationship because it seemed reasonable that bone tissue might employ osteocyte density networks that perceive regional non-uniformities in the strain environment (e.g., strain modes and magnitudes). However, correlations of osteocyte lacuna population density with strain history (modes and magnitudes) and marrow proximity have not been found in bones where these relationships should be obvious (J. G. Skedros, 2005; J. G. Skedros, Grunander, & Hamrick, 2005; J. G. Skedros, Weaver, & Doutré, 2016b). Studies of the regional prevalence of sealed secondary osteons, drifting osteons, and other osteon variants (e.g., elongated, zonal, dumbbell, and inter-connected), and concentrations of osteon cement line interfaces, have also shown that these characteristics are not helpful for detecting differences in load history (e.g., across a habitual bending region) or in reflecting aging or proximity to the medullary cavity of adult human and chimpanzee femora and in various other non-human limb bones (Finlinson, King, Henrie, Lohse, & Skedros, 2019; Henrie, Finlinson, Smith, Keenan, & Skedros, 2018; Knight, Keenan, Halley, & Skedros, 2012; J. G. Skedros, Henrie, Doutre, & Bloebaum, 2018; J. G. Skedros, Keenan, Halley, Knight, & Bloebaum, 2012; J. G. Skedros, Sorenson, & Jenson, 2007b).

5.8 Exceptions to the Rule, Important Caveats, and Consideration of "Tissue Types"

Although strain-mode-related variations in the patterns of CFO can be present in bones where remodeling does not occur (Mason, et al., 1995; J. G. Skedros & Hunt, 2004; J. G. Skedros & Kuo, 1999), there can be circumstances of growth and/or cases in some bones that poorly remodel (few secondary osteons form) where it is difficult or impossible to use this characteristic for deciphering load history. In this perspective, it is important to point out exceptional cases where predominant CFO can be misleading. We have shown the lack of expected strain-mode-specific (tension *vs.* compression) CFO patterns in skeletally mature Rhesus macaque ulnae and chicken tarsometatarsi even though they are considered to be loaded habitually in bending (Demes, et al., 2001; Judex, Gross, & Zernicke, 1997b;

J. G. Skedros, Demes, & Judex, 2003). In the case of the macaque ulnae that we examined, the histology in the "tension region" and "compression region" did not show significant differences in CFO and appeared to be comprising fairly recently deposited laminar bone without much remodeling (hence likely reflecting the sometimes difficult-to-interpret histology seen when skeletal maturity has been recently attained). The lack of expected regional tension *vs.* compression CFO patterns in the chicken tarsometatarsus also evades a similarly straight-forward maturation-related explanation, which might be based on high shear strains without significant neutral axis rotation (see footnote of **Table 5.2**). Thus it may be difficult, or impossible, to discern load history in terms of "expected" histomorphological correlates of functional adaptation in some bones that are growing, have recently stopped growing, and/or are from species that do not exhibit much osteonal remodeling, or other exceptional cases (J. G. Skedros & Doutré, 2019; J. G. Skedros, et al., 2004; J. G. Skedros, Sorenson, Hunt, et al., 2007).

Most of the unexpected cases mentioned above demonstrate that regional strain-mode-related variations in predominant CFO and/or associated osteon morphotypes, although seemingly having high validity/accuracy for differentiating habitual unidirectional bending from torsion/multi-directional loading, are not 100% sensitive and 100% specific in this context.

5.8.1 Regional Differences in Tissue Types and Growth Dynamics

Regional bone growth dynamics might be reflected in the seemingly patchy/irregular distribution of several primary histological "tissue types," as shown by McFarlin et al. (2008) in various non-human primate limb long bones spanning a range from immature to adult. In another study of non-human primate long bones, Warshaw et al. (2017) also argue that consideration must be given to the general type of histology that is also present (e.g., fibrolamellar, parallel-fibered, lamellar, compacted coarse cancellous, Sharpey fiber, and secondary osteonal; also see Prondvai et al. (2014)). This is important because most studies that have examined patterns of predominant CFO in bone cortices in the context of load history interpretations have focused on regionally "averaged" birefringence (image gray level) differences in polarized light and have provided little or nothing in terms of detailed descriptions of the histological type (see references in Warshaw et al.). However, Warshaw and co-worker's study also looked for regional variations of CFO in circularly polarized light but failed to find clear relationships. This reflects the probability that the bones that they studied typically experienced high-complexity loading.

We sought to determine if various primary bone tissue "types" also occur during growth of the relatively simply loaded sheep radius, and if regional strain-mode-specific ("cranial/tension" *vs.* "caudal/compression" cortices) differences in predominant CFO also occur in this case where secondary osteons are frequent (mature bones) or scarce or absent (Figure 5.26). Although in our ontogenetic sample of sheep radii (newborn to adult) we did not find the notably different tissue types shown by McFarlin et al. (2008) and Warshaw et al. (2017), the radii from immature (8–10 months old) animals did show statistically significant CFO differences in the primary bone (i.e., when relatively few secondary osteons have formed) between the "tension cortex" and "compression cortex" (J. G. Skedros & Kuo, 1999). These differences appeared to be strain-mode-specific. This, again, is an example showing that a bone can produce strain-mode-specific adaptation in CFO without secondary osteon formation.

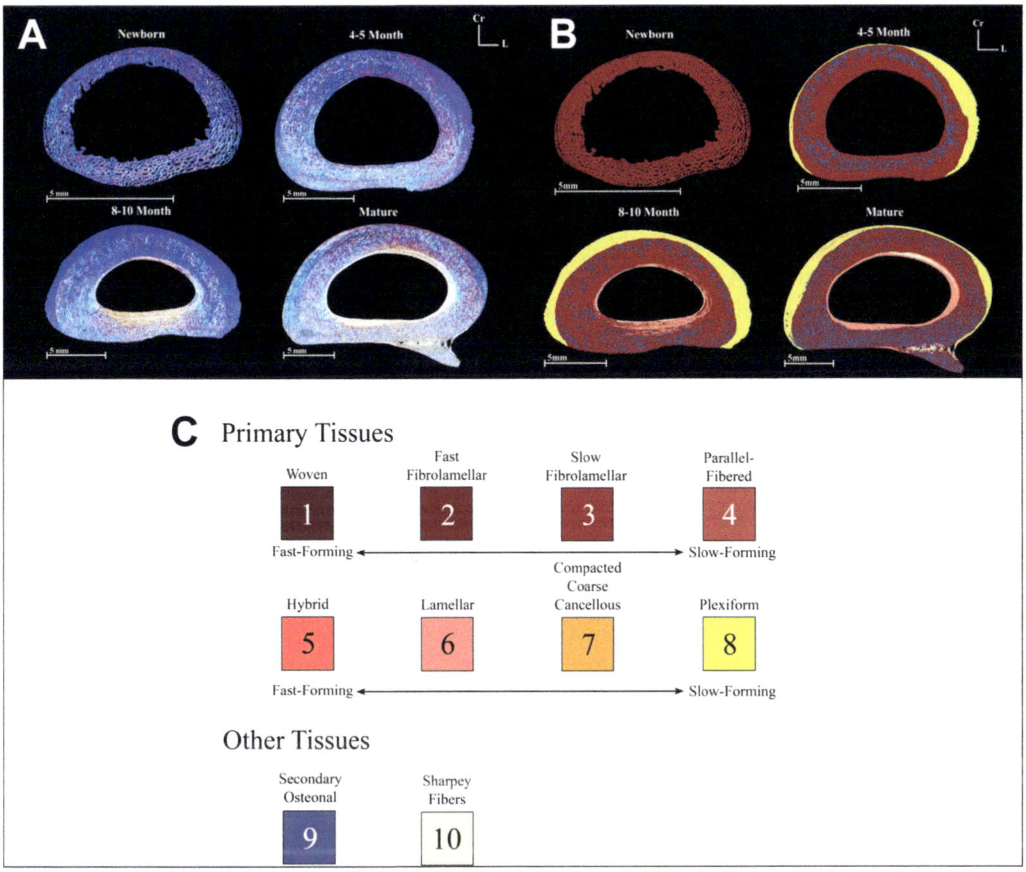

Figure 5.26 (**a**) Circularly polarized light (CPL) montages from a transversely sectioned mid diaphysis of one sheep radius from each of four age groups (note that the ulna is shown only in the mature group). The brighter/whiter gray levels are apparent in the caudal "compression cortices" in the older three groups. Notably, evidence of what appears to be strain-mode-related adaptation is found within the primary bone (i.e., in the non-osteonal bone) in the 4- to 5-month and 8- to 10-month age groups. The specimens were embedded in polymethyl methacrylate in an unstained and fully calcified state, and ultramilled to 100 microns thickness. All images were obtained in the same imaging session using the same illumination and 50x magnification. Montages of each cross-section shown here were created from many smaller microscopic images. Cr = cranial ("tension cortex"), L = lateral. (**b**) Pseudo colored montages of the images in (**a**). (**c**) The color scheme that was used to indicate the various histological "types" shown in (**b**) and include: dark red = slow fibrolamellar*, salmon = lamellar, cream = Sharpey's fibers, yellow = plexiform*, blue = secondary osteonal, and others. Cr = cranial, L = lateral. The pseudo coloring scheme that is used here is adapted from the scheme described by McFarlin et al. (2008). Color #8 was added to the scheme to account for the more "brick-like" plexiform bone as described by Goldman et al. (2009). *Note that plexiform bone is a type of fibrolamellar bone that is organized a specific way.

In (**a**) and (**b**) that the magnification has been adjusted (see scale bars). Also note that asymmetry in modeling activity is suggested by the color patterns in the 8- to 10-month and mature bones. Here it seems likely that more bone formation is occurring on the cranial, lateral, and medial cortical surfaces (and less on the caudal cortex), which is consistent with the regional cortical drift patterns described by Lanyon and Baggott (1976).

5.8.2 Seemingly Contradictory Predominant CFO in Cranial and Caudal Cortices of Sheep and Goat Radii

Our finding of statistically significant regional strain-mode-related CFO differences between the "cranial/tension" and "caudal/compression" cortices of radii of sub-adult and adult sheep (Figure 5.27) (J. G. Skedros & Kuo, 1999; J. G. Skedros, et al., 2009) starkly contrast with the lack of similar regional strain-mode-specific CFO differences in the radius of adult goats (R. P. Main, 2007). In an excellent study conducted in domesticated adult goats, Russell Main (R. P. Main, 2007) did not find the "expected" regional tension- and compression-adapted CFO patterns in the cranial and caudal cortices, respectively. These "unexpected" results in goats likely reflect their more variable locomotor behaviors when compared to adult sheep (Lloveras, Rissech, Davis, & Pares-Casanova, 2022; Moreno, et al., 2008), including climbing trees (Barr, 2020; Delibes, Castañeda, & Fedriani, 2017). This is why the goat radius is in the high-complexity load category (where regional strain-mode-specific material variations do not occur) and the sheep radius is in an intermediate-complexity load category (where regional strain-mode-specific material variations do occur) (Table 5.2).

5.8.3 Unexpected Regions of Bone Growth in the Exercised Sheep Tibia Model

The study of bone adaptation of the sheep tibia in skeletally immature animals by Wallace and co-workers (2014) is very important because it shows that considering influences of different load-complexity can be helpful in interpreting experimental results. They studied the sheep tibia diaphysis during treadmill exercise and examined mid-diaphyseal cross-sections to determine whether this activity promoted bone formation disproportionately in the direction of loading in diaphyseal regions that experience the highest strains. Hence, their study focused on strain magnitude and not strain mode. The treadmill exercise induced anterior-posterior (cranial-caudal) bending, which, as expected, placed the cranial bone surface in maximum tension strain and the caudal bone surface in maximum compression strain. Unexpectedly, new bone formation did not correspond to areas of maximal strains across the cranial-caudal axis. In fact, exercise-related osteogenesis occurred mostly along the neutral axis during the imposed exercise and also in some high strain regions. The authors concluded that this "...highlights the inherent problem with inferring loading history from bone shaft shape in the absence of *in vivo* data on how bones are loaded and how they actually respond to loading."

However, there is healthy skepticism in asking the question: What would be the outcome of an experiment where a bone region habitually subjected to high-complexity loading (i.e., diffuse shear strains from torsion with some bending) is then exposed to a tension/compression cranial/caudal loading regime (intermediate-complexity loading)? (Note that **Table 5.2** lists the sheep tibia in the high-complexity load category.) Perhaps the bone surfaces do not respond in a "conventionally expected" manner if the bone cells responsible for mechanosensation (osteocytes) and those responsible for bone formation (osteoblasts) are habitually attuned to a strain distribution (C. Rubin, Judex, & Hadjiargrou, 2002) that is quite different from the one that is experimentally imposed. This issue was considered nearly 40 years ago by Lanyon and co-workers (1982) in a study of bone formation and remodeling of the sheep radius diaphysis of skeletally mature animals with and without excision of a segment of the nearby ulna. They observed that although overstrain of the

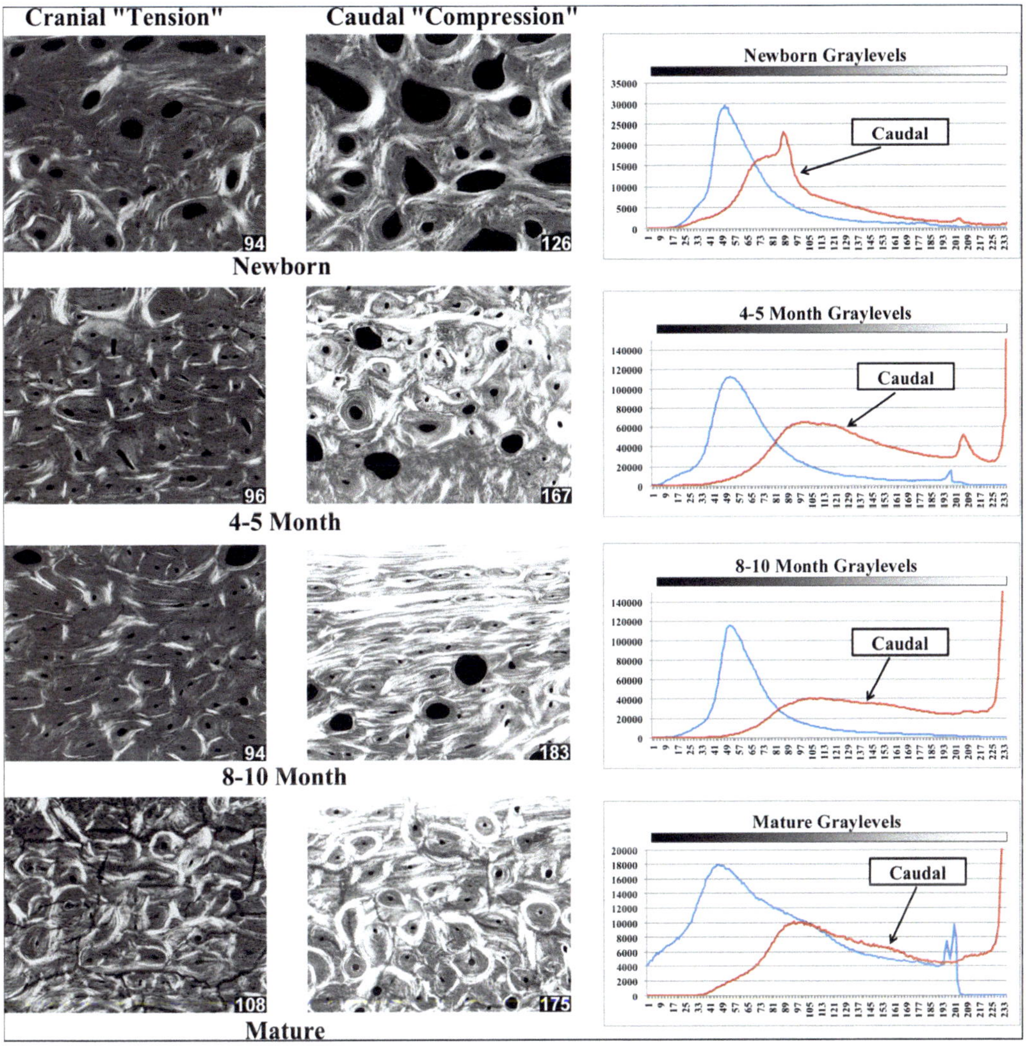

Figure 5.27 Circularly polarized light (CPL) images and corresponding gray level profiles (at **right**) of sheep radii from four age groups. The dark-to-bright gradient over each of the gray level profiles show how the gray level (i.e., predominant CFO) changes from left to right on the x axes of these profiles. The images are from the whole-bone montages Figure 5.26a. The pores have been darkened in order to help distinguish them from the mineralized bone. (The images in this figure were not obtained in the same session as the images obtained in Figure 5.24.) There are notable differences in brightness/contrast between the sheep radius images in these two figures. Brighter gray level in the caudal cortex represent relatively more oblique-to-transverse CFO (considered an adaptation for compression) than the relatively darker gray level in the cranial cortex (considered to be an adaptation for tension). Note that these differences in CPL birefringence appear prior to the extensive osteonal remodeling that is seen in the mature bones (i.e., two-year-old animals). The numerical mean gray level for each image is shown at the lower right. The specimens were embedded in polymethyl methacrylate in an unstained and fully calcified state, and ultramilled to 100 microns thickness. All images were obtained in the same imaging session using the same illumination and 50x magnification.

tension cortex of the radius was greater than that on the compression surface after the ulnar ostectomy (20% *vs.* 8%), most new bone formation occurred on the periosteal surface of the compression surface and adjacent to the missing ulna (and this new bone was highly remodeled with secondary osteons and the new bone on the tension side was not; CFO analysis was not done). This change on the radius appeared to be an attempt to replace the missing ulna by increasing the nearby cortical area of the radius. Similar to the findings of Wallace et al. (2014) in immature sheep tibiae, this response in the sheep radius is also not consistent with the idea that strain magnitude controlled the adaptive response, because the greatest amount of bone formation was on the surface where the increase in strain was the smallest. In this case the intermediate-complexity loading of the sheep radius simply changed with respect to the direction of bending (the change in load history was not as dramatic as in the sheep tibia experiment of Wallace et al.). Lanyon and co-workers suggested that alterations in "strain distribution" (see Figure 5.1 left side), or site-specific thresholds, may have some control of bone modeling and remodeling activities (Goodship, et al., 1979; Lanyon, et al., 1982). Additional controlled experimental studies are needed to further explore these possibilities and others, including relationships with strain mode in the contexts of the putative differential influences that strain magnitude and load frequency have on osteonal collagen/lamellar morphology and remodeling activity (Lad, et al., 2016; Lad, et al., 2023; Lad, et al., 2019).

5.9 Skeletal Immaturity and Precocial vs. Altricial Ambulation

5.9.1 Bone Histology in Precocial vs. Altricial Growth/Ambulation

Appendicular bones of some larger animals function well even though their osteonal remodeling is relatively sparse while other bones of the same limb skeleton are highly remodeled with secondary osteons. An example of this is shown in Figure 5.28. Bones that must grow fast for precocial ambulation, such as in limb skeletons of artiodactyls (e.g., sheep, cows, deer) and horses, often have primary bone (e.g., fibrolamellar or plexiform) earlier in their growth, which is a situation that is not encountered in anthropological studies of bone adaptation (J.G. Skedros, 2012). In many cases, osteons eventually form in these bones, probably reflecting the repair of microdamage that forms during normal use (J. G. Skedros, Sybrowsky, Parry, & Bloebaum, 2003a). By contrast, primates exhibit altricial ambulation and thus have appendicular bones that grow more slowly. For reasons that are unclear and do not necessarily appear attributable to slow growth, the relatively rapid onset of remodeling is often found in this situation.

5.9.2 The Bugaboo of Non-Primate/Inter-Species Comparisons and Extrapolations

The bugaboo in non-primate *vs.* primate comparisons, such as many that are presented in this chapter, is the idea that extrapolating bone adaptations in some species (like sheep, deer, and horses) to primates (like humans) can be invalid. While there seems to be few specific reasons that this might be so, this idea is pervasive in bone adaptation studies in anthropology. Fortunately, it has been shown that the adaptability of the cross-sectional morphology of limb bones is similar between primate and non-primate species of roughly similar size (Polk et al., 2000). But there is reason for concern about the adaptability of

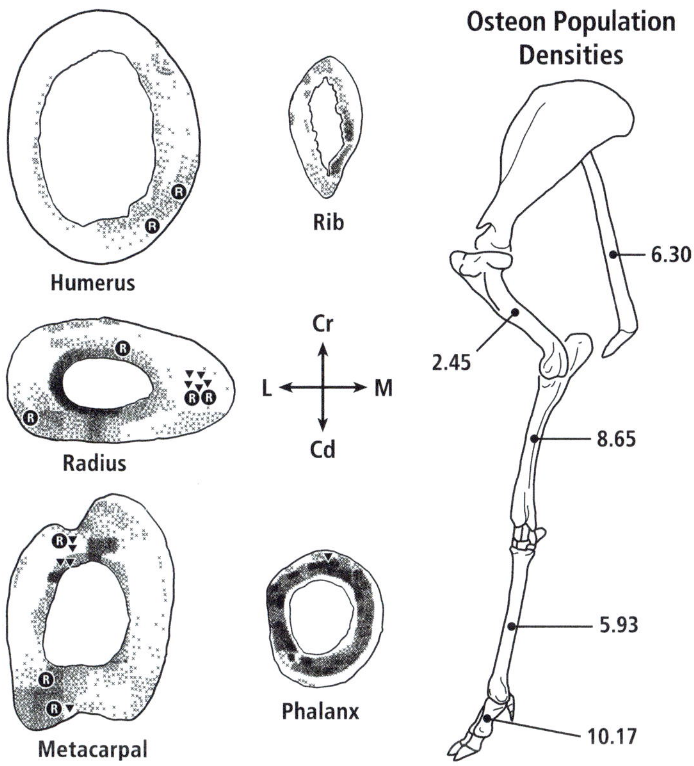

Figure 5.28 At left are transverse sections of the mid-diaphyses of the forelimb bones from one of the mature mule deer examined by Skedros et al. (2003b). **X** = secondary osteons, **R** = resorption spaces, and triangles (▼) = newly forming secondary osteons. Secondary osteons typically occupy approximately 30% or less of the mid-diaphyseal cortex of the principal metacarpal, radius, and humerus. The histological type of the non-osteonal bone is plexiform-type of bone (which is a type of fibrolamellar bone that is typically present in animals that grow relatively fast for precocial ambulation). At right is the side view of the forelimb skeleton of an adult Rocky Mountain mule deer showing the averaged values of secondary osteon population densities (no./mm^2) at mid-diaphysis from 11 adult animals.

primates vs. non-primates at the microstructural level. This is because there are dramatic differences between some species in terms of bone histology, especially if fibrolamellar/plexiform-type bone is prevalent—again, as seen in many animals that have rapid growth/precocial ambulation (Hillier & Bell, 2007). Conducting primate/non-primate comparative studies is feasible as long as: (1) the bones subject to comparison have the capacity to form significant densities of osteons (e.g., comparing rats to humans would be very problematic in this case (Lad, 2023)), and (2) the "volume effect" is not at play in some of the bones under scrutiny (Section 5.5).

5.10 Calibration of CFO for Quantitative Analysis

Although used successfully for many decades, the CPL method described in all aforementioned studies provides only relative differences in CFO—hence the data are qualitative because none of these studies used a calibration standard. Spiesz et al. (2011) have described a quantitative method for CFO analyses that uses linear birefringence to provide actual angular deviations from a reference orientation such as the long axis of a bone. Their method calibrates CFO using the mineralized turkey tendon model, where segments of the tendon are analyzed after being cut at a variety of angles with respect to the main tendon axis. This allows for quantitative CFO measurements, and they showed its utility in human secondary osteons. Studies are needed to: (1) make this calibrated methodology generally available and reproducible, and (2) explore its strengths and limitations when compared to conventional qualitative CFO analyses in various bone tissue types and load conditions.

5.11 Conclusions

A bone's adaptations for its load history can best be interpreted when using a systematic approach that includes the assessment of many structural and material characteristics (**Table 5.1**) and comprehensive evaluation provided by the checklist (**Table 5.3**). In addition to the load-complexity category hypothesis (**Table 5.2**), other important considerations include the: (1) shear resistance-priority hypothesis, (2) modeling/remodeling distinction and "division of labor," and (3) mechanical relevance of osteons, osteon morphotypes, and predominant CFO. The checklist fosters the accurate assessment of many of the issues that must be considered when interpreting a bone's load history in terms of characteristics of its structural and material organization. Among these myriad characteristics, regional variations in predominant CFO—that are often strongly correlated with population densities of secondary osteon collagen/lamellar morphotypes—appear to provide the strongest level of evidence for differentiating habitual bending from torsion/multi-directional loading when *in vivo* strain data are lacking or insufficient. The application of the checklist approach also facilitates making accurate comparisons between various bones, as well as within regions of the same bone. Identifying the load-complexity category of the bone or bone region that is being studied is of paramount importance and should be done prospectively for all studies of bone adaptation.

5.12 Appendix: Definitions

Adaptation: Adaptation in cortical bone commonly refers to either: (1) changes in bone structure and/or material organization in response to loading conditions outside a normal physiologic stress-strain range, distribution, and/or duration etc. (Martin & Burr, 1989; Schaffler, et al., 1985), or (2) the presence of regional differences in structural and/or material organization that are strongly influenced by normal functional stimuli occurring during normal development within or between bones (J. E. Bertram & Swartz, 1991; Martin & Burr, 1989; Riggs, et al., 1993). In this chapter the use of the term "adaptation" is not meant to imply an evolutionary adaptation within a species, but rather a specific adaptation within the lifetime of an individual. In other words, in this chapter "adaptations" are considered to be biomechanically relevant regional variations and temporal changes in cortical bone structural and material organization that are produced by the modeling and remodeling processes during normal skeletal development. In addition to being mediated by genetic and epigenetic influences, which are heritable, these processes can be influenced by non-heritable ("extra-genetic") stimuli such as regional variations in microdamage incidence. For a more in-depth discussion of "types" of bone adaptation, you might want to examine the excellent book chapters by Russell Main, Erin Simons, and Andrew Lee (R.P. Main, Simons, & Lee, 2021) and Ian Wallace, Brigette Demes, and Stefan Judex (I.J. Wallace, et al., 2017); they are especially useful because they astutely discuss bone adaptations in ontogenetic and phylogenetic contexts, and also with respect to myriad related issues are not addressed in this chapter.

This chapter primarily deals with identifying correlations between structure, material, function, and load history. These correlations might be produced by: (1) adaptation via natural selection, and (2) accommodation via epigenetic events. We have suggested that the non-uniform strain distribution experienced in the early development of most limb bones is proximate to the historical origin (i.e., mechanistic/causal) of the structure-function relationship (J. G. Skedros, Sorenson, Hunt, et al., 2007). Therefore, the word "adaptation" in this context is deemed appropriate *sensu stricto*. Although the word "accommodation" is also used in this chapter, some readers might argue that this is inappropriate because in most cases where it is used there has been no attempt to see how affecting the epigenetic signals that influence bone development would change structure or material (like Starck & Chinsamy, 2002).

Modeling: Modeling activities affect the formation and/or resorption of secondary or non-secondary bone (e.g., primary bone, and trabecular bone in some cases) on periosteal or endosteal surfaces. They are detected as changes and/or differences in a bone's curvature, cross-sectional shape, and/or regional cortical thickness. Consequently, modeling is a concept describing a combination of non-proximate, though coordinated, resorption and formation drifts whose net result is, typically, to change the distribution of bone. Such drifts are called macro-modeling in cortical bone and mini-modeling in cancellous bone. The re-alignment of trabecular tracts along the lines of stress would be a consequence of mini-modeling.

Remodeling: Remodeling activities affect the replacement of intracortical bone; this is achieved through the activation of basic multicellular units (BMUs = osteoclasts and osteoblasts) that create secondary osteons (Haversian systems) in cortical bone and secondary osteons or hemi-osteons in trabecular bone (Jee, Li, & Ke, 1991; Parfitt, Mundy, Roodman, Hughes, & Boyce, 1996). Readers that might be new to the field of bone adaptation should realize that "remodeling" is a polysemantic term that can mean bone turnover (i.e., secondary osteon formation as used in this chapter), bone functional adaptation in a general sense, fracture repair, and bone growth remodeling. Additionally, remodeling is also used to refer to a broad array of pathological skeletal changes (Wei & Cooper, 2023).

Toughness: Toughness refers to the amount of energy required to fracture a material; the more the overall amount of energy consumed, the "tougher" the material (Zioupos & Currey, 1998). Tough materials resist damage propagation but do not necessarily resist damage formation. Toughness tests typically involve propagating a crack in a controlled direction through a specimen machined into a specific shape for this test. Toughness measurements obtained from these "formal" toughness tests are different from the "toughness" that is often "informally" used to describe the energy absorbed by a specimen during a more conventional failure test. In this case, energy absorption is measured as the area under the stress-strain curve (C.H. Turner & Burr, 1993), which is an indirect measure of propagation toughness (Vashishth, 2004).

Notes

1 Mechanical strain is the change in length of a loaded structure as a percentage of its initial (unloaded) length. (In contrast, shear strain is not a change in length but a change in angle.) This dimensionless ratio is a measure of material or tissue deformation. 200 microstrain (με) in compression means a shortening of the structure by 0.02%; for example, from 100% to 99.98% of its original length. *In vivo* strain data from a variety of animals suggests that physiological strains are generally between 200 and 3,000 microstrain (i.e., between 0.02% and 0.30% change in length) in compression (Biewener, Thomason, Goodship, & Lanyon, 1983; Biewener, Thomason, & Lanyon, 1983; C. T. Rubin & Lanyon, 1985). The upper limit may be only 1,500 microstrain in tension (Fritton, McLeod, & Rubin, 2000). For an isotropic material loaded axially, stress (force per unit area) and strain are related by Hooke's law, which says that they are proportional to one another.

2 The complete record of forces imposed on a bone over a period of time is called the "load history" of the bone (Carter & Beaupré, 2001) (see also note 10). The functional adaptation of a bone to its load history is best explained or understood by considering the strains and stresses that are created by the applied loads.

3 'Material' characteristics include histomorphological characteristics (e.g., predominant collagen fiber orientation, secondary osteon collagen/lamellar morphotypes, and osteon size, shape, and population density) and histocompositional characteristics (e.g., regional variations in mineralization, specific collagen molecular crosslink densities, and noncollagenous proteins).

4 In this chapter, "remodeling" refers to secondary osteon (Haversian system) formation. Hence, it is redundant to say "osteonal remodeling"; but I still do this because some researchers use the term "remodeling" to mean both modeling and remodeling. As defined in the Appendix, remodeling in this chapter refers to the activation-resorption-formation (A-R-F) sequence of the basic multi-cellular unit (BMU) that forms either a secondary osteon within cortical bone or hemi-osteon on surfaces such as trabecular struts in cancellous bone.

5 There are types of microdamage (e.g., diffuse-type) that might be repaired by osteocytes (i.e., without evoking an osteon remodeling response) (Herman, Cardoso, Majeska, Jepsen, & Schaffler, 2010; Seref-Ferlengez, Basta-Pljakic, Kennedy, Philemon, & Schaffler, 2014). This non-osteon repair process might be at play in bones that do not remodel (i.e., lack the capacity for making biomechanically meaningful numbers of secondary osteons). In either case, osteocytes (that are believed to be mechanosensors), and possibly also osteoblasts (Jung, Best, & Akkus, 2015; Robling & Bonewald, 2020), appear to play an important role in detecting microdamage entities.

6 The three strain "modes" are tension (pulled), compression, and shear. The highest strain "magnitudes" are usually in compression (Lieberman, Polk, & Demes, 2004). Among various 'types' of stimuli, available data suggest that strain is the mechanical parameter that seems to be most directly involved in causally mediating bone adaptations (Ehrlich & Lanyon, 2002; Lanyon, 1987; C. T. Rubin & Lanyon, 1984; I.J. Wallace, Demes, & Judex, 2017). In bone, strain-induced fluid-flow dynamics that differ between tension, compression, and shear environments are likely the most proximate stimuli in effecting the cell activities that mediate functional adaptations in these different environments (Cowin, 2002; Fan, Pei, Lucas Lu, & Wang, 2016; Judex, Gross, & Zernicke, 1997; J.G. Skedros, Doutré, & Weaver, 2016; van Tol et al., 2020). Distinguishing between compression, tension and shear is also important because experimental evidence shows that bone cells can perceive and respond to different strain modes (Ei Hsu Hlaing et al., 2020; Kanzaki et al., 2019; Li, Zhang, Yang, Huo, & Zhang, 2015; Zhong, Zeng, Ni, & Huang, 2013), which can have clinical relevance even in adults (Rooney, McNeill, Ross, Bostrom, & van der Meulen, 2023).

7 For example, strain-mode-specific compression loading means that a region that is habitually loaded in compression is loaded (or mechanically tested) in compression (and similarly for strain-mode-specific tension loading).

8 All osteons referred to in this chapter are secondary osteons (Haversian systems).

9 Note that the correct nomenclature is, for example, "high-complexity load category" and not "high-load complexity category." The point here is that the magnitude of loading is irrelevant in most cases in this categorization scheme. Hence, a low-complexity loaded bone can habitually experience low magnitude loads or high magnitude loads.

10 In their review article, Ruff et al. (2006) emphasize that *in vivo* strain data available for most bones that have been studied generally consist of a few average cyclic strain parameters that are extracted from a short period of recordings while the animal performs a very restricted task. This could be construed to mean that the concept of "habitual load history" used in this chapter is excessively simplified, supporting the view that a much more complete record of strain history is required to relate bone biology and morphology to strain. But it is also important to realize that experimental studies have shown that very few loading cycles are sufficient for maintaining bone mass (C. T. Rubin, et al., 1996; C. T. Rubin, Gross, McLeod, & Bain, 1995). This might suggest that even brief amounts of loading are sufficient to produce regional strain-related remodeling variations between regions of the same cross-section of a limb bone diaphysis. Consequently, a "habitual" load history that produces regional variations in bone structural or material organization could, on a daily basis, be relatively brief. Prospective experimental studies are needed to better define "habitual load histories" for specific bones and circumstances (e.g., (Adams et al., 1997; Fritton, et al., 2000)) and how/when they are sufficient to produce mechanically important regional variations in bone histomorphology and/or composition.

11 Bones that resemble I-beams more closely such as the mandibular ramus of a wolf are very uncommon, perhaps because this design would preclude the various benefits of a marrow cavity and its contents for most animals. But in the case of manatees, it is clear that tubular morphology is not an absolute constraint of long bone design. In these animals the marrow cavities of the limb bones and ribs are essentially absent; they are filled with bone for buoyancy dynamics. In manatees, the primary site of hematopoiesis is in the vertebral bodies.

12 This chapter does not overview studies that have examined the relationship of long-bone structure (e.g., cross-sectional geometry and curvature of the diaphysis) with respect to ontogenetic changes in locomotor behaviors/patterns or as the result of different exercise regimes in skeletally immature or mature individuals (e.g., (Canington, Sylvester, Burgess, Junno, & Ruff, 2018; Gosman, et al., 2013; Macdonald, et al., 2009; Pantinople, McCabe, Henderson, Richards, & Milne, 2017; Sarringhaus, MacLatchy, & Mitani, 2016; Swan, Ives, Wilson, & Humphrey, 2020)). Such studies have great value because the adaptations that occur are typically retained in the adult animals. Histocompositional analyses make ontogenetic studies even more valuable (see Figure 28) (J. G. Skedros, et al., 2004; Zimmermann et al., 2019). See section 5.9 for mention of the biomechanical importance of the differences in microstructure between bones of species with altricial locomotion *vs.* species with precocial locomotion.

13 The CFO heterogeneity seen within the walls of hybrid osteon morphotypes (Figures 5.3 and 5.21) might reflect plasticity in the synthesis processes that occur during osteon infilling and in accordance with variability in the ambient fluid-flow/strain environment. Varga et al. (2013) have shown that changes in the matrix motifs across the wall of human osteons do occur. This suggests that during osteon infilling the osteoblasts can, for example, switch from synthesizing a twisted plywood motif to an oscillating plywood motif. This can result in biomechanically relevant CFO birefringent patterns within the osteonal wall (Katsamenis, et al., 2015) that are not seen in non-hybrid osteon morphotypes.

14 Two studies have reported nanoindentation data of individual secondary osteons that are revealing for the strain-mode-specific concept dealing with CFO. Stockhausen et al. (2021) examined individual secondary osteons that they considered to be tension and compression morphotypes from an adult middle-aged human femur at mid diaphysis. Nanoindentation and micro-mineralization analyses revealed that the tension/dark osteons had greater stiffness and the compression/bright osteons had higher energy absorption properties (tested axially = in the direction of the long axis of the bone). Similarly, in an adult equine radius, Rho et al. (2001)

showed that the tension/dark osteon morphotypes (from the cranial cortex) were much stiffer and more anisotropic than the compression/bright osteons (from the caudal cortex). These data from individual secondary osteons generally corroborate our findings in larger specimens.

15 In addition to remodeling-mediated adaptation, adaptive modeling-mediated adjustments in the cross-section shape of the third metacarpals can occur in racing Thoroughbreds. This is possible in racing Thoroughbreds because they are skeletally immature (D. M. Nunamaker, Butterweck, & Provost, 1989); hence, they still have the capacity to activate sub-periosteal/sub-endosteal bone growth.

References

Aamodt, A., Lund-Larsen, J., Eine, J., Andersen, E., Benum, P., & Schnell Husby, O. (1997). *In vivo* measurements show tensile axial strain in the proximal lateral aspect of the human femur. *J Ortho Res*, *15*(6), 927–931.

Abbott, S., Trinkaus, E., & Burr, D. B. (1996). Dynamic bone remodeling in later Pleistocene fossil hominids. *Am J Phys Anthropol*, *99*, 585–601.

Adams, D. J., Spirt, A. A., Brown, T. D., Fritton, S. P., Rubin, C. T., & Brand, R. A. (1997). Testing the daily stress stimulus theory of bone adaptation with natural and experimentally controlled strain histories. *J. Biomech.*, *30*(7), 671–678.

Aiello, B. R., Blob, R. W., & Butcher, M. T. (2013). Correlation of muscle function and bone strain in the hindlimb of the river cooter turtle (Pseudemys concinna). *J Morphol*, *274*(9), 1060–1069. doi: 10.1002/jmor.20162

Aiello, B. R., Iriarte-Diaz, J., Blob, R. W., Butcher, M. T., Carrano, M. T., Espinoza, N. R., et al. (2015). Bone strain magnitude is correlated with bone strain rate in tetrapods: implications for models of mechanotransduction. *Proc Biol Sci*, *282*(1810), 1–8. doi: 10.1098/rspb.2015.0321

Akkus, O., Knott, D. F., Jepsen, K. J., Davy, D. T., & Rimnac, C. M. (2003). Relationship between damage accumulation and mechanical property degradation in cortical bone: Microcrack orientation is important. *J Biomed Mater Res*, *65A*(4), 482–488.

Allahyari, P., Silani, M., Yaghoubi, V., Milovanovic, P., Schmidt, F. N., Busse, B., et al. (2023). On the fracture behavior of cortical bone microstructure: The effects of morphology and material characteristics of bone structural components. *J Mech Behav Biomed Mater*, *137*, 105530. doi: S1751-6161(22)00435-0 [pii]; doi: 10.1016/j.jmbbm.2022.105530

Ascenzi, A. (1988). The micromechanics versus the macromechanics of cortical bone--a comprehensive presentation. *J Biomech Eng*, *110*(4), 357–363.

Ascenzi, A., & Bonucci, E. (1967). The tensile properties of single osteons. *Anat Rec*, *158*(4), 375–386.

Ascenzi, A., & Bonucci, E. (1968). The compressive properties of single osteons. *Anat Rec*, *161*(3), 377–391.

Ascenzi, M. G., Ascenzi, A., Benvenuti, A., Burghammer, M., Panzavolta, S., & Bigi, A. (2003). Structural differences between "dark" and "bright" isolated human osteonic lamellae. *J Struct Biol*, *141*(1), 22–33.

Ascenzi, M. G., & Lomovtsev, A. (2006). Collagen orientation patterns in human secondary osteons, quantified in the radial direction by confocal microscopy. *J Struct Biol*, *153*(1), 14–30.

Banse, X., Sims, T. J., & Bailey, A. J. (2002). Mechanical properties of adult vertebral cancellous bone: correlation with collagen intermolecular cross-links. *J Bone Miner Res*, *17*(9), 1621–1628.

Barr, W. A. (2020). The morphology of the bovid calcaneus: function, phylogenetic signal, and allometric scaling. *Journal of Mammalian Evolution*, *27*(1), 111–121.

Barth, H. D., Launey, M. E., Macdowell, A. A., Ager, J. W., 3rd, & Ritchie, R. O. (2010). On the effect of X-ray irradiation on the deformation and fracture behavior of human cortical bone. *Bone*, *46*(6), 1475–1485. doi: S8756-3282(10)00485-0 [pii]; doi: 10.1016/j.bone.2010.02.025

Beck, T. J., Looker, A. C., Ruff, C. B., Sievanen, H., & Wahner, H. W. (2000). Structural trends in the aging femoral neck and proximal shaft: Analysis of the Third National Health and Nutrition Examination Survey dual-energy X-ray absorptiometry data. *J Bone Miner Res*, *15*(12), 2297–2304.

Beckstrom, A., Skedros, J., Kiser, C., & Keenan, K. (2010). Predominant collagen fiber orientation data support the multi-domain load hypothesis in the chimpanzee femur. *Am J Phys Anthropol, Supplement 50*, 63.

Beraudi, A., Stea, S., Bordini, B., Baleani, M., & Viceconti, M. (2010). Osteon classification in human fibular shaft by circularly polarized light. *Cells Tissues Organs*, *191*(3), 260–268. doi: 000240045 [pii]; doi: 10.1159/000240045

Bertram, J. E., & Swartz, S. M. (1991). The 'law of bone transformation': a case of crying Wolff? *Biol Rev Camb Philos Soc*, *66*(3), 245–273.

Bertram, J. E. A., & Biewener, A. A. (1988). Bone curvature: Sacrificing strength for load predictability? *J Theor Biol*, *131*, 75–92.

Biewener, A. A., & Bertram, J. E. A. (1993a). Mechanical loading and bone growth *in vivo*. In B. K. Hall (Ed.), *Bone, Volume 7, Bone Growth - B* (Vol. 7, pp. 1–36). Boca Raton, Florida, USA: CRC Press.

Biewener, A. A., & Bertram, J. E. A. (1993b). Skeletal strain patterns in relation to exercise training during growth. *J. Exp. Biol.*, *185*, 51–69.

Biewener, A. A., & Dial, K. P. (1995). *In vivo* strain in the humerus of pigeons (*Columba livia*) during flight. *J Morphol*, *225*, 61–75.

Biewener, A. A., Fazzalari, N. L., Konieczynski, D. D., & Baudinette, R. V. (1996). Adaptive changes in trabecular architecture in relation to functional strain patterns and disuse. *Bone*, *19*(1), 1–8.

Biewener, A. A., Swartz, S. M., & Bertram, J. E. A. (1986). Bone modeling during growth: Dynamic strain equilibrium in the chick tibiotarsus. *Calcif Tissue Int*, *39*, 390–395.

Biewener, A. A., Thomason, J., Goodship, A., & Lanyon, L. E. (1983a). Bone stress in the horse forelimb during locomotion at different gaits: A comparison of two experimental methods. *J Biomech*, *16*, 565–576.

Biewener, A. A., Thomason, J., & Lanyon, L. E. (1983b). Mechanics of locomotion and jumping in the forelimb of the horse (*Equus*): *In vivo* stress developed in the radius and metacarpus. *J Zool Lond*, *201*(1), 67–82.

Bigley, R. F., Gibeling, J. C., Stover, S. M., Hazelwood, S. J., Fyhrie, D. P., & Martin, R. B. (2007). Volume effects on fatigue life of equine cortical bone. *J Biomech*, *40*(16), 3548–3554.

Bigley, R. F., Gibeling, J. C., Stover, S. M., Hazelwood, S. J., Fyhrie, D. P., & Martin, R. B. (2008). Volume effects on yield strength of equine cortical bone. *J Mech Behav Biomed Mater*, *1*(4), 295–302. doi: S1751-6161(07)00035-5 [pii]; doi: 10.1016/j.jmbbm.2007.11.001

Bigley, R. F., Griffin, L. V., Christensen, L., & Vandenbosch, R. (2006). Osteonal interfacial strength and histomorphometry of equine cortical bone. *J Biomech*, *39*, 1629–1640.

Birkhold, A. I., Razi, H., Duda, G. N., Weinkamer, R., Checa, S., & Willie, B. M. (2016). The periosteal bone surface is less mechano-responsive than the endocortical. *Sci Reports*, *6*, 23480. doi: srep23480 [pii]; doi: 10.1038/srep23480

Blob, R. W., & Biewener, A. A. (1999). *In vivo* locomotor strain in the hindlimb bones of *Alligator mississippiensis* and *Iguana iguana*: Implications for the evolution of limb bone safety factor and non-sprawling limb posture. *J Exp Biol*, *202*, 1023–1046.

Blob, R. W., & Biewener, A. A. (2001). Mechanics of limb bone loading during terrestrial locomotion in the green iguana (Iguana iguana) and American alligator (Alligator mississippiensis). *J Exp Biol*, *204*(Pt 6), 1099–1122. doi: 10.1242/jeb.204.6.1099

Bouvier, M., & Hylander, W. L. (1984). *In vivo* bone strain on the dog tibia during locomotion. *Acta Anat (Basel)*, *118*(3), 187–192. doi: 10.1159/000145842

Boyce, T. M., Fyhrie, D. P., Glotkowski, M. C., Radin, E. L., & Schaffler, M. B. (1998). Damage type and strain mode associations in human compact bone bending fatigue. *J Orthop Res*, *16*, 322–329.

Britz, H. M., Thomas, C. D., Clement, J. G., & Cooper, D. M. (2009). The relation of femoral osteon geometry to age, sex, height and weight. *Bone*, *45*(1), 77–83.

Bromage, T. G., Goldman, H. M., McFarlin, S. C., Warshaw, J., Boyde, A., & Riggs, C. M. (2003). Circularly polarized light standards for investigations of collagen fiber orientation in bone. *Anat Rec, 274B*(1), 157–168.

Burger, E., Klein-Nulend, J., & Smit, T. (2003). Strain-derived canalicular fluid flow regulates osteoclast activity in a remodelling osteon--a proposal. *J Biomech, 36*(10), 1453–1459.

Burr, D. B. (2002). The contribution of the organic matrix to bone's material properties. *Bone, 31*(1), 8–11.

Burr, D. B., Milgrom, C., Fyhrie, D., Forwood, M., Nyska, M., Finestone, A., et al. (1996). *In vivo* measurement of human tibial strains during vigorous activity. *Bone, 18*(5), 405–410.

Burr, D. B., Ruff, C. B., & Thompson, D. D. (1990). Patterns of skeletal histologic change through time: Comparison of an archaic native American population with modern populations. *The Anat. Rec., 226*, 307–313.

Burr, D. B., Turner, C. H., Naick, P., Forwood, M. R., Ambrosius, W., Hasan, M. S., et al. (1998). Does microdamage accumulation affect the mechanical properties of bone? *J Biomech, 31*, 337–347.

Buss, D. J., Kroger, R., McKee, M. D., & Reznikov, N. (2022). Hierarchical organization of bone in three dimensions: a twist of twists. *J Struct Biol X, 6*, 100057. doi: S2590-1524(21)00014-3 [pii]; doi: 10.1016/j.yjsbx.2021.100057

Butcher, M. T., & Blob, R. W. (2008). Mechanics of limb bone loading during terrestrial locomotion in river cooter turtles (Pseudemys concinna). *J Exp Biol, 211*(Pt 8), 1187–1202. doi: 211/8/1187 [pii]; doi: 10.1242/jeb.012989

Butcher, M. T., White, B. J., Hudzik, N. B., Gosnell, W. C., Parrish, J. H., & Blob, R. W. (2011). *In vivo* strains in the femur of the Virginia opossum (*Didelphis virginiana*) during terrestrial locomotion: testing hypotheses of evolutionary shifts in mammalian bone loading and design. *J Exp Biol, 214*(Pt 15), 2631–2640. doi: 214/15/2631 [pii]; doi: 10.1242/jeb.049544

Canington, S. L., Sylvester, A. D., Burgess, M. L., Junno, J. A., & Ruff, C. B. (2018). Long bone diaphyseal shape follows different ontogenetic trajectories in captive and wild gorillas. *Am J Phys Anthropol, 167*(2), 366–376. doi: 10.1002/ajpa.23636

Carlson, K. J., Doran-Sheehy, D. M., Hunt, K. D., Nishida, T., Yamanaka, A., & Boesch, C. (2006). Locomotor behavior and long bone morphology in individual free-ranging chimpanzees. *Journal of Human Evolution, 50*, 394–404.

Carlson, K. J., & Judex, S. (2007). Increased non-linear locomotion alters diaphyseal bone shape. *J Exp Biol, 210*(Pt 17), 3117–3125. doi: 210/17/3117 [pii]; doi: 10.1242/jeb.006544

Carlson, K. J., Sumner, D. R., Morbeck, M. E., Nishida, T., Yamanaka, A., & Boesch, C. (2008). Role of nonbehavioral factors in adjusting long bone diaphyseal structure in free-ranging *Pan troglodytes. Int J Primatol, 29*(6), 1401–1420. doi: 10.1007/s10764-008-9297-y

Carrano, M. T., & Biewener, A. A. (1999). Experimental alteration of limb posture in the chicken (*Gallus gallus*) and its bearing on the use of birds as analogs for dinosaur locomotion. *J Morphol, 240*(3), 237–249.

Carrier, D. R. (1983). Postnatal ontogeny of the musculo-skeletal system in the black-tailed jack rabbit (*Lepus californicus*). *J Zool Lond, 201*, 27–55.

Carrier, D. R., & Leon, L. R. (1990). Skeletal growth and function in the California gull (*Larus californicus*). *J. Zool. Lond., 222*, 375–389.

Carter, D. R., & Beaupré, G. S. (2001). *Skeletal Function and Form. Mechanobiology of Skeletal Development, Aging, and Regeneration*. Cambridge, UK: University of Cambridge.

Carter, D. R., Caler, W. E., Spengler, D. M., & Frankel, V. H. (1981). Fatigue behavior of adult cortical bone: The influence of mean strain and strain range. *Acta Orthop Scand, 52*, 481–490.

Carter, D. R., Hayes, W. C., & Schurman, D. J. (1976). Fatigue life of compact bone-II. Effects of microstructure and density. *J Biomech, 9*, 211–218.

Carter, D. R., Smith, D. J., Spengler, D. M., Daly, C. H., & Frankel, V. H. (1980). Measurements and analysis of *in vivo* bone strains on the canine radius and ulna. *J. Biomech., 13*, 27–38.

Carter, D. R., & Spengler, D. M. (1982). Biomechanics of fracture. In G. Sumner-Smith (Ed.), *Bone in clinical orthopaedics: A study in comparative osteology* (pp. 305–334). Philadelphia, PA: W.B. Saunders Co.

Chang, B., & Liu, X. (2022). Osteon: structure, turnover, and regeneration. *Tissue Eng Part B Rev, 28*(2), 261–278. doi: 10.1089/ten.teb.2020.0322 [pii]; doi: 10.1089/ten.TEB.2020.0322

Cho, H., Stout, S. D., & Bishop, T. A. (2006). Cortical bone remodeling rates in a sample of African American and European American descent groups from the American Midwest: comparisons of age and sex in ribs. *Am J Phys Anthropol, 130*(2), 214–226.

Ciarelli, T. E., Tjhia, C., Rao, D. S., Qiu, S., Parfitt, A. M., & Fyhrie, D. P. (2009). Trabecular packet-level lamellar density patterns differ by fracture status and bone formation rate in white females. *Bone, 45*(5), 903–908. doi: 10.1016/j.bone.2009.07.002

Cochran, G. V. B., Zickel, R. E., & Fielding, J. W. (1980). Stress analysis of subtrochanteric fractures: effect of muscle forces and internal fixation devices. In H. K. Uhthoff (Ed.), *Current Concepts of Internal Fixation of Fractures* (pp. 211–227). New York: Springer-Verlag Berlin Heidelberg.

Coleman, J. C., Hart, R. T., Owan, I., Takano, Y., & Burr, D. B. (2002). Characterization of dynamic three-dimensional strain fields in the canine radius. *J Biomech, 35*(12), 1677–1683.

Copploe, J. V., 2nd, Blob, R. W., Parrish, J. H., & Butcher, M. T. (2015). *In vivo* strains in the femur of the nine-banded armadillo (*Dasypus novemcinctus*). *J Morphol, 276*(8), 889–899. doi: 10.1002/jmor.20387

Corondan, G., & Haworth, W. L. (1986). A fractographic study of human long bone. *J Biomech, 19*(3), 207–218.

Cosnefroy, Q., Marchal, F., Bellaiche, L., Carlier, R., Cazeau, C., Lamberton, F., et al. (2022). Do femoral biomechanical properties follow locomotor changes in primates? An ontogenetic study of olive baboons (*Papio anubis*). *Am J Biol Anthropol, 179*, 609–623.

Cowin, S. C. (1989). The mechanical properties of cortical bone tissue. In S. C. Cowin (Ed.), *Bone Mechanics* (pp. 97–127). Boca Raton, Florida, USA: CRC Press.

Cowin, S. C. (2001). *Mechanics of materials*. Boca Raton, FL: CRC Press.

Cowin, S. C. (2002). Mechanosensation and fluid transport in living bone. *J Musculoskelet Neuronal Interact, 2*(3), 256–260.

Currey, J. D. (2002). *Bones: Structure and Mechanics*. Princeton, NJ: Princeton University Press.

Currey, J. D., & Pond, C. M. (1989). Mechanical properties of very young bone in the axis deer (*Axis axis*) and humans. *J. Zool. Lond., 218*, 59–67.

Dakoulas, L., Keenan, K. E., & Skedros, J. G. (2011). Collagen fiber orientation heterogeneity (CFO-Het): evaluation of this new characteristic in antler, primate, and non-primate bones shows that it does not correlate well with load history. *Am J Phys Anthropol*, (Supp. 52), 120.

Delibes, M., Castañeda, I., & Fedriani, J. M. (2017). Tree-climbing goats disperse seeds during rumination. *Frontiers in Ecology and the Environment, 15*(4), 222–223.

Demes, B. (2007). *In vivo* bone strain and bone functional adaptation. *Am J Phys Anthropol, 133*(1), 717–722.

Demes, B., Qin, Y. X., Stern, J. T., Jr., Larson, S. G., & Rubin, C. T. (2001). Patterns of strain in the macaque tibia during functional activity. *Am J Phys Anthropol, 116*(4), 257–265.

Demes, B., Stern, J. T., Jr., Hausman, M. R., Larson, S. G., McLeod, K. J., & Rubin, C. T. (1998). Patterns of strain in the macaque ulna during functional activity. *Am J Phys Anthropol, 106*(1), 87–100.

de Vet, T., Jhirad, A., Pravato, L., & Wohl, G. R. (2021). Bone bioelectricity and bone-cell response to electrical stimulation: a review. *Crit Rev Biomed Eng, 49*(1), 1–19. doi: 76be2dbf552bfd42,49d20a9760885c4e [pii]; doi: 10.1615/CritRevBiomedEng.2021035327

Diab, T., & Vashishth, D. (2007). Morphology, localization and accumulation of *in vivo* microdamage in human cortical bone. *Bone, 40*(3), 612–618.

Drapeau, M. S., & Streeter, M. A. (2006). Modeling and remodeling responses to normal loading in the human lower limb. *Am J Phys Anthropol, 129*(3), 403–409.

Duda, G. N., Heller, M., Albinger, J., Schulz, O., Schneider, E., & Claes, L. (1998). Influence of muscle forces on femoral strain distribution. *J Biomech, 31*(9), 841–846.

Edwards, W. B., Gillette, J. C., Thomas, J. M., & Derrick, T. R. (2008). Internal femoral forces and moments during running: implications for stress fracture development. *Clin Biomech (Bristol,*

Avon), *23*(10), 1269–1278. doi: S0268-0033(08)00223-4 [pii]; doi: 10.1016/j.clinbiomech.2008.06.011

Ehrlich, P. J., & Lanyon, L. E. (2002). Mechanical strain and bone cell function: a review. *Osteoporos Int*, *13*(9), 688–700.

Ei Hsu Hlaing, E., Ishihara, Y., Odagaki, N., Wang, Z., Ikegame, M., & Kamioka, H. (2020). The expression and regulation of Wnt1 in tooth movement-initiated mechanotransduction. *Am J Orthod Dentofacial Orthop*, *158*(6), e151–e160. doi: S0889-5406(20)30565-5 [pii]; doi: 10.1016/j.ajodo.2020.08.006

Espinoza Orias, A. A., Deuerling, J. M., Landrigan, M. D., Renaud, J. E., & Roeder, R. K. (2009). Anatomic variation in the elastic anisotropy of cortical bone tissue in the human femur. *J Mech Behav Biomed Mater*, *2*(3), 255–263. doi: S1751-6161(08)00075-1 [pii]; doi: 10.1016/j.jmbbm.2008.08.005

Evans, F. G. (1958). Relations between the microscopic structure and tensile strength of human bone. *Acta Anat (Basel)*, *35*(4), 285–301.

Evans, F. G. (1976). Mechanical properties and histology of cortical bone from younger and older men. *Anatomical Record*, *185*, 1–12.

Evans, F. G., & Bang, S. (1966). Physical and histological differences between human fibular and femoral compact bone. In F. G. Evans (Ed.), *Studies on the anatomy and function of bone and joints* (pp. 142–155). New York: Springer.

Evans, F. G., & Bang, S. (1967). Differences and relationships between the physical properties and the microscopic structure of human femoral, tibial and fibular cortical bone. *Am J Anat*, *120*, 79–88.

Evans, F. G., & Vincentelli, R. (1974). Relations of the compressive properties of human cortical bone to histological structure and calcification. *J Biomech*, *7*(1), 1–10.

Fan, L., Pei, S., Lucas Lu, X., & Wang, L. (2016). A multiscale 3D finite element analysis of fluid/solute transport in mechanically loaded bone. *Bone Res*, *4*, 16032. doi: 10.1038/boneres.2016.32

Feng, X., & McDonald, J. M. (2011). Disorders of bone remodeling. *Annu Rev Pathol*, *6*, 121–145. doi: 10.1146/annurev-pathol-011110-130203

Fetto, J., Leali, A., & Moroz, A. (2002). Evolution of the Koch model of the biomechanics of the hip: clinical perspective. *J Orthop Sci*, *7*(6), 724–730.

Finlinson, E. D., King, B. W., Henrie, M. K., Lohse, Z. M., & Skedros, J. G. (2019). Evaluating osteon cement-line interfaces for advancing understanding of bone adaptation. *Am J. Physical Anthrop*, *168(S68)*, 74.

Francillon-Vieillot, H., de Buffrénil, V., Castanet, J., Géraudie, J., Meunier, F., Sire, J., et al. (1990). Microstructure and mineralization of vertebrate skeletal tissues. In J. Carter (Ed.), *Skeletal Biomineralization: Patterns, Processes and Evolutionary Trends* (Vol. 1, pp. 471–530). New York: Van Nostrand Reinhold.

Fritton, S. P., McLeod, K. J., & Rubin, C. T. (2000). Quantifying the strain history of bone: Spatial uniformity and self-similarity of low-magnitude strains. *J Biomech*, *33*(3), 317–325.

Gautier, E., Perren, S. M., & Cordey, J. (2000). Strain distribution in plated and unplated sheep tibia an *in vivo* experiment. *Injury*, *31 Suppl 3*, C37–44. doi: S0020-1383(00)80030-3 [pii]; doi: 10.1016/s0020-1383(00)80030-3

Gebhardt, W. (1906). Über funktionell wichtige anordnungsweisen der feineren und görberen bauelemente des wirbeltierknochens. II. Spezieller teil. Der bau der Haversschen lamellensysteme und seine funktionelle bedeutung. *Arch Entw Mech Org*, *20*, 187–322.

George, W. T., & Vashishth, D. (2005). Damage mechanisms and failure modes of cortical bone under components of physiological loading. *J Orthop Res*, *23*(5), 1047–1053.

Gibson, V. A., Stover, S. M., Gibeling, J. C., Hazelwood, S. J., & Martin, R. B. (2006). Osteonal effects on elastic modulus and fatigue life in equine bone. *J Biomech*, *39*(2), 217–225.

Goldman, H. M., Bromage, T. G., Thomas, C. D., & Clement, J. G. (2003). Preferred collagen fiber orientation in the human mid-shaft femur. *Anat Rec*, *272A*(1), 434–445.

Goldman, H. M., Hampson, N. A., Guth, J. J., Lin, D., & Jepsen, K. J. (2014). Intracortical remodeling parameters are associated with measures of bone robustness. *Anat Rec (Hoboken), 297*(10), 1817–1828. doi: 10.1002/ar.22962

Goldman, H. M., McFarlin, S. C., Cooper, D. M., Thomas, C. D., & Clement, J. G. (2009). Ontogenetic patterning of cortical bone microstructure and geometry at the human mid-shaft femur. *Anat Rec (Hoboken), 292*(1), 48–64.

Goodship, A. E., Lanyon, L. E., & McFie, H. (1979). Functional adaptation of bone to increased stress. An experimental study. *J Bone Joint Surg Am, 61*(4), 539–546.

Gosman, J. H., Hubbell, Z. R., Shaw, C. N., & Ryan, T. M. (2013). Development of cortical bone geometry in the human femoral and tibial diaphysis. *Anat Rec (Hoboken), 296*(5), 774–787. doi: 10.1002/ar.22688

Gosnell, W. C., Butcher, M. T., Maie, T., & Blob, R. W. (2011). Femoral loading mechanics in the Virginia opossum, Didelphis virginiana: torsion and mediolateral bending in mammalian locomotion. *J Exp Biol, 214*(Pt 20), 3455–3466. doi: 214/20/3455 [pii]; doi: 10.1242/jeb.060178

Gross, T. S., McLeod, K. J., & Rubin, C. T. (1992). Characterizing bone strain distribution *in vivo* using three triple rosette strain gauges. *J. Biomech., 25*, 1081–1087.

Haapasalo, H., Kontulainen, S., Sievanen, H., Kannus, P., Jarvinen, M., & Vuori, I. (2000). Exercise-induced bone gain is due to enlargement in bone size without a change in volumetric bone density: a peripheral quantitative computed tomography study of the upper arms of male tennis players. *Bone, 27*(3), 351–357. doi: S8756-3282(00)00331-8 [pii]

Hammond, A. S., Ning, J., Ward, C. V., & Ravosa, M. J. (2010). Mammalian limb loading and chondral modeling during ontogeny. *Anat Rec (Hoboken), 293*(4), 658–670. doi: 10.1002/ar.21136

Han, Z. H., Palnitkar, S., Rao, D. S., Nelson, D., & Parfitt, A. M. (1997). Effects of ethnicity and age or menopause on the remodeling and turnover of iliac bone: implications for mechanisms of bone loss. *J Bone Miner Res, 12*(4), 498–508. doi: 10.1359/jbmr.1997.12.4.498

Hansen, H. L., Bredbenner, T. L., Nicolella, D. P., Mahaney, M. C., & Havill, L. M. (2009). Cross-sectional geometry of the femoral midshaft in baboons is heritable. *Bone, 45*(5), 892–897. doi: S8756-3282(09)01624-X [pii]; doi: 10.1016/j.bone.2009.05.028

Havill, L. M., Allen, M. R., Harris, J. A., Levine, S. M., Coan, H. B., Mahaney, M. C., et al. (2013). Intracortical bone remodeling variation shows strong genetic effects. *Calcif Tissue Int, 93*(5), 472–480. doi: 10.1007/s00223-013-9775-x

Heaney, R. P., McCarron, D. A., Dawson-Hughes, B., Oparil, S., Berga, S. L., Stern, J. S., et al. (1999). Dietary changes favorably affect bone remodeling in older adults. *J Am Diet Assoc, 99*(10), 1228–1233. doi: S0002-8223(99)00302-8 [pii]; doi: 10.1016/S0002-8223(99)00302-8

Heinrich, R. E., Ruff, C. B., & Adamczewski, J. Z. (1999). Ontogenetic changes in mineralization and bone geometry in the femur of muskoxen (*Ovibos moschatus*). *J. Zool. Lond., 247*, 215–223.

Henrie, M. K., Finlinson, E. D., Smith, J. S., Keenan, K. E., & Skedros, J. G. (2018). Predominant collagen fiber orientation (CFO) is a stronger predictor of load history than drifting osteon prevalence or osteon population density (OPD of conventional osteons): an evaluation in the bending regions of adult human fibulae, femora, and chimpanzee femora. *Am J Phys Anthropol, 165(S68)*, 117.

Herman, B. C., Cardoso, L., Majeska, R. J., Jepsen, K. J., & Schaffler, M. B. (2010). Activation of bone remodeling after fatigue: differential response to linear microcracks and diffuse damage. *Bone, 47*(4), 766–772. doi: S8756-3282(10)01349-9 [pii]; 10.1016/j.bone.2010.07.006

Hiller, L. P., Stover, S. M., Gibson, V. A., Gibeling, J. C., Prater, C. S., Hazelwood, S. J., et al. (2003). Osteon pullout in the equine third metacarpal bone: effects of *ex vivo* fatigue. *J Orthop Res, 21*(3), 481–488.

Hillier, M., & Bell, L. (2007). Differentiating human bone from animal bone: a review of histological methods. *J Forensic Sci, 52*(2), 249–263.

Hsieh, Y.-F., Robling, A. G., Ambrosius, W. T., Burr, D. B., & Turner, C. H. (2001). Mechanical loading of diaphyseal bone *in vivo*: The strain threshold for an osteogenic response varies with location. *J Bone and Min Research, 16*(12), 2291–2297.

Jee, W. S. S., Li, X. J., & Ke, H. Z. (1991). The skeletal adaptation to mechanical usage in the rat. *Cells and Materials, 1*, 131–142.

Jepsen, K. J. (2011). Functional interactions among morphologic and tissue quality traits define bone quality. *Clin Orthop Relat Res, 469*(8), 2150–2159. doi: 10.1007/s11999-010-1706-9

Joo, W., Jepsen, K. J., & Davy, D. T. (2004). Complex cross-modal effects of damage on cortical bone properties. *Trans Orthop Res Soc, 29*, 515.

Judex, S., Gross, T. S., Bray, R. C., & Zernicke, R. F. (1997a). Adaptation of bone to physiological stimuli. *J Biomech, 30*, 421–429.

Judex, S., Gross, T. S., & Zernicke, R. F. (1997b). Strain gradients correlate with sites of exercise-induced bone-forming surfaces in adult skeleton. *J Bone Miner Res, 12*, 1737–1745.

Jung, H., Best, M., & Akkus, O. (2015). Microdamage induced calcium efflux from bone matrix activates intracellular calcium signaling in osteoblasts via L-type and T-type voltage-gated calcium channels. *Bone, 76*, 88–96. doi: S8756-3282(15)00092-7 [pii]; doi: 10.1016/j.bone.2015.03.014

Kalmey, J. K., & Lovejoy, C. O. (2002). Collagen fiber orientation in the femoral necks of apes and humans: Do their histological structures reflect differences in locomotor loading? *Bone, 31*, 327–332.

Kanzaki, H., Wada, S., Yamaguchi, Y., Katsumata, Y., Itohiya, K., Fukaya, S., et al. (2019). Compression and tension variably alter Osteoprotegerin expression via miR-3198 in periodontal ligament cells. *BMC Mol Cell Biol, 20*(1), 6. doi: 10.1186/s12860-019-0187-2; doi: 10.1186/s12860-019-0187-2 [pii]

Kassem, M., Melton, L. J., & Riggs, B. L. (1996). The Type I Type II mode for involutional osteoporosis. In R. Marcus, D. Feldman & J. Kelsay (Eds.), *Osteoporosis* (pp. 691–702). New York: Academic Press.

Katsamenis, O. L., Jenkins, T., & Thurner, P. J. (2015). Toughness and damage susceptibility in human cortical bone is proportional to mechanical inhomogeneity at the osteonal-level. *Bone, 76*, 158–168. doi: S8756-3282(15)00106-4 [pii]; doi: 10.1016/j.bone.2015.03.020

Keaveny, T. M., Morgan, E. F., Niebur, G. L., & Yeh, O. C. (2001). Biomechanics of trabecular bone. *Annu Rev Biomed Eng, 3*, 307–333.

Keenan, K., & Skedros, J. G. (2011). Collagen fiber orientation heterogeneity (CFO-Het): Does this new characteristic reflect habitual load history in the chimpanzee femur and does it corroborate CFO based on image gray levels? *Am J Phys Anthropol* (Supp. 52), 183.

Keenan, K. E., Mears, C. S., & Skedros, J. G. (2017). Utility of osteon circularity for determining species and interpreting load history in primates and nonprimates. *Am J Phys Anthropol, 162*(4), 657–681. doi: 10.1002/ajpa.23154

Kerley, E. R. (1965). The microscopic determination of age in human bone. *Am J Phys Anthropol, 23*(2), 149–163.

Knight, A. N., Keenan, K. E., Halley, J. A., & Skedros, J. G. (2012). Drifting osteons seem unlikely to be a mechanical adaptation in view of inconsistent regional distributions and overall low prevalence in adult chimpanzee and human femora. *Am J. Physical Anthrop, 147(Suppl. 54)*, 184.

Koch, J. C. (1917). The laws of bone architecture. *American Journal of Anatomy* (21), 177–298.

Kumar, N. C., Dantzig, J. A., Jasiuk, I. M., Robling, A. G., & Turner, C. H. (2010). Numerical modeling of long bone adaptation due to mechanical loading: correlation with experiments. *Annals of Biomedical Engineering*. doi: 10.1007/s10439-009-9861-4

Lad, S. E. (2023). Absence of secondary osteons in femora of aged rats: implications of lifespan on Haversian remodeling in mammals. *J. Morphology, 284*, e21600.

Lad, S. E., Daegling, D. J., & McGraw, W. S. (2016). Bone remodeling is reduced in high stress regions of the cercopithecoid mandible. *Am J Phys Anthropol, 161*(3), 426–435. doi: 10.1002/ajpa.23041

Lad, S. E., Kowalkowski, H., Liggio, D. F., Ding, H., & Ravosa, M. J. (2023). Prolonged cyclical loading induces Haversian remodeling in mandibles of growing rabbits. *J. Exp. Biol, in press*.

Lad, S. E., McGraw, W. S., & Daegling, D. J. (2019). Haversian remodeling corresponds to load frequency but not strain magnitude in the macaque (*Macaca fascicularis*) skeleton. *Bone, 127*, 571–576. doi: S8756-3282(19)30301-1 [pii]; doi: 10.1016/j.bone.2019.07.027

Lambert, K. L. (1971). The weight-bearing function of the fibula. A strain gauge study. *J Bone Joint Surg Am, 53*(3), 507–513.
Lanyon, L. E. (1974). Experimental support for the trajectorial theory of bone structure. *J Bone Joint Surg Br, 56*, 160–166.
Lanyon, L. E. (1987). Functional strain in bone tissue as an objective, and controlling stimulus for adaptive bone remodelling. *J Biomech, 20*, 1083–1093.
Lanyon, L. E., & Baggott, D. G. (1976). Mechanical function as an influence on the structure and form of bone. *J Bone Joint Surg, 58-B*, 436–443.
Lanyon, L. E., & Bourn, S. (1979). The influence of mechanical function on the development and remodeling of the tibia: an experimental study in sheep. *J Bone Joint Surg Am, 61*, 263–273.
Lanyon, L. E., Goodship, A. E., Pye, C. J., & MacFie, J. H. (1982). Mechanically adaptive bone remodeling. *J Biomech, 15*(3), 141–154.
Lanyon, L. E., Hampson, W. G., Goodship, A. E., & Shah, J. S. (1975). Bone deformation recorded *in vivo* from strain gauges attached to the human tibial shaft. *Acta Orthop Scand, 46*(2), 256–268.
Lazenby, R. (1986). Porosity-geometry interaction in the conversation of bone strength. *J Biomech, 19*(3), 257–258.
Lee, A. H. (2004). Histological organization and its relationship to function in the femur of *Alligator mississippiensis. J. Anat, 204*(3), 197–207.
Lee, T. C., Staines, A., & Taylor, D. (2002). Bone adaptation to load: microdamage as a stimulus for bone remodelling. *J Anat, 201*(6), 437–446.
Li, S., Zhang, H., Yang, Y., Huo, B., & Zhang, D. (2015). Connexin 43 and ERK regulate tension-induced signal transduction in human periodontal ligament fibroblasts. *J Orthop Res, 33*(7), 1008–1014. doi: 10.1002/jor.22830
Lieberman, D. E., Pearson, O. M., Polk, J. D., Demes, B., & Crompton, A. W. (2003). Optimization of bone growth and remodeling in response to loading in tapered mammalian limbs. *J Exp Biol, 206*, 3125–3138.
Lieberman, D. E., Polk, J. D., & Demes, B. (2004). Predicting long bone loading from cross-sectional geometry. *Am J Phys Anthropol, 123*, 156–171.
Liu, D., Wagner, H. D., & Weiner, S. (2000). Bending and fracture of compact circumferential and osteonal lamellar bone of the baboon tibia. *J Mater Sci Mater Med, 11*(1), 49–60. doi: 253932 [pii]
Lloveras, L., Rissech, C., Davis, S., & Pares-Casanova, P. M. (2022). Morphological differences between sheep and goat calcanea using two-dimensional geometric morphometrics. *Animals (Basel), 12*(21). doi: ani12212945 [pii]. animals-12-02945 [pii]; doi: 10.3390/ani12212945
Loundagin, L. L., Pohl, A. J., & Edwards, W. B. (2021). Stressed volume estimated by finite element analysis predicts the fatigue life of human cortical bone: The role of vascular canals as stress concentrators. *Bone, 143*, 115647. doi: S8756-3282(20)30427-0 [pii]; doi: 10.1016/j.bone.2020.115647
Lovejoy, C. O. (1988). Evolution of human walking. *Sci Am, 259*(5), 118–125.
Macdonald, H. M., Cooper, D. M., & McKay, H. A. (2009). Anterior-posterior bending strength at the tibial shaft increases with physical activity in boys: evidence for non-uniform geometric adaptation. *Osteoporos Int, 20*(1), 61–70. doi: 10.1007/s00198-008-0636-9
Main, R. P. (2007). Ontogenetic relationships between *in vivo* strain environment, bone histomorphometry and growth in the goat radius. *J Anat, 210*(3), 272–293.
Main, R. P., & Biewener, A. A. (2004). Ontogenetic patterns of limb loading, *in vivo* bone strains and growth in the goat radius. *J Exp Biol, 207*(Pt 15), 2577–2588.
Main, R. P., & Biewener, A. A. (2007). Skeletal strain patterns and growth in the emu hindlimb during ontogeny. *J Exp Biol, 210*(Pt 15), 2676–2690. doi: 210/15/2676 [pii]; doi: 10.1242/jeb.004580
Main, R. P., Simons, E. L. R., & Lee, A. H. (2021). Interpreting Mechanical Function in Extant and Fossil Long Bones. In A. J. D. R. V. De Buffrénil, L. Zylberberg, K. Padian (Ed.), *Vertebrate Skeletal Histology and Paleohistology* (pp. 688–723). Boca Raton, Florida, USA: CRC Press.

Manandhar, S., Song, H., Moshage, S. G., Craggette, J., Polk, J. D., & Kersh, M. E. (2023). Spatial variation in young ovine cortical bone properties. *J. Biomech Engineering, 145*(6), 061002.

Marotti, G. (1996). The structure of bone tissues and the cellular control of their deposition. *Ital J Anat Embryol, 101*, 25–79.

Martin, R. B. (2003). Fatigue microdamage as an essential element of bone mechanics and biology. *Calcif Tissue Int, 73*(2), 101–107.

Martin, R. B., & Boardman, D. L. (1993). The effects of collagen fiber orientation, porosity, density, and mineralization on bovine cortical bone bending properties. *J Biomech, 26*, 1047–1054.

Martin, R. B., & Burr, D. B. (1989). *Structure, Function and Adaptation of Compact Bone*. New York, New York: Raven Press.

Martin, R. B., Burr, D. B., & Sharkey, N. A. (1998). *Skeletal Tissue Mechanics*. New York, New York: Springer-Verlag.

Martin, R. B., Gibson, V. A., Stover, S. M., Gibeling, J. C., & Griffin, L. V. (1996b). Osteonal structure in the equine third metacarpus. *Bone, 19*(2), 165–171.

Martin, R. B., & Ishida, J. (1989). The relative effects of collagen fiber orientation, porosity, density, and mineralization on bone strength. *J Biomech, 22*(5), 419–426.

Martin, R. B., Mathews, P. V., Lau, S. T., Gibson, V. A., & Stover, S. M. (1996a). Collagen fiber organization is related to mechanical properties and remodeling in equine bone. A comparison of two methods. *J Biomech, 29*, 1515–1521.

Martiniaková, M., Grosskopf, B., Omelka, R., Vondraková, M., & Bauerová, M. (2006). Differences among species in compact bone tissue microstructure of mammalian skeleton: use of a discriminant function analysis for species identification. *J Forensic Sci, 51*(6), 1235–1239.

Mason, M. W., Skedros, J. G., & Bloebaum, R. D. (1995). Evidence of strain-mode-related cortical adaptation in the diaphysis of the horse radius. *Bone, 17*(3), 229–237.

Mayhew, P. M., Thomas, C. D., Clement, J. G., Loveridge, N., Beck, T. J., Bonfield, W., et al. (2005). Relation between age, femoral neck cortical stability, and hip fracture risk. *Lancet, 366*(9480), 129–135.

McFarlin, S. C., Terranova, C. J., Zihlman, A. L., Enlow, D. H., & Bromage, T. G. (2008). Regional variability in secondary remodeling within long bone cortices of catarrhine primates: the influence of bone growth history. *J Anat, 213*(3), 308–324.

Mears, C. S., Keenan, K. E., & Skedros, J. G. (2018). Regional variations in predominant collagen fiber orientation in the diaphysis and neck of sub-adult baboon femora resemble those in adult chimpanzee femora. Does this reflect similar load histories? *Am. J. Physical Anthrop., in press.*

Milgrom, C., Finestone, A., Simkin, A., Ekenman, I., Mendelson, S., Millgram, M., et al. (2000). *In-vivo* strain measurements to evaluate the strengthening potential of exercises on the tibial bone. *J Bone Joint Surg Br, 82*(4), 591–594.

Milne, N. (2016). Curved bones: an adaptation to habitual loading. *J Theor Biol, 407*, 18–24. doi: S0022-5193(16)30202-8 [pii]; doi: 10.1016/j.jtbi.2016.07.019

Moreno, C. A., Main, R. P., & Biewener, A. A. (2008). Variability in forelimb bone strains during non-steady locomotor activities in goats. *J Exp Biol, 211*(Pt 7), 1148–1162.

Morimoto, N., Ponce de Leon, M. S., & Zollikofer, C. P. (2011). Exploring femoral diaphyseal shape variation in wild and captive chimpanzees by means of morphometric mapping: a test of Wolff's law. *Anat Rec (Hoboken), 294*(4), 589–609. doi: 10.1002/ar.21346

Moyle, D. D., Welborn, J. W., 3rd, & Cooke, F. W. (1978). Work to fracture of canine femoral bone. *J Biomech, 11*(10–12), 435–440.

Muir, P., Johnson, K. A., & Ruaux-Mason, C. P. (1999). *In vivo* matrix microdamage in a naturally occurring canine fatigue fracture. *Bone, 25*(5), 571–576.

Mulhern, D. M., & Ubelaker, D. H. (2009). Bone microstructure in juvenile chimpanzees. *Am J Phys Anthropol, 140*(2), 368–375. doi: 10.1002/ajpa.20959

Niinimäki, S., Narra, N., Härkönen, L., Abe, S., Nikander, R., Hyttinen, J., et al. (2017). The relationship between loading history and proximal femoral diaphysis cross-sectional geometry. *Am J Hum Biol, 29*(4). doi: 10.1002/ajhb.22965

Nunamaker, D. (2001). Bucked shins in horses. In D. B. Burr & C. Milgrom (Eds.), *Musculoskeletal Fatigue and Stress Fractures* (pp. 203–219). Boca Raton, Florida, USA: CRC Press.
Nunamaker, D. M., Butterweck, D. M., & Provost, M. T. (1989). Some geometric properties of the third metacarpal bone: A comparison between the thoroughbred and standardbred racehorse. *J Biomech, 22*(2), 129–134.
Nyman, J. S., Roy, A., Tyler, J. H., Acuna, R. L., Gayle, H. J., & Wang, X. (2007). Age-related factors affecting the postyield energy dissipation of human cortical bone. *J Orthop Res, 25*(5), 646–655.
O'Brien, F., Taylor, D., & Lee, T. C. (2005). The effect of bone microstructure on the initiation and growth of microcracks. *J Orthop Res, 23*(2), 475–480.
O'Leary, T. J., Wardle, S. L., Gifford, R. M., Double, R. L., Reynolds, R. M., Woods, D. R., et al. (2021). Tibial macrostructure and microarchitecture adaptations in women during 44 weeks of arduous military training. *J Bone Miner Res, 36*(7), 1300–1315. doi: 10.1002/jbmr.4290
Page, A. E., Allan, C., Jasty, M., Harrigan, T. P., Bragdon, C. R., & Harris, W. H. (1993). Determination of loading parameters in the canine hip *in vivo*. *J Biomech, 26*(4–5), 571–579. doi: 0021-9290(93)90018-A [pii]; doi: 10.1016/0021-9290(93)90018-a
Pantinople, J., McCabe, K., Henderson, K., Richards, H. L., & Milne, N. (2017). The development of curvature in the porcine radioulna. *PeerJ, 5*, e3386. doi: 10.7717/peerj.3386 3386 [pii]
Parfitt, A. M., Mundy, G. R., Roodman, G. D., Hughes, D. E., & Boyce, B. F. (1996). A new model for the regulation of bone resorption, with particular reference to the effects of bisphosphonates. *J Bone Miner Res, 11*, 150–159.
Peacock, S. J., Coats, B. R., Kirkland, J. K., Tanner, C. A., Garland, T., Jr., & Middleton, K. M. (2018b). Predicting the bending properties of long bones: Insights from an experimental mouse model. *Am J Phys Anthropol, 165*(3), 457–470. doi: 10.1002/ajpa.23363
Peacock, S. J., Garland, T., Jr., & Middleton, K. M. (2018a). Reply to Ruff, Warden, and Karlson. *Am J Phys Anthropol, 167*(1), 190–193. doi: 10.1002/ajpa.23614
Pearson, O. M., & Lieberman, D. E. (2004). The aging of Wolff's law: Ontogeny and responses to mechanical loading in cortical bone. *Yearbook of Physical Anthropology, 47*, 63–99.
Peterman, M. M., Hamel, A. J., Cavanagh, P. R., Piazza, S. J., & Sharkey, N. A. (2001). *In vitro* modeling of human tibial strains during exercise in micro-gravity. *J Biomech, 34*(5), 693–698.
Pfeiffer, S., Crowder, C., Harrington, L., & Brown, M. (2006). Secondary osteons and Haversian canal dimensions as behavioral indicators. *Am J Phys Anthropol, 131*, 460–468.
Pidaparti, R. M. V., & Turner, C. H. (1997). Cancellous bone architecture: Advantages of nonorthogonal trabecular alignment under multidirectional loading. *J Biomech, 30*, 979–983.
Pinto, M., Jepsen, K. J., Terranova, C. J., & Buffenstein, R. (2010). Lack of sexual dimorphism in femora of the eusocial and hypogonadic naked mole-rat: a novel animal model for the study of delayed puberty on the skeletal system. *Bone, 46*(1), 112–120. doi: S8756-3282(09)01826-2 [pii]; doi: 10.1016/j.bone.2009.08.060
Polk, J. D., Demes, B., Jungers, W. L., Biknevicius, A. R., Heinrich, R. E., & Runestad, J. A. (2000). A comparison of primate, carnivoran and rodent limb bone cross-sectional properties: are primates really unique? *J Hum Evol, 39*(3), 297–325. doi: S0047-2484(00)90420-2 [pii]; doi: 10.1006/jhev.2000.0420
Portigliatti Barbos, M., Bianco, P., Ascenzi, A., & Boyde, A. (1984). Collagen orientation in compact bone: II. Distribution of lamellae in the whole of the human femoral shaft with reference to its mechanical properties. *Metab Bone Dis and Relat Res, 5*, 309–315.
Portigliatti Barbos, M., Carando, S., Ascenzi, A., & Boyde, A. (1987). On the structural symmetry of human femurs. *Bone, 8*, 165–169.
Prondvai, E., Stein, K. H. W., Ricqlès, A., & J., C. (2014). Development-based revision of bone tissue classification: the importance of semantics for science. *Biol J Linn Soc, 112*, 799–816.
Raisz, L. G. (1999). Physiology and pathophysiology of bone remodeling. *Clin Chem, 45*(8 Pt 2), 1353–1358.
Reilly, D. T., & Burstein, A. H. (1974). The mechanical properties of cortical bone. *J Bone Joint Surg, 56-A*, 1001–1022.

Reilly, D. T., & Burstein, A. H. (1975). The elastic and ultimate properties of compact bone tissue. *J Biomech, 8*, 393–405.

Reilly, G. C. (2000). Observations of microdamage around osteocyte lacunae in bone. *J Biomech, 33*, 1131–1134.

Reilly, G. C., & Currey, J. D. (1999). The development of microcracking and failure in bone depends on the loading mode to which it is adapted. *J Exp Biol, 202*, 543–552.

Reilly, G. C., & Currey, J. D. (2000). The effects of damage and microcracking on the impact strength of bone. *J Biomech, 33*, 337–343.

Reilly, G. C., Currey, J. D., & Goodship, A. E. (1997). Exercise of young thoroughbred horses increases impact strength of the third metacarpal bone. *J Orthop Res, 15*, 862–868.

Rho, J. Y., Currey, J. D., Zioupos, P., & Pharr, G. M. (2001). The anisotropic Young's modulus of equine secondary osteones and interstitial bone determined by nanoindentation. *J Exp Biol, 204*(Pt 10), 1775–1781. doi: 10.1242/jeb.204.10.1775

Riggs, C. M., Lanyon, L. E., & Boyde, A. (1993). Functional associations between collagen fibre orientation and locomotor strain direction in cortical bone of the equine radius. *Anat Embryol, 187*, 231–238.

Robling, A. G., & Bonewald, L. F. (2020). The osteocyte: new insights. *Annu Rev Physiol, 82*, 485–506. doi: 10.1146/annurev-physiol-021119-034332

Robling, A. G., & Stout, S. D. (2003). Histomorphology, geometry, and mechanical loading in past populations. In S. C. Agarwal & S. D. Stout (Eds.), *Bone Loss and Osteoporosis: An Anthropological Perspective* (pp. 189–205). New York: Kluwer Academic/Plenum Publishers.

Rooney, A. M., McNeill, T. J., Ross, F. P., Bostrom, M. P. G., & van der Meulen, M. C. H. (2023). PTH treatment increases cortical bone mass more in response to compression than tension in mice. *J Bone Mineral Res, 38*(1), 59–69.

Rubin, C., Judex, S., & Hadjiargrou, M. (2002). Skeletal adaptation to mechanical stimuli in the absence of formation or resorption of bone. *J Musculoskel Neuron Interact, 2*(3), 264–267.

Rubin, C. T., Fritton, S., Sun, Y. Q., & McLeod, K. J. (1996). Biomechanical parameters which stimulate bone formation: The ugly duckling of the skeletal growth factors. In Z. Davidovitch & L. A. Norton (Eds.), *Biological mechanisms of tooth movement and craniofacial adaptation* (pp. 51–59). Boston: Harvard Society for the Advancement of Orthodontics.

Rubin, C. T., Gross, T. S., McLeod, K. J., & Bain, S. D. (1995). Morphologic stages in lamellar bone formation stimulated by a potent mechanical stimulus. *J Bone Miner Res, 10*, 488–495.

Rubin, C. T., & Lanyon, L. E. (1982). Limb mechanics as a function of speed and gait: A study of functional strains in the radius and tibia of horse and dog. *J Exp Biol, 101*, 187–211.

Rubin, C. T., & Lanyon, L. E. (1984). Regulation of bone formation by applied dynamic loads. *J Bone Joint Surg Am, 66*(3), 397–402.

Rubin, C. T., & Lanyon, L. E. (1985). Regulation of bone mass by mechanical strain magnitude. *Calcif Tissue Int, 37*, 411–417.

Rubin, C. T., Seeherman, H., Qin, Y. X., & Gross, T. S. (2013). The mechanical consequences of load bearing in the equine third metacarpal across speed and gait: the nonuniform distributions of normal strain, shear strain, and strain energy density. *FASEB J, 27*(5), 1887–1894. doi: fj.12-216804 [pii]; doi:10.1096/fj.12-216804

Ruff, C., Holt, B., & Trinkaus, E. (2006). Who's afraid of the big bad Wolff?: "Wolff's law" and bone functional adaptation. *Am J Phys Anthropol, 129*(4), 484–498.

Ruff, C. B. (1981). *Structural changes in the lower limb bones with aging at Pecos Pueblo*. Philadelphia, PA: University of Pennsylvania.

Ruff, C. B., & Hayes, W. C. (1988). Sex differences in age-related remodeling of the femur and tibia. *J Orthop Res, 6*(6), 886–896.

Ruff, C. B., Warden, S. J., & Carlson, K. J. (2018). Of mice and men (and women): Comment on Peacock et al., 2018. *Am J Phys Anthropol, 167*(1), 185–189. doi: 10.1002/ajpa.23615

Sarringhaus, L. A., MacLatchy, L. M., & Mitani, J. C. (2016). Long bone cross-sectional properties reflect changes in locomotor behavior in developing chimpanzees. *Am J Phys Anthropol, 160*(1), 16–29. doi: 10.1002/ajpa.22930

Schaffler, M. B., Burr, D. B., Jungers, W. L., & Ruff, C. B. (1985). Structural and mechanical indicators of limb specialization in primates. *Folia Primatol, Basel* (45), 61–75.

Schatzker, J., Manley, P. A., & Sumner-Smith, G. (1980). *In vivo* strain gauge study of bone response to loading with and without internal fixation. In H. K. Uhthoff (Ed.), *Current Concepts of Internal Fixation of Fractures* (pp. 306–314). New York: Springer-Verlag Berlin Heidelberg.

Schneider, P., Stauber, M., Voide, R., Stampanoni, M., Donahue, L. R., & Muller, R. (2007). Ultrastructural properties in cortical bone vary greatly in two inbred strains of mice as assessed by synchrotron light based micro- and nano-CT. *J Bone Miner Res, 22*(10), 1557–1570. doi: 10.1359/jbmr.070703

Schneider, R. K., Milne, D. W., Gabel, A. A., Groom, J. J., & Bramlage, L. R. (1982). Multidirectional *in vivo* strain analysis of the equine radius and tibia during dynamic loading with and without a cast. *Am J Vet Res, 43*(9), 1541–1550.

Seref-Ferlengez, Z., Basta-Pljakic, J., Kennedy, O. D., Philemon, C. J., & Schaffler, M. B. (2014). Structural and mechanical repair of diffuse damage in cortical bone *in vivo*. *J Bone Miner Res, 29*(12), 2537–2544. doi: 10.1002/jbmr.2309

Shahar, R., Banks-Sills, L., & Eliasy, R. (2003). Stress and strain distribution in the intact canine femur: finite element analysis. *Med Eng Phys, 25*(5), 387–395. doi: S135045330300002X [pii]

Sheffield, K. M., Butcher, M. T., Shugart, S. K., Gander, J. C., & Blob, R. W. (2011). Locomotor loading mechanics in the hindlimbs of tegu lizards (*Tupinambis merianae*): comparative and evolutionary implications. *J Exp Biol, 214*(Pt 15), 2616–2630. doi: 214/15/2616 [pii]; doi: 10.1242/jeb.048801

Simkin, A., & Robin, G. (1974). Fracture formation in differing collagen fiber pattern of compact bone. *J Biomech, 7*(2), 183–188.

Sinclair, K. D., Farnsworth, R. W., Pham, T. X., Knight, A. N., Bloebaum, R. D., & Skedros, J. G. (2013). The artiodactyl calcaneus as a potential 'control bone' cautions against simple interpretations of trabecular bone adaptation in the anthropoid femoral neck. *J Hum Evol, 64*(5), 366–379. doi: S0047-2484(13)00008-0 [pii]; doi: 10.1016/j.jhevol.2013.01.003

Skedros, J. G. (2002). Use of predominant collagen fiber orientation for interpreting cortical bone loading history: Bending vs. torsion. *J. Bone Miner. Res., 17*(suppl 1), S307.

Skedros, J. G. (2005). Osteocyte lacuna population densities in sheep, elk and horse calcanei. *Cells Tissues Organs, 181*(1), 23–37.

Skedros, J. G. (2011). Collagen fiber orientation (CFO) variations in the hominid femoral neck are likely invalid for deciphering load history when cortical robusticity is low. *Am J Phys Anthropol, S52*, 275–276.

Skedros, J. G. (2012). When artiodactyls lead anthropologists astray: Important considerations, strengths, and limitations of comparing limb bone adaptation between artiodactyls and primates. *Am J Phys Anthropol, S54*, 270–271.

Skedros, J. G., & Baucom, S. L. (2007). Mathematical analysis of trabecular 'trajectories' in apparent trajectorial structures: the unfortunate historical emphasis on the human proximal femur. *J Theor Biol, 244*(1), 15–45.

Skedros, J. G., Beckstrom, A. B., Kiser, C. J., & Bloebaum, R. D. (2008). The importance of bipedality/bending in mediating morphological adaptation in the chimpanzee femoral neck might be overstated. *Am J Phys Anthropol, Suppl 46*, 195.

Skedros, J. G., Bloebaum, R. D., Mason, M. W., & Bramble, D. M. (1994). Analysis of a tension/compression skeletal system: possible strain-specific differences in the hierarchical organization of bone. *Anat Rec, 239*(4), 396–404. doi: 10.1002/ar.1092390406

Skedros, J. G., Cronin, J. T., Dayton, M. R., Bloebaum, R. D., & Bachus, K. N. (2023a). Exploration of the synergistic role of cortical thickness asymmetry ("trabecular eccentricity" concept) in reducing fracture risk in the human femoral neck and a control bone (artiodactyl calcaneus). *J. Theor Biol* (Apr 15:567), 111495.

Skedros, J. G., Dayton, M. R., Bachus, K. N., Bloebaum, R. D., & Cronin, J. T. (2023b). Strain-mode-specific mechanical test results can be misleading if load-sharing tension members are not considered in the deer calcaneus model. *J. Anatomy, in press.*

Skedros, J. G., Dayton, M. R., Sybrowsky, C. L., Bloebaum, R. D., & Bachus, K. N. (2006c). The influence of collagen fiber orientation and other histocompositional characteristics on the mechanical properties of equine cortical bone. *J Exp Biol, 209*, 3025–3042.

Skedros, J. G., Demes, B., & Judex, S. (2003). Limitations in the use of predominant collagen fiber orientation for inferring loading history in cortical bone. *Am J Phys Anthropol, Suppl 36*, 193.

Skedros, J. G., & Doutré, M. S. (2019). Collagen fiber orientation pattern, osteon morphology and distribution, and presence of laminar histology do not distinguish torsion from bending in bat and pigeon wing bones. *J. Anatomy, 234*(6), 748–763.

Skedros, J. G., Doutré, M. S., & Weaver, D. J. (2016a). Proximate mechanisms involved in the formation of secondary osteon morphotypes: Important considerations and a putative role of primary cilia of osteoblasts and osteocytes. *Osteologie, 25*(2), 101–112.

Skedros, J. G., Grunander, T. R., & Hamrick, M. W. (2005). Spatial distribution of osteocyte lacunae in equine radii and third metacarpals: considerations for cellular communication, microdamage detection and metabolism. *Cells, Tissues, Organs, 180*(4), 215–236.

Skedros, J. G., Henrie, T. R., Doutre, M. S., & Bloebaum, R. D. (2018). Sealed osteons in animals and humans: low prevalence and lack of relationship with age. *J Anat, 232*(5), 824–835. doi: 10.1111/joa.12786

Skedros, J. G., Hughes, P. E., Nelson, K., & Winet, H. (1999). Collagen fiber orientation in the proximal femur: challenging Wolff's tension/compression interpretation. *J Bone Miner Res, 14* (Suppl 1), S441.

Skedros, J. G., & Hunt, K. J. (2004). Does the degree of laminarity mediate site-specific differences in collagen fiber orientation in primary bone? An evaluation in the turkey ulna diaphysis. *J. Anat., 205*(2), 121–134.

Skedros, J. G., Hunt, K. J., & Bloebaum, R. D. (2004). Relationships of loading history and structural and material characteristics of bone: development of the mule deer calcaneus. *J Morphol, 259*, 281–307.

Skedros, J. G., & Keenan, K. E. (2016). Patterns of collagen fiber orientation in the human fibula middle-to-proximal diaphysis suggest a history of anterior-posterior bending and torsion consistent with "intermediate complexity" loading. *Am J Phys Anthropol, 159*(S62), 293.

Skedros, J. G., Keenan, K. E., Cooper, D. M., & Bloebaum, R. D. (2014a). Histocompositional organization and toughening mechanisms in antler. *J Struct Biol, 187*(2), 129–148. doi: S1047-8477(14)00140-3 [pii]; doi: 10.1016/j.jsb.2014.06.004

Skedros, J. G., Keenan, K. E., Halley, J. A., Knight, A. N., & Bloebaum, R. D. (2012). Osteon morphotypes and predominant collagen fiber orientation are adaptations for habitual medial-lateral bending in the human proximal diaphysis: implications for understanding the etiology of atypical fractures. *58th Annual Meeting of the Orthopaedic Research Society, 37*, 1512.

Skedros, J. G., Keenan, K. E., Mears, C. S., & Langston, T. D. (2015). The "shear resistance-priority hypothesis": a means for enhancing understanding of material adaptations in bones that habitually experience complex loading. *Am J Phys Anthropol, 154*(Suppl. 60), 290.

Skedros, J. G., Keenan, K. E., Williams, T. J., & Kiser, C. J. (2013a). Secondary osteon size and collagen/lamellar organization ("osteon morphotypes") are not coupled, but potentially adapt independently for local strain mode or magnitude. *J Struct Biol, 181*(2), 95–107. doi: S1047-8477(12)00295-X [pii]; doi: 10.1016/j.jsb.2012.10.013

Skedros, J. G., Kiser, C. J., Keenan, K. E., & Thomas, S. C. (2011b). Analysis of osteon morphotype scoring schemes for interpreting load history: evaluation in the chimpanzee femur. *J Anat, 218*(5), 480–499. doi: 10.1111/j.1469-7580.2011.01348.x

Skedros, J. G., Kiser, C. J., & Mendenhall, S. D. (2011c). A weighted osteon morphotype score outperforms regional osteon percent prevalence calculations for interpreting cortical bone adaptation. *Am J Phys Anthropol, 144*(1), 41–50.

Skedros, J. G., Knight, A. N., Clark, G. C., Crowder, C. M., Dominguez, V. M., Qiu, S., et al. (2013b). Scaling of Haversian canal surface area to secondary osteon bone volume in ribs and limb bones. *Am J Phys Anthropol, 151*(2), 230–244. doi: 10.1002/ajpa.22270

Skedros, J. G., & Kuo, T. Y. (1999). Ontogenetic changes in regional collagen fiber orientation suggest a role for variant strain stimuli in cortical bone construction. *J. Bone Miner. Res.*, *14*(Suppl. 1), S441.

Skedros, J. G., Mason, M. W., & Bloebaum, R. D. (2001a). Modeling and remodeling in a developing artiodactyl calcaneus: a model for evaluating Frost's Mechanostat hypothesis and its corollaries. *Anat Rec*, *263*(2), 167–185. doi: 10.1002/ar.1094 [pii]

Skedros, J. G., Mason, M. W., & Bloebaum, R. D. (2001b). Modeling and remodeling in a developing artiodactyl calcaneus: A model for evaluating Frost's mechanostat hypothesis and its corollaries. *Anat Rec*, *263*, 167–185.

Skedros, J. G., Mason, M. W., Nelson, M. C., & Bloebaum, R. D. (1996). Evidence of structural and material adaptation to specific strain features in cortical bone. *Anat Rec*, *246*(1), 47–63.

Skedros, J. G., Mears, C. S., Skedros, G. A., Henrie, T. R., Litton, S. M., & Keenan, K. E. (2014b). The mechanical relevance of secondary osteon "morphotypes": implications for accommodating non-uniform strains and the etiology of stress fractures. *60th Annual Meeting of the Orthopaedic Research Society*, 1293–1295.

Skedros, J. G., Mendenhall, S. D., Kiser, C. J., & Winet, H. (2009). Interpreting cortical bone adaptation and load history by quantifying osteon morphotypes in circularly polarized light images. *Bone*, *44*(3), 392–403.

Skedros, J. G., Sorenson, S. M., Hunt, K. J., & Holyoak, J. D. (2007a). Ontogenetic structural and material variations in ovine calcanei: a model for interpreting bone adaptation. *Anat Rec*, *290*(3), 284–300.

Skedros, J. G., Sorenson, S. M., & Jenson, N. H. (2007b). Are distributions of secondary osteon variants useful for interpreting load history in mammalian bones? *Cells Tissues Organs*, *185*(4), 285–307. doi: 000102176 [pii]; doi: 10.1159/000102176

Skedros, J. G., Sorenson, S. M., Takano, Y., & Turner, C. H. (2006b). Dissociation of mineral and collagen orientations may differentially adapt compact bone for regional loading environments: Results from acoustic velocity measurements in deer calcanei. *Bone*, *39*, 143–151.

Skedros, J. G., Su, S. C., & Bloebaum, R. D. (1997). Biomechanical implications of mineral content and microstructural variations in cortical bone of horse, elk, and sheep calcanei. *Anat Rec*, *249*(3), 297–316.

Skedros, J. G., Su, S. C., Knight, A. N., Bachus, K. N., & Bloebaum, R. D. (2019). Advancing the deer calcaneus model for bone adaptation studies: *ex vivo* strains obtained after transecting the tension members suggest an unrecognized important role for shear strains. *J. Anatomy*, *234*(1), 66–82.

Skedros, J. G., Sybrowsky, C. L., Anderson, W. E., & Chow, F. (2011a). Relationships between *in vivo* microdamage and the remarkable regional material and strain heterogeneity of cortical bone of adult deer, elk, sheep and horse calcanei. *J Anat*, *219*(6), 722–733. doi: 10.1111/j.1469-7580.2011.01428.x

Skedros, J. G., Sybrowsky, C. L., Bloebaum, R. D., Bachus, K. N., & Wang, X. (2006a). The relative influence of collagen crosslinks on the mechanical properties of equine cortical bone. *In: Transactions of the 52nd Annual Meeting of the Orthopaedic Research Society*, p. 1581.

Skedros, J. G., Sybrowsky, C. L., Parry, T. R., & Bloebaum, R. D. (2003a). Regional differences in cortical bone organization and microdamage prevalence in Rocky Mountain mule deer. *Anat Rec A Discov Mol Cell Evol Biol*, *274*(1), 837–850. doi: 10.1002/ar.a.10102

Skedros, J. G., Sybrowsky, C. L., Parry, T. R., & Bloebaum, R. D. (2003b). Regional differences in cortical bone organization and microdamage prevalence in Rocky Mountain mule deer. *Anat Rec*, *274A*(1), 837–850.

Skedros, J. G., Weaver, D. J., & Doutré, M. S. (2016b). Osteocyte size, shape, orientation, and population density: Scaling relationships, interpretation of load history, and mechanical consequences. *Osteologie*, *25*(2), 92–100.

Skerry, T. M. (2006). One mechanostat or many? Modifications of the site-specific response of bone to mechanical loading by nature and nurture. *J Musculoskelet Neuronal Interact*, *6*(2), 122–127.

Spiesz, E. M., Kaminsky, W., & Zysset, P. K. (2011). A quantitative collagen fibers orientation assessment using birefringence measurements: calibration and application to human osteons. *J Struct Biol, 176*(3), 302–306. doi: S1047-8477(11)00280-2 [pii]; doi: 10.1016/j.jsb.2011.09.009

Starck, J. M., & Chinsamy, A. (2002). Bone microstructure and developmental plasticity in birds and other dinosaurs. *J Morphol, 254*(3), 232–246.

Stockhausen, K. E., Qwamizadeh, M., Wolfel, E. M., Hemmatian, H., Fiedler, I. A. K., Flenner, S., et al. (2021). Collagen fiber orientation is coupled with specific nano-compositional patterns in dark and bright osteons modulating their biomechanical properties. *ACS Nano, 15*(1), 455–467. doi: 10.1021/acsnano.0c04786

Streeter, M., Stout, S., Trinkaus, E., & Burr, D. (2010). Brief communication: Bone remodeling rates in Pleistocene humans are not slower than the rates observed in modern populations: a reexamination of Abbott et al. (1996). *Am J Phys Anthropol, 141*(2), 315–318. doi: 10.1002/ajpa.21192

Su, S. C., Skedros, J. G., Bachus, K. N., & Bloebaum, R. D. (1999). Loading conditions and cortical bone construction of an artiodactyl calcaneus. *J Exp Biol, 202*(22), 3239–3254.

Swan, K. R., Ives, R., Wilson, L. A. B., & Humphrey, L. T. (2020). Ontogenetic changes in femoral cross-sectional geometry during childhood locomotor development. *Am J Phys Anthropol, 173*(1), 80–95. doi: 10.1002/ajpa.24080

Swartz, S. M., Bennett, M. B., & Carrier, D. R. (1992). Wing bone stresses in free flying bats and the evolution of skeletal design for flight. *Nature, 359*(6397), 726–729. doi: 10.1038/359726a0

Swartz, S. M., & Middleton, K. M. (2008). Biomechanics of the bat limb skeleton: scaling, material properties and mechanics. *Cells Tissues Organs, 187*(1), 59–84. doi: 000109964 [pii]; doi: 10.1159/000109964

Szivek, J. A., Johnson, E. M., & Magee, F. P. (1992). *In vivo* strain analysis of the greyhound femoral diaphysis. *J Invest Surg, 5*(2), 91–108.

Takano, Y., Turner, C. H., Owan, I., Martin, R. B., Lau, S. T., Forwood, M. R., et al. (1999). Elastic anisotropy and collagen orientation of osteonal bone are dependent on the mechanical strain distribution. *J Orthop Res, 17*, 59–66.

Tami, A. E., Nasser, P., Schaffler, M. B., & Knothe Tate, M. L. (2003). Noninvasive fatigue fracture model of the rat ulna. *J Orthop Res, 21*(6), 1018–1024. doi: S0736026603000998 [pii]; doi: 10.1016/S0736-0266(03)00099-8

Tang, T., Cripton, P. A., Guy, P., McKay, H. A., & Wang, R. (2018). Clinical hip fracture is accompanied by compression induced failure in the superior cortex of the femoral neck. *Bone, 108*, 121–131. doi: S8756-3282(17)30480-5 [pii]; doi: 10.1016/j.bone.2017.12.020

Taylor, D. (2000). Scaling effects in the fatigue strength of bones from different animals. *J Theor Biol, 206*, 299–306.

Thomas, K. A., Harris, M. B., Willis, M. C., Lu, Y., & MacEwen, G. D. (1995). The effects of the interosseous membrane and partial fibulectomy on loading of the tibia: a biomechanical study. *Orthopedics, 18*(4), 373–383.

Tommasini, S. M., Nasser, P., & Jepsen, K. J. (2007). Sexual dimorphism affects tibia size and shape but not tissue-level mechanical properties. *Bone, 40*(2), 498–505.

Tommasini, S. M., Nasser, P., Schaffler, M. B., & Jepsen, K. J. (2005). Relationship between bone morphology and bone quality in male tibias: implications for stress fracture risk. *J Bone Miner Res, 20*(8), 1372–1380.

Tovar-López, F. J., Domínguez-Hernández, V. M., Diez-García, M. D. P., & Araujo-Monsalvo, V. M. (2015). Finite-element analysis of the effect of basic hip movements on the mechanical stimulus within a proximal femur. *Revista de Investigación Clíni ca, 66*(Supl. 1), s32–s38.

Turner, A. S., Mills, E. J., & Gabel, A. A. (1975). *In vivo* measurement of bone strain in the horse. *Am J Vet Res, 36*(11), 1573–1579.

Turner, C. H. (1999). Toward a mathematical description of bone biology: The principle of cellular accommodation. *Calc Tissue Int, 65*, 466–471.

Turner, C. H., & Burr, D. B. (1993). Basic biomechanical measurements of bone: A tutorial. *Bone, 14*, 595–608.

Ural, A., & Vashishth, D. (2006). Interactions between microstructural and geometrical adaptation in human cortical bone. *J Orthop Res, 24*(7), 1489–1498.

Urbanová, P., & Novotný, V. (2005). Distinguishing between human and non-human bones: histometric method for forensic anthropology. *Anthropologie, XLIII*, 77–85.

Vainionpää, A., Korpelainen, R., Sievänen, H., Vihriälä, E., Leppäluoto, J., & Jämsä, T. (2007). Effect of impact exercise and its intensity on bone geometry at weight-bearing tibia and femur. *Bone, 40*(3), 604–611. doi: S8756-3282(06)00784-8 [pii]; doi: 10.1016/j.bone.2006.10.005

van der Meulen, M. C., Jepsen, K. J., & Mikić, B. (2001). Understanding bone strength: Size isn't everything. *Bone, 29*, 101–104.

van Tol, A. F., Schemenz, V., Wagermaier, W., Roschger, A., Razi, H., Vitienes, I., et al. (2020). The mechanoresponse of bone is closely related to the osteocyte lacunocanalicular network architecture. *Proc. National Academy Science, 117*, 32251–32259.

Varga, P., Pacureanu, A., Langer, M., Suhonen, H., Hesse, B., Grimal, Q., et al. (2013). Investigation of the three-dimensional orientation of mineralized collagen fibrils in human lamellar bone using synchrotron X-ray phase nano-tomography. *Acta Biomater, 9*(9), 8118–8127. doi: S1742-7061(13)00248-1 [pii]; doi: 10.1016/j.actbio.2013.05.015

Vashishth, D. (2004). Rising crack-growth-resistance behavior in cortical bone: implications for toughness measurements. *J Biomech, 37*(6), 943–946.

Vincentelli, R. (1978). Relation between collagen fiber orientation and age of osteon formation in human tibial compact bone. *Acta Anat (Basel), 100*(1), 120–128.

Vincentelli, R., & Evans, F. G. (1971). Relations among mechanical properties, collagen fibers, and calcification in adult human cortical bone. *J Biomech, 4*(3), 193–201.

Vincentelli, R., & Grigorov, M. (1985). The effect of Haversian remodeling on the tensile properties of human cortical bone. *J Biomech, 18*, 201–207.

Vitienes, I., Mikolajewicz, N., Hosseinitabatabaei, S., Bouchard, A., Julien, C., Graceffa, G., et al. (2023). Breed and loading history influence *in vivo* skeletal strain patterns in pre-pubertal female chickens. *Bone*, 116785. doi: S8756-3282(23)00118-7 [pii]; doi: 10.1016/j.bone.2023.116785

von Ebner, V. (1887). Über den feineren bau der skeletteile der kalkschwamme usw. *Wiener Stizber, 95*, 213–236.

Wainwright, S. A., Biggs, W. D., Currey, J. D., & Gosline, J. M. (1982). Elements of structural systems *Mechanical design in organisms*. (pp 254–258). Princeton, NJ: Princeton University Press.

Wallace, I. J., Demes, B., & Judex, S. (2017). Ontogenetic and genetic influences on bone's responsiveness to mechanical signals In: C.J. Percival and J.T. Reichtsmeier (Eds.), *Building Bones: Bone Formation and Development in Anthropology* (1st ed., pp. 233–253): Cambridge University Press.

Wallace, I. J., Demes, B., Mongle, C., Pearson, O. M., Polk, J. D., & Lieberman, D. E. (2014). Exercise-induced bone formation is poorly linked to local strain magnitude in the sheep tibia. *PLoS One, 9*(6), e99108. doi: PONE-D-14-08282 [pii]; doi: 10.1371/journal.pone.0099108

Wallace, I. J., Judex, S., & Demes, B. (2015). Effects of load-bearing exercise on skeletal structure and mechanics differ between outbred populations of mice. *Bone, 72*, 1–8. doi: S8756-3282(14)00421-9 [pii]; doi: 10.1016/j.bone.2014.11.013

Wallace, I. J., Middleton, K. M., Lublinsky, S., Kelly, S. A., Judex, S., Garland, T., Jr., et al. (2010). Functional significance of genetic variation underlying limb bone diaphyseal structure. *Am J Phys Anthropol*. doi: 10.1002/ajpa.21286

Wang, X. D., Masilamani, N. S., Mabrey, J. D., Alder, M. E., & Agrawal, C. M. (1998). Changes in the fracture toughness of bone may not be reflected in its mineral density, porosity, and tensile properties. *Bone, 23*(1), 67–72.

Warshaw, J., Bromage, T. G., Terranova, C. J., & Enlow, D. H. (2017). Collagen fiber orientation in primate long bones. *Anat Rec (Hoboken), 300*(7), 1189–1207. doi: 10.1002/ar.23571

Weaver, D. J., & Skedros, J. G. (2016). Do disparities in *ex vivo* strain data for the human fibula reflect heterogeneous load conditions or limitations of experimental designs? *Am J Phys Anthropol, 159(S62)*, 331.

Wei, X., & Cooper, D. M. L. (2023). The various meanings and uses of bone "remodeling" in biological anthropology: a review. *Am J Biol Anthropol*, 1–12. doi: 10.1002/ajpa.24825

Williams, J. L., Do, P. D., Eick, J. D., & Schmidt, T. L. (2001). Tensile properties of the physis vary with anatomic location, thickness, strain rate and age. *J Orthop Res*, *19*(6), 1043–1048.

Williams, J. L., Vani, J. N., Eick, J. D., Petersen, E. C., & Schmidt, T. L. (1999). Shear strength of the physis varies with anatomic location and is a function of modulus, inclination, and thickness. *J Orthop Res*, *17*(2), 214–222.

Yamada, H. (1970). *Strength of Biological Materials.* Baltimore, Maryland, USA: Williams and Wilkins Co.

Yamamoto, T., Domon, T., Takahashi, S., Islam, N., & Suzuki, R. (2000). Twisted plywood structure of an alternating lamellar pattern in cellular cementum of human teeth. *Anat Embryol (Berl)*, *202*(1), 25–30.

Yang, P. F., Sanno, M., Ganse, B., Koy, T., Bruggemann, G. P., Muller, L. P., et al. (2014). Torsion and antero-posterior bending in the *in vivo* human tibia loading regimes during walking and running. *PLoS One*, *9*(4), e94525. doi: 10.1371/journal.pone.0094525; PONE-D-13-23714 [pii]

Yeni, Y. N., Brown, C. U., Wang, Z., & Norman, T. L. (1997). The influence of bone morphology on fracture toughness of the human femur and tibia. *Bone*, *21*, 453–459.

Yeni, Y. N., & Norman, T. L. (2000). Calculation of porosity and osteonal cement line effects on the effective fracture toughness of cortical bone in longitudinal crack growth. *J. Biomed. Mater. Res.*, *51*, 504–509.

Zhong, Z., Zeng, X. L., Ni, J. H., & Huang, X. F. (2013). Comparison of the biological response of osteoblasts after tension and compression. *Eur J Orthod*, *35*(1), 59–65. doi: cjr016 [pii]; doi: 10.1093/ejo/cjr016

Zimmermann, E. A., Riedel, C., Schmidt, F. N., Stockhausen, K. E., Chushkin, Y., Schaible, E., et al. (2019). Mechanical competence and bone quality develop during skeletal growth. *J Bone Miner Res*, *34*(8), 1461–1472. doi: 10.1002/jbmr.3730

Zioupos, P., & Currey, J. D. (1998). Changes in the stiffness, strength, and toughness of human cortical bone with age. *Bone*, *22*, 57–66.

6 Histological Features of Dental Hard Tissues
Methodologies and Utilities in Forensic Anthropology and Bioarchaeology

CHRISTOPHER ARIS

6.1 Introduction

While bone is a robust material, its relative hardness pales in comparison to that of teeth. As a result, dentition is the most enduring material part of an individual, able to resist most erosion over time especially after death. On top of being robust to breakdown once the post-mortem period begins, teeth retain and preserve a wealth of an individual's information which can illuminate aspects of childhood, adulthood, and the later days of life before their death. This information is preserved across three hard dental tissues: enamel, dentine, and cementum (Figure 6.1); and is typically associated with internal macro- and microscopic features which can be observed through thin cross sections of teeth.

Within forensic analysis, long-standing methods using the observation of internal dental hard tissues can allow us to estimate age-at-death of individuals (Hillson, 1996) with some methods also able to help evaluate if infantile remains were stillborn (Whittaker and Richards, 1978; Adserias and Tejada, 2020; Canturk et al., 2014). In bioarchaeological analyses, dental histology has improved our understanding of factors that impact the growth and final shape of teeth, these include time (Aris et al., 2020a), location (McFarlane, et al., 2021b), diet (Le Luyer and Bayle, 2017; Aris et al., 2020b), lifestyle (Behie and Miszkiewicz, 2019), and pathology and disease (D'Ortenzio et al., 2018; Aris and Street, 2021; Snoddy et al., 2020, 2021). These methods may also possess utility to forensics in the future as methods for identifying habitual behaviours and afflictions with potential to modify dental structures improve through continued research. In this chapter, the focus will be on human dental histology and what intra-species analysis has revealed. However, the use of human dental hard tissue has also been key to illuminating evolutionary growth patterns and associated life histories, and experiences of physiological stress, in our hominin ancestors in the field of palaeoanthropology (see Dean, 2006 and Guatelli-Steinberg, 2008 for reviews respectively).

The uses of dental histology within the mixed fields of anthropology and archaeology stem from the retention of information within the internal structures of teeth across all three of its hard tissues. The primary aim of this chapter is therefore to provide a basic introduction to the anatomy of internal dental hard tissues, the methods by which we can produce dental cross sections, and how variations in the appearance of these tissues occur and what this means. The secondary aim will be to highlight how these methods can be applied in forensic and non-forensic (e.g. bioarchaeological) settings. While the purpose of

DOI: 10.4324/9781003385608-6

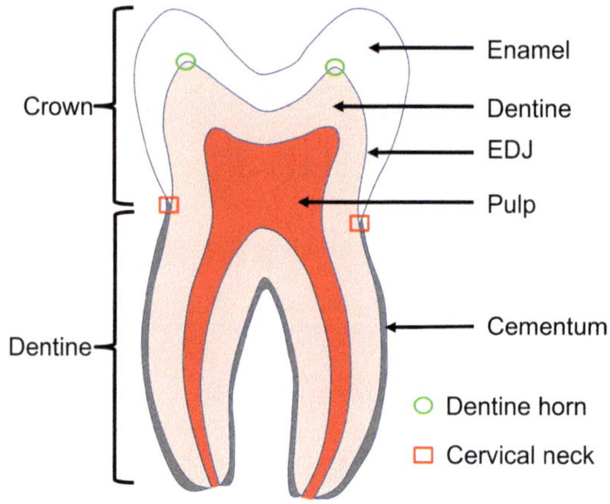

Figure 6.1 Diagram a longitudinal cross section of a molar tooth highlighting the crown and root regions as well as showing the primary tissues (hard tissues: enamel, dentine, and cementum; soft tissues: pulp).

the chapter is to provide an overview of all these factors, detailed explanations of research analyses and methodologies can be found in the literature cited throughout.

6.2 The Processes of Dental Histology

In order to understand the microstructure of teeth it is helpful to have a basic knowledge of how teeth are typically sectioned for examination. While methods can vary according to the context of the sectioning, they near-universally involve the key preparatory steps of: exfoliating (if necessary) teeth from alveolar bone (i.e. removing them from the socket), cleaning them, and embedding them in resin (e.g. Weale et al., 2023). Dental histological methods also involve comparable cutting and cleaning steps including: two cuts made using a slow speed wafering blade, mounting on glass slides, cleaning using alcohols, and cover slipping using ultra-thin glass. It is only recently that the processes of dental histology have become the focus of dedicated research (Aris, 2020). This recent work has highlighted the importance of ethical and preservative steps that can be included when sectioning teeth which include: conducting micro-CT scans where possible, producing resin casts of teeth before cutting/embedding, removing portions of roots for alternative future research before embedding, and digitising finished cross sections. After taking these steps final cross section will typically be 100–150 μm (0.01 cm) thick. By taking all the aforementioned steps dental histology allows for the viewing of microscopic internal structures of all three dental hard tissues (enamel, dentine, and cementum), the study of cell-sized features, and the development of methods using these features in research and casework. While this chapter will not go into detail on any of these steps specifically, reviews of each can be found in recent publications (Aris, 2020 and Weale et al., 2023).

6.3 Enamel: Growth Processes and Associated Histological Features

While teeth are harder than bone in general, hardness also varies within a tooth between its different parts. Enamel is by far the hardest of all these parts and the pale-white outer shell and is directly involved in breaking down food when chewing. Enamel's hardness is as a result of its microscopic structure which is inorganic, consisting of 96% mineralised hydroxyapatite in permanent, or fully formed teeth (Berkovitz et al., 2017). A further impact of enamel's composition is that it contains no living cells which, as toothpaste adverts like to remind us, mean enamel cannot repair like bone is able to (Ten Cate, 1998). While devoid of cells, enamel is packed with a variety of incrementally forming lines, artefacts of the growth process of the enamel cap in humans. Comparison can be made between the lines in enamel and tree rings, however the lines in human teeth represent much shorter periods of time – typically 1–12 days depending on the line (Berkovitz et al., 2017). Despite forming over such short periods of time, as different teeth grow at different times during early life (AlQahtani et al., 2010), these lines encase the chronology of enamel growth from the prenatal to late childhood period (Hillson, 1996). Due to the hardness of enamel, and teeth in general, these lines can be preserved for tens of thousands of years, and as a result readily examined for most of human history via histology (e.g. Smith et al., 2010).

Enamel formation (amelogenesis) is a pyramid-like shaped process, starting at the tip of the dentine horn where oral ectodermal cells begin to differentiate into enamel-forming cells (ameloblasts). This differentiation continues like a ripple downwards across the enamel dentine junction (EDJ) toward, and finishing at, the neck of the tooth (Berkovitz et al., 2017). Once differentiated, ameloblasts begin secreting enamel matrix as they migrate away from the EDJ towards what will eventually be the outer surface of the enamel cap. This process is known as the 'secretory' stage of enamel formation, and at this point enamel (by virtue of being an un-mineralised matrix) is only 27–30% mineralised (Berkovitz et al., 2017). Once all ameloblasts have reached the now established outer enamel surface, and thereby the full thickness of the enamel is formed, they reform. This sees the cells lose the features used for enamel matrix secretion, and gain features required for maturing enamel. The reformed ameloblasts then begin removing water and organic components from the secreted enamel while also injecting additional minerals (Berkovitz et al., 2017). Once this is complete enamel is fully mineralised and the ameloblasts go through programmed cell death (apoptosis; Nanci, 2007).

The secretion of enamel matrix occurs according to a number of incremental growth patterns and biorhythms which can be observed through the microscopic incremental line mentioned previously. The first of these marks the pathway of each ameloblast (Risnes, 1998) known as 'enamel rods' (Berkovitz et al., 2017) and/or 'enamel prisms' (Dean, 1987). Enamel rods can be mapped throughout the thickness of the enamel bar, using the small portions immediately adjacent to the outer enamel surface and EDJ. The diameter of enamel prisms restricts rhythmically along their length when viewed through a scanning electron microscope (SEM; Dean, 1987) – to again reference toothpaste, appearing almost like the paste when squeezed from its tube. These restrictions are believed to be the manifestation of cyclic daily (circadian) rhythms in ameloblast secretion, where regular interruptions to matrix secretion causes changes in local hydroxyapatite composition and CO_2 concentration which result in differential appearance when viewed under normal or polarised light (Antoine et al., 2009). These appearances show as lines known as 'cross striations', most easily observed through ground dental thin sections and light microscopy, which form sequentially perpendicular to enamel rods roughly every two to four microns (although this varies between individuals, see Aris, 2022 for a detailed review; see also Figure 6.2). While the volume of enamel between cross striations along prisms is known to vary within and between individuals, experimental studies from the 1930s to 1940s (see Dean, 1987 for a detailed review) and more recently in the late 1970s (Boyde, 1979) identified their formation rhythm. This was done by injecting fluorescent dyes into various mammals at know intervals to identify the circadian rhythm of cross striation formation.

Cross striations are occasionally referred to as 'short period lines' in order to differentiate them from 'long period lines'. Another incremental line forming as a result of enamel growth processes, long period lines, are more commonly known as 'Retzius lines', or 'striae of Retzius' (Dean, 2000). Unlike cross striations which form within single prisms, Retzius lines form across multiple enamel rods, appearing as dark lines when viewed under a microscope (Hillson, 2005) – in fact, in some histological slides particularly large Retzius lines can be viewed with the naked eye (see Figure 6.2). It is best to consider the anatomy and formation of Retzius lines on a three-dimensional level, as a single stria is in fact a series of growth layers formed as all ameloblasts along the enamel matrix-secreting front slow simultaneously during amelogenesis (Dean, 1987). Like cross striations however, Retzius lines form somewhat regularly throughout enamel. The regularity of these striae formation is measured through 'Retzius periodicity' which equals the number of cross striations separating each line. In humans Retzius periodicity has been observed to vary between 6 and 12 overall, most commonly 8–9 (Dean and Reid, 2001; Reid and Dean, 2006), and to be relatively consistent within individuals (Fitzgerald, 1998; McFarlane et al., 2021a). As a result of regular Retzius lines' consistent formation and relation to daily forming cross striations,

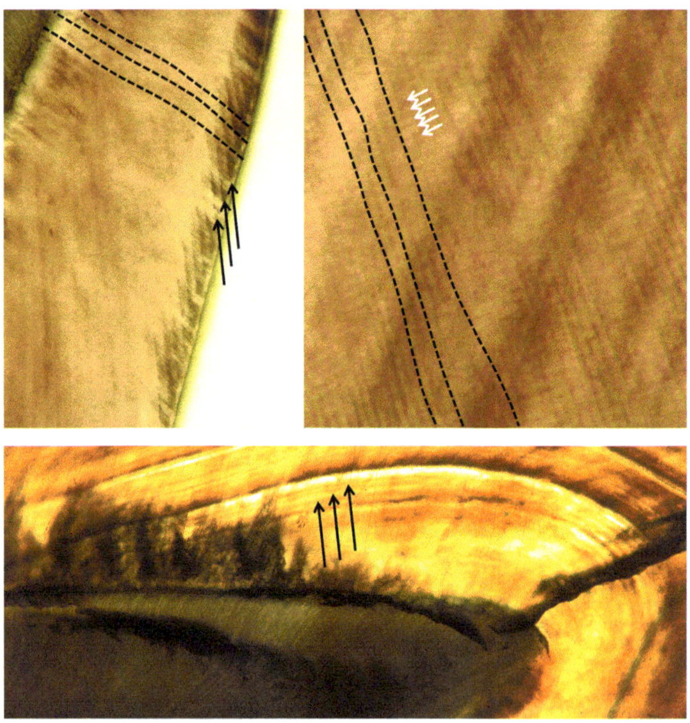

Figure 6.2 Top: longitudinal cross sections of an upper second premolar with dashed black lines highlighting enamel prism pathways. Left image is taken at 10x magnification with black oblique arrows pointing to striae of Retzius formed in lateral enamel. Right image is taken at 20x magnification with white arrows pointing to individual cross striations. Bottom: longitudinal cross section of a lower first permanent molar. Image is taken at 4x magnification with black arrows pointing to striae of Retzius formed in cuspal enamel. Image adapted from Weale et al., 2023.

Retzius periodicity is a useful tool for age-at-death and enamel cap formation time estimations (discussed in more detail later).

Until recently the exact cause of Retzius line formation was unknown. However, recent findings suggest that these lines form as a result of genetic and hormonal clocks causing oscillating biorhythms (Havers-Halberg Oscillation) which reflect like ripples in enamel growth (see Mahoney et al. 2022 for a detailed review). While this biorhythm has been studied extensively, it has not always been within the field of dental histology. However, the work which has been done using histological methods has used ground thin sections to link variation in Retzius periodicity to life history and body mass (e.g. Mahoney et al., 2017).

Within a tooth's earliest forming enamel at the tip of the enamel cap (cuspal or occlusal enamel) Retzius lines form like a stack of cones within the enamel and are thus not visible from the outside. In contrast, Retzius lines forming at the sides of the enamel cap (lateral enamel) reach the outer surface of enamel (see Figure 6.2). Where a line reaches the outer lateral enamel it forms as a slight ringed bevelling on the surface – these are called 'perikymata' (singular: perikyma; see Figure 6.3; Hillson, 1996). Perikymata relate to dental histology in that they can be used where such methods are not possible. In particular, as perikymata are external manifestations of Retzius lines, the periodicity of the two is the same. Thus, when viewed under reflected light the number of perikymata can be counted on a single tooth, average human Retzius periodicity be considered, and then an estimated range for the time the tooth's lateral enamel took to form be calculated (Guatelli-Steinberg and Reid, 2008).

Alongside the numerous cross striations and Retzius lines, there are two other kinds of notable linear developmental structures within enamel – the neonatal line (NNL) and accentuated striae (colloquially known as Wilson bands; Rose et al., 1978). The NNL is a prominent stria, appearing thicker and darker than other enamel striae, that forms in enamel as a result of birth, and roughly 7–10 days after the day of birth (Adserias-Garriga, & Visnapuu, 2019; Aka et al., 2015). As a result of this timing neonatal lines only form in deciduous (milk/baby) teeth and permanent (adult) first molars (Hillson, 1996; Figure 6.4). The cause of neonatal line formation is believed to be the result of the birth process disturbing ameloblast secretion

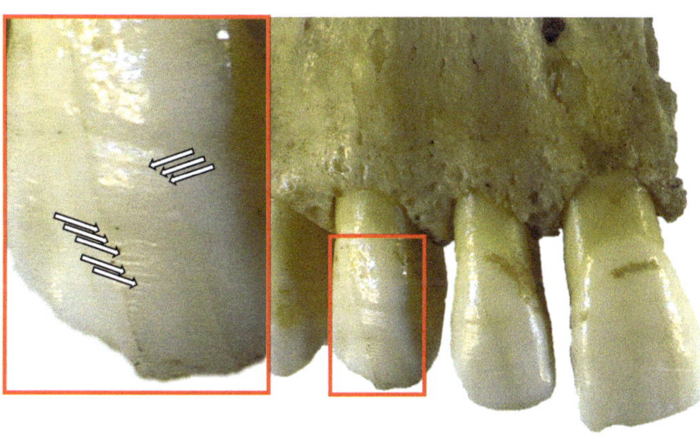

Figure 6.3 Image of right anterior permanent incisors and canine in situ within the maxilla. White arrows on superimposition point to surface perikymata lines. Photos courtesy of Megan Smith-Hands.

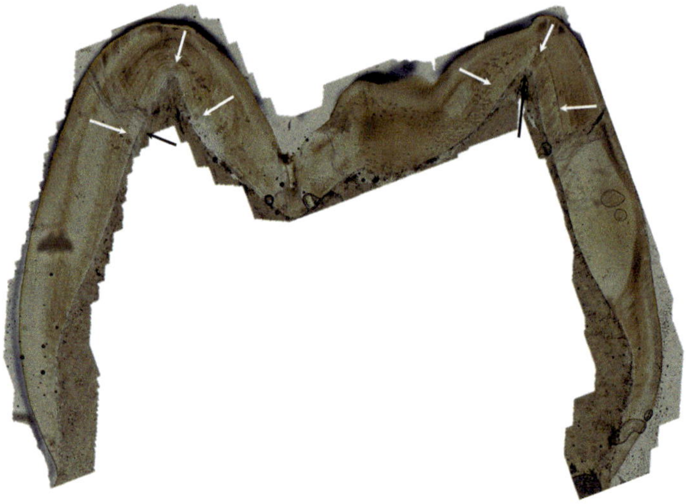

Figure 6.4 Longitudinal section of permanent upper first molar. White arrows point to neonatal lines across the enamel cusps. Black arrows point to matching neonatal lines in the dentine. Image produced using digital stitching under 20x magnification.

Figure 6.5 Longitudinal section of a permanent upper first incisor. Black arrows point to a number of Wilson Band accentuated striae. Image produced using digital stitching under 20x magnification.

(Whittaker and Richards, 1978) and/or changes in biochemical structure of pre- and postnatal enamel (Kurek et al., 2015).

Wilson bands form less consistently within enamel, appearing alongside or in-between striae of Retzius. Similar to the neonatal line, Wilson bands appear darker and thicker when viewed under a microscope (see Figure 6.5) and similar to Retzius lines can, in severe cases, manifest externally on the lateral enamel surface as deeper, hypoplastic defect, bevels known as 'linear enamel hypoplasia' lines (Goodman and Rose, 1990). Wilson bands develop as a result of physiological stress interrupting ameloblast secretion, typically nutritional deficiencies, disease, and physical stress/trauma (e.g. Fitzgerald and Saunders, 2005; Goodman, 1991). For example, the locations of accentuated striae have been found to correspond with the development of pathologies, based on documented evidence, in infants (see Teiven et al., 1996 for detail). As accentuated lines form along the forming front of enamel, it is possible to match sequences of accentuated striae across different enamel cap areas between different (but with chronologically overlapping development) teeth (e.g. Beynon et al., 1988; Boyde, 1963; Dean and Beynon, 1991; Reid et al., 1998). This has further implication for age-at-death estimations in subadult remains as explained in a later section.

6.4 Enamel: Implications for Forensic Anthropology and Paleopathology

Estimating age-at-death is an important component of the biological profile produced in forensic anthropological and bioarchaeological analyses. Age-at-death analysis from enamel structures was first conducted in the mid-late 1990s (Boyde, 1963, 1989, 1990). Boyde matched accentuated striae patterns from a child's only recently formed (at time of death) permanent incisor enamel cap to those of a first permanent molar. Using the neonatal line identified in the molar's cusp, the number of cross striations between the earliest forming post-natal molar enamel and latest forming incisor enamel could be counted to estimate age-at-death to a scale of days. This method arrived at an age estimation of 1,692 days. This method has since been tested for its accuracy, perhaps most prominently by Antoine et al. (1999; reviewed further in Antoine et al., 2009). They counted cross striations from the neonatal line to the last forming enamel in a child excavated from Christchurch, London, England – now a part of the well-known Spitalfields collection curated at the Natural History Museum in London. This collection is famous for being made up of all know age and sex individuals. Antoine and colleagues' age estimation using histological methods was three years and just over three months, nearly matching the known age of 3 years and 4 months.

Despite the utility of counting cross striations in forensic contexts, they are by the nature of their formation often difficult to count continuously. The enamel prisms along which cross striations form regularly weave around each other, especially in early forming cuspal enamel, and thus appear 'gnarled' when viewed and not aligning with the cut plane of a histological section (Hillson, 2005; also Figure 6.6). In the past methods have been proposed to account for this where the distance between consecutive visible cross striations is applied to the length of enamel prisms spanning the thickness of enamel and adjusted according to an enamel gnarling correction factor (Dean, 1998; Reid et al., 1998). However, recent research has found the distance between cross striations to vary significantly across different regions of the enamel cap within individuals (Aris et al., 2020a, 2020b) meaning the accuracy of this past method for use in forensics is questionable. What is consistent however is the Retzius periodicity of an individual. As a result, in lateral enamel where they are more visible, a count of Retzius lines can simply be multiplied by their periodicity to calculate lateral enamel formation time for an individual (e.g. Hillson, 2005). This is useful as it means that later forming enamel (i.e. lateral enamel) can remain easy to analyse within the range of accuracy necessary in forensics.

Figure 6.6 Longitudinal section of an upper third molar. Dashed white lines highlight a number of prism pathways within early forming gnarled enamel. Image taken at 10x magnification in the inner cuspal region of the buccal enamel cusp.

A number of dental charts which map both enamel and dentine development stages to age brackets have been produced over the years, displaying both internal and external dental shapes. These charts concisely demonstrate the utility of using dental growth to estimate age-at-death in young individuals using skeletal remains (Schour and Massler, 1941; Ubelaker, 1978; AlQahtani et al., 2010). These charts have been reviewed for modern use (AlQahtani et al., 2014) finding the 'London Atlas' dental chart to be the most accurate and therefor the most useful to forensic anthropologists (AlQahtani et al., 2010).

Accentuated striae in enamel are useful in forensics beyond just age estimation. Due to the nature of neonatal line formation, an infant must survive for a minimum of 7–10 days after birth for the line to be visible under a light microscope after histological sectioning. This is to allow a thin layer of enamel to form after the neonatal line so that it may be differentiated from the forming enamel front (e.g. Witzel, 2014; Behie and Miszkiewicz, 2019). This means that an absence of a neonatal line from dental remains indicates that an infant did not survive long after birth or was stillborn. Understanding the different ways in which the neonatal can be viewed through histological analysis is important to understand in forensic contexts - in particular the nature of birth and prenatal conditions can have a notable effect on neonatal line morphology. In some populations births with difficult deliveries have correlated with the formation of wider and darker lines, whereas infants born via Caesarean formed thinner lines by comparison (e.g. Eli et al., 1989; Canturk et al., 2014). In addition, maternal lifestyles including alcohol consumption (Behie and Miszkiewicz, 2019) or use of antispasmodic medications (Kurek et al., 2015) during pregnancy, and longer pregnancy generally (Zanolli et al., 2011) can result in formation of thicker neonatal lines in the associated infants.

Accentuated striae other than the neonatal line also have notable use to forensic investigations. In a well-known case Skinner and Anderson (1991) explored mapping Wilson band appearance and timing of known physiologically stressful events. Using the dental records and teeth of a Canadian child, they were able to match periods of stressful events/episodes documented in their medical records to the timing of accentuated band formation in their enamel. While the authors note that evidence of stress identified in enamel may be difficult to use to identify individuals from their teeth alone (Skinner and Anderson, 1991), it is still evidence of potential forensic relevance. Expanding this work and further exploring the potential use of Wilson bands in forensics is ongoing (e.g. Fitzgerald and Saunders, 2005; Gurian, 2021), with particular reference to more clearly defining how we can differentiate Wilson bands from Retzius lines, and narrowing down inter-population differences to refine their accuracy for use in forensic investigations.

Additional uses of accentuated lines in enamel have been used to explore physiological stress in relation to sudden and unexpected infant death (SUID) – a feature of importance in forensic cases involving subadult remains typically under one year of age-at-death (Mage and Donner, 2009; Horne, 2019). Teivens et al. (1996) analysed two samples of dental histological sections, examining the enamel of infants who died from known acute and/or chronic diseases as well as individuals with undetermined causes of death (sometimes classified as sudden infant death syndrome (SIDS)). SIDS is a 'diagnosis' of exclusion that includes autopsy findings, death scene investigation, family interviews, and review of clinical history. Teivens et al. (1996) noted that fewer accentuated striae were observed in the SIDS classified individuals. While this could suggest an individual dying at a young age with fewer accentuated striae was more likely to have died suddenly, via a lack of evidence of prolonged non-specific stress, it must be taken under strong consideration that a lack of accentuated striae in infant teeth could only be used to support a potential SIDS classification, rather than being the sole reason for such a classification in a medical forensic setting. It must also be noted that conducting histological analysis on subadult remains, especially those retaining soft tissue, is highly unlikely to be conducted as a part of a post-mortem examination based on the invasive and destructive nature of exfoliation and sectioning. However, the potential for future research and development of this application of dental histology remains, especially regarding skeletonised remains. However, it should also be noted that while SIDS assessments are widely used, research into the purpose and classification criteria used is limited (for an in-depth review: Shipstone et al., 2020). Because of this, dental histological applications to SIDS assessments could range from an opportunity for classification development, to an inappropriate inclusion until medical applications are refined.

Experimental approaches to enamel histology have also helped expand understanding of how different pathologies can impact dental development. For example, until recently enamel was believed to grow and mature at continually accelerating rates overtime through its development (Berkovitz et al., 2017). However, within analysis of a first permanent incisor with a Talon Cusp (a large accessory enamel cusp growth) slowing enamel growth patterns were identified in the accessory enamel (Aris and Street, 2021). Moreover, in recent and ongoing research the impact of nutritional vitamin D deficiency on enamel growth has been investigated. This has already been found to interrupt the development of dentine through the formation of inter-globular dentine (IGD) which can be observed in histological sections (discussed in detail later, also see: D'Ortenzio et al., 2018; Hemer and Verlinden, 2020). By comparing the enamel secretion rates via histology of individuals lacking IGD, to those of the same population with IGD, significant differences in enamel growth rates have been observed in both pre-medieval Welsh and industrial-era England (e.g. Aris et al., 2023). Those with IGD displayed greatly increased ranges in secretion rates, with the authors postulating that this may be the result of vitamin D deficiency being able to interrupt the regulated pattern of enamel growth as it does dentine, or evidence of catch-up growth occurring in teeth mirroring that which can occur in bone (e.g. Koo and Walyat, 2013).

6.5 Dentine: Growth and Histological Structures

Where enamel is the hard white outer surface of a tooth, dentine it the yellow-brown hard tissue that supports the enamel crown and forms the majority of dental roots. Dentine contains more organic material than enamel, consisting of roughly 70% mineralised material (primarily hydroxyapatite), 20% organic material (primarily collagen), and 10% water (Berkovtiz et al., 2017). The higher organic content allows dentine to be more elastic and thus support the brittle outer shell of enamel from fracturing as a result of masticatory stress.

Dentine is the first hard tissue of a tooth to begin formation, at the dental epithelium where cells differentiate into either ameloblasts which form enamel (as mentioned previously) or odontoblasts which form dentine (Berkovitz et al., 2017). Odontoblasts differentiate and begin matrix secretion following the same pattern as ameloblasts – starting at the dentine cusp apex and extending down the EDJ. Within this pattern odontoblast differentiation preludes that of ameloblasts (Hillson, 2005). Similar to the action of ameloblasts forming enamel, odontoblasts form dentine as they travel away from the EDJ, as well as the root surface, towards what will become the dental pulp chamber. As they travel odontoblasts trail behind long processes within structures called 'dentine tubules' (see Figure 6.7; e.g. Hillson, 2005). Both dentine tubules and the odontoblast processes remain for the lifespan of the dentine, with the tubules displaying the paths travelled by the original cells. These paths typically appear S-shaped near the dental crown, becoming straighter in the root, and span the entire thickness of a tooth's dentine (Berkovitz et al., 2017).

The material which forms between the dentine tubules is known as 'intertubular dentine' and that which forms within is called 'intratubular dentine' (Ten Cate, 1998). Intertubular dentine is formed by odontoblasts secreting an organic matrix (known a 'pre-dentine') and subsequently secreting matrix vesicles containing hydroxyapatite crystallites (Ten Cate, 1998). Intratubular dentine forms via a different process which results in it being 40% more mineralised than its intertubular counterpart (Ten Cate, 1998).

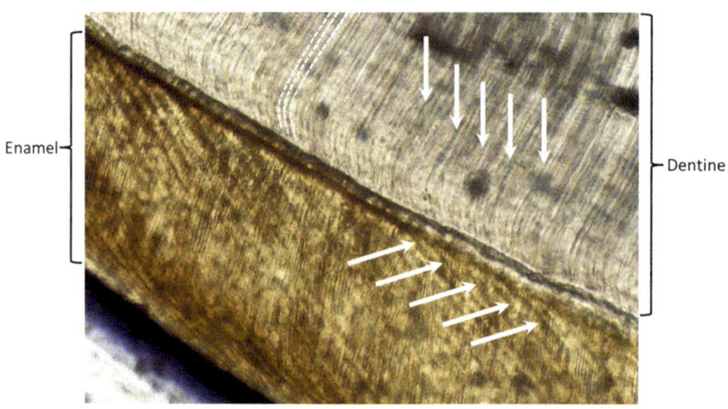

Figure 6.7 Longitudinal section of a permanent lower canine with dashed white lines highlighting adjacent dentine tubules. The vertical white arrows point to a number of contour lines of Owen in the dentine. The oblique white arrows point to a series of accentuated striae in the enamel. Interesting to note the slight offset between the dentine's contour lines and the enamel' striae along the EDJ which displays the offset of enamel formation which occurs briefly after the onset of dentine formation. Image captured at 20x magnification.

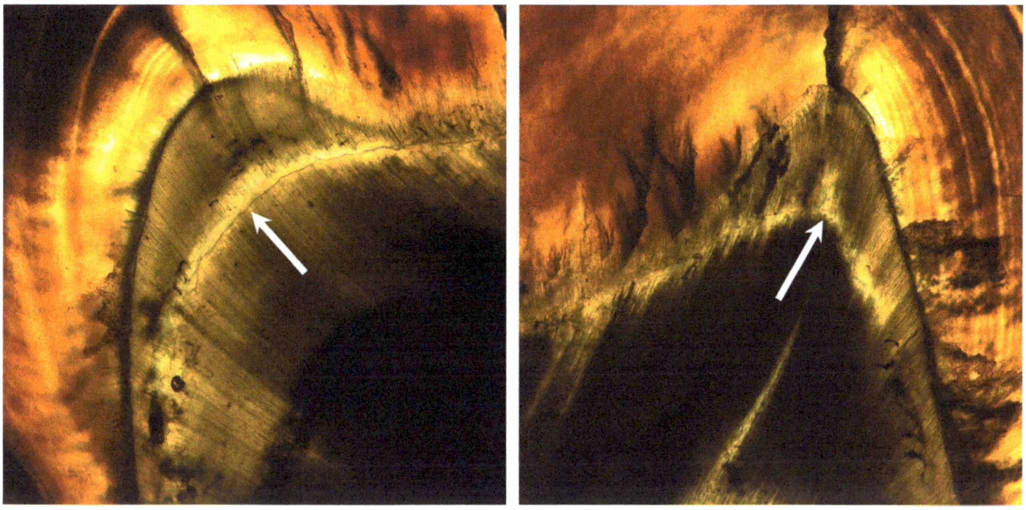

Figure 6.8 Longitudinal cross sections of two lower permanent first molars. Images captured at 4x magnification with white arrows pointing to the apices of the dentine neonatal lines. Images adapted from Weale 2023.

Dentine also forms incrementally like enamel, possessing its own short- and long-period lines. Dentine's short-period lines like enamel form daily – these are called 'von Eber lines' (Dean, 2000). The long-period lines in dentine also mirror those of enamel, these, known as 'Andresen Lines', have the same periodicity of Retzius lines within individuals and are thus thought to form from the same growth biorhythm (Dean, 2000). Dentine can also form a third prominent line with similarities to Wilson bands in enamel, 'contour lines of Owen', which typically form parallel to Andresen lines as a likely result of stressful periods interrupting dentine formation (Hillson, 2005). Dentine, according to its tooth type, may also present a particularly accentuated contour line which corresponds to an individual's day of birth – as with enamel this line is also known as a neonatal line (see Figure 6.8; Ten Cate, 1998). Much like Wilson bands, as a result of the causation of contour lines of Owen their position maps the formation front of dentine (e.g. Macho et al., 1996). It must be noted however, that as enamel formation is slightly delayed compared to dentine formation, Wilson bands form more incisally compared to their dentinal counterparts (Hillson, 2005).

Dentine is commonly described by its two component parts which can be viewed via histology, known as 'primary' and 'secondary' dentine. Primary dentine forms the majority of a tooth and is the dentine which develops during the formation of dental crown and root(s), and thereby in tandem with enamel (Arana-Chavez and Massa, 2004). By contrast secondary dentine is that which is deposited after the whole tooth has formed, specifically dentine which forms at the borders of the root canal(s) and dental pulp (Arana-Chavez and Massa, 2004). While secondary dentine forms after the enamel cap is formed, it is not reparative. Instead 'tertiary dentine' is that which is deposited in reaction to destructive process caused by external stimuli, most commonly caries (Ten Cate, 1998). Dentine can also be produced within dentine tubules in response to carie development which acts to prevent continued infection of the tooth (Arnold et al., 2001). Similar processes of dentine depositions within tubules also occur with age, with incremental deposition over time casing 'sclerotic' dentine to form (Ten Cate, 1998). This process of complete mineralisation of dentine tubules can be observed through histology as it impacts the refractive index of a tooth (i.e. what it looks like with light shone through) and can thus be used to estimate age of an individual (e.g. Singhal et al., 2010). This is explained in more detail later.

6.6 Cementum: Histological Features

As its name alludes to, cementum covers the root(s) of a tooth and acts to 'cement' the roots to the alveolar bone of the jaw (Berkovitz et al., 2017). Specifically, cementum adheres teeth within the jaw by providing a broad attachment site for periodontal ligaments (PDLs) found on the outer surface of bony tooth sockets (Hillson, 2005). Where the cementum adheres to the dentine adjacent to the outer EDJ, the ring of cementum that touches the enamel is known as the cemento-enamel junction. (White et al., 2011). Cementum is more akin to bone than other more mineralised dental tissues, consisting of 45–50% inorganic material and 50–55% organic material (mostly collagen) and water (Berkovitz et al., 2017). It is the collagen within cementum which compromises the extrinsic fibres of the attaching PDL (Hillson, 2005).

Where enamel is formed by ameloblasts, and dentine odontoblasts, cementum is formed by cementoblast cells. Cementoblasts differentiate in response to the deposition of primary root dentine (Berkovitz et al., 2017) and then deposit organic matrix around the extrinsic PDL fibres before mineralising in a similar fashion as dentine (Ten Cate, 1998). The overall formation of cementum can be split into two phases. Phase one sees the deposition of 'primary cementum' which covers the first two thirds of outer forming dentine, that being the region nearest the dental crown (see earlier regarding dentine formation); primary cementum thus forms as a tooth erupts (Ten Cate, 1998). In phase two 'secondary cementum' is deposited more rapidly at the apical third of roots after a tooth has erupted (Ten Cate, 1994). As secondary cementum is deposited the cementoblasts are tapped within its internal spaces (lacunae), at which point the cementoblasts are known as 'cementocytes' creating pockets of cellular tissue (Hillson, 2005; Ten Cate, 1998). Once embedded cementocytes then continue to produce cementum throughout a person's lifetime and as discussed later, creates the opportunity for ageing methods which remains an ongoing area of research (discussed in detail later). However, where abnormally high quantities of cementum, particularly at root apices, are produced this can impact related age-at-death estimation methods by thickening regional cementum (hypercementosis). In such cases this is likely a response to root inflammation (such as from infection) and/or where a tooth's enamel (and even dentine in extreme cases) undergoes excessive wear (Hillson, 2005).

Despite being especially thin and low on volume relative to enamel and dentine, linear formations are visible within cementum through histological analysis. Through viewing cementum in thin sections layered alternating bands of dark and light, known as 'contour lines of Salter', can be observed (see Figure 6.9). These visually distinct bands are believed to represent accelerating and decelerating cementum deposition, as more rapidly formed cementum traps more cementocytes and extrinsic collagen fibres and is thus less mineralised and possesses a different refractive index (Hillson, 2005). Conversely, slower forming cementum is more mineralised, with fewer different compromising features which causes is to scatter less transmitted light, and thus appear darker during microscopy (Hillson, 2005). As a result of the banded contour lines mapping variations in cementum deposition rates, these can be used within dental-based ageing methods (discussed in more detail next).

Figure 6.9 Light dark bands running the length, from top to bottom, on the image show cementum layers observed under transmitted light. Image taken at 20x magnification from a contemporary modern upper third molar.

6.7 Dentine and Cementum: Implications for Forensic Anthropology and Bioarchaeology

One of the most commonly analysed features of dentine, which can assist in biological analysis of remains through histology, is the formation of interglobular dentine (IGD). Interglobular dentine is a growth defect and appears as areas or bands of black patches typically within primary dentine (see Figure 6.10; e.g. D'Ortenzio et al., 2018; Hemer and Verlinden, 2020; Snoddy et al., 2021). The observation of IGD within a tooth is thought to be the result of vitamin D deficiency and/or other nutritional stress during early life while dentine is forming (dentinogenesis; e.g. Kagayama et al., 1997; Tsuchiya et al., 2002). Simply by producing histological cross sections of teeth we can therefore investigate whether individuals, or populations, suffered periods of nutritional stress, a potentially useful factor in both bioarchaeological and forensic investigations. Due to the incremental nature of dentine formation, we can go one step further, and examine IGD formation across the volume of dentine with the potential to identify multiple periods of nutritional stress through the presence of multiple IGD bands, thus expanding our ability to comment on individual early life history and health (e.g. Jayawardena et al., 2009; Weale, 2023). Research into IGD is

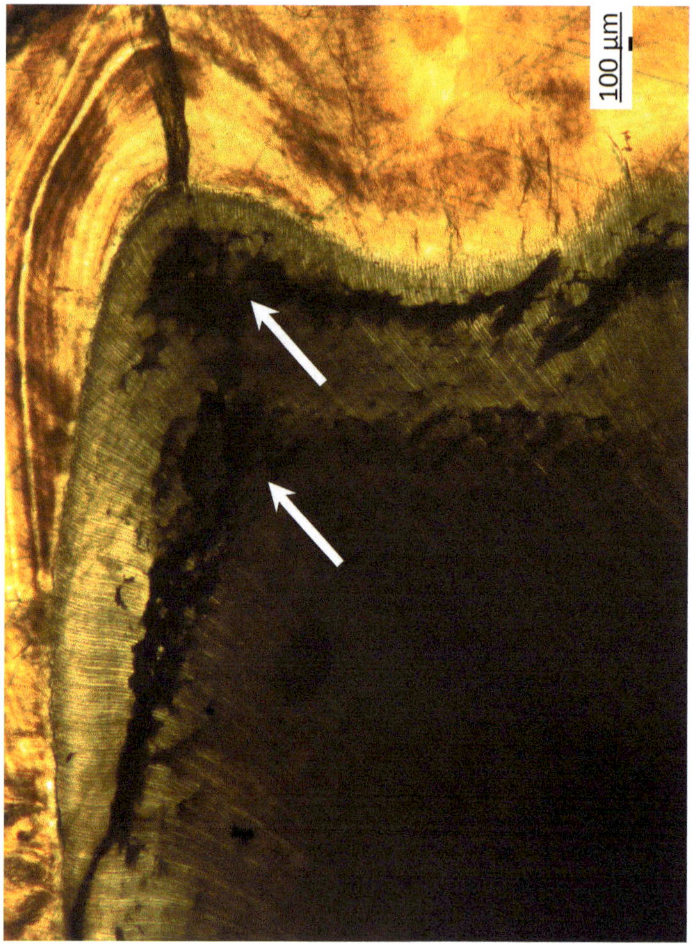

Figure 6.10 Longitudinal cross section of a lower first permanent molar. Image captured at 4x magnification with white arrows highlighting the apices of two separate bands of IGD. The top arrow points to the IGD band that formed earlier in childhood during root development. Image adapted from Weale 2023.

ongoing, however the analyses already conducted has helped inform our understanding of how tooth formation can be affected by an individual's environment, and is thus continuing to highlight the utility of dental histology within all fields of skeletal analysis.

Where IGD is an irregular forming defect, dentine possesses a number of regular forming incremental formation lines (Andresen lines and contour lines of Owen), thus it could be assumed that these lines can be suitably used to estimate age-at-death in both forensic and bioarchaeological settings. Dentine also possessing a neonatal line in certain teeth further makes the use of dentine for age-at-death estimation, at the surface, comparable in utility to enamel. In forensic anthropology however, accuracy and clarity of visual analyses must be maximised. As a result, with the incremental lines in enamel being more clearly seen than in dentine through histological analysis (e.g. Dean and Scandrett, 1996), it is recommended that enamel be used (via methods discussed previously) for age-at-death estimation where possible in tandem with dentine-based estimations (e.g. Hillson, 2005). However, when producing the biological profile of an individual with complete enamel crowns with corresponding incomplete roots, incremental dentine lines are fully recommended for use. In these cases, root extension rates (i.e. the rate at which a root is extending via growth) can be calculated using average daily formation rates of dentine and observations of the orientation of clearer Andresen lines and/or contour lines of Owen (see Dean and Vesey, 2008 for full detail). By subsequently measuring the maximum root length of the same tooth the extension rate can be used to estimate the time taken for the incomplete root to form (e.g. Dean and Cole, 2013). By then calculating crown formation time of the tooth's enamel after the formation of its neonatal line, the formation time of the incomplete root can be added to provide a relatively accurate estimation of an individual's age-at-death. While a long process, this provides methodological options to forensic and bioarchaeological practitioners when working with sub-adult and adolescent remains.

Where working with remains presenting complete dental roots, there are still methods involving histology of dentine that can assist with age-at-death estimation. In particular observations of the development of sclerotic dentine in dental roots have been found to strongly correlate with age (Hillson, 1996). Sclerotic dentine, as mentioned previously, is more mineralised than standard dentine and thus appears more translucent in histological samples due to its lower organic content (Ten Cate, 1994; Hillson, 2005). Formation of sclerotic dentine begins as early as the late teens at the apex of dental roots (e.g. Hillson, 1996) and thus methods utilising it can be applied to the majority of individuals where the incomplete root growth method cannot be applied. The correlation between sclerotic dentine formation and age has been quantified for methodological use on multiple occasions. In one of the earliest and most prominently examples, 926 extracted teeth spanning 265 individuals of known age were analysed (Bang and Ramm, 1970). Results identified notable correlations between age-at-extraction (note not age-at-death) and root dentine transparency, from which regression formula were developed which were found to estimate age within ±5 years, in a select sample of individuals, up to the age of 60 years.

A string of studies have worked to improve on, or have commented on, Bang and Ramm's (1970) initial study, with results suggesting improved accuracies of age estimation up to ±3 years in individuals up to the age of 50 years (e.g. Vasilidas et al., 1983a, 1983b; Hillson, 2005). One of the more recent studies continuing this work was conducted in India and looked at a sample of 50 extracted teeth, with a near-even split of male and female individuals, and representative samples from ages ranging from 15–75 years of age (Sabarad and Kalburge, 2014). Their findings further supported the use of dentine transparency, noting the usefulness of dentine sclerosis development being robust to environmental factors which otherwise can influence skeletal elements – as a result making the method more accurate and useful to forensic anthropology in particular. Multiple studies have been able to replicate these results as recently as 2023, with particular success found using ground histological sections (e.g. Alghonemy, 2023).

Much like in dentine, variation in the formation of cementum in terms of mineralisation can be used, via histology, to estimate age-at-death. The basis of the initial methodological theory comes from early analysis of marine and terrestrial mammals, which identified a seasonal growth cycle in dental cementum (see Hillson, 2005 for detailed review). In terrestrial mammals this growth cycle appears dependent on food resources, as evidenced by Mitchell (1963, 1967). They found that histological dental cross sections of red deer (*Cervus elaphus*) showed darker cementum bands (signalling more organic cementum and interrupted growth) on their outermost layer when killed in leaner winter months, in contrast to brighter outermost bands when killed in resource abundant periods. On the logical basis that paired dark and bright band represented one year of cementum growth Mitchell managed to accurately estimate age-at-death from the dental slides compared to the known age of the dear (Mitchell, 1963, 1967). While showing promising results, the variation between human and non-human dental development and internal morphology is known (e.g. Dean, 1998; Berkovitz and Shellis, 2018), and therefore at face value there are potential issues with applying the method to human samples in bioarchaeological contexts, and especially in forensic anthropological contexts where accuracy is a necessity.

Some research has been conducted testing the viability of applying the method to human samples. Initially Stott et al. (1982) found positive results when adding age of tooth eruption to extra time calculated looking from counting cementum layers from three human cadavers. However, these results were not replicated by a subsequent study with an expanded data set of 100 extracted teeth (Miller et al., 1988). Mixed results were subsequently identified in the early 2000s with analysis of a sample of 363 extracted teeth yielding results as accurate at ±2.5 years of known age (Wittwer-Backofen et al., 2004), but with a stark decrease in accuracy when tested on a sample of 121 early medieval teeth (Wittwer-Backofen et al., 2008). This loss in accuracy was suggested to be the result of high inter-observer variability in the application of the cementum layer counting method, and the impact of quality of dental sections with accuracy only likely to be obtained from the clearest of histological sections (Wittwer-Backofen et al., 2008).

Alongside the mixed findings discussed, additional research has continued to identify high inter-observer error when estimating age from cementum (Huffman, 2008; Roksandic et al., 2009) and has commented on the issues surrounding the non-standardisation of the method (Wittwer-Backofen, 2012). In addition, diagenetic changes, hypercementosis, and even the location of cementum analysed have all be found to impact the ability to accurately produce age estimations based on cementum layer annulation (Roksandic et al., 2009; Huffman 2008; Jackes, 2009). In addition, recent research on a large number of teeth from multiple modern Czech samples has managed to estimate age from cementum, on average, to within ±1.7 years (Zazvonilová et al., 2022). However, within the samples ranges for individuals were still found to vary up to ±10.8 years, and as a result the authors highlighted an urgent need to standardise cementum-based techniques for age estimation (Zazvonilová et al., 2022). Promise has also been found in estimation made from cementum thickness, rather than cementum layers, observed through horizontal transverse sections of roots rather than vertical sections through the root and crown (Birimiša et al., 2021). In their study of 57 modern donated teeth, positive correlation was found between cementum thickness and donor age, although variation was noted between location of histological section, sex, crown condition, and number of roots of the sectioned tooth (Birimiša et al., 2021). Overall, then, while there is potential utility in estimating age from cementum bands in bioarchaeology with further research, the limitations in accuracy noted across multiple studies effectively rules out its utility in forensic analyses. Moreover, when compared to the promise shown by dentine-based histological ageing methods and considering the fact it would be rare to have a histological slide with only one of cementum or dentine present for analysis, dentine analysis is likely to continue to have more utility in both the bioarchaeological and forensic fields.

Recent research has discovered new features within cementum, via histology, that have more promising implications. These features are primarily 'stress lines', commonly known as 'irregular incremental lines', form similarly to accentuated striae in enamel forming parallel to typical cementum bands but with darker and more localised appearance as a result of interrupted mineralisation (see Mani-Caplazi et al., 2019). As cementum forms more regularly throughout life (as previously detailed) any ability to glean information regarding life history has clear value to bioarchaeological and forensic analysis of dental remains. This is highlighted in recent research by Mani-Caplazi et al. (2017 and 2019) who identified possible correlations between irregular incremental lines in cementum and pregnancy and/or disease through analysing both modern extracted teeth and archaeological teeth. Alongside this correlation, specific work has been conducted to confirm the replicability of the method and standardise its use (see Mani-Caplazi et al., 2019 for detailed summary). With both identifiable life history aspects now identifiable in cementum, with supported detailed, replicable, and standardised techniques available, histological cementum analysis has a clear place in analysis of modern and archaeological teeth despite limitations on age estimation methods.

6.8 Conclusions

The fact that all three primary hard tissues of teeth, enamel, dentine, and cementum, all contain physiological features associated with their growth and development gives them clear utility via histological analysis in bioarchaeological, palaeopathological, and forensic contexts. Where issues lie are in the variability of development between these hard tissues, and the limitation of accuracy of methods which utilise their analysis after completion of development – for example, loss of age-at-death estimation accuracy after completed enamel and root formation (e.g. Antoine et al., 1999 and 2009). However, there is existing and developing promise in the combination of concurrent analysis of multiple hard tissues via histology. In particular, in subadult individuals where enamel has completed formation within a tooth containing a neonatal line, but tooth formation is incomplete, crown formation time can be calculated and added to the existing root formation time give an accurate age estimation (e.g. Hillson 2005, Dean and Cole, 2013). In addition, recent analysis of the impact of pathologies and physiological stressors on dentition has identified concurrent impacts on dentine and enamel formation within single individuals (Aris and Street, 2021; Aris et al., 2023). These multi-hard tissue inclusive methods and studies highlight dental histology's utility to bioarchaeology and forensic anthropology, and the great potential analysing multiple dental tissues at once has to methodological developments.

Enamel presents some of the greatest advantages to skeletal analyses as well as some of the most unavoidable limitations. The formation of cross striations allows investigators to analyse enamel formation to a daily level of accuracy, with further utility in age estimation of younger individuals (Hillson, 2005; Antoine et al., 2009). Moreover, the consistent pattern in formation of Retzius striae allows for accurate formation time estimations for the entire enamel cap, and the formation of similar accentuated lines allows for preliminary analysis of physical and physiological stress in early life including the day of birth (e.g. Skinner and Anderson, 1991; Behie and Miszkiewicz, 2019). However, once enamel has formed it is highly mineralised and unable to undergo repair and reformation (Berkovitz et al., 2017). It is therefore an extremely useful time capsule of an individual's early life, with extensive value to bioarchaeology, as well as to forensic anthropology in select circumstances – in particular it will be fascinating to see how enamel research improves our ability to investigate the quality of life of children before death in forensic cases.

Dentine, much like enamel, contains a series of formation lines related to development including Andresen lines and contour lines of Owen which can be used for age estimation techniques (Dean and Vesey, 2008). In addition, the dentine compromising roots forming at the time of birth can also possess a neonatal line alongside evidence of stress, particularly stress as a result of dietary deficiency (Hillson, 2005; Hemer and Verlinden, 2020). Dentine therefore in theory has near identical uses to bioarchaeology and forensics as enamel. However, the clarity of its interior lines are less clear and harder to analyse via histology (Dean and Scandrett, 1996) thereby making enamel the more commonly selected hard tissue for histological research. However unlike enamel, dentine undergoes changes after it completes developing, giving it an extended timespan of utility when analysing life history into adulthood. In particular, the formation of sclerotic dentine throughout life and the resulting loss of transparency of dentine in ground histological sections, has been found to be able to provide accurate age-at death estimations with refinement of this method continually being made (e.g. Sabarad and Kalburge, 2014; Alghonemy, 2023).

Cementum in comparison to dentine and enamel appears on the surface as having more limited utility within bioarchaeology and forensic histological analyses, but recent research highlights how this may not be the case as previously believed. Like other dental hard tissues cementum possess lines that form over time with potential viability for use in age estimation, however their use has been linked to inconsistent accuracy and high inter-observer error (Huffman, 2008; Roksandic et al., 2009) with strong calls for associated methods to be standardised (Zazvonilová et al., 2022). Moreover, diagenetic changes and pathology in individuals have also been found to impact the clarity of internal cementum lines, further limiting their use in estimating age (Roksandic et al., 2009; Huffman, 2008; Jackes, 2009). However, the impact of pathology on cementum physiology has opened up new lines of investigation into different aspects of biological profiling. Also like the other hard tissues of dentition, cementum can respond to stress by forming irregular incremental lines. These lines in cementum have been linked to pregnancy and disease in ways unique to this dental hard tissue (Mani-Caplazi et al., 2017 and 2019). This recent research therefore highlights the persisting utility of cementum to both bioarchaeology and forensic anthropology, and it would be interesting to know if the stress lines forming in cementum correlated with stress lines forming elsewhere within a tooth.

Acknowledgments

I would like to thank Dr Marie C. Weale for her feedback and comments on the manuscript as well her permission to use, and help in producing, a number of images in this chapter. Thanks also go to Dr Julia Beaumont for her advice and supportive comments, and to Megan Smith-Hands for their help with figure imaging. Finally, I would like to express my gratitude to Dr Debbie Guatelli-Steinberg, and to Dr Christian Crowder and Professor Sam Stout, for the recommendation and invitation respectively to contribute to this volume.

References

Adserias-Garriga, J., & Tejada, J. (2020). Dental aging methods and population variation. *Forensic Science and Humanitarian Action: Interacting with the Dead and the Living, 10*(1), 219–233.

Adserias-Garriga, J., & Visnapuu, V. (2019). The neonatal line as evidence of live birth. In *Age Estimation*, 161–168. Academic Press.

Aka, P. S., Yagan, M., Canturk, N., & Dagalp, R. (2015). *Primary Tooth Development in Infancy: A Text and Atlas*. CRC Press.

Alghonemy, W. (2023). Dental Age Estimation Using Radicular Dentin Translucency for Egyptian Adults. *Egyptian Dental Journal, 69*(1), 197–205.

AlQahtani, S. J., Hector, M. P., & Liversidge, H. M. (2010). Brief communication: The London atlas of human tooth development and eruption. *American Journal of Physical Anthropology, 142*(3), 481–490.

AlQahtani, S. J., Hector, M. P., & Liversidge, H. M. (2014). Accuracy of dental age estimation charts: Schour and Massler, Ubelaker and the London Atlas. *American Journal of Physical Anthropology, 154*(1), 70–78.

Antoine, D., Dean, C., & Hillson, S. (1999). The periodicity of incremental structures in dental enamel based on the developing dentition of post-medieval known-age children. In *Dental Morphology'98: Proceedings of the 11th International Symposium on Dental Morphology* (pp. 48–55). Oulu University Press.

Antoine, D., Hillson, S., & Dean, M. C. (2009). The developmental clock of dental enamel: a test for the periodicity of prism cross-striations in modern humans and an evaluation of the most likely sources of error in histological studies of this kind. *Journal of Anatomy, 214*(1), 45–55.

Arana-Chavez, V. E., & Massa, L. F. (2004). Odontoblasts: the cells forming and maintaining dentine. *The International Journal of Biochemistry & Cell Biology, 36*(8), 1367–1373.

Arnold, W. H., Konopka, S., & Gaengler, P. (2001). Qualitative and quantitative assessment of intratubular dentin formation in human natural carious lesions. *Calcified Tissue International, 69*(5).

Aris, C. (2020). The histological paradox: Methodology and efficacy of dental sectioning. *Papers from the Institute of Archaeology, 29*(1).

Aris, C. (2022). Enamel growth rate variation of inner, mid, and outer enamel regions between select permanent tooth types across five temporally distinct British samples. *Archives of Oral Biology, 137*, 105394.

Aris, C., Hemer, K., Weale, M., Street, E., & Collings, J. A. (2023). Enamel growth in pathological samples and its potential used as a diagnostic tool in biological and forensic anthropology. *American Journal of Biological Anthropology*. In press.

Aris, C., Mahoney, P., O'Hara, M. C., & Deter, C. (2020a). Enamel thickness and growth rates in modern human permanent first molars over a 2000 year period in Britain. *American Journal of Physical Anthropology, 173*(1), 141–157.

Aris, C., Mahoney, P., O'Hara, M. C., & Deter, C. (2020b). Enamel thickness and growth rates in modern human permanent first molars over a 2000 year period in Britain. *American Journal of Physical Anthropology, 173*(1), 141–157.

Aris, C., & Street, E. (2021). Growth rates of accessory human enamel: a histological case study of a modern-day incisor from Northern England. *Dental Anthropology Journal, 34*(1), 3–12.

Bang, G., & Ramm, E. (1970). Determination of age in humans from root dentin transparency. *Acta Odontologica Scandinavica, 28*(1), 3–35.

Behie, A. M., & Miszkiewicz, J. J. (2019). Enamel neonatal line thickness in deciduous teeth of Australian children from known maternal health and pregnancy conditions. *Early Human Development, 137*, 104821.

Berkovitz, B. K., Holland, G. R., & Moxham, B. J. (2017). *Oral anatomy, histology and embryology E-book*. Elsevier Health Sciences.

Berkovitz, B. K., & Shellis, P. (2018). *The teeth of mammalian vertebrates*. Academic Press.

Beynon, A. D., Clayton, C. B., Rozzi, F. V. R., & Reid, D. J. (1998). Radiographic and histological methodologies in estimating the chronology of crown development in modern humans and great apes: a review, with some applications for studies on juvenile hominids. *Journal of Human Evolution, 35*(4–5), 351–370.

Birimiša, M., Dumančić, J., Vodanović, M., Milošević, S. A., Marić, M., & Brkić, H. (2021). Forensic determination of dental age by cementum thickness of human teeth. *The Journal of Forensic Odonto-Stomatology, 39*(3), 41.

Boyde A (1963). Estimation of age at death of young human skeletal remains from incremental lines in dental enamel. London: Third International Meeting in Forensic Immunology, Medicine, Pathology, and Toxicology: Plenary Session 11A.

Boyde A (1979). Carbonate concentration, crystal centres, core dissolution, caries, cross striations, circadian rhythms, and compositional contrast in the SEM. *J Dent Res (special issue)*, 58, 981–983.

Boyde A (1989). Enamel. In: Berkovitz BKB, Boyde A, Frank RM, Höhling HJ, Moxham BJ, Nalbandian J, Tonge CH (eds) *Teeth: Handbook of Microscopic Anatomy*. Springer, New York.

Boyde A (1990). Developmental interpretations of dental microstructure. In: DeRousseaus CJ (ed) *Primate Life History and Evolution: Monographs in Primatology*, 14th vol. Wiley-Liss, New York.

Canturk, N., Atsu, S. S., Aka, P. S., & Dagalp, R. (2014). Neonatal line on fetus and infant teeth: An indicator of live birth and mode of delivery. *Early Human Development*, *90*(8), 393–397.

Dean MC (1987). Growth layers and incremental markings in hard tissues; a review of the literature and some preliminary observations about enamel structure in Paranthropus boisei. Journal of Human Evolution, 16, 157–172.

Dean, M. C. (1998). Comparative observations on the spacing of short-period (von Ebner's) lines in dentine. *Archives of Oral Biology*, *43*(12), 1009–1021.

Dean, M. C. (2000). Incremental markings in enamel and dentine: what they can tell us about the way teeth grow. *Development, Function and Evolution of Teeth*, *1*(1), 119–130.

Dean, M. (2006). Tooth microstructure tracks the pace of human life-history evolution. *Proceedings of the Royal Society B: Biological Sciences*, *273*(1603), 2799–2808.

Dean, M. C., & Beynon, A. D. (1991). Histological reconstruction of crown formation times and initial root formation times in a modern human child. *American Journal of Physical Anthropology*, *86*(2), 215–228.

Dean, M. C., & Cole, T. J. (2013). Human life history evolution explains dissociation between the timing of tooth eruption and peak rates of root growth. *PLoS One*, *8*(1), e54534.

Dean, M. C., & Reid, D. J. (2001). Perikymata spacing and distribution on hominid anterior teeth. *American Journal of Physical Anthropology*, *116*(3), 209–215.

Dean, M. C., & Scandrett, A. E. (1996). The relation between long-period incremental markings in dentine and daily cross-striations in enamel in human teeth. *Archives of Oral Biology*, *41*(3), 233–241.

Dean, M. C., & Vesey, P. (2008). Preliminary observations on increasing root length during the eruptive phase of tooth development in modern humans and great apes. *Journal of Human Evolution*, *54*(2), 258–271.

D'Ortenzio, L., Kahlon, B., Peacock, T., Salahuddin, H., & Brickley, M. (2018). The rachitic tooth: Refining the use of interglobular dentine in diagnosing vitamin D deficiency. *International Journal of Paleopathology*, *22*, 101–108.

Eli I, Sarnat H, & Talmi E (1989). Effects of the birth process on the neonatal line in the primary tooth enamel. *Pediatric Dentistry*, *11*, 220–223.

FitzGerald, C. M. (1998). Do enamel microstructures have regular time dependency? Conclusions from the literature and a large-scale study. *Journal of Human Evolution*, *35*(4–5), 371–386.

FitzGerald, C. M., & Saunders, S. R. (2005). Test of histological methods of determining chronology of accentuated striae in deciduous teeth. *American Journal of Physical Anthropology*, *127*(3), 277–290.

Goodman, A. H. (1991). Stress, adaptation, and enamel developmental defects. *Human Paleopathology: Current Syntheses and Future Options*, *1*(1), 280–287.

Goodman, A. H., & Rose, J. C. (1990). Assessment of systemic physiological perturbations from dental enamel hypoplasias and associated histological structures. *American Journal of Physical Anthropology*, *33*(S11), 59–110.

Guatelli-Steinberg, D. (2008). Using perikymata to estimate the duration of growth disruptions in fossil hominin teeth: issues of methodology and interpretation. *Cambridge Studies in Biological and Evolutionary Anthropology*, *1*(53), 71–86.

Guatelli-Steinberg, D., & Reid, D. J. (2008). What molars contribute to an emerging understanding of lateral enamel formation in Neandertals vs. modern humans. *Journal of Human Evolution, 54*(2), 236–250.

Gurian, K. N. (2021). *What accentuated striae in tooth enamel reveal about developmental stress in two groups of disparate socioeconomic status in Ohio* (Doctoral dissertation, The Ohio State University).

Hemer, K. A., & Verlinden, P. (2020). Vitamin D deficiency rickets in early medieval Wales: a multi-methodological case study. *Childhood in the Past, 13*(1), 20–37.

Hillson, S. (1996). *Dental anthropology*. Cambridge University Press.

Hillson S (2005) i. Cambridge University Press, Cambridge.

Horne, R. S. (2019). Sudden infant death syndrome: current perspectives. *Internal Medicine Journal, 49*(4), 433–438.

Huffman M (2008). Determining the effects of preparatory methods when observing cementum layering: exploration of techniques and microscopy. Unpublished MSc dissertation, University of London.

Jackes, M. (2009). Teeth and the past in Portugal and the Mesolithic-Neolithic transition. In *Frontiers of Oral Biology: Comparative Dental Morphology. 14th International Symposium on Dental Morphology, 13th vol.*

Jayawardena, C., Nandasena, T., Abeywardena, A., & Nanayakkara, D. (2009). Regional distribution of interglobular dentine in human teeth. *Archives of Oral Biology, 54*(11), 1016–1021.

Kagayama M, Zhu JX, Sasano Y, Sato H, Mayanagi H. (1997). Development of interglobular dentine in rat molars and its relation to maturation of enamel. *Anatomy and Embryology (Berlin)* 196(6): 477–483.

Koo, W., & Walyat, N. (2013). Vitamin D and skeletal growth and development. *Current Osteoporosis Reports, 11*, 188–193.

Kurek, M., Żądzińska, E., Sitek, A., Borowska-Strugińska, B., Rosset, I., & Lorkiewicz, W. (2015). Prenatal factors associated with the neonatal line thickness in human deciduous incisors. *Homo, 66*(3), 251–263.

Le Luyer, M., & Bayle, P. (2017). Microevolution of outer and inner structures of upper molars in Late Pleistocene and Early Holocene humans. *Comptes Rendus Palevol, 16*(5–6), 632–644.

Macho GA, Reid DJ, Leakey MG, Jablonski H, & Beynon AD (1996). Climatic effects on dental development of eropithecus oswaldi from Koobi Fora and Olorgesailie. Jounral of Human Evolution. *30*(1), 57–70.

Mage, D. T., & Donner, M. (2009). A unifying theory for SIDS. *International Journal Of Pediatrics, 2009*, 2–10.

Mahoney, P., McFarlane, G., Loch, C., White, S., Floyd, B., Dunn, E. C., & Guatelli-Steinberg, D. (2022). Dental biorhythm is associated with adolescent weight gain. *Communications Medicine, 2*(1), 99.

Mahoney, P., Miszkiewicz, J. J., Pitfield, R., Deter, C., & Guatelli-Steinberg, D. (2017). Enamel biorhythms of humans and great apes: the Havers-Halberg Oscillation hypothesis reconsidered. *Journal of Anatomy, 230*(2), 272–281.

Mani-Caplazi, G., Hotz, G., Wittwer-Backofen, U., & Vach, W. (2019). Measuring incremental line width and appearance in the tooth cementum of recent and archaeological human teeth to identify irregularities: First insights using a standardized protocol. *International Journal of Paleopathology, 27*, 24–37.

Mani-Caplazi, G., Schulz, G., Deyhle, H., Hotz, G., Vach, W., Wittwer-Backofen, U., & Müller, B. (2017). Imaging of the human tooth cementum ultrastructure of archeological teeth, using hard x-ray microtomography to determine age-at-death and stress periods. In *Developments in X-Ray tomography XI* (Vol. 10391, pp. 195–202). SPIE.

McFarlane, G., Guatelli-Steinberg, D., Loch, C., White, S., Bayle, P., Floyd, B., & Mahoney, P. (2021a). An inconstant biorhythm: the changing pace of Retzius periodicity in human permanent teeth. *American Journal of Physical Anthropology, 175*(1), 172–186.

McFarlane, G., Loch, C., Guatelli-Steinberg, D., Bayle, P., Le Luyer, M., Sabel, N., & Mahoney, P. (2021b). Enamel daily secretion rates of deciduous molars from a global sample of children. *Archives of Oral Biology, 132*, 105290.

Miller CS, Dove SB, Cottone JA (1988). Failure of use of cemental annulations in teeth to determine the age of humans. *Journal of Forensic Sciences*, 33, 137–143.

Mitchell, B. (1963). Determination of age in Scottish red deer from growth layers in dental cementum. *Nature*, 198, 350–351.

Mitchell B (1967). Growth layers in dental cementum for determining the age of red deer (Cervus elaphus). *The Journal of Animal Ecology* 36, 279–293.

Nanci A (2007). Ten Cate's Oral Histology, 7th ed. Elsevier Science, Amsterdam.

Reid DJ, Beynon AD, Ramirez Rozzi F. 1998. Histological reconstruction of dental development in four individuals from a mediaeval site in Picardie, France. *Journal of Human Evolution*, 35, 463–479.

Reid, D. J., & Dean, M. C. (2006). Variation in modern human enamel formation times. *Journal of Human Evolution, 50*(3), 329–346.

Risnes, S. (1998). Growth tracks in dental enamel. *Journal of Human Evolution, 35*(4–5), 331–350.

Roksandic, M., Vlak, D., Schillaci, M. A., & Voicu, D. (2009). Applicability of tooth cementum annulation to an archaeological population. *American Journal of Physical Anthropology, 140*(3), 583–588.

Rose, J. C., Armelagos, G. J., & Lallo, J. W. (1978). Histological enamel indicator of childhood stress in prehistoric skeletal samples. *American Journal of Physical Anthropology, 49*(4), 511–516.

Sabarad, P., & Kalburge, J. (2014). Chronological age estimation using transparent root dentin: A stereomicroscopic study. *Journal of Advanced Medical and Dental Sciences Research, 2*(2).

Schour, I., & Massler, M., (1941). *Development of human dentition chart*, 2nd ed. Chicago: American Dental Association.

Shipstone, R. A., Young, J., Thompson, J. M., & Byard, R. W. (2020). An evaluation of pathologists' application of the diagnostic criteria from the San Diego definition of SIDS and unclassified sudden infant death. *International Journal of Legal Medicine, 134*(3), 1015–1021.

Singhal, A., Ramesh, V., & Balamurali, P. D. (2010). A comparative analysis of root dentin transparency with known age. *Journal of Forensic Dental Sciences*, 18–21.

Skinner, M., & Anderson, G. S. (1991). Individualization and enamel histology: a case report in forensic anthropology. *Journal of Forensic Science, 36*(3), 939–948.

Smith, T. M., Tafforeau, P., Reid, D. J., Pouech, J., Lazzari, V., Zermeno, J. P., & Hublin, J. J. (2010). Dental evidence for ontogenetic differences between modern humans and Neanderthals. *Proceedings of the National Academy of Sciences, 107*(49), 20923–20928.

Snoddy, A. M., Buckley, H., King, C., Kinaston, R., Nowell, G., Gröcke, D., & Petchey, P. (2020). 'Captain of all these men of death': an integrated case study of tuberculosis in nineteenth-century Otago, New Zealand. *Bioarchaeology International, 3*(4), 217–237.

Snoddy, A. M. E., Miszkiewicz, J. J., Loch, C., Tromp, M., & Buckley, H. R. (2021). An image analysis protocol for the quantification of interglobular dentine in anthropological tooth sections. *American Journal of Physical Anthropology, 174*(1), 144–148.

Stott GG, Sis RF, Levy BM (1982). Cementum annulations as an age criterion in forensic dentistry. *Journal of Dental Research*, 61, 814–817.

Tsuchiya M., Sasano Y, Kagayama M, Watanabe M (2002). The extent of odontoblast processes in the dentine is distinct between cusp and cervical regions during development and aging. *Archives of Histology and Cytology* 65: 179–88.

Ten Cate, J. M. (1994). In situ models, physico-chemical aspects. *Advances in dental research, 8*(2), 125–133.

Ten Cate, A. R. (1998). *Oral histology: development, structure, and function*. Mosby, St. Louis; Toronto.

Teivens, A., Mörnstad, H., Norén, J. G., & Gidlund, E. (1996). Enamel incremental lines as recorders for disease in infancy and their relation to the diagnosis of SIDS. *Forensic Science International, 81*(2–3), 175–183.

Ubelaker, D.H. (1978). *Human skeletal remains: excavation, analysis, interpretation*. Aldine Publishing Co. Inc., Chicago.

Vasiliadis, L., Darling, A. I., & Levers, B. G. H. (1983a). The amount and distribution of sclerotic human root dentine. *Archives of Oral Biology*, *28*(7), 645–649.

Vasiliadis, L., Darling, A. I., & Levers, B. G. H. (1983b). The histology of sclerotic human root dentine. *Archives of Oral Biology*, *28*(8), 693–700.

Weale, M. C. (2023). The Life course of Vitamin D: the risk of deficiency in British Archaeological Populations. PhD Thesis. University of Bradford (United Kingdom).

Weale, M. C., Koon, H. Faber, E. W., Aris, C., Buckberry, J. (2023). Half of a whole: Optimising sample preparation of histological thin sections for the study of archaeological human enamel and dentine. *Dental Anthropology Journal*, in review.

White, T. D., Black, M. T., & Folkens, P. A. (2011). *Human osteology*. Academic Press, London, UK.

Whittaker, D. K., & Richards, D. (1978). Scanning electron microscopy of the neonatal line in human enamel. *Archives of Oral Biology*, *23*(1), 45–50.

Wittwer-Backofen, U. (2012). Age estimation using tooth cementum annulation. *Forensic Microscopy for Skeletal Tissues: Methods and Protocols*, 129–143.

Wittwer-Backofen, U., Buckberry, J., Czarnctzki, A., Doppler, S., Grupe, G., Hotz, G., & Weise, S. (2008). Basics in paleodemography: a comparison of age indicators applied to the early medieval skeletal sample of Lauchheim. *American Journal of Physical Anthropology*, *137*(4), 384–396.

Wittwer-Backofen, U., Gampe, J., & Vaupel, J. W. (2004). Tooth cementum annulation for age estimation: Results from a large known-age validation study. *American Journal of Physical Anthropology*, *123*(2), 119–129.

Witzel, C. (2014). Echoes from birth–Mutual benefits for physical and forensic anthropology by applying increment counts in enamel of deciduous teeth for aging. *Anthropologischer Anzeiger*, *71*(1–2), 87–103.

Zanolli, C., Bondioli, L., Manni, F., Rossi, P., & Macchiarelli, R. (2011). Gestation length, mode of delivery, and neonatal line-thickness variation. *Human Biology*, *83*(6), 695–713.

Zazvonilová, E., Velemínský, P., Černíková, A., Danielisová, A., & Brůžek, J. (2022). Protocol matters: A need for standardized procedure in cementochronology. *Forensic Science International, 340*, 111439.

Human versus Non-Human
Bone Microstructure in Species Differentiation and Experimental Research

7

VICTORIA M. DOMINGUEZ

7.1 Introduction

The ability to distinguish between human and non-human bone plays a fundamental role in forensic casework and in bioarchaeological analyses (Porto, Flavel, Maggio, & Franklin, 2023), both of which depend on accurate identification to interpret a scene and/or extrapolate past events. For a skilled osteologist, the presence of whole bones or even large bone fragments makes species identification, or at least the differentiation of humans from other species, a relatively straightforward affair. However, in cases of high fragmentation, commingling, and/or other taphonomic alteration, gross morphology alone is often insufficient to make a reliable determination. The addition of histological analysis of bone's microstructure in many cases resolves this issue in a fairly quick and cost-effective way. Additionally, non-human skeletal material is often selected as a proxy for human bone in experimental testing in anthropological studies, where considerations like composition differences across species may significantly impact study outcomes. While non-human samples provide invaluable information, understanding the material and structural composition of the samples utilized in testing is essential to ensure that the results of these studies actually apply to humans.

Chapters such as this one are often a litany of the distinct or not so distinct features of cortical bone microscopically observed in different species. While absolutely vital information, bone's microstructure in any species does not change as quickly as articles are published or books updated. Therefore, rather than providing another such list of characteristics by species, something which has been done both more thoroughly and eloquently by others (Enlow & Brown, 1956, 1957, 1958; D. M. Mulhern & Ubelaker, 2011), this chapter takes a slightly different approach. The goals here are as follows: (1) to briefly lay out the evolutionary history and origins of skeletal tissues as they are currently understood, (2) to provide an introduction to bone microstructure in the context of human and non-human growth and development, (3) to present an overview of existing methods and related considerations for distinguishing between human and non-human bone at the microstructural level, and (4) to illustrate the importance of microstructure for selecting appropriate non-human models for experimental research.

7.2 The Origins of Skeletal Tissue

Mineralized tissues exist in many forms across the animal kingdom, as well as more specifically among vertebrates. To understand how and why tissues like enamel, dentin, and bone demonstrate both substantial material similarity and significant structural variation

 DOI: 10.4324/9781003385608-7

in extant vertebrates, it helps to look back at their earliest forms in the fossil record. Biomineralization likely arose independently on several occasions in different animal lineages by co-opting a variety of gene sets and regulatory networks involved in metabolism from a soft-bodied metazoan ancestor during the Cambrian Explosion (Murdock, 2020). A variety of hypotheses based on either physiology or function have been proposed to explain why this widespread and sudden uptick in mineralized skeletons occurred during this time, but in vertebrates the evidence points to the rise of mineralization being tied to feeding in conodonts (Doherty, Ghalambor, & Donahue, 2015; Donoghue & Sansom, 2002). Mineralized tissues and eventually mineralized skeletons would later be co-opted in turn for other purposes, such as protection and structural support and mobility, especially in vertebrates. This may be tied to a compositional difference in vertebrate versus invertebrate mineralization. In vertebrates, unlike our invertebrate fellows who use calcium carbonate, the inorganic component that comprises bone is calcium phosphate in the form of hydroxyapatite. Calcium phosphate is more insoluble and thus better adapted to the acidic environment produced by vertebrate muscle activity, as well as creating a phosphate store that could improve energy metabolism overall (Ruben & Bennett, 1987).

Bone as we know it today likely originated from aspidin found in jawless fish with dermal scales, as the mineralized skeleton in vertebrates predates jawed fish (Donoghue & Sansom, 2002). Though the nature of aspidin has long been debated, three-dimensional analysis recently established aspidin as acellular, intramembranously derived bone due to morphology that indicated the presence of collagen fiber bundles that acted as a scaffold for mineralization (Keating, Marquart, Marone, & Donoghue, 2018). This suggests that the full range of bone cell and tissue types were present prior to the divergence of skeletonizing vertebrates and that the pleisomorphic condition for vertebrates is acellular bone (Keating et al., 2018). Why then is cellular bone the majority in extant vertebrates? Cellular bone, meaning bone with osteocytes, arose later and may have been favored because it established a mechanism to aid in mineral homeostasis via osteocytic osteolysis, a form of demineralization that could meet the physiological demands for phosphorous by withdrawing it from the skeleton (Haridy et al., 2021). Additionally, evidence suggests that mineralized tissue repair, specifically for bone, aspidin, and dentin, is also ancestral for vertebrates and that there is a link between bone's healing and remodeling abilities (see below for further discussion) (Herbst, Doube, Smithson, Clack, & Hutchinson, 2019).

7.3 Bone Structure, Growth, and Development

Given the long evolutionary history of our hard tissues, it should come as no surprise that among extant vertebrates the processes that form and alter bone are generally comparable. Three types of bone cell control most of bone's activity, though we will only touch on them briefly here. Osteoblasts, derived from mesenchymal stem cells, are bone depositing cells that lay down osteoid that eventually ossifies to become bone. A portion of these osteoblasts become embedded within the matrix they deposit, entombing themselves in a lacunar space and growing dendritic processes that extend out through the bone's fine canalicular network. These embedded cells are called osteocytes and they communicate with each other via their dendritic extensions to respond to mechanosensory signals and to maintain the skeleton. Finally, derived from hematopoietic stem cells, osteoclasts are multinucleated, phagocytotic cells that remove bone, either altering bone's overall shape or clearing out damaged or otherwise degraded tissue to make way for osteoblasts to lay new bone. These cell types work in concert to grow and maintain the skeleton throughout life in vertebrate species. However, the structural organization of bone tissue that results from these coordinated efforts varies substantially depending on the demands of the organism.

Bone is generally classified as either cortical or trabecular, distinctions based on structural organization since the materials composing these two types are essentially the same. Cortical bone is the dense tissue that forms the outer surfaces of bones, making up the majority of the skeleton and carrying the brunt of the mechanical load. Trabecular bone, sometimes referred to as spongy bone because of its lattice-like appearance, exists within the surrounding cortical bone and is most prevalent in areas like the vertebral bodies, between the cortical layers of the flat bones, and at the joint ends of the long bones, organized into supportive struts for the mechanical loads placed on the body. Though cortical and trabecular bone is found in all species, particularly in the diaphyses of long bones, humans tend to have more trabecular bone than non-humans. With this in mind, for the rest of this chapter "bone" will refer to cortical bone, which has been the focal point of nearly all the research completed to date comparing humans and non-humans.

In humans, fetal and infant bone structure is composed largely of primary woven bone. Characterized by a highly disorganized collagen matrix, woven bone's haphazard schema results in a material that is structurally weak, but has the advantage of being very rapidly laid down. This allows for the very quick growth seen in human infants with limited risk to their survival since humans require a high level and relatively long period of parental care. Infants are regularly carried and fed for most of the first year of life, abating the immediate need for a skeleton able to support ambulation and mobility. Instead, woven bone forms the majority of the initial skeleton and establishes a scaffold upon which the more structurally sound, but also more time and energy intensive primary lamellar bone builds, forming the majority of the body's cortical bone by the time a child is 2–3 years of age (Shapiro & Wu, 2019). Woven bone also appears in the adult skeleton, but usually in specific instances of trauma or pathology. In fracture healing, woven bone replaces the fibrocartilaginous callus, further stabilizing the injury site and allowing a relatively quick return to mobility; while in pathological conditions, proliferative lesions often form from woven bone.

Primary lamellar bone replaces woven bone via remodeling, creating materially and structurally more sound skeletal tissue. The primary distinction between woven and lamellar bone lies in the deposition and organization of the bone's collagen component. In lamellar bone, collagen deposition occurs in organized layers laid at right angles to each other that confer increased mechanical strength to the tissue. Vascularization in this more structured tissue occurs as primary vascular channels and/or primary osteons, the latter of which form when circumferential lamellae are deposited around a blood vessel in woven bone (Shapiro & Wu, 2019). Primary lamellar bone forms the majority of the juvenile skeleton, shaping the bones of the body through apposition and cortical drift. Lamellar deposition via these processes is most active during growth, but primary lamellar deposition continues reshaping the skeleton in a reduced capacity throughout life. As such, areas of lamellar apposition can be seen in regions of the periosteal surface of long bones well into adulthood and old age.

Growth patterns in non-human species follow a similar trajectory, though depending on the size of the species in question, the tissue replacing primary woven bone may differ significantly in structure. Many large-bodied species, such as artiodactyls and perissodactyls, form bone that is both woven and lamellar resulting in a woven-parallel complex, also called a fibrolamellar complex (de Buffrénil & Quilhac, 2021). The terminology and associated descriptions surrounding this bone tissue are complicated, but in this chapter, we opt to use the term "fibrolamellar bone" for consistency with most current forensic and bioarchaeological research. Fibrolamellar bone has several distinct and identifiable vascularization patterns thought to reflect the quicker bone formation rates of these species (de Buffrénil & Quilhac, 2021) following Amprino's rule (Amprino, 1947) that suggests that primary bone microstructure reflects formation rate. These specific patterns result in qualitative differences described later in this chapter as a means of differentiating human from non-human bone. This is because fibrolamellar bone in the human skeleton is vanishingly rare, disappearing if and when present within the first year of life to be replaced by the lamellar bone more characteristic of humans (Caccia et al., 2016). But what makes these other species and humans so different, enough for substantial differences in tissue types to arise? Unlike humans, a great many non-human species need to be mobile relatively soon, if not immediately, after birth. In such cases, rapid skeletal deposition must be matched with a structure capable of supporting and levering the body's musculature. The fibrolamellar complex achieves this feat.

In either lamellar or fibrolamellar bone, as individuals age, lamellar bone in the form of secondary osteons replaces these primary tissues, first in specific regions and eventually throughout individual bone cortices and the skeleton as a whole. Secondary osteons, also called Haversian systems, are tubular structures composed of several layers of lamellar tissue that surround a central Haversian canal through which passes a blood vessel. These concentric layers of lamellae form when osteoblasts deposit and bury themselves in osteoid about one cell thick, with subsequent layers infilling the tunnels left by osteoclasts removing damaged tissue. Essentially, these concentric layers nest one inside the other, leaving open a central canal. They can be distinguished from primary osteons by the presence of a scalloped reversal line, indicative of the osteoclastic activity that initiated their formation. Osteons form mostly parallel to the long axis of the bone, appearing as complex branching structures when observed in three-dimensions (Maggiano et al., 2016) but as a series of circular structures when observed in transverse, two-dimensional cross-sections. Remodeling in humans begins early in life, but constitutes only a limited portion of the total bone mass of the skeleton in childhood during which primary lamellar bone dominates. With age, as growth slows and modeling activity diminishes, remodeling begins to occupy more and more cortical bone, replacing the primary lamellar bone and eventually even other remodeling events, becoming the predominant bone type in most regions of the skeleton and known as dense Haversian bone (see Figure 7.1). Remodeling also occurs in species with woven-parallel bone, but often is less pervasive than in humans.

The structural arrangement of bone tissues as briefly described here varies not only from species to species, but from element to element within an individual and from area to area of the same bone. For this reason, a thorough understanding of micromorphological variation needs to consider factors such as element, age, and even life history when possible, in order to provide useful data for species determination. Also, it should be noted that what has been described in this section as the "human" patterns of growth and development more accurately describe primate patterns of growth and development. Because forensic and bioarchaeological work by their nature are usually human-centric, researchers often default to describing these factors as human or non-human. But our common lineage with

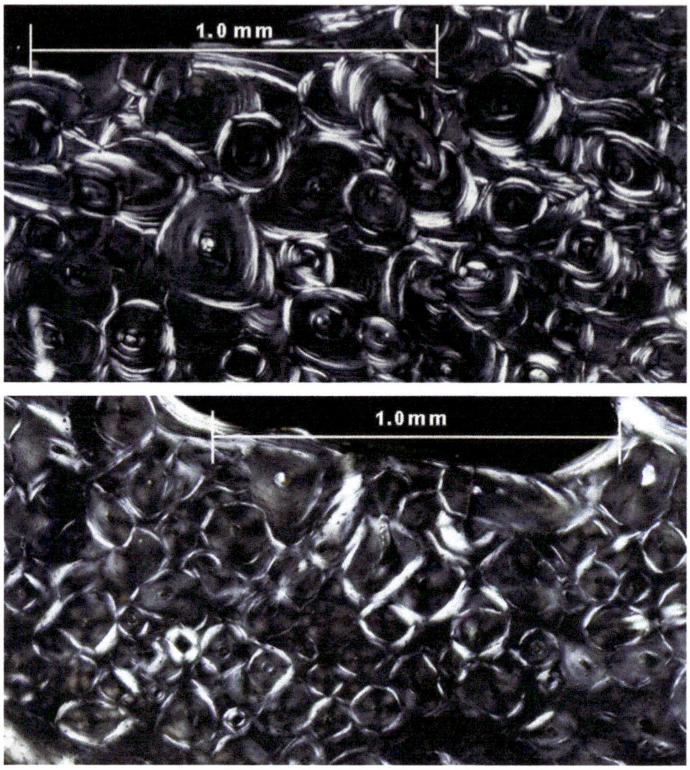

Figure 7.1 Dense Haversian bone showing a crowded cortex filled with remodeling events from a human rib (upper image) and from a deer rib (lower image).

non-human primates and the fact that a significant proportion of bone structure is determined by phylogeny means that both macroscopically and microscopically all primates are more similar to each other than to any other species. Further discussion of the similarities and differences between humans and other primates can be found in the "Experimental Studies" section later in this chapter.

7.4 Species Identification in Forensics and Bioarchaeology

The practical differentiation of human from non-human skeletal remains contributes to a major proportion of the work carried out both in forensic cases and in bioarchaeological settings. In forensic cases, identification of remains to at least to the level of human versus non-human composes a significant portion of the practitioner's casework, as much as 25% in some offices in the United States (C. M. Crowder, Wiersema, Adams, Austin, & Love, 2016). For archaeological settings, distinguishing not only human from non-human, but making identification to specific faunal species often plays a major role in being able to properly interpret site context (French, Crowder, & Crabtree, 2022). While experienced osteologists face little trouble making such determinations when presented with intact skeletal elements, cases of high fragmentation and/or other modifications (such as burning) benefit from more advanced techniques.

Methods such as DNA and/or protein analyses can provide excellent results (Lowenstein et al., 2006), but the inherent issues with these approaches are well known. As destructive techniques, they can often consume whole samples when applied to fragmented material. Furthermore, the expenses both in terms of time and cost are high, making such approaches unfeasible, particularly in large bioarchaeological assemblages. Finally, material preservation plays a huge role in the utility of these approaches, as degradation of bone's organic components means that in some cases, there will be no results even if such testing can be arranged. Histology shares some of these difficulties. It too is destructive and subject to preservation challenges, but sectioning rarely consumes the whole sample and even in cases of poor preservation or diagenetic change, something often remains observable. In addition, histological methods cost less and generate quicker results. As such, histological studies of bone microstructure form a significant portion of the literature on species differentiation.

The remainder of this section provides an overview of the methods and techniques described in these works. Though by no means exhaustive, both commonly applied and validated approaches for identification of human versus non-human bone are presented, as well as a brief consideration of the underpinning biology. Methods tend to fall into one of two categories: qualitative and quantitative. Qualitative methods depend on the observation of bone's microarchitecture, specifically looking for features that are unique or rare to particular species/groups. Quantitative methods measure microstructural features common across multiple species and use mathematics to identify samples to particular species/groups. Finally, the section concludes with specific considerations that may influence method selection and utility depending on the samples in question.

7.4.1 Qualitative Methods

Both early and contemporary methods for distinguishing species rely upon, at least in part, understanding what constitutes typical bone microstructure in each species. Comparative analysis of multiple species led to qualitative descriptions and corresponding illustrations that created a framework for describing bone tissue types and facilitating species comparisons (see "Bone Structure, Growth, and Development" above). Among the earliest comparative reviews of differing non-human microstructures, Quekett gave a paper generally comparing histological bone sections of mammals, birds, reptiles, and fish in 1846 (Quekett, 1849). In 1916, James Foote published one of the broadest comparisons of bone microstructure across species, building on work he began several years earlier. Interested in bone's structural variation and its significance, Foote examined approximately 600 cross-sectional slides of the femur from dozens of species, including amphibians, birds, mammals, and reptiles. These efforts culminated in the publication of a monograph that included 467 illustrations and their corresponding descriptions. Several decades later, Amprino and Godina (1947) took a similar comparative approach, examining about 350 samples of vertebrates from amphibians to humans, and noted that while adult bone structures vary considerably, the processes by which bone grows and restructures with age are comparable across many species. And just a few years later, Enlow and Brown (1956, 1957, 1958) proposed a system for the classification of major bone structural patterns and published a histological survey using this system to describe the major vertebrate groups, but also expanded on the work of their predecessors by including both fossil and modern specimens. These expansive studies illustrate that while bone's component parts are largely consistent across the animal kingdom, their structural organization is anything but, leading to the identification of certain diagnostic features useful for distinguishing species, particularly human from non-human.

While being able to identify each individual species would be ideal, considering the multitudes that make up the subphylum Vertebrata, it is not surprising anthropologists, most of who are interested in human remains, focus their efforts largely on comparisons with other mammalian species. Similar evolutionary histories give rise to more similar skeletal organization. Furthermore, in most environments where humans live it is the remains of other large-bodied mammals that are the most likely to be mistaken for humans, especially in circumstances of commingling. Domesticated species, such as farm animals like bovids and equines and pets like canids and felids, inspire the lion's share of this research. Their long coexistence with humans means that they frequently occur in bioarchaeological contexts, from burials to hearths, and provide valuable information not only about animal domestication and husbandry, but the activities, nutrition, and even beliefs of past peoples. In forensic settings, these same species compose a large percentage of human versus non-human casework for the same reason. Methods focused on non-domesticated species tend to vary by region, reflecting variation in local wildlife by environment. For example, deer and other cervids overlap with humans in many wooded areas in North America, making them a group of particular interest. All things considered, however, the number of defining qualitative traits for species identification is limited.

As mentioned earlier, bone microstructure reflects evolutionary history and growth trajectory. The most commonly cited qualitative trait for distinguishing human from non-human is the presence of fibrolamellar bone. As described above, fibrolamellar bone presents with a number of different vascular patterns. Reticular bone possesses primary osteons with no particular orientation, while in radial bone primary osteons run perpendicular to the long axis of the bone (de Buffrénil & Quilhac, 2021). Laminar bone forms with circumferentially oriented vascular channels providing the appearance of several stacked layers and is most often found in large land mammals (Hillier & Bell, 2007). Finally, there is plexiform bone. Plexiform strongly resembles laminar bone, but with a far greater number of radial anastomoses (de Buffrénil & Quilhac, 2021) that create a characteristic "brick-wall" appearance (see Figure 7.2). Resulting from the dual need for strong but rapidly laid down bone, plexiform comprises the majority of long bone tissue in large bodied species. Plexiform also occurs in carnivores but has only been rarely observed in humans, all of whom were infants (Caccia et al., 2016; Enlow, 1963). As a result, the presence of plexiform bone in any bone or bone fragments is widely accepted as exclusionary criteria for the identification of a sample as human. Other than the plexiform vascularization pattern of fibrolamellar tissue, there are currently no other qualitative primary bone features accepted to definitively discriminate humans and non-humans, though there are two secondary features that can aid in distinguishing between the dense Haversian bone of species.

Osteon banding refers to a distinct row formed by at least five primary and/or secondary osteons (D. M. Mulhern & Ubelaker, 2001) (see Figures 7.3 & 7.4). In non-humans, osteon bands appear in rows with multiple bands of differing lengths that are sometimes surrounded by layers of primary lamellar bone (D. M. Mulhern & Ubelaker, 2001). Osteon bands can and do occur in humans as well, though previous research indicates they tend to be isolated and shorter than those seen in non-humans (Cummaudo et al., 2018; D. M. Mulhern & Ubelaker, 2001). The only three-dimensional study of osteon banding to date used high resolution synchrotron radiation-based micro-CT to look at 43 areas in

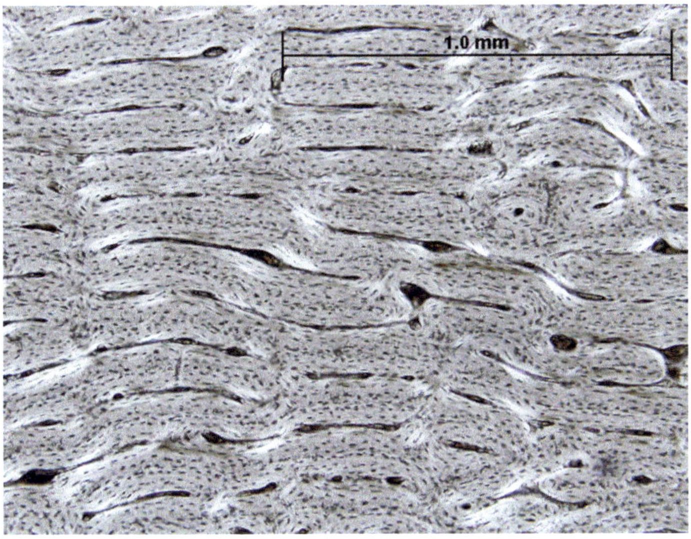

Figure 7.2 Plexiform bone from a pig humerus, illustrating the brick-wall appearance created by the vasculature in this fibrolamellar bone type.

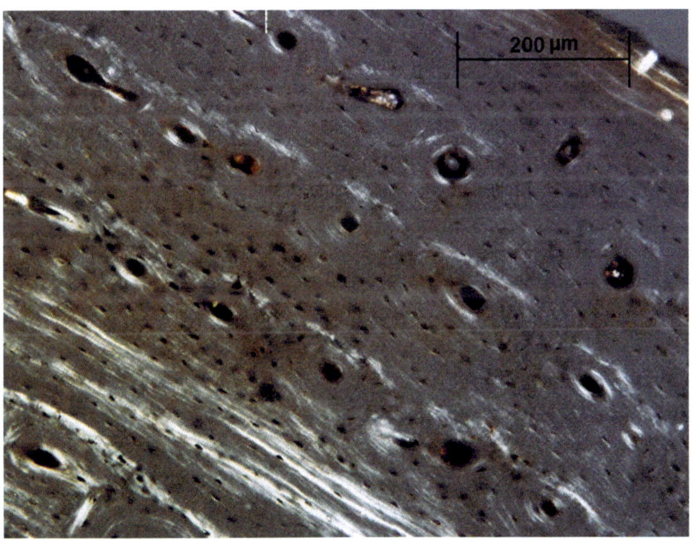

Figure 7.3 Two osteon bands formed by primary osteons in the femoral cortex of a dog.

three human subjects and found that linear osteonal structures do occur throughout the skeleton (Andronowski, Pratt, & Cooper, 2017). While the authors of this study caution against using osteon banding alone as a diagnostic non-human trait, their data generally support the existing consensus that multiple, long bands in humans are uncommon. None of their sample displayed more than three osteon bands, with the majority displaying only one or two, and most bands were composed of 5–7 osteons each, with just one band from a temporal bone breaking into the double digits with 12 osteons (Andronowski et al., 2017). Therefore, the presence of multiple osteon bands, particularly longer ones, remains a relatively reliable indication that a sample is non-human. And this greater tendency toward

Figure 7.4 One osteon band formed by secondary osteons and surrounded by lamellar tissue in the humeral cortex of a dog. The secondary osteons are distinguished from the primary osteons by the presence of a reversal line around each remodeling event.

multiple, long bands of osteon in non-human species likely arises from the remodeling of fibrolamellar bone, particularly laminar and plexiform bone types, that possesses vascular patterns distinctly different than that observed in humans in most cases.

Besides the presence of osteon bands, in rib or long bone samples composed of Haversian bone, osteon variants may also serve as species indicators. Recent work proposes the presence of drifting, or waltzing, osteons as a qualitative signifier of human bone (French, Mavroudas, & Dominguez, submitted). Of the limited research that examines osteon variants in non-humans, drifting osteons are noted as rare (Skedros, Sorenson, & Jenson, 2007), unlike in human bone, where they have been frequently observed throughout the skeleton (Cummaudo et al., 2018; Epker & Frost, 1965; Robling & Stout, 1999; Streeter, 2010). Drifting osteons are secondary osteons that appear to move through the cortex, having a circular Haversian canal and forming a long, hemicyclical tail behind them (C. Crowder, Dominguez, Heinrich, Pinto, & Mavroudas, 2022) (see Figure 7.4). Though highly variable in their morphology (Robling & Stout, 1999), for the purposes of species identification drifting osteons have a tail that is at least twice as long as the maximum diameter of the osteon itself. Following a decision tree, positive identification of human bone occurs if: (1) the sample is from a rib or long bone, (2) the sample is unlikely to be from a non-human primate given context, (3) more than two drifting osteons are noted, and (4) the prevalence of drifting osteons is >= 1% of the total remodeling observed (French et al., submitted). Unlike other qualitative traits in common use, the presence of multiple drifting osteons indicates a human rather than a non-human origin. However, what drifting osteons really indicate are bones of primate origin, as non-human primates also display high numbers of drifters (French et al., submitted). Though the etiology of drifting osteons remains unclear, with arguments for mechanical (van Oers, Ruimerman, van Rietbergen, Hilbers, & Huiskes, 2008) and/or metabolic (Coutelier, 1976) impetuses, their high incidence across primates compared to other mammalian orders suggests that further study may provide some evolutionary evidence for either or both arguments.

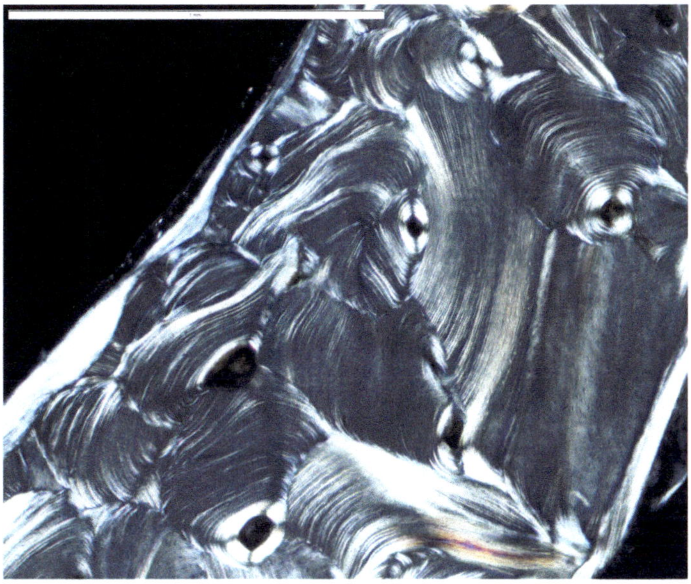

Figure 7.5 Several drifting osteons in the cortex of a human rib. Drifting osteons are distinguished from a typical secondary osteon by the presence of a long, hemicyclical trail that creates the appearance that they are moving through the bone.

7.4.2 Quantitative Methods

As described above, Haversian bone forms when remodeling replaces primary bone with secondary osteons. Remodeling processes in humans and non-humans are essentially the same and result in similar microstructure. Histomorphometric approaches to species differentiation arose primarily to deal with non-distinct, dense Haversian bone samples lacking the obvious non-human indication of plexiform tissue. Though osteon banding and drifting osteons may occur, neither trait is currently as widely accepted as definitive for species identification as plexiform bone, highlighting the need for further validation studies. Furthermore, it is common for large proportions of fragmented material to lack any of these distinct features, leaving only a field of non-descript Haversian systems from which to determine species origin.

As the most visible features of bone's microstructure, secondary osteons and their Haversian canals became and remain the focus of the greatest share of histomorphometric research. Most often, sections are imaged for data collection, which often includes measurements such as area, perimeter, and min/max diameter in microns. Preliminary studies demonstrated that differences in osteon and canal dimensions between species do exist (Jowsey, 1966; Singh, Tonna, & Gandel, 1974), as well as the fact that body mass influences osteon dimensions so that smaller species have smaller osteons than bigger species (Jowsey, 1968). More recent work suggests these differences in osteon and canal sizes are allometric, with measurements in large species being absolutely bigger but relatively smaller than in other species when controlling for differences in body mass (Felder et al., 2017). Considering that remodeling creates transient porosity when an osteoclast tunnels through the cortex, limiting maximum osteon size in small cortices reduces the risk of fracture, whereas osteon size in larger species may be limited by osteocytic demands. With these size differences in mind, several researchers have proposed methods that apply discriminant function analysis to osteon and Haversian canal metrics to differentiate species with varying degrees of success (Cattaneo et al., 1999; Martiniaková, Grosskopf, Omelka, Vondráková, & Bauerová, 2006; Urbanova & Novotny, 2005). Cattaneo et al.'s (1999) method specifically deals with burned bone, which can alter histomorphometry and thus makes this approach less reliable for unburned material (see "Special Considerations" below for further discussion on this matter). Urbanova and Novotny (2005) presented their method with a 100% correct classification, but only when incorporating the additional metric of cortical thickness, a variable demonstrated to vary across species, but that proved insufficient on its own to make reliable species identifications (Croker, Reed, & Donlon, 2016). Martiniaková et al. (2006) used only osteon and Haversian canal metrics, producing a correct overall classification rate of 76% of samples to one of five specific species; correct classification rates for the human subsample was 100%. A recent validation study found that Martiniaková et al.'s (2006) method had only a 32% correct classification rate due largely to the fact that everything tested classified as human, however this may be due to issues of method application, as well as the fact that several species used in the validation were not included in the original study (Lagace, Schmitt, Martrille, Benassi, & Adalian, 2020). Finally, it should be noted that though distinguishable, considerable overlap exists in terms of osteon size, particularly in larger-bodied species, with several studies finding that there are no significant differences for this metric in species as large as or larger than humans (Jowsey, 1968; Zedda & Babosova, 2021).

In addition to size, another variable often considered in species identification is shape. The rise of digital imaging and analysis greatly facilitated shape analysis of osteons, which first drew interest as a potential indicator/estimator of age for humans in the forensic context (Britz, Thomas, Clement, & Cooper, 2009; Currey, 1964; Goliath, Stewart, & Stout, 2016). The most often utilized shape descriptor is circularity, which calculates how near or far the dimensions of each osteon are relative to a perfect circle. By tracing the reversal line of a secondary osteon, its shape can be described as either more round or more elongated. While circularity appears to have some discriminating power between humans and non-humans (Crescimanno & Stout, 2012), shape alone does not appear sufficient to consistently differentiate among non-human species (Dominguez & Crowder, 2012; Keenan, Mears, & Skedros, 2017). As stated by Keenan et al. (2017), this indicates that species, as well as strain-mode distribution and load-complexity category, do not explain osteon shape variation and some other factor(s) must be involved. Though we lack a full understanding of what drives shape differences in the osteons of humans and non-humans, combining circularity with other histomorphometric variables may improve the robusticity of equations and predictive outcomes in cases where the goal is simply to identify human remains from a mixed assemblage. Dominguez and Crowder (2012) carried out discriminant function analyses that utilized both osteon area and circularity. Data illustrated that osteon's size and shape are largely independent of each other, with area alone correctly classifying samples in 92% of cases and circularity alone in 76%. However, when combined, the variables differentiated human from non-human in their sample in all cases. Though the study examined only three species (deer, dog, and human), a recent validation with a larger variety of non-humans still demonstrated correct classification to human and non-human groups in 86% of cases (Lagace et al., 2020). The misclassified non-humans were species not included in the original study's sample, cow and horse, which happen to be species with considerable histomorphometric overlap with humans (Lagace et al., 2020).

In addition to histomorphometry of secondary osteon and Haversian canals, a few other quantitative methods using these and other variables merit consideration. Similar to the histomorphometric approaches already described, a preliminary study comparing the osteocytic lacuna of humans and pigs found pigs to have smaller and more circular lacunae than humans, suggesting that lacunar data could further aid in species identification (Cummaudo et al., 2019). Furthermore, remodeling density, which is essentially a count of secondary osteons in a given area, may aid in distinguishing human from some non-human species in a general sense. Most vertebrates other than birds and mammals tend not to remodel (de Buffrénil & Quilhac, 2021), and even among these two exceptions, body size plays an important role. Small birds and mammals, such as rodents, typically show no or very limited intracortical remodeling activity (Taguchi & Lopez, 2021). Large-bodied birds and mammals regularly undergo much greater amounts of bone remodeling, and while particularly in cases of fragmentation this can lead to confusion about origin, groups like the ungulates tend to have lower remodeling densities over the entirety of their bone cross-sections than members of Carnivora or Primates.

7.4.3 Special Considerations

When presented with a bone fragment of unknown origin, an osteologist's first step is to attempt identification using macroscopic methods. We have attempted to establish that if that fails, histological methods provide an alternative course of action that begins with assessing the overall microstructure of the fragment. Depending on the observable features, either qualitative or quantitative methods may be applied to make a determination, though most often using these approaches in conjunction with one another provides the best results (Hillier & Bell, 2007). As with any kind of analysis, however, there are some important factors to consider to ensure that the methods selected are appropriate to the samples in question and thus will provide accurate determinations. In this section, we review specific factors that may influence the methods and comparative samples selected, as well as any interpretation when attempting to make a species determination.

As alluded to earlier, a major consideration for selecting the appropriate method for species determination, particularly when using histomorphometric techniques, is an awareness of local fauna. In both bioarchaeological and forensic contexts, familiarity with species endemic to where skeletal remains are recovered narrows the initial pool of animals that need to be considered when making a species determination. Furthermore, there exist so many species that it is impossible and impractical for any one method to adequately sample all possible fauna. Instead, most studies concentrate on common domesticated species (Martiniaková et al., 2006), wildlife native to specific areas (Hidaka, Matsumoto, Ohsako, Toyoshima, & Nishinakagawa, 1998; Owsley, Mires, & Keith, 1985), or whatever species are available (Singh et al., 1974). Having some sense of what species are most likely to be found in an area can make it easier to find appropriate comparative descriptions and data, if available. Even when things can be narrowed to a single species, variation within a species can be a confounding factor in analysis. Some animals may have very large ranges that include various environments and others may consist of several sub-species, both such conditions warranting care and requiring validation that data for individuals from one area apply to those from another because existing variation has been adequately sampled.

Where a sample comes from within the body matters nearly as much as its geographic origins. All of the methods described in this chapter examine cortical bone, largely because trabecular bone preserves poorly and is generally viewed as less informative in this context. Not all cortical bone is created equal, however. Significant microstructural variation exists not only between different types of skeletal elements, for example between long bones and the flat bones of the skull, but also between regions of the same element. Both mechanics and metabolism lead to variegated responses that alter microarchitecture across the hierarchical levels of the skeleton. Because of the ubiquity of long bone fragments, which are often the hardest thing to place to the species level, many existing methods are based on data from long bones. The likelihood of misclassifying cranial fragments when using long bone derived data also increases because these elements have a very different cortical structure and merit more individual attention. Additionally, while some elements of the human skeleton have been the focus of many years of histological research, for example the ribs and femur, many other regions of the skeleton have been overlooked. Long bones other than the femur, tibia, and humerus have received relatively limited attention in the histological literature, and the same can be said of most elements in the hands and feet and the vertebral column. Even in well-researched elements like the ribs, much of the existing data looks only at mid-thoracic elements and focus on the midshaft portion of the bones. This underrepresents the variation within elements as multiple locations of the same bone demonstrate significant differences in microstructure (Dominguez, Harden, Wascher, & Agnew, 2020). Clearly there is more of a gap in our understanding of the range of variation within human bone microstructure then might be anticipated. In the majority of non-human species, even among other primates, this gap is a screaming chasm.

Another major consideration is age. As laid out above, growth and development of the skeleton play a determining role in the appearance of bone's microstructure. What the body demands from our skeletons in infancy and childhood differs markedly from the demands of adulthood. Generally, younger individuals of any species have higher proportions of primary bone tissue types that gradually become replaced with Haversian bone. And Haversian bone itself is not a "one and done" situation, as remodeling continues throughout life, with new bone replacing pre-existing osteons repeatedly. Evidence suggests that secondary osteons in humans become smaller and rounder as we age (Currey, 1964), throwing a wrench in the effectiveness of some quantitative methods as human microstructure becomes more similar to that of other species. Additionally, several studies demonstrate that plexiform bone exists in the skeleton of infant humans, and that other fibrolamellar types are observable within the first five years of life, corresponding to when humans experience their most rapid bone growth (Caccia et al., 2016). Lastly, while we know a lot about how the human skeleton changes throughout life, relatively little data about the state of non-human microstructural changes with age exist at all.

Lastly, in terms of inherent variation, factors such as sex differences, genetic influence, and pathology still require further study. The existence of sex-based differences in bone microstructure, at least in some aspects, remains uncertain. For example, differences in cross-sectional geometry and in cortical area between females and males are well known in humans (Seeman, 2001), but the relationship between sex and histomorphometric variables is less clear, with studies indicating both significant (Britz et al., 2009) and non-significant (Dominguez & Agnew, 2016) differences. The role of genetic factors in features of micromorphology, such as osteon size and shape, are only recently beginning to be explored in humans and non-humans alike (Bjornerem et al., 2015). And last but not least, pathological alterations of bone microstructure resulting from metabolic or other systemic disruptions can complicate assessments by changing typical bone structure. While there is evidence that metabolic disruption can alter remodeling rates, either decreasing (Fazeli & Klibanski, 2018) or increasing (Roodman & Windle, 2005) activity, the influence of pathology on bone microstructure merits further investigation. Overall, the field needs more comprehensive, systematic, and regional studies of microstructural variation within species that account for inherent factors to improve methods for differentiating between species.

Extrinsic factors also influence microstructural analysis and by extension successful species identification, in particular taphonomic processes that can substantially influence bone's microstructure. Unfortunately, as any bone histologist will tell you, pristine macro-morphology does not guarantee usable or diagnostic micromorphology. Diagenesis caused by bacteria and fungi can wreak havoc on and obscure bone microstructure (Hackett, 1981; Hollund et al., 2012). This creates difficulties for analysis if qualitative differences are lost due to physical alterations in the cortex, such as with microbial tunneling or mineralization changes that affect birefringence, as well as for metric methods that usually require quantification of a minimum bone area or number of osteons for reliable results.

Finally, the influence of burning on bone microstructure has received a lot of attention. Fire damage, whether in archaeological cremations or forensic house fires, often results in high fragmentation and commingling of skeletal material, and depending on the temperature and duration of the burn can damage and/or destroy the organic matter useful for DNA/protein analyses. In other words, this is the situation for which histological methods of species differentiation were built. While burns can sometimes obscure microstructure, in many cases, even when samples are completely calcined visible microstructure remains. Research demonstrates that qualitative features such as the presence of fibrolamellar bone remain equally reliable for distinguishing human from non-human bone in cases of burning (Simmons, Goodburn, & Singhrao, 2016). However, osteon and Haversian canal measurements demonstrate significant differences in burned and unburned samples, though there appears to be no pattern in terms of element or temperature (Mavroudas, Meckel, Gocha, Goldstein, & Garza, 2022). This indicates that the use of methods based on unburned tissue is not reliable for quantitative species assessment in burned tissue and vice versa. This appears to have been corroborated by a recent validation study that examined the work of Cattaneo et al. (1999), one of the first papers to apply discriminant function analysis for species identification but which actually used a burned sample to do so. The values presented in Cattaneo et al.'s (1999) original study for osteon and Haversian canals were found to be considerably smaller than those reported in the validation and other studies of unburned material (Lagace et al., 2020). Even allowing for the fact that these are different samples taken in different geographic areas and decades apart, the reported differences can reasonably be argued to reflect taphonomic alteration resulting from burning. The loss of bone's organic component through thermal alteration is the most likely culprit for these morphometric changes. As the extracellular matrix and the collagen component that give bone its elasticity evaporate under heating, the remaining mineral component experiences expansion and/or contraction resulting in permanent dimensional changes to the microstructure. Further research on microstructure response to post-mortem factors is warranted, but for the present analysts and technicians applying quantitative approaches should be careful when selecting methods and comparative samples for determining species.

As with any method, the more information available, the better the choices that can be made about what techniques to apply and how to interpret data. Though identification to species level is not always possible, the ability to distinguish human from non-human bone meets the needs of most anthropologists. Qualitative and quantitative approaches independently work well, but ultimately, combining these approaches leads to the greatest success in terms of accurate species identification. Furthermore, while many existing methods demonstrate good accuracy and significant promise, there is a major need for validation studies, particularly for methods intended for use in forensics that must meet higher evidentiary standards. Additionally, as laid out in this section, variation both within and across species is very high but underrepresented in the current literature. Expansion of existing methods and/or new studies that include a much wider representation of age, elements, origins, etc., particularly for non-human species, are a potential area of research worth exploring.

7.5 Non-Human Models in Experimental Studies

Though not as often directly addressed, anthropologists have a vested interest in non-human species that extends beyond our ability to differentiate them from humans for purposes of identification. Non-human animals play an essential role in a staggering proportion of experimental research in biological anthropology, particularly in forensic fields. Having long relied heavily on case studies and anecdotes (Dempsey & Blau, 2020), forensic anthropology increasingly faces pressure to meet evidentiary standards of admissibility that are scrutinizing the science and research applied to casework (Christensen & Crowder, 2009). This has led to a rise in controlled experiments, many of which use non-human models as a proxy for human bone. This occurs for a variety of reasons, most often due either to difficulty in acquiring or working with humans or human material for analysis and/or the nature of the questions being asked. Animal models in biomedical research date back millennia and remain a mainstay of current clinical research (Ericsson, Crim, & Franklin, 2013). In anthropology, despite shifting social attitudes in many areas of the United States and around the world that have resulted in an increase of voluntary human body donations for research (Habicht, Kiessling, & Winkelmann, 2018), the choice to use animal proxies both in in vivo and ex vivo studies stems from several factors, including but not limited to ethical considerations, a lack of appropriate facilities, demographic data, element availability, sample sizes, and cost. In many cases, the use of non-human models provides valuable and essential data, but in recent years, questions have arisen about the validity of such proxies for clinical studies, where human physiology might differ substantially (Robinson et al., 2019). The intent of this section is not to argue against the use of animal models or even to provide an overview of their use in anthropology. Rather, the objective is to highlight some areas of research that require an understanding of bone microstructure for both selecting appropriate experimental proxies and improving the interpretation of these outcomes.

Theoretically, it is well understood that bone's microstructure underpins and reflects its myriad functions, and as clearly demonstrated above, decades of research have established both similarities and differences between human and non-human bone. Practically these considerations are often overlooked or given only perfunctory mention when selecting animal models or interpreting experimental outcomes. But as demonstrated by biomedical and clinical research, extrapolating findings about one species to another should be a major concern. Though there are many instances when accounting for microstructure can improve research outcomes, here we provide just a few specific examples to illustrate how incorporating microstructural considerations can improve the basic science in some major areas of anthropological research.

We begin by looking at in vivo studies of bone remodeling and the necessary considerations for selecting appropriate animal models. Remodeling studies provide key data for methods of species determination and age-at-death estimation, but also inform interpretations of skeletal and/or metabolic health. Current methods for assessing remodeling via histology cannot be carried out in living subjects, so the use of human models for in vivo studies remains out of the question until technology improves or new approaches develop. This makes animal models essential for such research. While ethical and practical considerations also play massively important roles in determining research models, here we concentrate on the underlying biology for selecting non-human proxies.

In vivo studies, especially in biomedical and clinical settings, often use mice or other rodents because of their small size, quick reproductive cycle, and limited genetic variance, which allow for excellent intrinsic controls and longitudinal studies. But, while intracortical remodeling can be induced in rodents, they rarely experience spontaneous remodeling typical of humans (Taguchi & Lopez, 2021), making them poor models for understanding remodeling and remodeling rates in humans. Much of the early work on this topic relied instead on canine, whose remodeling dynamics are thought to more closely resemble humans, demonstrating very high levels of secondary remodeling that convert most of their cortices to dense Haversian bone with time (Jaworski & Lok, 1972). More recently, rabbit models have been utilized to study the effects of opioid use on remodeling rates, because their cortical bone dynamics more closely resemble humans than rodent models (Andronowski et al., 2021). In these instances, knowledge of human and non-human microstructure results in a more informative model since the skeletal processes of two species that spontaneously and extensively remodel are likely more similar to each other than to a species that does not.

In ex vivo bone trauma research, bone's mechanical properties are determined by material and structural factors, including microstructure. Cortical and trabecular bone vary primarily in structural organization and are often studied as distinct entities, but little if any consideration is usually given to the microarchitecture of the cortex, even though as discussed above, this also reflects adaptation to mechanical requirements. Many animal models commonly used for ex vivo trauma experiments (i.e. pig, deer, sheep/goat) have cortical bone composed largely of fibrolamellar tissue, unlike the primary lamellar or dense Haversian bone found in humans. When dense Haversian bone does occur in these large mammals, it often occupies a much smaller percentage of the cortex then in humans, which may lead to a differential bone mechanical response because the presence of extensive secondary osteons has been suggested to attenuate fracture propagation in humans (O'Brien, Taylor, & Clive Lee, 2005).

Not only microstructure, but many other variables of bone composition, density, and quality display significant interspecies variation and merit further consideration in discussions of model selection for trauma studies (Aerssens, Boonen, Lowet, & Dequeker, 1998). Knowledge of the underlying microstructure can also improve interpretations of tested material. For example, microstructure may explain the appearance and prevalence of specific features observed in fractographic analysis. Lacking data on this question, current guidelines for fractographic analysis recommend the use of human over non-human bone in experimental research (Christensen et al., 2022). This seems like a good recommendation for many ex vivo bone trauma studies, particularly those looking at questions of fracture propagation, for example, where microstructural variation in vasculature can substantially alter outcomes.

In post-mortem decomposition studies, microstructure plays a role in bone's response to environmental factors. The effects of fire on microstructure were described above and highlighted the need to analyze both human and non-human remains in comparable experimental conditions. This is true also in histotaphonomic studies of diagenesis, where the majority of experimental work looking at the role of microbial agents has been carried out in non-human species or in archaeological material lacking controls and decompositional data. In fact, a recent study looking at human remains in the context of histotaphonomy suggests that microstructural differences in bone vasculature such as that seen in fibrolamellar versus dense Haversian bone may influence microbial activity and thus histotaphonomy (Mavroudas, Alfsdotter, Bricking, & Madgwick, 2023). In this instance, knowledge of bone microstructure can establish an additional intrinsic control (microstructure itself) for experimental models.

Finally, non-human primates (NHPs) deserve some special consideration in terms of their relationship to humans and their role in experimental studies since we share a long evolutionary history and thus they are more similar to us in many respects than most other species. Because of our close phylogenetic ties, the use of NHPs for research is not new. In any number of fields, not just within anthropology, NHPs have for a long time been used as analogues for humans (Phillips et al., 2014). In terms of microstructure, NHPs demonstrate similar bone types to humans, lacking fibrolamellar bone and demonstrating high rates of remodeling similarly influenced by age, sex, and skeletal maturity (Havill, 2004; McFarlin, Terranova, Zihlman, Enlow, & Bromage, 2008; Dawn M. Mulhern & Ubelaker, 2009). As indicated above, differentiating human from non-human primates based on microstructure is not possible based on any currently available methods. However, the ethics of in vivo studies using NHPs is beyond the scope of this paper and from a practical perspective, it is probably easier to ethically source human donors than NHPs for most trauma and taphonomic studies.

Where NHPs may play an important and currently unparalleled role is in untangling the influence of genetics and inheritance on remodeling. Some research carried out in human twins demonstrates a fair amount of genetic control on bone remodeling (Bjornerem et al., 2015), but longitudinal and more extensive data is required to fully explore these relationships. For humans, the necessary longitudinal and skeletal data for such analyses is impossible to collect. For NHPs though, well-documented, closed populations like the Cayo Santiago rhesus macaques for whom genetic and skeletal data over several generations exists and continues to be collected, provide an incredible opportunity to study the potential links between microstructure and genetics (Widdig et al., 2016).

This section briefly illustrates some ways in which increased consideration of bone microstructure can benefit experimental research models. None of this implies that non-human models should never be used, but rather like age, sex, and mechanical loading that microstructure should also factor into decisions about and interpretations of these models. There is no universally appropriate animal model for research (Matuszewski et al., 2020), rather models should be based on the proposed questions (Taguchi & Lopez, 2021). Finally, non-human models play an essential role in developing methods and establishing proof of concept for new techniques, but for many questions, testing with human tissue remains essential to bridge the gap between species and to help meet evidentiary standards.

7.6 Conclusions

Though bioanthropologists place a considerable amount of effort and attention on the human skeleton, a thorough knowledge of non-human bone structure plays a pivotal role in casework and experimental studies. The common origin of mineralized tissue in the vertebrate line explains the significant overlap and resulting difficulties in differentiating human from non-human bone, but also establishes the appropriateness of non-human tissue as a proxy for humans in some experimental studies. Differing mechanical and metabolic requirements lead to differences in skeletal microstructure among different species that are best assessed using a combination of qualitative and quantitative methods, when possible. Even when identification to the species level is not possible, exclusions and/or the confident determination of "human" provide valuable information in both forensic and bioarchaeological contexts. However, many of the currently available methods would benefit from expansion of their test samples to include a wider array of species, elements, and ages. Even more important, validation studies of these methods would increase confidence in their application and are fundamental to meet minimum evidentiary standards in forensic casework. Finally, the experimental methods that provide the basic knowledge essential for such varied tasks as interpreting skeletal trauma and histotaphonomy or providing the biological reasoning that underpins methods of histological age estimation rely in large part on appropriate non-human proxies. As such, a comprehensive knowledge of bone microstructure in humans and our living and extinct brethren is fundamental to improving current and developing methods across many aspects of anthropological analysis.

References

Aerssens, J., Boonen, S., Lowet, G., & Dequeker, J. (1998). Interspecies differences in bone composition, density, and quality: potential implications for in vivo bone research. *Endocrinology*, 139(2), 663–670. doi:10.1210/endo.139.2.5751

Amprino, R. (1947). La structure du tissue osseux envisagée comme expression de différences dans la vitesse de l'accroissement. 315–330.

Amprino, R., & Godina, G. (1947). La struttura delle ossa nei vertebrati: ricerche comparative negli anfibi e negli amnioti: na.

Andronowski, J. M., Pratt, I. V., & Cooper, D. M. L. (2017). Occurrence of osteon banding in adult human cortical bone. *American Journal of Physical Anthropology*, 164(3), 635–642. doi:10.1002/ajpa.23297

Andronowski, J. M., Schuller, A. J., Cole, M. E., LaMarca, A. R., Davis, R. A., & Tubo, G. R. (2021). Rabbits (Oryctolagus cuniculus) as a model system for longitudinal experimental opioid treatments: implications for orthopedic and biomedical research. *Osteology*, 1(4), 225–237. Retrieved from https://www.mdpi.com/2673-4036/1/4/21

Bjornerem, A., Bui, M., Wang, X., Ghasem-Zadeh, A., Hopper, J. L., Zebaze, R., & Seeman, E. (2015). Genetic and environmental variances of bone microarchitecture and bone remodeling markers: a twin study. *Journal of Bone and Mineral Research*, 30(3), 519–527. doi:10.1002/jbmr.2365

Britz, H. M., Thomas, C. D., Clement, J. G., & Cooper, D. M. (2009). The relation of femoral osteon geometry to age, sex, height and weight. *Bone*, 45(1), 77–83. doi:10.1016/j.bone.2009.03.654

Caccia, G., Magli, F., Tagi, V. M., Porta, D. G., Cummaudo, M., Marquez-Grant, N., & Cattaneo, C. (2016). Histological determination of the human origin from dry bone: a cautionary note for subadults. *International Journal of Legal Medicine*, 130(1), 299–307. doi:10.1007/s00414-015-1271-6

Cattaneo, C., DiMartino, S., Scali, S., Craig, O. E., Grandi, M., & Sokol, R. J. (1999). Determining the human origin of fragments of burnt bone: a comparative study of histological, immunological and DNA techniques. *Forensic Science International*, 102(2–3), 181–191. doi:10.1016/s0379-0738(99)00059-6

Christensen, A. M., Berryman, H. E., Isa, M., Saginor, I. S., Smith, M. A., Webb, J. B., & Hefner, J. T. (2022). *A Guide to Forensic Fractography of Bone: Zenodo.*

Christensen, A. M., & Crowder, C. M. (2009). Evidentiary standards for forensic anthropology. *Journal of Forensic Sciences*, 54(6), 1211–1216. doi:10.1111/j.1556-4029.2009.01176.x

Coutelier, L. (1976). [The internal remodeling of compact bone in children]. *Bull Assoc Anat (Nancy)*, 60(168), 95–110. Retrieved from https://www.ncbi.nlm.nih.gov/pubmed/1016750

Crescimanno, A., & Stout, S. D. (2012). Differentiating fragmented human and nonhuman long bone using osteon circularity. *Journal of Forensic Sciences*, 57(2), 287–294. doi:10.1111/j.1556-4029.2011.01973.x

Croker, S. L., Reed, W., & Donlon, D. (2016). Comparative cortical bone thickness between the long bones of humans and five common non-human mammal taxa. *Forensic Sci Int*, 260, 104 e101–104 e117. doi:10.1016/j.forsciint.2015.12.022

Crowder, C., Dominguez, V. M., Heinrich, J., Pinto, D., & Mavroudas, S. (2022). Analysis of histomorphometric variables: Proposal and validation of osteon definitions. *Journal of Forensic Sciences*, 67(1), 80–91. doi:10.1111/1556-4029.14949

Crowder, C. M., Wiersema, J. M., Adams, B. J., Austin, D. E., & Love, J. C. (2016). The Utility of Forensic Anthropology in the Medical Examiner's Office. *Acad Forensic Pathol*, 6(3), 349–360. doi:10.23907/2016.037

Cummaudo, M., Cappella, A., Biraghi, M., Raffone, C., Marquez-Grant, N., & Cattaneo, C. (2018). Histomorphological analysis of the variability of the human skeleton: forensic implications. *International Journal of Legal Medicine*, 132(5), 1493–1503. doi:10.1007/s00414-018-1781-0

Cummaudo, M., Cappella, A., Giacomini, F., Raffone, C., Marquez-Grant, N., & Cattaneo, C. (2019). Histomorphometric analysis of osteocyte lacunae in human and pig: exploring its potential for species discrimination. International *Journal of Legal Medicine*, 133(3), 711–718. doi:10.1007/s00414-018-01989-9

Currey, J. D. (1964). Some effects of ageing in human haversian systems. *J Anat*, 98, 69–75. Retrieved from https://www.ncbi.nlm.nih.gov/pubmed/14109815

de Buffrénil, V., & Quilhac, A. (2021). Bone tissue types: A brief account of currently used categories. In V. de Buffrénil, A. J. de Ricqlès, L. Zylberberg, & K. Padian (Eds.), *Vertebrate Skeletal Histology and Paleohistology* (pp. 147–182). Boca Raton: CRC Press.

Dempsey, N., & Blau, S. (2020). Evaluating the evidentiary value of the analysis of skeletal trauma in forensic research: A review of research and practice. *Forensic Science International*, 307, 110140. doi:10.1016/j.forsciint.2020.110140

Doherty, A. H., Ghalambor, C. K., & Donahue, S. W. (2015). Evolutionary physiology of bone: bone metabolism in changing environments. *Physiology*, 30(1), 17–29. doi:10.1152/physiol.00022.2014

Dominguez, V. M., & Agnew, A. M. (2016). Examination of Factors potentially influencing osteon size in the human Rib. *Anat Rec (Hoboken)*, 299(3), 313–324. doi:10.1002/ar.23305

Dominguez, V. M., & Crowder, C. M. (2012). The utility of osteon shape and circularity for differentiating human and non-human Haversian bone. *American Journal of Physical Anthropology*, 149(1), 84–91. doi:10.1002/ajpa.22097

Dominguez, V. M., Harden, A. L., Wascher, M., & Agnew, A. M. (2020). Rib variation at multiple locations and implications for histological age estimation. *Journal of Forensic Sciences*, 65(6), 2108–2111. doi:10.1111/1556-4029.14520

Donoghue, P. C. J., & Sansom, I. J. (2002). Origin and early evolution of vertebrate skeletonization. *Microscopy Research and Technique*, 59(5), 352–372. doi:10.1002/jemt.10217

Enlow, D. H. (1963). *Principles of bone remodeling; an account of post-natal growth and remodeling processes in long bones and the mandible*. Springfield, Ill,: Thomas.

Enlow, D. H., & Brown, S. O. (1956). A comparative histological study of fossil and recent bone tissues. *Part I. The Texas Journal of Science*, 8, 405–412.

Enlow, D. H., & Brown, S. O. (1957). A comparative histological study of fossil and recent bone tissues. *Part II. The Texas Journal of Science*, 9, 186–214.

Enlow, D. H., & Brown, S. O. (1958). A comparative histological study of fossil and recent bone tissues. *Part III. The Texas Journal of Science*, 10, 187–230.

Epker, B. N., & Frost, H. M. (1965). The direction of transverse drift of actively forming osteons in human rib cortex. *The Journal of Bone and Joint Surgery. American Volume*, 47(6), 1211–1215. Retrieved from https://www.ncbi.nlm.nih.gov/pubmed/14337779

Ericsson, A. C., Crim, M. J., & Franklin, C. L. (2013). A brief history of animal modeling. *Missouri Medicine*, 110(3), 201–205.

Fazeli, P. K., & Klibanski, A. (2018). Effects of anorexia nervosa on bone metabolism. *Endocrine Reviews*, 39(6), 895–910. doi:10.1210/er.2018-00063

Felder, A. A., Phillips, C., Cornish, H., Cooke, M., Hutchinson, J. R., & Doube, M. (2017). Secondary osteons scale allometrically in mammalian humerus and femur. *Royal Society Open Science*, 4(11), 170431. doi:10.1098/rsos.170431

French, K. M., Crowder, C., & Crabtree, P. J. (2022). Integrating histology in the analysis of multispecies cremations: A case study from early medieval England. *International Journal of Osteoarchaeology*, 32(6), 1253–1263. doi:10.1002/oa.3160

French, K. M., Mavroudas, S. R., & Dominguez, V. M. (submitted). Prevalence of drifting osteons distinguishes human bone fragments. *PLoS One*.

Goliath, J. R., Stewart, M. C., & Stout, S. D. (2016). Variation in osteon histomorphometrics and their impact on age-at-death estimation in older individuals. *Forensic Science International*, 262, 282 e281–e286. doi:10.1016/j.forsciint.2016.02.053

Habicht, J. L., Kiessling, C., & Winkelmann, A. (2018). Bodies for anatomy education in medical schools: an overview of the sources of cadavers worldwide. *Academic Medicine*, 93(9), 1293–1300. doi:10.1097/acm.0000000000002227

Hackett, C. J. (1981). Microscopical focal destruction (tunnels) in exhumed human bones. *Medicine, Science and the Law*, 21(4), 243–265. doi:10.1177/002580248102100403

Haridy, Y., Osenberg, M., Hilger, A., Manke, I., Davesne, D., & Witzmann, F. (2021). Bone metabolism and evolutionary origin of osteocytes: Novel application of FIB-SEM tomography. *Science Advances*, 7(14), eabb9113. doi: 10.1126/sciadv.abb9113

Havill, L. M. (2004). Osteon remodeling dynamics in Macaca mulatta: normal variation with regard to age, sex, and skeletal maturity. *Calcified Tissue International*, 74(1), 95–102. doi:10.1007/s00223-003-9038-3

Herbst, E. C., Doube, M., Smithson, T. R., Clack, J. A., & Hutchinson, J. R. (2019). Bony lesions in early tetrapods and the evolution of mineralized tissue repair. *Paleobiology*, 45(4), 676–697, 622. doi: 10.1017/pab.2019.31

Hidaka, S., Matsumoto, M., Ohsako, S., Toyoshima, Y., & Nishinakagawa, H. (1998). A histometrical study on the long bones of raccoon dogs, Nyctereutes procyonoides and badgers, Meles meles. *The Journal of Veterinary Medical Science*, 60(3), 323–326. doi:10.1292/jvms.60.323

Hillier, M. L., & Bell, L. S. (2007). Differentiating human bone from animal bone: a review of histological methods. *Journal of Forensic Sciences*, 52(2), 249–263. doi:10.1111/j.1556-4029.2006.00368.x

Hollund, H. I., Jans, M. M. E., Collins, M. J., Kars, H., Joosten, I., & Kars, S. M. (2012). What happened here? Bone histology as a tool in decoding the postmortem histories of archaeological bone from castricum, The Netherlands. *International Journal of Osteoarchaeology*, 22(5), 537–548. doi:10.1002/oa.1273

Jaworski, Z. F., & Lok, E. (1972). The rate of osteoclastic bone erosion in Haversian remodeling sites of adult dog's rib. *Calcified Tissue Research*, 10(1), 103–112. doi:10.1007/BF02012540

Jowsey, J. (1966). Studies of Haversian systems in man and some animals. *Journal of Anatomy*, 100(Pt 4), 857–864.

Jowsey, J. (1968). Symposium on equine bone and joint diseases. Age and species differences in bone. *The Cornell Veterinarian*, 58(Journal Article), Suppl:74–94. Retrieved from https://www.ncbi.nlm.nih.gov/pubmed/4295746

Keating, J. N., Marquart, C. L., Marone, F., & Donoghue, P. C. J. (2018). The nature of aspidin and the evolutionary origin of bone. *Nature Ecology & Evolution*, 2(9), 1501–1506. doi:10.1038/s41559-018-0624-1

Keenan, K. E., Mears, C. S., & Skedros, J. G. (2017). Utility of osteon circularity for determining species and interpreting load history in primates and nonprimates. *American Journal of Physical Anthropology*, 162(4), 657–681. doi:10.1002/ajpa.23154.

Lagace, F., Schmitt, A., Martrille, L., Benassi, J., & Adalian, P. (2020). Using histomorphometry for human and nonhuman distinction: A test of four methods on fresh and archaeological fragmented bones. *Forensic Science International*, 313, 110369. doi:10.1016/j.forsciint.2020.110369

Lowenstein, J. M., Reuther, J. D., Hood, D. G., Scheuenstuhl, G., Gerlach, S. C., & Ubelaker, D. H. (2006). Identification of animal species by protein radioimmunoassay of bone fragments and bloodstained stone tools. *Forensic Science International*, 159(2–3), 182–188. doi:10.1016/j.forsciint.2005.08.007

Maggiano, I. S., Maggiano, C. M., Clement, J. G., Thomas, C. D., Carter, Y., & Cooper, D. M. (2016). Three-dimensional reconstruction of Haversian systems in human cortical bone using synchrotron radiation-based micro-CT: morphology and quantification of branching and transverse connections across age. *Journal of Anatomy*, 228(5), 719–732. doi:10.1111/joa.12430

Martiniaková, M., Grosskopf, B., Omelka, R., Vondráková, M., & Bauerová, M. (2006). Differences among species in compact bone tissue microstructure of mammalian skeleton: use of a discriminant function analysis for species identification. *Journal of Forensic Sciences*, 51(6), 1235–1239. doi: 10.1111/j.1556-4029.2006.00260.x

Matuszewski, S., Hall, M. J. R., Moreau, G., Schoenly, K. G., Tarone, A. M., & Villet, M. H. (2020). Pigs vs people: the use of pigs as analogues for humans in forensic entomology and taphonomy research. *International Journal of Legal Medicine*, 134(2), 793–810. doi:10.1007/s00414-019-02074-5

Mavroudas, S. R., Alfsdotter, C., Bricking, A., & Madgwick, R. (2023). Experimental investigation of histotaphonomic changes in human bone from whole-body donors demonstrates limited effects of early post-mortem change in bone. *Journal of Archaeological Science*, 154, 105789. doi:10.1016/j.jas.2023.105789

Mavroudas, S. R., Meckel, L. A., Gocha, T. P., Goldstein, J. Z., & Garza, S. L. (2022). The effects of experimental whole-body burning on histological age-at-death estimation from human cortical bone and dental cementum. *Biology (Basel)*, 11(11). doi:10.3390/biology11111569

McFarlin, S. C., Terranova, C. J., Zihlman, A. L., Enlow, D. H., & Bromage, T. G. (2008). Regional variability in secondary remodeling within long bone cortices of catarrhine primates: the influence of bone growth history. *Journal of Anatomy*, 213(3), 308–324. doi:10.1111/j.1469-7580.2008.00947.x

Mulhern, D. M., & Ubelaker, D. H. (2001). Differences in osteon banding between human and nonhuman bone. *Journal of Forensic Sciences*, 46(2), 220–222.

Mulhern, D. M., & Ubelaker, D. H. (2009). Bone microstructure in juvenile chimpanzees. *American Journal of Physical Anthropology*, 140(2), 368–375. doi:10.1002/ajpa.20959

Mulhern, D. M., & Ubelaker, D. H. (2011). Differentiating Human from Nonhuman Bone Microstructure. In C. Crowder & S. Stout (Eds.), *Bone Histology* (pp. 109–134). Boca Raton: CRC Press.

Murdock, D. J. E. (2020). The 'biomineralization toolkit' and the origin of animal skeletons. *Biological Reviews*, 95(5), 1372–1392. doi:10.1111/brv.12614

O'Brien, F. J., Taylor, D., & Clive Lee, T. (2005). The effect of bone microstructure on the initiation and growth of microcracks. *Journal of Orthopaedic Research*, 23(2), 475–480. doi:10.1016/j.orthres.2004.08.005

Owsley, D. W., Mires, A. M., & Keith, M. S. (1985). Case involving differentiation of deer and human bone fragments. *Journal of Forensic Sciences*, 30(2), 572–578. Retrieved from https://www.ncbi.nlm.nih.gov/pubmed/3998705

Phillips, K. A., Bales, K. L., Capitanio, J. P., Conley, A., Czoty, P. W., 't Hart, B. A., … Voytko, M. L. (2014). Why primate models matter. *American Journal of Primatology*, 76(9), 801–827. doi:10.1002/ajp.22281

Porto, S., Flavel, A., Maggio, A., & Franklin, D. (2023). Differentiating human from non-human bone fragments through histomorphological assessment of remains from Camposanto cemetery, Italy. *Archaeometry*, 65(1), 213–229. doi:10.1111/arcm.12801

Quekett, J. (1849). On the intimate structure of bone as composing the skeleton in the four great classes of animals: mammals, birds, reptiles, and fishes, with some remarks on the great value of the knowledge of such structure in determining the affinities of minute fragments of organic remains. *Transactions of the Microscopical Society of London*, 2, 46–64.

Robinson, N. B., Krieger, K., Khan, F. M., Huffman, W., Chang, M., Naik, A., … Gaudino, M. (2019). The current state of animal models in research: A review. *International Journal of Surgery*, 72, 9–13. doi:10.1016/j.ijsu.2019.10.015

Robling, A. G., & Stout, S. D. (1999). Morphology of the drifting osteon. *Cells, Tissues, Organs*, 164(4), 192–204. doi:10.1159/000016659

Roodman, G. D., & Windle, J. J. (2005). Paget disease of bone. *The Journal of Clinical Investigation*, 115(2), 200–208. doi:10.1172/jci24281

Ruben, J. A., & Bennett, A. A. (1987). The Evolution of Bone. *Evolution*, 41(6), 1187–1197. doi:10.2307/2409087

Seeman, E. (2001). Sexual dimorphism in skeletal size, density, and strength. *The Journal of Clinical Endocrinology & Metabolism*, 86(10), 4576–4584. doi:10.1210/jcem.86.10.7960

Shapiro, F., & Wu, J. Y. (2019). Woven bone overview: Structural classification based on its integral role in developmental, repair and pathological bone formation throughout vertebrate groups. *European Cells and Materials*, 38, 137–167.

Simmons, T., Goodburn, B., & Singhrao, S. K. (2016). Decision tree analysis as a supplementary tool to enhance histomorphological differentiation when distinguishing human from non-human cranial bone in both burnt and unburnt states: A feasibility study. *Medicine, Science, and the Law*, 56(1), 36–45. doi:10.1177/0025802415589776

Singh, I. J., Tonna, E. A., & Gandel, C. P. (1974). A comparative histological study of mammalian bone. *Journal of Morphology*, 144(4), 421–437. doi:10.1002/jmor.1051440404

Skedros, J. G., Sorenson, S. M., & Jenson, N. H. (2007). Are distributions of secondary osteon variants useful for interpreting load history in mammalian bones? *Cells, Tissues, Organs*, 185(4), 285–307. doi:10.1159/000102176

Streeter, M. (2010). A four-stage method of age at death estimation for use in the subadult rib cortex. *Journal of Forensic Sciences*, 55(4), 1019–1024. doi:10.1111/j.1556-4029.2010.01396.x

Taguchi, T., & Lopez, M. J. (2021). An overview of de novo bone generation in animal models. *Journal of Orthopaedic Research*, 39(1), 7–21. doi:10.1002/jor.24852

Urbanova, P., & Novotny, V. (2005). Distinguishing between human and non-human bones: Histometric method for forensic anthropology (Vol. 43). Brno, TCHEQUE, REPUBLIQUE: Moravské muzeum.

van Oers, R. F., Ruimerman, R., van Rietbergen, B., Hilbers, P. A., & Huiskes, R. (2008). Relating osteon diameter to strain. *Bone*, 43(3), 476–482. doi:10.1016/j.bone.2008.05.015

Widdig, A., Kessler, M. J., Bercovitch, F. B., Berard, J. D., Duggleby, C., Nürnberg, P., … Schmidtke, J. (2016). Genetic studies on the Cayo Santiago rhesus macaques: A review of 40 years of research. *American Journal of Primatology*, 78(1), 44–62. doi:10.1002/ajp.22424

Zedda, M., & Babosova, R. (2021). Does the osteon morphology depend on the body mass? A scaling study on macroscopic and histomorphometric differences between cow (Bos taurus) and sheep (Ovis aries). *Zoomorphology*, 140(1), 169–181. doi:10.1007/s00435-021-00516-6

Age-at-Death Estimation from Bone Microstructure

8

TIMOTHY P. GOCHA,
SOPHIA R. MAVROUDAS, AND
JUSTIN Z. GOLDSTEIN

8.1 Introduction

Estimation of age-at-death is a vital first step for biological anthropologists when examining skeletal remains from various contexts, often providing baseline information from which to apply other interpretive lenses. For example, in paleoanthropology comparative skeletal analysis between potentially divergent species first requires an understanding of relative age (e.g., whether skeletal maturity has been reached) in order to compare skeletal morphologies. In bioarchaeology, age estimation is critical for paleodemography and efforts to reconstruct population structure and life expectancy, as well as mortality and fertility rates of past populations. It is also of great importance in forensic anthropology, where accurate and precise age estimates can help narrow down a missing persons list for comparison to an unidentified decedent. Regardless of context, anthropologists must make observations of the available skeletal material, weigh their options for estimating age-at-death, and ensure use of the most informative and appropriate methods. Such methods rely on the established relationship between chronological age (i.e., how many years someone has lived) and biological age (i.e., how old tissues appear to be) in skeletal tissues. Importantly, though, biological age is affected by intrinsic factors such as genetics and individual behavior, as well as extrinsic factors such as environmental stressors and taphonomic forces after death. While age-at-death estimation is possible because biological and chronological age are often well correlated, they are rarely, if ever, synonymous (Kemkes-Grottenthaler, 2002).

When evaluating skeletal age, anthropologists often assume that traditional, gross morphological observations are the most reliable and appropriate means for age-at-death estimation. In subadults, gross morphological age estimation focuses on growth and development of skeletal elements including the dentition, resulting in relatively accurate and narrow age estimates that are explainable by known developmental processes. This reliability allows researchers to predict how methods will perform across populations and in response to various stressors. Macroscopic age estimation for adult skeletal material, however, is inherently more difficult (Garvin and Passalacqua, 2012), most often relying on a qualitative assessment of "wear-and-tear" degeneration of macroscopic skeletal structures (Baccino and Schmitt, 2006). Such changes are determined less by underlying developmental processes and more often related to idiosyncratic use over time. As a result, age ranges provided by macroscopic aging methods are often so broad that they have somewhat limited utility in forensic and bioarchaeological investigations.

Alternatively, or in addition to macroscopic methods, analysts can consider skeletal histology. Age estimation from bone histology has a historical time depth equal to or greater

 DOI: 10.4324/9781003385608-8

than most macroscopic methods for age estimation. Since 1911, the medicolegal community has recognized that microscopic changes in bone correlate with age (Balthazard and Lebrun, 1911), and established age-at-death estimation methods have been available since 1965 (Kerley). Most histological age-at-death estimation has primarily focused on intracortical remodeling of long-bones and ribs. To date, many elements have been investigated for histological age-related changes including long bones of the arm (e.g., humerus, radius, ulna), the thorax (rib), long bones of the leg (femur, tibia, fibula), metacarpals, cranial bone, and the mandible. Each element has been examined for histological changes that correlate with advancing skeletal age, though most use a unique combination of histological variables. While many methods have been developed on these elements, most have used small sample sizes and linear regression analysis which assumes a linear relationship on histological changes that do not necessarily progress linearly through time. All skeletal elements have patterns of development that could be explored, qualified, and quantified for age-at-death estimation purposes. As a result, there are many opportunities in histological research for anthropologists to explore age-related histological skeletal changes and experiment with new statistical approaches that accurately reflect the underlying biological changes.

8.2 Methods for Age Estimation from Bone Microstructure

This section surveys some of the major developments in methods for age estimation through histological investigation of cortical bone microstructure. A chronological approach is taken to offer a temporal context to advances in the field as they arose. For a summary of the variety of methodologies, see Appendix A.

8.2.1 Balthazard and Lebrun: Accidental Origins for Age Estimation from Histomorphometry

More than a century ago, Balthazard and Lebrun (1911) were the first to report using skeletal histomorphometry to estimate age-at-death, though this was not their original intent. They initially set out to investigate the utility of Haversian canal diameter to determine if bone fragments were human or non-human—a question of great interest to the medicolegal community in the early 20th century and one that is still being explored. Their human samples included the anterior region of the midshaft tibia from 65 individuals aged from full-term fetus to 84 years at death. From each individual they measured 100–200 randomly selected Haversian canals from an area spanning from the periosteal to endosteal borders. Because of the amount of regional variation observed, they emphasized the importance of collecting data across the whole expanse of a bone. Measurements were taken directly through the microscope using an ocular micrometer. Their results reported that Haversian canal diameter is sufficient in most cases to adequately distinguish human from non-human skeletal material, with humans having larger Haversian canals, though bone fragments from non-human primates might still be similar enough to be confused. However, the wide age range of samples also allowed Balthazard and Lebrun to observe that average Haversian canal diameter increases with age. They plotted Haversian canal diameter against age-at-death to provide forensic practitioners the first method for age estimation using histomorphometry, reporting that from 10 years of age and into “extreme old age” Haversian canal size is useful to estimate age with error of <5 years in two-thirds of cases and with no error >15 years between known and estimated age. Because of the amount of regional variation observed, they emphasized the importance of collecting data across the whole expanse of a bone. However, subsequent research called these results into question.

8.2.2 Twentieth-Century Anthropological Exploration and Applications

Skeletal histology for age estimation within the field of biological anthropology began in earnest in the 1960s. In the decades that followed there was a period of exploration with a proliferation of research exploring various skeletal elements, region of interest (ROI) placement and size, and predicting variables. Histological variables were primarily quantified directly through microscopic observation.

8.2.2.1 Kerley: First Applications in Anthropology

The earliest application of skeletal histology for estimation of age-at-death in the anthropological literature was by Ellis Kerley with his 1961 dissertation (published in 1965). Interestingly, Kerley was seemingly unaware of earlier work by Balthazard and Lebrun (1911), as their research was not cited in either his dissertation or his 1965 work. Instead, Kerley cited inspiration from Amprino and Bairati (1936) and Jowsey (1960), who both studied midshaft diaphyses of various long bones from dozens of individuals across the age spectrum from neonates to octogenarians and described general changes in bone microarchitecture with age, though neither focused specifically on estimation of age-at-death.

With a sample of 126 individuals aged 0–95 years, Kerley examined cross-sections from the midshaft of femora, tibiae, and fibulae. The femoral sample included 67 individuals (43 male, 17 female, 7 of unknown sex); the tibial sample included 33 individuals (24 male, 8 female, 1 of unknown sex); and the fibular sample included only 25 individuals (19 male, 5 female, 1 of unknown sex). Four ROIs were examined on each skeletal element: anterior, posterior, medial, and lateral. Histological variables were quantified along the periosteal border, with data collected directly through a microscope using a circular field reported to be 1.25 mm in diameter. Four variables were examined and quantified at each location: (1) number of intact secondary osteons (defined as an osteon with ≥80% of its original area unremodeled and with an intact Haversian canal); (2) number of fragmentary osteons (defined as an osteon with <80% of its area intact, with or without an identifiable Haversian canal); (3) percentage of circumferential lamellar bone; and 4) number of primary osteons. Regression analyses showed that each variable significantly correlated with age, the most significant of which was osteon fragments in the fibula. A subsequent validation test was conducted on an additional 56 specimens, yielding 87.3% of estimates within + 5 years of actual age and all estimates within + 10 years.

Kerley's original work was later modified by Kerley and Ubelaker in 1978. In a study designed to directly compare Kerley's (1965) method and that of Ahlqvist and Damsten (1969), Bouvier and Ubelaker (1977) found that a field size of 1.25 mm in diameter, as originally reported by Kerley, yielded age estimates far too young, so Kerley and Ubelaker (1978) re-analyzed Kerley's original samples. To obtain results concordant with his original research, Kerley and Ubelaker concluded that the original field size must have been 1.62 mm in diameter rather than 1.25 mm. Further, mathematical errors were noted for some regressions in Kerley's 1965 publication, and thus Kerley and Ubelaker (1978) produced new regression formulae; therefore, this reference should be consulted when applying the Kerley method.

8.2.2.2 Ahlqvist and Damsten: Introducing Percentage of Remodeled bone

In 1969, Ahlqvist and Damsten published a modification of Kerley's original method for age estimation from femoral histology, citing a desire to simplify the method. The authors pointed out that in his original publication, Kerley did not describe how the percent of circumferential lamellar bone was measured and claimed that the seemingly arbitrary value of 80% to distinguish intact from fragmentary osteons could lead to inconsistent results. Further, Ahlqvist and Damsten argued that the posterior ROI in Kerley's method, located at the linea aspera, could be a source of interpersonal variation due to its function as a major muscle attachment site and variability in individual activity levels. In the modified method, ROIs were shifted to anteromedial, posteromedial, posterolateral, and anterolateral to avoid the linea aspera. In a sample of 20 femoral midshaft sections obtained at autopsy (no age range or sex ratio provided), each ROI was examined using a reticle in a microscope eyepiece, superimposing a 1-mm square grid consisting of 100 square cells to visualize the percentage of remodeled bone. If an individual square was more than 50% filled with either intact and/or fragmentary osteons it was counted as remodeled, and the sum of all grid squares was used to determine the percentage of remodeled bone in that ROI. The percentage for each sampling region was determined, and these were averaged to determine the relationship between percent remodeled bone and age. Ahlqvist and Damsten rightfully argued the resulting method is easier to use because it does not require quantification of four separate variables or distinguishing between intact and fragmentary osteons. Further, they argued their method is fairly reliable for estimating age, as linear regression analyses of percent remodeled bone plotted against known age yielded a standard error of the estimate of 6.71 years. However, later tests of the method demonstrated that the reported standard error was likely artificially low due to the small sample sizes used in method generation and that Kerley's method actually produced more accurate and reliable results (Bouvier and Ubelaker, 1977; Stout and Gehlert 1980).

8.2.2.3 Singh and Gunberg: The First Non–Long Bone Method

Noting that macroscopic age estimation methods were unreliable for older individuals and often required substantial portions of the skeleton, Singh and Gunberg (1970) examined the femur, tibia, and the first non–long bone element for histological aging, the mandible. Their sample was 52 male dissecting room cadavers, aged 39-87 years. Samples were taken from all three elements for 33 of the individuals; these samples were decalcified and stained for histological investigation. From the remaining 19 individuals only the mandible was available, samples from which were not decalcified nor stained. To minimize destructive sampling, sections measuring ~1 x 1 cm were taken from the anterior midshaft of the femur and tibia, and the posterior border of the mandibular ramus, opposite the lingula. Using a circular microscopic field measuring 2.0 mm in diameter, the authors examined two randomly selected viewing fields from the periosteal third of each element and quantified the total number of intact osteons in a field, average number of lamellae per osteon, and average Haversian canal diameter directly through the microscope.

For all three skeletal elements investigated, the authors reported that the number of lamellae per osteon increases with age but has the lowest correlation of any variable. Not surprisingly, the number of intact osteons also increases with age; however, Haversian canal diameter decreases with age, contrary to Balthazard and Lebrun's (1911) findings. Using multiple regression analyses, Singh and Gunberg provide age predicting equations for the femur, tibia, and mandible, though they report the mandible to be the most reliable for age estimation (likely a product of having a larger sample size. Using the mandible, Singh and Gunberg concluded that individuals aged 40–80 years could be estimated to within + 2.58 years in 67% of cases and + 5.16 years in 95% of cases. Interestingly, the authors note some nonlinearity in their data, especially at older ages, but admit to ignoring it and using linear models for age estimation because previous studies had concluded accurate age estimation was possible using linear models.

8.2.2.4 Thompson: Minimizing Destructive Sampling through Bone Cores

In his 1978 dissertation, Thompson noted that the "principal shortcoming" of existing methods for histological age estimation was the need for complete cross-sections of diaphyseal bone. He argued that to examine whether population-specific histological methods were necessary, anthropologists needed access to large series of both archaeological and forensically derived skeletal material but that obtaining complete cross-sections from many such collections would be prohibitive. To address this issue, in 1979 Thompson published a new method to minimize destructive sampling termed the "core technique." Specially constructed bone corers were attached to a rotary (e.g., Dremel) saw to obtain cores 0.4 cm in diameter from the diaphysis of long bones. Cores were taken from the anterior midshaft of the femur, medial midshaft of the tibia, medial midshaft of the humerus, and lateral distal ulna. Their total sample included 116 individuals (64 male, 52 female), though cores from femora and tibiae were only collected from 113 cadavers, while cores from humeri and ulnae were taken from 31 individuals. The sample came from dissecting room cadavers and thus included individuals who died of acute and chronic diseases, subdivided into "nonpathological" and "pathological" groups, respectively.

Before histological slides were prepared from the sample, intact bone cores were analyzed for macroscopic variables. After histological preparation, four adjacent fields each measuring 0.992 mm^2 were analyzed from the periosteal border of each bone core. Histological data were collected directly through the microscope using a 10 x 10 grid eyepiece disk micrometer. Data were collected for 19 variables: cortical thickness, core weight, cortical bone density, mineral content, mineral index, aggregate osteon lamellae area, aggregate Haversian canal area, osteon area, secondary osteon number, Haversian canal number, individual osteon lamellae area, individual Haversian canal area, aggregate osteon perimeter, aggregate Haversian canal perimeter, individual osteon perimeter, individual Haversian canal perimeter, ratio of aggregate Haversian canal area to aggregate secondary osteon lamellae area, ratio of aggregate Haversian canal perimeter to aggregate osteon perimeter, and ratio of individual Haversian canal perimeter to individual osteon perimeter.

Stepwise linear regression was used to analyze all 19 variables, resulting in numerous equations for estimating age-at-death, including equations generated from the entire sample, male- and female-specific equations, nonpathological group-specific equations, as well as separate equations for left- and right-sided elements. Osteon area was reported to be the single best variable for predicting age-at-death, though up to five variables were present in various equations. Standard errors of the estimate for equations ranged 6.2–10.6 years.

This study was later shown to be reliable for histological age-at-death estimation of the femur by Crowder (2005) and was important in development of the field of histological age-at-death estimation because it allowed researchers to examine methods that incorporated less cortical area of bone.

8.2.2.5 Hauser: Looking beyond the Periosteal Border

In 1980, work by Hauser and colleagues was the first to look beyond the periosteal border of the cortex in both the femur and tibia. Their sample came from modern European cadavers, from which 96 femoral samples were collected from individuals aged 21–87 years and 33 tibial samples from individuals aged 18–88 years (no sex ratio was provided for either sample). In both skeletal elements, researchers quantified the population density of osteons, mean minimum diameter of Haversian canals, and percentage of the surface occupied by Haversian canals. Quantification occurred directly through the microscope using circular fields measuring 1.16 mm in diameter. In the femur, these variables were quantified in three ROIs each at the anterior, medial, and lateral periosteal borders, as well as medial, lateral, and posterior endosteal borders. In the tibia, the same variables were quantified in the anterior cortex, along both the periosteal and endosteal borders, as well as a line running the depth of the cortex between them. Variables were subjected to linear regression, and separate equations were provided for periosteal and endosteal sampling regions. Equations for age estimation from the femur yielded standard crrors of 10.7–11.4 years, while for the tibia these estimates were 13.5–16.0 years.

While results based on the standard errors provided are not the most promising from histological aging methods, inclusion of endosteal predictors of age could be useful in situations where taphonomic processes or diagenesis has damaged the periosteal surface of bone.

8.2.2.6 Drusini: The Importance of Spatial Variation

In 1987, Drusini conducted research originally designed to test the methods proposed by Kerley (1965; Kerley and Ubelaker, 1978) and Thompson (1979). The sample included 24 individuals (15 males, 9 females) aged 19–50 years from an osteological collection with known demographics in Italy. Full femoral cross-sections were obtained from 20 individuals. Drusini followed ROI size and placement of the original publications, although he continued beyond merely applying the published methods and examined ROIs around the entire perimeter of three femora. Drusini concluded that there is considerable spatial variation in remodeling distribution around the circumference of the femur and that sampling only the four ROIs as required in each method would be insufficient. Drusini thus abandoned his method test and instead quantified only the number of secondary osteons per square millimeter in 10–20 fields, each 1 mm^2, around the perimeter of the femur, excluding the linea aspera. He used linear regression to produce a formula for age estimation with a standard error of only 3.92 years, though it should be noted that the sample size was small and did not include any individuals older than middle-aged. While the work used a small sample size with a heavily skewed age distribution, it was the first time research highlighted spatial variation in remodeling density circumferentially around the femoral midshaft.

8.2.2.7 Samson and Branigan: Haversian Canal Size in Taphonomically Altered Bone

Samson (1983) reported difficulty in applying previously published histological aging methods to archaeological skeletal material. As a result of taphonomic forces, important histological structures such as osteons were difficult to distinguish, although Haversian canals themselves were still well-preserved. Noting that the femur is among the bones most likely to survive in archaeological contexts, Samson and Branigan (1987) investigated use of Haversian canal size in femora for age estimation using a sample of 58 modern Caucasian individuals aged 16–91 years (31 males, 27 females). A complete cross-section was taken from the midshaft of each femur, and mean cortical thickness was derived by measuring at eight equally spaced locations around the cortex, avoiding the linea aspera. A square piece of bone measuring 1 x 1 cm was subsequently removed from both the anterolateral and anteromedial area of the cortex, and histological slides were prepared for each square section. The minimum diameter of 30 Haversian canals was measured for both the anterolateral and anteromedial sections, from which a mean minimum Haversian canal was calculated.

Of the three original variables collected by Samson and Branigan, linear regression indicated only the number of Haversian canals per unit area significantly correlates with age. The authors subsequently created a new variable they termed "morphological character" by multiplying the mean number of Haversian canals per unit area by the mean minimum Haversian canal diameter. Linear regression with this variable yielded more accurate results, with standard errors of the estimate of 6 years for males and 16 years for females. Unfortunately, the authors did not provide any of the derived equations, precluding further application of their method.

8.2.2.8 Ericksen: Minimizing Destructive Sampling through Anterior Wedges

Similar to Thompson (1979), Ericksen (1991) expressed concern over histological methods that required complete femoral cross-sections. To address this, Ericksen devised a technique using only a wedge of bone from the anterior surface of the femur using a large sample of 328 individuals (174 males, 154 females) aged 14–97 years from the United States, Dominican Republic, and Chile. From each anterior wedge, five fields along the periosteal border each measuring 0.886 mm^2 were analyzed directly through the microscope. Data were collected on printed photographs corresponding to each field, allowing the observer to move the viewing plane to focus on peripheral structures without "losing" the original field. From each ROI the following variables were quantified: number of secondary intact osteons (defined as having an intact Haversian canal), number of type II or embedded osteons, number of fragmentary osteons (having a Haversian canal disturbed or absent due to subsequent remodeling), number of resorption spaces (indicating initiation of a new remodeling event), and number of non-Haversian or primary vascular canals.

The sum of each variable count was divided by the total area sampled (4.43 mm^2) to yield an average number of variables per square millimeter. Ericksen also superimposed a 100-space grid (similar to Ahlqvist and Damsten, 1969) to quantify the percentage of unremodeled circumferential lamellar bone, percentage of intact osteonal bone, and percentage of bone containing fragmentary osteons. The percentage for each variable was averaged for the five fields to provide an overall average for age estimation. Stepwise regression analyses yielded an overall standard error of the estimate of 10.1–12.2 years. Sex-specific equations were also determined, yielding results of 10.1–12.0 years for males and 10.0–11.6 years for females. Although these standard errors were larger than previously published methods, the significantly larger sample size was likely a contributing factor, in that Ericksen's sample captured more naturally occurring human variation in remodeling.

8.2.2.9 Stout and Paine: The First Use of Thoracic Skeletal Elements

Building on years of biomedical and clinical research on age-related changes in the human rib conducted by Harold Frost and his Orthopaedic Research Laboratory at the Henry Ford Hospital (e.g., Epker and Frost 1964; Sedlin et al., 1963a, 1963b; Frost 1969), in 1986,

Stout published the first use of histology from bones of the thorax to estimate age in a bioarchaeolgoical setting. Specifically, he published a case study on histological analysis of archaeological remains purported to be those of Francisco Pizarro. Analyzing a complete cross-section from the middle of the left sixth rib, Stout estimated the age of the remains using an unpublished formula developed in his lab.

In 1992, Stout and Paine expanded on this earlier work and put forth a new method for histological age estimation using either a sixth rib or clavicle, or both bones. The authors noted that because remodeling, at least to some degree, stems from microdamage caused by loading, long bones and even the mandible might exhibit significant variation due to lifestyle factors affecting loading on those bones. The ribs and clavicles, however, were argued to undergo minimal load bearing and thus believed to have a more static remodeling rate. Additionally, as cross-sections of these bones are much smaller than long bones, not only could entire cross-sections be analyzed to reduce risk of reading a potentially abnormal field, but multiple serial sections can be analyzed in an efficient manner. Stout and Paine recommend sampling at least two entire cross-sections for ribs and clavicles.

Their sample came from 40 dissecting room cadavers (32 males, 7 females, 1 unknown sex) aged 13–62 years. Quantification occurred directly through the microscope using a Merz reticle and counting every other or alternating fields from each cross-section. Stout and Paine measured cortical area, intact osteon density, and fragmentary osteon density, which they used to determine osteon population density (OPD)—the sum of intact and fragmentary osteon density divided by cortical area. Regression analyses were conducted for the rib and clavicle separately, as well as combined. A validation test on 19 further cadaver specimens yielded a mean absolute difference of 3.9 + 0.9 years for the rib, 5.91 + 2.67 years for the clavicle, and 3.8 + 1.91 for the rib and clavicle combined.

In 1994, Stout teamed up with İşcan and Loth, who pioneered use of sternal rib end morphology to examine the potential combination of histomorphometry and macroscopic indicators for age-at-death. Using the sternal end of the fourth rib for 59 white individuals aged 11–88 years, histological quantification was performed for intact osteon density, fragmentary osteon density, and their combination to generate OPD. Macroscopic phase scores were also assigned according to the methods of İşcan et al. (1984, 1985). An equation for estimating age from histology of the fourth rib alone yielded a standard error of the estimate of 10.43 years, though a separate equation combining histological variables and the assigned phase score reduced standard error to 7.18 years. This led Stout et al. (1994) to suggest that use of both histology and macroscopic morphology is preferable to either alone.

In 1996, Stout and colleagues revisited earlier work on the clavicle, expanding the sample to include 83 known-age individuals from a Swiss archaeological cemetery in addition to the original cadaver sample of 40. OPD was measured in the same way as Stout and Paine (1992), and the team conducted new regression analyses and generated a new prediction formula. This yielded age estimates with a mean absolute difference of only 5.5 years, making the revised formula preferable over the original.

Collectively, the work of Stout and colleagues demonstrated for the first time that skeletal elements from the thorax could be used for histological age-at-death estimation Subsequent work of Stout and colleagues that incorporated both gross morphological and histological variables into age prediction equations was also an important first step toward holistic age-at-death estimation.

8.2.2.10 *Kimura: Exploring Use of Metacarpals*

Citing Stout's (1989) observation that non-weight-bearing bones should be further investigated because biomechanical forces affect skeletal remodeling, Kimura (1992) introduced a new element for histological age estimation—the midshaft of the second metacarpal. Samples came from 227 Japanese cadavers (114 male, 113 female) aged 30–98 years. Kimura measured cortical thickness and cortical bone density using radiographs of the metacarpals. Microradiographs were used to quantify intact and fragmentary osteons, as well as intact and fragmentary osteons divided by cortical area [essentially the same as Stout and Paine's OPD, though Kimura termed this total osteon density (TOD)]. Stepwise linear regression was used to calculate a series of equations for age estimation from the total combined sample, for males and females separately, and for a younger age group (30–65 years) and an older age group (≥65 years). Standard errors ranged 5.51–14.82 years. In the younger age group, TOD was the best predictor of age, while for the older age group cortical thickness was best. Overall, sex-specific equations yielded better results than for the sexes combined, and results for the younger age group were more promising than the older group. Figures provided in Kimura's 1992 manuscript indicate that OPD seems to hit an asymptote in the second metacarpal at 50–60 years of age, which likely explains why it is not as useful to estimate age in older individuals.

This method expands the utility of histological age-at-death estimation beyond traditional long bone elements and provides an alternative method for material from less-complete skeletal remains. Future research may improve on this method by exploring additional two- or three-dimensional histological variables that could increase the utility of metacarpals for age-at-death estimation beyond 60 years of age.

8.2.2.11 *Cool and Colleagues: Investigating Cranial Bone*

Noting that it is not uncommon for forensic investigators to have only an edentulous cranium for examination, in 1995 Cool and colleagues introduced the first and still only published method for age estimation from the cranium via skeletal histology. Researchers extracted a 12-mm-diameter core from the occipital bone in 17 white males aged 21–70 years. In both ectocranial and endocranial tables, three random fields 0.85 x 0.83 mm were examined directly through the microscope, and four variables were quantified: fractional volume of primary osteons, intact and fragmentary osteons, and lamellar bone. Regression analyses indicated the best results were obtained from fragmentary osteons alone ($r^2 = 0.44$), though multivariate analyses for a combination of variables did not significantly improve results. Overall, Cool and colleagues reported that despite some observable, age-related histological changes in human occipital bone, random variation in the variables examined preclude their use for accurate age estimation.

8.2.2.12 *Watanabe and Colleagues: Utilizing the Middle of the Femoral Cortex*

Similar to many researchers before them, Watanabe et al. (1998) examined the anterior femur for histological age estimation, though unlike previous research that mostly focused on the periosteal or endosteal border, their research was first to examine the middle third of the cortex. In a sample of 88 Japanese individuals (67 males, 21 females) aged 20–92 years, the researchers examined three square fields each 0.708 mm^2 and quantified the area, maximum and minimum diameters for intact osteons and Haversian canals, and number of type II (embedded) osteons and fragmentary osteons. Stepwise linear regression selected only osteon perimeter, Haversian canal diameter, and number of fragmentary osteons, resulting in an age estimation formula with a standard error of 6.39 years. Similar to the work of Hauser et al. (1980), Watanabe and colleagues' exploration of the middle third of the femoral cortex could be useful in situations where taphonomic processes or diagenesis has damaged the periosteal surface of bone.

8.2.3 Histological Age Estimation in the 21st Century

Moving into the 21st century, there appears to be less proliferation of new methods for histological age estimation, though some general research trends do appear. These include larger sample sizes, less reliance on direct observation through the microscope in favor of newer technologies, examination of population variation, and more sophisticated statistical analyses.

8.2.3.1 Thomas and Colleagues: Attempts at Automation

In 2000, Thomas and colleagues were the first to attempt age estimation from examining the entirety of the femoral midshaft rather than small ROIs. Their sample came from modern Australian forensic cases and included 96 individuals (50 males, 46 females) aged 21–92 years. Only individuals with a normal body mass index were included to avoid possible confounding variables such as malnutrition or morbid obesity, and no individuals with medical conditions known to affect bone remodeling were included.

The authors collected several macroscopic measurements from each femoral section including total subperiosteal area, periosteal perimeter, endosteal perimeter, cortical bone area, and medullary area. Additionally, the authors calculated second moments of area to investigate potential age-related patterns in bone biomechanics. The most novel component of their research, however, was use of an automated histological data collection system. The authors devised a video system that could automatically detect pores >400 μm^2 and did so across the whole femoral cross-section. Because of the detected pore size, the authors were essentially quantifying Haversian canals (i.e., intact osteons) and larger pores, likely age-related intracortical porosity, with no ability to quantify fragmentary osteons, which represent evidence of past remodeling events which is the basis for most histological age estimation methods for cortical bone.

Multiple regression analyses were used to examine all variables. Interestingly, while total subperiosteal area and periosteal perimeter were independent predictors of age, the number of pores in a cortex was not a significant predictor of age. Age prediction equations were then applied to samples in the research and the authors report that their automated method could not predict age at a "clinically acceptable level," with more than half of samples having a difference of >8 years between estimated age and known age.

8.2.3.2 Cho and Colleagues: Revisiting the Rib for Population Variation

In 2002, Cho and colleagues argued that a major deficit in histological age-at-death estimation was the lack of methods based on genetically heterogeneous populations. In response, they published population-specific histological age-at-death estimation equations for African American and European American samples using the mid thoracic rib. The researchers used a sample of 154 individuals split into developmental and testing sets to calculate new equations using forward stepwise regression with variables of OPD, osteon area, and relative cortical area. The authors reported that population affinity played a significant role in age prediction and thus developed population-specific equations. They also produced an equation for individuals of unknown ancestry and an equation for incomplete cross-sections (when relative cortical area could not be determined). Data collection for this method was done through direct observation of the cross-section with a light microscope and counting reticle.

The work of Cho et al. (2002) is important because it included diverse samples to provide accurate age-at-death estimates for a broader population, specifically in the United States. As the authors point out in the article, however, little research had been done at the time on whether population affinity affects bone remodeling, which remains an open question to this day. Although population-specific methods have been developed for individuals in Africa, America, Asia, and Europe, age mimicry of the original sample data may have more effect on the interpopulational utility of methods rather than underlying genetic/environmental differences in population themselves. The issue with age mimicry limiting the utility of age estimation methods on interpopulation samples is discussed in detail by Konigsberg et al. (2008), and examples of the effect of age mimicry in archaeological samples is discussed by Navitainuck et al. (2022).

8.2.3.3 *Maat and Colleagues: Investigating the Percentage of Non-Remodeled Bone*

In 2006, Maat and colleagues introduced another new method for age-at-death estimation from the anterior femur, one that is reportedly uncomplicated and minimally invasive. The study examined 162 individuals (86 males, 76 females) aged 15–96 years, all from either modern forensic cases or anatomy dissection rooms in Western Europe. The authors excluded individuals with chronic diseases that affect bone metabolism (e.g., diabetes). After removing an anterior wedge from the femoral midshaft, three ROIs were investigated: one directly anterior, one 25° medial to anterior, and one 25° lateral to anterior, each located along the periosteal border and 1 mm^2 in size. Using a 10 x 10 grid projected into the microscope eyepiece, the authors quantified the percentage of non-remodeled bone in each ROI. As expected, the percentage of non-remodeled bone decreased with increasing age; however, the authors report when non-remodeled bone occupied 80%–100% of the anterior femur there was no strong relationship with age, limiting applicability in younger individuals. Curvilinear regression was used to generate age estimation formulae from either a single ROI or various combinations of ROIs, yielding standard errors of the estimate of 10.60–14.79 years. Although such results are not as promising as some earlier methods, the authors make the case that determining the percentage of non-remodeled bone avoids the complexity of distinguishing and counting intact versus fragmentary osteons.

8.2.3.4 *Streeter: Subadult Age Estimation*

Streeter (2010) developed a qualitative method for estimating subadult age using histomorphology of human mid thoracic ribs. Her method used 72 subadult human ribs from forensic contexts to develop four age phases: individuals <5 years old, 5- to 9-year-olds, 10- to 17-year-olds, and 18- to 21-year-olds. The method involved direct observation of the cross-section with both bright field and polarized light. Streeter's method used general descriptions of osteon morphotypes and observation of spatial placement of those morphotypes on specific cortices to assign a phase for age estimation, rather than quantification of osteons as in most histological aging methods.

Descriptions of phases included the presence or absence of specific bone types and histological structures, including woven bone, primary lamellar bone, primary vascular canals, concentric lamellar bone, Volkmann's canals, type I osteons, and drifting osteons. While there is no validation of the method in the original article, Streeter (2010) does mention unsuccessful application of this method to archaeological material and suggests that the age ranges for phases used to estimate age are population-specific.

An age-at-death estimation method for subadult rib material is beneficial to anthropologists with no other age indicators at their disposal, although the phase-based nature of this method perpetuates the problem of age mimicry in estimation methodology and could benefit from reanalysis with more advanced statistical approaches. For example, mirroring work done on phase-based gross morphological age-at-death indicators like the pubic symphysis, it may be possible to improve the utility of Streeter's (2010) method through application of Bayesian transition analysis (e.g., Konigsberg et al., 2008).

8.2.3.5 Crowder and Dominguez: Looking beyond the Periosteal Border

Noting that previous methods often applied varied and inconsistent definitions of histological variables and quantified histological changes only along the periosteal border, Crowder and Dominguez (2012, Crowder, 2013) developed a method looking at the anterior femur across the cortex from the periosteal to endosteal border using newly validated definitions of remodeling events (Heinrich et al., 2012; Crowder et al., 2022). Their total sample size was 319 individuals (169 males, 150 females) aged 15–97 years, with a majority of samples from the Ericksen femur collection (n = 279) and other samples from Kerley's original work (n = 14) as well as modern forensic samples from the Office of Chief Medical Examiner in New York City (n = 26). Using a 5-mm-wide ROI separated into 10 columns, data were collected directly through the microscope using a Merz reticle by counting every other field in a checkerboard pattern (after Iwaniec et al., 1998). A total of eight variables were quantified and analyzed: surface area, number of intact secondary osteons, number of fragmentary secondary osteons, intact secondary osteon density, fragmentary osteon density, OPD, mean osteon cross-sectional area, and mean anterior cortical width. Moderate to strong Pearson correlations were found between each variable and age, with the exception of intact osteon density, leading Crowder and Dominguez to conclude intact and fragmentary osteons should be treated statistically separate rather than combining into OPD. Stepwise linear regression models were generated for males, females, and combined sexes for when biological sex cannot be estimated from skeletal remains, yielding standard errors of the estimate of 11.13 years, 9.77 years, and 10.70 years, respectively.

An important contribution of this method is that it provides strong evidence that separating OPD into its constituent parts may provide better information on skeletal age and contribute to more accurate age predictions. While there is still some debate as to whether fragmentary osteons can be a reliable variable with an acceptable level of interobserver error, new osteon definitions with less ambiguous criteria (Crowder et al., 2022) can improve error rates.

8.2.3.6 Gocha: Geographic Information Systems Software for Spatial Patterning

In 2014, Gocha noted that despite nearly half a century of research on histological age estimation from the femoral midshaft, there was still no consensus on where ROIs should be placed, how many ROIs should be investigated, and how large the ROIs should be.

Admittedly, answering such questions would have been nearly impossible decades earlier through direct microscopic observation. However, in 2012, Rose and colleagues demonstrated that geographic information systems (GIS) software could be used to digitally map and spatially analyze histological variables. Building on that, Gocha conducted a pilot study using GIS to map all remodeling events across the entirety of the femoral midshaft on a sample of 30 individuals (15 male, 15 female) aged 21–97 years.

After digitally marking/counting all remodeling events, density maps generated in GIS revealed spatial heterogeneity of remodeling events across the cortex, where after ~35 years of age the lateral and anterolateral regions of the femur generally show the greatest remodeling event density (Figure 8.1; see Gocha and Agnew, 2016 for biomechanical interpretations). The flexibility of GIS allowed application of 10 segmentation techniques to investigate the entire cortex, separating the cortex into quadrants and octants at different orientations, as well as to investigate quadrants and octants by separating them into periosteal, middle, and endosteal thirds, yielding a total of 100 ROIs of varying size and location. Remodeling counts were divided by the area of each ROI to generate OPD values, which were investigated for their relationship with age-at-death.

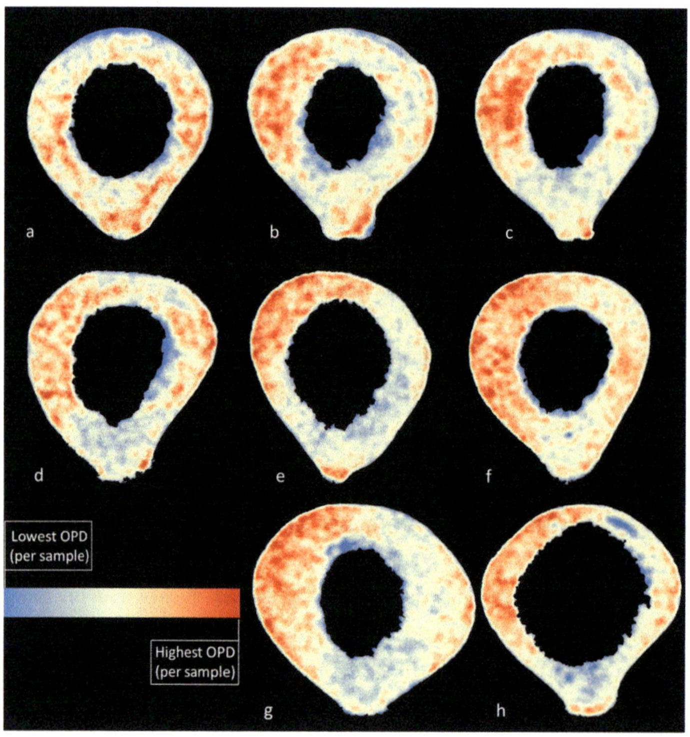

Figure 8.1 Relative density maps of OPD. Red indicates highest observed OPD in a sample, and blue indicates lowest OPD in a sample, with all values between stretched along the color palate. (a) 21-year-old male, high OPD = 25.0; (b) 35-year-old female, high OPD = 32.0; (c) 44-year-old female, high OPD = 42.0; (d) 57-year-old female, high OPD = 38.0; (e) 69-year-old female, high OPD = 45.0; (f) 76-year-old male, high OPD = 48.0; (g) 84-year-old male, high OPD = 48.0; (h) 92-year-old female, high OPD = 52.0. Lowest observed OPD for each sample was 0.0. All samples are oriented with anterior at the top and medial to the right. Modified from Gocha and Agnew 2016, *Journal of Anatomy*; used with permission.

OPD from each of the 100 ROIs positively and significantly correlated with age-at-death, though strength of that correlation varied depending on the ROI. Figure 8.2 shows an example of two segmentation techniques (APML quadrants by thirds, and APML octants by thirds) along with Pearson correlation coefficients for OPD and age-at-death for each ROI (all $p < 0.05$). Generally, a trend is evident wherein remodeling in the anterior and lateral regions of the femur, particularly in the periosteal one-third, correlates more strongly with age than remodeling in the medial or posterior regions of the femur or areas deeper in the cortex.

Stepwise linear regression was used to determine which ROIs from each of the 10 segmentation methods could be combined to predict age with greatest accuracy. Each regression model indicated a robust relationship between OPD and its ability to estimate age-at-death. The two most promising models, as judged by adjusted R^2 values and standard errors, were APML quadrants by thirds and APML octants by thirds (Figure 8.2), with both models explaining >90% of age variability and standard errors ranging 5.82–6.73 years. To evaluate model performance, the sample set was subjected to jackknife leave-one-out cross-validation, which demonstrated the models worked equally well in young and old individuals, even into the tenth decade of life. For example, a female individual who was 92 years old at death was estimated to be 92.4 years old by the AMPL quadrants by thirds model and 92.1 years old by the APML octants by thirds model.

Importantly, Gocha suggests that the femur, compared to bones with smaller cross-sections, is less likely to demonstrate an OPD asymptote if large enough ROI's are investigated and the bone is non-pathological. An OPD asymptote occurs when new osteons remove evidence of previous ones creating an asymptotic value. Frost (1987) hypothesized that the femur would hit an asymptote at an OPD of 50/mm^2, and although Gocha found OPD values as high as 55/mm^2, they only existed in small pockets of 1–2 mm^2 in the antero-lateral or lateral femur, and none were evident until the eighth decade of life. This suggests that if ROIs are large enough to encompass sufficient variation in remodeling density, such small asymptotic areas are of little concern.

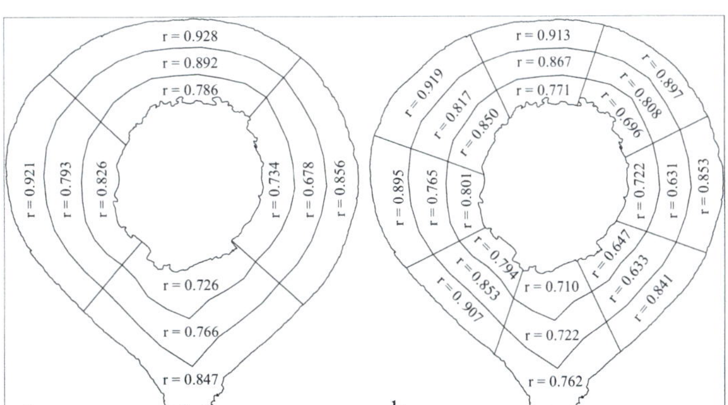

Figure 8.2 Schematic of two femoral segmentation methods from Gocha 2014, with Pearson correlation coefficients between OPD and age-at-death listed within each ROI. (a) APML quadrants by thirds sampling strategy; (b) APML octants by thirds sampling strategy. Both samples are oriented with anterior at the top and medial to the right. Figure by T. Gocha.

Despite the promise of such results, the sample for this pilot research included only 30 individuals and thus falls short of a truly applicable method in medicolegal/forensic casework. To remedy this, Gocha and colleagues are currently expanding this research with an ongoing National Institute of Justice grant that plans to examine >300 individuals from a relatively diverse sample and apply more advanced statistical modeling such as random forest rather than linear regression.

8.2.3.7 Goliath and Colleagues: Combining Skeletal Elements for Age-at-Death Estimation

To explore the potential link between osteon shape and size as well as their association with age-at-death estimation, Goliath et al. (2016) developed age-at-death estimation equations using a femur and rib sample from 27 individuals from dissecting room cadavers from the University of Missouri and Washington University St. Louis. There is no explicit description of the data collection procedures, but both direct observation through the microscope and digital images were used to collect variables of OPD, osteon circularity, and osteon area. From these data, stepwise linear regression equations are proposed for the rib and femur elements separately and together, yielding standard errors of the estimate of 6.059 years, 5.919 years, and 5.614 years, respectively.

The importance of this study is that it provides a histological age-at-death estimation method for combined skeletal elements and explores an additional histomorphometric variable for age-at-death estimation (osteon circularity), though expanding the research to a larger sample size with a greater age range should be explored.

8.2.3.8 Pfeiffer and Colleagues: Population-Specific Equations for Age-at-Death Estimation

To examine the necessity of population-specific histological age-at-death estimation methods, Pfeiffer et al. (2016) used 213 rib samples from the Kirsten Skeletal Collection in South Africa to develop regression equations for subpopulations in the collection ("Colored," "Black," and "White"). This was done using direct observation of the cross-section through a counting reticle in bright field light. Their results indicated that population-specific equations did not improve age-at-death estimates.

Pfeiffer et al. also included a pointed critique on the use of linear regression analysis for histological age-at-death estimation. As the authors note, the nature of the linear regression approach is to model accumulation of osteons at the same rate throughout the adult lifespan of an individual, although biologically this relationship is not linear. This mismatch between model choice and the underlying biological reality of histological senescence results in consistent overestimation of younger individuals and underestimation of older individuals. In their discussion, the authors describe that when linear regression is used to examine the relationship between OPD and age, an OPD value of 40/mm^2 would be needed to precisely estimate the age of an 80-year-old from their sample, where the highest recorded OPD value was 34.79/mm^2. Similarly, at the lower end of their sample age range, a 20-year-old individual would need an OPD value of 2.5/mm^2, which is smaller than their smallest observed value (7/mm^2). Pfeiffer and colleagues thus call for a new statistical approach to histological age-at-death estimation and specifically suggest a maximum likelihood approach, which still has not been explored. However, it should be noted that larger skeletal elements such as the femur allow researchers to observe higher OPD values, suggesting they may be better suited for estimating age in older individuals.

8.2.3.9 *Michener and Colleagues: Three-Dimensional Histomorphometry*

Although three-dimensional imaging of bone microstructure has become more commonplace in recent decades through synchrotron and micro-computed tomography (e.g., Cooper et al., 2006; Cooper et al., 2007; see Chapter 12 in this volume), rarely has such three-dimensional data been applied to age estimation. However, Michener et al. (2020) expanded use of GIS software in researching bone microstructure using serial sections from a single femoral midshaft to push beyond two-dimensional and into three-dimensional histomorphometry. Using three serial sections from a single femur, the authors identified >1000 osteons that were connected across the sections and extrapolated the volume of each osteon within GIS. Through their serial sections the authors noted the area of an osteon can vary between sections, and therefore osteon volume may be a more promising variable for age estimation than osteon area as used in some previous methods. Unfortunately, because only a single individual of unknown age was used in this research, the relationship of osteon volume with age is still unclear. Yet this research remains an important proof-of-concept for three-dimensional use of GIS for histomorphometry.

8.2.3.10 *Kenyhercz and Colleagues: Random Forest Regression Modeling*

Kenyhercz et al. (2020) reanalyzed the data from Crowder and Dominguez (2012) using machine learning, specifically random forest regression analysis, which improved correlation values between estimated and known ages and reduced bias compared to earlier 2013 equations. Further, to make this method more readily usable for practitioners, a graphical user interface (GUI) named keRley was developed. The GUI uses the random forest package for R to allow practitioners to input any combination of variables among intact OPD, fragmentary OPD, anterior width of the femur, and osteon area to produce a prediction interval based on a summary of quantiles (2.5 and 97.5) from 800 random forest trees. The interface also produces a histogram summarizing the results from each tree as well as a mean, median, and mode age generated from the forest. While the random forest regression approach would normally provide a single point age estimate, in order to be applicable to forensic scenarios where age intervals are best practice the method provides a 95% prediction interval based on the range of estimates from the forest for the unknown sample. While this method has not formally been published, it was limitedly tested on burned and unburned femora from a modern sample and shown to accurately predict age-at-death (Mavroudas et al., 2022). The GUI is currently available online at anthropologyapps.com.

Development of this random forest model approach for histological age-at-death estimation is an important step to improving estimation methodology and warrants further exploration using more diverse geographic populations and more skeletal elements. This is the first nonlinear regression approach in histological age-at-death estimation methodology that addresses the issues with sample size outlined by Hennig and Cooper (2011) and issues with age mimicry outlined by Konigsberg et al. (2008).

8.2.3.11 *García-Donas and Colleagues: Interpopulation Differences in the Mediterranean*

To further explore the effect of population variation on histological age-at-death estimation of the human mid thoracic rib, García-Donas et al. (2022) used a sample of 88 individuals from Greece and Cyprus to produce 12 models for histological age-at-death estimation using general linear regression models with linear and multiple regression analyses. The best performing model was based on use of OPD, osteon perimeter, and osteon circularity. Data were collected using digital imaging with ImageJ software and direct observation through a light microscope with polarized light.

The results of this study indicate that there may be interpopulation differences between Greek and Greek-Cypriot individuals, which merit further exploration in larger samples. The study also shows that in this Mediterranean sample osteon circularity is a useful predictor variable for age-at-death estimation, as shown previously by Goliath et al. (2016). However, the utility of osteon circularity in these studies is potentially related to the small sizes and age distribution of the overall sample, as subsequent studies examining the utility of osteon circularity in the human rib for age estimation have shown weak correlations (Mavroudas, 2023).

8.2.3.12 *Mavroudas: Random Forest Modeling and Use of Multiple Osteon Morphotypes*

Building off the Kenyhercz et al. (2020) femur method, Mavroudas (2023) applied random forest regression modeling to age-at-death estimation of the human mid thoracic rib using diverse variables including osteon morphotypes. This was done via direct observation through the microscope combined with digital image analysis using Image J software (Schneider et al., 2012) and a digitizing tablet for the entire rib cross-section. Variables included intact OPD, fragmentary OPD, drifting OPD, type II OPD, primary OPD, osteon area, and relative cortical area. These variables were used in multiple combinations to explore the utility and relative importance of each for age-at-death estimation. The total sample consisted of 246 individuals aged 16–102 years primarily from the United States and Mexico, although taphonomic alterations and differential preservation did not allow for all data types to be collected from each sample.

Using leave-one-out validation to maximize age distribution of the developmental sample, the results showed accurate age-at-death estimates (known age within the 95% predicted age interval), with the highest accuracy for runs that included multiple osteon morphotype variables. As predicted, due to the nature of random forest regression modeling, inaccuracy and bias values were greatly reduced when compared to linear regression methods (specifically the Cho et al., 2002 method) and comparable to gross morphological methods.

This method also directly answered the call of Pfeiffer et al. (2016) to explore nonlinear regression-based statistical approaches to avoid overfitting data to a linear model. The benefit of random forest modeling generally is that the ensemble approach significantly reduces overall bias, but practitioners should remember that the method cannot extrapolate data. Consequently, it cannot predict an age that is beyond the sample data age distribution. For example, if the developmental data used to produce a random forest model consisted of individuals aged 30–102 years, the model could never produce an age-at-death estimation of <30 years. To avoid this issue, it is especially important for histological age-at-death estimation methods applying random forest modeling to be explicit in their sample composition and for practitioners to take this limitation into account when applying the method. Both Kenyhercz et al. (2020) and Mavroudas (2023) provide these data.

The importance of this study is that it expanded the utility of random forest modeling for histological age-at-death estimation to an additional skeletal element (the rib), demonstrated the value of using non-traditional osteon morphotype data as predictor variables, and expanded the utility of histological age-at-death estimation of the human rib into the ninth decade of life.

8.3 Discussion and a Path Forward

Although the methods surveyed above do not cover every histological age-at-death estimation method ever developed, they do represent most major developments in the field as well as investigations that provide cautionary tales. Despite more than a century of investigations into using skeletal histology to estimate age-at-death, there is significant variability in published approaches. Methods differ in key areas including sample construction, skeletal element(s) investigated, variables included, sampling/ROI size and location, and the statistical methods used.

Constructing a sample to develop histological age estimation methods is often not straightforward. Despite development and growth of several modern documented and ethically sourced skeletal collections worldwide (Alves-Cardoso et al., 2022), most such collections have restrictions on destructive sampling, so collections that do allow sampling offer a crucial resource to researchers. Another valuable resource is the handful of documented collections designed specifically to study bone microstructure, including those that consist primarily of bone sections taken for further sampling rather than whole skeletons (see Chapter 11 in this volume). Even when permissions are granted to sample from traditional or histological skeletal collections, these collections are often heavily skewed in their age distributions and are rarely balanced between biological sexes. These factors are evident in most of the 20th-century methods that are often plagued by small sample sizes, age distributions biased heavily toward the elderly, and almost universally samples with more males than females. Although many early methods for histological age estimation report no differences between male and female data (e.g., Kerley 1965, Stout and Paine 1992), these methods were often constructed with relatively small sample sizes that had a large imbalance between the number of males and females in their sample. Conversely, methods with much larger sample sizes that are more balanced between biological males and females have indeed reported significant differences between male and female data and report sex-specific equations to be more accurate (e.g., Ericksen 1991; Kenyhercz et al., 2020). Considering differences in remodeling between biological males and females, especially later in life, it is imperative to have a reasonable number of both male and female samples available for method construction and validation. Sex-specific equations should be developed; however, pooled data should also be examined and non-sex-specific age estimation formulae be developed since biological sex might be unknown in numerous forensic cases and/or bioarchaeological contexts, especially in cases of incomplete recoveries or with taphonomic alterations. Fortunately, most newer methods have tried to remedy these issues with larger samples sizes that are more balanced between the sexes and with more equal age distributions.

The health status of individuals included in method development also varies significantly. Some researchers have preferred to include only individuals who were "healthy" at the time of death (e.g., Thomas et al., 2000; Maat et al., 2006), whereas others have divided their sample into pathological and nonpathological groups (e.g., Thompson, 1979). While these samples undoubtedly avoid the confounding nature of disease processes that can affect skeletal remodeling (e.g., diabetes, metabolic disorders), they represent an unrealistic, idealized reference population rather than being representative of real-world populations. Moreover, this omission raises concerns about the applicability of such methods in forensic contexts, which frequently involve decedents from vulnerable and/or marginalized groups. Since there is no a priori way to know if an unknown individual had any such diseases or disorders, especially if represented only by fragmentary skeletal remains, to compare those remains only to a "healthy" reference population lacking sufficient variation of the human condition is not justified. Although it might increase error rates, methods developed using samples that include both pathological and nonpathological individuals are expected to be more realistic in their ability to estimate age from the general population. A 2006 study by Paine and Brenton highlights this well. The authors set out to test the Stout and Paine (1992) rib equation for age estimation on 26 individuals from South Africa who were known to suffer from malnutrition or the niacin deficiency disease pellagra at the time of their death. Results revealed that these individuals were underaged by an average of 29.2 years, and the authors rightfully point out that methods developed on 'healthy' individuals are likely of limited utility to individuals of poor health who are regularly encountered in forensic anthropology casework.

Another source of methodological variability is the skeletal element(s) used by various researchers. Long bones, particularly weight-bearing long bones, have been favored by researchers. However, their large size usually prevents analyzing much of the cortex in a timely manner. Further, weight-bearing long bones are subject to significant mechanical demands from locomotion, which impacts the spatial variation of remodeling events within bone cortices (Gocha and Agnew, 2016). As such, biomechanical demand on long bones may result in disparate remodeling rates between individuals with varied activity habits and body sizes. Though this has yet to be equivocally demonstrated, research has shown that the cross-sectional properties of diaphyses vary in archaeological populations that exhibited different subsistence strategies, as well as those that traversed different terrain types (Ruff, 2018). Moreover, biomechanical loading has been shown to impact histological structures including but not limited to collagen fiber orientation, which may help optimally dissipate energy through the cortex (see Chapter 5 in this volume). The growth of documented skeletal collections with data being collected on mobility patterns and habitual activities may help more decidedly answer questions related to activity patterns, though weight restrictions in most body donation programs will largely hamper the evaluation of the effects of body size. For many of these reasons, smaller, non-weight bearing bones such as the ribs, clavicle, and metacarpals have been suggested for age-at-death estimation. The smaller size allows for greater topographical sampling of these elements, allowing analysts to often assess the entire cross-section to estimate age-at-death. Unfortunately, their smaller cross-sections also mean they are more likely to reach an OPD asymptote at an earlier age, although research on the rib has documented the OPD asymptote at >70 years of age (Beresheim et al., 2018; Mavroudas, 2023). Overall, the majority of methods have focused on the femur and rib, although more elements should be thoroughly investigated to better understand remodeling patterns throughout the skeleton and how they might be useful in age estimation. Following the example of Goliath et al. (2016), incorporating multiple elements to construct a holistic age-at-death interval may provide important insights into intra-skeletal variation.

A great deal of variation also exists in the variables that have been examined in method development. Some methods have incorporated structural variables such as cortical thickness (e.g., Thomas et al., 2000; Crowder and Dominguez, 2012), but these measures have rarely appreciably increased the accuracy of age estimates (see Mavroudas et al., 2022). Further, the number of variables investigated varies widely, ranging from a single variable (e.g., Maat et al., 2006) up to 19 (Thompson, 1979), though including additional variables also does not seem to necessarily improve model performance for traditional regression methods. When only a single variable is investigated, it is usually either a percentage of remodeled or non-remodeled bone (e.g., Ahlqvist and Damsten, 1969; Maat et al., 2006) or OPD. Arguably, since calculating the percentage of remodeled or non-remodeled bone does not require differentiating between intact and fragmentary osteons, data collection for this variable is easier. However, because osteon size decreases with age, the same percentage of remodeled bone could belong to a younger individual with fewer but larger osteons or an older individual with more numerous but smaller osteons. In the only research to compare these two variables directly, Stout and Stanley (1991) quantified OPD and percentage of remodeled bone in the radius, tibia, and fibula of 36 cadavers and found osteon counts are a significantly better predictor of age. Although distinguishing between intact and fragmentary osteons is sometimes a source of consternation for researchers (Lynnerup et al., 1998), the authors of this chapter firmly believe it is a worthwhile endeavor. Simply put, the biological process of intracortical remodeling correlates well with chronological age, and as such it should be remodeling events themselves, both intact and fragmentary, that are quantified to estimate age. In fact, recent research demonstrates that fragmentary OPD correlates more strongly with age than intact OPD in both the femur (Crowder and Dominguez, 2012) and rib (García-Donas et al., 2022; Mavroudas, 2023). When some of these data are systematically ignored, for example only quantifying intact osteons to the exclusion of fragmentary osteons (e.g., Thomas et al., 2000), the results are significantly less promising.

Another source of methodological variation is the size and location(s) of ROIs used in the cross-section of larger skeletal elements such as the femur (Figure 8.3). Due largely to technological limitations, most of the earliest methods used a static-sized ROI for quantification (e.g., Kerley, 1965) assessed directly through the microscope. In practice, this was and still is difficult to operationalize because the size of a viewing field varies based on the technical specifications of different microscopes. To remedy this, use of correction factors has been suggested wherein one collects data using the field size obtainable by a particular microscope and then adjusts osteon counts proportionally. However, research by Stout and Gehlert (1982) demonstrated that correction factors are of limited value due to spatial variation in remodeling event distribution. Admittedly, this is less of a concern in the 21st century with the relative ubiquity of digital microscopic imaging and accompanying software. If one wanted to apply the Kerley method to a modern sample, it would be relatively easy to construct a circular ROI of exactly 1.62 mm in diameter using programs such as ImageJ or GIS software.

Nevertheless, nearly 60 years after Kerley first introduced a method for histological age estimation from the femur, the optimal size for ROI investigation remains an open question. As Frost (1969) noted, smaller ROIs can fail to capture variation in a cross-section. To remedy this, he recommended investigation of a total area of ≥50 mm^2 to avoid spatial variation that negatively impacts interpretation. The earliest methods typically used small ROIs, usually <2 mm in diameter, but more recent methods have used much larger ROIs. For example, Crowder and Dominguez (2012) used a variable-sized ROI at the anterior femur, fixed at 5-mm-wide and spanning the depth of the cortex from the periosteum to the endosteum, resulting in an average ROI of ~19 mm^2. Gocha (2014) examined anatomically defined octants and quadrants that varied in size, with the anterior octant averaging ~42 mm^2 and the anterior quadrant averaging ~90 mm^2. Considering both Crowder and Dominguez (2012) and Gocha (2014) primarily used the Ericksen collection for their research, Ingvoldstad et al. (2017) performed a meta-analysis comparing ROI size and the

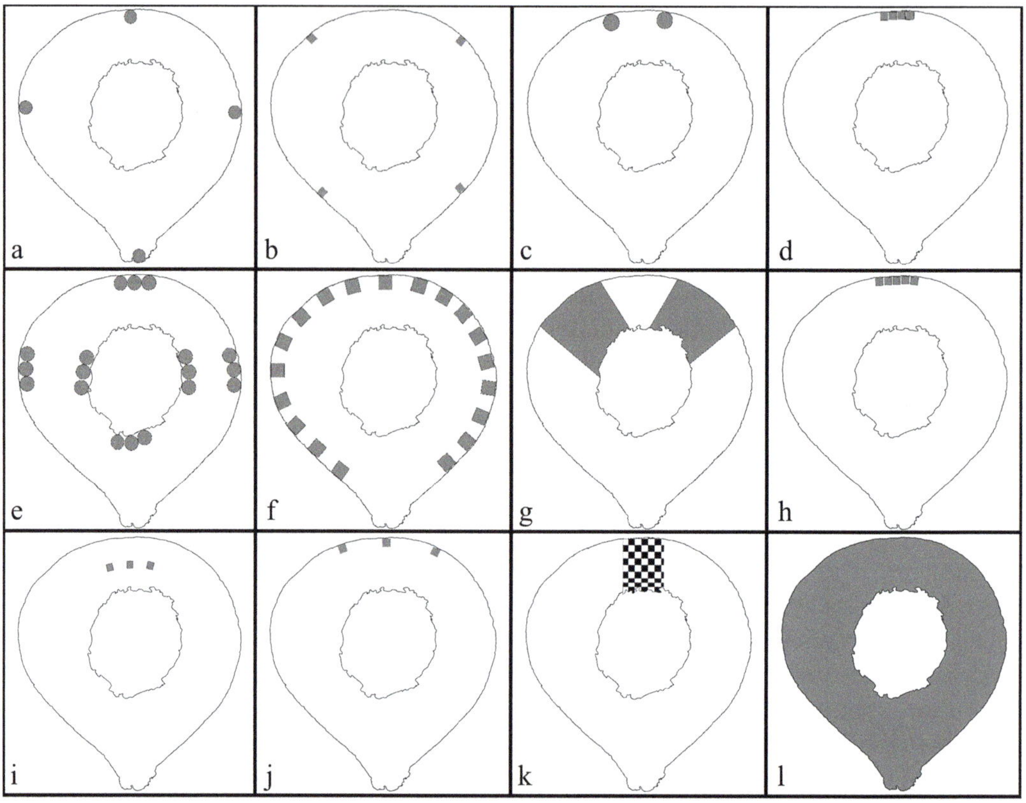

Figure 8.3 Areas sampled for quantification of variables for histological age estimation from the femur. All ROIs are to scale with reference to the cortex. (a) Kerley, 1965; (b) Ahlqvist and Damsten, 1969; (c) Singh and Gunberg, 1970; (d) Thompson, 1979; (e) Hauser et al., 1980; (f) Drusini, 1987; (g) Samson and Branigan, 1987; (h) Ericksen, 1991; (i) Watanabe et al., 1998; (j) Maat et al., 2006; (k) Crowder and Dominguez, 2012, and Kenyhercz et al., 2020; (l) Gocha, 2014. Figure by T. Gocha.

relationship of OPD and age-at-death. Simple correlation analysis revealed that OPD and age-at-death were positively correlated for both methods and across all ROI sizes, although the correlation coefficient increased as ROI size increased. For Crowder and Dominguez's ROI (average size: ~19 mm^2) the correlation coefficient was $r = 0.681$; for Gocha's octant (average size: ~42 mm^2) the correlation coefficient was $r = 0.907$; for Gocha's quadrant (average size: ~90 mm^2) the correlation coefficient was $r = 0.918$. Admittedly, this meta-analysis examined only OPD, not the constituent variables of intact and fragmentary OPD, and it did not account for possible sex differences. Still, the results strongly suggest examination of small, isolated ROIs is not recommended. Ingvoldstad and colleagues concluded future studies should consider larger ROIs (although exact size recommendations were not provided), even at the expense of sample size.

Another significant consideration in development of histological age estimation methods is the statistical approaches used for model generation. While gross morphological approaches have been incorporating more rigorous statistical methods into skeletal age-at-death estimation since the beginning of the 21st century (Konigsberg et al., 2008;

DiGangi et al., 2009; Slice and Algee-Hewitt, 2015; Dudzik and Langley, 2015; Boldsen et al., 2022; Stull et al., 2022), histological approaches have been less ambitious with the statistical modeling used to develop age estimates. To date, statistical interpretation of bone histology for age-at-death estimation has largely relied on regression techniques. Although these methods have proven accurate, it is worth considering that the limitation of current histological methods may be surpassed by using more advanced statistical modeling. With the exception of Kenyhercz et al. (2020) for the femur and Mavroudas (2023) for the rib, there has been no rigorous exploration of machine learning methods or maximum-likelihood approaches to histological age-at-death estimation. The success of machine learning methods to improve age-at-death estimates while incorporating multiple and complex variable types such as deep neural network (Navega et al., 2022) or mixed cumulative probit methods (Stull et al., 2022) is promising and indicates that these methods could be applied to histological data as well. Another technique worth exploring is structural equation modeling. Although structural equation modeling has yet to be applied in histological age-at-death estimation, ongoing research is exploring this potential.

Ideally, future researchers will have multifactorial gross and histological data available from the same individuals to incorporate both data types into a comprehensive, accurate, and useful age-at-death estimation method for unknown skeletal material. Application of advanced statistical modeling to combined gross and histological data from the same individuals will also allow future researchers to evaluate patterns of skeletal senescence within an individual and across populations to better understand which environmental and genetic factors impact skeletal age indicators.

8.3.1 Use of Histological Age-at-death Estimation in Applied Contexts

There are myriad ways in which histological age-at-death estimation methods can be advantageous to examine skeletal senescence in applied contexts. When limited skeletal material is available for examination, histological techniques allow for analysis of even fragmentary remains. This is not limited to age-at-death analyses, as microscopic analysis of cortical bone has proven efficient for distinguishing between human and non-human material (see Chapter 7 in this volume). Additionally, taphonomic alterations including thermal alterations can often affect the efficacy of gross morphological age-at-death estimation indicators, whereas histological age-at-death estimation has been shown in recent research to still be applicable to thermally altered remains (Mavroudas et al., 2022).

Another notable benefit is the relatively small amount of bone required for histological age-at-death estimation techniques. When analyzing fully fleshed individuals, it may be impractical to macerate an entire individual to perform age-at-death estimations. In these scenarios, sampling a few select skeletal elements for histological analysis is less invasive than cleaning an entire skeleton. Even in settings where gross indicators of age can be easily removed from fully fleshed individuals (e.g., medial clavicles, fourth sternal rib ends, pubic symphyses), also removing histological samples would be a simple addition and allow for more holistic age estimates. Particularly considering humanitarian contexts in which multiple human remains are recovered at once and/or there are time constraints on field analyses, histological sampling is a fast and easy approach to age-at-death estimation from many individuals at once. Here, incorporation of histological estimation into a sampling strategy could reduce the time from sample acquisition to age estimation and provide holistic age estimates for identification and documentation purposes. Even in archaeological contexts, which may require analysis of fragmented, taphonomically altered, and/or commingled remains in time-sensitive settings, histological techniques provide effective tools to mitigate such technical and logistical difficulties.

8.3.2 The Case for Using Histological Age-at-Death Estimates

Biological anthropologists most often rely on macroscopic methods for age estimation. In a survey of practicing forensic anthropologists, Garvin and Passalacqua (2012) report that preferred methods are all macroscopic in nature, but that the most common response was histology in regard to reliable, yet underutilized, skeletal methods. Many respondents justified the under-use of histology by noting it requires training, experience, and specialized equipment and is destructive in nature. The authors of this chapter contend that some shortcomings of macroscopic methods are often overlooked, the critiques of histology are perhaps exaggerated, and there are some instances in which histological methods are preferable for use.

The myriad ways in which various populations use their skeletons lead to vast differences in degeneration rates that contribute not only to large age-at-death estimation ranges (sometimes >50 years) but also difficulty predicting changes in degenerative rates between populations. This population-use dependency is also exacerbated by the effect of body size differences on gross morphological age estimates, wherein ages for larger and obese individuals are often underestimated (Merritt, 2015, 2017; Wescott and Drew, 2015). As obesity rates continue to climb in recent decades, this problem will only increase for forensic anthropologists. Alternatively, from a bone biology perspective, population variation in histological structures should be predictable based on genetically constrained features such as osteon area (Havill et al., 2013), and body size seems to have little effect on age-at-death estimation variables (Britz et al., 2009; Beresheim et al., 2018; Mavroudas, 2023). As such, histological methods may be more reliable for age estimation in individuals of larger body size than macroscopic methods, though this should be more fully researched.

Age estimation in the elderly is another area where macroscopic methods historically leave much to be desired. For decades, anthropologists have noted that while traditional macroscopic methods perform acceptably for younger adults, they offer little precision for individuals older than middle-aged. Indeed, age ranges produced from some pubic symphysis or auricular surface methods are >50 years, often leading practitioners to truncate an age estimate for an unidentified set of remains to merely ≥50 years old at the time of death. Although more recent macroscopic methods such as transition analysis may offer some hope in this area, it is uncertain how far into the adult age spectrum this method can provide accurate and precise age estimates. This is an area where recent histological research shows much promise. Using larger ROIs in the femur, or using various osteon morphotypes in the rib, histological methods offer the ability to estimate age accurately and precisely into the ninth and tenth decades of life (Crowder and Dominguez, 2012; Gocha, 2014; Mavroudas, 2023).

Additionally, many biological anthropologists might assume the need for specialized equipment is cost-prohibitive to undertake histological analysis. While most histological laboratories have specialized saws with diamond blades for cutting bone, grinders/polishers for thinning bone sections, and transmitted light microscopes for analysis of slides—which together typically cost tens of thousands of dollars (US)—there are economical alternatives. In 1958 Frost introduced a method for rapid, manual preparation of undecalcified bone sections for histological analysis that requires no specialized equipment. More recently, Maat et al. (2001) updated Frost's method to improve ease of application and point out that it is simple, extremely cheap, rapid, and reliable. Such methods eliminate the need for a specialized saw and grinder/polisher. As an alternative to traditional transmitted light microscopes, stereoscopic microscopes can be used for histological analysis and are already a fixture in many forensic anthropology laboratories for trauma analysis. Figure 8.4 depicts a full cross-sectional image from a sixth rib taken with a stereoscope at the Forensic Anthropology Center at Texas State. Figures 8.5 and 8.6 both depict a zoomed-in view of the inferior pleural surface of this rib, one image taken with a traditional transmitted light microscope (Figure 8.5) and the other with a stereoscope (Figure 8.6), demonstrating quantification of histological variables is quite achievable even with only a stereoscope.

Overall, histological analysis of bone provides alternative means of estimating age-at-death from skeletal remains that can be used in all contexts including fragmented or taphonomically altered material. Beyond forensic applications, examination of microscopic age-related changes in human bone throughout the skeleton and between geographic populations can provide insight into macroevolutionary processes of skeletal senescence. Incorporating histological age-at-death estimation methods in addition to gross morphological indicators also provides a more holistic age estimate when compared to gross indicators alone (Stout et al., 1994), and future research should explore such possibilities. Although technological demands have constrained research questions and methods development in the past, newer approaches to histological analysis have embraced readily available software and accessible GUI statistical interfaces, making it more practical than ever to incorporate histological techniques.

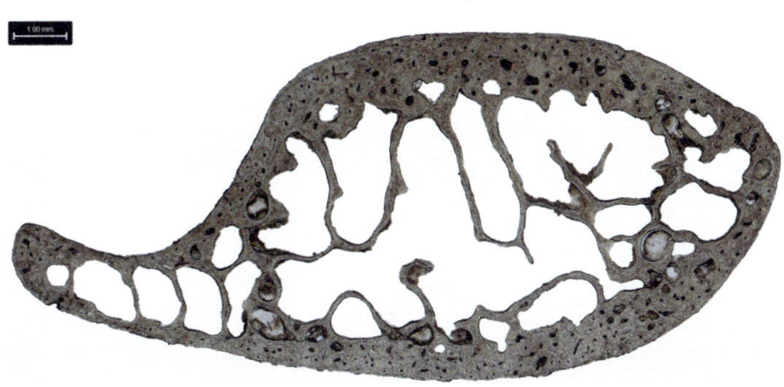

Figure 8.4 Cross-sectional image from a sixth rib taken with a Leica M205C stereoscope. Image by S. Mavroudas.

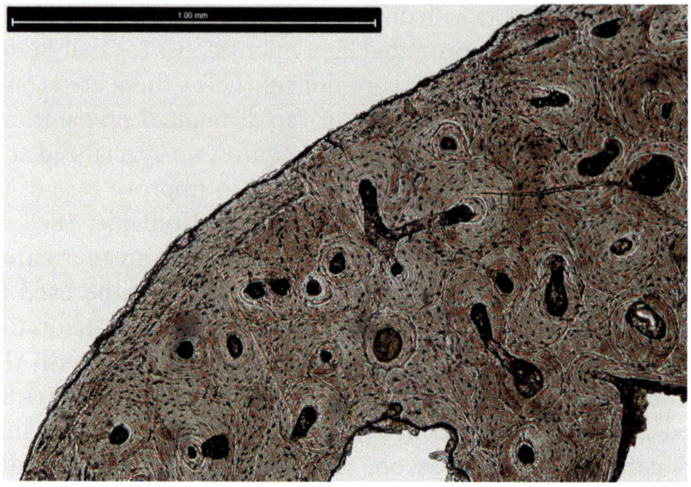

Figure 8.5 Zoomed-in view of the inferior pleural surface of the same rib in Figure 8.4. Image taken with a Leica DM6M traditional transmitted light microscope. Image by S. Mavroudas.

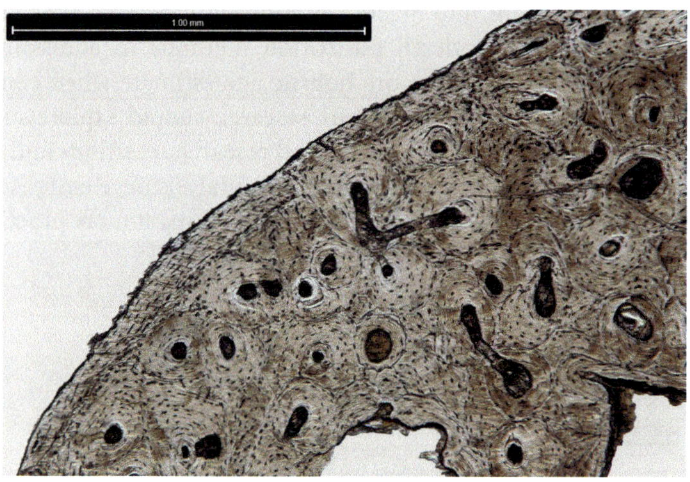

Figure 8.6 The same zoomed-in view of the inferior pleural surface as in Figure 8.5. However, this image was taken with a Leica M205C stereoscope. Image by S. Mavroudas.

Appendix A: Summary of selected histological age-estimation methods[a]

Method	Bone	Specimen Sampling Technique	Sample Parameters: (N); Age Range {$\bar{x}$}, Sex [♂:♀:?] and Ancestry	Variables Used In Analysis	Statistical Methods Used In Analysis	Reported Accuracy
Balthazard and Lebrun, 1911	Tibia	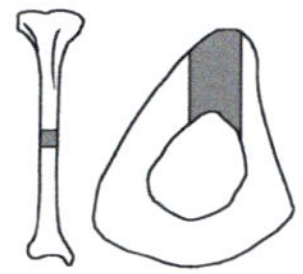	(65); 0–84 {39.2} [?:?:?] ? ? ? ?	1) Mean diameter of (100–200) Haversian canals	Not specified	67% of sample within ± 5 years
Kerley, 1965 & Kerley and Ubelaker, 1978	Femu	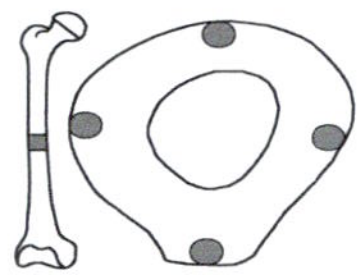	(67); 0–95 {41.6} [43:17:7] White and Black; distribution in sample not provided	Number of: 1) intact osteons, 2) fragmentary osteons, 3) non-Haversian canals, 4) % unremodeled bone	Kerley (1965): Linear and curvilinear regression Kerley & Ubelaker (1978): polynomial regression	Kerley (1965): S.E.E. = 9.39–13.85 Kerley & Ubelaker (1978) S.E.E. = 6.98–12.52
	Tibiar	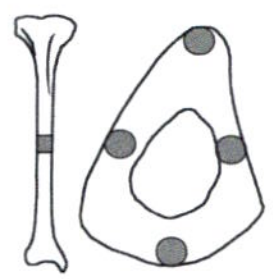	(33); 0–85 {?} [24:8:1] White and Black; distribution in sample not provided			Kerley (1965): S.E.E. = 6.69–13.62 Kerley & Ubelaker (1978): S.E.E. = 8.42–14.28
	Fibula	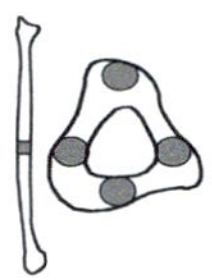	(25); 0–83 {34.5} [19:5:1] White and Black; distribution in sample not provided			Kerley (1965): S.E.E. = 5.27–10.85 S.E.E. = 3.66–14.62
Ahlqvist and Damsten, 1969	Femur	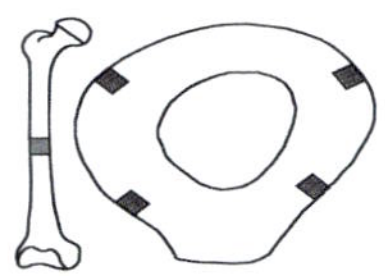	(20); ?-? {55} [?:?:?] Autopsy specimens; no ancestry data provided	1) % remodeled bone	Linear regression	S.E.E. = 6.71

(Continued)

Method	Bone	Specimen Sampling Technique	Sample Parameters: (N); Age Range {$\bar{x}$}, Sex [♂:♀:?] and Ancestry	Variables Used In Analysis	Statistical Methods Used In Analysis	Reported Accuracy
Singh and Gunberg 1970	Femur		(33); 39–87 {62.3} [33:0:0] Cadavers from US; no ancestry data provided	1) total number of osteons 2) mean number of lamellae per osteon 3) mean Haversian canal diameter	Multiple linear regression	S.E.E. = 3.24–5.01
	Tibia	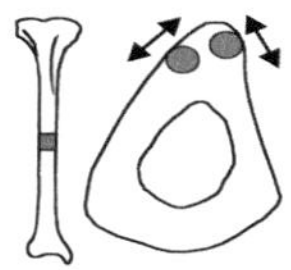				S.E.E. = 3.02–4.59
	Mandible	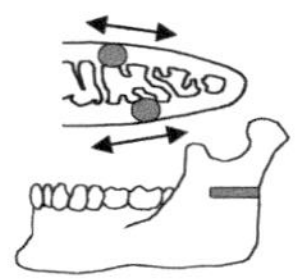	(52); 39–87 {64} [52:0:0] Cadavers from US; no ancestry data provided			S.E.E. = 2.55–3.83 95% of sample within ± 5 yrs. 67% of sample within ± 3 yrs.
Thompson, 1979	Femur	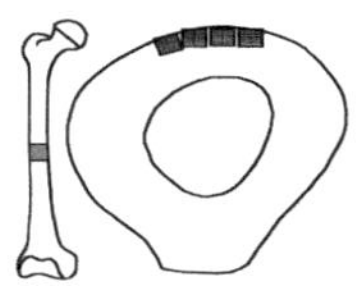	(113); 30–97 {72.1} [63:50:0] New England whites	19 variables analyzed. See original article for variable descriptions	Stepwise linear regression	S.E.E. = 7.1–8.6 (pooled sex); 6.4–8.2 (males) 7.2–9.4 (females)
	Tibia	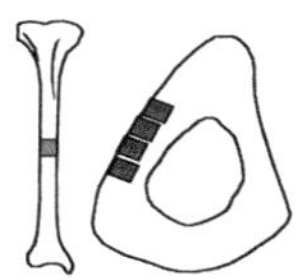	(113); 30–97 {72.1} [63:50:0] New England whites			S.E.E. = 7.6–9.5 (pooled sex); 7.4–9.3 (males) 7.0–10.0 (females)
	Humerus	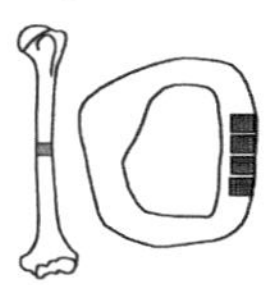	(31); ?–? {67.8} [?:?:?] New England whites			S.E.E. = 6.2–9.5
	Ulna	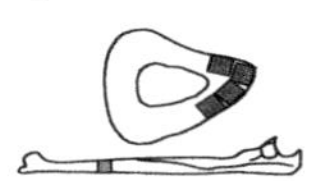	(31); ?–? {68.8} [?:?:?] New England whites			S.E.E. = 7.9–10.6

Hauser et al., 1980	Femur	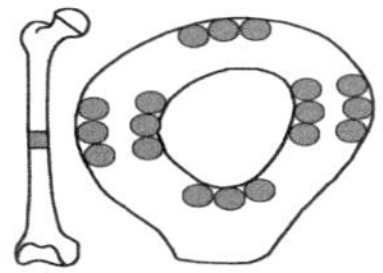	(96); 21–87 {?} [?:?:?] Modern Europeans	1) total number of osteons 2) mean Haversian canal diameter 3) % of ROI occupied by Haversian canals	Linear regression	S.E.E = 10.7–11.4
	Tibia	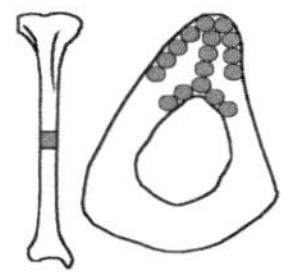	(31); 18–88 {?} [?:?:?] Modern Europeans			S.E.E. = 13.5–16.
Drusini, 1987	Femur	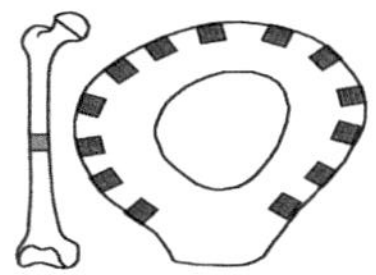	(20) 19–50 {28.8} [?:?:?] Modern Italians	1) number of secondary osteons per mm^2	Linear regression	S.E.E. = 3.92
Samson and Branigan, 1987	Femur	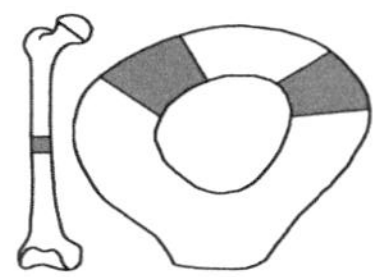	(58) 16–91 {?} [31:27:0] Caucasians	2) number of Haversian canals per mm^2 3) Haversian canal diameter 4) mean cortical thickness excluding linea aspera	Linear regression	S.E.E = 6 for males; 16 for females
Ericksen, 1991	Femur	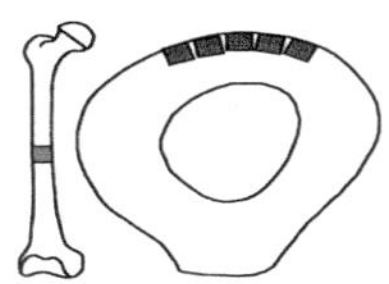	(328) 14–97 {62.8} [174:154:0] 251 US whites, 1 US Asian, 12 US blacks, 6 Chilean Hispanics 58 Domin. Repub. Blacks	Density of: 1) intact osteons, 2) type II osteons, 3) fragmentary osteons, 4) resorption spaces, and 5) non-Haversian canals. Also, mean % of 6) unremodeled, 7) osteonal and 8) fragmental bone.	Stepwise linear regression	S.E.E. = 10.1–12.2 (pooled sex); 10.1–12.0 (males) 10.0–11.6 (females)

(Continued)

Method	Bone	Specimen Sampling Technique	Sample Parameters: (N); Age Range {$\bar{x}$}, Sex [♂:♀:?] and Ancestry	Variables Used In Analysis	Statistical Methods Used In Analysis	Reported Accuracy
Kimura, 1992	2nd Metacarpal	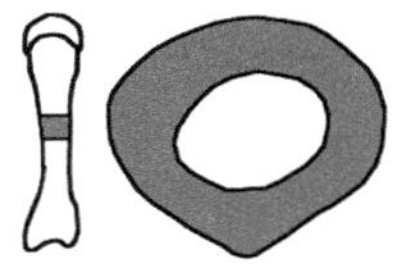	(227); 30–98 {68} [114:113:0] Modern Japanese	1) cortical bone density 2) cortical thickness 3) total number of osteons and osteon fragments in section 4) total osteon density (OPD)	Stepwise linear regression	S.E.E. = 9.99–14.82 (for (total sample); 5.51–8.64 (for 30–65 year old group); 5.64–7.68 (for 66–98 year old group)
Stout and Paine, 1992	6th Rib		(40) 13–62 {28.6} [32:7:1] 32 US whites, 4 US blacks, 4 unknown	1) sum of intact and fragmentary osteons per mm^2 (OPD)	Linear regression	mean absoulte difference between estimates and actual age (rib only) = 3.9
	Clavicle	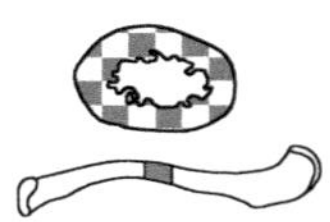				mean absoulte difference between estimates and actual age (rib & clavicle combined) = 3.8 mean absoulte difference between estimates and actual age (clav. only) = 5.9
Stout et al. 1994	4th Rib	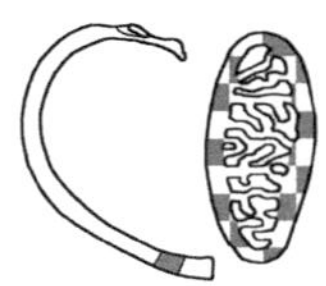	(59) 11–88 {39.2} [?:?:?] US whites	1) sum of intact and fragmentary osteons per mm^2 (OPD) 2) sternal rib phase	Histology only: linear regression Histology & rib phase: mulitple regression	S.E.E. = 10.43 for rib histology alone; S.E.E. = 7.18 for combined rib histology and sternal rib end phase

Cool et al., 1995	Occipital	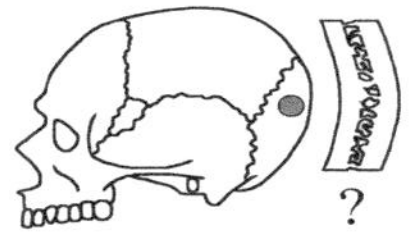	(17); 21–70 {46.8} [17:0:0] Whites	Fractional volume of: 1) primary osteons 2) secondary osteons 3) secondary osteon fragments 4) unremodeled lamellar bone	Linear regression & Canonical correlation analysis	r^2 = .10 (secondary osteons) r^2 = .44 (osteon fragments) r^2 = .32 (lamellar bone)
Stout et al. 1996	Clavicle	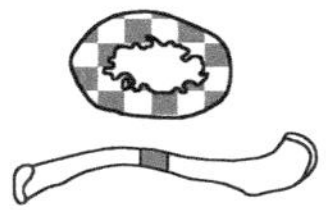	(123) 13–75 {34.0} [73:49:1] 32 US whites; 4 US blacks; 4 unknown; 83 19th cent. Swiss	1) sum of intact and fragmentary osteons per mm^2 (OPD)	Linear regression	r^2 = .85
Watanabe et al., 1998	Femur	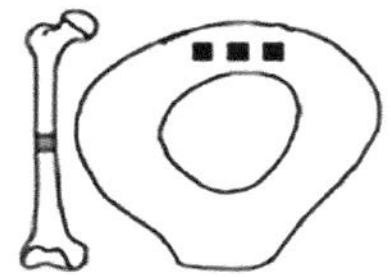	(93); 20–92 {49.6} [72:26:0] Modern Japanese	1) osteon perimeter 2) Haversian canal length 3) number of osteon fragments	Stepwise linear regression	S.E.E. = 6.39
Thomas et al., 2000	Femur	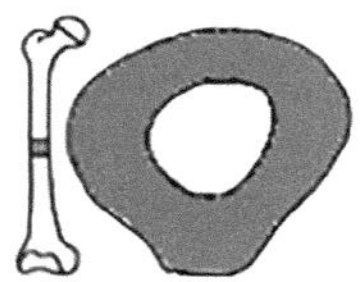	(96); 21–92 {?} [50:46:0] Modern Australians	Numerous macroscopic variables, as well as number size, and density of pores >400 µm. See original text for detailed descriptions.	Multiple linear regression	Macroscopic: adj. r^2 = 0.39 Microscopic: adj. r^2 = 0.33 Macroscopic & microscopic: adj. r^2 = 0.47
Cho et al., 2002	6th Rib		(103) 17–95 {50.4} [?:?:?] 69 historic U.S. blacks, 34 modern U.S. whites	1) OPD (sum of intact and fragmentary osteons per mm^2) 2) osteon area 3) relative cortical area 4) Grouping variable	Stepwise linear regression	50% of estimates within + 11.63 years

(*Continued*)

Method	Bone	Specimen Sampling Technique	Sample Parameters: (N); Age Range {$\bar{x}$}, Sex [♂:♀:?] and Ancestry	Variables Used In Analysis	Statistical Methods Used In Analysis	Reported Accuracy
Maat et al., 2006	Femur	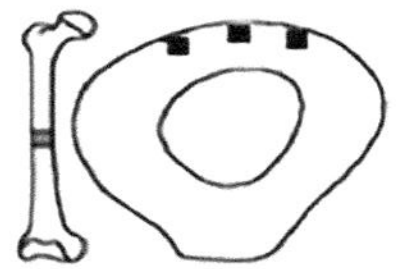	(162); 15–96 {?} [86:76:0] Modern European whites	1) % nonremodeled bone	Quadratic regression & Stepwise linear regression	$r^2 = 0.608–0.799$ S.D. of prediction error = 9.16–14.79 years
Crowder and Dominguez, 2012 & Crowder, 2013	Femur	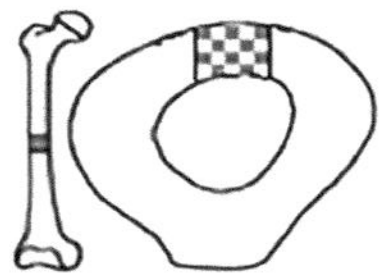	(319); 15–97 {♂63.4, ♀69.5} [169:150:0] 289 U.S. whites, 18 U.S. blacks, 1 asian, 5 hispanic, 7 not reported	1) OPD 2) OPD of intacts (OPD(I)) 3) OPD of fragments (OPD(F)) 4) Osteon area 5) Anterior cortical width	Stepwise linear regression	Pooled sex: $r^2 = 0.587$; S.E.E. = 11.2 Males: $r^2 = 0.51$; S.E.E = 11.95 Females: $r^2 = 0.68$; S.E.E = 9.6
Gocha, 2014 & Gocha et al., 2016	Femur	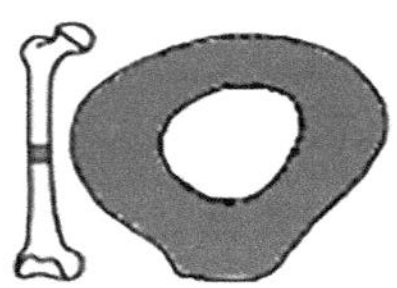	(30); 21–97 {58.9} [15:15:0] 29 U.S. whites, 1 U.S. black	1) sum of intact and fragmentary osteons per mm^2 (OPD)	Stepwise linear regression	Adj. $r^2 = 0.814–0.931$ S.E.E. = 5.82–9.55 95% C.I. = + 2.51–+ 3.57
Goliath et al., 2016	6th Rib		(27) 39–82 {62} [11:16:0] U.S. Whites	1) sum of intact and fragmentary osteons per mm^2 (OPD) 2) Osteon area 3) Osteon circularity	Stepwise linear regression	Rib only: adj. $r^2 = 0.648$ S.E.E. = 6.059
	Femur	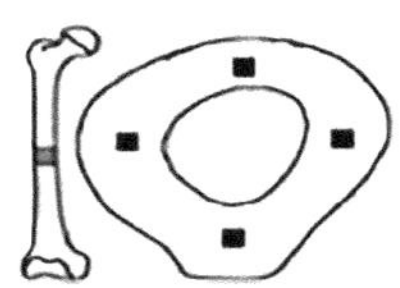				Rib & Femur combined: adj. $r^2 = 0.698$ S.E.E. = 5.614 Femur only: adj. $r^2 = 0.664$ S.E.E. = 5.919

Kenyhercz et al., 2020	Femur	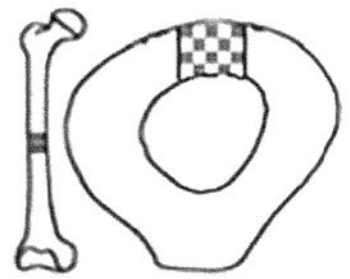	(236); ?-? {?}, subsample from Crowder & Dominguez 2012; no ancestry data reported	1) OPD 2) OPD of intacts (OPD(I)) 3) OPD of fragments (OPD(F)) 4) Osteon area 5) Anterior cortical width	Random forest regression	Pooled sex: $r^2 = 0.719$ Males: $r^2 = 0.692$ Females: $r^2 = 0.796$
García-Donas et al., 2022	Mid-thoracic Rib		(88); 19–100 {60.33}, [40:48:0] 47 Greek-Cypriot; 41 Cretan	13 variables analyzed. See original article for variable descriptions.	Multiple linear regression	Adj. $r^2 = 0.448–0.654$ S.E.E. = 10.71–13.37
Mavroudas 2023	Mid-thoracic Rib		(246) 16–102 {54.6} [128:77:41]; 183 from U.S., 53 from Mexico, 10 from Central America. Mostly modern, some historic.	1) OPD, 2) osteon area, 3) relative cortical area, 4) OPD of intacts (iOPD), 5) OPD of fragments (fOPD), 6) OPD of type II osteons (tIIOPD), 7) OPD of drifting osteons (dOPD), 8) OPD of primary osteons (pOPD)	Random forest regression	$r^2 = 0.31–0.50$

a Modified from Gocha et al., 2018, used with permission. Predicting equations and variable definitions are not provided in the table. In order to use a method properly, the original article should be consulted.

References

Ahlqvist, J. F., & Damsten, O. (1969). A modification of Kerley's method for the microscopic determination of age in human bone. *Journal of Forensic Sciences, 14*(2), 205–212.

Alves-Cardoso, F., Campanacho, V., & Plens, C. R. (2022). Topical collection "the rise of forensic anthropology and documented human osteological collections". *Forensic Sciences, 2*(3), 551–555. https://doi.org/10.3390/forensicsci2030039

Amprino, R., & Bairati, A. (1936). Processi di ricostruzione e di riassorbimento nella sostanza compatta delle ossa dell'uomo. Ricerche su cento soggetti dalla nascita fino a tarda eta. *Zeitschrift fur Zellforschung und mikroskopische Anatomie, 24*, 439–511.

Baccino, E., & Schmitt, A. (2006). Determination of adult age at death in the forensic context. In A. Schmitt, E. Cunha, & J. Pinheiro (Eds.), *Forensic anthropology and medicine: Complementary sciences from recovery to cause of death* (pp. 259–280). Towtowa: Humana Press.

Balthazard, & Lebrun. (1911). Les canaux de Havers de l'os humain aux differents ages. *Annales d'Hygiene Publíque et de Medicine Legale, 15*, 144–152.

Beresheim, A. C., Pfeiffer, S. K., & Alblas, A. (2018). The Influence of Body Size and Bone Mass on Cortical Bone Histomorphometry in Human Ribs. *The Anatomical Record (Hoboken), 301*(10), 1788–1796. https://doi.org/10.1002/ar.23933

Boldsen, J. L., Milner, G. R., & Ousley, S. D. (2022). Paleodemography: From archaeology and skeletal age estimation to life in the past. *American Journal of Biological Anthropology, 178*(S74), 115–150. https://doi.org/10.1002/ajpa.24462

Bouvier, M., & Ubelaker, D. H. (1977). A comparison of two methods for the microscopic determination of age at death. *American Journal of Physical Anthropology, 46*(3), 391–394. https://doi.org/10.1002/ajpa.1330460303

Britz, H. M., Thomas, C. D. L., Clement, J. G., & Cooper, D. M. L. (2009). The relation of femoral osteon geometry to age, sex, height and weight. *Bone, 45*(1), 77–83.

Cho, H., Stout, S. D., Madsen, R. W., & Streeter, M. A. (2002). Population-specific histological age-estimating method: a model for known African-American and European-American skeletal remains. *Journal of Forensic Sciences, 47*(1), 12–18.

Cool, S. M., Hendrikz, J. K., & Wood, W. B. (1995). Microscopic age changes in the human occipital bone. *Journal of Forensic Sciences, 40*(5), 789–796.

Cooper, D. M. L., Thomas, C. D. L., Clement, J. G., Hallgrímsson, B. (2006). Three-dimensional microcomputed tomography imaging of basic multicellular unit-related resorption spaces in human cortical bone. *Anatomical Record Part A: Discoveries in Molecular, Cellular, and Evolutionary Biology* 288A: 806–816. https://doi.org/10.1002/ar.a.20344

Cooper, D. M. L., Thomas, C. D. L., Clement, J. G., Turinsky, A. L., Sensen, C. W., Hallgrímsson, B. (2007). Age-dependent change in the 3D structure of cortical porosity at the human femoral midshaft. *Bone, 40*(4), 957–965. https://doi.org/10.1016/j.bone.2006.11.011

Crowder, C. (2005). Evaluating The Use of Quantitative Bone Histology To Estimate Adult Age At Death [Dissertation] University of Toronto.

Crowder, C. (2013). *Estimation of Age at Death Using Cortical Bone Histomorphometry*. National Institute of Justice. https://nij.ojp.gov/library/publications/estimation-age-death-using-cortical-bone-histomorphometry

Crowder, C., & Dominguez, V. M. (2012). A new method for histological age estimation of the femur. *Proceedings of the American Academy of Forensic Sciences 64th Annual Meeting*, American Academy of Forensic Sciences 64th Annual Meeting in Atlanta, GA.

Crowder, C., Dominguez, V. M., Heinrich, J., Pinto, D., & Mavroudas, S. (2022). Analysis of histomorphometric variables: Proposal and validation of osteon definitions. *Journal of Forensic Sciences, 67*(1), 80–91. https://doi.org/10.1111/1556-4029.14949

DiGangi, E. A., Bethard, J. D., Kimmerle, E. H., & Konigsberg, L. W. (2009). A new method for estimating age-at-death from the first rib. *American Journal of Physical Anthropology, 138*(2), 164–176. https://doi.org/10.1002/ajpa.20916

Drusini, A. (1987). Refinements of two methods for the histomorphometric determination of age in human bone. *Zeitschrift Für Morphologie Und Anthropologie*, *77*(2), 167–176. JSTOR.

Dudzik, B., & Langley, N. R. (2015). Estimating age from the pubic symphysis: A new component-based system. *Forensic Science International*, *257*, 98–105. https://doi.org/10.1016/j.forsciint.2015.07.047

Epker, B. N., Frost, H. M. (1964). The Parabolic Index: a Proposed Index of the Degree of Osteoporosis in Ribs. *Journal of Gerontology*, *19*(4), 469–473. https://doi.org/10.1093/geronj/19.4.469

Ericksen, M. F. (1991). Histologic estimation of age at death using the anterior cortex of the femur. *American Journal of Physical Anthropology*, *84*(2), 171–179.

Frost, H. M. (1958). Preparation of thin undecalcified bone sections by rapid manual method. *Stain Technology*, *33*(6), 273–277.

Frost, H. M. (1969). Tetracycline-based histological analysis of bone remodeling. *Calcified Tissue Research*, *3*(3), 211–237.

Frost, H. M. (1987). Secondary osteon population densities: An algorithm for estimating the missing osteons. *American Journal of Physical Anthropology*, *30*(S8), 239–254. https://doi.org/10.1002/ajpa.1330300513

García-Donas, J. G., Paine, R. R., Bonicelli, A., & Kranioti, E. F. (2022). Age estimation for two Mediterranean populations: rib histomorphometry applied to forensic identification and bone remodeling research. *International Journal of Legal Medicine*, *136*(5), 1469–1481. https://doi.org/10.1007/s00414-022-02812-2

Garvin, H. M., & Passalacqua, N. V. (2012). Current Practices by Forensic Anthropologists in Adult Skeletal Age Estimation. *Journal of Forensic Sciences*, *57*(2), 427–433. https://doi.org/10.1111/j.1556-4029.2011.01979.x

Gocha, T. P. (2014). Mapping spatial patterns in cortical remodeling from femoral midshaft using geographic information systems software: implications for age estimation from adult human skeletal remains [Dissertation]. The Ohio State University.

Gocha, T. P., & Agnew, A. M. (2016). Spatial variation in osteon population density at the human femoral midshaft: histomorphometric adaptations to habitual load environment. *Journal of Anatomy*, *228*(5), 733–745. https://doi.org/10.1111/joa.12433

Gocha, T.P., Robling, A. G., & Stout, S. D. (2018). Histomorphometry of human cortical bone: applications to age estimation. In M. A. Katzenberg & A. L. Grauer (Eds.), *Biological anthropology of the human skeleton* 3rd ed. (pp. 145–187). Hoboken, NJ: John Wiley & Sons, Inc.

Goliath, J. R., Stewart, M. C., & Stout, S. D. (2016). Variation in osteon histomorphometrics and their impact on age-at-death estimation in older individuals. *Forensic Science International*, *262*, 282 e1–6. https://doi.org/10.1016/j.forsciint.2016.02.053

Hauser, R., Barres, D., Durigon, M., & Derobert, L. (1980). Identification using histomorphometry of the femur and tibia. *Acta Medicinae Legalis Et Socialis*, *30*(2), 91–97.

Havill, L. M., Allen, M. R., Harris, J. A. K., Levine, S. M., Coan, H. B., Mahaney, M. C., & Nicolella, D. P. (2013). Intracortical Bone Remodeling Variation Shows Strong Genetic Effects. *Calcified Tissue International*, *93*(5), 472–480. https://doi.org/10.1007/s00223-013-9775-x

Heinrich, J. T., Crowder, C. M., Pinto, D. C. (2012). Proposal and validation of definitions for intact and fragmented osteons. *American Journal of Physical Anthropology* 147(S54): 163.

Hennig, C., & Cooper, D. (2011). Brief communication: The relation between standard error of the estimate and sample size of histomorphometric aging methods. *American Journal of Physical Anthropology*, *145*(4), 658–664. https://doi.org/10.1002/ajpa.21540

Ingvoldstad, M. E., Gocha, T. P., Crowder, C. M., & Dominguez, V. D. (2017). Examining the effect of ROI size on the ability to accurately estimate age at death from osteon population density. *Proceedings of the American Academy of Forensic Sciences 69th Annual Meeting. American Academy of Forensic Sciences 69th Annual Meeting in New Orleans, LA.*

İşcan, M. Y., Loth, S. R., & Wright, R. K. (1984). Metamorphosis at the sternal rib end: A new method to estimate age at death in white males. *American Journal of Physical Anthropology*, *65*(2), 147–156. https://doi.org/10.1002/ajpa.1330650206

Işcan, M. Y., Loth, S. R., & Wright, R. K. (1985). Age estimation from the rib by phase analysis: white females. *Journal of Forensic Sciences*. https://doi.org/10.1520/jfs11776j

Iwaniec, U. T., Crenshaw, T. D., Schoeninger, M. J., Stout, S. D., & Ericksen, M. F. (1998). Methods for improving the efficiency of estimating total osteon density in the human anterior mid-diaphyseal femur. *American Journal of Physical Anthropology*, *107*(1), 13–24.

Jowsey, J. (1960). Age changes in human bone. *Clinical Orthopedics*, *17*, 210–217.

Kemkes-Grottenthaler, A. (2002). Ageing through the ages: historical perspectives on age indicator methods. In R. Hoppa & J. Vaupel (Eds.), *Paleodemography: Age distributions from skeletal samples*. (pp. 48-72). Cambridge University Press.

Kenyhercz, M., Crowder, C. M., & Dominguez, V. M. (2020). Histological Age Estimation of the Femur Using Random Forest Regression. *Proceedings of the American Academy of Forensic Sciences 72nd Annual Meeting. American Academy of Forensic Sciences 72nd Annual Meeting in Anaheim, CA*.

Kerley, E. R. (1965). The microscopic determination of age in human bone. *American Journal of Physical Anthropology*, *23*, 149–164.

Kerley, E. R., & Ubelaker, D. H. (1978). Revisions in the microscopic method of estimating age at death in human cortical bone. *American Journal of Physical Anthropology*, *49*(4), 545–546. https://doi.org/10.1002/ajpa.1330490414

Kimura, K. (1992). Estimation of age at death from second metacarpals. *Zeitschrift für Morphologie und Anthropologie*, *79*(2), 169–181.

Konigsberg, L. W., Herrmann, N. P., Wescott, D. J., & Kimmerle, E. H. (2008). Estimation and Evidence in Forensic Anthropology: Age-at-Death. *Journal of Forensic Sciences*, *53*(3), 541–557. https://doi.org/10.1111/j.1556-4029.2008.00710.x

Lynnerup, N., Thomsen, J. L., & Frohlich, B. (1998). Intra- and inter-observer variation in histological criteria used in age at death determination based on femoral cortical bone. *Forensic Science International*, *91*(3), 219–230. https://doi.org/10.1016/S0379-0738(97)00197-7

Maat, G. J. R., Maes, A., Aarents, M. J., & Nagelkerke, N. J. D. (2006). Histological Age Prediction from the Femur in a Contemporary Dutch Sample. The decrease of nonremodeled bone in the anterior cortex. *Journal of Forensic Sciences*, *51*(2), 230–237. https://doi.org/10.1111/j.1556-4029.2006.00062.x

Maat, G. J. R., Van Den Bos, R. P. M., & Aarents, M. (Job). (2001). Manual preparation of ground sections for the microscopy of natural bone tissue: update and modification of Frost's rapid manual method. *International Journal of Osteoarchaeology*, *11*(5), 366–374. https://doi.org/10.1002/oa.578

Mavroudas, S. R. (2023). An exploration of histological variation in the human rib: applications for age estimation methods in applied anthropology [Dissertation]. Texas State University.

Mavroudas, S. R., Meckel, L. A., Gocha, T. P., Goldstein, J. Z., & Garza, S. L. (2022). The Effects of Experimental Whole-Body Burning on Histological Age-at-Death Estimation from Human Cortical Bone and Dental Cementum. *Biology*, *11*(11), 1569. https://doi.org/10.3390/biology11111569

Merritt, C. E. (2015). The influence of body size on adult skeletal age estimation methods. *American Journal of Physical Anthropology*, *156*(1), 35–57. https://doi.org/10.1002/ajpa.22626

Merritt, C. E. (2017). Inaccuracy and bias in adult skeletal age estimation: Assessing the reliability of eight methods on individuals of varying body sizes. *Forensic Science International*, *275*, 315.e1-315.e11. https://doi.org/10.1016/j.forsciint.2017.03.003

Michener, S., Bell, L. S., Schuurman, N. C., & Swanlund, D. (2020). A Method to Interpolate Osteon Volume Designed for Histological Age Estimation Research. *Journal of Forensic Sciences*, *65*(4), 1247–1259. https://doi.org/10.1111/1556-4029.14305

Navega, D., Costa, E., & Cunha, E. (2022). Adult Skeletal Age-at-Death Estimation through Deep Random Neural Networks: A New Method and Its Computational Analysis. *Biology*, *11*(4), 532. https://doi.org/10.3390/biology11040532

Navitainuck, D. U., Vach, W., Pichler, S. L., & Alt, K. W. (2022). Age-at-death estimation in archaeological samples: Differences in population means resulting from different aging methods can be predicted from the mean ages of method-specific reference samples. *International Journal of Osteoarchaeology*, *32*(6), 1226–1237. https://doi.org/10.1002/oa.3157

Paine, R. R., Brenton, B. P. (2006). Dietary health does affect histological age assessment: an evaluation of the Stout and Paine (1992) age estimation equation using secondary osteons from the rib. *Journal of Forensic Sciences*, *51*(3), 489–492. https://doi.org/10.1111/j.1556-4029.2006.00118.x

Pfeiffer, S., Heinrich, J., Beresheim, A., & Alblas, M. (2016). Cortical bone histomorphology of known-age skeletons from the Kirsten collection, Stellenbosch university, South Africa. *American Journal of Physical Anthropology*, *160*(1), 137–147. https://doi.org/10.1002/ajpa.22951

Rose, D. C., Agnew, A. M., Gocha, T. P., Stout, S. D., & Field, J. S. (2012). Technical note: The use of geographical information systems software for the spatial analysis of bone microstructure. *American Journal of Physical Anthropology*, *148*(4), 648–654. https://doi.org/10.1002/ajpa.22099

Ruff, C. B. (2018). Biomechanical analyses of archaeological human skeletons. In M. A. Katzenberg & A. L. Grauer (Eds.), *Biological anthropology of the human skeleton* 3rd ed. (pp. 189–224). Hoboken, NJ: John Wiley & Sons, Inc.

Samson, C. (1983). The determination of age at death by histological analysis of human bone [Dissertation]. University of Sheffield.

Samson, C., & Branigan, K. (1987). A new method of estimating age at death from fragmentary and weathered bone. In A. Boddington, A. N. Garland, & R. C. Janaway (Eds.), *Death, decay, and reconstruction: approaches to archaeology and forensic science*. Manchester University Press.

Schneider, C., Rasband, W. & Eliceiri, K. NIH Image to ImageJ: 25 years of image analysis. *Nature Methods* 9, 671–675 (2012). https://doi.org/10.1038/nmeth.2089

Sedlin, E. D., Frost, H. M., & Villanueva, A.R. (1963a). Variations in cross-section area of rib cortex with age. *Journal of Gerontology*, *18*(1), 9–13. https://doi.org/10.1093/geronj18.1.9

Sedlin, E. D., Frost, H. M., & Villanueva, A.R. (1963b). Age changes in resorption in human rib cortex. *Journal of Gerontology*, *18*(4), 345–349. https://doi.org/10.1093/geronj/18.4.345

Singh, I. J., & Gunberg, D. L. (1970). Estimation of age at death in human males from quantitative histology of bone fragments. *American Journal of Physical Anthropology*, *33*(3), 373–381. https://doi.org/10.1002/ajpa.1330330311

Slice, D. E., & Algee-Hewitt, B. F. B. (2015). Modeling Bone Surface Morphology: A Fully Quantitative Method for Age-at-Death Estimation Using the Pubic Symphysis. *Journal of Forensic Sciences*, *60*(4), 835–843. https://doi.org/10.1111/1556-4029.12778

Stout, S. D. (1986). The use of bone histomorphometry in skeletal identification: the case of Francisco Pizarro. *Journal of Forensic Sciences*, *31*(1), 296–300.

Stout, S. D. (1989). Histomorphometric analysis of human skeletal remains. In M. Y. Iscan & K. A. R. Kennedy (Eds.), *Reconstruction of Life from the Skeleton* (pp. 41–52). Alan R. Liss.

Stout, S. D., Dietze, W. H., Iscan, M. Y., & Loth, S. R. (1994). Estimation of age at death using cortical histomorphometry of the sternal end of the fourth rib. *Journal of Forensic Sciences*, *39*(3), 778–784.

Stout, S. D., & Gehlert, S. J. (1980). The relative accuracy and reliability of histological aging method. *Forensic Science International* *15*(3), 181–190. https://doi.org/10.1016/0379-0738(80)90132-2

Stout, S. D., & Gehlert, S. J. (1982). Effects of field size when using Kerley's histological method for determination of age at death. *American Journal of Physical Anthropology*, *58*(2), 123–125. https://doi.org/10.1002/ajpa.1330580203

Stout, S. D., & Paine, R. R. (1992). Brief communication: histological age estimation using rib and clavicle. *American Journal of Physical Anthropology*, *87*(1), 111–115. https://doi.org/10.1002/ajpa.1330870110

Stout, S. D., & Paine, R. R. (1994). Bone remodeling rates: A test of an algorithm for estimating missing osteons. *American Journal of Physical Anthropology*, *93*(1), 123–129. https://doi.org/10.1002/ajpa.1330930109

Stout, S. D., Porro, M. A., & Perotti, B. (1996). Brief communication: a test and correction of the clavicle method of Stout and Paine for histological age estimation of skeletal remains. *American Journal of Physical Anthropology*, *100*(1), 139–142. https://doi.org/10.1002/(SICI)1096-8644(199605)100:1<139::AID-AJPA12>3.0.CO;2-1

Stout, S. D., & Stanley, S. C. (1991). Percent osteonal bone versus osteon counts: The variable of choice for estimating age at death. *American Journal of Physical Anthropology*, *86*(4), 515–519. https://doi.org/10.1002/ajpa.1330860407

Streeter, M. (2010). A four-stage method of age at death estimation for use in the subadult rib cortex. *Journal of Forensic Sciences*, *55*(4), 1019–1024. https://doi.org/10.1111/j.1556-4029.2010.01396.x

Stull, K. E., Chu, E. Y., Corron, L. K., & Price, M. H. (2022). Subadult Age Estimation Using the Mixed Cumulative Probit and a Contemporary United States Population. *Forensic Sciences*, *2*(4), 741–779. https://doi.org/10.3390/forensicsci2040055

Thomas, C. D. L., Stein, M. S., Feik, S. A., Wark, J. D., & Clement, J. G. (2000). Determination of age at death using combined morphology and histology of the femur. *Journal of Anatomy*, *196*(3), 463–471. https://doi.org/10.1046/j.1469-7580.2000.19630463.x

Thompson, D. D. (1978). Age related changes in osteon remodeling and bone mineralization [Dissertation]. University of Connecticut.

Thompson, D. D. (1979). The core technique in the determination of age at death of skeletons. *Journal of Forensic Sciences*, *24*(4), 902–915.

Watanabe, Y., Konishi, M., Shimada, M., Ohara, H., & Iwamoto, S. (1998). Estimation of age from the femur of Japanese cadavers. *Forensic Sciences International*, *98*(1), 55–65. https://doi.org/10.1016/S0379-0738(98)00136-4

Wescott, D. J., & Drew, J. L. (2015). Effect of obesity on the reliability of age-at-death indicators of the pelvis. *American Journal of Physical Anthropology*, *156*(4), 595–605. https://doi.org/10.1002/ajpa.22674

The Histopathological Analysis of Human Skeletal Remains

9

SANDRA ASSIS, JULIETA GÓMEZ GARCÍA-DONAS, AND HANS H. DE BOER

9.1 Introduction

The skeleton provides support and protection to the body; stores and metabolizes elements and minerals; houses hematopoietic tissue; and regulates specific endocrine processes. Due to this variety of functions, bone tissue is intricately connected to many metabolic and physiologic processes, and all major organ systems. It also means that skeletal tissue is exposed to a wide variety of pathological insults such as trauma; physiologic, metabolic, and endocrine derangement; hematopoietic disorders; infections; and tumors (Horvai, 2015; Ragsdale & Lehmer, 2012).

The microscopic examination of bone tissue can provide valuable information about bone structure, composition, and properties. Histology has therefore long been a valuable source of information for those interested in musculoskeletal pathology, and clinical pathologists play an important role in the diagnosis of a wide variety of musculoskeletal diseases in living patients. Especially when combined with other diagnostic information, such as the results of radiological, biochemical, hematological and clinical analysis, bone histology can either definitively diagnose skeletal disease or narrow differential diagnosis considerably.

Notwithstanding the established benefits in a clinical setting, histology is far less often used as a tool to diagnose disease in skeletonized human remains. This may be due to various reasons. First of all, the invasive, destructive nature of histological analysis may preclude its use on museal or otherwise significant specimens. Second, practitioners may struggle to find the correct procedure to process skeletonized tissue into histological specimens, or to choose the best method of their visualization. Third, the interpretation of the specimens may be challenging, due to absence of soft tissue and the obscuring effect of diagenesis. And lastly, practitioners may be unsure of the diagnostic potential of histology within the context of a particular case. These limitations are, however, manageable, and the histopathological analysis of dry bone can still add essential information to the diagnostic process.

In an attempt to assist practitioners interested in the histopathology of skeletal human remains, this chapter aims to provide a comprehensive overview of dry bone pathophysiology, and the limitations, challenges and opportunities that are associated with it.

9.2 Basic Pathophysiology of Bone

Bone growth and remodeling depend on the activity of the three principal types of bone cells: osteoblast, osteocytes, and osteoclasts. Also in disease, these cells are responsible for

DOI: 10.4324/9781003385608-9

the formation, regulation, and resorption of bone tissue, and therefore for skeletal (histo) morphology. Understanding the histomorphological features of bone disease thus requires a short recapitulation of the cellular components of bone tissue, their reaction to stress, and how this translates to dry bone histomorphology. For a more detailed reiteration of bone histology the reader is referred to Chapter 1 of this volume.

Osteoblasts are the principal bone forming cells, derived from the mesenchymal stem cell lineage. All bone surfaces, either periosteal or endosteal, are covered by (usually dormant) osteoblasts. When activated, they synthesize an extracellular matrix (osteoid), consisting of collagen and various bone-specific proteins, that will form a scaffold for mineralization, and hence for the composite material that is bone. The orientation of the deposited collagen fibers is one of the key elements that provide bone with its tensile strength. Mineralization of the osteoid with hydroxyapatite, also produced by osteoblasts, will add structural strength. In clinical specimens, dormant osteoblasts are bland, elongated, spindle-shaped cells, while activated osteoblasts are plump and cuboidal. In dry bone specimens, osteoblasts and osteoid are no longer present, and only their mineralized extracellular product remains.

As osteoid accumulates around the osteoblast, some will become encased and differentiate into bone-residing cells: osteocytes. Osteocytes reside in a lacuna (plural: lacunae) and communicate with other osteocytes and surrounding tissues through a fine network of small cellular processes, housed in canaliculi. Osteocyte lacunae and canaliculi are visible in dry bone tissue as empty spaces, which – due to birefringence – may appear dark in undecalcified histological slides. Osteocytes are by no means quiescent cells (Tresguerres et al., 2020). They are pivotal in mineral homeostasis, while their mechanosensory capabilities inform bone remodeling and thereby the adaptation of bone architecture to its mechanical loading (also referred to as 'Wolff's law' (Frost, 2004). Although osteocytes play a central role in bone metabolism and therefore in many pathological processes, their effect on dry bone histomorphology is mostly indirect, through inducing osteoblastic or osteoclastic activity. It is believed that in certain conditions, osteocytes can resorb surrounding bone tissue in a process called osteocytic osteolysis (Heveran & Boerckel, 2023), histologically resulting in enlarged osteocyte lacunae. This increase of osteocyte lacunae size is, however, too limited and variable to be considered a helpful diagnostic feature in dry bone histology (Figure 9.1).

During bone formation, the orientation of the collagen fibers within the osteoid dictates whether the bone will be of the 'woven' or the 'lamellar' type. In woven bone, the collagen fibers are apparently deposited haphazardly, resulting in an irregular, relatively hypercellular microstructure. Like the collagen fibers, osteocyte lacunae are oriented irregularly, while the residing osteocytes (and hence their lacunae) are larger. Due to its irregular microstructure, woven bone is weaker, less rigid, and more flexible than lamellar bone (Mills, 2012). Woven bone indicates rapid bone formation. In lamellar bone, the collagen fibers are deposited in distinct layers, giving bone tissue a highly organized lamellar microarchitecture. In normal conditions, the fully developed skeleton consists entirely of lamellar bone. The formation of lamellar bone is much slower than woven bone, but its microstructure lends greater rigidity and tensile strength. Lamellar bone can be deposited primarily, i.e., without a precursor, or may be formed after the remodeling of woven bone.

In terms of bone formation, a distinction is also made between intramembranous bone formation and endochondral ossification. In intramembranous bone formation, the tissue on the site of future bone tissue consists of a fibrous membrane. This type of bone formation occurs, for instance, in the periosteal and endosteal envelopes, and forms, for instance, the

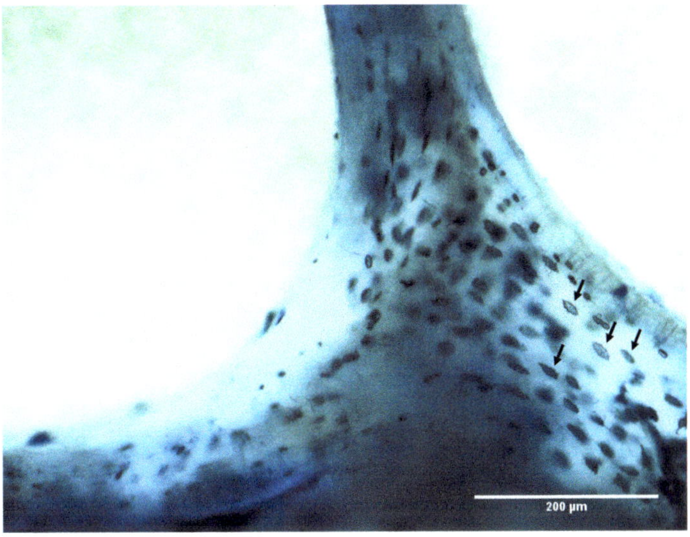

Figure 9.1 Sternal end of a right rib from a 75-year-old female, undecalcified specimen viewed with polarized light. Note the multiple osteocyte lacunae within the trabeculae, visible as either optically empty or opaque oval structures (black arrows). Specimen from the 19th- to 20th-century Luís Lopes Anthropological Collection. ULisboa-MUHNAC.

flat bones of the skull. The deposited bone can be of the woven or the lamellar type, depending on the speed of deposition. In endochondral ossification, first a scaffold of cartilage is formed (the 'anlage') which is then replaced by bone tissue. The first deposited bone in endochondral ossification is of the woven type, which is secondarily remodeled in lamellar bone. Endochondral ossification is, for instance, found at the epiphyseal growth plates in subadults, and during fracture repair.

Woven and lamellar bones are readily recognizable in non-skeletonized bone tissue and in dry bone remains, but the use of a polarization filter enhances their morphological difference (Figures 9.2 and 9.3). Woven bone, in a fully mature skeleton, always indicates an aberration from normal bone formation and therefore provides an important diagnostic clue. It can, for instance, indicate the presence of previous trauma (fracture repair, ossification of periosteal hematoma), infection (involucrum formation), or primary or secondary bone tumors. Lamellar bone on the other hand indicates a slow, more benign form of bone formation. It is therefore less often indicative of pathology. When present in pathological lesions it could, for instance, favor a benign nature of a tumorous growth, or another form of slowly developing lesion. When found within areas of woven bone, it signals remodeling, and therefore chronicity.

Osteoclasts are the principal bone resorbing cells, derived from the monocyte cell lineage. They are solitary, large, multinucleated cells that – once activated – make contact with the bone tissue surface with their unilateral 'brush border'. Bone resorption then follows due to the excretion of lytic substances, creating resorption pits (Howship's lacunae). These Howship's lacunae remain visible in skeletonized tissue (Figure 9.4). Due to ongoing remodeling, Howship's lacunae are normally present in bone tissue. On the surface of trabecular bone, or the periosteal and endosteal surfaces of cortical bone, they are visible as scattered trabecular indentations. Within cortical bone, grouped Howship's lacunae can be found in the bone multicellular units (BMUs) which are responsible for normal cortical

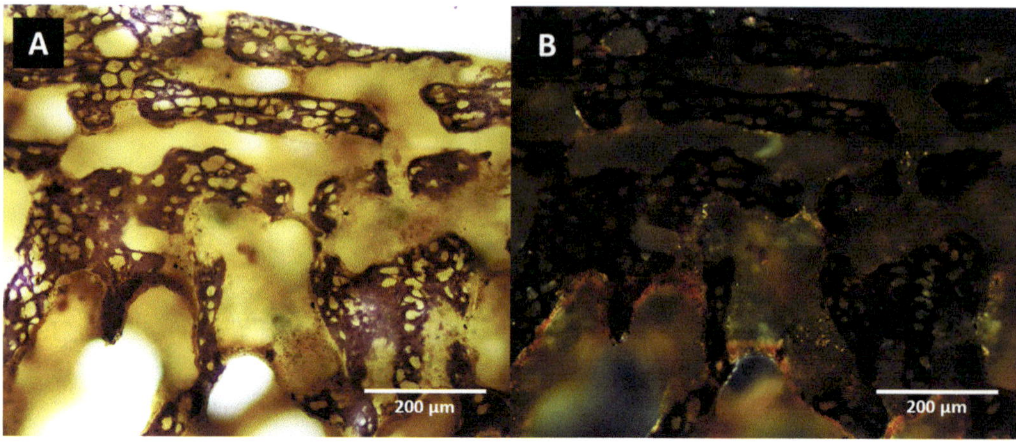

Figure 9.2 Callus of a fractured rib of a 23-year-old male, undecalcified specimen, stained with hematoxylin. Panels A and B show woven bone, viewed with bright light and polarization filter, respectively. In panel A, the large osteocyte lacunae can be appreciated. The bone is haphazardly arranged and lacks a lamellar structure. Panel B highlights the lack of birefringence, although its lack should be interpreted cautiously in specimen with marked taphonomical alteration. Specimen from the histological collection of the Amsterdam University Medical Centre.

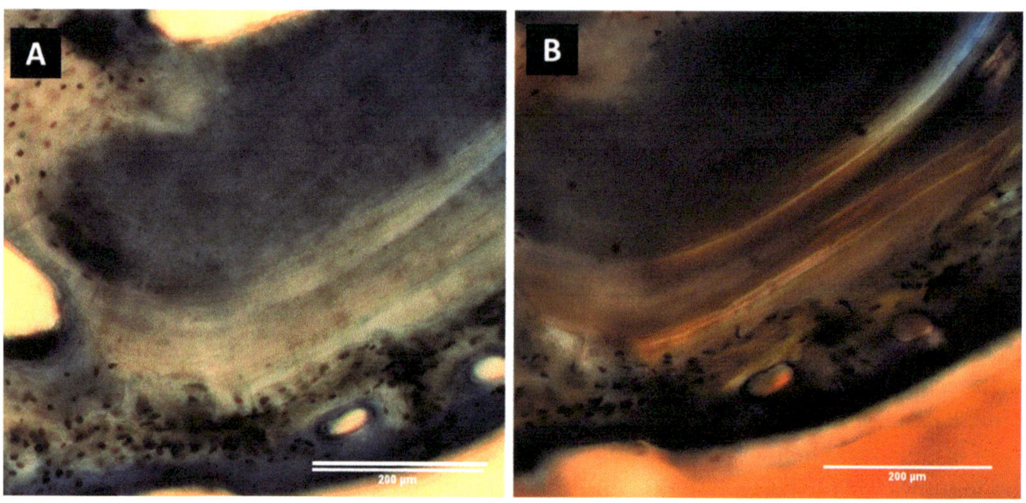

Figure 9.3 Vertebral portion of the 11th right rib from a 9-year-old female, undecalcified specimen. Panels A and B show lamellar bone, viewed with bright light and polarization filter, respectively. The orientation of the lamellae is enhanced by the use of polarized light. Specimen from the 19th- to 20th-century Luís Lopes Anthropological Collection. ULisboa-MUHNAC.

remodeling (Kenkre & Bassett, 2018). In a pathological state, the number of osteoclasts / Howship's lacunae can increase dramatically, with resultant localized focal osteolysis (e.g., as seen in osteomyelitis or osteolytic metastases), or a general/systemic paucity of bone tissue (e.g., as seen in osteopenia or osteoporosis).

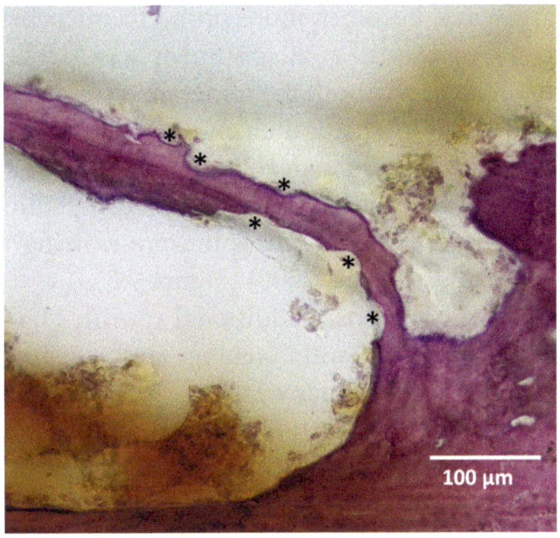

Figure 9.4 Endosteal surface of a fractured rib of an middle-aged male, undecalcified specimen, stained with hematoxylin. Note the conspicuous bone resorption, indicated by numerous Howship's lacunae (black asterisks). Specimen from the histological collection of the Amsterdam University Medical Centre.

9.3 Limitations in Dry Bone Histopathology

The histopathological examination of skeletonized remains is challenging for a variety of reasons, which often coexist. First of all, not all diseases affect the skeleton or only do so to a limited extent. Also, the circumstantial information that is often pivotal in a clinical setting is regularly scarce and/or unreliable when dealing with skeletonized remains. Decomposition and diagenesis present further challenges. No tissues are immune to the effects of decomposition, but soft tissues and cells are far more prone to the destructive forces of autolysis and putrefaction (Janaway et al., 2009; Kendall et al., 2018). In the remaining skeletonized tissue, the cytonuclear and soft tissue features that clinical orthopedic pathologists rely on for a diagnosis are no longer present. At the same time, the effect of disease on dry bone histomorphology is quite limited in terms of its variety. From a dry bone histopathology perspective, a disease can either be appositional (bone forming/osteoblastic), lytic (bone resorbing/osteoclastic), or a combination. These basic categories provide an important starting point for differential diagnosis but also underpin why dry bone histomorphology often lacks specificity (Bell & Piper, 2000).

Although bone is relatively resistant to decomposition, even skeletonized tissue eventually decays. Any postmortem change of bone tissue (also: diagenesis) directly impacts on the preservation of bone tissue, and thus on the quality and quantity of information that can be obtained. The speed and extent of diagenesis is dependent on a wide variety of factors, including physical, chemical and biological agents (Assis et al., 2015; Bertoglio et al., 2021; Delannoy et al., 2018; Hedges, 2002; Jans et al., 2004; Lander et al., 2023; Turner-Walker, 2007). Dissolution of the mineralized fraction of bone is accelerated in free-draining and acidic soils, particularly in environments with frequent wetting and drying cycles (Hedges, 2002). Loss of collagen is accelerated in demineralized bone, as well as environments with extreme pH values and high temperatures. This loss of collagen can affect the application of some histological techniques, as it may preclude bone decalcification. As the diagenetic alteration of bone progresses, it becomes more prone to microbial attack (Collins et al., 2002; Hedges & Millard, 1995; Kendall et al., 2018; Nielsen-Marsh et al., 2000). Microbial activity in turn may cause depletion of both the mineralized and organic components, while also producing conspicuous microstructural changes. This microfocal destruction (MFD) may obscure or erase evidence of pathological bone response (Hackett, 1981; Jans, 2008). Diagenesis can furthermore induce dissolution and/or redistribution of apatite crystallites, known as recrystallization, which also obscures the pre-existing bone microarchitecture (Hedges et al., 1995; Nielsen-Marsh et al., 2000) (Figure 9.5).

Thermally altered (burnt) bone poses additional specific challenges. Changes due to heat include color changes, shrinking and cracking, and warping. The extent and severity of such changes are dependent on many factors, including temperature, duration of exposure, and circumstances of combustion (Fairgrieve, 2008; Mayne Correia, 1997; Thompson, 2004; Thompson et al., 2017). The impact of thermal changes on the bone microarchitecture is relatively well documented (e.g., Absolonova et al., 2013; Carroll & Squires, 2020; Cattaneo et al., 1999; Fernández Castillo et al., 2013; Forbes, 1941; Squires et al., 2011). Changes in the mineral structure (e.g., polyhedral and square-shaped crystalline formations) are visible at temperatures of at least 300°C and 400°C, and usually increase above temperatures of 500°C and 600°C. Haversian and Volkmann's canals remain visible at temperatures between 300°C and 600°C, despite bone cracking displaying. Beyond this temperature, the organic content of the bone is destroyed (no lamellar systems visible), and the bone recrystallizes (Fairgrieve, 2008). The loss of the organic component and recrystallization also makes bone more brittle and consequently more difficult to process (Thompson et al., 2017).

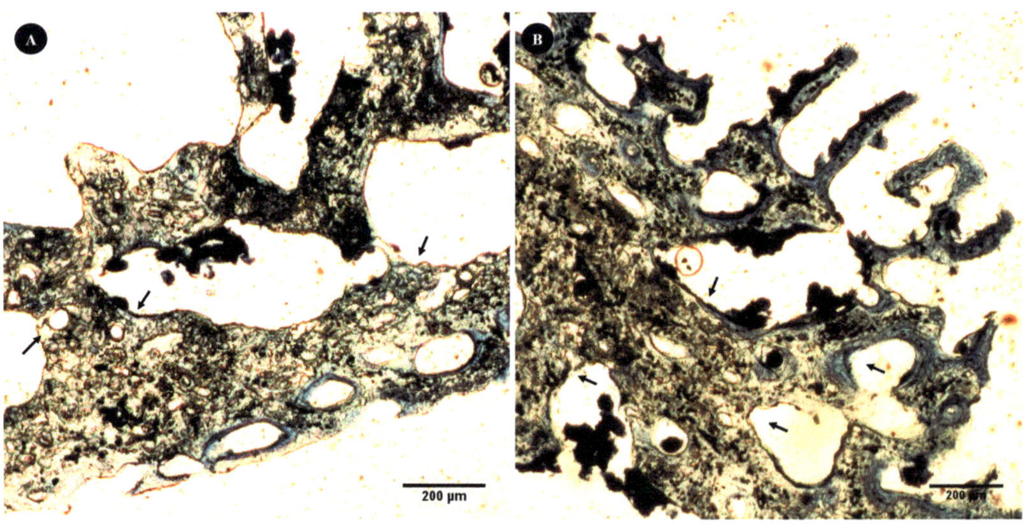

Figure 9.5 Distal end of the right fibula of a 2-year-old male, undecalcified specimen. Panel A shows the presence of some relatively well-defined cortical Haversian canals surrounded by microscopic focal destruction (MFD) and inclusions, visible as black/opaque areas (black arrows). In Panel B, the microscopic focal destruction is mainly situated at the endosteal surface of the bone (black arrows). Specimen from the 19th- to 20th-century Luís Lopes Anthropological Collection. ULisboa-MUHNAC.

9.4 Methods of Examination

Different research questions and types of material (e.g., recent or archaeological) require different techniques for sample processing and visualization, and the following section briefly covers the available methods. Further literature can be consulted for more detailed information on methods, protocols and techniques (e.g., An & Martin, 2003; Bell, 2012; De Boer & Maat, 2012; Maat et al., 2001).

9.4.1 Qualitative and Quantitative Light Microscopy

Light microscopy has been the mainstay of dry bone histopathology (De Boer & Maat, 2012; Schultz, 2001; Stout et al., 2019). It consists of the assessment of bone histology in a two-dimensional plane using an optical microscope (Stout & Crowder, 2012). Light microscopy requires translucent or reflective histological specimens, and therefore quite elaborate sample preparation. Guidelines for the processing of undecalcified and decalcified bone tissue, and for the embedding and staining of specimens are available (e.g., De Boer et al., 2012; Piper & Valentine, 2012). The best approach will vary depending on the nature of the sample and time available, as for example embedding will be preferred for handling fragile samples but is time-consuming, while undecalcified sections can be used for mummified remains (Schultz, 2012). Bright field visualization can be complemented by the use of polarization filters. Such filters absorb light waves in one direction, which enhances the visibility of the orientation of collagen and bone tissue orientation. This can readily demonstrate key features in pathologies such as Paget's disease (Aaron et al., 1992). Histochemical staining has been used both on fresh and dry bone, and can, for instance, enhance the visibility of a lamellar structure, or the presence of Howship's lacunae (De Boer & Van der Merwe, 2016).

As implied by the name, qualitative light microscopy relies on qualitative descriptions of characteristics, such as 'trabecular tunneling' as seen in hyperparathyroidism or the 'radiating spongy bone' as seen in subperiosteal hematomas (Van der Merwe et al., 2010; Zink et al., 2005). Qualitative light microscopy thus requires clear-cut definitions with information on the distribution, morphology, and orientation of the described features (Turner-Walker & Mays, 2007). The inter-observer variability associated with many qualitative features is noted. Quantitative light microscopy can refer to dynamic or static histomorphometry (Stout & Crowder, 2012), with the latter being commonly applied in biological and forensic anthropology as well as archaeological and paleoanthropological research. This approach entails the quantitative assessment of histological parameters most often those related to bone remodeling rates (Parfitt et al., 1983; Skingle & Compston, 2011). Different software can be used for collecting histomorphometric data such as ImageJ, with several plugins being useful for the assessment of bone quality and quantity (Cole et al., 2022; Schneider et al., 2012).

9.4.2 Scanning Electron Microscopy

Scanning electron microscopy (SEM) produces images based on the interaction between a focused beam of electrons and the bone specimen surface. It provides a highly detailed spatial resolution, high depth and wide field of view, and allows for focus on specific tissue components (Boyde et al., 1986; Shah et al., 2019). SEM requires specific sample preparation techniques. Traditional high vacuum SEM requires sample cleaning, drying and conductive fixation methods, and can produce images with a resolution of 3 nm (Jones, 2012). Low vacuum and environmental SEM may provide a lower resolution, but with the advantage of requiring minimum preparation (Shah et al., 2019). Backscattered scanning electron microscope (BSE-SEM) allows the observation of bone mineral content through high energy electrons reflecting on the bone surface, providing high quality images of full cross-sections, with quantitative backscattered electron imaging (qBEI) enabling quantification of bone mineral density distribution (Beresheim et al., 2019; Busse et al., 2009; Shah et al., 2019).

SEM images emphasize the delineations between osteonal structures and remaining cortical microstructures as well as trabecular compartments boundaries, with new techniques (e.g., topographical BSE) being available to assess complex and very porous trabecular tissue such as associated with osteoporosis (Boyde, 2003). BSE-SEM images can also facilitate the observation of bone surface topography indicating resorbing or incomplete mineralization as seen in primary hyperparathyroidism (Mays et al., 2007). Moreover, diagenetic changes and pathological alterations can be differentiated based on appearance and distribution of the defects (Van Der Merwe et al., 2018).

9.4.3 Other Approaches

Other histological approaches include quantitative computer tomography (qCT), micro-computer tomography (micro-CT), microradiography, and magnetic resonance microimaging (micro-MRI). Micro-CT has been used for the observation of bone microstructures in a three-dimensional level and allows for the assessment of complex microarchitectural patterns (e.g., Cole et al., 2019). It does not require invasive/destructive biopsy of specimens. Initial micro-CT studies focused on trabecular bone architecture (Engekel et al., 2008), with parameters such as trabecular thickness and bone volume used in the assessment of osteoporosis. Other research focused on microscopic assessment of fracture healing (Körmendi et al., 2020). Paleopathological evidence of tuberculosis on ribs have been observed through micro-CT as well as 2D histology and suggested that the combination of the two approaches can assist in better understanding the disease process (Nicklisch et al., 2012). The automated analysis of images using software applications such as Dragonfly or ImageJ is an advantage of micro-CT, although specific features such as woven and lamellar bone differentiation would still require histological assessment by a specialist (Rühli et al., 2007). For a further discussion, see Aksoy et al. (2020) and Kuhn et al. (2007).

9.5 Integrating Histology in the Diagnostic Workup

Despite its various limitations and challenges, literature has shown that the microscopic examination of bone can still be of considerable diagnostic value (De Boer et al., 2013; Schultz, 2001). However, the same literature illustrates that the extent in which histology will inform a definitive diagnosis is entirely dependent on the research question, and how histology is integrated in the diagnostic process. As in clinical orthopedic pathology, histological examination should be one of the final steps of the examination, within a much larger diagnostic workup. Such a workup ensures that histological examination is applied appropriately, and prevents disappointing results after sometimes lengthy and painstaking laboratory preparations. Essential elements of such a diagnostic workup, prior to considering histology, are epidemiological consideration, macroscopic analysis, and radiological examination

If the human remains allow for reliable estimation of biological parameters (e.g., sex or age), the first step of the diagnostic process should be based on epidemiological considerations. For instance, some metabolic bone diseases, primary bone tumors, or hematological disorders/malignancies correlate well with particular age groups. Also, the incidence of metastatic bone disease is age-related, and often sex-related as well. Consider, for instance, the difference between breast and prostate cancer. Lastly, some orthopedic diseases are related to geographical areas (e.g., malaria, treponemal diseases) or are more prevalent in particular socio-economic groups.

Macroscopic examination will, among others, determine location, extent, and type (blastic, lytic, mixed) of the lesion. Subsequent radiological imaging should at least include plain X-rays, but the addition of CT-imaging is preferred. Radiology is not only vital to enhance the interpretation of macroscopically visible lesions, but also to visualize pathologies that would otherwise remain undetected. The macroscopic and radiological findings are preferably discussed in a multidisciplinary team, with input from at least a skeletal radiologist and an orthopedic pathologist. Depending on the differential diagnosis, a clinical specialist may be needed, such as a rheumatologist, endocrinologist, or orthopedic surgeon. In all forensic cases, a forensic pathologist should be included. None of these specialists may have specific knowledge on the interpretation of dry bone histology, but their expertise will nonetheless help to narrow down the differential diagnosis, and may even render histological examination superfluous.

The results of multidisciplinary discussion will determine if histological examination is an appropriate next step. At this moment, usually one of three common scenarios presents itself. First, it may still be uncertain if the observed lesion represents true pathology or is in fact artifactual ('pseudo-pathology'). Due to its ability to visualize bone deposition and resorption, histology is an apt tool to differentiate antemortem processes from postmortem artifacts. Therefore, assuming that preservation is sufficient to allow for histological examination, microscopy would be considered an appropriate next step.

The second scenario represents a situation in which the lesion is deemed to be pathological, but no definitive diagnosis can be established. Naturally, the diagnostic value of histology will then depend on the histomorphological differences between the entities that are considered. This requires knowledge of clinical orthopedic pathology, but even when diseases are histologically distinct in clinical specimens, their dry bone histomorphology may turn out to be quite similar (e.g., De Boer & Maat, 2018; De Boer & Van der Merwe, 2016). In other words, as discussed later in this chapter, there are only a limited number of bone diseases that retain specific (pathognomonic) diagnostic features once skeletonized. This problem of nonspecificity can be further compounded by the obscuring effect of diagenesis.

A last, less common scenario presents itself when a particular diagnosis is established, but histology is considered as a means to determine the extent or stage of disease. This would only be considered relevant in a minority of cases, for instance, when one requires knowledge on the age of a lesion, or the health status of an individual. Histology can diagnose remodeling at an earlier stage than radiology, and therefore certainly holds diagnostic potential for this particular use. This is further illustrated by various papers on the features of (healing) scurvy or the severity of vitamin D deficiency, or on fracture age estimation (Cappella et al., 2019; Mays et al., 2015; Welsh et al., 2020).

9.6 Dry Bone Histomorphology of Specific Diseases

9.6.1 Metabolic Bone Diseases

Metabolic bone disease refers to disorders caused by an abnormal chemical milieu in the body, and entails complex multifactorial influences, manifestations, and factors (McCarthy, 2010). The following sections discuss the histological changes that may be seen in osteoporosis, vitamin D and C deficiency, and hyperparathyroidism. It is not uncommon for these diseases to co-exists, and their histological features overlap to a certain extent. In line with convention, Paget's disease of bone is also discussed in this section, although its etiology is not completely understood.

9.6.1.1 Osteoporosis

Osteoporosis is defined as a pathologically decrease in bone mass, and affects bone mass and strength through macrostructural and microstructural changes (Nih Consensus Development Panel On Osteoporosis Prevention, Diagnosis, And Therapy, 2001; Raisz et al., 2008). A division is made between primary osteoporosis (Type I and Type II) and secondary osteoporosis, which arises from pathological conditions other than menopause or aging (Marcus & Bouxsein, 2008; Mirza & Canalis, 2015). The essence of osteoporosis is an excessive bone resorption not accompanied by an appropriate increase in osteoblast activity (Raisz, 2005; Raisz et al., 2008). The diagnosis is difficult to make on histology alone, but if attempted, a quantitative approach is preferred. Such an approach should account for sex-related differences and other conditions that manifest with osteoporosis (Agarwal et al., 2004; De Boer et al., 2013; Raisz, 2005).

In trabecular bone, osteoporotic bone shows a reduction in trabecular volume, number, size and connectivity, with an increase in trabecular microarchitectural space indicating an increased perforation of the bony trabeculae (Figure 9.6) (Chen et al., 2008; Stauber & Müller, 2006). The remodeling BMU hemicones in trabecular plates perforate the surface more deeply, with rapid resorption or remodeling foci producing a complete removal of structural elements (Brickley et al., 2020c; Parfitt, 2008). Healing microfractures showing callus and woven bone are also associated features (Roberts & Wakely, 1992; Vernon-Roberts & Pirie, 1973). These microfractures alone are suggestive, but not specific of osteoporosis (Agarwal, 2018; Brickley & Agarwal, 2003).

In cortical bone, resorption occurs at the endosteal surface, as well as intra-cortically at the Haversian canals, and within the osteocyte lacunae (Brickley et al., 2020c; Thomas et al., 2006). Bone loss in the endocortical envelope or medullary cavity produces a 'trabecularized' appearance of the cortex, with a higher resorption depth reflecting a deeper perforation by BMUs at the endocortical interface (Figures 9.7 and 9.8) (Parfitt, 1989; Power et al., 2003; Tanizawa et al., 1999). Cortical area reduction due to aging has been shown through 2D histomorphometric studies (e.g., Cho et al., 2002; García-Donas et al., 2022; Lee et al., 2014). Relative cortical area can provide further insights into bone mass loss, e.g., for sex comparisons, as this parameter is unrelated to size, although intra-skeletal and inter-population variability should be considered (García-Donas et al., 2021; Kim et al., 2007; Stewart et al., 2015). Other measurements such as bone formation rates and bone activation frequency can offer information about general skeletal health status, with different rates being observed between osteoporotic and healthy individuals (Wu et al., 1970). See also Stout et al. (2019) for further details.

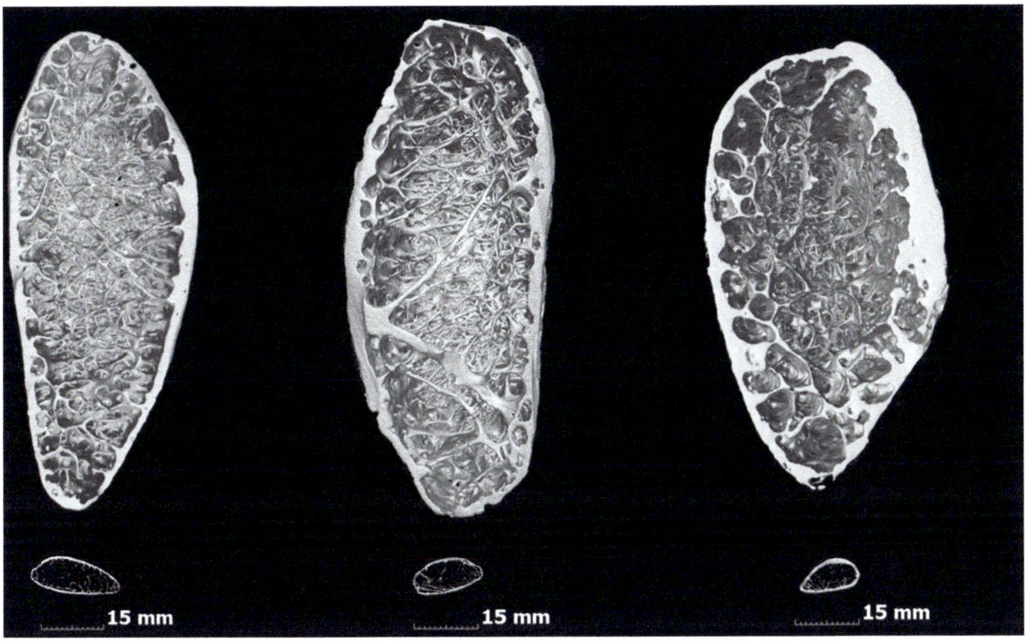

Figure 9.6 Micro-CT 3D (top) and 2D (bottom) reconstructions of the sternal end of three 3rd ribs, showing a reduction of bone volume and increase in trabecular separation with increasing age. Left to right: male 48 years old, male 68 years old, male 84 years old. Specimen from the histology collection of the Centre for Anatomy and Human Identification (CAHID).

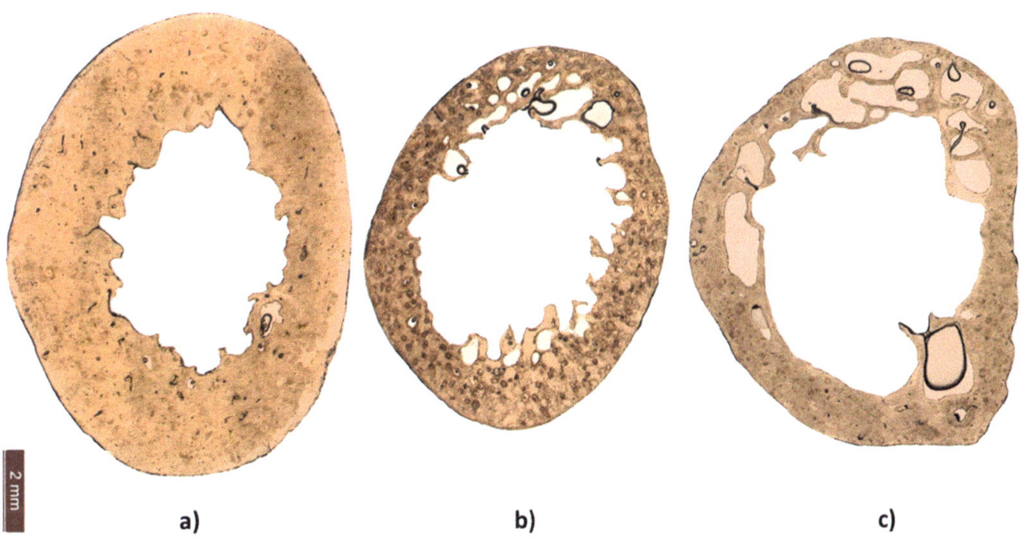

Figure 9.7 Cross-sections of the medial segments of three left clavicles, undecalcified specimen. (A) Clavicle of a 55-year-old female with normal cortical thickness and preservation of the endosteal envelope. (B) Clavicle of a 78-year-old female with signs of thinning and trabecularization on the superior endosteal envelope. (C) Clavicle of a 97-year-old female showing further cortical area thinning and trabeculation of the cortex. Note the differences in size due to inter-individual variation. Specimen from the histology collection of the Centre for Anatomy and Human Identification (CAHID).

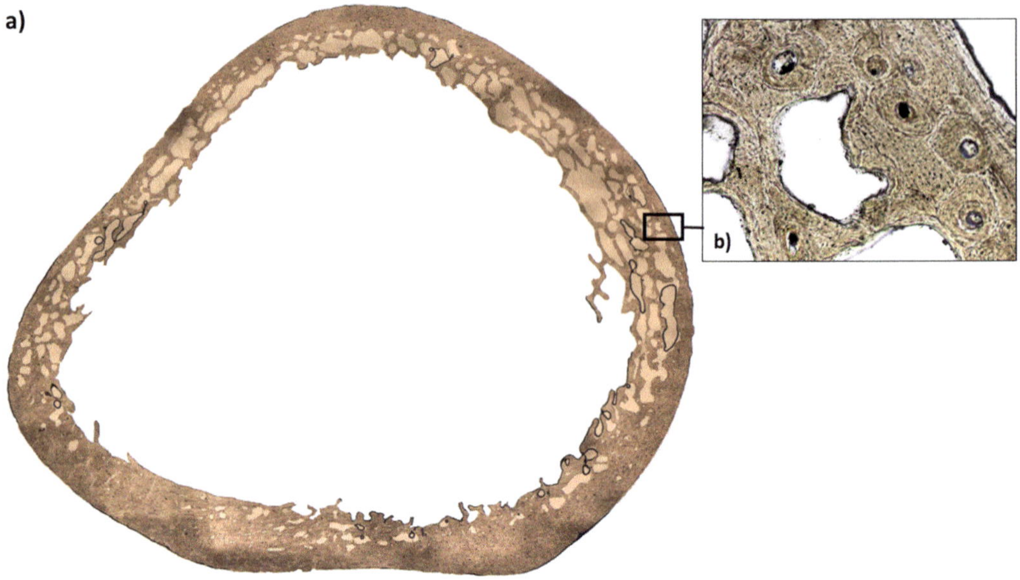

Figure 9.8 Cross-section of an undecalcified distal femur of a 78-year-old female, showing extreme trabeculation of the cortex (panel A). The detail in panel B shows increased intracortical porosity due to resorption of the cortical bone. Specimen from the histology collection of the Centre for Anatomy and Human Identification (CAHID).

9.6.1.2 Scurvy

Scurvy is caused by a deficiency of vitamin C (ascorbic acid) (Brickley et al., 2020a; Maxfield & Crane, 2023). In vitamin C deficiency, osteoblasts activity is depressed, provoking an insufficient or absence of osteoid formation. Vitamin C is furthermore essential for collagen production, resulting in a weak organic bone component. The defective collagen in other connective tissues such as blood vessel walls will predispose to hemorrhage (Fain, 2005; Klaus, 2017). If hemorrhage occurs in joints or subperiosteally, subperiosteal bone formation as well as porosity might be observed (Mays, 2014; Ortner et al., 1999).

Microscopic assessment can confirm the diagnosis of macroscopic scorbutic manifestations, although histological changes are not specific for scurvy (De Boer et al., 2013; Follis et al., 1940). Other histological lesions include bone osteopenia, metaphyseal fractures and epiphyseal plate trabecular micro-fractures (Brickley et al., 2020a; Schultz, 2003). Most seen in subadults, subperiosteal bleeding with or without previous trauma can be demonstrated due to the subperiosteal new bone formation it incites (Schultz, 2001). The remodeled subperiosteal hematoma is often visible as a demarcated spongy bone with thick and short bridges arranged radially, perpendicular to and on the top of the cortical surface with preservation of the periosteal circumferential lamellae (Figure 9.9) (Schultz, 2001, 2003; Van der Merwe et al., 2010). The trabeculae orientation depends on the size and amount of bleeding and forces applied from the surrounding muscles, with appositional layers within the hematoma suggesting repetitive episodes of healed scurvy (De Boer et al., 2013; Maat, 2004; Schultz, 2003).

Different stages of remodeling can be differentiated (for further details see Van der Merwe et al., 2010). In chronic scurvy, subperiosteal new bone formation as a result of mineralized hemorrhage can appears on limbs shafts and skull and this process (periostosis)

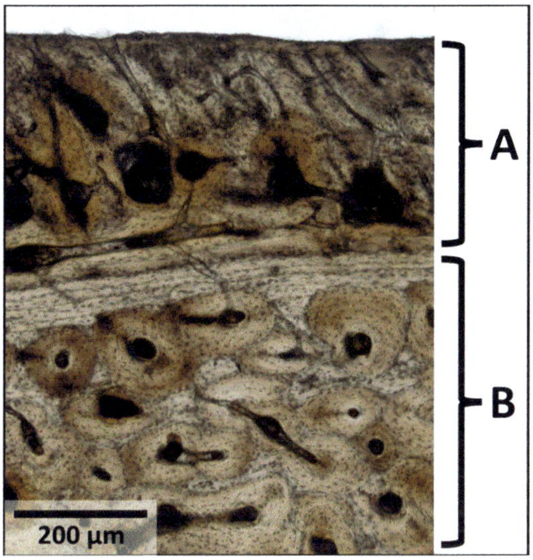

Figure 9.9 Ossified hematoma secondary to scurvy, observed in the skeletonized remains of an adult male (20–25 years old). Note the orientation of the lamellae and vascular canals in the ossified hematoma (A). The periosteal circumferential lamellae of the pre-existing cortical bone tissue (B) are intact. Photo courtesy of Dr. Lida van der Merwe, Amsterdam University Medical Centre.

might produce microscopic features similar to other inflammatory processes and infection (Schultz, 2001, 2012). The cause of an ossified hematoma (e.g., scurvy or trauma) cannot be determined on the basis of histology alone either (Van der Merwe et al., 2010). Orbital microscopic features such as tangential trabeculae reflecting the organized hematoma and cortex preservation can indicate scorbutic features (Lovász et al., 2013; Schultz, 2012), but are also seen in disorders such as anemia and rickets (Klaus, 2014; Schultz, 2003; Zuckerman et al., 2014). Although histological changes alone cannot provide a confirmation of scurvy, they can assist in reaching a differential diagnosis, if for example, microscopic lesions are in line with key gross-anatomical scorbutic manifestations (De Boer & Van der Merwe, 2016; Stark, 2014; Thompson et al., 2022).

9.6.1.3 Vitamin D Deficiency

Vitamin D is essential for calcium homeostasis (Holick, 2003), and a deficiency causes poor osteoid mineralization during growth and remodeling (Imel et al., 2014; Lips et al., 2013). Depending on the age of the patient, vitamin D deficiency results in either rickets (in subadults) or osteomalacia (in adults) (Bikle, 2012; Brickley et al., 2007; Zhang et al., 2016). Microscopic changes can support the diagnosis of rickets and osteomalacia along with radiography and macroscopic observation (e.g., Brickley et al., 2007; Mays et al., 2007; Schamall et al., 2003; Van Der Merwe et al., 2018).

Early phases of vitamin D deficiency can be observed microscopically without macroscopic lesions (Brickley et al., 2020b). Mineralization defects occurring through poorly mineralized bone are indicative of osteomalacia (Van Der Merwe et al., 2018). Although the osteoid is lost postmortem, traces of defective mineralization can be seen as poorly or incompletely areas of mineralized bone separated from existing bone near cement lines (Brickley et al., 2020b; Mankin, 1974; Mays et al., 2007; Schamall et al., 2003). These

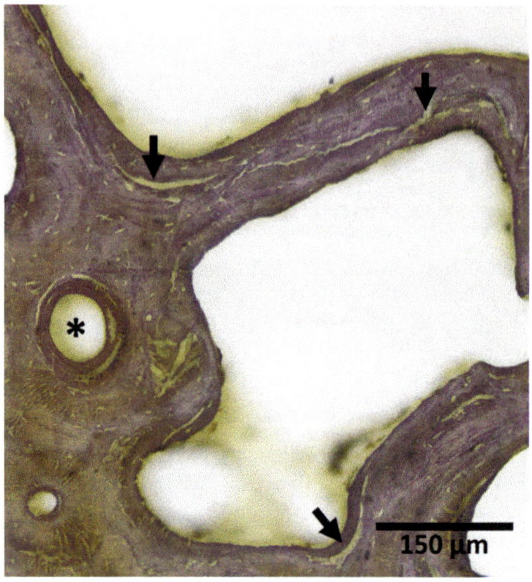

Figure 9.10 Detail of a rib of an individual with osteomalacia (age unknown), undecalcified specimen, hematoxylin stained. Note the 'defective cement lines': the areas of poorly or incompletely mineralized bone separated from existing bone. These can be seen either around Haversian systems (asterisk) or within bone trabeculae (arrows). Photo courtesy of Dr. Lida van der Merwe of the Amsterdam University Medical Centre.

'defective cement lines' are believed to be the result of bone mineralization failure causing a disconnect between the cement line and later deposited bone (Figure 9.10) (Boyde et al., 1986; Brickley et al., 2007). Van Der Merwe et al. (2018) confirmed the presence of bone mineralization defects appearing along the cement lines, near the osteon canals or around the osteocyte lacunae, basing a differentiation from postmortem alterations on its location/distribution and their fine granular appearance. Mineralization defects can be easily seen using BSE-SEM and lesions should be observed consistently throughout the skeleton to reach a diagnosis (Brickley et al., 2020b).

An increase of Howship's lacunae and enlarged osteocyte lacunae (osteocytic osteolysis) indicates increased resorption (Schnitzler & Pettifor, 2019; Van Der Merwe et al., 2018). In extreme cases of osteomalacia, osteoclastic tunneling might be present and the mineralized bone underneath the osteoid seams shows an irregular outline reflecting previous resorption episodes (Marie & Glorieux, 1982; Skingle & Compston, 2011). As vitamin D deficiency can cause secondary hyperparathyroidism, histological features from both conditions might co-exist (Mays et al., 2007; Schnitzler & Pettifor, 2019).

9.6.1.4 Hyperparathyroidism

Parathyroid hormone (PTH) is responsible for the levels of ionized calcium concentration in the blood and affects bone metabolism (Bellido & Hill Gallant, 2014). Elevated levels of PTH induce bone resorption through the induction of osteoclastic activity (Bellido & Hill Gallant, 2014; Silva & Bilezikian, 2015; Unnanuntana et al., 2011). Abnormal activity of the glands is subdivided in hypo- or hyperparathyroidism, with the latter being subdivided in primary, secondary, and tertiary hyperparathyroidism, depending on its cause (Rubin & Levine, 2013; Tominaga, 2021).

Trabecular tunneling is a pattern of bone resorption that is believed to be diagnostic of hyperparathyroidism. It describes the distinctive features of osteoclastic trabecular resorption in which a central channel is formed within the trabeculae (McCarthy, 2010, 2016; Vigorita & Ghelman, 1999). This resorption is ordinarily compensated by deposition of fibrous tissue, but this will not be present in skeletonized remains. The distinctive trabecular tunneling or 'dissecting osteitis', is believed to be still visible in well-preserved specimens (De Boer et al., 2013). This tunneling, along with 'cookie-bite' defects in areas of high lytic activity, are common signs of hyperparathyroidism (Cook et al., 1988; Mays et al., 2007; Zink et al., 2005). Additionally, intracortical porosity is significantly increased with an uneven sub-periosteal surface presenting tunneling through joined large, irregular, and deep canals, while the endosteal envelope presents trabecularization and pronounced Howship's lacunae (Christiansen et al., 1993; Cook et al., 1988; Fadda et al., 1990; Vu et al., 2013).

Osteocytic osteolysis can be present with a high number of osteocytes showing irregular lacunae as well as enlarged canaliculi (Bonucci et al., 1978; Stout et al., 2019). This feature is, however, not limited to hyperparathyroidism, and can also be seen in, for instance, glucocorticoid-induced osteoporosis or tumors (Tsourdi et al., 2018). Its recognition in skeletonized remains is challenging, if not impossible.

Histology can assist in differentiating hyperparathyroidism and osteoporosis, as although both conditions produce bone loss, osteoclastic and osteoblastic activity is more balanced in the latter. The presence of dissecting osteitis favors a diagnosis of hyperparathyroidism (Cook et al., 1988; McCarthy, 2010, 2016). The disease might easily be missed if the pathological assessment of a skeleton is limited to a macroscopic analysis (De Boer et al., 2013).

9.6.1.5 Paget's Disease

Paget's disease of bone (PDB) is characterized by an increase of bone remodeling occurring at one or more circumscript foci (Khurana & Fitzpatrick, 2009; McCarthy, 2010). Combined radiographic and histological assessment is essential for diagnosis, as gross-anatomical observation might lead to mis- or overdiagnosis (Garnett & Lewis, 2022; Roches et al., 2002). Three phases of the condition are recognized physiologically and histologically (Brickley et al., 2020d; Jaffe, 1977; Khurana & Fitzpatrick, 2009; Singer & Roodman, 2008).

The first phase consists mostly of lytic activity in which multiple Howship's lacunae as well as thinning of trabecular elements occur (Mirra et al., 1995; Ortner & Putschar, 1981). Other diseases associated with high bone turnover will produce a similar effect (De Boer & Van der Merwe, 2016). Irregularly shaped portions of lamellar bone can be present both in PDB and in hyperparathyroidism, although the absence of dissecting osteitis in PDB and the lack of osteoblastic reaction alterations in hyperparathyroidism along with radiographic features should be considered (Brickley et al., 2020d; Burrell et al., 2019; Ortner, 2003).

In the second phase, a mixed osteoblastic and osteoclastic activity occurs, at rates up to 20 times faster than seen during normal remodeling (Mundy, 1999). The typical histological Pagetic bone appearance, along with trabecular and cortical thickening, starts to appear at the end of this phase (Jaffe, 1977). Both osteoclastic and osteoblastic processes will occur simultaneously producing a microscopic 'mosaic pattern' in trabecular and cortical bone (Aaron et al., 1992; De Boer & Van der Merwe, 2016; Roches et al., 2002). In general, osteons will appear incomplete with scalloped *polycyclic* cement lines and with irregular shapes (Meunier et al., 1980; Rossetti et al., 2018). This bone eventually consists of a 'patchwork' of lamellar and woven bone separated by multiple interconnecting haphazard, scalloped, and disrupted cement lines representing consecutive episodes of repair (Brickley et al., 2020d; McCarthy, 2010; Singer & Roodman, 2008). Osteoclastic-osteoblastic forms of bone metastasis (especially prostate carcinoma) can also present this mosaic pattern, but radiological and morphological differences will help to establish a definitive diagnosis (De Boer et al., 2013; Roches et al., 2002). Multiple erratic cement lines can be also present in processes such as chronic osteomyelitis, but again morphological and radiological manifestations of each condition will assist in differential diagnosis (Rossetti et al., 2018).

The third 'sclerotic' or 'burnt-out' phase relates predominantly to osteoblastic activity resulting in overall sclerotic, but structurally weaker bone (Brickley et al., 2020d).

9.6.2 Tumors

Many tumors and tumor-like lesions can affect the skeleton (Resnick & Kransdorf, 2005). Depending on the tissue of origin and malignancy, tumors are either primary or secondary, and either benign or malignant. Primary bone tumors originate from tissues in which they are found, and may be either benign (without metastatic potential) or malignant (with metastatic potential). Secondary tumors are malignant by definition as they originate from elsewhere in the body, and therefore have metastasized (Marques, 2019; Waldron, 2008). Primary bone tumors are rare in comparison to metastatic and hematopoietic tumors. Benign bone tumors greatly outnumber the malignant forms. For a detailed review of tumors that may affect bone see the World Health Organization (WHO) classification of soft tissue and bone tumors (WHO sand IARC, 2020).

The diagnosis of bone tumors is complex and is mainly based on the morphology of the lesion, the site or distribution of skeletal involvement, and the individual's age. The morphological characterization of lesions (e.g., pattern of bone destruction, size, shape, and margin of the lesions, and type of periosteal new bone formation) is useful to ascertain the aggressive or nonaggressive behavior of lesions. The anatomical location and age of the individual are also important, since there is a propensity of specific tumors to affect certain age groups and locations. For instance, many bone tumors have a peak incidence during the first decades of life and affect predominantly the long bones of the extremities (Horvai, 2015; Resnick & Kransdorf, 2005).

Several cases of benign and malignant tumors have been identified in dry bone remains (for a review, see, for instance: Hunt et al., 2018). Their analysis is often limited to a macroscopic examination, occasionally complemented with imaging techniques, and less often aided by microscopic analysis (Brothwell, 2011). Even though the differential diagnosis of some tumors may be challenging, as different tumors share similar histological features, the use of microscopy can add information, for instance, on the nature of the tumors or to exclude other, non-pathological entities (so-called 'pseudo pathologies') (De Boer & Maat, 2018; Resnick & Kransdorf, 2005).

9.6.3 Bone-Forming Tumors

Bone-forming tumors are characterized by the production of unmineralized osteoid or mineralized woven and/or lamellar bone. They may range from small benign tumor and tumor-like lesions (e.g., osteoid osteoma, osteoblastoma, and osteoma), to aggressive malignant processes (e.g., osteosarcoma) (Horvai, 2015).

9.6.3.1 Osteoid Osteoma and Osteoblastoma

Osteoid osteoma and osteoblastoma are benign tumors that share histological features, including the presence of interconnecting networks of osteoid and trabeculae of woven bone rimmed by active osteoblasts (Horvai, 2015; Rezeanu, 2016). Osteoblastomas usually present a less organized microstructure and wider trabeculae than osteoid osteomas, and may potentially progress to aggressive forms. Osteoid osteomas and osteoblastomas are rarely reported in skeletonized remains (e.g., Bews et al., 2023; Riccomi et al., 2018). The histological description of a case of osteoblastoma of the frontal sinus revealed a microstructure composed of an external calcified shell with the underlying tissue showing wide vascular spaces and no mature lamellar bone (Riccomi et al., 2018).

9.6.3.2 Osteomas

Osteomas are benign, slow-growing lesions composed of dense mature compact or cancellous bone with either a parosteal or periosteal origin. They have been relatively often reported in skeletal human remains (e.g., Capasso, 1997; Eshed et al., 2002; Moghaddam et al., 2013; Shin et al., 2015; Valentina et al., 2022). Histologically, they present dense compact bone with intraosseous Haversian systems. Parosteal osteomas can have a trabecular component. Bone lesions may display multiple parallel cement lines, which is indicative of a slow growth process (Horvai, 2015; Rezeanu, 2016). Comparable histological features have been reported in the study of dry bone remains. A few studies on the histology of dry bone osteomas noticed a microstructure composed of regular and/or irregular non-lamellar trabecular bone, at some points, covered by osteocyte lacunae, and without signs of osteoblastic activity. The presence of wide trabeculae with reduced intertrabecular space was also observed (Capasso, 1997; Strouhal et al., 1996). A mature, well-organized lamellar bone with homogeneous thickness and an ordered formation of trabeculae was reported in further cases of osteoma, in addition to erratic osteocyte lacunae, but without absence of immature or fibrous bone components or lytic lesions (Eshed et al., 2002; Moghaddam et al., 2013).

9.6.3.3 Osteosarcomas

Osteosarcomas, particularly conventional osteosarcomas, are the most common primary malignant tumor of bone (for a complete overview, see, for instance: Resnick & Kransdorf, 2005). They are characterized by the presence of malignant osteoid-producing cells and the presence of immature/woven bone. The pattern of bone involvement in conventional osteosarcomas is variable, however, the co-occurrence of osteolytic and osteosclerotic foci is most common. The tumor is invasive, and frequently breaks through the cortex and lifts the periosteum, leading to exuberant periosteal reactions, radiologically recognizable as a 'Codman's triangle' or a 'sunburst' appearance (Horvai, 2015; Resnick & Kransdorf, 2005). The histological analysis of dry bone samples reported a mixture of cortical and cancellous bone defects entwined with new bone deposition, mostly at the periosteal level. The new bone formation was usually of the woven type. Its microarchitecture varied between an amorphous, poorly organized tissue without osteon remodeling, to an irregular, trabecular, or sponge-like architecture, also displaying slim bony spiculae formation oriented more or less perpendicularly to the cortical bone. Numerous Howship's lacunae associated with rapid resorption process were also reported (e.g., Alt et al., 2002; Aufderheide et al., 1997; De Boer & Maat, 2018; Molnár et al., 2017; Strouhal et al., 1997; Suzuki, 1987).

9.6.3.4 *Cartilage-Forming Tumors*

Cartilage-forming tumors, of either the benign or malignant types, are the most frequent primary bone tumors, being characterized by the formation of hyaline or myxoid cartilage. This cartilage may undergo endochondral ossification (Horvai, 2015). Osteochondromas and a possible case of chondrosarcoma, have been examined histologically in skeletonized human remains (e.g., De Boer & Maat, 2018; Laffranchi et al., 2021; Varotto et al., 2021; Vyhnánek et al., 1999).

9.6.3.5 *Osteochondromas*

Osteochondromas are cartilage-covered bony outgrowths (pedunculated or sessile) that arise from the surface of bones, and are attached to the underlying bone by bony stalks. Osteochondromas may be solitary, which account for 85% of the cases, or multiple. The latter develop as part of a multiple hereditary exostosis syndrome, which is an autosomal dominant hereditary disease (Adler, 2000; Horvai, 2015; Resnick & Kransdorf, 2005). In osteochondromas, the cartilaginous cap is the tumorous component, and the bone is the result of its endochondral ossification. The bone often exhibits an internal wide-meshed and irregular trabecular framework, outlined by a thin layer of cortical bone. The new bone close to the cartilaginous cap is usually immature without a Haversian structure. Bone trabeculae show lamellar layering with smooth outer borders, in which scattered osteocytes (osteocyte lacunae) and reversal lines can be visible. Signs of bone remodeling (e.g., Howship's lacunae) are less frequent (Adler, 2000; De Boer & Maat, 2018; Vyhnánek et al., 1999).

9.6.3.6 *Chondrosarcomas*

Chondrosarcomas are malignant cartilage-producing tumors. Conventional chondrosarcoma is a relatively common type of malignant primary bone tumor that develops either *de novo* or can arise in previously benign cartilaginous lesions (e.g., osteochondroma). They may have an intramedullary origin (central), representing circa 90% of chondrosarcomas, or a juxtacortical (peripheral) origin (Horvai, 2015; Resnick & Kransdorf, 2005). Due to calcification/ossification, chondrosarcoma may produce spotty calcifications (Resnick & Kransdorf, 2005), as well as extensive bone resorption with abundant Howship's lacunae and new bone formation of the woven type. Remodeled lamellar bone is rare (De Boer & Maat, 2018; Horvai, 2015).

9.6.3.7 *Other Mesenchymal Tumors of Bone*

A wide variety of other mesenchymal tumors can originate from the bone. In this section, angiosarcomas and multiple myeloma are discussed. Fibro-osseous lesions are strictly speaking not all neoplasms but do present as tumorous lesions, and are therefore also discussed in this section.

9.6.3.8 *Angiosarcoma*

Angiosarcoma is a rare malignant vascular tumor characterized by the proliferation of malignant endothelial cells. Skeletal localizations of this tumor are characterized by marked bone osteolysis with multifocal involvement, uncommonly accompanied by osteosclerosis. The osteolytic foci are variable in size and may present well-defined or irregular contours. Irregular defects at cortical and trabecular level with multiple Howship's lacunae have also been noticed in the study of dry bone samples (De Boer & Maat, 2018). Cortical thinning and extensive periosteal bone formations may occur, the latter associated with more aggressive tumors (Resnick & Kransdorf, 2005). Small deposits of new bone, mostly of the woven type, encircling the osteolytic foci have also been reported (De Boer & Maat, 2018).

9.6.3.9 Multiple Myeloma

Multiple myeloma is a tumor of plasma cells that usually develops in the bone marrow. It produces small lytic lesions with uniform diameter (0.5–2 mm) and spheroid morphology (Aufderheide & Rodríguez Martín, 1998; Rothschild et al., 1998; Rothschild & Rothschild, 1995). The osteolytic lesions are sharply defined with a 'punched-out' appearance and scalloped contours, and with no signs of osteoblastic activity (Ortner, 2003; Rothschild et al., 1998). Several cases of multiple myeloma have been identified in dry bone remains (for a review see: (Riccomi et al., 2019)). The histological analysis of some cases revealed lesions with a microarchitecture composed of spongiotic trabeculae without signs of osteoclastic resorption or with minimal Howship's lacunae. No evidence of osteoblastic response was noticed (e.g., Strouhal et al., 1996; Wakely et al., 1998).

9.6.3.10 Fibro-osseous Lesions

Fibrous dysplasia (or fibrocartilaginous dysplasia) is a developmental anomaly characterized by a failure to form mature lamellar bone, particularly at the trabecular level (Adler, 2000). Bone tumor lesions are characterized by a disorganized mixture of fibrous and osseous elements in the medullary region (Garcia et al., 2012; López-Terrada & Hicks, 2009). They often display 'fibrous and osseous component with bland spindle cells and irregular curvilinear trabeculae of woven bone' (López-Terrada & Hicks, 2009: 99).

Ossifying fibroma is a closely related condition which displays a central fibro-osseous lesion with a coarse network of fibro-osseous trabeculae with varying widths and lengths (Resnick and Kransdorf, 2005). Bone trabeculae may be mineralized to varying extents, and present osteocytes lacunae. Mature lesions may show lamellar bone trabeculae (Adler, 2000).

Osteofibrous dysplasia may cause cortical destruction and periosteal new bone formation (López-Terrada & Hicks, 2009). Fibrous-osseous processes documented in dry bone remains usually display core lesions composed of networks of thin trabeculae, mainly of the woven type, in some cases, surrounded by a well-defined cortex (e.g., Colard et al., 2008; Gresky et al., 2018).

9.6.3.11 Secondary Bone Tumors (Metastases)

Metastatic carcinomas are the most common form of skeletal malignancy, significantly outnumbering primary bone malignancies. By definition, they are metastatic and invasive, and are characterized by an uncontrolled proliferation of abnormal tumor cells that may reach bone through direct extension, lymphatic or hematogenous dissemination, or via intraspinal seeding (e.g., via the Batson plexus of veins) (Horvai, 2015). The malignant carcinomas that most often spread to bone, particularly in adults, are those of the prostate, breast, kidney, and lung. Bone metastases are usually but not exclusively multifocal (Horvai, 2015; Resnick & Kransdorf, 2005). Histologically, most metastatic lesions are predominantly resorptive, with an irregular lytic appearance in both cortical and spongy tissues. There are several reports of its histological appearance in skeletonized remains (e.g., Assis & Keenleyside, 2019; De Boer & Maat, 2018; De La Rúa et al., 1995; Luna et al., 2008; Mays et al., 1996; McCarthy, 2016; Schultz et al., 2007; Šefčáková et al., 2001). In these reports, the core of the lesions was located in the medullary cavity and a significant number of Howship's lacunae were seen bordering the primary foci. Multiple destructive cavities (Howship's lacunae)

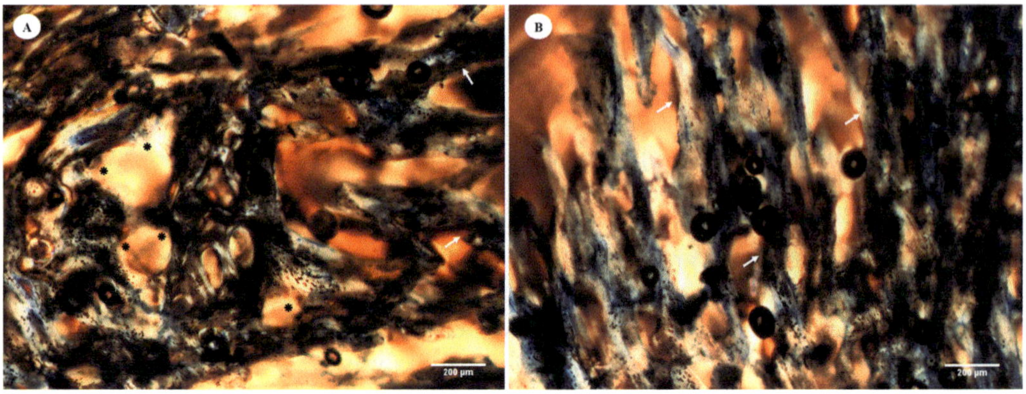

Figure 9.11 Location of a colon carcinoma metastasis in a 5th left rib from a 55-year-old male, undecalcified specimen, viewed with polarized light. The malignancy is no longer visible due to decomposition, but the disorderly mixed osteoclastic and osteoblastic microarchitecture associated with the bone metastasis can still be appreciated. Panel A demonstrates cortical osteoclastic resorption (black asterisk) and deposition of immature, woven bone (white arrows). Panel B demonstrates multiple parallel spiculae of immature bone (white arrows). Specimen from the 19th- to 20th-century Luís Lopes Anthropological Collection. ULisboa-MUHNAC.

were also seen on the trabeculae of the spongy bone. In some cases, the preexisting bone was substituted by irregular trabeculae, indicating an osteoblastic response. When present, newly formed bone displayed an immature structure (woven bone) and was located at periosteal level. The arrangement of the newly built bone varied between elongated spiculae perpendicular to the cortex to bony bars or more smooth formations (Figure 9.11).

9.6.4 Infectious Diseases

Bone infections usually comprise osteomyelitis, i.e., an infection of bone and marrow cavity. Histologically, acute osteomyelitis is identified by necrotic bone with empty and enlarged lacunae and bone scalloping, while chronic osteomyelitis presents bone formation (involucrum formation and 'creeping substitution') (Bruder et al., 2009; De Boer & Van der Merwe, 2016; Keel, 2015). In dry bone, osteolytic activity through the presence of multiple Howship's lacunae and reactive bone formation can be observed (De Boer & Van der Merwe, 2016; Schultz, 2001). Bone formation and resorption is, however, nonspecific of osteomyelitis, appearing in conditions such as malignant primary bone tumors or osteolytic metastases, so further assessment is needed to reach a definitive diagnosis (De Boer & Van der Merwe, 2016). Additionally, a bone pattern with multiple irregular cement lines with pagetoid appearance can be present in primary chronic osteomyelitis (Baltensperger et al., 2004; Bruder et al., 2009). Infectious bone formation can be differentiated by bone apposition due to an ossified hematoma. In infections, the pre-existing cortical surface will be irregular/remodeled, where in ossified hematomas it will be unchanged (Schultz, 2012; Van der Merwe et al., 2010).

A wide variety of micro-organisms can cause osteomyelitis, although most cases are due to infection by *Staphylococcus aureus*. Infections may be localized or systemic (Lew & Waldvogel, 2004; Sia & Berbari, 2006). There are some reports that discuss the dry bone histology of specific forms of osteomyelitis. These reports suggest that nonspecific osteomyelitis (generally a high-grade infection) produces a more irregular appearance with large osteolytic holes and connecting canals (Schultz & Schmidt-Schultz, 2015). Low-grade infections, such as tuberculosis, leprosy, and treponemal disease are said to produce a more chronic histological image, with padding ('polster') and band-like stripes ('grenzstreifen') reported in treponemal diseases (see Schultz, 2001, 2003). However, the specificity of these features has been debated (Schultz, 2001; Von Hunnius et al., 2006; Weston, 2008).

Although histology cannot identify the infectious pathogen by dry bone histopathology, it can help to assess the degree, location and chronicity of infection, which assists in narrowing down differential diagnosis or corroborate a diagnosis (De Boer & Van der Merwe, 2016; Santos & Suby, 2015; Schultz & Schmidt-Schultz, 2015; Spekker et al., 2022; Traversari et al., 2019; Wilbur et al., 2009).

9.7 Trauma

Since biopsy interferes with healing, the histological analysis of bone trauma is not a common occurrence in clinical practice. This, however, does not mean that in the histological analysis of trauma is without merit. Especially for specific questions relating to fractures or amputations, microscopy can provide valuable information in addition to the results of macroscopic and radiological analysis. For instance, histological analysis can help to determine whether a trauma actually occurred. For acute trauma, this diagnostic challenge mainly presents itself in pediatric cases, where distinguishing between developmental anomalies and traumatic injury can be challenging. Differentiating between accessory skull sutures and skull fractures has been discussed in multiple publications (e.g. Wiedijk et al., 2016) and histological analysis is indicated in those cases in which radiology is inconclusive. In the case of old, well-healed fractures, the macroscopic and radiological examination may be unable to determine whether the angulation of a bone is due to previous fracture. In these cases, histology may show changes in the cortical microarchitecture, indeed confirming the presence of an old fracture with extensive remodeling.

Another valuable addition of microscopy to the diagnostic toolkit, however, relates to the temporal positioning of the trauma, i.e., when the trauma occurred. Fractures that show evidence of healing by definition occurred during life, and histological analysis of skeletal tissue can detect features of healing at an earlier stage than gross examination and radiology. Resorption of fracture ends, indicated by the presence of Howship's lacunae (Figure 9.12), can occur in the first days following trauma, and may be the only available evidence that the trauma occurred during life. The deposition of osseous callus on the periosteal surface (subperiosteal new bone formation) is another healing feature that can be recognized at an earlier moment histologically, if compared to radiology. Therefore, histological analysis of fractures is advised when antemortem trauma is considered, but no macroscopic or radiological healing features are present (Figure 9.13).

When it is decided that a trauma occurred antemortem, histological examination can aid in the estimation of the posttraumatic time, i.e., the time lapsed between trauma and death. Although fracture healing is a complex, incompletely understood process that is dependent on a large amount of intrinsic and extrinsic variables (Giannoudis et al., 2007; Marsell & Einhorn, 2011), it follows a strict sequential process. The various phases of the healing process (the inflammatory phase, reparative phase, and remodeling phase) all have distinct histological features that are still detectable in skeletonized remains (De Boer et al., 2015). Since each phase or histological features requires a minimum amount of time to develop, these features can be used to estimate the minimum

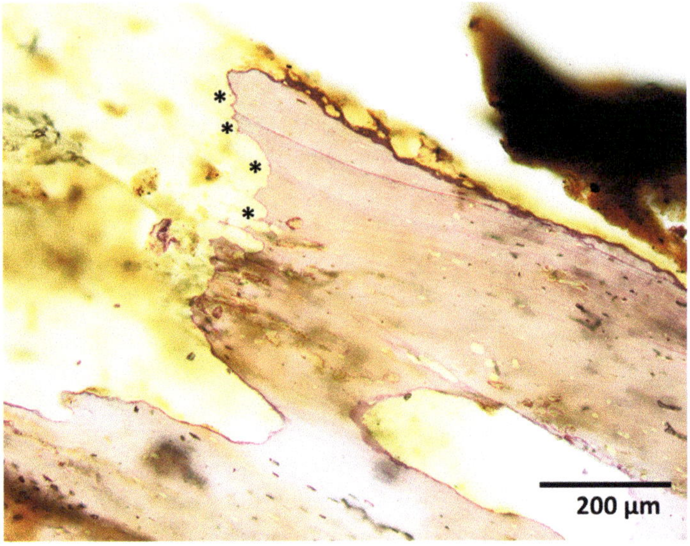

Figure 9.12 Fracture edge of a fractured 1st metacarpal bone, undecalcified section, stained with hematoxylin. Note the numerous Howship's lacunae at the fracture edge (asterisks), indicating osteoclastic activity. Specimen from the histological collection of the Amsterdam University Medical Centre.

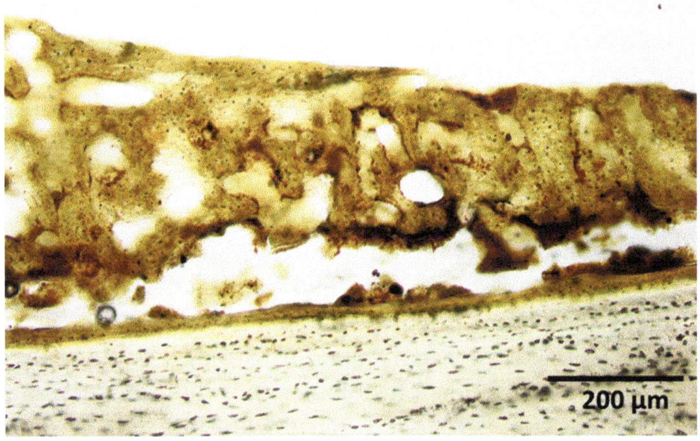

Figure 9.13 Early callus formation on the periosteal surface of a fractured 5th rib, undecalcified section. The callus consists entirely of woven bone and is affected by a considerable degree of diagenesis. Compare with the lamellar appearance of the underlying cortical bone. Specimen from the histological collection of the Amsterdam University Medical Centre.

amount of time between trauma and death (Cappella et al., 2019; Steyn et al., 2014). Various publications list the relation between fracture histology and post-traumatic time (Klotzbach et al., 2003; Naqvi et al., 2019), with more large cohort data expected in the near future (see: Isaac et al., 2022). However, at the time of publication of this volume, the histological estimation of post-traumatic time should still be considered an inexact science with a

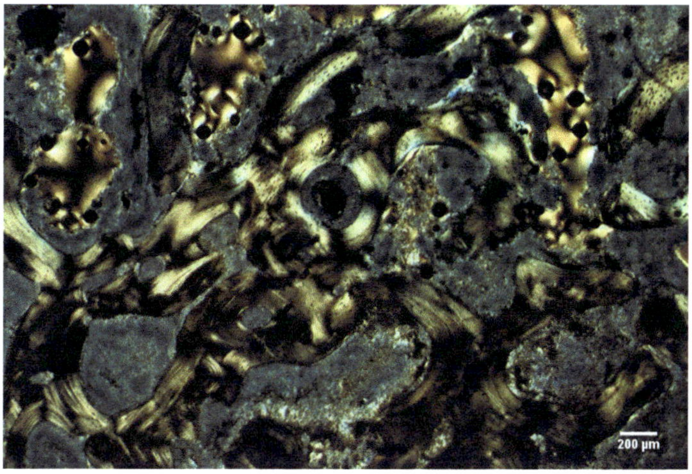

Figure 9.14 Largely remodeled endosteal callus of the right fibula of a 68-year-old male, undecalcified specimen, viewed with polarized light. The callus shows large areas of lamellar bone, indicating advanced remodeling and hence a substantial post-traumatic survival time. Specimen from the 19th- to 20th-century Luís Lopes Anthropological Collection. ULisboa-MUHNAC.

considerable error margin. Opinions on post-traumatic time should therefore be provided with the greatest caution (Figure 9.14).

The histological differentiation between perimortem trauma and postmortem breakage is equally challenging. Despite various publications on this particular subject, mostly related to microcrack patterns (Cappella & Cattaneo, 2019; Cattaneo & Cappella, 2017; Ebacher et al., 2012; Hanaue et al., 2007; Pechníková et al., 2011; Winter-Buchwalder et al., 2022), it is currently unresolved if histological analysis of dry bones can reliably differentiate between perimortem and postmortem fractures.

9.8 Conclusion

The histopathological examination of dry bone tissue may be associated with challenges, but the appropriate application of histological techniques can still add significantly to the diagnostic processes. It can help to differentiate between disease processes and pseudo pathologies. In a small subset of pathologies it may be possible to definitively diagnose a specific disease. In other cases, it may help to narrow down the differential diagnosis, assess the extent of disease, or determine the progression of healing. Dry bone histology always requires a robust embedding within a larger diagnostic framework, and its added value will be heavily dependent on the case circumstances.

Acknowledgments

The authors thank Dr. Lida van der Merwe of the Dept. of Medical Biology of the Amsterdam University Medical Centre, and the CAHID histology lab members for their assistance with some of the figures. We would like to thank the Luís Lopes Collection - Museu Nacional de História Natural e da Ciência, Universidade de Lisboa (MUHNAC) and its former curators for granting permission to assess bone samples from this repository.

References

Aaron, J. E., Rogers, J., & Kanis, J. A. (1992). Paleohistology of Paget's disease in two medieval skeletons. *American Journal of Physical Anthropology*, *89*(3), 325–331. https://doi.org/10.1002/ajpa.1330890306

Absolonova, K., Veleminsky, P., Dobisikova, M., Beran, M., & Zocova, J. (2013). Histological Estimation of Age at Death from the Compact Bone of Burned and Unburned Human Ribs. *Journal of Forensic Sciences*, *58*, S135–S145. https://doi.org/10.1111/j.1556-4029.2012.02303.x

Adler, C.-P. (2000). *Bone Diseases: Macroscopic, Histological, and Radiological Diagnosis of Structural Changes in the Skeleton*. Springer Berlin Heidelberg. https://doi.org/10.1007/978-3-662-04088-1

Agarwal, S. C. (2018). Understanding bone aging, loss, and osteoporosis in the past. In M. A. Katzenberg & A. L. Grauer (Eds.), *Biological Anthropology of the Human Skeleton* (pp. 385–414). John Wiley & Sons, Inc. https://doi.org/10.1002/9781119151647.ch11

Agarwal, S. C., Dumitriu, M., Tomlinson, G. A., & Grynpas, M. D. (2004). Medieval trabecular bone architecture: The influence of age, sex, and lifestyle. *American Journal of Physical Anthropology*, *124*(1), 33–44. https://doi.org/10.1002/ajpa.10335

Aksoy, U., Özkayalar, H., & Orhan, K. (2020). Micro-CT in Comparison with Histology in the Qualitative Assessment of Bone and Pathologies. In K. Orhan (Ed.), *Micro-computed Tomography (micro-CT) in Medicine and Engineering* (pp. 109–124). Springer International Publishing. https://doi.org/10.1007/978-3-030-16641-0_8

Alt, K. W., Adler, C. P., Buitrago-Téllez, C. H., & Lohrke, B. (2002). Infant osteosarcoma: Infant Osteosarcoma. *International Journal of Osteoarchaeology*, *12*(6), 442–448. https://doi.org/10.1002/oa.647

An, Y. H., & Martin, K. L. (Eds.). (2003). *Handbook of Histology Methods for Bone and Cartilage*. Humana Press. https://doi.org/10.1007/978-1-59259-417-7

Assis, S., & Keenleyside, A. (2019). The macroscopic and histomorphological properties of periosteal rib lesions and its relation with disease duration: Evidence from the Luis Lopes Skeletal Collection (Lisbon, Portugal). *Journal of Anatomy*, *234*(4), 480–501. https://doi.org/10.1111/joa.12936

Assis, S., Keenleyside, A., Santos, A. L., & Cardoso, F. A. (2015). Bone Diagenesis and its Implication for Disease Diagnosis: The Relevance of Bone Microstructure Analysis for the Study of Past Human Remains. *Microscopy and Microanalysis*, *21*(4), 805–825. https://doi.org/10.1017/S1431927615000768

Aufderheide, A. C., Ragsdale, B. D., Buikstra, J. E., Ekberg, F., & Vinh, T. (1997). Structure of the radiological "sunburst" pattern as revealed in an ancient osteosarcoma. *Journal of Paleopathology*, *9*(2), 101–106.

Aufderheide, A. C., & Rodríguez Martín, C. (1998). *The Cambridge encyclopedia of human paleopathology* [First paperback edition). Cambridge University Press.

Baltensperger, M., Grätz, K., Bruder, E., Lebeda, R., Makek, M., & Eyrich, G. (2004). Is primary chronic osteomyelitis a uniform disease? Proposal of a classification based on a retrospective analysis of patients treated in the past 30 years. *Journal of Cranio-Maxillofacial Surgery*, *32*(1), 43–50. https://doi.org/10.1016/j.jcms.2003.07.008

Bell, L. S. (Ed.). (2012). *Forensic Microscopy for Skeletal Tissues: Methods and Protocols* (Vol. 915). Humana Press. https://doi.org/10.1007/978-1-61779-977-8

Bell, L. S., & Piper, K. (2000). An introduction to palaeohistopathology. In M. Cox & S. Mays (Eds.), *Human osteology in archaeology and forensic sciences* (pp. 255–274). Greenwich Medical Media Limited.

Bellido, T., & Hill Gallant, K. M. (2014). Hormonal Effects on Bone Cells. In *Basic and Applied Bone Biology* (pp. 299–314). Elsevier. https://doi.org/10.1016/B978-0-12-416015-6.00015-0

Beresheim, A. C., Pfeiffer, S. K., Grynpas, M. D., & Alblas, A. (2019). Use of backscattered scanning electron microscopy to quantify the bone tissues of mid-thoracic human ribs. *American Journal of Physical Anthropology*, *168*(2), 262–278. https://doi.org/10.1002/ajpa.23716

Bertoglio, B., Messina, C., Cappella, A., Maderna, E., Mazzarelli, D., Lucheschi, S., Sardanelli, F., Sconfienza, L. M., Sforza, C., & Cattaneo, C. (2021). Bone tissue preservation in seawater environment: A preliminary comparative analysis of bones with different post-mortem histories through anthropological and radiological perspectives. *International Journal of Legal Medicine*, *135*(6), 2581–2594. https://doi.org/10.1007/s00414-021-02636-6

Bews, E. A., Aytek, A. I., Yavuz, A. Y., Kaya, E. H., Savran, G., Kalata, M., & Bethard, J. D. (2023). Differential diagnosis of an osseous cranial tumor from Hellenistic Muğla, Turkey. *International Journal of Paleopathology*, *40*, 103–108. https://doi.org/10.1016/j.ijpp.2023.01.003

Bikle, D. D. (2012). Vitamin D and Bone. *Current Osteoporosis Reports*, *10*(2), 151–159. https://doi.org/10.1007/s11914-012-0098-z

Bonucci, E., Cascio, V., Adami, S., Cominacini, L., Galvanini, G., & Scuro, A. (1978). The ultrastructure of bone cells and bone matrix in human primary hyperparathyroidism. *Virchows Archiv A Pathological Anatomy and Histology*, *379*(1), 11–23. https://doi.org/10.1007/BF00432779

Boyde, A. (2003). Improved digital SEM of cancellous bone: Scanning direction of detection, through focus for in-focus and sample orientation. *Journal of Anatomy*, *202*(2), 183–194. https://doi.org/10.1046/j.1469-7580.2003.00146.x

Boyde, A., Maconnachie, E., Reid, S. A., Delling, G., & Mundy, G. R. (1986). Scanning electron microscopy in bone pathology: Review of methods, potential and applications. *Scanning Electron Microscopy*, *Pt 4*, 1537–1554.

Brickley, M. B., & Agarwal, S. C. (2003). Techniques for the Investigation of Age-Related Bone Loss and Osteoporosis in Archaeological Bone. In S. C. Agarwal & S. D. Stout (Eds.), *Bone Loss and Osteoporosis: An Anthropological Perspective* (pp. 157–172). Springer US. https://doi.org/10.1007/978-1-4419-8891-1_10

Brickley, M. B., Ives, R., & Mays, S. (2020a). 4—Vitamin C deficiency, scurvy. In M. B. Brickley, R. Ives, & S. Mays (Eds.), *The Bioarchaeology of Metabolic Bone Disease (Second Edition)* (pp. 43–74). Academic Press. https://doi.org/10.1016/B978-0-08-101020-4.00004-5

Brickley, M. B., Ives, R., & Mays, S. (2020b). 5—Vitamin D deficiency. In M. B. Brickley, R. Ives, & S. Mays (Eds.), *The Bioarchaeology of Metabolic Bone Disease (Second Edition)* (pp. 75–128). Academic Press. https://doi.org/10.1016/B978-0-08-101020-4.00005-7

Brickley, M. B., Ives, R., & Mays, S. (2020c). 6—Age-related bone loss and osteoporosis. In M. B. Brickley, R. Ives, & S. Mays (Eds.), *The Bioarchaeology of Metabolic Bone Disease (Second Edition)* (pp. 129–163). Academic Press. https://doi.org/10.1016/B978-0-08-101020-4.00006-9

Brickley, M. B., Ives, R., & Mays, S. (2020d). 8—Paget's disease of bone. In M. B. Brickley, R. Ives, & S. Mays (Eds.), *The Bioarchaeology of Metabolic Bone Disease (Second Edition)* (pp. 179–200). Academic Press. https://doi.org/10.1016/B978-0-08-101020-4.00008-2

Brickley, M. B., Mays, S., & Ives, R. (2007). An investigation of skeletal indicators of vitamin D deficiency in adults: Effective markers for interpreting past living conditions and pollution levels in 18th and 19th century Birmingham, England. *American Journal of Physical Anthropology*, *132*(1), 67–79. https://doi.org/10.1002/ajpa.20491

Brothwell, D. (2011). Tumors: Problems of Differential Diagnosis in Paleopathology. In *A Companion to Paleopathology* (pp. 420–433). John Wiley & Sons, Ltd. https://doi.org/10.1002/9781444345940.ch23

Bruder, E., Jundt, G., & Eyrich, G. (2009). Pathology of Osteomyelitis. In M. M. Baltensperger & G. K. H. Eyrich (Eds.), *Osteomyelitis of the Jaws* (pp. 121–133). Springer. https://doi.org/10.1007/978-3-540-28766-7_6

Burrell, C. L., Emery, M. M., & Gonzalez, S. (2019). Paget's disease of bone in two medieval skeletons from Poulton Chapel, Cheshire, UK. *International Journal of Osteoarchaeology*, *29*(6), 922–933. https://doi.org/10.1002/oa.2807

Busse, B., Hahn, M., Soltau, M., Zustin, J., Püschel, K., Duda, G. N., & Amling, M. (2009). Increased calcium content and inhomogeneity of mineralization render bone toughness in osteoporosis: Mineralization, morphology and biomechanics of human single trabeculae. *Bone*, *45*(6), 1034–1043. https://doi.org/10.1016/j.bone.2009.08.002

Capasso, L. (1997). Osteoma: Palaeopathology and phylogeny. *International Journal of Osteoarchaeology*, *7*(6), 615–620. https://doi.org/10.1002/(SICI)1099-1212(199711/12)7:6<615::AID-OA370>3.0.CO;2-1

Cappella, A., & Cattaneo, C. (2019). Exiting the limbo of perimortem trauma: A brief review of microscopic markers of hemorrhaging and early healing signs in bone. *Forensic Science International*, *302*, 109856. https://doi.org/10.1016/j.forsciint.2019.06.014

Cappella, A., De Boer, H. H., Cammilli, P., De Angelis, D., Messina, C., Sconfienza, L. M., Sardanelli, F., Sforza, C., & Cattaneo, C. (2019). Histologic and radiological analysis on bone fractures: Estimation of posttraumatic survival time in skeletal trauma. *Forensic Science International*, *302*, 109909. https://doi.org/10.1016/j.forsciint.2019.109909

Carroll, E. L., & Squires, K. E. (2020). Burning by numbers: A pilot study using quantitative petrography in the analysis of heat-induced alteration in burned bone. *International Journal of Osteoarchaeology*, *30*(5), 691–699. https://doi.org/10.1002/oa.2902

Cattaneo, C., & Cappella, A. (2017). Distinguishing between Peri- and Post-Mortem Trauma on Bone. In *Taphonomy of Human Remains: Forensic Analysis of the Dead and the Depositional Environment* (pp. 352–368). John Wiley & Sons, Ltd. https://doi.org/10.1002/9781118953358.ch23

Cattaneo, C., DiMartino, S., Scali, S., Craig, O. E., Grandi, M., & Sokol, R. J. (1999). Determining the human origin of fragments of burnt bone: A comparative study of histological, immunological and DNA techniques. *Forensic Science International*, *102*(2–3), 181–191. https://doi.org/10.1016/S0379-0738(99)00059-6

Chen, H., Shoumura, S., Emura, S., & Bunai, Y. (2008). Regional variations of vertebral trabecular bone microstructure with age and gender. *Osteoporosis International*, *19*(10), 1473–1483. https://doi.org/10.1007/s00198-008-0593-3

Cho, H., Stout, S. D., Madsen, R. W., & Streeter, M. A. (2002). Population-specific histological age-estimating method: A model for known African-American and European-American skeletal remains. *Journal of Forensic Sciences*, *47*(1), 12–18.

Christiansen, P., Steiniche, T., Brockstedt, H., Mosekilde, L., Hessov, I., & Melsen, F. (1993). Primary hyperparathyroidism: Iliac crest cortical thickness, structure, and remodeling evaluated by histomorphometric methods. *Bone*, *14*(5), 755–762. https://doi.org/10.1016/8756-3282(93)90207-Q

Colard, T., Gabart, N., & Blondiaux, J. (2008). A palaeopathological case of a right maxilla's cemento-ossifying fibroma. *International Journal of Osteoarchaeology*, *18*(2), 195–201. https://doi.org/10.1002/oa.936

Cole, M. E., Stout, S., & Agnew, A. M. (2019). Image processing techniques for extracting complex three-dimensional cortical pore networks from high-resolution micro-computed tomography (micro-CT) images of the human femoral neck and rib. *American Journal of Physical Anthropology*, *168*(S68), 46.

Cole, M. E., Stout, S. D., Dominguez, V. M., & Agnew, A. M. (2022). Pore Extractor 2D: An ImageJ toolkit for quantifying cortical pore morphometry on histological bone images, with application to intraskeletal and regional patterning. *American Journal of Biological Anthropology*, *179*(3), 365–385. https://doi.org/10.1002/ajpa.24618

Collins, M. J., Nielsen-Marsh, C. M., Hiller, J., Smith, C. I., Roberts, J. P., Prigodich, R. V., Wess, T. J., Csapo, J., Millard, A. R., & Turner-Walker, G. (2002). The survival of organic matter in bone: A review. *Archaeometry*, *44*(3), 383–394. https://doi.org/10.1111/1475-4754.t01-1-00071

Cook, M., Molto, E., & Anderson, C. (1988). Possible case of hyperparathyroidism in a roman period skeleton from the Dakhleh Oasis, Egypt, diagnosed using bone histomorphometry. *American Journal of Physical Anthropology*, *75*(1), 23–30. https://doi.org/10.1002/ajpa.1330750104

De Boer, H. H., Aarents, M. J., & Maat, G. J. R. (2012). Staining ground sections of natural dry bone tissue for microscopy. *International Journal of Osteoarchaeology*, *22*(4), 379–386. https://doi.org/10.1002/oa.1208

De Boer, H. H., & Maat, G. J. R. (2012). The histology of human dry bone (a review). *Cuadernos de Prehistoria y Arqueología de la Universidad de Granada*, *22*, 49–65.

De Boer, H. H., & Maat, G. J. R. (George). (2018). Dry bone histology of bone tumours. *International Journal of Paleopathology*, *21*, 56–63. https://doi.org/10.1016/j.ijpp.2016.11.005

De Boer, H. H., van der Merwe, A. E., Hammer, S., Steyn, M., & Maat, G. J. R. (2015). Assessing Post-traumatic Time Interval in Human Dry Bone. *International Journal of Osteoarchaeology*, *25*(1), 98–109. https://doi.org/10.1002/oa.2267

De Boer, H. H., & Van der Merwe, A. E. (Lida). (2016). Diagnostic dry bone histology in human paleopathology. *Clinical Anatomy*, *29*(7), 831–843. https://doi.org/10.1002/ca.22753

De Boer, H. H., Van Der Merwe, A. E., & Maat, G. J. R. (2013). The diagnostic value of microscopy in dry bone palaeopathology: A review. *International Journal of Paleopathology*, *3*(2), 113–121. https://doi.org/10.1016/j.ijpp.2013.03.004

De La Rúa, C., Baraybar, J. P., & Etxeberria, F. (1995). Neolithic case of metastasizing carcinoma: Multiple approaches to differential diagnosis. *International Journal of Osteoarchaeology*, *5*(3), 254–264. https://doi.org/10.1002/oa.1390050304

Delannoy, Y., Colard, T., Cannet, C., Mesli, V., Hédouin, V., Penel, G., & Ludes, B. (2018). Characterization of bone diagenesis by histology in forensic contexts: A human taphonomic study. *International Journal of Legal Medicine*, *132*(1), 219–227. https://doi.org/10.1007/s00414-017-1699-y

Ebacher, V., Guy, P., Oxland, T. R., & Wang, R. (2012). Sub-lamellar microcracking and roles of canaliculi in human cortical bone. *Acta Biomaterialia*, *8*(3), 1093–1100. https://doi.org/10.1016/j.actbio.2011.11.013

Engekel, K., Prevrhal, S., & Genant, H. (2008). Macro- and Microimaging of Bone Architecture. In J. Bilezikian, L. Raisz, & T. Martin (Eds.), *Principles of Bone Biology* (pp. 1905–1942). Academic Press.

Eshed, V., Latimer, B., Greenwald, C. M., Jellema, L. M., Rothschild, B. M., Wish-Baratz, S., & Hershkovitz, I. (2002). Button osteoma: Its etiology and pathophysiology. *American Journal of Physical Anthropology*, *118*(3), 217–230. https://doi.org/10.1002/ajpa.10087

Fadda, M., Zirattu, G., & Delrio, A. N. (1990). [Osteosis due to primary hyperparathyroidism. The SEM ultrastructural features of the periosteal and endosteal surfaces]. *Archivio "Putti" Di Chirurgia Degli Organi Di Movimento*, *38*(1), 155–162.

Fain, O. (2005). Musculoskeletal manifestations of scurvy. *Joint Bone Spine*, *72*(2), 124–128. https://doi.org/10.1016/j.jbspin.2004.01.007

Fairgrieve, S. I. (2008). *Forensic cremation: Recovery and analysis*. CRC Press.

Fernández Castillo, R., Ubelaker, D. H., Acosta, J. A. L., De La Rosa, R. J. E., & Garcia, I. G. (2013). Effect of Temperature on Bone Tissue: Histological Changes. *Journal of Forensic Sciences*, *58*(3), 578–582. https://doi.org/10.1111/1556-4029.12093

Follis, R., Jackson, D., & Park, E. (1940). The problem of the association of rickets and scurvy. *The American Journal of Diseases of Children*, *14*(1), 50–60.

Forbes, G. (1941). The Effects of Heat on the Histological Structure of Bone. *The Police Journal: Theory, Practice and Principles*, *14*(1), 50–60. https://doi.org/10.1177/0032258X4101400108

Frost, H. M. (2004). A 2003 update of bone physiology and Wolff's Law for clinicians. *The Angle Orthodontist*, *74*(1), 3–15. https://doi.org/10.1043/0003-3219(2004)074<0003:AUOBPA>2.0.CO;2

Garcia, R., Klein, M., & Schiller, A. (2012). Bones and Joints. In Rubin R, & D. Strayer (Eds.), *Essentials of Rubin's Pathology* (pp. 1199–1272). Wolters Kluwer Health/Lippincott Williams & Wilkins.

García-Donas, J. G., Bonicelli, A., Scholl, A. R., Lill, C., Paine, R. R., & Kranioti, E. F. (2021). Rib histomorphometry: A reliability and validation study with a critical review of histological techniques for forensic age estimation. *Legal Medicine*, *49*, 101827. https://doi.org/10.1016/j.legalmed.2020.101827

García-Donas, J. G., Paine, R. R., Bonicelli, A., & Kranioti, E. F. (2022). Age estimation for two Mediterranean populations: Rib histomorphometry applied to forensic identification and bone remodelling research. *International Journal of Legal Medicine*, *136*(5), 1469–1481. https://doi.org/10.1007/s00414-022-02812-2

Garnett, E., & Lewis, M. (2022). The potential for over diagnosis of Paget's disease of bone using macroscopic analysis. *International Journal of Paleopathology, 38*, 55–63. https://doi.org/10.1016/j.ijpp.2022.07.001

Giannoudis, P. V., Einhorn, T. A., & Marsh, D. (2007). Fracture healing: The diamond concept. *Injury, 38*, S3–S6. https://doi.org/10.1016/S0020-1383(08)70003-2

Gresky, J., Kalmykov, A., & Berezina, N. (2018). Benign fibro-osseous lesion of the mandible in a Middle Bronze Age skeleton from Southern Russia. *International Journal of Paleopathology, 20*, 90–97. https://doi.org/10.1016/j.ijpp.2017.09.001

Hackett, C. J. (1981). Microscopical Focal Destruction (Tunnels) in Exhumed Human Bones. *Medicine, Science and the Law, 21*(4), 243–265. https://doi.org/10.1177/002580248102100403

Hanaue, K., Katakura, A., Kasahara, K., Kamiyama, I., Takaki, T., Shibahara, T., Abe, S., & Ide, Y. (2007). Course of fracture line in sagittal splitting of human mandible. *The Bulletin of Tokyo Dental College, 48*(4), 163–170. https://doi.org/10.2209/tdcpublication.48.163

Hedges, R. E. M. (2002). Bone diagenesis: An overview of processes. *Archaeometry, 44*(3), 319–328. https://doi.org/10.1111/1475-4754.00064

Hedges, R. E. M., & Millard, A. R. (1995). Bones and groundwater: Towards the modelling of diagenetic processes. *Journal of Archaeological Science, 22*(2), 155–164. https://doi.org/10.1006/jasc.1995.0017

Hedges, R. E. M., Millard, A. R., & Pike, A. W. G. (1995). Measurements and Relationships of Diagenetic Alteration of Bone from Three Archaeological Sites. *Journal of Archaeological Science, 22*(2), 201–209. https://doi.org/10.1006/jasc.1995.0022

Heveran, C. M., & Boerckel, J. D. (2023). Osteocyte Remodeling of the Lacunar-Canalicular System: What's in a Name? *Current Osteoporosis Reports, 21*(1), 11–20. https://doi.org/10.1007/s11914-022-00766-3

Holick, M. F. (2003). Vitamin D: A millenium perspective: Vitamin D. *Journal of Cellular Biochemistry, 88*(2), 296–307. https://doi.org/10.1002/jcb.10338

Horvai, A. (2015). Bones, Joints, and Soft Tissue Tumors. In Kumar V, Abbas AK (Ed.), *Robbins and Cotran pathologic basis of disease* (pp. 1179–1226). Elsevier Saunders.

Hunt, K. J., Roberts, C., & Kirkpatrick, C. (2018). Taking stock: A systematic review of archaeological evidence of cancers in human and early hominin remains. *International Journal of Paleopathology, 21*, 12–26. https://doi.org/10.1016/j.ijpp.2018.03.002

Imel, E., DiMeglio, L., Burr, D., & Allen, M. (2014). Metabolic Bone Diseases. In D. Burr (Ed.), *Basic and Applied Bone Biology* (pp. 317–344). Elsevier. https://doi.org/10.1016/B978-0-12-416015-6.00015-0

Isaac, C. V., Cornelison, J. B., Prahlow, J. A., Devota, C. J., & Christensen, E. R. (2022). The Repository of Antemortem Injury Response (REPAIR): An online database for skeletal injuries of known ages. *International Journal of Legal Medicine, 136*(4), 1189–1196. https://doi.org/10.1007/s00414-021-02756-z

Jaffe, H. L. (1977). The Classic Paget's Disease of Bone. *Clinical Orthopaedics and Related Research®, 127*, 4.

Janaway, R. C., Percival, S. L., & Wilson, A. S. (2009). Decomposition of Human Remains. In S. L. Percival (Ed.), *Microbiology and Aging* (pp. 313–334). Humana Press. https://doi.org/10.1007/978-1-59745-327-1_14

Jans, M. M. E. (2008). Microbial bioerosion of bone – a review. In M. Wisshak & L. Tapanila (Eds.), *Current Developments in Bioerosion* (pp. 397–413). Springer Berlin Heidelberg. https://doi.org/10.1007/978-3-540-77598-0_20

Jans, M. M. E., Nielsen-Marsh, C. M., Smith, C. I., Collins, M. J., & Kars, H. (2004). Characterisation of microbial attack on archaeological bone. *Journal of Archaeological Science, 31*(1), 87–95. https://doi.org/10.1016/j.jas.2003.07.007

Jones, C. G. (2012). Scanning Electron Microscopy: Preparation and Imaging for SEM. In L. S. Bell (Ed.), *Forensic Microscopy for Skeletal Tissues: Methods and Protocols* (pp. 1–20). Humana Press. https://doi.org/10.1007/978-1-61779-977-8_1

Keel, S. B. (2015). Pathologic Diagnosis of Osteomyelitis. In T. J. Boffeli (Ed.), *Osteomyelitis of the Foot and Ankle* (pp. 49–53). Springer International Publishing. https://doi.org/10.1007/978-3-319-18926-0_5

Kendall, C., Eriksen, A. M. H., Kontopoulos, I., Collins, M. J., & Turner-Walker, G. (2018). Diagenesis of archaeological bone and tooth. *Palaeogeography, Palaeoclimatology, Palaeoecology*, *491*, 21–37. https://doi.org/10.1016/j.palaeo.2017.11.041

Kenkre, J., & Bassett, J. (2018). The bone remodelling cycle. *Annals of Clinical Biochemistry: International Journal of Laboratory Medicine*, *55*(3), 308–327. https://doi.org/10.1177/0004563218759371

Khurana, J. S., & Fitzpatrick, L. A. (2009). Osteoporosis and Metabolic Bone Disease. In J. S. Khurana (Ed.), *Bone Pathology* (pp. 217–237). Humana Press. https://doi.org/10.1007/978-1-59745-347-9_14

Kim, Y.-S., Kim, D.-I., Park, D.-K., Lee, J.-H., Chung, N.-E., Lee, W.-T., & Han, S.-H. (2007). Assessment of Histomorphological Features of the Sternal End of the Fourth Rib for Age Estimation in Koreans*: HISTOLOGICAL AGE ESTIMATION FROM THE RIB. *Journal of Forensic Sciences*, *52*(6), 1237–1242. https://doi.org/10.1111/j.1556-4029.2007.00566.x

Klaus, H. D. (2014). Subadult scurvy in Andean South America: Evidence of vitamin C deficiency in the late pre-Hispanic and Colonial Lambayeque Valley, Peru. *International Journal of Paleopathology*, *5*, 34–45. https://doi.org/10.1016/j.ijpp.2013.09.002

Klaus, H. D. (2017). Paleopathological rigor and differential diagnosis: Case studies involving terminology, description, and diagnostic frameworks for scurvy in skeletal remains. *International Journal of Paleopathology*, *19*, 96–110. https://doi.org/10.1016/j.ijpp.2015.10.002

Klotzbach, H., Delling, G., Richter, E., Sperhake, J., & Püschel, K. (2003). Post-mortem diagnosis and age estimation of infants' fractures. *International Journal of Legal Medicine*, *117*(2), 82–89. https://doi.org/10.1007/s00414-002-0338-3

Körmendi, S., Vecsei, B., Orhan, K., & Dobó-Nagy, C. (2020). Micro-CT in Osteoporosis Research. In K. Orhan (Ed.), *Micro-computed Tomography (micro-CT) in Medicine and Engineering* (pp. 87–107). Springer International Publishing. https://doi.org/10.1007/978-3-030-16641-0_7

Kuhn, G., Schultz, M., Müller, R., & Rühli, F. J. (2007). Diagnostic value of micro-CT in comparison with histology in the qualitative assessment of historical human postcranial bone pathologies. *HOMO*, *58*(2), 97–115. https://doi.org/10.1016/j.jchb.2006.11.002

Laffranchi, Z., Milella, M., Lombardo, P., Langer, R., & Lösch, S. (2021). Co-occurrence of malignant neoplasm and Hyperostosis Frontalis Interna in an Iron Age individual from Münsingen-Rain (Switzerland): A multi-diagnostic study. *International Journal of Paleopathology*, *32*, 1–8. https://doi.org/10.1016/j.ijpp.2020.10.008

Lander, S. L., Hosie, M., & Brits, D. (2023). The Diagenetic Alterations of Historic Skeletons from the Crown Mines Cemetery, South Africa. *Biology*, *12*(3), 378. https://doi.org/10.3390/biology12030378

Lee, U.-Y., Jung, G.-U., Choi, S.-G., & Kim, Y.-S. (2014). Anthropological Age Estimation with Bone Histomorphometry from the Human Clavicle. *The Anthropologist*, *17*(3), 929–936. https://doi.org/10.1080/09720073.2014.11891509

Lew, D. P., & Waldvogel, F. A. (2004). Osteomyelitis. *The Lancet*, *364*(9431), 369–379. https://doi.org/10.1016/S0140-6736(04)16727-5

Lips, P., van Schoor, N., & Bravenboer, N. (2013). Vitamin D defiency-related disorders. In C. Rosen (Ed.), *Primer on the Metabolic Bone Diseases and Disorders of Mineral Metabolism* (pp. 613–623). Wiley.

López-Terrada, D., & Hicks, J. M. (2009). Genetics and Molecular Biology of Bone and Soft Tissue Tumors. In J. S. Khurana (Ed.), *Bone Pathology* (pp. 91–124). Humana Press. https://doi.org/10.1007/978-1-59745-347-9_6

Lovász, G., Schultz, M., Gödde, J., Bereczki, Z., Pálfi, G., Marcsik, A., & Molnár, E. (2013). Skeletal manifestations of infantile scurvy in a late medieval anthropological series from Hungary. *Anthropological Science*, *121*(3), 173–185. https://doi.org/10.1537/ase.130905

Luna, L. H., Aranda, C. M., Bosio, L. A., & Beron, M. A. (2008). A case of multiple metastasis in Late Holocene hunter-gatherers from the Argentine Pampean region. *International Journal of Osteoarchaeology*, *18*(5), 492–506. https://doi.org/10.1002/oa.950

Maat, G. J. R. (2004). Scurvy in adults and youngsters: The Dutch experience. A review of the history and pathology of a disregarded disease. *International Journal of Osteoarchaeology*, *14*(2), 77–81. https://doi.org/10.1002/oa.708

Maat, G. J. R., Van Den Bos, R. P. M., & Aarents, M. (Job). (2001). Manual preparation of ground sections for the microscopy of natural bone tissue: Update and modification of Frost's ?rapid manual method? *International Journal of Osteoarchaeology*, *11*(5), 366–374. https://doi.org/10.1002/oa.578

Mankin, H. J. (1974). Rickets, osteomalacia, and renal osteodystrophy. Part II. *The Journal of Bone and Joint Surgery. American Volume*, *56*(2), 352–386.

Marcus, R., & Bouxsein, M. (2008). The Nature of Osteoporosis. In Marcus R, D. Feldman, D. Nelson, & C. Rosen (Eds.), *Osteoporosis* (pp. 27–36). Elsevier, Academic Press.

Marie, P. J., & Glorieux, F. H. (1982). Bone histomorphometry in asymptomatic adults with hereditary hypophosphatemic vitamin D-resistant osteomalacia. *Metabolic Bone Disease and Related Research*, *4*(4), 249–253. https://doi.org/10.1016/0221-8747(82)90035-2

Marques, C. (2019). Chapter 19—Tumors of Bone. In J. E. Buikstra (Ed.), *Ortner's Identification of Pathological Conditions in Human Skeletal Remains (Third Edition)* (pp. 639–717). Academic Press. https://doi.org/10.1016/B978-0-12-809738-0.00019-3

Marsell, R., & Einhorn, T. A. (2011). The biology of fracture healing. *Injury*, *42*(6), 551–555. https://doi.org/10.1016/j.injury.2011.03.031

Maxfield, L., & Crane, J. S. (2023). Vitamin C Deficiency. In *StatPearls*. StatPearls Publishing. http://www.ncbi.nlm.nih.gov/books/NBK493187/

Mayne Correia, P. M. (1997). Fire modification of bone: A review of the literature. In W. D. Haglund & M. H. Sorg (Eds.), *Forensic Taphonomy: The postmortem fate of human remains* (pp. 275–293). CRC Press.

Mays, S. (2014). The palaeopathology of scurvy in Europe. *International Journal of Paleopathology*, *5*, 55–62. https://doi.org/10.1016/j.ijpp.2013.09.001

Mays, S., Brickley, M., & Ives, R. (2007). Skeletal evidence for hyperparathyroidism in a 19th century child with rickets. *International Journal of Osteoarchaeology*, *17*(1), 73–81. https://doi.org/10.1002/oa.854

Mays, S., Maat, G. J. R., & de Boer, H. H. (2015). Scurvy as a factor in the loss of the 1845 Franklin expedition to the Arctic: A reconsideration. *International Journal of Osteoarchaeology*, *25*(3), 334–344. https://doi.org/10.1002/oa.2305

Mays, S., Strouhal, E., Vyhnánek, L., & Němečková A. A. (1996). A case of metastatic carcinoma of medieval date from Wharram Percy, England. *Journal of Paleopathology*, *8*(1), 33–42.

McCarthy, E. F. (2010). Metabolic Bone Disorders. In J. S. Khurana, E. F. McCarthy, & P. J. Zhang (Eds.), *Essentials in Bone and Soft-Tissue Pathology* (pp. 61–68). Springer US. https://doi.org/10.1007/978-0-387-89845-2_5

McCarthy, E. F. (2016). The histology of metabolic bone disease. *Diagnostic Histopathology*, *22*(10), 378–383. https://doi.org/10.1016/j.mpdhp.2016.09.004

Meunier, P. J., Coindre, J. M., Edouard, C. M., & Arlot, M. E. (1980). Bone histomorphometry in Paget's disease. Quantitative and dynamic analysis of pagetic and nonpagetic bone tissue. *Arthritis and Rheumatism*, *23*(10), 1095–1103. https://doi.org/10.1002/art.1780231005

Mills, S. E. (2012). Bone. In *Histology for Pathologists* (4th ed., p. 91). Lippincott Williams & Wilkins.

Mirra, JosephM., Brien, EarlW., & Tehranzadeh, J. (1995). Paget's disease of bone: Review with emphasis on radiologic features, part II. *Skeletal Radiology*, *24*(3). https://doi.org/10.1007/BF00228919

Mirza, F., & Canalis, E. (2015). MANAGEMENT OF ENDOCRINE DISEASE: Secondary osteoporosis: pathophysiology and management. *European Journal of Endocrinology*, *173*(3), R131–R151. https://doi.org/10.1530/EJE-15-0118

Moghaddam, N., Langer, R., Ross, S., Nielsen, E., & Lösch, S. (2013). Multiple osteosclerotic lesions in an Iron Age skull from Switzerland (320–250 BC) – an unusual case. *Swiss Medical Weekly*. https://doi.org/10.4414/smw.2013.13819

Molnár, E., Schultz, M., Schmidt-Schultz, T. H., Marcsik, A., Buczkó, K., Zádori, P., Biró, G., Bernert, Z., Baumhoer, D., & Hajdu, T. (2017). Rare Case of an Ancient Craniofacial Osteosarcoma with Probable Surgical Intervention. *Pathology & Oncology Research*, *23*(3), 583–587. https://doi.org/10.1007/s12253-016-0153-7

Mundy, G. R. (1999). *Bone remodeling and its disorders* (2. ed). Martin Dunitz.

Naqvi, A., Raynor, E., & Freemont, A. J. (2019). Histological ageing of fractures in infants: A practical algorithm for assessing infants suspected of accidental or non-accidental injury. *Histopathology*, *75*(1), 74–80. https://doi.org/10.1111/his.13850

Nicklisch, N., Maixner, F., Ganslmeier, R., Friederich, S., Dresely, V., Meller, H., Zink, A., & Alt, K. W. (2012). Rib lesions in skeletons from early neolithic sites in Central Germany: On the trail of tuberculosis at the onset of agriculture. *American Journal of Physical Anthropology*, *149*(3), 391–404. https://doi.org/10.1002/ajpa.22137

Nielsen-Marsh, C., Gernaey, A., Turner-Walker, G., Hedges, R., Pike, A., & Collins, M. (2000). The chemical degradation of bone. In M. Cox & S. Mays (Eds.), *Human Osteology: In Archaeology and Forensic Science* (pp. 439–452). Greenwich Medical Media Limited.

Nih Consensus Development Panel On Osteoporosis Prevention, Diagnosis, And Therapy. (2001). Osteoporosis Prevention, Diagnosis, and Therapy. *JAMA: The Journal of the American Medical Association*, *285*(6), 785–795. https://doi.org/10.1001/jama.285.6.785

Ortner, D. (2003). *Identification of Pathological Conditions in Human Skeletal Remains*. Elsevier. https://doi.org/10.1016/B978-0-12-528628-2.X5037-6

Ortner, D., Kimmerle, E. H., & Diez, M. (1999). Probable evidence of scurvy in subadults from archeological sites in Peru. *American Journal of Physical Anthropology*, *108*(3), 321–331. https://doi.org/10.1002/(SICI)1096-8644(199903)108:3<321::AID-AJPA7>3.0.CO;2-7

Ortner, D., & Putschar, W. (1981). *Identification of pathological conditions in human skeletal remains. Smithsonian Contributions to Anthropology 28*. Smithsonian Institution Press.

Parfitt, A. M. (1989). Surface specific bone remodeling in health and disease. In M. Kleerekoper & S. Krane (Eds.), *Clinical Disorders of Bone and Mineral Metabolism* (pp. 7–14). Mary Ann Liebert.

Parfitt, A. M. (2008). Skeletal Heterogeneity and the Purposes of Bone Remodelling. In R. Marcus, D. Feldman, D. Nelson, & C. Rosen (Eds.), *Osteoporosis* (pp. 855–872). Academic Press.

Parfitt, A. M., Mathews, C. H., Villanueva, A. R., Kleerekoper, M., Frame, B., & Rao, D. S. (1983). Relationships between surface, volume, and thickness of iliac trabecular bone in aging and in osteoporosis. Implications for the microanatomic and cellular mechanisms of bone loss. *Journal of Clinical Investigation*, *72*(4), 1396–1409. https://doi.org/10.1172/JCI111096

Pechníková, M., Porta, D., & Cattaneo, C. (2011). Distinguishing between perimortem and post-mortem fractures: Are osteons of any help? *International Journal of Legal Medicine*, *125*(4), 591–595. https://doi.org/10.1007/s00414-011-0570-9

Piper, K., & Valentine, G. (2012). Bone Pathology. In L. S. Bell (Ed.), *Forensic Microscopy for Skeletal Tissues: Methods and Protocols* (pp. 51–88). Humana Press. https://doi.org/10.1007/978-1-61779-977-8_4

Power, J., Loveridge, N., Lyon, A., Rushton, N., Parker, M., & Reeve, J. (2003). Bone Remodeling at the Endocortical Surface of the Human Femoral Neck: A Mechanism for Regional Cortical Thinning in Cases of Hip Fracture. *Journal of Bone and Mineral Research*, *18*(10), 1775–1780. https://doi.org/10.1359/jbmr.2003.18.10.1775

Ragsdale, B. D., & Lehmer, L. M. (2012). A Knowledge of Bone at the Cellular (Histological) Level is Essential to Paleopathology. In A. L. Grauer (Ed.), *A Companion to Paleopathology* (pp. 225–249). Wiley-Blackwell. https://doi.org/10.1002/9781444345940.ch13

Raisz, L. G. (2005). Pathogenesis of osteoporosis: Concepts, conflicts, and prospects. *Journal of Clinical Investigation*, *115*(12), 3318–3325. https://doi.org/10.1172/JCI27071

Raisz, L. G., Bilezikian JP, J., & Martin, T. (2008). Pathophysiology of Osteoporosis. In J. Bilezikian, L. G. Raisz, & T. Martin (Eds.), *Principles of Bone Biology* (pp. 1635–1647). Academic Press.

Resnick, D., & Kransdorf, M. J. (2005). *Bone and joint imaging* (3rd ed). Saunders.

Rezeanu, L. (2016). Lesions Characterized by Osteoid Deposition and Nonaggressive Radiology. In A. Deyrup & G. Siegal (Eds.), *Practical Orthopedic Pathology: A Diagnostic Approach* (pp. 67–83). Elsevier Inc.

Riccomi, G., Fornaciari, G., & Giuffra, V. (2019). Multiple myeloma in paleopathology: A critical review. *International Journal of Paleopathology*, *24*, 201–212. https://doi.org/10.1016/j.ijpp.2018.12.001

Riccomi, G., Fornaciari, G., Minozzi, S., Aringhieri, G., & Giuffra, V. (2018). A rare case of osteoblastoma from medieval Tuscany. *The Lancet Oncology*, *19*(1), 26. https://doi.org/10.1016/S1470-2045(17)30917-8

Roberts, C., & Wakely, J. (1992). Microscopical findings associated with the diagnosis of osteoporosis in palaeopathology. *International Journal of Osteoarchaeology*, *2*(1), 23–30. https://doi.org/10.1002/oa.1390020105

Roches, E., Blondiaux, J., Cotten, A., Chastanet, P., & Flipo, R.-M. (2002). Microscopic evidence for Paget's disease in two osteoarchaeological samples from early northern France. *International Journal of Osteoarchaeology*, *12*(4), 229–234. https://doi.org/10.1002/oa.617

Rossetti, C., Pasquinelli, L., Verzeletti, A., Armocida, G., Licata, M., & Fulcheri, E. (2018). A case of Paget from a Northern Italy medieval necropolis. *International Journal of Paleopathology*, *20*, 104–107. https://doi.org/10.1016/j.ijpp.2017.11.003

Rothschild, B. M., Hershkovitz, I., & Dutour, O. (1998). Clues potentially distinguishing lytic lesions of multiple myeloma from those of metastatic carcinoma. *American Journal of Physical Anthropology*, *105*(2), 241–250. https://doi.org/10.1002/(SICI)1096-8644(199802)105:2<241::AID-AJPA10>3.0.CO;2-0

Rothschild, B. M., & Rothschild, C. (1995). Comparison of radiologic and gross examination for detection of cancer in defleshed skeletons. *American Journal of Physical Anthropology*, *96*(4), 357–363. https://doi.org/10.1002/ajpa.1330960404

Rubin, M., & Levine, M. (2013). Hypoparathyroidism and Pseudohypoparathyroidism. In C.J. Rosen (Ed.), *Primer on the Metabolic Bone Diseases and Disorders of Mineral Metabolism* (pp. 579–589). John Willey.

Rühli, F. J., Kuhn, G., Evison, R., Müller, R., & Schultz, M. (2007). Diagnostic value of micro-CT in comparison with histology in the qualitative assessment of historical human skull bone pathologies. *American Journal of Physical Anthropology*, *133*(4), 1099–1111. https://doi.org/10.1002/ajpa.20611

Santos, A. L., & Suby, J. A. (2015). Skeletal and Surgical Evidence for Acute Osteomyelitis in Non-Adult Individuals: Acute Osteomyelitis in Non-Adults Individuals. *International Journal of Osteoarchaeology*, *25*(1), 110–118. https://doi.org/10.1002/oa.2276

Schamall, D., Teschler-Nicola, M., Kainberger, F., Tangl, S., Brandstätter, F., Patzak, B., Muhsil, J., & Plenk, H. (2003). Changes in trabecular bone structure in rickets and osteomalacia: The potential of a medico-historical collection: Changes in Trabecular Bone Structure in Rickets and Osteomalacia. *International Journal of Osteoarchaeology*, *13*(5), 283–288. https://doi.org/10.1002/oa.688

Schneider, C. A., Rasband, W. S., & Eliceiri, K. W. (2012). NIH Image to ImageJ: 25 years of image analysis. *Nature Methods*, *9*(7), Article 7. https://doi.org/10.1038/nmeth.2089

Schnitzler, C. M., & Pettifor, J. M. (2019). Calcium Deficiency Rickets in African Adolescents: Cortical Bone Histomorphometry. *JBMR Plus*, *3*(6). https://doi.org/10.1002/jbm4.10169

Schultz, M. (2001). Paleohistopathology of bone: A new approach to the study of ancient diseases. *American Journal of Physical Anthropology*, *116*(S33), 106–147. https://doi.org/10.1002/ajpa.10024

Schultz, M. (2003). Light microscopic analysis in skeletal paleopathology. In D. Ortner (Ed.), *Identification of Pathological Conditions in Human Skeletal Remains* (pp. 99–146). Academic Press.

Schultz, M. (2012). Light microscopic analysis of macerated pathologically changed bones. In S. D. Stout & C. Crowder (Eds.), *Bone Histology: An Anthropological Perspective* (pp. 253–296). CRC Press.

Schultz, M., Parzinger, H., Posdnjakov, D. V., Chikisheva, T. A., & Schmidt-Schultz, T. H. (2007). Oldest known case of metastasizing prostate carcinoma diagnosed in the skeleton of a 2,700-year-old Scythian king from Arzhan (Siberia, Russia). *International Journal of Cancer, 121*(12), 2591–2595. https://doi.org/10.1002/ijc.23073

Schultz, M., & Schmidt-Schultz, T. H. (2015). Is it possible to diagnose TB in ancient bone using microscopy? *Tuberculosis, 95*, S80–S86. https://doi.org/10.1016/j.tube.2015.02.035

Šefčáková, A., Strouhal, E., Němečková, A., Thurzo, M., & Stassíková-Stukovská, D. (2001). Case of metastatic carcinoma from end of the 8th-Early 9th century Slovakia. *American Journal of Physical Anthropology, 116*(3), 216–229. https://doi.org/10.1002/ajpa.1117

Shah, F. A., Ruscsák, K., & Palmquist, A. (2019). 50 years of scanning electron microscopy of bone—A comprehensive overview of the important discoveries made and insights gained into bone material properties in health, disease, and taphonomy. *Bone Research, 7*(1), 15. https://doi.org/10.1038/s41413-019-0053-z

Shin, D. H., Oh, C. S., Kim, Y.-S., Kim, Y., Oh, S. W., Park, J. B., Lee, I. S., & Kim, M. J. (2015). Bone tumors in pre-modern skulls from human skeletal series of Joseon Dynasty. *Anatomy & Cell Biology, 48*(3), 213. https://doi.org/10.5115/acb.2015.48.3.213

Sia, I. G., & Berbari, E. F. (2006). Osteomyelitis. *Best Practice & Research Clinical Rheumatology, 20*(6), 1065–1081. https://doi.org/10.1016/j.berh.2006.08.014

Silva, B. C., & Bilezikian, J. P. (2015). Parathyroid hormone: Anabolic and catabolic actions on the skeleton. *Current Opinion in Pharmacology, 22*, 41–50. https://doi.org/10.1016/j.coph.2015.03.005

Singer, F., & Roodman, G. (2008). Paget's disease of Bone. In Bilezikian JP, Raisz LG, Martin TJ (Ed.), *Principles of Bone Biology* (pp. 1599–1609). Academic Press.

Skingle, L., & Compston, J. (2011). Bone Histomorphometry. In D. Feldman, J. Pike, & J. Adams (Eds.), *Vitamin D* (3rd edition, (pp. 845–859). Academic Press.

Spekker, O., Váradi, O. A., Szekeres, A., Jäger, H. Y., Zink, A., Berner, M., Pany-Kucera, D., Strondl, L., Klostermann, P., Samu, L., Király, K., Bereczki, Z., Molnár, E., Pálfi, G., & Tihanyi, B. (2022). A rare case of calvarial tuberculosis from the Avar Age (8th century CE) cemetery of Kaba–Bitózug (Hajdú-Bihar county, Hungary) – Pathogenesis and differential diagnostic aspects. *Tuberculosis, 135*, 102226. https://doi.org/10.1016/j.tube.2022.102226

Squires, K. E., Thompson, T. J. U., Islam, M., & Chamberlain, A. (2011). The application of histomorphometry and Fourier Transform Infrared Spectroscopy to the analysis of early Anglo-Saxon burned bone. *Journal of Archaeological Science, 38*(9), 2399–2409. https://doi.org/10.1016/j.jas.2011.04.025

Stark, R. J. (2014). A proposed framework for the study of paleopathological cases of subadult scurvy. *International Journal of Paleopathology, 5*, 18–26. https://doi.org/10.1016/j.ijpp.2014.01.005

Stauber, M., & Müller, R. (2006). Age-related changes in trabecular bone microstructures: Global and local morphometry. *Osteoporosis International, 17*(4), 616–626. https://doi.org/10.1007/s00198-005-0025-6

Stewart, M. C., Goliath, J. R., Stout, S. D., & Hubbe, M. (2015). Intraskeletal Variability of Relative Cortical Area in Humans: Variability of Relative Cortical Area. *The Anatomical Record, 298*(9), 1635–1643. https://doi.org/10.1002/ar.23181

Steyn, M., De Boer, H. H., & Van Der Merwe, A. E. (2014). Cranial trauma and the assessment of post-traumatic survival time. *Forensic Science International, 244*, e25–e29. https://doi.org/10.1016/j.forsciint.2014.08.021

Stout, S. D., & Crowder, C. (2012). Bone Remodeling, Histomorphology, and Histomorphometry. In S. D. Stout & C. Crowder (Eds.), *Bone Histology: An anthropological perspective* (pp. 1–17). CRC Press.

Stout, S. D., Mary E. Cole, & Amanda M. Agnew. (2019). Chapter 6 - Histomorphology: Deciphering the Metabolic Record. In Jane E. Buikstra (Ed.), *Ortner's Identification of Pathological Conditions in Human Skeletal Remains (Third Edition)* (pp. 91–167). Academic Press. https://doi.org/10.1016/B978-0-12-809738-0.00006-5

Strouhal, E., Vhynánek, L., Horáčková, L., Benešová, L., & Němečková, A. (1996). Two Unusual Benign Tumours in Skulls from the Ossuary at Křtiny (Czech Republic). *International*

Journal of Osteoarchaeology, *6*(3), 289–299. https://doi.org/10.1002/(SICI)1099-1212(199606)6:3<289::AID-OA272>3.0.CO;2-R

Strouhal, E., Vyhnánek, L., Horáčková, L., Benešová, L., & Němečková, A. (1997). A Case of Osteosarcoma in a Late Medieval–Early Modern Skull from Kyjov (Czech Republic). *International Journal of Osteoarchaeology*, *7*(1), 82–90. https://doi.org/10.1002/(SICI)1099-1212(199701)7:1<82::AID-OA320>3.0.CO;2-J

Suzuki, T. (1987). Paleopathological study on a case of osteosarcoma. *American Journal of Physical Anthropology*, *74*(3), 309–318. https://doi.org/10.1002/ajpa.1330740305

Tanizawa, T., Itoh, A., Uchiyama, T., Zhang, L., & Yamamoto, N. (1999). Changes in cortical width with bone turnover in the three different endosteal envelopes of the ilium in postmenopausal osteoporosis. *Bone*, *25*(4), 493–499. https://doi.org/10.1016/S8756-3282(99)00183-0

Thomas, C. D. L., Feik, S. A., & Clement, J. G. (2006). Increase in pore area, and not pore density, is the main determinant in the development of porosity in human cortical bone. *Journal of Anatomy*, *209*(2), 219–230. https://doi.org/10.1111/j.1469-7580.2006.00589.x

Thompson, J., Power, R. K., Mercieca-Spiteri, B., Magnussen, J. S., Pardey, M., Buck, L. T., Stock, J. T., McLaughlin, T. R., Stoddart, S., & Malone, C. (2022). Analysis of periosteal lesions from commingled human remains at the Xagħra Circle hypogeum reveals the first case of probable scurvy from Neolithic Malta. *International Journal of Osteoarchaeology*, *32*(1), 18–37. https://doi.org/10.1002/oa.3040

Thompson, T. (2004). Recent advances in the study of burned bone and their implications for forensic anthropology. *Forensic Science International*, *146*, S203–S205. https://doi.org/10.1016/j.forsciint.2004.09.063

Thompson, T., Gonçalves, D., Squires, K., & Ulguim, P. (2017). Thermal Alteration to the Body. In E. M. J. Schotsmans, N. Márquez-Grant, & S. L. Forbes (Eds.), *Taphonomy of Human Remains: Forensic Analysis of the Dead and the Depositional Environment* (pp. 318–334). John Wiley & Sons, Ltd. https://doi.org/10.1002/9781118953358.ch21

Tominaga, Y. (2021). 61—Surgical Management of Secondary and Tertiary Hyperparathyroidism. In G. W. Randolph (Ed.), *Surgery of the Thyroid and Parathyroid Glands (Third Edition)* (pp. 564–575. e4). Elsevier. https://doi.org/10.1016/B978-0-323-66127-0.00061-2

Traversari, M., Serrangeli, M. C., Catalano, G., Petrella, E., Piciucchi, S., Feletti, F., Oxilia, G., Cristiani, E., Vazzana, A., Sorrentino, R., De Fanti, S., Luiselli, D., Calcagnile, L., Saragoni, L., Feeney, R. N. M., Gruppioni, G., Cilli, E., & Benazzi, S. (2019). Multi-analytic study of a probable case of fibrous dysplasia (FD) from certosa monumental cemetery (Bologna, Italy). *International Journal of Paleopathology*, *25*, 1–8. https://doi.org/10.1016/j.ijpp.2019.03.003

Tresguerres, F. G. F., Torres, J., López-Quiles, J., Hernández, G., Vega, J. A., & Tresguerres, I. F. (2020). The osteocyte: A multifunctional cell within the bone. *Annals of Anatomy - Anatomischer Anzeiger*, *227*, 151422. https://doi.org/10.1016/j.aanat.2019.151422

Tsourdi, E., Jähn, K., Rauner, M., Busse, B., & Bonewald, L. F. (2018). Physiological and pathological osteocytic osteolysis. *Journal of Musculoskeletal & Neuronal Interactions*, *18*(3), 292–303.

Turner-Walker, G. (2007). The Chemical and Microbial Degradation of Bones and Teeth. In R. Pinhasi & S. Mays (Eds.), *Advances in Human Palaeopathology* (pp. 3–29). John Wiley & Sons, Ltd. https://doi.org/10.1002/9780470724187.ch1

Turner-Walker, G., & Mays, S. (2007). Histological Studies on Ancient Bone. In *Advances in Human Palaeopathology* (pp. 121–146). John Wiley & Sons, Ltd. https://doi.org/10.1002/9780470724187.ch7

Unnanuntana, A., Rebolledo, B. J., Khair, M. M., DiCarlo, E. F., & Lane, J. M. (2011). Diseases Affecting Bone Quality: Beyond Osteoporosis. *Clinical Orthopaedics & Related Research*, *469*(8), 2194–2206. https://doi.org/10.1007/s11999-010-1694-9

Valentina, G., Luigi, R., Ileana, B., & Marco, M. (2022). Osteoma of the zygomatic bone from 16th century Sardinia (Italy) in the framework of a paleoepidemiological study. *International Journal of Paleopathology*, *39*, 70–74. https://doi.org/10.1016/j.ijpp.2022.10.003

Van der Merwe, A. E., Maat, G. J. R., & Steyn, M. (2010). Ossified haematomas and infectious bone changes on the anterior tibia: Histomorphological features as an aid for accurate diagnosis. *International Journal of Osteoarchaeology, 20*(2), 227–239. https://doi.org/10.1002/oa.1026

Van Der Merwe, A. E., Veselka, B., Van Veen, H. A., Van Rijn, R. R., Colman, K. L., & De Boer, H. H. (2018). Four possible cases of osteomalacia: The value of a multidisciplinary diagnostic approach. *International Journal of Paleopathology, 23*, 15–25. https://doi.org/10.1016/j.ijpp.2018.03.004

Varotto, E., Militello, P. M., Platania, E., Sferrazza, P., & Galassi, F. M. (2021). Paleopathological study of a podal osteochondroma from the prehistoric Hypogeum of Calaforno (Sicily). *Clinical Anatomy, 34*(1), 19–23. https://doi.org/10.1002/ca.23603

Vernon-Roberts, B., & Pirie, C. J. (1973). Healing trabecular microfractures in the bodies of lumbar vertebrae. *Annals of the Rheumatic Diseases, 32*(5), 406–412. https://doi.org/10.1136/ard.32.5.406

Vigorita, V. J., & Ghelman, B. (1999). *Orthopaedic pathology*. Lippincott Williams & Wilkins.

Von Hunnius, T. E., Roberts, C. A., Boylston, A., & Saunders, S. R. (2006). Histological identification of syphilis in pre-Columbian England. *American Journal of Physical Anthropology, 129*(4), 559–566. https://doi.org/10.1002/ajpa.20335

Vu, T. D. T., Wang, X. F., Wang, Q., Cusano, N. E., Irani, D., Silva, B. C., Ghasem-Zadeh, A., Udesky, J., Romano, M. E., Zebaze, R., Jerums, G., Boutroy, S., Bilezikian, J. P., & Seeman, E. (2013). New insights into the effects of primary hyperparathyroidism on the cortical and trabecular compartments of bone. *Bone, 55*(1), 57–63. https://doi.org/10.1016/j.bone.2013.03.009

Vyhnánek, L., Strouhal, E., & Němečková, A. (1999). 'Kissing' osteochondroma: A case from Ancient Egypt. *International Journal of Osteoarchaeology, 9*(5), 361–368. https://doi.org/10.1002/(SICI)1099-1212(199909/10)9:5<361::AID-OA469>3.0.CO;2-A

Wakely, J., Strouhal, E., Vyhnánek, L., & Němečková, A. (1998). Case of a Malignant Tumour from Abingdon, Oxfordshire, England. *Journal of Archaeological Science, 25*(10), 949–955. https://doi.org/10.1006/jasc.1997.0228

Waldron, T. (2008). *Palaeopathology* (1st ed.). Cambridge University Press. https://doi.org/10.1017/CBO9780511812569

Welsh, H., Nelson, A. J., Van Der Merwe, A. E., De Boer, H. H., & Brickley, M. B. (2020). An Investigation of Micro-CT Analysis of Bone as a New Diagnostic Method for Paleopathological Cases of Osteomalacia. *International Journal of Paleopathology, 31*, 23–33. https://doi.org/10.1016/j.ijpp.2020.08.004

Weston, D. A. (2008). Investigating the specificity of periosteal reactions in pathology museum specimens. *American Journal of Physical Anthropology, 137*(1), 48–59. https://doi.org/10.1002/ajpa.20839

World Health Organisation (WHO) and International Associatio Cancer Research (Eds.). (2020). *Soft tissue and bone tumours* (5th ed). OMS.

Wiedijk, J. E. F., Soerdjbalie-Maikoe, V., Maat, G. J. R., Maes, A., Van Rijn, R. R., & De Boer, H. H. (2016). An accessory skull suture mimicking a skull fracture. *Forensic Science International, 260*, e11–e13. https://doi.org/10.1016/j.forsciint.2016.01.025

Wilbur, A. K., Bouwman, A. S., Stone, A. C., Roberts, C. A., Pfister, L.-A., Buikstra, J. E., & Brown, T. A. (2009). Deficiencies and challenges in the study of ancient tuberculosis DNA. *Journal of Archaeological Science, 36*(9), 1990–1997. https://doi.org/10.1016/j.jas.2009.05.020

Winter-Buchwalder, M., Schwab, N., Galtés, I., Ortega-Sánchez, M., Scheirs, S., & Jordana, X. (2022). Microcracking pattern in fractured bones: New approach for distinguishing between peri- and postmortem fractures. *International Journal of Legal Medicine*. https://doi.org/10.1007/s00414-022-02875-1

Wu, K., Schubeck, K. E., Frost, H. M., & Villanueva, A. (1970). Haversian bone formation rates determined by a new method in a mastodon, and in human diabetes mellitus and osteoporosis. *Calcified Tissue Research, 6*(1), 204–219. https://doi.org/10.1007/BF02196201

Zhang, M., Shen, F., Petryk, A., Tang, J., Chen, X., & Sergi, C. (2016). "English Disease": Historical Notes on Rickets, the Bone–Lung Link and Child Neglect Issues. *Nutrients*, *8*(11), 722. https://doi.org/10.3390/nu8110722

Zink, A. R., Panzer, S., Fesq-Martin, M., Burger-Heinrich, E., Wahl, J., & Nerlich, A. G. (2005). Evidence for a 7000-year-old case of primary hyperparathyroidism. *JAMA*, *293*(1), 40–42. https://doi.org/10.1001/jama.293.1.40-c

Zuckerman, M. K., Garofalo, E. M., Frohlich, B., & Ortner, D. J. (2014). Anemia or scurvy: A pilot study on differential diagnosis of porous and hyperostotic lesions using differential cranial vault thickness in subadult humans. *International Journal of Paleopathology*, *5*, 27–33. https://doi.org/10.1016/j.ijpp.2014.02.001

Histological Analyses of Human Bone from Archaeological Contexts

10

SUSAN PFEIFFER AND
DEBORRAH PINTO

This chapter will focus on the histological study of bone tissue of archaeological origin, a topic that inevitably extends into the worlds of paleoanthropology and forensic anthropology. Bioarchaeology is the study of human remains from archaeological sites, understood and interpreted by considering the context from which they derive (Larsen, 1997; Pfeiffer, 2022; Roberts, 2009). In this context, bone histology can provide helpful information by indicating whether small tissue fragments are human, the age of a human at time of death, and can sometimes provide information about the person's health prior to death. Histological research of archaeologically derived bone tissue is part of a methodological continuum that includes various tissues other than bone (such as teeth and desiccated soft tissue) and applications that go beyond, but are complementary to, histomorphological characterization such as microradiography and micro-CT (or X-ray microtomography). The focus here will primarily be on transmitted light microscopy as applied to bone. Early studies of bone histology were exploratory, establishing the potential for this approach (Day & Molleson, 1973; Dubois, 1937; Graf, 1949). Given its long history and its relatively simple laboratory requirements, this approach has considerable history, and is a normal part of many analytical protocols. Guides regarding the methods of sample preparation for transmitted light microscopy are available elsewhere in this volume, and beyond (data from Robling & Stout, 2008; Schultz, 1997a, 1997b; Turner-Walker & Mays, 2008; Ubelaker, 1989). There is considerable variation in methodological approaches, so a newcomer to this field should read broadly. Definitions and labels of structures are standardized, however, and all researchers are urged to follow established terminology for histomorphometry (Crowder et al., 2022; Parfitt et al., 1987).

10.1 Taphonomy and Diagenesis

An archaeological context means that bone tissue will have changed in some ways from its fresh state, i.e., grossly and microscopically. The nature and extent of the post-mortem changes will determine the feasibility of histological research. While even brief post-mortem intervals can alter tissue (Haglund & Sorg, 1997), longer time periods can introduce diverse factors. Taphonomy, based on *Taphos*, the Greek word for burial or tomb, is the study of how organisms decay or become preserved through fossilization. Physical changes from weathering and ravaging are examples of taphonomic factors that affect the amount and condition of bone from archaeological contexts that is available for histological study. Chemical and physical changes, referred to as diagenesis, are commonly studied aspects of taphonomy

DOI: 10.4324/9781003385608-10

Table 10.1 The Oxford Histological Index (OHI), after Millard (2001)

OHI	Approx. % of intact bone	Description
5	>95	Very well preserved, indistinguishable from modern bone
4	>85	Only minor amounts of destructive foci, otherwise generally well preserved.
3	>67	Clear preservation of some osteocyte lacunae
2	<33	Clear lamellate structures preserved between destructive foci
1	<15	Small areas of well-preserved bone, or some lamellar structures preserved by pattern of destructive foci
0	<5	No original features identifiable other than Haversian canals

(Pfeiffer, 2000; Pfeiffer & Varney, 2000; Turner-Walker & Mays, 2008) and can include organic decomposition, recrystallization, mineral replacement and increases in porosity (Kendall et al., 2018). The extent of diagenetic change to bone tissue is linked to soil and moisture conditions, as the agents of change can be various combinations of bacteria, fungi, algae, plants, worms, and insects. Topics of current interest include the question of whether microbial agents come primarily from inside or outside the intact deceased body (Mavroudas et al., 2023) and how the breakdown of mineralized tissue progresses (Turner-Walker 2019).

The burial environments most conducive to bone tissue preservation are dry and sterile, although some bacteria are inherent to the body itself, especially in the gut. Research suggests that an environment with fluctuating water levels in and around the bone may be the least conducive to preservation. Bodies enclosed in coffins are often particularly vulnerable to diagenetic destruction (Kontopoulos et al., 2022). Warm environments are also deleterious. Bacterial and fungal tunneling is associated with loss of collagen and redeposition/replacement of bone mineral (Jans et al., 2004). An increase in crystallinity has been used as a parameter to characterize the extent of diagenetic alteration of bone with analogous changes in dental tissues as well. The relatively porous, less mineralized bone of infants and children is more vulnerable to diagenetic destruction (Caruso et al., 2021).

The mechanisms by which a corpse turns into a skeleton and that skeleton is subsequently altered has been termed bioerosion (Jans, 2008; Bronnimann et al., 2018). A scheme for characterizing the extent of microbial bioerosion as well as other sources of diagenetic change has been proposed (Hedges et al., 1995), known as the Oxford Histological Index (OHI) (Table 10.1). It is a tool for quantifying the extent of tissue preservation, an overall assessment of the amount of altered versus unaltered microstructure without reference to the cause(s) of the destruction. It has proven to be a helpful tool to standardize language as research into diagenesis expands. Modification to include non-biological factors like cracking and staining has been suggested (Hollund et al., 2012) as has separate evaluation of periosteal, endosteal and mid-cortical sectors (Jans, 2008).

10.2 Micro-Computed Tomography (Micro-CT) in Archaeological Contexts

While transmitted light microscopy has provided the basis for observations of diagenetic processes, researchers have expanded the scope of exploration through newer methods, including scanning electron microscopy (SEM) (Morales et al., 2018), synchrotron radiation

Table 10.2 The Virtual Histological Index (VHI) based on micro-CT, after Mandl et al. (2022); MFD = microscopic focal destruction. The stages are designed to parallel those of the Oxford Histological Index (OHI, Table 10.1)

VHI	Approx. % of intact bone	Description
5	>95	Very well preserved, indistinguishable from modern bone: canals and lamellae are clear; grey values are consistent within and between slices
4	>85	Only minor amounts of destructive foci, otherwise generally well preserved; small patches of MFD present (darker grey values). The canals and lamellae are easily distinguished throughout the stack.
3	>67	Destructive foci evident throughout the stack but more than half of the bone remains unaffected; bioerosion spreads beyond lamellae in patches but the foci are still visible.
2	<33	Destructive foci may still be present but bioerosion now primarily presents in wide swathes of dark grey values in which lamellae are nearly indistinguishable; in small unaffected or less affected regions, lamellae remain distinguishable; there is more affected bone than non-affected.
1	<15	Most of the stack is filled with bioerosion; only the smallest patches of MFD are distinguishable within a wide swathe of dark grey values; in most regions the lamellae have disappeared or are only vaguely distinguishable.
0	<5	Erosion is pervasive throughout and hardly any unaffected bone remains; lamellae are indistinguishable; only the slightest outline remains within the erosion; a small number of canals may still be present or may be obliterated.

micro-CT (SR-micro-CT) (Caruso et al., 2020), and lab-based micro-CT (or XRM) (Dal Sasso et al., 2014). While micro-CT slices do not provide as much detail, nor can they detect staining, they are less time-consuming to prepare, they are somewhat less destructive, and they can provide information about bioerosion that is comparable to that from microscopy.

Compared to traditional histology and microradiography, micro-CT allows less destructive imaging of bone at a microscopic level; however, the samples are trimmed to fit into the micro-CT scanner and the resolution may not be optimal with loss of details in 3D reconstructions (Carew et al., 2022). Micro-CT has been successfully used to study trauma, pathology, and biomechanics in historic and prehistoric samples (Appleby et al., 2015; Beaudet et al., 2020; Beresheim et al., 2019; Dittmar et al., 2021; Doershuk et al., 2019; Figus et al., 2023; Mitchell et al., 2022; Nicklisch et al., 2022; Perchalski et al., 2018; Ruff et al., 2022; Ryan & Shaw, 2015). Anticipating the continued use of micro-CT, a set of benchmarks has been proposed in which the rankings are analogous to those of the OHI (Table 10.2). For a more detailed discussion on micro-CT, see Chapter 12 of this volume.

10.3 A Component of Archaeological Science

A relatively new use of bone histology is in the context of assessing tissue quality prior to paleogenomic study (Bell et al., 2008; Guarino et al., 2006; Kendall et al., 2018; Roberts, 2009). The preservation of histological structures is seen as a strong indicator of probable success in extracting ancient DNA (aDNA) from bone tissue (Dash & Das, 2018). Particular attention has been directed toward understanding the preservation of anatomical regions, like the inner ear, at which collagen is particularly well preserved (Kontopoulos et al., 2019: Kontopoulos et al., 2022).

The extent of bone preservation is also relevant to studies of past lives through quantification of stable isotope values extracted from bone, hair/nail, and dental tissue. Stable isotope analysis of human tissue can provide valuable life history information including the composition of an individual's diet and geolocation. It also has applications in sorting commingled remains (Chesson & Berg 2021). We refer here to research on variation in dietary isotopes like those of carbon and nitrogen, or isotopes, like strontium, that reflect the geology of the environment within which a person's growth took place. The microbes responsible for bone destruction may influence biogenic isotope values through selective feeding (Grupe et al., 2000). Histologically well-preserved bone tissue is also a necessary pre-condition for isotope studies using bone density fractionation. This approach seeks to distinguish bone tissue linked to different periods of formation and mineralization, thereby exploring a person's life history (Bell et al., 2001).

A newer application of this technology has been the use of isotope analysis in the personal identification of unknown human remains. The United States Department of Defense's Defense POW/MIA Account Agency (DPAA) has employed its use to assist in repatriating remains of service personnel who died during military conflicts, typically on foreign soil. These analyses have aided in locating the geographic origin of a set of remains or separating commingled remains (Chesson & Berg 2021). Bartelink et al. (2014) demonstrated the use of this type of analysis in distinguishing the remains of U.S. servicemen from native inhabitants of Asia and the Pacific. Isotope analysis is particularly important in the identification of unknown human remains when dental records are unavailable, and DNA is inconclusive. These examples illustrate why archaeological science training should include some introduction to the histological appearance of normal cortical and cancellous human bone structures as it can guide the analyst in selecting appropriate methods for research design.

10.4 Reconstructing Funerary Practices

Histological approaches have been used to ascertain whether funerary treatments were immediate or delayed, or whether treatments differed between socioeconomic groups (Booth & Bruck, 2020; Bronnimann et al., 2018; Miszkiewicz & Mahoney, 2016). Haddow et al. (2023) examined bone samples from 57 individuals from a Neolithic site (Catalhoyuk), using light SEM imaging techniques, to determine whether specific funerary treatments could be associated with specific patterns of microstructural preservation, generating inconclusive results. Histological ascertainment of bioerosion in cremated bone may suggest delayed incineration, as the microbial action would have preceded the burning (Lemmers et al., 2020; Vegh et al., 2021). The hypothesis that gut microbes are an important source of bioerosion has led to the suggestion that neonatal and stillborn infants may show minimal bone diagenesis, since they had not yet developed their gut microbiomes. Histological exploration of this hypothesis has generated considerable interest (Booth, 2020; Booth et al., 2016; Kontopoulos et al., 2022).

Another area of interest is that of the potential effect of iron oxides on bone diagenesis, especially red ochre as used in funerary rituals. While the application of red ochre in body adornment and funerary ritual is justifiably characterized as symbolic (Watts, 2009), the substance has been shown to have antiseptic, antiparasitic, insect repellent, and sun protection properties (Havenga et al., 2022; Huntley, 2018; Teklay et al., 2023). It has been suggested that when a paste containing ochre was applied to the deceased, as among the ancestral Maya, that coating could deter bioerosion, especially if it blocked orifices through which insects normally enter the body (Mandl & Carlson, In Prep; Mandl & Grassberger, 2021). By delaying putrefaction, this could prolong the time available for funerary rituals. The question of whether it would also deter exogenous microbial action can be explored through bone histology.

10.5 Reconstructing Past Lives

Histological study of archaeologically derived material can contribute to research questions arising in forensic and biomedical fields. Archaeological sites, including historic cemeteries, can provide access to large numbers of skeletons, a higher proportion of non-adult skeletons than are found in modern settings, and instances of bone that has been modified by diseases that were untreated or were treated in inefficacious ways in the past. There are challenges to doing histology on such samples, however. Even if there is headstone or cultural information provided by artifacts, we still have incomplete knowledge of each person in life.

The histological study of bone tissue is inevitably destructive, although the extent of alteration to a specimen may be rather slight, like the removal of less than a centimeter of bone from an already broken rib, or the removal of a small core from an intact long bone. Guidelines for sampling have been developed by various institutions, and by research groups. Most current and future work in archaeologically derived bone histology will likely occur in contexts where human remains are being disinterred briefly (e.g., historic cemeteries), or descendants have authorized specific research activities. A basic principle for such projects has been proposed: "The scientific question being addressed should be important enough to justify invasive sampling of hominid remains and should not be answerable by any other means" (Hublin et al., 2008: 756).

Most histological research is undertaken to answer questions in four categories, these being the differentiation of human from non-human, the estimation of age at death, the reconstruction of habitual activities in life, and the diagnosis of pathological or traumatic conditions.

10.5.1 Is the Bone Tissue from *Homo sapiens*?

There is extensive historical literature on the differentiation of adult human cortical bone from that of other species (Enlow & Brown, 1956, 1957, 1958; Foote, 1916). There can be uncertainty when the bone tissue is cancellous or diploic in structure, or when the tissue is from an infant or juvenile of various mammalian species. The usual identified types of bone tissue are woven bone ("bundle bone"), primary bone that includes two types (fibrolamellar bone, and primary lamellar bone with primary osteons), and secondary bone. Secondary bone includes structures known as secondary osteons (Haversian systems), as well as interstitial and circumferential lamellar bone. Non-human secondary osteons tend to be more rectangular, more regular in size within a cortical sample, are substantially smaller than those of humans, although with exceptions (Hillier & Bell, 2007), and they can form bands (Mulhern & Ubelaker 2001). Some sources will assert that the plexiform pattern of fibrolamellar bone is not found in humans, yet others demonstrate that it does occur in bones of human infants under the age of 1 year (Caccia et al., 2016). However, there is lack of clarity regarding whether the bone tissue seen in human infant bone tissue includes the plexiform type (Pfeiffer, 2006). The human infant bone structures are probably primary woven bone or lamellar bone rather than fibrolamellar, but diagenesis is especially common in the porous, small, partially mineralized bones of infants, so clarity is unlikely. Nevertheless, through the histological study of autopsy specimens, the processes of bone modeling and remodeling are becoming more clearly understood (Streeter, 2005; Streeter & Stout, 2003).

Encouraging developments in this field continue to arise from the combined assessment of shape and size of structures. For example, while study samples remain small, secondary osteons in human juveniles appear to be of "adult size" from first appearance (Pfeiffer, 2006). Hence, human secondary structures at all ages are more than twice as large in area as the secondary osteons of deer, dog, or pig (Hillier & Bell, 2007; Morris, 2007). Non-destructive methods, like micro-CT, are being standardized (Carlson et al., 2022; Mandl et al., 2022). They can be used to distinguish human bone structure in contexts of fragmentary or fossilized remains (Bennett et al., 2019; Carlson et al., 2022; Mandl et al., 2022). For more information regarding differentiating human from nonhuman bone, see Chapter 7 in this volume.

10.5.2 Age at Death

The methods of estimating adult age at death from cortical bone tissue are addressed in this volume (Chapter 8) and elsewhere (Crowder et al., 2009; Crowder et al., 2012). The methods have been developed from known-age samples derived from dissection and forensic contexts, although research using historic cemetery samples has allowed researchers to refine their approaches (Cho et al., 2002). Archaeological samples of this type provide more genetic diversity and a different health regimen in life, as compared to modern samples. Crowder, examining femoral and rib samples from the large Spitalfields crypt sample (Molleson & Cox, 1993), demonstrated that the most accurate histological age determination methods show the same bias as the results of aging methods based on gross morphology. There is a tendency for the youngest adults to be over-aged, and the oldest adults to be under-aged. The error for the best histological methods is not greater than that associated with more established methods, like those based on pubic symphyses and sternal rib ends (Crowder, 2005; Crowder & Pfeiffer, 2009a).

During the first decades since Kerley's publications stimulated interest in the histological age estimation (Kerley, 1965, 1969, 1970; Kerley & Ubelaker, 1978), there were some attempts to use histological age estimates as a basis for paleodemography of archaeologically derived cemeteries (cf. Pfeiffer, 1980). However, aspiring researchers faced several challenges. From burial environments that are frequently moist and organically active comes bone tissue that has been diagenetically altered, leading to relatively low research yields relative to the effort expended. The necessary permission for destructive testing must be based on a strong promise of definitive results and this is difficult to assess *a priori*. In addition, these same decades saw ongoing dialogue about the relative value of paleodemographic reconstructions, as compared to other forms of bioarchaeological analyses (Bocquet-Appel & Masset, 1982; Hoppa & Vaupel, 2002; Wood et al., 1992). This approach is now rarely pursued, but remains enticing (Boldsen et al., 2022).

Histological methods may be most helpful when only an isolated bone or unique burial is discovered and there is substantial interest in estimating age-at-death. Histomorphometric analyses were used to produce age estimates of Neanderthals from Shanidar (Thompson & Trinkaus, 1981; Trinkaus & Thompson, 1987) revealing some of the biologically oldest

Neanderthals recovered. Streeter and colleagues were able to provide an age estimate for a Middle Pleistocene tibia from Boxgrove (U.K.) from a core of cortical bone (Streeter et al., 2001). The estimate produced by microscopic techniques was higher than the macroscopic estimates and showed that the Boxgrove 1 individual was most likely in his/her fourth or fifth decade of life. Pfeiffer and Zehr estimated age-at-death of the humerus shaft from Border Cave (southern Africa, putatively late Pleistocene) from a flake of bone that spalled off a broken surface (Pfeiffer & Zehr, 1996).

Some studies have generated age estimates for archaeological samples to test the accuracy of methods developed on forensic or cadaveric populations. A commonly used method is that developed by Thompson (Thompson, 1979, 1981; Thompson & Galvin, 1983) using small cores from the anterior mid-diaphysis of various major long bones. In 1981, Thompson and colleagues reported that histological analyses of femoral samples of Eskimo skeletons from St. Lawrence Island, Kodiak Island, Baffin Island, and Southampton Island using the Thompson method demonstrated consistently higher age estimates than morphological methods (Thompson & Gunness-Hey, 1981; Thompson et al., 1981). This same analytical approach was tested on an archaeologically derived set of adults from historic burials from southern Ontario (ca. AD 1825–1920), with generally good results (Pfeiffer, 1992). Pfeiffer noted that post-mortem erosion of the periosteal surface led to a bias toward older ages and cautioned that tissue preservation influences both accuracy (including bias) and precision.

The Thompson method has been popular in bone histological analyses of archaeological material because it requires removal of only a small core of bone. In part to avoid even this amount of destruction to large bones, researchers were motivated to explore other possible sampling sites such as the (frequently broken) rib midshaft. An archaeologically derived sample, putatively the skeleton of Francisco Pizarro, was used to explore a new regression equation for estimating age from rib fragments (Stout, 1986). Stout was able to acquire an age range in the early sixties consistent with Pizarro's known age at death. Assessment of the rib as described by Cho, Stout, Madsen, and Streeter (2002) generates age estimates with about the same range of error as methods based on gross morphology (Crowder & Pfeiffer, 2009b; Pfeiffer et al., 2016).

Age at death in the archaeological context is an important piece of the biological profile; however, histological methods are rarely pursued (Falys & Lewis, 2011). The reasons for this lack of use include the destructive nature of histological analyses, poor preservation of important microscopic structures, the possibility of population-specific variations, and the specific training and equipment required for these analyses. In methods using the periosteal boundary such as those developed by Kerley (1965) and Thompson (1979), preservation of the outer surface is of utmost importance. Deterioration of these areas can produce increasingly erroneous results so that the results are not as useful as those of macroscopic methods. Nevertheless, in cases where macroscopic preservation of key features like the pelvis and skull is poor, yet tissue quality is good at histological sampling sites, histological analyses can provide acceptable estimates of age at death. In this sense, microscopic methods cannot be deemed better or worse than macroscopic methods; each complements the other in producing the biological profiles of unknown individuals. Histological methods may be especially useful in contexts where other methods are known to be biased, like the underestimation of age-at-death of small, gracile adults when gross anatomical indicators are used (Merritt, 2015). A study by Pfeiffer and colleagues demonstrated that in that context the histological estimates are notably older and more plausible, more consistent with normal human life histories (Pfeiffer et al., 2019).

10.5.3 Habitual Activity and Diet

There is a rich and dynamic scholarly literature exploring the way that bone tissue reflects a norm of reaction, responding to environmental factors within a genetically established range (Carter & Beaupre, 2001; Garn, 1970; Martin & Burr, 1989). This literature forms the basis for reconstructions of past behavior based on the biomechanical properties of human long bones (Ruff, 2008; Ruff et al., 2006; Ruff et al., 1993; Stock & Pfeiffer, 2001; Stock & Pfeiffer, 2004; Stock et al., 2005). A logical extrapolation from this line of investigation is that the structures that comprise the bone tissue should also be responsive, since it is through these structures that bone is gained or lost. Researchers continue to investigate the relationship between bone structure and intrinsic as well as extrinsic factors.

The proposition that the relatively small secondary osteons of Late Pleistocene humans reflected their morphological robusticity (Abbott et al., 1996) provided the catalyst for various subsequent research approaches. Using the measure of osteon area (On.Ar, including both osteon and Haversian canal), the Pleistocene average of 27.7 $(mm^2)(10^{-3})$ is substantially smaller than the value from contemporary humans of around 40 $(mm^2)(10^{-3})$ that was known at that time, suggesting that the pattern is behavioral, rather than genetic. The pattern of smaller than modern secondary osteon areas has been found in other putatively ancient specimens (Pfeiffer & Zehr, 1996), but not all of them (Sawada et al., 2004).

If variation in the size or shape of secondary osteons could be linked with the load history of bones, this would be helpful for interpreting the past. One challenge has been to characterize the range of normal variability in human secondary osteon dimensions across a range of behavioral settings (Burr et al., 1990; Mulhern, 2000; Mulhern & Van Gerven, 1997; Pfeiffer, 1998; Pfeiffer et al., 2006). It has been demonstrated that average secondary osteon size varies from one sample to another. Quite consistently, the secondary osteons from ribs are smaller than those from femoral midshafts (the two most studied sampling sites), suggesting the influence of some biomechanical factor. Results from a study of a large, well-documented cadaveral sample indicate significant contributions to variance in femoral osteon size due to age, body weight and sex, but not stature (Britz et al., 2009). However, few individuals in any of these studies show osteon sizes as small as those observed in the ancient humans. Femoral midshafts from diverse recent groups yielded median On.Ar as low or lower than 27.7$(mm^2)(10^{-3})$ in four of twenty 19th-century Canadian settlers (St. Thomas', three females and one male), and two of fourteen hunter-gatherers (South African Later Stone Age, both males), but none of twenty 18th-century Huguenots (Spitalfields) (data from Pfeiffer et al., 2006). None of the 87 femora studied by Britz et al. (2009) had median sizes this small. The current challenge is to ascertain what behavioral features might distinguish individuals who have particularly large or small secondary osteons.

Not only the size of osteons, but also features such as double zoning, type II osteons and drifting have been quantified in experimental studies of various quadrupedal mammals and humans. That research has demonstrated the absence of a consistent relationship between secondary osteon morphological characteristics and loading history (Skedros et al., 2007). In an archaeological sample of ribs from 45 adults, double zoned and type II variants were observed in 30, with presence corresponding best with age at death (Pfeiffer et al., 2019) (Figure 10.1). On the other hand, examples from archaeologically derived contexts demonstrate the effect of habitual behaviors on secondary osteon density. Amputation is one such context. A young man from the turn of the 20th century had lived with a homemade below the

(a)

(b)

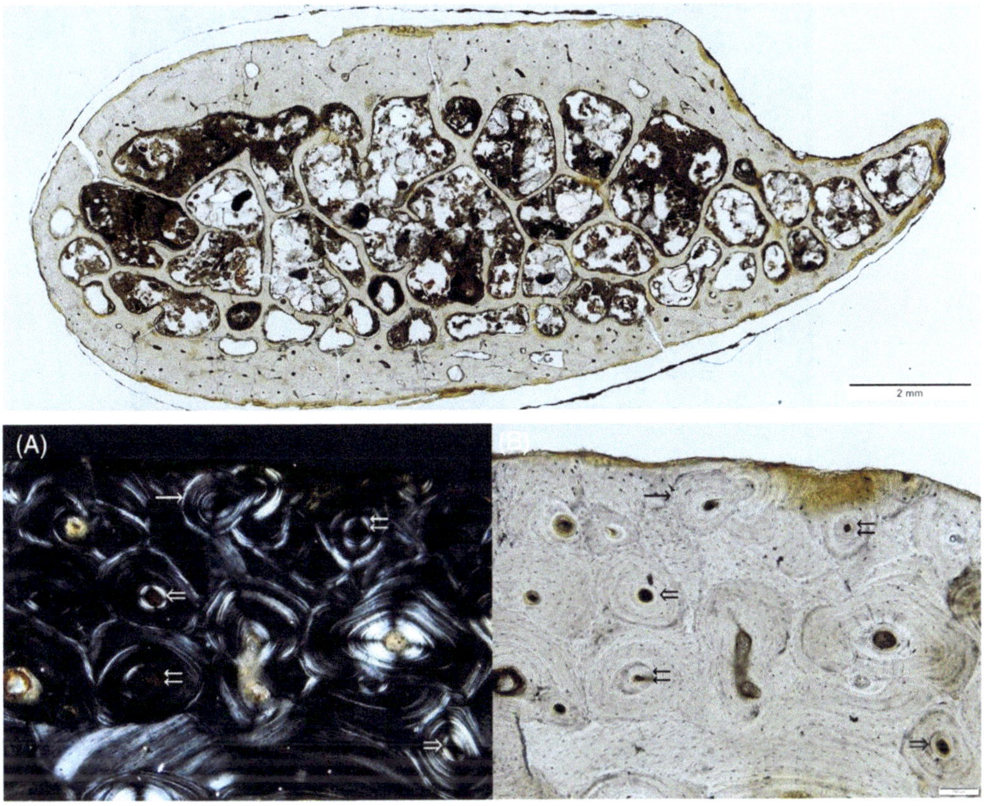

Figure 10.1 (a) Image in brightfield of rib cross-section (Mostert G8 H8) illustrating some surface erosion, cracks and other characteristics of archeologically derived bone tissue; (b) Images in partially polarized light (A) and bright field (B) of rib bone tissue showing osteon variants: Type I (single arrow), Type II or embedded (double arrows) and double zonal (double line arrow). Credit for images: A. Beresheim (Pfeiffer et al., 2019:134).

knee prosthesis for several years. The femoral midshaft from his intact leg showed 28% remodeled bone, but the contralateral femur was 61% remodeled (Figure 10.2) (Lazenby & Pfeiffer, 1993). Another study on the effects of biomechanical loading on bone histomorphology, Schlecht et al. (2012), investigated the effects of disuse in long bones of two cadaveric quadriplegic individuals. They compared multiple variables, including On.Ar, from the quadriplegic individuals to 28 mobile individuals of similar age and found lower osteon density and larger On.Ar. in the paraplegics. Thus, while age, weight, and sex may explain some portion of normal variability, some behavioral factors can also affect remodeling rate. A well substantiated framework for interpreting variation in secondary osteon structures is not yet available.

In the histological study of ribs from 45 adults mentioned previously, Pfeiffer and colleagues found that Holocene foragers from one region of South African coast died at older ages and had higher trophic-level diets, as determined from nitrogen isotope ratios, compared to foragers from other parts of the coast (Pfeiffer et al., 2019). This suggested that a higher quality diet could enhance longevity. Using a sample from a prehistoric village in Paloma, Peru, Robling and Stout (2003) compared the relationships between bone histology, bone geometric properties, and mechanical loading. They demonstrated that both bone

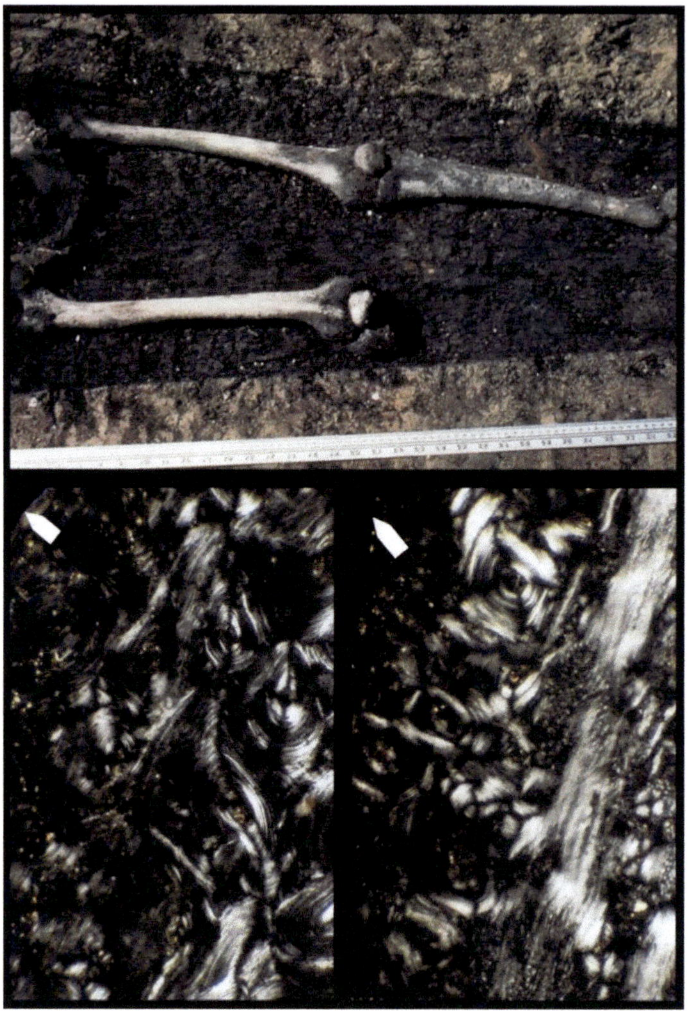

Figure 10.2 The grave (top) of a young man who used a home-made prosthesis to compensate for a below-knee amputation that occurred a considerable time prior to his death, AD 1899; femoral midshafts show proportionally more remodeled bone (secondary osteons) on the leg that sustained the amputation (left) than on the intact leg (right). Photo of grave courtesy of M.W. Spence; photos of histology courtesy of R.A. Lazenby.

histology and bone geometry provided useful information regarding mechanical loading. Therefore, when specimens are degraded due to diagenesis and microanatomical features are obscured, bone geometric properties can be used to study habitual behavior.

Histological exploration of archaeologically derived human bone can lead to novel discoveries, for example, the report of tetracycline-like fluorescence in bone from Sudanese Nubia (Bassett et al., 1980). It was argued that ingestion occurred through bacterially contaminated grain or ale, and therapeutic health effects were postulated (Keith & Armelagos, 1983). Thereafter, similar fluorescence was reported from other sites in the region, signaling a continuance of this pattern for 1,200 years (Cook et al., 1989). An alternate interpretation, that of postmortem diagenesis as a source of the fluorescence, was debated (Keith & Armelagos, 1988; Piepenbrink, 1986), but scrutiny with improved analytical methods appears to verify the original interpretation (Maggiano et al., 2006). To date, this phenomenon seems to be limited to one region and time.

Micro-CT has been employed to investigate bone architecture such as trabecular connectivity and other microanatomical features as they relate to biomechanics. Several studies have addressed the effects of biomechanics on bone, in contexts of growth and development as well as subsistence practices. Using archaeologically derived samples, Doershuk and colleagues (2018) and Ryan and Shaw (2015) demonstrated that bone microarchitecture of highly mobile people are different from more sedentary ones, thereby providing evidence that activity patterns influence bone structure. The more active populations, i.e., foragers, tend to have thicker and less connected trabeculae, lower bone to surface ratio, and higher bone volume fraction. Since modern pediatric samples are not easily accessible to study bone development, researchers often focus on archaeological pediatric samples for their studies. They have demonstrated that bone microarchitecture, particularly of the appendicular skeleton, changes from birth to adulthood, accommodating biomechanical needs (Figus et al., 2023; Gosman & Ketcham, 2009; Perchalski et al., 2018). These studies show that trabecular connectivity decreases, and trabecular thickness increases from birth to maturity. Using a sample from Spitalfields Christ Church in London, UK, Beresheim et al. (2019) explored differences related to growth and maturation in cortical and trabecular bone microstructure (Figures 10.3 and 10.4). They found that unfilled resorption spaces had a greater effect on overall bone porosity than Haversian canals.

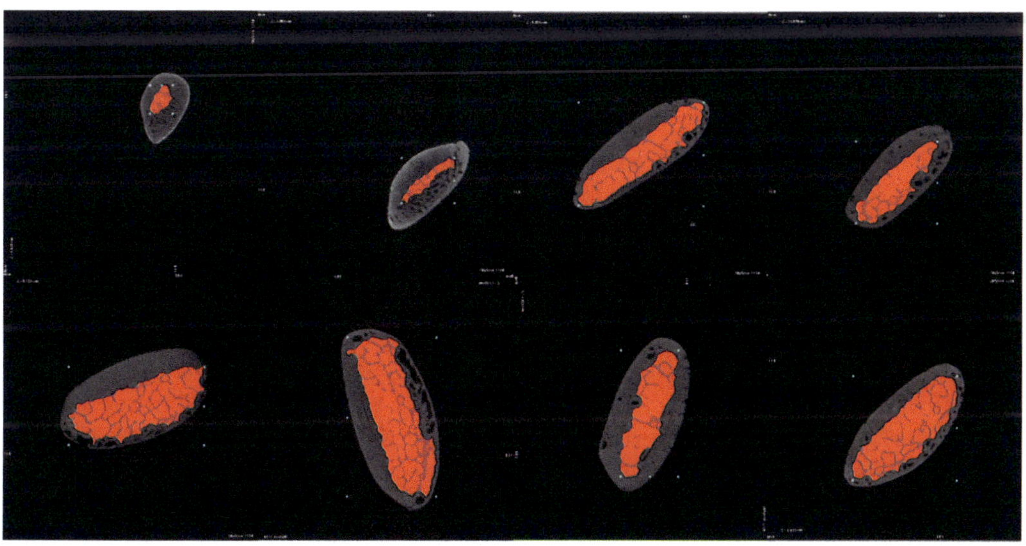

Figure 10.3 Ontogenetic series of ribs from known-age girls and women from the Spitalfields community, U.K. Each image represents a single 2-D slice at midpoint of the micro-CT image stack (figure 4, Beresheim et al., 2019:10).

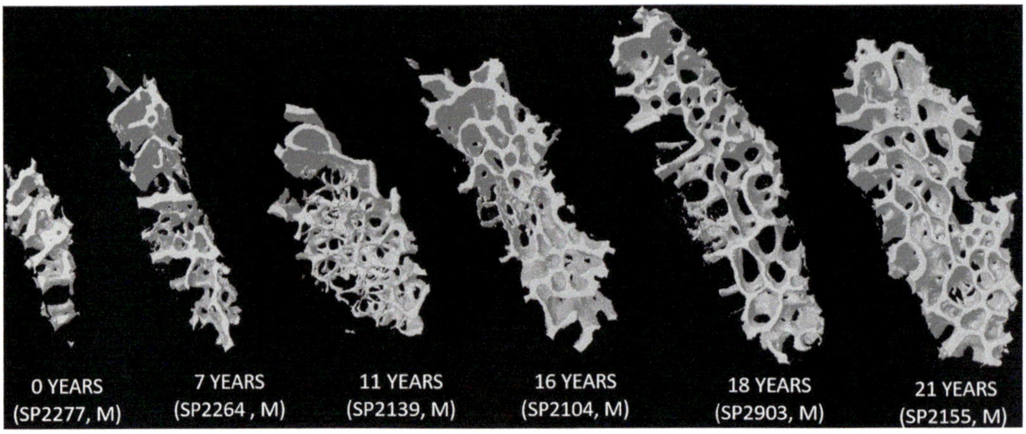

Figure 10.4 Three-dimensional trabecular bone reconstructions of thoracic rib midshafts in six boys and men from the Spitalfields community, U.K. (figure 5, Beresheim et al., 2019:10).

10.5.4 Pathological Conditions

Histological investigation can help to establish a probable etiology for abnormal bone conditions found in archaeological contexts and can differentiate normal from pathological features (Welsh & Brickley n.d.). Schultz noted that paleohistology is centrally important to paleopathological diagnoses (Schultz, 2001, 2003). He argued that clinically established diagnostic features can be re-interpreted in the context of ancient remains in which many of the cells are absent and only their sequelae remain. A benefit from this interest in paleohistology is that it necessitates a thorough understanding of tissue response among researchers. There are relatively few conditions in which histological structures are definitive, but histological characterization may contribute useful information in many contexts. Because it is destructive, it has not been considered a routine procedure (Ortner & Putschar, 1981). Imaging methods that provide a cross-sectional image of cortical bone with minimal sample preparation, like micro-CT scanning, can facilitate analysis when there is a need to differentiate between etiologies; for example, osteoporosis (osteopenia with thinned cortex) and hyperparathyroidism (osteopenia with porosity of the cortex). Although Chapter 10 of this volume discusses histopathological analysis in detail, here we provide examples of its use for investigating archaeological remains.

Changes to bone tissue arising from pathogens are often non-specific, since the bone modeling unit is limited to only two responses, namely resorption and deposition. The goal of preparing a thin section or a surface for SEM is often that of making the apparently simple distinction of whether bone deposition or resorption was occurring at the time of death. Smooth surfaces and edges comprising well-mineralized bone, often lamellar in structure, suggest that deposition had been well underway at the site. Scalloped edges along the bone surfaces (such as Howship's lacunae) can indicate sites at which osteoclasts were active at the time of death, signaling resorption. However, the bone deposition process begins with a non-mineralized template that is only gradually mineralized. Hence, an apparently resorbing surface may have been in the early stages of deposition (Mosekilde, 1990). The interpretation of slides from archaeological abnormal bone must be approached from a well-informed perspective, always attentive to the absent soft tissue. Such information can be useful for understanding human lives in the distant past, and in the forensic context when an assessment of timing of injury is important.

There are relatively few pathological conditions that modify bone in ways that are distinctive enough that histological results can contribute to a definitive diagnosis. For example, most bone tumors, with the exception of osteosarcomas, display non-specific histological features (de Boer & Maat, 2018). One of the liveliest areas of research focuses on treponemal infections, in part because of a sustained interest in the natural history of venereal diseases in the "Old" and "New" Worlds (Dutour et al., 1994; Powell & Cook, 2005). In paleopathology, a proposed etiology of syphilis is often based on changes to one or more long bones, especially the tibiae. The tibia is often affected in treponemal diseases but is affected in ways that are grossly similar by hematogenous osteomyelitis. Schultz and colleagues have identified two histological features that appear to be characteristic of the former, but not of the latter (Schultz, 2003). Polsters ("padding") and grenzstreifen ("border stripes") can be seen in thin sections because of the relatively slow progress of the treponemal diseases (venereal syphilis, endemic syphilis, yaws), whereas osteomyelitis and several other infectious diseases progress more quickly, leaving a very different histological signature. A study of bone sections from individuals with documented venereal syphilis concluded that polsters and grenzstreifen are useful as a "rough guide for identifying syphilis in bone sections" (Von Hunnius, Roberts, Saunders, & Boylston, 2006:564). Interpretive complications arise from evidence that leprosy, too, can lead to polsters and perhaps to grenzstreifen (Blondiaux et al., 2002; Schultz & Roberts, 2002).

Another example of a paleopathological topic in which histology plays a role is that of Paget's disease, *osteitis deformans*. Its etiology is unclear, but it may arise from a low-grade viral infection. In modern times, it affects older Caucasians in all parts of the world, at frequencies of 10–15% by the ninth decade of life (Roberts & Manchester, 1995). To better understand the disease, there is an interest in finding it among archaeological collections, especially those representing non-Caucasians. In the disruption of bone metabolism caused by Paget's disease, both deposition and resorption are heightened. This results in a mosaic of cortical bone structures, marked by numerous cement lines, which is the histological signature that distinguishes it from osteomalacia and hyperparathyroidism. Although rare, various instances of the disease have been documented from European contexts. The diagnostic histological features have been well demonstrated in a case from medieval Ipswich, U.K. (Turner-Walker & Mays, 2008). It has been argued that in the absence of characteristic histological features, a diagnosis of Paget's disease is unwarranted (Pinto & Stout, 2009) and concern has been expressed that there is a potential for over-diagnosis (Garnett & Lewis, 2022). A particular challenge arises because the high bone turnover and bone "softening" that characterize this and other metabolic diseases of bone can exacerbate diagenetic deterioration of the bone tissue. This can make it more difficult to see the histological features.

Occasionally, fossils with pathological features are preserved sufficiently well for histological analysis. This approach has been applied to hominid fossils found in South Africa, exploring the antiquity of degenerative joint disease (Odes et al., 2017) and cancerous conditions (Odes et al., 2016). Pathological long bone fragments from Koobi Fora, Kenya, KNM-ER 1808, come from the femur of a partial female skeleton initially considered to be *Homo erectus* from ca. 1.7 million years ago (https://humanorigins.si.edu/evidence/human-fossils/fossils/knm-er-1808). Enveloping the normal cortex of the shaft is a substantial region of woven bone, causing an irregular and expanded bone surface. Researchers suggest that she suffered from an overdose of vitamin A possibly caused by her eating the liver of a large predator, like a lion or leopard (Walker & Shipman, 1996; Walker, Zimmerman, & Leakey, 1982). It has been argued that the advanced nature of the condition suggests that these *Homo sapiens* precursors practiced compassion toward group members (Walker & Shipman, 1996).

Micro-CT has also been successfully used to identify pathological conditions and trauma on archaeological samples. Mitchell et al. (2022) found that micro-CT had better sensitivity for identifying cancer in a medieval European sample, as compared to visual inspection or radiography. Using a sample from the same era, Dittmar et al. (2021) were able to demonstrate the relationship between the presence of gout and the social status of the decedents. Like analysis of pathological conditions, assessing trauma requires the ability to examine the bone surface. Typically, trauma is assessed by gross visual examination of the injured skeletal elements which is then complemented by microscopy techniques to investigate evidence of healing. Micro-CT offers the possibility to perform the gross and microscopic examination with a single analytical tool. Appleby et al. (2015) used this approach to examine the excavated remains of King Richard III of England. They identified several perimortem incised and penetrating injuries with no evidence of healing. Similarly, Nicklisch et al. (2022) were able to identify perimortem sharp force trauma in skeletal remains from an Early Bronze Age tomb.

10.6 Conclusions

There is potential for considerable progress in studies of histomorphology of cortical and cancellous bone using tissue from archaeological contexts. Despite the challenges presented by variable tissue preservation and the reluctance to apply destructive methods, there are times when histological investigation can provide information that would be otherwise unavailable. This may pertain to the species represented by fragmentary remains, the age at death of deceased humans, distinctive behaviors, and some features of health during life. As discussed here, there are several areas where a full understanding of bone structural phenomena is just beyond reach. There are symbiotic relationships between research undertaken on archaeologically derived material and forensic material, and between histological and micro-CT methods. Each can inform the other, and future research questions will arise from both domains. Various aspects of the research reviewed here would benefit from tighter calibration. Our knowledge will be furthered as researchers consult with one another and commit to defining variables and structural features in the same manner.

References

Abbott, S., Trinkaus, E., & Burr, D. B. (1996). Dynamic bone remodeling in later Pleistocene fossil hominids. *American Journal of Physical Anthropology*, *99*, 585–601.

Appleby, J., Rutty, G. N., Hainsworth, S. V., Woosnam-Savage, R. C., Morgan, B., Brough, A, Earp R.W., Robinson, C., King, T.E., Morris, M., Buckley, R. (2015). Perimortem trauma in King Richard III: a skeletal analysis. *Lancet*, *385*(9964), 253–259. doi:10.1016/s0140-6736(14)60804-7

Bartelink, E.J., Berg, G.E., Beasley, M.M., Chesson, L.A. (2014). Application of stable isotope forensics for predicting region of origin of human remains from past wars and conflicts. *Annals of Anthropological Practice*, *38.1*, 124–136.

Bassett, E., Keith, M., Armelagos, G., Martin, D., & Villaneuva, A. (1980). Tetracycline-labeled human bone from ancient Sudanese Nubia (A.D. 350). *Science, 209*, 1532–1534.

Beaudet, A., Clarke, R. J., Heaton, J. L., Pickering, T. R., Carlson, K. J., Crompton, R. H., Jashashvili, T., Bruxelles, L., Jakata., K., Bam, L., Van Hoorebeke, L., Kuman, K., Stratford, D. (2020). The atlas of StW 573 and the late emergence of human-like head mobility and brain metabolism. *Scientific Reports, 10*(1). doi:10.1038/s41598-020-60837-2

Bell, L. S., Cox, G., & Sealy, J. C. (2001). Determining life history trajectories using bone density fractionation and stable light isotope analysis: a new approach. *American Journal of Physical Anthropology, 166*, 66–79.

Bell, L. S., Kayser, M., & Jones, C. (2008). The mineralized osteocyte: A living fossil. *American Journal of Physical Anthropology, 137*(4), 449–456.

Bennett, E. A., Crevecoeur, I., Viola, B., Derevianko, A. P., Shunkov, M. V., Grange, T., Maureille, B., Geigl, E. M. (2019). Morphology of the Denisovan phalanx closer to modern humans than to Neanderthals. *Science Advances, 5*(9). doi:10.1126/sciadv.aaw3950

Beresheim, A. C., Pfeiffer, S., & Grynpas, M. (2019). Ontogenetic changes to bone microstructure in an archaeologically derived sample of human ribs. *Journal of Anatomy, 236*, 448–462. doi:10.1111/joa.13116

Blondiaux, J., Durr, J., Khouchaf, L., & Eisenberg, L. E. (2002). Microscopic study and X-ray analysis of t[wo] 5th-century cases of leprosy. In C. A. Roberts, M. E. Lewis, & K. Manchester (Eds.), *The Past and Present of Leprosy: Archaeological, Historical, Palaeopathological and Clinical Approaches. Proceedings of the International Congress on the Evolution and palaeoepidemiology of the infectious diseases 3 (ICEPID), University of Bradford, 26–31 July, 1999* (pp. 105–110). Oxford: Archaeopress.

Bocquet-Appel, J., & Masset, C. (1982). Farewell to Paleodemograqhy. *J.Hum.Evo., 11*, 321–333.

Boldsen, J. L., Milner, G. R., & Ousley, S. D. (2022). Paleodemography: From archaeology and skeletal age estimation to life in the past. *American Journal of Biological Anthropology, 178*, 115–150. doi:10.1002/ajpa.24462

Booth, T. J. (2020). *Using Bone Histology to Identify Stillborn Infants in the Archaeological Record.*

Booth, T. J., & Bruck, J. (2020). Death is not the end: radiocarbon and histo-taphonomic evidence for the curation and excarnation of human remains in Bronze Age Britain. *Antiquity, 94*(377), 1186–1203. doi:10.15184/aqy.2020.152

Booth, T. J., Redfern, R. C., & Gowland, R. L. (2016). Immaculate conceptions: Micro-CT analysis of diagenesis in Romano-British infant skeletons. *Journal of Archaeological Science, 74*, 124–134. doi:10.1016/j.jas.2016.08.007

Britz, H. M., Thomas, C. D. L., Clement, J. G., & Cooper, D. M. L. (2009). The relation of femoral osteon geometry to age, sex, height and weight. *Bone, 45*, 77–83.

Bronnimann, D., Portmann, C., Pichler, S. L., Booth, T. Rderoder, B., Vach, W., Schibler, J., Rentzel, P. (2018). Contextualising the d–ad - Combining geoarchaeology and osteo-anthropology in a new multi-focus approach in bone histotaphonomy. *Journal of Archaeological Science, 98*, 45–58. doi:10.1016/j.jas.2018.08.005

Burr, D. B., Ruff, C., & Thompson, D. D. (1990). Patterns of skeletal histologic change through time: comparison of an Archaic Native American population with modern populations. *Anatomical Record, 226*, 613–616.

Caccia, G, Magli F, Tagi, V. M., Porta, D. G. A., Cummaudo, M., Márquez-Grant, N., & Cattaneo, C. (2016). Histological determination of the human origin from dry bone: a cautionary note for subadults. *International Journal of Legal Medicine, 130*(1), 299–307.

Carew, R. M., Iacoviello, F., Rando, C., Moss, R. M., Speller, R., French, J., & Morgan, R. M. (2022). A multi-method assessment of 3D printed micromorphological osteological features. *International Journal of Legal Medicine, 136*(5), 1391–1406. doi:10.1007/s00414-022-02789-y

Carlson, K. S. D., Mandl, K., McCall, A., Bronnimann, D., Teschler-Nicola, M., Weiss-Krejci, E., & Metscher, B. (2022). 3D visualization of bioerosion in archaeological bone. *Journal of Archaeological Science, 145*. doi:10.1016/j.jas.2022.105646

Carter, D. R., & Beaupre, G. S. (2001). *Skeletal Function and Form*. Cambridge: Cambridge University Press.

Caruso, V., Marinoni, N., Diella, V., Berna, F., Cantaluppi, M., Mancini, L., Trombino, L., Cattaneo, C., Pastero, L., & Pavese, A. (2020). Bone diagenesis in archaeological and contemporary human remains: an investigation of bone 3D microstructure and minero-chemical assessment. *Archaeological and Anthropological Sciences, 12*(8). doi:10.1007/s12520-020-01090-6

Caruso, V., Marinoni, N., Diella, V., Possenti, E., Mancini, L., Cantaluppi, M., Berna, F., Cattneo, C., & Pavese, A. (2021). Diagenesis of juvenile skeletal remains: A multimodal and multiscale approach to examine the post-mortem decay of chil'ren's bones. *Journal of Archaeological Science, 135*. doi:10.1016/j.jas.2021.105477

Chesson, L. A., Berg, G. E. (2021). The use of stable isotopes in postconflict forensic identification. *Wiley Interdisciplinary Reviews: Forensic Science*, e1439. https://doi.org/10.1002/wfs2.1439

Cho, H., Stout, S. D., Madsen, R. W., & Streeter, M. (2002). Population-specific histological age-estimating method: A model for known African-American and European-American skeletal remains. *Journal of Forensic Sciences, 47*(1), 12–18.

Cook, M., Molto, E., & Anderson, C. (1989). Fluorochrome labeling in Roman Period skeletons from Dakhleh Oasis, Egypt. *American Journal of Physical Anthropology, 80*, 137–143.

Crowder, C. (2005). *Evaluating the use of quantitative bone histology to estimate adult age at death*. [Doctorate Dissertation, University of Toronto].

Crowder, C., Dominguez, V. M., Heinrich, J., Pinto, D., & Mavroudas, S. (2022). Analysis of histomorphometric variables: Proposal and validation of osteon definitions. *Journal of Forensic Sciences, 67*(1), 80–91. doi:10.1111/1556-4029.14949

Crowder, C., Heinrich, J., & Dominquez, V. (2009). Histological age estimation. In S. Blau & D. Ubelaker (Eds.), *Handobook of Forensic Anthropology and Archaeology* (pp. 222–235). Walnut Creek, California: Left Coast Press.

Crowder, C., Heinrich, J., & Stout, S. D. (2012). Rib histomorphometry for adult age estimation. In L. S. Bell (Ed.), *Forensic Microscopy for Skeletal Tissues: Methods and Protocols* (Vol. Methods in Molecular Biology, vol. 915, pp. 109–127). New York City: Humana Press.

Crowder, C., & Pfeiffer, S. (2009a). The application of cortical bone histomorphometry to estimate age-at-death. In K. E. Latham & M. Finnegan (Eds.), *Age Estimation from the Human Skeleton* (pp. 193–215), Springfield, Ill.: CC Thomas.

Crowder, C., & Pfeiffer, S. (2009b). The application of cortical bone histomorphometry to estimate age-at-death. In K. E. Latham & M. Finnegan (Eds.), *Age Estimation from the Human Skeleton* (pp. 193–215). Springfield, Ill.: CC Thomas.

Dal Sasso, G., Maritan, L., Usai, D., Angelini, I., & Artioli, G. (2014). Bone diagenesis at the micro-scale: Bone alteration patterns during multiple burial phases at Al Khiday (Khartoum, Sudan) between the Early Holocene and the II century AD. *Palaeogeography Palaeoclimatology Palaeoecology, 416*, 30–42. doi:10.1016/j.palaeo.2014.06.034

Dash, H. R., & Das, S. (2018). Microbial Degradation of Forensic Samples of Biological Origin: Potential Threat to Human DNA Typing. *Molecular Biotechnology, 60*(2), 141–153. doi:10.1007/s12033-017-0052-5

Day, M. H., & Molleson, T. I. (1973). The Trinil femora. In M. H. Day (Ed.), *Human Evolution* (pp. 127–154). London: Taylor and Francis.

De Boer, H. H., & Maat, G. J. R. (2018). Dry bone histology of bone tumors. *International Journal of Paleopathology, 21*, 56–63.

Dittmar, J. M., Mitchell, P. D., Jones, P. M., Mulder, B., Inskip, S. A., Cessford, C., & Robb, J. E. (2021). Gout and 'Podagra' in medieval Cambridge, England. *International Journal of Paleopathology, 33*, 170–181. doi:10.1016/j.ijpp.2021.04.007

Doershuk, L. J., Saers, J. P. P., Shaw, C. N., Jashashvili, T., Carlson, K. J., Stock, J. T., & Ryan, T. M. (2019). Complex variation of trabecular bone structure in the proximal humerus and femur of five modern human populations. *American Journal of Physical Anthropology, 168*(1), 104–118. doi:10.1002/ajpa.23725

Dubois, E. (1937). The osteone arrangement of the thigh-bone compacta of Man identical with that, first found, of Pithecanthropus. *K.ned.Akad.Wet., 40*, 864–870.

Dutour, O., Palfi, G., Berato, J., & Brun, J.-P. (1994). *L'origine de la syphilis en Europe: avant ou apres 1493?* Toulon, France: Editions Errance.

Enlow, D. H., & Brown, S. O. (1956). A comparative histological study of fossil and recent bone tissues. Part 1. *Texas Journal of Science, 8*, 405–443.

Enlow, D. H., & Brown, S. O. (1957). A comparative histological study of fossil and recent bone tissues. Part 2. *Texas Journal of Science, 9*, 186–214.

Enlow, D. H., & Brown, S. O. (1958). A comparative histological study of fossil and recent bone tissues. Part 3. *Texas Journal of Science, 10*, 187–230.

Falys, C. G., & Lewis, M. E. (2011). Proposing a way forward: a review of tandardization in the use of age categories and aging techniques in osteological analysis (2004–2009). *International Journal of Osteoarchaeology, 21*, 704–716.

Figus, C., Stephens, N. B., Sorrentino, R., Bortolini, E., Arrighi, S., Higgins, O. A., Lugli, Fl, Marciani, G., Oxilia, G., Romandini, M., Silvestrini, S., Baruffaldi, F., Belcastro, M.G., Bernardini, F., Festa, A., Hajdu, T., Mateovics-László, O., Pap, I., Szeniczey, T., Tuniz, Cl, Ryan, T.M., Benazzi, S. (2023). Morphologies in-between: The impact of the first steps on the human talus. *Anatomical Record-Advances in Integrative Anatomy and Evolutionary Biology, 306*(1), 124–142. doi:10.1002/ar.25010

Foote, J. S. (1916). *A Contribution to the Comparative Histology of the Femur* (Vol. 35). Washington, D.C.: Smithsonian Institution.

Garn, S. M. (1970). *The Earlier Gain and Later Loss of Cortical Bone, in Nutritional Perspective.* Springfield, Illinois: C.C. Thomas.

Garnett, E. M., & Lewis, M. E. (2022). The potential for over diagnosis of Paget's disease of bone using microscopic analysis. *International Journal of Paleopathology, 38*, 55–63.

Gosman, J. H., & Ketcham, R. A. (2009). Patterns in ontogeny of human trabecular bone from Sunwatch Village in the Prehistoric Ohio Valley: General features of microarchitectural change. *American Journal of Physical Anthropology, 138*(3), 318–332. doi:10.1002/ajpa.20931

Graf, W. (1949). Preserved histological structures in Egyptian mummy tissues and ancient Swedish skeletons. *Acta Anatomie, 8*(3), 236–250.

Grupe, G., Balzer, A., & Turban-Just, S. (2000). Modeling protein diagenesis in ancient bone: towards a validation of stable isotope data. In *Biogeochemical Approaches to Paleodietary Analysis* (Vol. 5, pp. 173–187). Boston, MA: Springer US.

Guarino, F. M., Angelini, F., Vollono, C., & Orefice, C. (2006). Bone preservation in human remains from the Terme del Sarno at Pompeii using light microscopy and scanning electron microscopy. *Journal of Archaeological Science, 33*, 513–520.

Haddow, S. D., Mazzucato, C., Mangaloglu-Votruba, S., Yagci, B., Booth, T., Schotsmans, E. M. J., & Knusel, C. J. (2023). Scratching the surface? A histotaphonomic study of human remains at Neolithic catalhoyuk. *Archaeological and Anthropological Sciences, 15*(6). doi:10.1007/s12520-023-01756-x

Haglund, A. H., & Sorg, M. (Eds.). (1997). *Forensic Taphonomy: the postmortem fate of human remains.* Boca Raton: CRC Press.

Havenga, D., Akoba, R., Menzi, L., Azizi, S., Sackey, J., Swanepoel, N., Gibaud, A., Maaza, M. (2022). From Himba indigenous knowledge to engineered Fe2O3 UV-blocking green nanocosmetics. *Scientific Reports, 12*(1). doi:10.1038/s41598-021-04663-0

Hedges, R. E. M., Millard, A. R., & Pike, A. W. G. (1995). Measurements and relationships of diagenetic alteration of bone from three archaeological sites. *Journal of Archaeological Science, 22*(2), 201–209.

Hillier, M., & Bell, L. S. (2007). Differentiating human bone from animal bone: a review of histological methods. *Journal of Forensic Science, 52*(2), 249–263.

Hollund, H. I., Jans, M. M. E., Collins, M. J., Kars, H., Joosten, I., & Kars, S. M. (2012). What Happened Here? Bone Histology as a Tool in Decoding the Postmortem Histories of Archaeological Bone from Castricum, The Netherlands. *International Journal of Osteoarchaeology, 22*(5), 537–548. doi:10.1002/oa.1273

Hoppa, R. D., & Vaupel, J. W. (Eds.). (2002). *Paleodemography: age distributions from skeletal samples.* Cambridge, U.K.: Cambridge University Press.

Hublin, J.-J., Pääbo, S., Derevianko, A.P., Doronichev, V.B., Golovanova, L.V., Friess, M., Froment, Al, Hoffman, Al, Kachache, N.E.J., Kullmer, O., Lordkipanidze, D., Moncel, M.-H., Potts, R., Radovcic, J., Rak, Y.Z., Richards, M., Rodriguez Mendez, J., Rosas, A., Schmauder, M., Schmitz, R.W., Semal, P, Smith, T., Tafuri, M.A., Tattersall, I., Tournepiche, J.-F., Toussaint, M., Vassiliev, S., Vilet, A., White, T., and Ziegler, R. (2008). Suggested guidelines for invasive sampling of hominid remains. *Journal of Human Evolution, 55*(4), 756–757.

Huntley, J. (2018). Australian indigenous ochres: use, sourcing and exchange. In I. J. McNiven & B. David (Eds.), *The Oxford Handbook of the Archeology of Indigenous Australia and New Guinea.* Oxford: Oxford University Press.

Jans, M. M. E. (2008). Microbial bioerosion of bone – a review. In M. Wisshak & L. Tapanila (Eds.), *Current Developments in Bioerosion* (pp. 397–413). Berlin: Springer.

Jans, M. M. E., Nielsen-Marsh, C. M., Smith, C. I., Collins, M. J., & Kars, H. (2004). Characterisation of microbial attack on archaeological bone. *Journal of Archaeological Science, 31,* 87–95.

Keith, M., & Armelagos, G. (1983). Naturally occurring antibiotics and human health. In L. Romanucci-Ross, D. Moerman, & L. Tancredi (Eds.), *The Anthropology of Medicine: From culture to method* (pp. 68–79). New York: Praeger.

Keith, M., & Armelagos, G. (1988). An example of in vivo tetracycline labeling: reply to Piepenbrink. *Journal of Archaeological Science, 15,* 595–601.

Kendall, C., Eriksen, A. M. H., Kontopoulos, I., Collins, M. J., & Turner-Walker, G. (2018). Diagenesis of archaeological bone and tooth. *Palaeogeography Palaeoclimatology Palaeoecology, 491,* 21–37. doi:10.1016/j.palaeo.2017.11.041

Kerley, E. R. (1965). The Microscopic Determination of Age in Human Bone. *American Journal of Physical Anthropology, 23,* 149–164.

Kerley, E. R. (1969). Age Determination of Bone Fragments. *Journal of Forensic Sciences, 14*(1), 59–67.

Kerley, E. R. (1970). Estimation of Skeletal Age: After About Age 30. In T. D. Stewart (Ed.), *Personal Identification in Mass Disasters* (pp. 57–70): Smithsonian Institute.

Kerley, E. R., & Ubelaker, D. H. (1978). Revisions in the Microscopic Method of Estimating Age at Death in Human Cortical Bone. *American Journal of Physical Anthropology, 49*(4), 545–546.

Kontopoulos, I., Penkman, K., McAllister, G. D., Lynnerup, N., Damgaard, P. B., Hansen, H. B., … Collins, M. J. (2019). Petrous bone diagenesis: a multi-analytical approach. *Palaeogeography Palaeoclimatology Palaeoecology, 518,* 143–154. doi:10.1016/j.palaeo.2019.01.005

Kontopoulos, I., Van de Vijver, K., Robberechts, B., von Tersch, M., Turner-Walker, G., Penkman, K., & Collins, M. J. (2022) Histological and stable isotope analysis of archeological bones from St. Rombout's cemetery (Mechelen, Belgium): Intrasite, intraindividual, and intrabone variability. *International Journal of Osteoarchaeology.* doi:10.1002/oa.3145

Larsen, C. S. (1997). *Bioarchaeology: Interpreting Behavior from the Human Skeleton.* New York: Cambridge University Press.

Lazenby, R. A., & Pfeiffer, S. (1993). Effects of a nineteenth century below-knee amputation and prosthesis on femoral morphology. *International Journal of Osteoarchaeology, 3,* 19–28.

Lemmers, S. A. M., Goncalves, D., Cunha, E., Vassalo, A. R., & Appleby, J. (2020). Burned Fleshed or Dry? The Potential of Bioerosion to Determine the Pre-Burning Condition of Human Remains. *Journal of Archaeological Method and Theory, 27*(4), 972–991. doi:10.1007/s10816-020-09446-x

Maggiano, C., Dupras, T., Schultz, M., & Biggerstaff, J. (2006). Spectral and photobleaching analysis using confocal laser scanning microscopy: a comparison of modern and archaeological bone fluorescence. *Molecular and Cellular Probes, 20,* 154–162.

Mandl, K., & Carlson, K. J. (In Prep). Preliminary experiment into the effects of red ochre on skeletonization rate and perinate bone microstructure in *Sus scrofa domesticus. TBA.*

Mandl, K., Carlson, K. S. D., Bronnimann, D., McCall, A., Grassberger, M., Teschler-Nicola, M., Weiss-Krejci, E., & Metscher, B. (2022). Substantiating microCT for diagnosing bioerosion in archaeological bone using a new Virtual Histological Index (VHI). *Archaeological and Anthropological Sciences, 14*(6). doi:10.1007/s12520-022-01563-w

Mandl, K., & Grassberger, M. (2021). As red as blood. Treatment of the dead with ochre: Ritual or deliberate manipulation? In C. Mayer & G. Langer (Eds.), *Experimentelle Archäeologie/ Experimental Archaeology* (pp. 70–73): WMB, Weinviertal Museum Betriber GmbH.

Martin, R. B., & Burr, D. B. (1989). *Structure, function, and adaptation of compact bone.* New York: Raven Press.

Mavroudas, S. R., Alfsdotter, C., Bricking, A., & Madgwick, R. (2023). Experimental investigation of histotaphonomic changes in human bone from whole-body donors demonstrates limited effects of early post-mortem change in bone. *Journal of Archaeological Science, 154.* doi:10.1016/j.jas.2023.105789

Merritt, C. E. (2015). The Influence of body size on adult skeletal age estimation methods. *American Journal of Physical Anthropology, 156,* 35–57.

Millard, A. R. (2001). The deterioration of bone. In D. R. Brothwell & A. M. Pollard (Eds.), *Handbook of Archaeological Sciences* (pp. 637–674). New York: Wiley.

Miszkiewicz, J. J., & Mahoney, P. (2016). Ancient Human Bone Microstructure in Medieval England: Comparisons between Two Socio-Economic Groups. *Anatomical Record-Advances in Integrative Anatomy and Evolutionary Biology, 299*(1), 42–59. doi:10.1002/ar.23285

Mitchell, P. D., Dittmar, J. M., Mulder, B., Inskip, S., Littlewood, A., Cessford, C., & Robb, J. E. (2022). Assessing the relative benefits of imaging with plain radiographs and microCT scanning to diagnose cancer in past populations. *International Journal of Paleopathology, 36,* 24–29. doi:10.1016/j.ijpp.2021.12.001

Molleson, T., & Cox, M. (1993). *The Spitalfields Project, Volume 2: The Anthropology, The Middling Sort.* York: Council for British Archaeology.

Morales, N. S., Catella, L., Oliva, F., Sarmiento, P. L., & Barrientos, G. (2018). A SEM-based assessment of bioerosion in Late Holocene faunal bone assemblages from the southern Pampas of Argentina. *Journal of Archaeological Science-Reports, 18,* 782–791. doi:10.1016/j.jasrep.2017.07.012

Morris, H. (2007). *Quantitative and Spatial Analysis of the Microscopic Bone Structures of Deer, Dog and Pig.* (M.A.), Louisiana State University.

Mosekilde, L. (1990). Consequences of the remodelling process for vertebral trabecular bone structure: a scanning electron microscope study (uncoupling of unloaded structures). *Bone and Mineral, 10,* 13–35.

Mulhern, D. M. (2000). Rib remodeling dynamics in a skeletal population from Kulubnarti, Nubia. *American Journal of Physical Anthropology, 111,* 519–530.

Mulhern D.M., Ubelaker, D.H. (2001). Differences in osteon banding between human and nonhuman bone. *Journal Forensic Sciences,* 46:220–2.

Mulhern, D. M., & Van Gerven, D. P. (1997). Patterns of Femoral Bone Remodeling Dynamics in a Medieval Nubian Population. *American Journal of Physical Anthropology, 104,* 133–146.

Nicklisch, N., Ramsthaler, F., Bunnefeld, J. H., Schulz, G., Friedrich, R., Alt, K. W., & Meller, H. (2022). Bioarchaeological investigations of the princely grave at Helmsdorf attesting to the violent death of an Early Bronze Age leader. *Scientific Reports, 12*(1). doi:10.1038/s41598-022-20720-8

Odes, E. J., Parkinson, A. H., Randolph-Quinney, P. S., Zipfel, B., Jakata, K., Bonney, H., & Berger, L. R. (2017). Osteopathology and insect traces in the Australopithecus africanus skeleton StW 431. *South African Journal of Science, 113*(1–2), 56–62. doi:10.17159/sajs.2017/20160143

Odes, E. J., Randolph-Quinney, P. S., Steyn, M., Throckmorton, Z., Smilg, J. S., Zipfel, B., Augustine, T.N., de Beer, F., Joffman, J.W., Franklin, R.D., & Berger, L. R. (2016). Earliest hominin cancer: 1.7-million-year-old osteosarcoma from Swartkrans Cave, South Africa. *South African Journal of Science, 112*(7–8), 100–104. doi:10.17159/sajs.2016/20150471

Ortner, D. J., & Putschar, W. G. J. (1981). *Identification of Pathological Conditions in Human Skeletal Remains* (Vol. 28). Washington, D.C.

Parfitt, A. M., Drezner, M. K., Glorieux, F. H., Kanis, J. A., Malluche, H., Meunier, P. J., Ott, S.M., & Recker, R. R. (1987). Bone histomorphometry: Standardization of nomenclature, symbols, and units. *Journal of Bone and Mineral Research, 2*(6), 595–610.

Perchalski, B., Placke, A., Sukhdeo, S. M., Shaw, C. N., Gosman, J. H., Raichlen, D. A., & Ryan, T. M. (2018). Asymmetry in the Cortical and Trabecular Bone of the Human Humerus During Development. *Anatomical Record-Advances in Integrative Anatomy and Evolutionary Biology, 301*(6), 1012–1025. doi:10.1002/ar.23705

Pfeiffer, S. (1980). Bone remodeling age estimates compared with estimates obtained by other techniques. *Current Anthropology, 21*(6), 793.

Pfeiffer, S. (1992). Cortical bone age estimates from historically known adults. *Zetischrift für Morphologie und Anthropologie, 79*, 1–10.

Pfeiffer, S. (1998). Variability in osteon size in recent human populations. *American Journal of Physical Anthropology, 106*(2), 219–227.

Pfeiffer, S. (2000). Palaeohistology: Health and Disease. In M. A. Katzenberg & S. R. Saunders (Eds.), *Biological Anthropology of the Human Skeleton* (pp. 281–296). New York: Wiley-Liss.

Pfeiffer, S. (2006). Cortical bone histology in juveniles. In G. Grupe & J. Peters (Eds.), *Microscopic Examinations of Bioarchaeological Remains: Keeping a close eye on ancient tissues* (pp. 15–28). Rahden/Westf: Verlag Marie Leidorf GmbH.

Pfeiffer, S. (2022). *Osteobiographies: The Discovery, Interpretation and Repatriation of Human Remains*. New York: Academic Press.

Pfeiffer, S., Cameron, M. E., Sealy, J., & Beresheim, A.C. (2019). Diet and adult age-a-death among mobile foragers: a synthesis of bioarchaeological methods. *American Journal of Physical Anthropology, 170*, 131–147.

Pfeiffer, S., Crowder, C., Harrington, L., & Brown, M. (2006). Secondary osteon and Haversian canal dimensions as behavioral indicators. *American Journal of Physical Anthropology, 131*, 460–468.

Pfeiffer, S., Heinrich, J., Beresheim, A., & Alblas, M. (2016). Cortical bone histomorphology of known-age skeletons from the Kirsten Collection, Stellenbosch University, South Africa *American Journal of Physical Anthropology, 160*, 137–147.

Pfeiffer, S., & Varney, T. (2000). Quantifying histological and chemical preservation in archaeological bone. In S. H. Ambrose & M. A. Katenberg (Eds.), *Biogeochemical Approaches to Paleodietary Analysis* (pp. 141–158). New York: Kluwer Academic/ Plenum Publishers.

Pfeiffer, S., & Zehr, M. (1996). A morphological and histological study of the human humerus from Border Cave. *Journal of Human Evolution, 31*, 49–59.

Piepenbrink, H. (1986). 2 examples of biogenous dead bone decomposition and their consequences for taphonomic interpretation. *Journal of Archaeological Science, 13*(5), 417–430.

Pinto, D. C., & Stout, S. D. (2009). P'get's disease in pre-contact Florida? Revisiting the Briarwoods site in Gulf coast Florida. *International Journal of Osteoarchaeology, early view*.

Powell, M. L., & Cook, D. C. (Eds.). (2005). *The myth of Syphillis: The natural history of Trepanematosis in North America*. Gainesville, Florida: University Press of Florida.

Roberts, C., & Manchester, K. (1995). *The Archaeology of Disease, 2nd Edition*. Ithaca, New York: Cornell University Press.

Roberts, C. A. (2009). *Human Remains in Archaeology: a handbook* (Vol. 19). York: Council for British Archaeology.

Robling, A. G., & Stout, S. D. (2003). Histomorphology, Geometry, and Mechanical Loading in Past Populations. In S.C. Agarwal and S.D. Stout (Eds.), *Bone Loss and Osteoporosis: An Anthropological Perspective* (pp. 189–206), New York: Kluwer Academic/Plenum Publishers.

Robling, A. G., & Stout, S. D. (2008). Histomorphometry of Human Cortical Bone: Applications to Age Estimation. In M. A. Katzenberg & S. R. Saunders (Eds.), *Biological Anthropology of the Human Skeleton* (second ed., pp. 149–182). New York: Wiley-Liss.

Ruff, C. B. (2008). Biomechanical analyses of archaeological human skeletons. In M. A. Katenberg & S. R. Saunders (Eds.), *Biological Anthropology of the Human Skeleton* (Second ed.). New York: Wiley-Liss.

Ruff, C. B., Holt, B., & Trinkaus, E. (2006). Who's afraid of the big bad Wolff? Wolff's law and functional adaptation. *American Journal of Physical Anthropology, 129*(4), 484–498.

Ruff, C. B., Sylvester, A. D., Rahmawati, N. T., Suriyanto, R. A., Storm, P., Aubert, M., Joannes-Boyau, R., Berghuis, H., Pop, E., Joost Battenburg, K., Coban, S.B., Kostenko, A., Noerwidi, S., Renema, W., Adhityatam, S., &. Joordens, J. C. (2022). *Two Late Pleistocene human femora from Trinil*, Indonesia: Implications for body size and behavior in Southeast Asia. *Journal of Human Evolution, 172*. doi:10.1016/j.jhevol.2022.103252

Ruff, C. B., Trinkaus, E., Walker, A., & Larsen, C. S. (1993). Postcranial robusticity in Homo. I: Temporal trends and mechanical interpretation. *American Journal of Physical Anthropology, 91*, 21–53.

Ryan, T. M., & Shaw, C. N. (2015). Gracility of the modern Homo sapiens skeleton is the result of decreased biomechanical loading. *Proceedings of the National Academy of Sciences of the United States of America, 112*(2), 372–377. doi:10.1073/pnas.1418646112

Sawada, J., Kondo, O., Nara, T., Dodo, Y., & Akazawa, T. (2004). Bone histomorphology of the Dederiyeh Neanderthal child. *Anthropological Science.*

Schlecht, S. H., Pinto, D. C., Agnew, A. M., & Stout, S. D. (2012). Brief communication: The effects of disuse on the mechanical properties of bone: What unloading tells us about the adaptive nature of skeletal tissue. *American Journal of Biological Anthropology, 149*(4), 599–605.

Schultz, M. (1997a). Microscopic investigation of excavated skeletal remains: A contribution to paleopathology and forensic medicine. In A. H. Haglund & M. Sorg (Eds.), *Forensic Taphonomy* (pp. 201–222).

Schultz, M. (1997b). Microscopic structure of bone. In W. D. Haglund & M. H. Sorg (Eds.), *Forensic Taphonomy: The Postmortem Fate of Human Remains* (pp. 187–199). Boca Raton, Florida: CRC Press.

Schultz, M. (2001). Paleohistology of bone: a new approach to the study of ancient diseases. *Yearbook of Physical Anthropology, 44*, 106–147.

Schultz, M. (2003). Light microscopic analysis in skeletal palaeopathology. In D. J. Ortner (Ed.), *Identification of Pathological Conditions in Human Skeletal Remains* (second ed., pp. 73–107). New York: Elsevier.

Schultz, M., & Roberts, C. A. (2002). Diagnosis of leprosy in skeletons from an English later medieval hsopital using histological analysis. In C. A. Roberts, M. E. Lewis, & K. Manchester (Eds.), *The Past and Present of Leprosy: Archaeological, Historical, Palaeopathological and Clinical Approaches. Proceedings of the International Congress on the Evolution and palaeoepidemiology of the infectious diseases 3 (ICEPID), University of Bradford, 26–31 July, 1999* (Vol. 1054, pp. 89–104). Oxford: Archaeopress.

Skedros, J. G., Sorenson, S. M., & Jenson, N. H. (2007). Are distributions of secondary osteon variants useful for interpreting load history in mammalian bones? *Cells, Tissues, Organs, 185*(4), 285–307.

Stock, J., & Pfeiffer, S. (2001). Linking structural variability in long bone diaphyses to habitual behaviors: foragers from the southern African Later Stone Age and the Andaman Islands. *American Journal of Physical Anthropology, 115*(4), 337–348.

Stock, J., & Pfeiffer, S. K. (2004). Long bone robusticity and subsistence behaviour among Later Stone Age foragers of the forest and fynbos biomes of South Africa. *Journal of Archaeological Science, 31*, 999–1013.

Stock, J. T., Pfeiffer, S. K., Chazan, M., & Janetski, J. (2005). F-81 skeleton from Wadi Mataha, Jordan, and its bearing on human variability in the Epipaleolithic of the Levant. *American Journal of Physical Anthropology, 128*, 453–465.

Stout, S. D. (1986). The use of bone histomorphometry in skeletal identification: the case of Francisco Pizarro. *Journal of Forensic Sciences, 31*, 296–300.

Streeter, M. (2005). *Histomorphometric characteristics of the subadult rib cortex: normal patterns of dynamic bone modeling and remodeling during growth and development.* (Ph.D.), University of Missouri - Columbia.

Streeter, M., & Stout, S. D. (2003). The histomorphometry of the subadult rib: age associated changes in bone mass and the creation of peak bone mass. In S. C. Agarwal & S. D. Stout (Eds.), *Bone Loss and Osteoporosis: An Anthropological Perspective* (pp. 91–101). New York: Kluwer Academic/Plenum.

Streeter, M., Stout, S. D., Trinkaus, E., Roberts, M. B., & Parfitt, S. A. (2001). Histomorphometric age assessment of the Boxgrove I tibial diaphysis. *Journal of Human Evolution, 40*, 331–338.

Teklay, M., Thole, J. T., Ndumbu, N., Vries, J., & Mezger, K. (2023). Mineralogical and chemical characterization of ochres used by the Himba and Nama people of Namibia. *Journal of Archaeological Science-Reports, 47.* doi:10.1016/j.jasrep.2022.103690

Thompson, D. D. (1979). The Core Technique in the Determination of Age at Death in Skeletons. *Journal of Forensic Sciences, 24*, 902–915.

Thompson, D. D. (1981). Microscopic Determination of Age at Death in an Autopsy Series. *Journal of Forensic Science*, 470–475.

Thompson, D. D., & Galvin, C. A. (1983). Estimation of age at death by tibial osteon remodeling in an autopsy series. *Forensic Science International, 22*, 203–211.

Thompson, D. D., & Gunness-Hey, M. (1981). Bone mineral-osteon analysis of Yupik-Inupiaq skeletons. *American Journal of Physical Anthropology, 55*, 1–7.

Thompson, D. D., Salter, E. M., & Laughlin, W. S. (1981). Bone Core Analysis of Baffin Island Skeletons. *Arctic Anthropology, 28*(1), 87–96.

Thompson, D. D., & Trinkaus, E. (1981). Age Determination for the Shanidar 3 Neanderthal. *Science, 212*, 575–577.

Trinkaus, E., & Thompson, D. D. (1987). Femoral diaphyseal histomorphometric age determinations for the Shanidar 3, 4, 5, and 6 Neandertals and Neandertal longevity. *American Journal of Physical Anthropology, 72*, 123–129.

Turner-Walker, G. (2019). Light at the end of the tunnels? The origins of microbial bioerosion in mineralised collagen. *Palaeogeography Palaeoclimatology Palaeoecology, 529*, 24–38. doi:10.1016/j.palaeo.2019.05.020

Turner-Walker, G., & Mays, S. (2008). Histological studies on ancient bone. In R. Pinhasi & S. Mays (Eds.), *Advances in Human Palaeopathology* (pp. 121–146). Chichester, U.K.: John Wiley & Sons.

Ubelaker, D. H. (1989). *Human Skeletal Remains: Excavation, Analysis, Interpretation* (2nd ed.). Washington, D.C.: Taraxacum.

Vegh, E. I., Czermak, A., Marquez-Grant, N., & Schulting, R. J. (2021). Assessing the reliability of microbial bioerosion features in burnt bones: A novel approach using feature-labelling in histotaphonomical analysis. *Journal of Archaeological Science-Reports, 37.* doi:10.1016/j.jasrep.2021.102906

Von Hunnius, T. E., Roberts, C. A., Saunders, S. R., & Boylston, A. (2006). Histological identification of syphilis in pre-Columbian England. *American Journal of Physical Anthropology, 129*, 559–566.

Walker, A., & Shipman, P. (1996). *The Wisdom of the Bones*. New York: Alfred Knopf.

Walker, A., Zimmerman, M. R., & Leakey, R. E. F. (1982). A possible case of hypervitaminosis A in Homo erectus. *Nature, 296*(18 March), 248–250.

Watts, I. (2009). Red ochre, body painting, and language: interpreting the Blombos ochre. In R. Botha & C. Knight (Eds.), *The Cradle of Language* (pp. 62–92). Oxford: Oxford University Press.

Welsh, H., & Brickley, M. B. (n.d.) Pathology or expected morphology? Investigating patterns of cortical porosity and trabecularization during infancy and early childhood. *Anatomical Record-Advances in Integrative Anatomy and Evolutionary Biology*. doi:10.1002/ar.25081

Wood, J. W., Milner, G. R., Harpending, H. C., & Weiss, K. M. (1992). The osteological paradox: Problems of inferring prehistoric health from skeletal samples. *Current Anthropology, 33*, 343–370.

11 Skeletal Collections for Histological and Imaging Research

JANNA M. ANDRONOWSKI,
JOSHUA T. TAYLOR,
BRIAN F. SPATOLA,
RITA HARDIMAN, AND
NATALIE A. SIMS

11.1 Introduction

The rise of biological anthropology as a discipline can be intimately tied to the establishment of skeletal collections (Alves-Cardoso & Campanacho, 2022). Such collections have proven invaluable for scientific research, teaching, and training. The first well-documented skeletal archives include the Hamann-Todd Human Osteological Collection (Işcan, 1992), Huntington Collection (Campanacho et al., 2021), and Robert J. Terry Anatomical Human Skeletal Collection (Hunt & Albanese, 2005), which were procured in the early 20th century. Since their initiation, these collections (among others) have served as the basis for numerous theses, dissertations, and manuscripts related to biological profile development, underlining their immense importance to the field. A recent study reported 376 publications utilizing skeletal collections published in 92 journals between 1969 and 2021 (Alves-Cardoso & Campanacho, 2022). It is essential, however, for well-documented skeletal collections to represent living populations as this is the demographic from which forensic cases are derived. Such historical collections no longer represent modern life circumstances (e.g., occupational and medical standards) – resulting in discussions about the validity and applicability of the data gleaned from these collections to modern and diverse groups.

Inherent biases exist when procuring specimens from anatomical body donation programs and surgical procedures, as they mostly contain older individuals with pathological or age-associated changes impacting skeletal physiology. Others, including the Hamann-Todd Collection, grew significantly in part through receiving unclaimed decedents from Medical Examiners' Offices (Cobb, 1959). Similarly, the Terry collection was developed via the acquisition of cadavers from neighboring hospitals and morgues and typically reflected unclaimed individuals and those of lower socioeconomic status (Hunt & Albanese, 2005). Current procurement protocols surrounding osteological collections are inconsistent, varying based on the institution and region. It is generally understood that donors have to provide informed consent (Nuremberg Code of 1947), the donors must be anonymized, and researchers should follow strict ethical guidelines (Belcastro et al., 2022; Campanacho et al., 2021; Mendes et al., 2022; Plens et al., 2022; Winburn et al., 2022). Several researchers suggested similar international guidelines for the ethical acquisition of skeletal assemblages (Mendes et al., 2022). With increasing attention and recent discussions regarding the repatriation of historical collections and the ethical acquisition of skeletal assemblages,

DOI: 10.4324/9781003385608-11

researchers are calling for detailed assessments of museum collections to move toward a more ethical future (Stantis et al., 2023). In response to these concerns, an ethical committee of over 60 anthropologists, archaeologists, curators, and geneticists convened in 2021 to discuss ethical human DNA research (Alpaslan-Roodenberg et al., 2021).

Resourcing for bone research remains critical, however, for further understanding natural age-related processes, biological variation in microstructural parameters, and guiding future clinical preventative interventions for common bone diseases (e.g., osteoporosis). Skeletal collections developed specifically for histological and imaging research remain scarce worldwide. The most complete well-documented collection specifically for this purpose, the Andronowski Skeletal Collection for Histological Research (ASCHR; Canada), was recently introduced (Andronowski & Taylor, 2022). This unique skeletal archive, initiated in 2017, offers a modern anatomical resource explicitly developed for hypothesis-driven research related to histological and imaging analyses. Additional established collections include the Melbourne Femur Research Collection (MFRC; Australia), skeletal tissue collections and orthopedic pathology slides at the National Museum of Health and Medicine (NMHM; USA), and the Ericksen collection (USA). These collections were instrumental in advancing scientific knowledge related to human growth and development (Carter et al., 2013; Scheiner et al., 2016), normal bone microarchitectural anatomy (Andronowski et al., 2023; Carter et al., 2014; Maggiano et al., 2016), histomorphometry (Andronowski et al., 2023; Hennig et al., 2015), bone and joint pathology (Buenzli et al., 2013; Wang et al., 2013), and comparative anatomy (Andronowski, Davis, et al., 2019).

The primary aims of our chapter are to (1) profile specific bone histology skeletal collections, including their backgrounds, ethical considerations, bone procurement processes, demographic compositions, and existing imaging and histological data available, (2) illustrate the emergence of digital repositories for data sharing retrieved from skeletal collections reference banks, and (3) encourage national and international collaboration with researchers interested in human skeletal biology, biological anthropology, and related fields (e.g., anatomical sciences, archaeology).

11.2 Skeletal Collections for Histological and Imaging Research

11.2.1 The Andronowski Skeletal Collection for Histological Research

11.2.1.1 Establishment and Ethical Considerations

Dr. Janna M. Andronowski is a researcher, clinical anatomy educator, forensic anthropologist, and bone imaging expert in the field of skeletal biology. Her research focuses on the high-resolution 3D imaging of bone microstructure and the related study of bone adaptation, aging, and quality and fragility associated with substance use. To encourage collaboration, digital data sharing, and access to skeletal material specifically for histological and imaging research in North America, Dr. Andronowski founded the Andronowski Skeletal Collection for Histological Research (ASCHR).

Initiated in 2017, this unique modern skeletal archive provides an ethically sourced resource for studying skeletal microarchitectural variability with advancing age and between the sexes. Protocols for tissue procurement and anatomical specimen requests were prepared for various institutions/organizations according to research protocols and reviewed and approved by Medical Advisory Boards and/or Ethics Advisory Panels. Professional relationships with researchers, anatomists, organ donation nonprofit operations, and US and Canadian Medical Examiners' offices subsequently allowed for routine collection of cadaveric bone specimens in accordance with current medical research guidelines. All samples are collected with strict oversight and informed consent provided by the patients themselves or their next of kin. For organ donors, advanced procurement coordinators, responsible for securing organ and tissue collection authorizations (e.g., heart valves, bone, corneas), administer questionnaires to obtain clinically relevant data (e.g., substance use history, past medical history, infection history, genetic defects, etc.) in addition to serology, urinalysis, and toxicology reports, culminating in hundreds of pages of documentation. The study of the skeletal material was ethically cleared by The University of Akron Institutional Review Board for the Protection of Human Subjects and the Newfoundland and Labrador Health Research Ethics Board (Protocol Reference #2020.308) (Andronowski & Taylor, 2022).

The collection currently boasts skeletal elements from over **700 individuals**, with bone procurement primarily focused on two sites used in histological age-at-death estimation in biological anthropology: the midshaft sixth rib and femoral midshaft. Specimens in the collection have been procured from donors ranging from 15 to 105 years of age-at-death, with a roughly equal sex distribution. Demographic information, including age, sex, ancestry, and primary and secondary cause of death, is known for each individual. Only 4% of the collection is missing demographic information in these categories. The primary objective of this collection is to (1) provide researchers with access to an invaluable extensive collection of physical samples for potentially destructive analyses, and (2) elucidate data repositories consisting of all sample data collected from the collection.

In terms of physical samples, ASCHR is the most extensive skeletal collection of its kind in Canada – and now the world – recently surpassing the Melbourne Femur Research Collection in Australia. The collection is securely housed in Dr. Andronowski's laboratory at Memorial University of Newfoundland (MUN) and is available to domestic and international researchers via research requests. Routine histologic thin sections and digital data sets have further been prepared for use, with permission. Research collaborations and associated tissue loans are ongoing in Canada and abroad.

11.2.1.2 Demographics and Specimen Acquisition and Preparation

The ASCHR represents the largest modern, well-documented, and fastest-growing collection developed specifically for bone histological and imaging research. At the time of this writing, the assemblage contains over 1,382 skeletal elements from 739 donors (Figure 11.1). It consists of 451 left femora, 525 left sixth ribs, 406 cranial samples (including frontal, parietal, temporal, and occipital), and additional hard tissue specimens (e.g., metacarpals, mandibles, maxillae, teeth, and cremains) (Figure 11.2). Tissues from fresh and embalmed donors are included. All samples are labeled with an anonymized donor identification number, photographed, and routinely maintained to ensure integrity, longevity, and maximal utilization.

All skeletal elements are procured following a stringent protocol to ensure the consistent removal of tissue samples from specific skeletal regions. Femoral and rib samples are collected by first dissecting soft tissues (e.g., musculature, neurovasculature) from the anterior midshaft and employing a bone saw (e.g., oscillating or reciprocating) or rib shears to extract approximately 3-inch bone blocks. Cranial samples are extracted via an oscillating

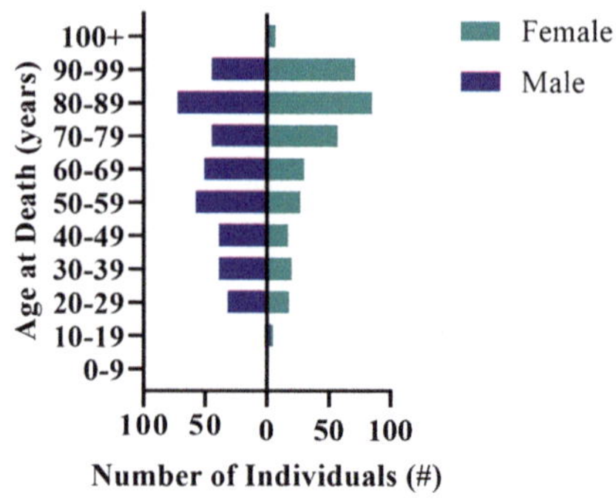

Figure 11.1 Sex and age distribution of individuals in the ASCHR collection.

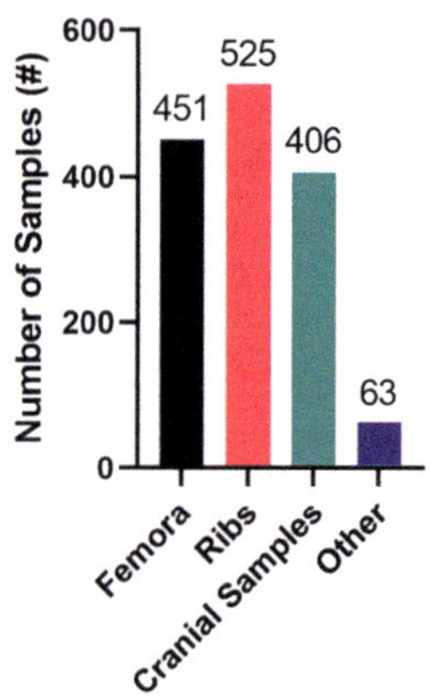

Figure 11.2 The number of samples per skeletal element in ASCHR.

saw to produce 2x2 cm sections. The samples are placed in individual collection tubes/specimen containers labeled with the donor identification number, sex, age, ancestry, skeletal element, and cause of death. Specimens are securely wrapped in sterile saline-soaked gauze and safely stored in a sample container in a –20°C freezer upon return to the lab.

Following an established protocol for histological specimen preparation (Crowder, 2012), embalmed cadaveric specimens are processed under low heat in a Tergazyme© (Sigma-Aldrich), and water mixture. Periosteal soft tissues are subsequently removed, and marrow cavities are flushed with a water flosser and/or via an ultrasonicator. Samples are submerged in 70% ethanol for 24 hours for fixation and air-dried for an additional 24 hours.

Specimens are labeled with an anonymized identification number using fine-tip archival ink pens, fixed with clear nail polish, and left to air dry for 24 hours. All specimens are photographed pre- and post-processing and from all angles on a black photo tarp with a scale bar using a Nikon D5600 DSLR camera with a standard 18–55 mm lens. Each sample is placed in an individually labeled bag (with desiccant packets) and stored in a designated museum-grade cardboard box lined with adhesive neoprene or polyurethane foam. Alternative cardboard boxes are used for adequate storage depending on the number of samples and their size. The entire collection is checked monthly for any irregularities (cracks, mold, and/or the accumulation of grease). Further avenues (e.g., reagents) are being experimentally tested to sufficiently and effectively address grease removal.

11.2.1.3 Digital Data

Most ASCHR specimens have associated histological sections and/or bone cores, as well as imaging data. Current availability includes 597 one-millimeter-thick sections and 511 five-millimeter-thick femoral and rib sections for research loans. Such specimens are especially beneficial to researchers who may not have the infrastructure and/or consumables to prepare bone histological sections in-house.

Over 700 ASCHR specimens have been visualized and quantitatively analyzed using various 3D imaging modalities, resulting in over 55 terabytes of data, including Synchrotron Radiation micro-Computed Tomography (SRμCT), laboratory micro-Computed Tomography (μCT), confocal microscopy (brightfield, polarized light, and fluorescence), and X-Ray Photoelectron Spectroscopy (XPS) (Figure 11.3). The collection includes 642 total SRμCT datasets (femora – 443, ribs – 199) imaged at the Canadian Light Source (CLS) synchrotron on the BioMedical Imaging and Therapy (BMIT) beamlines. Specific samples were further imaged at various resolutions using a Skyscan 1172 laboratory μCT (Bruker) and Skycan 1076 *in vivo* laboratory μCT. A Kratos AXIS Supra XPS system has been further employed to evaluate the surface chemistry of various ASCHR specimens. The imaging technology depends on the targeted region of interest and the resolution required to distinguish the specimens' desired microstructural features. All data are stored on internal SATA hard drives with zipped backups in a secure Google Drive as a backup.

11.2.1.4 Synchrotron Radiation Micro-Computed Tomography

Synchrotron Radiation micro-Computed Tomography (SRμCT) uses brilliant breaking radiation (or bremsstrahlung) produced by a synchrotron facility to procure high-resolution 3D data sets and subsequent visualizations (Figure 11.3A). All ASCHR specimens thus far have been imaged at the Canadian Light Source (CLS) national synchrotron facility in Saskatchewan, Canada. The Andronowski Lab is pioneering the use of SRμCT to address novel forensic anthropological queries. Recent projects utilizing SRμCT technology include (1) identifying the effects of opioid abuse on cortical bone microstructure in 2D and 3D (Andronowski, Davis, & Cole, 2020), (2) analyzing bone microarchitectural parameters to address differences in nuclear DNA yield rates among bone tissue types (Andronowski, Mundorff, et al., 2017), (3) assessing characteristics of osteon banding in human bone for determining species origin (Andronowski, Pratt, et al., 2017), (4) isolating sex- and age-related changes in cortical bone of the human anterior femur (Andronowski et al., 2023). A significant selection of ASCHR cranial specimens (e.g., frontal, parietal, temporal, occipital) are further being deployed to analyze osteocyte lacunar and vascular pore parameters associated with intramembranous bone development.

The Andronowski Lab routinely uses the BioMedical Imaging and Therapy (BMIT) Insertion Device (ID) and Bending Magnet (BM) beamlines at CLS. In total, 642 ASCHR and 66 ZSCHR (nonhuman skeletal collection; see ZSCHR description below) specimens have been scanned and reconstructed. Our experimental setup employs the BM beamline, which is equipped with a white beam microscope, 5x objective, photon energy of 20 keV, 5 cm sample-detector distance, and 1.5 μm resolution. Three thousand raw tomographic images were captured for each specimen at a step angle of 0.06° for a total of 180° rotation. The majority of raw image files were reconstructed with *tofu ez* (Faragó et al., 2022), with certain data sets reconstructed using NRecon (Bruker). Image visualization and analysis routinely occurs via CTAn (Bruker), ORS Dragonfly (http://www.theobjects.com/dragonfly/ (accessed on 1 May 2023)), and Amira-Avizo (FEI, www.thermofisher.com/ (accessed on 1 May 2023)).

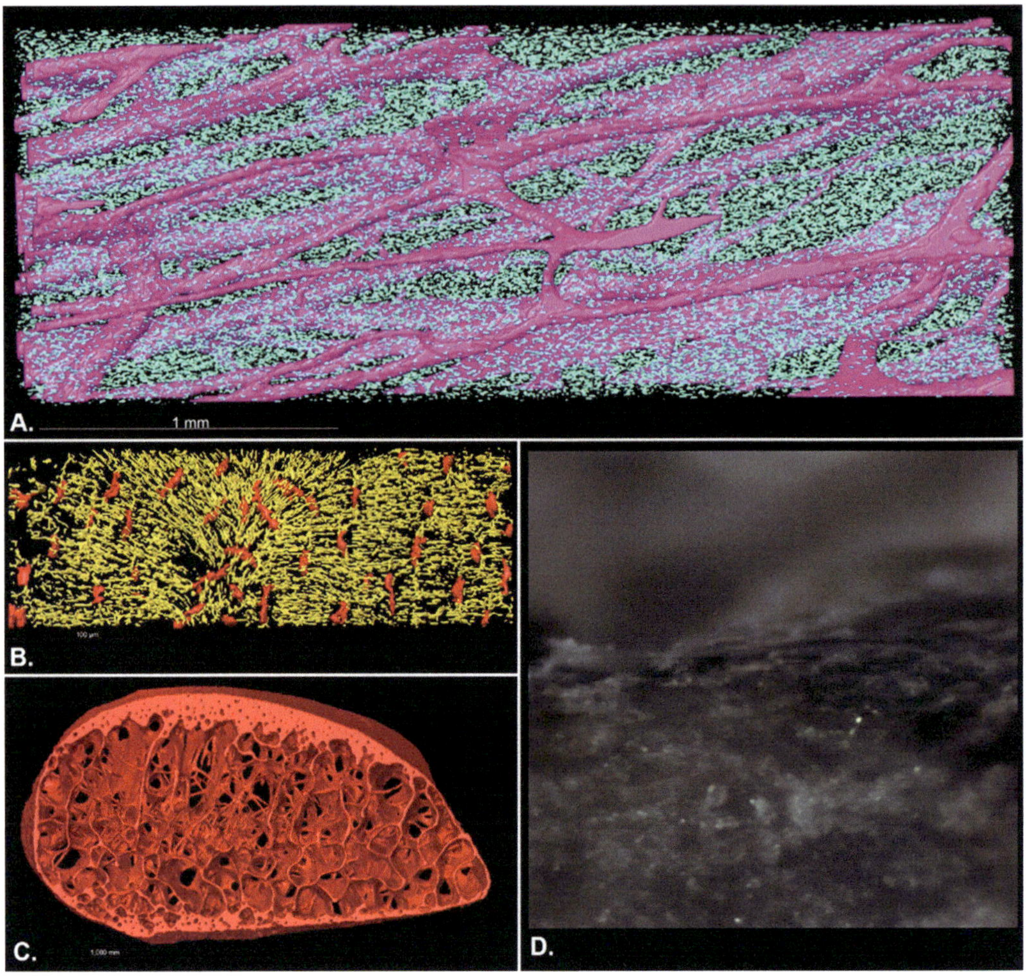

Figure 11.3 Examples of digital data present in ASCHR's data repository. A 3D render of an SRµCT dataset (**A**) depicting the vascular pores (magenta) and osteocyte lacunae (cyan). Scale bar = 1 mm. Confocal Laser Scanning Microscopy (CLSM) 3D render from a human midshaft femur (**B**) depicting the lacunae (red) and canaliculi (yellow) that compose the lacuno-canalicular network (LCN). Scale bar = 100 µm. A 3D laboratory µCT reconstruction (**C**) of a human midshaft sixth rib including cortical and cancellous bone. (**D**) An X-ray photoelectron spectroscopy image showing the surface of cortical bone.

11.2.1.5 Confocal Laser Scanning Microscopy

Confocal laser scanning microscopy (CLSM) employs a scanning laser that targets and excites fluorophores within the sample. Only focused light is received as the pinhole blocks out unfocused light. CLSM offers a significantly higher resolution than SRµCT, allowing for the visualization of the canaliculi that contribute to the lacuno-canalicular network (Figure 11.3B). The Andronowski Lab has employed a Leica TCS SPE CLSM with a motorized Z-Galvo stage (Leica Microsystems, Wetzlar, Germany) to visualize the LCN in the anterior intracortical femoral diaphysis of 40 individuals (Andronowski et al., 2023). This specific instrument had a 63x oil immersion lens with a laser wavelength of 488 nm and a 32.5% intensity with a 485 – 585 nm spectral window. ASCHR specimens were included in a recent manuscript for *The Anatomical Record* (Andronowski et al., 2023).

11.2.1.6 Laboratory Micro-Computed Tomography

Laboratory micro-Computed Tomography (μCT) is more commercially available than SRμCT, however, it lacks the resolution and the efficiency to achieve higher resolutions (e.g., 1.5 μm). With the capability for a larger field of view, visualization of a larger sample is possible. Thus, cross-sectional and macroscopic parameters can be tested (e.g., tissue volume, canal morphology). SkyScan 1172 (Bruker) and 1076 (Bruker) images have been used to image 45 ASCHR (Figure 11.3C) and 47 ZSCHR specimens ranging from 5.49 to 18 μm. Laboratory μCT has allowed us to address several anthropological and bone biological questions, including (1) sex- and age-related changes in bone microarchitecture of the left sixth midshaft rib (Hicks et al., 2019), (2) characteristics and progression of hyperostosis frontalis interna (White, 2019), (3) microstructural comparisons of human and black bear metacarpal and metatarsals for human vs. nonhuman bone identification (Andronowski, Davis, et al., 2019; Stephen, 2018), and (4) characterization of micromorphology of the harp seal baculum and humerus (Hande, 2023).

11.2.1.7 Significance

Skeletal collections offer unique insight into bone biology, pharmacology, bone cellular biology, medicine, and forensic anthropology. Currently, most well-documented collections are not of modern origin (e.g., Hamann-Todd), reflecting demographics of the past and without consideration of secular change.

The ASCHR thus provides a unique opportunity to continue the expansion of an ethically sourced and modern well-documented osteological collection in North America. The collection has grown substantially in only six short years, allowing a unique perspective to investigate factors affecting a modern and diverse population. Since its inception, it has become the most extensive and fastest growing, amassing approximately 123 samples annually.

The Andronowski Lab has studied ASCHR specimens in collaboration with colleagues of the (1) Memorial University of Newfoundland Department of Biology; (2) Department of Fisheries and Oceans Canada; (3) Departments of Anatomy, Physiology, and Pharmacology and Chemistry, the University of Saskatchewan; (4) Departments of Surgery and Mechanical Engineering, the University of Alberta; (5) Department of Anthropology, the University of Tennessee, Knoxville; (6) Forensic Anthropology Division at the New York City Office of Chief Medical Examiner (NYC-OCME); (7) Divisions of Physical Anthropology and Vertebrate Mammals, Cleveland Museum of Natural History; (8) Lifebanc; and (9) the Cuyahoga County Medical Examiner's Office locally, nationally, and internationally. Additionally, there are local collaborations within the Division of BioMedical Sciences, Faculty of Medicine, Memorial University of Newfoundland, and the Office of the Chief Medical Examiner in St. John's, NL.

The ASCHR has contributed to various theses, dissertations, and manuscripts (as described above) (Andronowski, Davis, et al., 2019; Andronowski, Davis, & Holyoke, 2020; Andronowski et al., 2023; Andronowski, Mundorff, et al., 2019) and offers promise to be expansive, leading to novel and exciting discoveries. ASCHR and its partner ZSCHR have further been implemented into additional published research and other projects that are currently under development:

- Microarchitectural differences between human and bear metapodials for forensic identification using laboratory μCT (Andronowski, Davis, et al., 2019; Stephen, 2018).

- An accurate and precise cortical bone core sampling methodology for synchrotron μCT-based imaging (Andronowski, Davis, & Holyoke, 2020).
- The influence of age and sex on bone microarchitecture utilizing confocal laser scanning microscopy and Synchrotron radiation-based micro-Computed Tomography (Andronowski et al., 2023).
- Investigating the impacts of prolonged opioid exposure on 2D microstructural parameters (e.g., osteon population density, cortical area) and 3D measures (e.g., intracortical porosity, osteocyte lacunar density) (Andronowski, Davis, & Cole, 2020; Andronowski et al., 2021, 2022).
- Development of deep learning models to accurately and precisely segment μCT imaging data (Hassan et al., 2023).
- Development of a novel method for extraction of degraded fats from bone in curated skeletal assemblages.
- Analyzing variation and patterns of vascular porosity among human cranial elements (e.g., frontal, parietal, temporal, occipital).
- SRμCT and μCT imaging of harp seal bacula to deduce biomechanical parameters related to bone strength over the lifespan (Hande, 2023).

11.2.2 Zoological Skeletal Collection for Histological Research

Alongside ASCHR, the Zoological Skeletal Collection for Histological Research (ZSCHR) was founded in 2021. Newfoundland and Labrador contain unique zoological species exclusive to this most easterly region of North America. Andronowski is leading new research initiatives to include investigating the microarchitecture of various animal species with high-resolution imaging modalities. A current project, for example, works to infer the species origin of various Newfoundland mammals through virtual histology to aid forensic identification efforts in the province.

Though in its infancy, the collection currently possesses 88 different skeletal specimens from eight species, including harp seals (*Pagophilus groenlandicus*), moose (*Alces alces*), northern gannets (*Morus bassanus*), black bear (*Ursus americanus*), rat (*Rattus rattus*), white-tailed deer (*Odocoileus virginianus*), lynx (*Lynx canadensis*), and eastern coyote (*Canis latrans*). The collection is poised to expand further as Andronowski continues forging fruitful professional collaborations with local ethical hunters, museums, and researchers. Like ASCHR, the collection has associated high-resolution SRμCT imaging data for 67 specimens and μCT data for 36 individuals.

11.2.2.1 Collections Management and Access

The ASCHR and ZSCHR collections are securely stored in a temperature-controlled laboratory within the Faculty of Medicine at Memorial University of Newfoundland, St. John's, Newfoundland and Labrador. The collections are open for researchers in anthropology, biomedicine, biology, and other fields. Interested parties should contact Dr. Janna M. Andronowski and complete a Research Request Form on the Andronowski Lab website (https://www.andronowskilab.com/andronowski-skeletal-collection; accessed 18 April 2023). Scientific merit and the proposed research impacts will be heavily considered with each request.

Submission of additional supporting materials, including research grant proposals, ethical approval documents, equipment training documentation, and other relevant documentation (e.g., experience with experimental protocol) are encouraged. Student researchers interested in the collection must submit a signed letter from their primary supervisor in addition to the research request form. No fees are associated with using ASCHR materials, but shipping and handling costs may apply.

Researchers may document their research endeavors using various imaging modalities but must provide copies of all images and associated data to the Andronowski Lab. It is required that ASCHR and Dr. Andronowski are acknowledged in any publications and presentations resulting from the use of the collection.

11.2.2.2 Conclusions

The Andronowski Skeletal Collection for Histological Research is a novel osteological collection offering an avenue for research specifically related to histological and imaging research. The collection provides unique opportunities to conceive novel regression formulae for age-at-death estimation and address questions concerning age-related microarchitectural changes and biomechanical processes. It serves as an open skeletal reference database for researchers from various disciplines.

11.2.3 Melbourne Femur Collection

11.2.3.1 Introduction

The Melbourne Femur Research Collection was established a little over three decades ago. The collection was founded by Professor John Clement (Figure 11.4), who was the consultant forensic odontologist in Victoria, Australia, at the time, together with a small scientific

Figure 11.4 Professor John Clement inspecting a manually tiled cross-section microradiograph of the femoral midshaft cortex.

group based at the School of Dental Science at the University of Melbourne. Its original purpose was to answer questions about how bone changes could be used to estimate the age-at-death of unknown individuals: one of the important facts required to help establish an identity. Professor Clement was a keen scientist, whose research interests spanned mineralized tissues and how their characteristics could answer related scientific and medico-legal questions. It was Professor Clement's ultimate vision to nurture the collection into a world-renowned resource for the study of age changes in bone. He recognized the value and necessity of building a well-provenanced and documented collection to support excellent research. The collection now is a vast repository, not only of specimens, but also of new knowledge and data, with a large digital component. The collection continues to support important research throughout the world.

None of this work would have been possible without the generosity and kindness of the next-of-kin of those whose bone samples and imaging data make up the Melbourne Femur Research Collection, and without the authorities who gave their time and expertise to ensure proper ethical oversight and management of the collection.

11.2.3.2 Establishing the Collection

The Melbourne Femur Research Collection began as an attempt to determine age-at-death from histomorphometric analysis of sections of the femoral midshaft cortex. The original group of investigators set out to gather a sample of femoral midshaft cortical bone from each year of the human lifespan. These bone samples were collected at the Victorian Institute of Forensic Medicine in Melbourne, Australia, and formed the kernel of the collection.

An important aspect of forensic odontology is establishing the identity of unknown human remains. Odontologically, this can be done by comparing postmortem dentition with antemortem dental records for example. Forensic odontology can also support the establishment of identity and assist forensic anthropologists with constructing the biological profile through the evaluation of dental development, eruption, and morphology to estimate biological parameters: sex, population affinity, and age. Age-at-death can be established with reasonable accuracy in people up to the time that their teeth have finished forming. In adults, however, an estimation of age-at-death from skeletal elements is much less accurate, varying by up to ten years either side of the true age-at-death. It was this problem that the group establishing the Melbourne Femur Research Collection was trying to solve: to improve the histological aging methods for forensic cases. Anthropological literature available at the time (Amprino, 1963; Kerley, 1965; Singh & Gunberg, 1970) suggested that remodeling of cortical bone over the lifespan would provide a measurable pattern that could be used to estimate the age-at-death of an individual: with an increase in spatial density of osteons over time as bone is remodeled, and an increase in the number of osteon fragments, caused by establishment of new cutting cones through the preexisting osteons, as an individual ages.

The team in Melbourne sought to improve the accuracy of age-at-death estimation using the histomorphometry of human femoral cortical bone by studying whole cross-sections of femoral midshaft cortical bone, and developed an automated system to record and measure features of interest from microradiographs made from cortical bone cross-sections. As the research project moved forward, it became increasingly apparent that there was a poor relationship between osteon density (as measured from microradiographs) and age-at-death (Bertelsen et al., 1995). By this stage, the group had gathered a large amount of data about osteon distribution and other variables in femoral midshaft cortical bone.

As with many research endeavors, the team had generated many questions during the course of the research. In this case, the questions were about the processes of bone tissue adaptation and aging. New projects were developed to study the remodeling process and the mechanisms of age-related changes in bone tissue.

The driving force at this time (in the mid-1990s) was the threat to population health of age-related diseases such as osteoporosis. At the time, the cost to the Australian community of osteoporotic hip fractures was AU$400 million per annum. More recently (2019), it has been estimated that Australia spends AU$960 million on hip fracture care (Tatangelo et al., 2019). The direct cost of osteoporosis annually in Australia is estimated at US$2.6 billion (Tarrrant et al., 2020). These figures are of course replicated in many other parts of the world and increase with population growth and population aging. Age-related fractures remain an important threat to the health of the population, and this is still the main purpose for the MFRC: to support good research into age-related changes in bone, leading to prevention of and treatments for health-impacting changes that come with advanced age.

11.2.3.3 Composition/Demographics

Accurate and relevant research data are created from pristine and sensibly sourced collections which include comprehensive information about the samples. The Melbourne Femur Research Collection consists of bone samples, imaging data, and demographic/biometric data from individuals who died suddenly with no known mobility constraints and no known diseases or conditions that affect bone metabolism. The information collected at the time of bone sample collection includes age at time of death, cause of death, biological sex, height, and weight. Some were collected together with a blood sample or with a health and medical questionnaire completed by the next-of-kin. The bone samples were collected with informed consent and ethics approval from individuals undergoing medicolegal autopsy at the Victorian Institute of Forensic Medicine, with informed consent from the next-of-kin and ethics approval from the VIFM.

The MFRC has grown in size and scope since its early days in the early 1990s, adding new data and samples in novel formats as technological advances were adopted. The composition of the collection reflects the changing research focus (from forensic age-determination from microscopic features in femoral cortical bone in the early 1990s to the study of age-related changes in bone tissue since the mid to late 1990s) as well as advancements in data capturing technology. Many of the later samples are digital representations of the lower limb, so the composition of the collection is diverse in a number of aspects. Diversity is found in the sample population demographics in terms of age distribution and representation of biological sex, but also in sample formats in terms of physical specimens, scans, medical/lifestyle questionnaires, and a number of other imaging modalities.

Currently, the collection represents data from 904 individuals (439 biologically male, 436 biologically female, with the remainder of unknown biological sex), with a number of different sample types and a range of data and associated materials including microradiographs, pQCT imaging, clinical CT, pelvic radiographs, medical and lifestyle questionnaires and hard ground sections (Table 11.1). Some specimens are preserved in fluid at room temperature; others are deep frozen at −70°C. Still other samples are thin sections (hard ground, undecalcified). The collection represents a wide range of ages in the human lifespan, with the original intention in setting up the collection to study at least one sample from every year of the human lifespan.

Table 11.1 Current composition of the Melbourne Femur Research Collection

Specimen type	2 cm midshaft block		Osteoarthritic femoral head		Proximal femur		Short stack CT		Total
Total (n)	230 (1 u/k sex)		122 (14 u/k sex)		166 (1 u/k sex)		385 (18 u/k sex)		903 (34 u/k)
Total by sex	♀	♂	♀	♂	♀	♂	♀	♂	
Specimen type total by sex	**109**	**120**	**66**	**42**	**69**	**96**	**204**	**163**	♀: 436 ♂: 439
			Additional material/documentation/digital data						
Blood samples	15	9	0	0	49	81	0	0	154
Microradiographs	103	113	0	0	9	0	0	0	225
Hard ground section	83	90	0	0	9	0	0	0	182
2-inch glass plates	32	37	2	5	4	1	0	0	81
Mounted specimen	15	11	0	0	4	7	0	0	37
Plane radiograph, pelvis	0	0	0	0	47	70	0	0	117
Clinical CT scan	0	0	0	0	16	18	0	0	34
pQCT scan	0	0	0	0	35	11	0	0	46
Micro-CT scan	21	17	0	0	17	18	0	0	73
Autopsy report	109	120	0	0	69	93	0	0	396

11.2.3.4 Research To-Date

The earliest studies of the initial cross-section specimens in the collection using the semi-automated tiling and thresholding mechanism attempted to determine age-at-death using Haversian canal number (Bertelsen et al., 1995), which proved to be unsuccessful due to the variation in intracortical porosity. Researchers then set about studying the age-related changes in cortical porosity, including a comparison between bone samples from both biological sexes (Feik et al., 1996, 1997). This showed an age-associated increase in subperiosteal area, polar moment of inertia, cortical area, and medullary area, with women showing a steeper increase in these variables around the age range of menopause compared with men across the same age range. For example, total subperiosteal porosity increased greatly in women by their 50s: in men this level of porosity was not reached until they were in their 80s. A later study combined morphological data with porosity data, again to attempt to determine age-at-death. Total subperiosteal area, peri- and endo-steal perimeter, cortical bone area, moments of area, and histological data such as the number and size of intracortical pores were analyzed with respect to age. The predicted age-at-death in this study was found to be inaccurate by over eight years in more than half the individuals (Thomas et al., 2000). When the early results of this study were emerging, the researchers concluded that regional variation in the cortical porosity might have affected the results and that a measure of periosteal shape at the femoral midshaft might aid age-at-death determination. The available technology at the time limited most studies to histomorphometry in the form of digitized microradiographs or light microscopy and image analysis using partially automated thresholding. At this time, the collection supported research into:

- Age changes in porosity and mineralization throughout the midshaft cortex and their effect on bone mechanical strength,
- Bone size and mechanics at the femoral midshaft (Stein et al., 1998),
- Preferred collagen fiber orientation in the femoral midshaft cortex (H. Goldman et al., 1999),
- Regional variation of cortical bone features (Feik et al., 2000; H. M. Goldman et al., 2003), and
- Remodeling clusters (super-osteons) (Bell et al., 2001; Loveridge et al., 2001).

With the addition to the collection of larger portions of the femur (proximal 5/8th), it was now also possible to study femoral neck cortical bone morphology and age-related changes at that location (Mayhew et al., 2004). A three-dimensional approach was added once this technology became available and accessible through synchrotron facilities throughout the world, and also desktop micro-computed tomography units.

With the development of higher-resolution 3D imaging (which is also nondestructive), attention turned to:

- The three-dimensional microstructure of the Haversian bone in the midshaft cortex, including how the canals are connected, osteocyte lacunar density, and how these variables change with age throughout the human lifespan, and differ between the sexes (Cooper et al., 2007),
- Visualizing the three-dimensional morphology of osteons (Arhatari et al., 2011; Cooper et al., 2011), Haversian systems (Maggiano et al., 2016), and osteocyte lacunar density (Carter et al., 2013), and
- Comparisons of imaging modalities measuring bone features (Hennig et al., 2015).

All these studies were designed to understand the remodeling process of human bone, and how this might be manipulated to improve or prevent conditions, such as age-related osteoporosis. Other related research has included investigations into the mineral phase of bone (Dicken et al., 2015) and gross morphological changes in the femoral cortex (Zhang et al., 2016).

Currently, research continues to reuse digital data already created to explore the relationship between sex and proximal femur volume, cross-sectional morphometry and muscle volume and comparisons in histomorphometry of modern samples from the MFRC and archaeological samples, as well as projects investigating the age changes in mineral accrual and remodeling.

11.2.3.5 Significance of the Collection

The MFRC remains one of the world's best documented and curated collections of contemporary human bone samples. The establishment, growth, and successful management of this collection relied on the early establishment of a close and trusting relationship between the Melbourne Dental School, Victorian Institute of Forensic Medicine and the next-of-kin of each individual from whom a sample of bone was removed for research. Many skeletal collections are historical or archaeological in nature, and consist of desiccated bones or bone samples which have been subjected to the elements and local environmental conditions since death. There is often little information about living conditions and disease status, e.g., metabolic diseases having an effect on the skeletal system, which may be detectable in bone, but are not able to be confirmed. The MFRC has only collected bone from individuals who had no known diseases or conditions affecting mineralized tissue, with one exception: the subset of femoral heads removed as part of hip replacement surgery for osteoarthritis. These also form an important sub-collection to study the changes in bone tissue as a result of osteoarthritis.

11.2.3.6 Ethics and Digital Data Management/Security

All projects involving the MFRC must now apply for ethics approval at the University of Melbourne (if researchers are staff or students at the University). Alternatively, local ethics approval would be appropriate, as long as the local institution at which the research is being conducted has ethical approval standards equivalent to the University of Melbourne. All research is conducted with adherence to the National Statement on Ethical Conduct in Human Research (NHMRC, 2018).

Throughout the life of the MFRC, there has been one change to the body responsible for ethical oversight of the collection. Originally, ethical approval for research into "Age Related Changes in the Skeleton" was granted by the Victorian Institute of Forensic Medicine (VIFM). This was transferred to the University of Melbourne in 2011 and continues to be administered by the university.

The issues arising with analysis of digital data add another layer of requirements for researchers and the custodians of the MFRC. While transferring and sharing of digital data to be analyzed for research projects has become easier (providing a download link, or remote access to database), this means that security issues become more prominent: what happens to the data that has been downloaded remotely once the research project is complete? How will the security of the data be maintained at the remote location? These issues are dealt with as best as we can through the ethics approval process, and through agreements with researchers when applying for access to the MFRC. This means that original data files downloaded from the MFRC are destroyed upon completion of research projects. Collections management, digital and data security are increasingly important considerations for anyone maintaining a research collection or biobank.

In any research collection that is to benefit society over a long period, care must be taken to ensure that consistent custodianship can be carried out beyond the career-span of an individual. In its early days, the collection was managed by the group of researchers originally involved in collecting the samples. As the collaborations grew, and the life of the collection spanned more than a single career-generation, a committee of management was established. This committee includes members involved in medical research, basic research, and experience in human research ethics and collections management. The committee mainly oversees and assesses applications to access the collection for research.

Access has become easier (depending on the type of samples required) since the storage of the digital data component of the collection was transferred to dedicated server storage at the University of Melbourne. All digital data, as well as a master catalogue, are stored on a Distributed and Reflective Informatics system (DaRIS) database, developed at the University of Melbourne. Storage is currently in the range of 5TB, and is managed under the University of Melbourne's Petascale Campus initiative. This biomedical imaging data management application is primarily used for the storage, processing, and curation of data from imaging scanners like MRI and PET/CT (source: University of Melbourne Research Computing Services).

The MFRC's management committee aims to adhere to the FAIR principles (Barker et al., 2022) (originally developed for research software). We strive to make the MFRC Findable, Accessible, Interoperable (in terms of digital assets in the collection) and Reusable (where appropriate and possible).

To improve the findability of the collection, it has webpages on the Melbourne Dental School's website (https://dental.unimelb.edu.au/research/melbourne-femur-research-collection), an entry in the Melbourne Academic Centre for Health's Biobank registry (mach-australia.org) and an entry in the Research Data Australia registry (https://researchdata.edu.au/melbourne-femur-collection/671966). To improve the reusability of the collection, an entry was created in the University of Melbourne's FigShare (figshare.com) account, to allow citation of the collection when it is used in research.

All applications for access require the applicant to produce a short proposal outlining the research. This proposal is then used to determine if the proposed research conforms to the purpose of the collection, is feasible to carry out, and is original and has not already been addressed in some way by previous research on the collection. Any costs incurred by transport (to and from the location of the research) must be borne by the applicant, and all ethical and other regulations for the storage and research on human tissue samples must be adhered to. Special attention is paid to research proposals that intend to carry out destructive testing on the samples. If there is a way to avoid this and still answer the research questions in the proposal, this alternative is presented to the applicant group.

Any physical specimens remaining (for example, after 3D volumetric scanning) are returned to the MFRC. Results data is requested to be returned to the collection to enable future interpretation or reuse of data to increase the sustainability of the MFRC. In this way, the MFRC will continue to help answer research questions about the aging of human bone for a long time to come.

11.2.3.7 Conclusions

The establishment, growth, and management through the decades have shown this collection to be a great example of how to manage an active research collection involving contemporary human tissue and digital assets.

The Melbourne Femur Research Collection has grown in size and diversity over the last 30 years, with the addition of digital assets mirroring the development of new digital technologies and data storage and management systems.

The philosophy of the management committee is that this collection has no ownership, but a group of custodians who must ensure that the collection and its use follow the FAIR principle: Findable, Accessible, Interdisciplinary [Interoperable in the FAIR Principle for Research Software guidelines] and Reusable wherever possible, and that any research is carried out to high ethical standards.

11.2.4 National Museum of Health and Medicine Collections

11.2.4.1 Bone Histology Collections at the NMHM

The National Museum of Health and Medicine's (NMHM) Anatomical Division curates several large and unique collections of histological slides of skeletal tissues containing more than 10,000 slides of both stained and undecalcified bone and joint specimens ranging in size from standard 1x3" to 5x7" whole-mount slides. The collections span the subjects of human growth and development, normal anatomy, bone and joint disease, tumor pathology, and comparative anatomy.

The collections began as the research archives of the Orthopedic Pathology Department of the now-defunct Armed Forces Institute of Pathology (AFIP). Just after World War II, the department was developed and led by Lent Johnson, an indefatigable and influential pathologist, researcher, educator, and international collaborator in the field of bone biology. Johnson was particularly influential in the field of diagnostic classification of bone tumors (Johnson, 1953; Spatola et al., 2012).

Under the auspices of the AFIP Registry of Bone Pathology, Johnson selected illustrative and novel cases of bone pathology from among the submissions to the Institute coming from military and civilian hospitals around the world for inclusion in the department's research archive.

In his capacity as head of the Registry Johnson preserved several historically important collections by having them transferred to his department from other organizations when they could no longer be cared for properly by their respective owners. The transfers included the Codman Registry of Bone Sarcoma, a portion of Dr. Dallas Phemister's whole-mount slide collection from the University of Chicago and the Richard H Follis Collection from his work on nutritional pathology at Johns Hopkins and the AFIP. Research material from the work of Dr. Ellis Kerley is also present from a period when he worked for the Veteran's Administration and was assigned to work under Dr. Johnson at the AFIP. The museum also has a massive unprocessed archive of Johnson's papers and case materials.

Overall, the collections are a longitudinal representation of refinements in the classification and understanding of bone tumors, beginning with the Registry of Bone Sarcoma, initiated by Ernest Codman in 1920 (Codman, 1922). Many diagnoses in the Codman collection were made by the pathologists who first described and named tumors during their involvement with the registry (e.g., Ewing's sarcoma, Codman's chondroblastoma, Phemister's chondrosarcoma). When paired with the Johnson-Sweet material, nearly 100 years of development in the understanding of bone tumors is available for study. The rest of the materials exemplify broad areas of research in bone biology and comparative anatomy.

Lent Johnson was also deeply interested in the biology of bone growth and development. Kerley built upon Johnson's prior research to develop applied techniques for histological aging of bone to apply to estimate the age of death in unidentified skeletons. Material from both of their research is present in the collections. Researchers interested in this material should be aware that it was not intended for continued study and remains somewhat disorganized.

11.2.4.2 *Johnson-Sweet Collection*

The Johnson-Sweet Whole-Mount Collection began as the research archives of the Armed Forces Institute of Pathology (AFIP) Division of Orthopedic Pathology. The collection was developed from material drawn from the clinical, collaborative, and educational activities of the department following World War II. It was initiated by the singular influence of Lent Johnson who was Chief of the Division from 1946 to 1980. Dr. Donald Sweet (Chief, 1980–2004) carried on the tradition of Johnson while the latter remained on staff as an advisor and senior investigator until his death in 1998. The Johnson-Sweet Collection is composed of research and diagnosed case material that often includes substantial documentation of case history, radiologic correlations, treatment, and in some cases follow-up. All categories of bone disease are well represented in the collections and include extremely rare conditions. The latter are available since diagnostically difficult tumors and osseous infections from around the world were regularly sent to the department for expert opinion.

The pathological material consists primarily of whole-mount glass slides from the histology laboratory of the AFIP Orthopedic Pathology Department from over 4,500 diagnosed cases of bone disease (Figure 11.5). Multiple cuts from each large tissue block were rendered in hematoxylin and eosin (H&E) and various special stains. Normal and comparative zoological reference material is also well represented in large format.

The most common diagnoses in the Johnson-Sweet Collection, are osteosarcoma (n = 379), chondrosarcoma (n = 301), giant cell tumor (n = 240), osteochondroma (n = 184), carcinoma (n = 146), fibrosarcoma (n = 121), enchondroma (n = 119), osteomyelitis (n = 117), fracture (n = 115), Ewing's sarcoma (n = 86), osteoarthritis (n = 84), Paget's disease (n = 80), fibrous dysplasia (n = 79), chondroma (n = 68), and myeloma (n = 64).

Many specimens have detailed case histories, clinical and specimen radiology images for comparison, micro X-rays, photomicrographs, and archived embedded gross tissue. Pathological material is supplemented by sections of developing human bone and normal adult bone as well as a substantial collection of comparative nonhuman material derived in part from material donated to the AFIP by the nearby Smithsonian National Zoological Park.

The Johnson-Sweet slides are derived from thousands of surgical and autopsy gross specimens and supporting material contributed by outside pathologists seeking consultative opinions from AFIP staff pathologists. Johnson began creating large format slides sometime after 1946 using a celloidin embedding technique originally developed for brain sections. Johnson learned the technique as a student at the University of Chicago from Dr. Dallas Phemister. This original process would take months to obtain a usable section. Soon after his arrival at AFIP, Johnson and the histology staff developed a paraffin embedding technique that made it possible to get a usable specimen within four days (Dowding, 1951; Johnson, 1993).

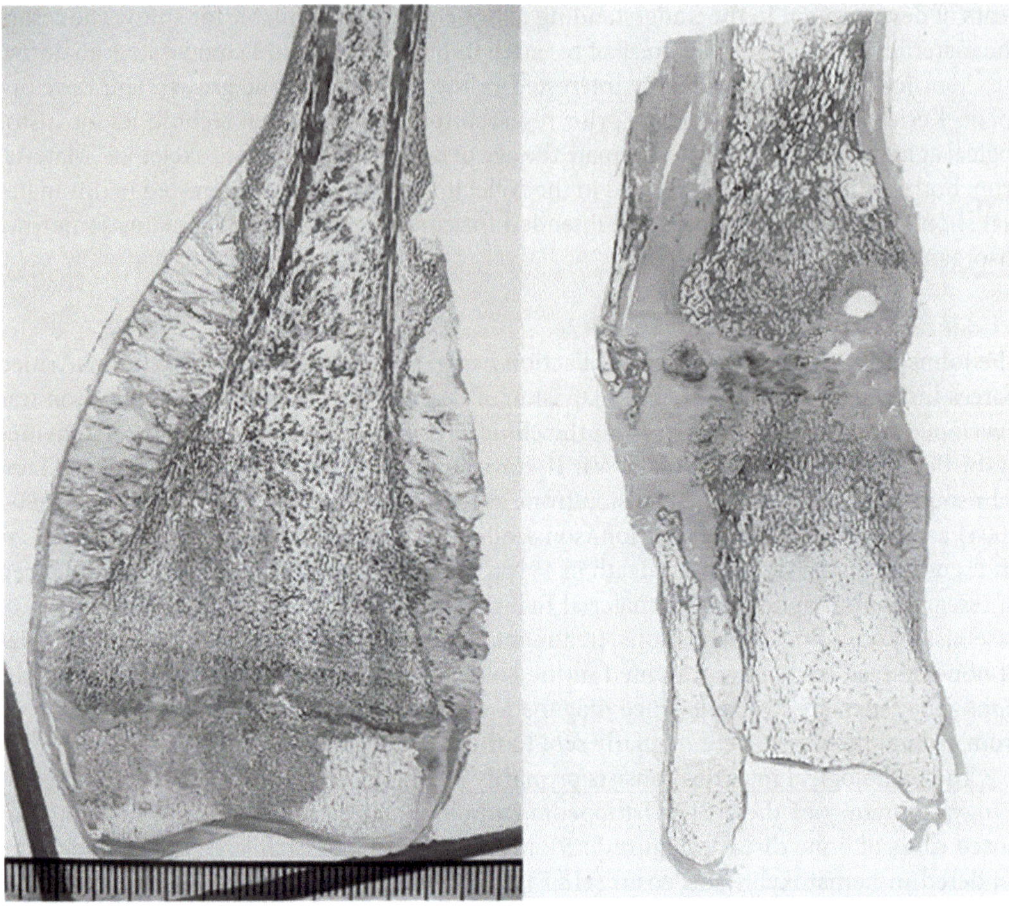

Figure 11.5 Examples from the Johnson-Sweet Collection: Osteosarcoma in a distal femur (Left) and a traumatic injury to the distal tibia / fibula with pseudarthrosis (right).

The primary methods for creating whole-mount slides are outlined in the out-of-print 1968 AFIP Histological Staining Methods Manual (Luna, 1968). Some details are summarized here for historical purposes. H&E stains predominate in the whole-mount collection, though Masson's trichrome and other special stains were commonly used. Undecalcified sections in smaller format were also created, as were hand-ground sections for demonstrating the structure of bone in microradiographs. All processing of large bone specimens was done by hand, whereas automatic processors handled the smaller pieces. A circa 1927 Sartorius microtome manufactured for cutting brains was being used to slice gross specimens of bone and joint as early as 1949 (AFIP, 1949).

For Johnson and the AFIP staff pathologists, context was fundamental to understanding and diagnosing bone disease. Large format slides enabled the systematic study of tumors with a greater appreciation of the histological transition from normal regions to the pathological in an individual specimen, a perspective that could not be obtained with small format sections of biopsied tissue.

Countless collaborations and special studies carried out by Johnson and AFIP staff pathologists led to an accumulation of more than 4,500 cases over 50 years of study and diagnosis primarily between the late 1940s through the 1980s. In addition, sections were made from many of the museum's specimens of traumatic gunshot wounds and infections from the Civil War and World War I.

11.2.4.3 Codman Collection

The Codman Registry of Bone Sarcoma is the oldest tumor registry in the country (Henry, 1964). The registry began in 1920 prompted by hospital standardization pioneer Ernest. A. Codman's (1869–1940) interest in refining diagnostic criteria and classification of bone tumors to improve knowledge of the outcomes of treatment. The following year the project achieved sponsorship by the American College of Surgeons. The material from the Codman collection was the basis for early classification nomenclature for bone tumors (Codman, 1925) and subsequent revisions (Ewing, 1939). At the time, depending on their education, pathologists employed a variety of terms for the same pathological conditions (Codman, 1922), which caused significant confusion and hindered the systematic study of bone tumors. The committee took on the work of standardizing nomenclature and pathobiology of the various entities they encountered.

The registry was a longitudinal study of a large series of cases of bone tumors. The study was carried out between the years 1920 and 1940 and included long-term patient follow-up. The collection consists of over 2,600 bone tumors or suspected tumors submitted by pathologists from across the United States. The slides were shuttled in trunks across the country to be evaluated by a committee of eminent surgeons and pathologists of the day along with all available histological, clinical, radiographic, and other case data. Among the prominent committee members were Joseph Bloodgood (1867–1935), William Coley (1862–1936), James Ewing (1866–1943), and Frank B. Mallory (1862–1941).

The Codman material was donated to the museum by the American College of Surgeons in 1953 as adjunct to the AFIPs orthopedic pathology registry. Digitally scanned case files include the original case files and handwritten diagnoses of participating doctors often with commentary (sometimes extensive) as well as photos of gross tissue and radiographs (Figure 11.6).

The most common diagnoses in the Codman Collection are osteosarcoma (n=1083), giant cell tumor (n=415), Ewing sarcoma (n=273), chondrosarcoma (n=132), myeloma (n=105), chondroma (n=80), carcinoma (n=62), fibrosarcoma (n=60), and lymphosarcoma (n=30).

11.2.4.4 Phemister Collection

Dr. Dallas B. Phemister was the second chair of the Registry of Bone Sarcoma (Codman being the first) and a participant from its inception. He was also a mentor of Johnson at the University of Chicago. The collection includes Phemister's collection of approximately 550 whole-mount orthopedic pathology slides produced from the late 1910s to the early 1930s. Phemister brought the large slide technique to the United States from Europe, where he had learned the technique from pathologist Jakob Erdheim (Johnson, 1993). This collection includes a unique 5 3/8" X 15 2/8" whole-mount slide of osteosarcoma arising from Paget's disease of an entire tibia. The case diagnoses are contained in a ledger and at the time of publication is uncataloged.

11.2.4.5 Kerley Collection

The museum's collection of nonpathological glass slides of decalcified (stained and unstained) and undecalcified human and nonhuman bone tissue consists of Dr. Ellis R. Kerley's research

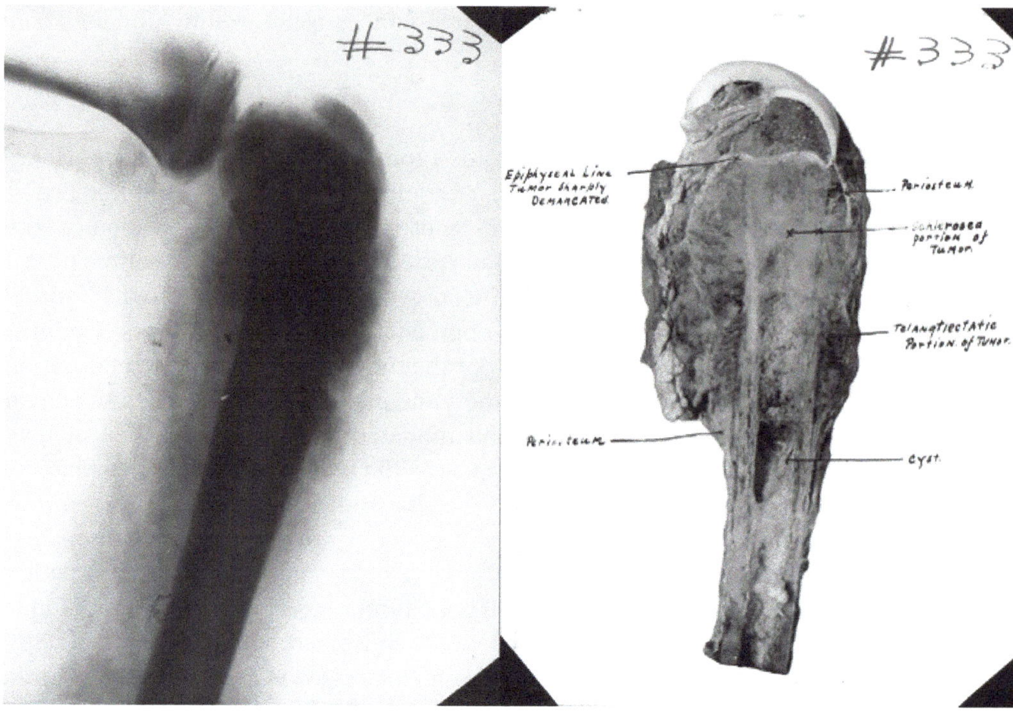

Figure 11.6 Radiograph and labeled soft tissue image of an osteosarcoma of the distal femur. Case 333. Codman Registry of Bone Sarcoma.

on microscopic age changes in bone structure of humans and chimpanzees. This research was conducted during his eight-year tenure with the AFIP. In 1957, Kerley took a position with the Veterans Administration as a physical anthropologist at the AFIP where he worked under the direction of Lent Johnson (Sledzik, 2001). Through his affiliation with the AFIP, and particularly Johnson, Kerley expanded his professional interests to include bone histology and pathology, amassing a collection of whole-mount human and nonhuman bone slides. Today, these slides along with accompanying documentation and original notes are curated by the Anatomical Division of the NMHM.

The Kerley human bone slides collection was created during the late 1950s and early 1960s and is associated with Kerley's AFIP research on age-related changes of bone histomorphology (Kerley, 1965, 1966). With these slides and associated data, Kerley explored normal human variation in four characteristics of bone microstructure: the frequency of osteons, osteon fragments, non-Haversian canals (primary vasculature canals) and the percentage of circumferential lamellar bone. This work resulted in his 1962 dissertation and subsequent publication of his research methods (Kerley, 1965).

The collection includes over 600 bone sections mounted on 1-by-3 inch and 2-by-3 inch glass slides, representing approximately 170 individuals. Slides from 126 of these individuals were used in Kerley's 1965 publication. Specimens originating from AFIP accessioned material are accompanied by antemortem clinical data. A small subsample of individuals lacks detailed antemortem data owing to either an incomplete record or specimens originating from an archaeological context. The basic demographic profile for the specimens that have a recorded age-at-death (n = 136) comprises individuals from 0 to 95 years of age, and a mean and standard deviation of 41 ± 24 years. Both males (n = 91) and females (n = 37) are represented. Due to their age, many slides show evidence of air infiltration in the coverslip medium making portions of the sample opaque. Limited in-house conservation may be available to make these slides available for research. The slides comprise midshaft cross-sections of femur, tibia, and fibula. Some sections are decalcified and stained with H&E, while others are undecalcified hand-ground sections.

The nonhuman primate bone sections were collected in the early 1960s and are related to research projects conducted at the AFIP by Johnson and Kerley on somatotype, skeletal biology, and comparative skeletal histomorphology. Kerley and Johnson initiated studies at the AFIP using the chimpanzee and other primates as research models of normal growth and development with the goal of equating primate skeletal age with human skeletal age and exploring the notion that skeletal type can predict body type. The nonhuman primates were obtained from two regional primate facilities: the Yerkes Laboratory of Primate Biology, Orange Park, Florida, and the Delta Regional Primate Research Center, Tulane University, Covington, Louisiana.

The nonhuman primate collection includes histological slides and associated records that document standard anthropometric data taken prior to death at the respective facilities. Necropsy data, such as radiographs and organ weights, were also recoded. By April 1966, 182 primates had been received by Johnson and Kerley though whole-mount histological slides were not prepared from every animal. Today, the histological slide collection of nonhuman primates includes 67 specimens. Major Primate genera represented include Pan, Hylobates, Erythrocebus, Saimiri, Papio, and Macaca. From each specimen, histological cross-sections and longitudinal sections were made. Midshaft cross-sections were taken from femur, tibia, fibula, humerus, radius, ulna, clavicle, third metacarpal, second metatarsal, and first lumbar vertebra. Metaphyseal and epiphyseal longitudinal sections were made from the proximal humerus, proximal and distal femur, proximal tibia, sterna, clavicle, second metatarsal, twelfth thoracic, seventh and third cervical, and the ninth costochondral rib. The slides are mostly large format, whole-mount glass slides of decalcified, H&E-stained bone tissue.

11.2.4.6 Follis-Park Collection

The Follis-Park collections consist of histology slides, X-rays, and archival records created by Richard H Follis and Edward A Park for the study of nutritional growth disturbances in infants and children in Baltimore between the years 1926 and 1942. The collection consists of histology slides and postmortem radiographs from 1263 cases ranging in age from birth to 11 years. Most cases are between birth and 2 years of age (n= 1248) with 517 dying before the first month. The remaining 15 cases died between 3 and 11 years of age. The slides are predominantly those taken from the cross sections of the growth plates of middle ribs where disturbances in bone metabolism are most easily visualized. The ends of various long bones are often also present. The diagnoses of the cases breakdown as 792 normal and 471 exhibiting signs of rickets (n=364), scurvy (n=52), congenital syphilis (n=54), and lead poisoning (n=8). Seventy-one (71) cases show more than one pathological condition. Clinical radiographs are available for approximately 140 cases.

Ninety-eight cases have the entire case file, and all have index cards with diagnosis, age, sex, and ancestry. Five hundred and forty (540) are listed as female and 721 as male. Seven hundred and twenty-six (726) are documented as black and 536 as white. Dr. Park's extensive notes and data sheets are also available along with Dr. Follis' notes and logs. While the study focused on the detection of growth disturbances in the skeleton, the actual clinical diagnoses and cause of death for each case span the gamut of childhood diseases and the findings related to growth disturbance should not be construed as necessarily related directly to the cause of death.

For preparation, the ribs were sawn in 3 to 5 mm slabs, decalcified, fixed in formaldehyde, embedded in celloidin, and cut in either 14-micron (decalcified with formic acid prior to 1934) or 20-micron (decalcified with nitric acid after 1934) sections. The predominant dyes used are hematoxylin and eosin (Follis et al., 1943).

11.2.4.7 Imaging Lab and Access

There are a variety of imaging tools available at NMHM for visiting researchers. There are Epson 11"x14" flatbed scanners for scanning large slides and documents and 5 Nikon light microscopes with automated stages for digital microscopy. A Huron TissueScope LE slide scanner is also available upon appointment; it can scan up to six bright field 2"×3" slides, or any size slide up to 6"×8".

In addition to imaging tools, several 3-D modeling software packages are available for registration and pre-analysis of digital microscopy data, NextGen data, CT and MRI scans, and models. Digital data and images in a wide variety of formats made from anatomical and developmental specimens are available upon request.

11.2.5 Ericksen Collection

11.2.5.1 Overview

Dr. Mary F. Ericksen established the Ericksen Collection at George Washington University Medical School. The sample primarily comprised femoral sections from cadaveric dissection room specimens. Additional samples were collected from Chilean autopsy specimens and Dominican Republic Cemeteries. Dr. Ericksen collected 328 femoral samples with an age range of 14–97 (Ericksen, 1991). The sex distribution for the collection is 47% female (154) and 53% male (174), and primarily comprised individuals with European ancestry (95%). The Ericksen Collection served as a source for bone blocks and thin sections that are currently located at The Ohio State University (Columbus, Ohio), Texas State University (San Marcos, Texas), and the Tarrant County Medical Examiner's Office (Fort Worth, TX). Research utilizing the Ericksen collection has produced several academic contributions (Andronowski & Taylor, 2022; C Crowder, 2013; Dominguez & Crowder, 2012; Ericksen, 1991; Andronowski & Crowder, 2014).

11.2.5.2 Digital Repositories and Data Dissemination

The COVID-19 pandemic influenced global policies and dramatically disrupted academic research, funding opportunities, networking, and student training. The major shift to virtual learning and collaboration resulted in the increased availability and use of digital repositories and accessible data dissemination. As such, histological imaging files provide unique opportunities for remote data sharing and collaboration. The New Mexico Decedent Image Database (NMDID) is one of the largest examples, with over 275 researchers accessing a database of CT images from over 15,000 decedents (Berry & Edgar, 2021).

A significant number of skeletal collections have a digital component, including the William M. Bass Donated Skeletal Collection (Campanacho et al., 2021; Shirley et al., 2011),

Forensic Isotopes Nation Database (Campanacho et al., 2021), the University of Bologna's skeletal collection (Belcastro et al., 2022), and the Subadult Virtual Anthropology Database (SVAD) (Stull & Corron, 2022). Further, SVAD collaborates with NMDID and has its own secure database of imaging files while acting as a host for an online Zenodo community to share R packages relevant to data processing (Stull & Corron, 2022). These online databases will revolutionize collaborations by allowing for sample conservation and reducing the need for travel or shipment of samples.

11.3 Conclusions

Our chapter explored well-documented bone histology skeletal collections available for study, highlighting their developmental trajectories, ethical considerations, bone procurement processes, demographic compositions, and existing imaging and histological data available. We further illustrated the emergence of digital repositories for data sharing retrieved from skeletal collections reference banks. The unique skeletal archives detailed here provide invaluable opportunities to answer questions concerning age-related microarchitectural changes and biomechanical processes, devise novel regression formulae for histological age-at-death estimation, and beyond. Our primary objective moving forward is to encourage national and international collaboration with researchers interested in human skeletal biology, biological anthropology, and related fields (e.g., anatomical sciences, archaeology).

Acknowledgments

None of this work and development of new knowledge would have been possible without the altruistic and generous attitudes of the donors or next-of-kin of each of the individuals represented in these collections. Their goodwill and trust have resulted in a wealth of investigations to answer questions relating to bone health.

The development of the ASCHR collection involved the efforts and support of many people and institutions. The authors thank Beth Dalzell from The University of Toledo College of Medicine and Life Sciences, the Department of Anatomy and Neurobiology at Northeast Ohio Medical University (NEOMED), Wright State University Boonshoft School of Medicine, and the SORC team of Lifebanc for access to cadaveric bone specimens. Certain specimen processing, data visualization, and collection curation (both skeletal and digital) would not have been possible without the support and assistance of current and past Andronowski Lab members from all training levels.

Additionally, certain research described in this paper was performed at the BMIT facility at the Canadian Light Source, which is supported by the Canada Foundation for Innovation, Natural Sciences and Engineering Research Council of Canada, the University of Saskatchewan, the Government of Saskatchewan, Western Economic Diversification Canada, the National Research Council Canada, and the Canadian Institutes of Health Research. The authors would like to thank the beamline scientists at the Canadian Light Source, particularly Drs. Adam Webb, Sergey Gasilov, Arash Panahifar, and Ning Zhu for the assistance in setup and troubleshooting of the SkyScan SRμCT and white beam microscope systems. The authors further wish to thank Drs. Christine Dengler-Crish and Matthew Smith of NEOMED for access to their confocal laser scanning microscope, and Dr. Mike Marsh of Objects Research Systems Inc. for assistance with troubleshooting our CLSM Dragonfly workflow.

References

AFIP. (1949). Armed Forces Institute of Pathology (AFIP_Washington DC annual report. *AFIP Historical Archives, National Museum of Health and Medicine, Washington, DC.*

Alpaslan-Roodenberg, S., Anthony, D., Babiker, H., Bánffy, E., Booth, T., Capone, P., Deshpande-Mukherjee, A., Eisenmann, S., Fehren-Schmitz, L., Frachetti, M., Fujita, R., Frieman, C. J., Fu, Q., Gibbon, V., Haak, W., Hajdinjak, M., Hofmann, K. P., Holguin, B., Inomata, T., ... Zahir, M. (2021). Ethics of DNA research on human remains: Five globally applicable guidelines. *Nature, 599*(7883), 41–46. doi:10.1038/s41586-021-04008-x

Alves-Cardoso, F., & Campanacho, V. (2022). The scientific profiles of documented collections via publication data: past, present, and future directions in forensic anthropology. *Forensic Sciences, 2*(1), 37–56. doi:10.3390/forensicsci2010004

Amprino, R. (1963). On the growth of cortical bone and the mechanism of osteon formation. *Cells Tissues Organs, 52*(3), Article 3. doi:10.1159/000142349

Andronowski, J. M., Cole, M. E., Davis, R. A., Schuller, A. J., Tubo, G. R., LaMarca, A. R., & Taylor, J. T. (2022). Longitudinal effects of prolonged opioid use on cortical bone remodeling in a rabbit model: Part II—intraskeletal histological variability and regional differences. *Proceedings of the American Academy of Forensic Sciences, 28.*

Andronowski, J. M., Cole, M. E., Davis, R. A., Tubo, G. R., Taylor, J. T., & Cooper, D. M. L. (2023). A multimodal 3D imaging approach of pore networks in the human femur to assess age-associated vascular expansion and Lacuno-Canalicular reduction. *The Anatomical Record*, ar.25089. doi:10.1002/ar.25089

Andronowski, J. M., Cole, M. E., Schuller, A. J., Davis, R. A., & Tubo, G. R. (2021). Longitudinal effects of prolonged opioid use on cortical bone remodeling in a rabbit model: Part I - intraskeletal variability and regional differences detected via micro-CT. *Roceedings of the American Academy of Forensic Sciences, 27*, 62.

Andronowski, J.M., & Crowder, C. (2014). Evaluating Age-related Bone Loss in the Ericksen Femur Collection. *Proceedings of the American Association of Physical Anthropologists, 83rd Annual Meeting, Calgary, AB*, 67.

Andronowski, J. M., Davis, R. A., & Cole, M. E. (2020). Investigating the Impact of Opioid Abuse on Intracortical Porosity and Bone Cellular Density: A Synchrotron-Radiation Micro-Computed Tomography (SRμCT) Approach. *Proceedings of the American Academy of Forensic Sciences, 72nd Annual Scientific Meeting, Anaheim, CA*, A59.

Andronowski, J. M., Davis, R. A., & Holyoke, C. W. (2020). A sectioning, coring, and image processing guide for high-throughput cortical bone sample procurement and analysis for synchrotron micro-CT. *Journal of Visualized Experiments, 160*, Article 160. doi:10.3791/61081

Andronowski, J. M., Davis, R. A., & Stephen, H. E. (2019). Inferring bone attribution to species through micro-Computed Tomography: A comparison of third metapodials from Homo sapiens and Ursus americanus. *Journal of Forensic Radiology and Imaging, 18*, 11–17. doi:10.1016/j.jofri.2019.08.001

Andronowski, J. M., Mundorff, A. Z., Pratt, I. V., Davoren, J. M., & Cooper, D. M. L. (2017). Evaluating differential nuclear DNA yield rates and osteocyte numbers among human bone tissue types: A synchrotron radiation micro-CT approach. *Forensic Science International: Genetics, 28*, 211–218. doi:10.1016/j.fsigen.2017.03.002

Andronowski, J. M., Mundorff, A. Z., Davis, R. A., & Price, E. (2019). Application of X-ray photoelectron spectroscopy to examine surface chemistry of cancellous bone and medullary contents to refine bone sample selection for nuclear DNA analysis. *Journal of Analytical Atomic Spectrometry, 34*(10), 2074–2082. doi:10.1039/C9JA00203K

Andronowski, J. M., Pratt, I. V., & Cooper, D. M. L. (2017). Occurrence of osteon banding in adult human cortical bone. *American Journal of Physical Anthropology, 164*(3), 635–642. doi:10.1002/ajpa.23297

Andronowski, J. M., & Taylor, J. T. (2022). The andronowski skeletal collection for histological research: A modern anatomical contribution. *Forensic Sciences, 2*(1), 175–189. doi:10.3390/forensicsci2010014

Arhatari, B. D., Cooper, D. M. L., Thomas, C. D. L., Clement, J. G., & Peele, A. G. (2011). Imaging the 3D structure of secondary osteons in human cortical bone using phase-retrieval tomography. *Physics in Medicine and Biology, 56*(16), Article 16. doi:10.1088/0031-9155/56/16/012

Barker, M., Chue Hong, N. P., Katz, D. S., Lamprecht, A.-L., Martinez-Ortiz, C., Psomopoulos, F., Harrow, J., Castro, L. J., Gruenpeter, M., Martinez, P. A., & Honeyman, T. (2022). Introducing the FAIR principles for research software. *Scientific Data, 9*(1), Article 1. doi:10.1038/s41597-022-01710-x

Belcastro, M. G., Pietrobelli, A., Nicolosi, T., Milella, M., & Mariotti, V. (2022). Scientific and ethical aspects of identified skeletal series: the case of the documented human osteological collections of the University of Bologna (Northern Italy). *Forensic Sciences, 2*(2), 349–361. doi:10.3390/forensicsci2020025

Bell, K. L., Loveridge, N., Reeve, J., Thomas, C. D. L., Feik, S. A., & Clement, J. G. (2001). Super-osteons (remodeling clusters) in the cortex of the femoral shaft: Influence of age and gender. *The Anatomical Record, 264*(4), Article 4. doi:10.1002/ar.10014

Berry, S. D., & Edgar, H. J. (2021). Announcement: The New Mexico decedent image database. *Forensic Imaging, 24*, 200436. doi:10.1016/j.fri.2021.200436

Bertelsen, P. K., Clement, J. G., & Thomas, C. D. L. (1995). A morphometric study of the cortex of the human femur from early childhood to advanced old age. *Forensic Science International, 74*(1–2), Article 1–2. doi:10.1016/0379-0738(95)01738-5

Buenzli, P. R., Thomas, C. D. L., Clement, J. G., & Pivonka, P. (2013). Endocortical bone loss in osteoporosis: The role of bone surface availability. *International Journal for Numerical Methods in Biomedical Engineering, 29*(12), 1307–1322. doi:10.1002/cnm.2567

Campanacho, V., Alves Cardoso, F., & Ubelaker, D. H. (2021). Documented skeletal collections and their importance in forensic anthropology in the United States. *Forensic Sciences, 1*(3), 228–239. doi:10.3390/forensicsci1030021

Carter, Y., Suchorab, J. L., Thomas, C. D. L., Clement, J. G., & Cooper, D. M. L. (2014). Normal variation in cortical osteocyte lacunar parameters in healthy young males. *Journal of Anatomy, 225*(3), 328–336. doi:10.1111/joa.12213

Carter, Y., Thomas, C. D. L., Clement, J. G., & Cooper, D. M. L. (2013). Femoral osteocyte lacunar density, volume and morphology in women across the lifespan. *Journal of Structural Biology, 183*(3), 519–526. doi:10.1016/j.jsb.2013.07.004

Cobb, W. M. (1959). Thomas Wingate Todd; 1885–1938. *Journal of the National Medical Association, 51*(3), 233–246.

Codman, E. (1922). The registry of cases of bone sarcoma. *Surg Gynecol Obstet, 34*, 338.

Codman, E. (1925). The nomeclature used by the Registry of Bone Sarcoma. *Am J Roentgenol Radium Ther, 13*, 105–126.

Cooper, D. M. L., Erickson, B., Peele, A. G., Hannah, K., Thomas, C. D. L., & Clement, J. G. (2011). Visualization of 3D osteon morphology by synchrotron radiation micro-CT: 3D osteon morphology. *Journal of Anatomy, 219*(4), Article 4. doi:10.1111/j.1469-7580.2011.01398.x

Cooper, D. M. L., Thomas, C. D. L., Clement, J. G., Turinsky, A. L., Sensen, C. W., & Hallgrímsson, B. (2007). Age-dependent change in the 3D structure of cortical porosity at the human femoral midshaft. *Bone, 40*(4), Article 4. doi:10.1016/j.bone.2006.11.011

Crowder, C. (2012). Rib Histomorphometry for Adult Age Estimation. In L.S. Bell (Ed.), *Forensic microscopy for skeletal tissues: Methods and protocols* (pp. 109–127). Humana Press; Springer.

C Crowder. (2013). *Estimation of Age at Death Using Cortical Bone Histomorphometry* (Final Technical Report No. 2010-DN-BX-K035; pp. 1–87). National Institute of Justice, U.S. Department of Justice. https://nij.ojp.gov/library/publications/estimation-age-death-using-cortical-bone-histomorphometry

Dicken, A. J., Evans, J. P. O., Rogers, K. D., Stone, N., Greenwood, C., Godber, S. X., Prokopiou, D., Clement, J. G., Lyburn, I. D., Martin, R. M., & Zioupos, P. (2015). X-ray diffraction from bone employing annular and semi-annular beams. *Physics in Medicine and Biology*, *60*(15), Article 15. doi:10.1088/0031-9155/60/15/5803

Dominguez, V. M., & Crowder, C. M. (2012). The utility of osteon shape and circularity for differentiating human and non-human Haversian bone. *American Journal of Physical Anthropology*, *149*(1), Article 1. doi:10.1002/ajpa.22097

Dowding, G. (1951). Rapid sectioning of whole bones. *Bull Int Assoc Med Museums XXXII.*

Ericksen, M. F. (1991). Histologic estimation of age at death using the anterior cortex of the femur. *American Journal of Physical Anthropology*, *84*(2), 171–179. doi:10.1002/ajpa.1330840207

Ewing, J. (1939). A review of the classification of bone tumors. *Surg Gynecol Obstet*, *68*, 971–976.

Faragó, T., Gasilov, S., Emslie, I., Zuber, M., Helfen, L., Vogelgesang, M., & Baumbach, T. (2022). *Tofu*: A fast, versatile and user-friendly image processing toolkit for computed tomography. *Journal of Synchrotron Radiation*, *29*(3), 916–927. doi:10.1107/S160057752200282X

Feik, S. A., Thomas, C. D. L., Bruns, R., & Clement, J. G. (2000). Regional variations in cortical modeling in the femoral mid-shaft: Sex and age differences. *American Journal of Physical Anthropology*, *112*(2), Article 2. doi:10.1002/(SICI)1096-8644(2000)112:2<191::AID-AJPA6>3.0.CO;2-3

Feik, S. A., Thomas, C. D. L., & Clement, J. G. (1996). Age trends in remodeling of the femoral mid-shaft differ between the sexes. *Journal of Orthopaedic Research*, *14*(4), Article 4. doi:10.1002/jor.1100140413

Feik, S. A., Thomas, C. D. L., & Clement, J. G. (1997). Age-related changes in cortical porosity of the midshaft of the human femur. *Journal of Anatomy*, *191*(3), Article 3. doi:10.1046/j.1469-7580.1997.19130407.x

Follis, R. H., Jackson, D., Eliot, M. M., & Park, E. A. (1943). Prevalence of Rickets in Children Between Two And Fourteen Years of Age. *Archives of Pediatrics and Adolescent Medicine*, *66*(1), 1–11. doi:10.1001/archpedi.1943.02010190008001

Goldman, H., Bromage, T., Bruns, R., Clement, J., Thomas, C., & Feik, S. (1999). Preferred collagen fiber orientation in the human mid-shaft femur. *American Journal of Physical Anthropology*, 136–137.

Goldman, H. M., Bromage, T. G., Boyde, A., Thomas, C. D. L., & Clement, J. G. (2003). Intrapopulation variability in mineralization density at the human femoral mid-shaft. *Journal of Anatomy*, *203*(2), Article 2. doi:10.1046/j.1469-7580.2003.00212.x

Hande, S. (2023). *The Microarchitecture of the Penis Bone (Baculum) of the Pagophilus Groenlandicus.* Memorial University of Newfoundland.

Hassan, M., Taylor, J. T., & Andronowski, J. M. (2023, February). *An Automated Deep Learning Approach to the Segmentation of Vascular Pore Systems and Osteocyte Lacunae in Virtual and Traditional Histologic Datasets* [Poster]. American Academy of Forensic Sciences, Orlando, Florida.

Hennig, C., Thomas, C. D. L., Clement, J. G., & Cooper, D. M. L. (2015). Does 3D orientation account for variation in osteon morphology assessed by 2D histology? *Journal of Anatomy*, *227*(4), 497–505. doi:10.1111/joa.12357

Henry, R. (1964). *The Armed Forces Institute of Pathology: Its first century, 1862–1962.* Washington, DC: Government Printing Office.

Hicks, T., Andronowski, J. M., & Davis, R. (2019). Age-related changes to bone microarchitecture in a non-weight bearing bone. *Williams Honors College, Honors Research Projects*, *871*. https://ideaexchange.uakron.edu/honors_research_projects/871

Hunt, D. R., & Albanese, J. (2005). History and demographic composition of the Robert J. Terry anatomical collection. *American Journal of Physical Anthropology*, *127*(4), 406–417. doi:10.1002/ajpa.20135

İşcan, M. (1992). A comparison of the hamann–todd and terry collections. *Anthropologie (1962-), 30*(1), 35–40. https://www.jstor.org/stable/26295719

Johnson, L. C. (1953). A general theory of bone tumors. *Bulletin of the New York Academy of Medicine, 29*(2), 164–171.

Johnson, L. C. (1993). *Transcript: Interview by Charles Stewart Kennedy* [Interview]. AFIP Oral History Program. AFIP Historical Archives, National Museum of Health and Medicine, Washington, DC.

Kerley, E. (1965). The microscopic determination of age in human bone. *American Journal of Physical Anthropology, 23*(2), 149–163. doi:10.1002/ajpa.1330230215

Kerley, E. (1966). Skeletal age changes in the chimpanzee. *Tulane Stud Zool, 13*, 71–82.

Loveridge, N., Bell, K., Reeve, J., Feik, S., & Clement, J. (2001). Cortical porosity and its relationship to remodelling clusters (super-osteons). *Journal of Bone and Mineral Research, 16*(6), Article 6. doi:10.1002/jbmr.2001.16.6.1157

Luna, L. (1968). *Manual of histologic staining methodsof the Armed Forces Institute of Pathology* (3rd ed.). McGraw-Hill.

Maggiano, I. S., Maggiano, C. M., Clement, J. G., Thomas, C. D. L., Carter, Y., & Cooper, D. M. L. (2016). Three-dimensional reconstruction of Haversian systems in human cortical bone using synchrotron radiation-based micro-CT: Morphology and quantification of branching and transverse connections across age. *Journal of Anatomy, 228*(5), 719–732. doi:10.1111/joa.12430

Mayhew, P., Thomas, C., Clement, J., Loveridge, N., Bonfield, W., Beck, T., & Reeve, J. (2004). Walking tall Is poor long-term Protection against Hip Fracture. *Journal of Bone and Mineral Research, 19*(S1), Article S1. doi:10.1002/jbmr.5650191305

Mendes, P. V., Silva, H. P., Bastos, M., Bittar, V., Reis, S., & Rodrigues-Carvalho, C. (2022). Osteological collections of the national museum in Brazil: challenges and new perspectives for a historical collection. *Forensic Sciences, 2*(2), 287–301. doi:10.3390/forensicsci2020022

NHMRC. (2018). *National statement on ethical conduct in human research 2007 (updated 2018).* National Health and Medical Research Council.

Plens, C. R., Górka, K., & Lopez Quintero, Y. A. (2022). The identified osteological collections of South America and their ethical dimensions. *Forensic Sciences, 2*(1), 238–252. doi:10.3390/forensicsci2010018

Scheiner, S., Pivonka, P., & Hellmich, C. (2016). Poromicromechanics reveals that physiological bone strains induce osteocyte-stimulating lacunar pressure. *Biomechanics and Modeling in Mechanobiology, 15*(1), 9–28. doi:10.1007/s10237-015-0704-y

Shirley, N. R., Wilson, R. J., & Jantz, L. M. (2011). Cadaver use at the University of Tennessee's anthropological research facility. *Clinical Anatomy, 24*(3), 372–380. doi:10.1002/ca.21154

Singh, I. J., & Gunberg, D. L. (1970). Estimation of age at death in human males from quantitative histology of bone fragments. *American Journal of Physical Anthropology, 33*(3), Article 3. doi:10.1002/ajpa.1330330311

Sledzik, P. S. (2001). A career takes form: Ellis kerley's tenure at the armed forces institute of pathology (1957–1966). *Journal of Forensic Sciences, 46*(4), 15044J. doi:10.1520/JFS15044J

Spatola, B. F., Damann, F. E., & Ragsdale, B. D. (2012). Bone histology collections of the national museum of health and medicine. In *Bone Histology: An Anthropological Perspective* (1st ed.). CRC Press, Taylor Francis Group.

Stantis, C., de la Cova, C., Lippert, D., & Sholts, S. B. (2023). Biological anthropology must reassess museum collections for a more ethical future. *Nature Ecology & Evolution.* doi:10.1038/s41559-023-02036-6

Stein, M. S., Thomas, C. D. L., Feik, S. A., Wark, J. D., & Clement, J. G. (1998). Bone size and mechanics at the femoral diaphysis across age and sex. *Journal of Biomechanics, 31*(12), Article 12. doi:10.1016/S0021-9290(98)00127-4

Stephen, H. (2018). Inferring species origin through virtual histology: A comparison of third metapodials from Homo sapiens and Ursus americanus using Micro-computed Tomography.

Williams Honors College, Honors Research Projects, 786. https://ideaexchange.uakron.edu/honors_research_projects/786 Download 112 DOWNLOADS Since November 29, 2018 PlumX Metrics Included in Biology Commons Share.

Stull, K. E., & Corron, L. K. (2022). The Subadult Virtual Anthropology Database (SVAD): An accessible repository of contemporary subadult reference data. *Forensic Sciences*, *2*(1), 20–36. doi:10.3390/forensicsci2010003

Tarrrant, S. M., Ajgaonkar, A., Babhulkar, S., Cui, Z., Harris, I. A., Kulkarni, S., Minehara, H., Miyamoto, T., Oppy, A., Shigemoto, K., Tian, Y., & Balogh, Z. J. (2020). Hip fracture care and national systems: Australia and Asia. *OTA International: The Open Access Journal of Orthopaedic Trauma*, *3*(1), Article 1. doi:10.1097/OI9.0000000000000058

Tatangelo, G., Watts, J., Lim, K., Connaughton, C., Abimanyi-Ochom, J., Borgström, F., Nicholson, G. C., Shore-Lorenti, C., Stuart, A. L., Iuliano-Burns, S., Seeman, E., Prince, R., March, L., Cross, M., Winzenberg, T., Laslett, L. L., Duque, G., Ebeling, P. R., & Sanders, K. M. (2019). The cost of osteoporosis, osteopenia, and associated fractures in Australia in 2017: Cost of OSTEOPOROSIS, OSTEOPENIA, AND FRACTURES. *Journal of Bone and Mineral Research*, *34*(4), Article 4. doi:10.1002/jbmr.3640

Thomas, C. D. L., Stein, M. S., Feik, S. A., Wark, J. D., & Clement, J. G. (2000). Determination of age at death using combined morphology and histology of the femur. *Journal of Anatomy*, *196*(3), Article 3. doi:10.1046/j.1469-7580.2000.19630463.x

Wang, J., Zhou, B., Parkinson, I., Thomas, C. D. L., Clement, J. G., Fazzalari, N., & Guo, X. E. (2013). Trabecular plate loss and deteriorating elastic modulus of femoral trabecular bone in intertrochanteric hip fractures. *Bone Research*, *1*(1), 346–354. doi:10.4248/BR201304005

White, B. (2019). Evaluating the impact of hyperostosis frontalis interna on nonmetric cranial traits for sex estimation. *Williams Honors College, Honors Research Projects*.

Winburn, A., Jennings, A., Steadman, D., & DiGangi, E. (2022). Ancestral diversity in skeletal collections: perspectives on african american body donation. *Forensic Anthropology*, *5*(2), 141–152. doi:10.5744/fa.2020.1023

Zhang, J., Hislop-Jambrich, J., & Besier, T. F. (2016). Predictive statistical models of baseline variations in 3-D femoral cortex morphology. *Medical Engineering & Physics*, *38*(5), Article 5. doi:10.1016/j.medengphy.2016.02.003

12 Three-Dimensional Microstructural Imaging of Bone

Technological Developments and Anthropological Applications

DAVID M. L. COOPER, JANNA M. ANDRONOWSKI, XUAN WEI, AND JOSHUA T. TAYLOR

12.1 Introduction

Bone has a complex, hierarchically arranged three-dimensional (3D) microstructure that has the capacity for change (growth, modeling, and remodeling) over time – making it a fundamentally four-dimensional (4D) tissue. It is this dynamic nature that makes the study of bone so intriguing and useful in numerous subdisciplines of biological anthropology – including the study of past and present populations. That said, our view on the microstructural dynamics of bone has generally been through a two-dimensional (2D) window, as most histological techniques rely upon the assessment of cut or ground sections. To be certain, 2D histological methods (including those presented in other chapters of this volume) have and will continue to provide a wealth of information about bone microstructure and ongoing advances provide increasingly detailed insights into bone at the microstructural, cellular, and even molecular levels. This includes the field of spatial-omics where molecular activity can be mapped down to the level of individual bone cells (Rai et al., 2021; Zhang et al., 2022). Whether directly or indirectly, biological anthropology will continue to benefit from the growing body of knowledge in the larger field of bone biology and 2D histological techniques have a central place in the advancement of the field. That said, a tremendous potential exists to complement 2D analyses of bone microstructure with 3D and even 4D approaches, thereby expanding our understanding of the spatiotemporal dynamics of this tissue. Imaging technology has been central to expanding the dimensions of bone microstructural analysis with X-ray micro-computed tomography (micro-CT) being at the center of these developments over recent decades. This chapter provides an overview of the development of 3D analytical techniques for the study of bone microstructure and their applications in biological anthropology, a more in-depth review of anthropological applications of 3D micro-CT and synchrotron radiation (SR) micro-CT, and a discussion of associated methodological issues and future uses of this technology in the field.

DOI: 10.4324/9781003385608-12

12.2 Reflections on the Past Chapter

This chapter is intended to be an update and expansion of that from approximately a decade ago (Cooper et al., 2012). While this previous chapter focused specifically on cortical bone, here we have expanded our scope to include trabecular bone. The previous chapter was also limited in its discussion of anthropological applications as the overall use of microstructural imaging in the field, and particularly that of cortical bone, was still in its infancy. Here, after a further decade, which has seen growing availability of high-resolution imaging technology, combined with a broader perspective that includes trabecular bone, we strive to provide a more fulsome exploration of 3D microstructural analysis of bone in biological anthropology, situated within the technological landscape that has brought us to this exciting point for the field. Of course, an exhaustive accounting of all topics we cover here is not possible within a single chapter. As such, our goal is to provide a foundational understanding, examples of how microstructural imaging has been employed in biological anthropology, and, perhaps more importantly, a gateway to the literature of the larger interdisciplinary field.

12.3 Technological Developments

12.3.1 Historical Perspectives on 3D Analysis

Bone was among the first tissues to be described by early pioneers of microscopy over 300 years ago (Havers, 1691; Leeuwenhoek, 1677). These descriptions noted the internal porosity ("pipes" in Leeuwenhoek's words) within cortical bone and even today Clopton Havers' name remains associated with the longitudinal vascular canals within secondary osteons (i.e., *Haversian* canals within *Haversian* systems). With respect to 3D assessment, van Leeuwenhoek's hand drawing of the microstructure of the bone of an ox is undoubtedly the first published 3D rendering of cortical bone microarchitecture (Figure 12.1a). The general arrangement of trabecular bone can be appreciated, to a degree, without high magnification, with a classic example being the work of Julius Wolff on carefully prepared sections of human proximal femora, where he linked the organization observed with stresses incurred during loading (Wolff, 1892). Thus, it is critical to recognize that researchers have long appreciated that bone has a complex and functionally significant internal microstructure. That said, our focus here is the direct pursuit of 3D methodologies for visualization and analysis which have been limited due to methodological hurdles faced in the assessment of all tissues but exacerbated by the mineralized matrix of bone. In broad terms, prior to the advent of high-resolution X-ray micro-CT, three general approaches enabled the evaluation of bone microstructure in 3D: (1) staining or casting the cortical canals, (2) serial reconstruction, and (3) confocal microscopy.

12.4 General Approaches I 3D Microstructural Analysis

12.4.1 Staining and Casting

The application of ink to abraded bone surfaces has been used to map generalized patterns of the superficial cortical vasculature (Hert et al., 1994; Petrtyl et al., 1996; Saito, 1984), while introduction of ink/stain followed by clearing of the bone tissue (Figure 12.1b) has been

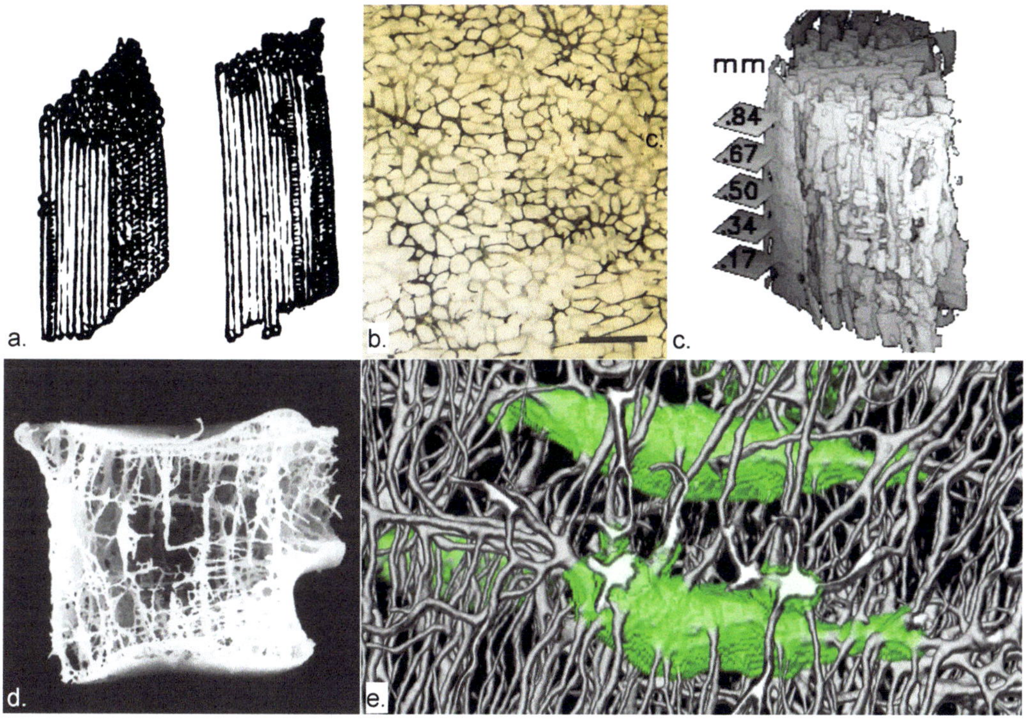

Figure 12.1 (a) Leeuwenhoek's first 3D render of the cortical bone of an ox (adapted from Leeuwenhoek, 1677); (b) ink-injected vascular canals in bovine humerus, scale bar 1 mm (Locke, 2004); (c) 3D rendering of cortical bone serial dataset from a dog femur (Stout et al., 1999); (d) human second lumbar vertebra rendered from automated serial sectioning/block face imaging (Odgaard et al., 1994); (e) CLSM-based 3D rendering of osteocyte lacunae (green) and canaliculi (gray) within human femoral bone utilizing FIT-C stain (image credit to Lorraine Hart).

used to evaluate vascular canals (Albu et al., 1984; Albu & Georgia, 1984; W. T. Dempster & Enlow, 1959; Georgia et al., 1982; Georgia & Albu, 1988; Hert et al., 1994; Locke, 2004; Ruth, 1947; Scheffer & Jansen, 1987) and remodeling spaces (Marotti & Zallone, 1980; Vasciaveo & Bartoli, 1961) deeper in the cortex of humans and other species. Casting of the spaces within bone followed by removal of the surrounding matrix, an approach known as **corrosion casting**, also allows detailed visualization of vascular canals (Arsenault, 1990), and even the lacuno-canalicular network (Pazzaglia & Congiu, 2013). Collectively, these approaches enable visualization, qualitative inspection, and even quantification of some parameters (e.g., canal orientations); however, they are inherently destructive in that the sample is fundamentally altered (grinding, staining, clearing, and even complete dissolution). Furthermore, quantitative analysis in 3D is not straightforward and, therefore, generally limited. With respect to biological anthropology, initial applications of these approaches by Hĕrt, Petrtýl (and their co-author, Fiala) utilized human skeletal material from an "ossuary in southern Bohemia" (Hert et al., 1994) including specimens exhibiting altered forms associated with pathology (attributed by the authors to inactivity and/or illness, including possibly poliomyelitis earlier in life). Beyond these specific examples, however, we are not aware of staining or corrosion casting approaches being applied in anthropological contexts.

12.4.2 Serial Reconstruction by Sectioning or Milling

Detailed assessment of bone microstructural features in 3D beyond the vascular canals of the cortex has traditionally required highly destructive and tedious preparation of **serial sections** which can subsequently be assembled into 3D datasets. Early applications of such approaches are truly remarkable given the challenge of making and assessing physical 3D models without the aid of computers (Amprino, 1948; Amstutz & Sissons, 1969; Cohen & Harris, 1958; Kragstrup & Melsen, 1983; Tappen, 1977). While the innovation of virtual reconstruction was a significant advance (Schnapper et al., 2002; Stout et al., 1999) enabled by the rise of microcomputers (Figure 12.1c), 3D reconstruction of physical sections remains challenging even today and thus is applied in a limited number of studies and generally to a limited number of specimens (Andreasen et al., 2023; Lassen et al., 2017; Mohsin et al., 2002). A key advantage of serial sectioning is that it can be combined with intra-vital fluorochrome labeling (tetracycline, calcein, etc.), classic histological staining approaches, and even molecular methods like *in situ* hybridization. Perhaps the most challenging issue for serial sectioning is its tedious nature that, despite early optimism about automation (DeHoff, 1983), remains limited to this day. With respect to anthropological applications, serial sectioning also comes with the complication of decalcification, which is generally used to facilitate the process, and this may not be suitable for fragile archaeological bone (Schultz, 2001) or taphonomically compromised forensic samples, let alone fossilized materials. **Sequential sectioning** (Robling & Stout, 1999) with a saw avoids the need for decalcification and fragile specimens can be consolidated within an embedding medium, but this approach yields incomplete 3D information (e.g., material is lost due to destruction by the saw blade, or *kerf*). A variant of these approaches is serial imaging of a sample block surface after either removing a section with a microtome (Figure 12.1d) (Odgaard et al., 1990, 1994) or milling/grinding away a layer of precise thickness (Beck et al., 1997). A downside to these approaches is that they entirely destroy the sample in order to acquire the information necessary for 3D reconstruction. On the other hand, they can be highly automated and imaging a block surface means the serial images are inherently registered (which is a complicating factor for serial/sequential sectioning). A further advantage is that fluorescence can be simultaneously captured (Kazakia et al., 2007) to enhance the detection of microstructural features. **Serial milling**, for example, has enabled 3D assessment of microdamage in trabecular bone (Bigley et al., 2008), high-resolution assessment of trabecular bone surface texture (Slyfield et al., 2009), and even visualization of individual trabecular resorption cavities (Tkachenko et al., 2009) and formation events (Slyfield et al., 2012). Taking the concept of serial milling to an even higher resolution, **focused ion beam scanning electron microscopy (FIB-SEM)** uses an ion beam to serially mill and image a surface with SEM, and is capable of resolving 3D structure of the lacuno-canalicular network and even the ultrastructural arrangement of collagen fibers (Shah et al., 2019). Given the highly destructive nature of these approaches, they have not, to our knowledge, seen application in anthropological contexts.

12.4.3 Confocal Microscopy

Confocal microscopy, most commonly implemented as **confocal laser scanning microscopy (CLSM)**, achieves optical sectioning of a sample by rejecting scattered and out-of-plane fluorescent light (A. D. Elliott, 2020). The creation of serial optical sections at varied depths (z-dimension) within the sample serves as the basis for 3D visualization and analysis (Feng et al., 2007), while preserving the physical sample. These optical sections have a thickness and thus the data are described as consisting of *volume elements* or **voxels**, an extension of 2D *picture elements* (**pixels**) into the third dimension. By stacking up serial 2D optical sections, 3D volumetric analysis can be achieved. An important point to consider, however, is that the scale of the voxels in the image plane (x- and y-dimensions) is generally smaller (reflecting higher resolution) than the thickness (z-dimension). As such, these voxels are referred to as **anisotropic**. Conversely, the term **isotropic** refers to equality of scale in all dimensions. Digitally reconstructing a 3D object with anisotropic voxels is analogous to building a model with rectangular blocks whereas isotropic voxels would be comparable to using cubic blocks. The latter will capture 3D structure with equal fidelity in all three dimensions. Thus, the capability of confocal microscopy to generate 3D data is tempered by the fact that resolution is not equal in all dimensions. In practical terms for bone, this means that while geometric (topologic) properties are well captured (e.g., the interconnectivity of lacunae and canaliculi – see Figure 12.1e), volumetric measurements (e.g., the volumes of lacunae and canaliculi) are distorted by the larger scale of the z-dimension (Dallas & Moore, 2020). Another important caveat to consider is the limited depth of optical sectioning – generally on the order of 100 μm or less, although multiphoton approaches can push this to 1 mm or more (Dallas & Moore, 2020).

Fluorescence in bone can either be naturally occurring (endogenous) or introduced (exogenous). Naturally occurring endogenous fluorescence in bone has primarily been attributed to collagen (Bachman & Ellis, 1965), enabling the differentiation of microstructural features including osteons of differing tissue age (Pilolli et al., 2008; Prentice, 1967). Another source of endogenous fluorescence can come from experimentally introduced fluorochrome labels. Examples of exogenous dyes applied *ex vivo*, after sections are prepared, include Texas red-X-conjugated phalloidin (Sugawara et al., 2005) and Fluorescein Isothiocyanate (FIT-C) (Andronowski et al., 2023; Ashique et al., 2017) both used to enhance visualization of the lacuno-canalicular network. For a more detailed overview of confocal applications for osteocyte and lacuno-canalicular visualization, readers are encouraged to read the recent review by Dallas and Moore (Dallas & Moore, 2020).

Optical sectioning of a sample is an advantage for anthropological applications as it enables analysis of thicker sections that are simpler to prepare (e.g., high surface polishing is not necessary), which is particularly beneficial for fragile archaeological samples. The limited penetration depth, however, means that confocal approaches are generally still destructive, requiring the preparation of sections, even if they can be thicker than those used for conventional light microscopy. A notable exception is the application of confocal techniques to subperiosteal microstructures, as demonstrated by the noninvasive visualization of lacunae in an *Australopithecus afarensis* femur (Bromage et al., 2009). Notwithstanding this exciting work of Bromage and colleagues, anthropological applications of confocal microscopy have been somewhat limited. An early application, for example, was the assessment of a hominid cranial bone which revealed vascular canals and osteocyte lacunae (Bartsiokas, 2002). That said, there is growing interest in capitalizing on the advantages of confocal approaches for anthropological applications. For CLSM, these include improved resolution in the z-plane compared with light microscopy, the ability to assess 3D structure, and the potential of spectral analysis (e.g., emission peak, maximum intensity and fluorescence range of a fluorochrome); however, in 2006, Maggiano and colleagues observed the method had seen limited application in the field (2006). Subsequent reviews of the potential of CLSM for anthropological applications have similarly expressed a lack of uptake of the technology in the field and strived to further highlight its potentials and thereby encourage its wider adoption (C. Maggiano et al., 2009; Papageorgopoulou et al., 2010). With confocal microscopes having become standard equipment for many biological disciplines, the continued and growing use of this technology in biological anthropology seems certain. Indeed, a recent study used confocal topography to assess cortical porosity and trabecularization in skeletal remains from early Pacific Islanders (Miszkiewicz et al., 2021), and there is also growing interest in applying confocal microscopy in forensic contexts as means of differentiating skeletal remains of medicolegal interest from older archaeological materials (Capasso et al., 2017) or to assess diagenetic loss of fluorescence over time (Smith et al., 2022).

12.5 X-Ray-Based Imaging of Bone Microstructure

Röntgen discovered X-rays in 1895 (1895) and was subsequently awarded the first Nobel Prize in Physics in 1901. The great potential of X-rays for the study of bone was immediately obvious, as demonstrated by the iconic image of Roentgen's wife's hand (Figure 12.2a). Compared to the soft-tissues of the body, the attenuation (loss of intensity due to absorption and scatter) of X-rays by the mineralized matrix of bones and teeth creates excellent contrast (the relative difference in attenuation). While X-rays enable the detection of internal structures within otherwise opaque objects, a limitation is the superimposition of 3D structures within a single 2D planar view. Here we use the term **radiograph** to refer to such images. Moving beyond qualitative inspection, quantifying parameters such as bone density from radiographs is complicated by varying amounts of overlying soft tissues. To overcome this, multi-energy energy (aka spectral) imaging has been developed to enable the assessment of mineralized and soft-tissue components more precisely. Dual photon absorptiometry (**DPA**), utilizing radionuclide sources, was developed first and has subsequently been replaced by dual energy X-ray absorptiometry (**DEXA** or, more commonly, **DXA**). DXA, colloquially referred to as a *bone density scan*, became commercially available in 1987 and was rapidly adopted clinically (Genant et al., 1996) as it is quicker and higher resolution than DPA. As these techniques remain 2D in nature, they provide *areal* measures of bone mineral density (**aBMD**; g/cm^2) and are unable to directly detect microstructural features, nor segment trabecular and cortical bone. Both DPA (Perzigian, 1973) and DXA have been applied to archaeological bone, including recent studies for the latter (Brødholt et al., 2022); however, it has long been recognized that such densitometric techniques cannot account for potential impacts of diagenesis (Kneissel et al., 1994; Stout & Simmons, 1979).

12.5.1 Microradiography

With respect to bone microstructure, a means of mitigating the limitations of 3D superimposition in 2D images is to make radiographs of thinly ground undecalcified sections (e.g., 50–100 μm), a technique known as **microradiography**. Microradiographs reveal differing X-ray attenuation within a section enabling the visualization of variable degrees of mineralization and, thereby, the detection of secondary osteons and other features of different tissue age (Figure 12.2b). Microradiography has been extensively used in anthropological contexts, reaching back to at least the 1950s (C. A. Baud & Morgenthaler, 1956) and

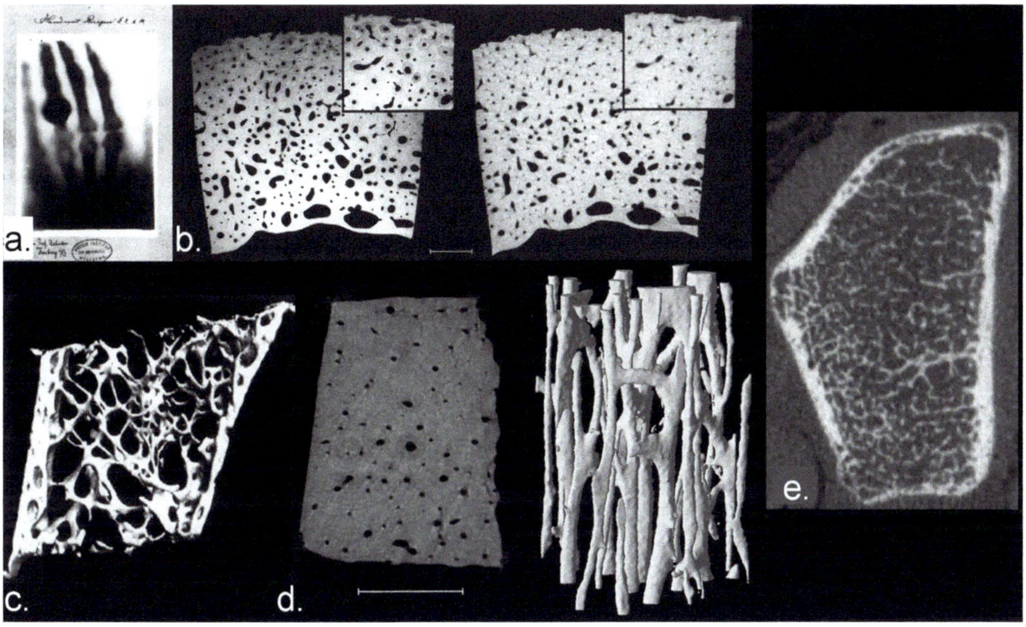

Figure 12.2 (a) Radiograph of Roentgen's wife's hand (image is in the public domain, source: wikipedia.org); (b) microradiograph vs. micro-CT of human cortical bone (Cooper et al., 2004) scale bar = 1 mm; (c) 3D Micro-CT human iliac crest biopsy (Borah et al., 2001); (d) 2D and 3D micro-CT images of human cortical bone porosity (Cooper et al., 2003) scale bar = 1 mm; (e) HR-pQCT (single 2D slice) of a human wrist (distal radius) (Soltan et al., 2019).

a comprehensive review of this topic is beyond the scope of this chapter. Briefly, anthropological applications have included evaluating mineralization differences associated with diagenetic change in ancient and archaeological bone (Lagier et al., 1982; Piepenbrink, 1986; Sergi et al., 1972; Stout, 1978), pathological change (e.g., lesion characteristics) (C.-A. Baud et al., 1978; Lagier et al., 1982; Marchiafava et al., 1974), and microscopic structures (e.g., to differentiate osteons of various types) used in age-at-death estimation (Drusini, 1987; Jowsey, 1960; Yoshino et al., 1994). A strength of microradiography is the high contrast between the bone tissue and voids within it, particularly the vascular porosity (i.e., Haversian/osteonal canals). This creates a very efficient platform for automatic segmentation (Liu et al., 1996, 1999), and this approach was extensively applied, for example, to the study of the Melbourne Femur Research Collection (Bertelsen et al., 1995; Feik et al., 1997, 2000; Stein et al., 1999; Thomas et al., 2000, 2005, 2006) (*see* Chapter 10 of this volume). The drawbacks of microradiography include that it remains inherently 2D, it is destructive (e.g., sections have to be prepared) and superimposition is not entirely eliminated due to the limited section thickness. Regarding the latter point, the structures detected depend, in part, on the section thickness and also the resolution of the imaging system and thus may be reduced relative to light microscopy. That said, ground sections prepared for microradiography can also undergo additional microscopic analyses, including light and confocal microscopy, since the X-rays leave the sample intact. The technique has also become increasingly accessible/straightforward due to the availability of digital systems that eliminate the need for films. Ultimately, with respect to the primary focus of this chapter, microradiography remains a 2D technique.

12.5.2 X-Ray Computed Tomography (CT)

To move into the third dimension, radiographic images taken from multiple angles (e.g., anterior-posterior and medial-lateral) are required to determine the 3D position of internal structures. An important caveat is that these structures must have sufficient contrast to be detectable in these projections. Taking an increasing number of angular views provides increasingly refined positional information. Building upon this principle, in 1973 Hounsfield published his foundational paper on computerized transverse axial tomography, a technology which we now simply refer to as **computed tomography** (**CT**) (Hounsfield, 1973). The term tomography is derived from the Greek term *tomos*, referring to a section or slice which is the same root for the word microtome – a device used in histology to cut sections. CT provides a noninvasive means of mapping the localized X-ray attenuation (linear attenuation coefficient) within voxels comprising a cross-section, yielding greatly improved spatial and density resolution. In practical terms this means that some structures which are poorly detected 2D radiographs (e.g., soft tissue features) can be better detected by CT. Scaling of the linear attenuation coefficients relative to water yields a standardized scale, measured in **Hounsfield Units (HU)**. As discussed above with respect to confocal microscopy, the tomograms (also commonly referred to as slices) generated by CT have a thickness and thus can be stacked to enable 3D visualization and analysis. Despite this potential, early CT applications generated a limited number of slices which were utilized primarily as 2D data from which 3D anatomy could be inferred, rather than directly volumetrically rendered. While CT has undergone advances spanning a number of technological "generations", the most common configuration today involves rotating the X-ray source and detector around the patient/target to collect the necessary angular views – such scanners are known as rotate-rotate or third generation machines (Schulz et al., 2021).

The advent of CT represented a revolutionary advance; however, its limitations include the increased radiation dose due to the need for many angular views, as well as resolution. The key point of relevance for this chapter is that CT, as applied in the clinic, has generally lacked the resolution necessary to directly resolve individual bone microstructural features, including trabeculae and cortical vascular pores. Indeed the size of the voxels generated by clinical scanners is on the order of hundreds of microns to millimeters. A particularly limiting factor has been the slice thickness (z-dimension), which can be considerably larger in scale than the in-plane (x and y) voxel dimensions – similar to that discussed above for confocal microscopy. It should be noted, however, that CT can be utilized to gain an appreciation of the overall distribution of trabeculae, and this has potential for anthropological applications such as age estimation (de Froidmont et al., 2013). While the individual microstructural features of bone may not be directly resolved, the volumetric attenuation data CT provides (i.e., the Hounsfield Units) can be calibrated into *volumetric* bone mineral density (**vBMD**; g/cm^3) using density standards (phantoms) – an approach known as **Quantitative CT (QCT)**. While not employed for clinical purposes, a variant focused on imaging of the limbs (primarily radius and tibia), **peripheral QCT** (**pQCT**) has been widely used for research on living human subjects. The outcomes of vBMD for bone are strongly correlated with the amount of mineralized tissue within the target voxels. For example, Bousson and colleagues reported that cortical porosity (assessed by microradiography) accounted for 71.6% of the variance in cortical vBMD in a sample derived from 163 human femora (Bousson et al., 2000). Similarly, in trabecular bone samples from human vertebrae vBMD explained 89% of the variation in apparent density (McBroom et al., 1985).

Perfect correlations between vBMD and cortical porosity/trabecular apparent density are not expected as tissue-level bone mineral density can vary to a degree and the porosity of the lacuno-canalicular network is unaccounted in a direct manner. This highlights an issue associated with some anthropological applications, as diagenesis may alter the material density of bone while leaving some microstructural features relatively intact – thus caution must be used when using approaches that do not directly resolve the target structures (e.g., DXA and QCT) but rather rely upon coarser-level density.

There have been continued innovations related to CT with notable areas of advance over the last 30 years including achieving **isotropic voxels** (equal x-, y-, and z-dimensions), improved temporal resolution (reduced scan time) and material decomposition through spectral (multi-energy) imaging employing the same principles applied with DPA and DXA. For more information on the past, present, and future of clinical CT technology, readers are directed to the excellent review by Hsieh and Flohr (Hsieh & Flohr, 2021). Due to its nondestructive nature and increasing accessibility, CT has seen extensive use in anthropological inquiry (Weber, 2015); however, these pursuits have necessarily been generally macroscopic in nature and thus fall outside the specific scope of this chapter. Again, however, just as with microradiography, the development of CT served as an important stepping stone toward higher resolution implementation which ultimately has enabled 3D microstructural analysis.

12.5.3 X-Ray Micro-Computed Tomography (Micro-CT)

To directly visualize and analyze individual microstructural features of bone, higher resolution than that achievable using clinical CT is needed. Here we use the term **micro-computed tomography (micro-CT)** to describe the parallel technology that was developed during the 1980s and 1990s, although it should be noted that alternative terms, including X-ray microtomography, X-ray tomographic microscopy (XTM), and computed microtomography, as well as different abbreviations (e.g., μCT), have all been employed by different studies to describe what is essentially the same technology. The common element in these terms, "micro", refers to imaging with voxel sizes on the order of single to tens of micrometers (microns, μm). By extension, the term **nano-CT** has been used to refer to tomographic imaging with sub-micron (nanometer) voxel sizes (Schneider et al., 2007). Elliott and Dover were pioneers of micro-CT, describing a system capable of 12 μm x 12 μm in-plane (x and y) voxel size within a 0.5 mm field of view in 1982 (J. C. Elliott & Dover, 1982). This approach, which utilized a highly collimated (pin-hole) laboratory X-ray source, was soon applied to a human femoral sample, demonstrating the internal porosity as well as some limited detail regarding density variation around Haversian canals (J. C. Elliott & Dover, 1984), likely related to differing levels of mineralization (bone tissue age), as has been well demonstrated by microradiography. As micro-CT was originally developed to target small, ideally cylindrical, samples, from a practical perspective these systems have utilized stationary X-ray sources and detectors with the target rotated on a stage placed between them. An exception to this is *in vivo* micro-CT systems for animal imaging (discussed in future directions later in this chapter) that generally utilize the same rotation-rotation configuration employed in clinical systems.

While the early applications of micro-CT were remarkable given their temporal context and capacity to resolve individual microstructural features of bone (trabeculae and vascular canals), they presented only 2D slices and, while 3D inferences could be made by assessment of adjacently acquired images, there was no explicit 3D rendering/analysis. Similar to early clinical scanners, each scan generated a single slice (generated by rastering the source or target relative to one another) and thus acquiring 3D datasets was time consuming. Combined advances in X-ray detector technology, computing power, and reconstruction algorithms were necessary to realize direct acquisition of volumetric datasets and to render them in 3D. In the late 1980s Feldkamp and colleagues described a micro-CT system that utilized an area detector (vs. a single point or even line detector) and a cone-beam reconstruction algorithm that meant entire volumes, with isotropic voxels, could be captured in a single scan (Feldkamp et al., 1989). The cone-beam approach also afforded the advantage of magnification via the divergence of the X-rays from a micro-focus source, meaning the effective voxel size could be smaller than the physical size of the detector elements. This combination of improved resolution, 3D **field of view** (**FOV**) and the computational means to directly process the data (along with the ever-increasing availability of computing power) meant that use of micro-CT technology exploded in the late 1990s and early 2000s, rapidly becoming the new gold standard for many bone morphometry applications – particularly for the 3D quantitative study trabecular bone microstructure *ex vivo* (Figure 12.2c). The ability of micro-CT, operating at isotropic voxels sizes at or below 10 μm, was also soon recognized as having great potential for *ex vivo* 3D visualization and quantification of cortical bone porosity (Cooper et al., 2003) (Figure 12.2d), although additional features related to density, including osteons of varying tissue age (Cooper et al., 2004), are either undetected or only occasionally observed (Figure 12.2b).

12.5.4 High Resolution peripheral Quantitative Computed Tomography (HR-pQCT)

With the advent of micro-CT technology capable of detecting bone microstructural features in 3D, the question of what is possible in living humans naturally arises (*animal imaging is discussed briefly below in future directions*). Ultimately, the limitations imposed by radiation dose impose the limits we see in terms of human imaging, with **High Resolution peripheral Quantitative Computed Tomography** (**HR-pQCT**) representing the most advanced technology currently available for assessment of human bone microstructure *in vivo*. First described in 2005, HR-pQCT at the time had isotropic 82 μm voxels and is capable of resolving individual trabeculae in the wrists (radii) and ankles (tibiae) of living humans (Boutroy et al., 2005) (Figure 12.2e). It was also not long before cortical bone porosity was being assessed from HR-pQCT scans (Burghardt et al., 2010; Nishiyama et al., 2009), although given the limited resolution, it has been demonstrated that only the largest of pores are resolved when comparisons are made with higher resolution techniques (Soltan et al., 2019). Nonetheless, this technology has seen growing application in the study of cortical porosity in living human subjects and the second generation (**HR-pQCT II**) introduced improved resolution (61 μm isotropic voxels) (Manske et al., 2017).

12.5.5 Synchrotron Radiation (SR) Micro-CT

A **synchrotron** is a circular particle accelerator which generates intense beams of light (including the X-ray portion of the electromagnetic spectrum) tangential to an electron storage ring either by the bending magnets, which create the ring itself, or insertion devices (comprised of clustered magnets), placed between the bending magnets (Figure 12.3a).

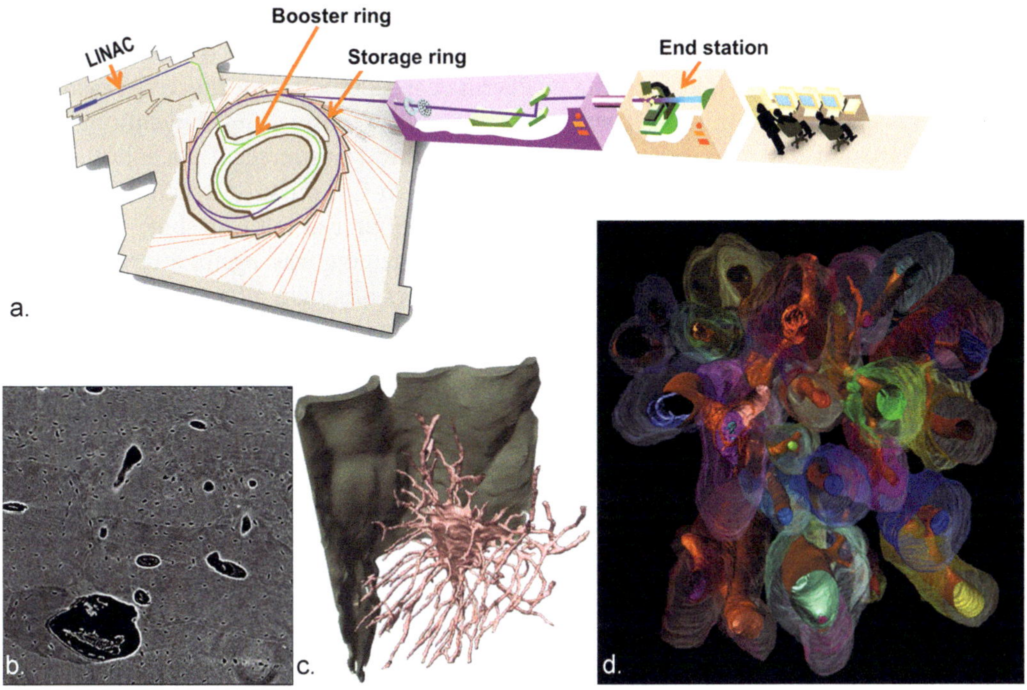

Figure 12.3 (a) Schematic showing the major components of a synchrotron radiation facility including the linear accelerator (LINAC) which feeds electrons into the booster ring and, ultimately, the storage ring. Lateral accelerations of the elections in the storage ring, by bending magnets or insertion devices, create tangential beams of intense radiation (beamlines) which are exploited at experimental end stations (image courtesy of the Canadian Light Source; www.lightsource.ca); (b) SR micro-CT of human femoral cortical bone showing both vascular canals and lacunae with bright white phase contrast enhancement at surfaces as well as soft-tissue remnants within the large remodeling space (bottom left) (Cooper et al., 2011); (c) SR micro-CT 3D rendering of an osteocyte and canaliculi (Langer & Peyrin, 2016); (d) SR micro-CT 3D rendering of osteon borders and vascular canals within them (I. S. Maggiano et al., 2016).

The deflection of the electrons by the magnetic fields results in the generation of the radiation. A single facility can thus simultaneously operate numerous tangentially arranged beam lines with associated imaging end-stations, each focused on exploiting a specific portion of the energy spectrum. What is key to understand about synchrotron radiation in the current context is that the X-ray beams generated are potentially a millionfold (Flannery & Roberge, 1987) more intense (e.g., greater brilliance) than those created by conventional X-ray tubes. This advantage, combined with some additional features of synchrotron radiation create unique imaging capabilities, which we discuss in turn below.

12.5.6 Resolution

As already discussed, achieving high resolution was a driving force behind the development of micro-CT in general. A limitation of imaging at high resolution is the need to pass enough X-rays through the target volume from multiple angles to create a high-quality image. If one envisions the voxel as the target volume, then, as the voxel size decreases (e.g., resolution increases), with all else equal, exposure time needs to increase to maintain image quality. For CT, this relation is nonlinear and increasing voxel size by a factor of 4 requires 64 times more acquisition time to achieve equivalent image quality if the intensity of the beam (X-ray/photon flux) is kept constant (Peyrin et al., 2000). It is for this very reason that the early development of micro-CT was pursued simultaneously using conventional X-ray sources (discussed above) as well as more intense synchrotron sources. Thus, Elliott, Dover, and colleagues, already discussed as pioneers of micro-CT, pursued synchrotron radiation (SR) micro-CT using the Synchrotron Radiation Source at the Daresbury Laboratory (UK; operational from 1981 to 2008). Cortical bone imaging was, again, a target in these early pursuits which achieved resolutions on the order of single microns, although only the detection of Haversian canals in human femoral samples is evident in their 2D images (Bowen et al., 1986; J. C. Elliott et al., 1987). Similar to the overall pattern observed for conventional micro-CT, 3D synchrotron applications initially focused on trabecular bone, although it was also recognized early on that higher resolution features, like osteocyte lacunae, were within reach (Peyrin et al., 1998, 2000) (Figure 12.3b). Visualization of cortical bone vascular-level porosity followed for human (Bousson et al., 2004) and rat (Matsumoto et al., 2006, 2007) bone and, soon thereafter, analysis specifically targeting osteocyte lacunae were reported for mouse (Schneider et al., 2007) and, subsequently, human bone (Hannah et al., 2010). Even the LCN has come within reach of SR-based nano-CT, including individual canaliculi (Langer et al., 2012; Pacureanu et al., 2012) (Figure 12.3c). That said, a general caveat is that as resolution increases, FOV decreases and thus studies of osteocyte lacunae are generally limited to specimens of a few millimeters in diameter (voxel sizes of 1–2 μm) and the groundbreaking LCN studies were limited to assessing samples of approximately 0.5 mm diameter (voxel size of 0.280 μm). While conventional micro/nano-CT is capable of increasingly impressive resolutions, scan times can be protracted and thus the combination of resolution and experimental throughput represent a core advantage of SR micro-CT. Scan times, depending on the resolution can take only minutes, seconds or even less with SR micro-CT and impressive automated experimental pipelines, such as that developed by Mader and colleagues to assess osteocyte lacunae in 1300 mouse femoral samples in a single study, are truly remarkable (Mader et al., 2015).

12.5.7 Tunability

An additional advantage derived from the increased intensity of SR is tunability, referring to the capacity to select a very narrow range of imaging energy, resulting in what is referred to as a **monochromatic** beam. This is contrasted with imaging using conventional X-ray sources which utilize a wider range of energies and thus are generally referred to as **polychromatic**. In essence, the greater intensity of an SR beam means that after selecting a narrow range of energies (via a crystal monochromator and/or filters), enough intensity remains to make imaging feasible in terms of time. The use of a monochromatic beam reduces artifacts such as "beam-hardening" which occurs in polychromatic imaging systems whereby lower energy (softer) X-rays are preferentially absorbed as the beam traverses the target leading to a greater proportion of higher-energy (harder) X-rays (Brooks & Chiro, 1976). In practical terms, this artifact creates artificial density gradients, thus monochromatic SR micro-CT has the potential to more precisely map localized bone mineral density than conventional micro-CT. SR micro-CT thus has the potential to facilitate better delineation of low-contrast features such as osteons of varying tissue age/density in cortical bone (Cooper et al., 2011; Dalstra et al., 2004, 2006; Peter et al., 2007) (Figure 12.3d). That said, the consistent segmentation of osteons remains challenging and has been aided by phase contrast, discussed next.

12.5.8 Phase Contrast

Thus far, the X-ray-based imaging discussed in this chapter has been based upon the principle of contrast derived from attenuation (*primarily absorption*) of X-rays, an approach well suited to bone imaging due to its high (mineralized) density. **Phase contrast** represents an alternative mechanism whereby differences associated with X-ray refractive index and thickness of the target are exploited to create imaging contrast (Wilkins et al., 1996). The brilliant and coherent sources of modern (third-generation) synchrotrons facilitate the straightforward implementation of phase contrast imaging – creating novel research opportunities. An X-ray source is considered coherent if it generates interference fringes associated with the refractive index of the sample and these fringes can, in the simplest configuration, be optimized by changing the distance between the target and the detector (Lengeler, 2001). This is commonly referred to as **propagation-based imaging (PBI)**. While phase contrast perhaps has its greatest potential in the detection of low-absorbing targets, such as soft-tissues, the interfaces within bone between mineral and soft-tissue are particularly enhanced by interference fringes. Phase contrast can exist within images as edge enhancements in absorption images improving creating bright fringes at the surfaces of vascular pores, lacunae (see Figure 12.3b) and even soft-tissue remnants within canals/marrow spaces, as well as enhancing of bone-bone interfaces such as cement lines and circumferential lamellae (Andronowski, Mundorff, et al., 2017; Cooper et al., 2011). Alternatively, phase (refractive index) itself can be computationally retrieved (Paganin et al., 2002), again enhancing the general detection of newly formed osteons and features such as cement lines (Gauthier et al., 2018, 2019). Of course, the primary advantage of these approaches is the capacity to generate 3D reconstructions of osteons and thereby generate new insights into spatiotemporal regulation of remodeling via the structures it generates (Cooper et al., 2011; Gauthier et al., 2019; I. S. Maggiano et al., 2016, 2021) (Figure 12.3d). We would be remiss to not note that phase retrieval and segmentation of osteons has also been achieved using a conventional (polychromatic) micro-CT system (Arhatari et al., 2011). Given the importance of remodeling and osteon-level features in many anthropological applications, further advancing noninvasive 3D analysis of these features presents an important area for further research and refinement of techniques. Enhanced contrast without increased radiation dose also conveys advantages for *in vivo* imaging (discussed below).

12.6 3D Bone Histomorphometry

Being able to resolve and render microstructural features of bone using (SR) micro-CT may provide important qualitative insights alone, as has been widely employed in anthropological inquiry; however, a key driving force behind the development and wider adoption of these technologies has been the capacity to facilitate efficient quantitative 3D morphometrics. Morphometry of bone has entailed a battery of different measures related to the amount of bone present and its spatial organization, with the community striving for standardization to improve communication within and outside the field of bone biology (D. W. Dempster et al., 2013; Parfitt et al., 1987). The field has long grappled with the challenging issue of 2D sampling of 3D structure, setting aside, for the moment, the fourth dimension of time. Measurements made on sections, in the most direct sense, are limited to parameters such as area, length (e.g., perimeter), distance between points/lines, and counts. While extrapolation of areal fraction (e.g., bone area/total area) to volume fraction (bone volume/total volume) may be straightforward, other parameters (e.g., connectivity, orientation, trabecular thickness) are not. The field of **stereology** explores the extrapolation of 3D structural parameters from 2D measures. Three-dimensional stereological measures rely upon models or theorems, generally based upon a number of assumptions about sampling (randomized orientation of sections, isotropic nature of the targeted structure, etc.) that are rarely met in practical terms (D. W. Dempster et al., 2013). As such, it is critical to interpret measures derived from stereological extrapolations (and indeed all histomorphometric measures, regardless of their 2D or 3D nature) within the specific context of what they actually measure and their underlying assumptions. Odgaard provides an excellent review of the challenges of extrapolating 3D trabecular bone parameters such as connectivity, orientation, and thickness from 2D measures (Odgaard, 1997). Volumetric visualization by micro-CT has provided an accessible platform for direct, model-independent measures of 3D trabecular microstructural parameters with an excellent example being trabecular thickness. Two-dimensional assessment of trabecular thickness relies upon assumptions, while direct 3D assessment using distance transforms – which can be conceptualized as fitting maximal diameter spheres throughout the trabeculae – provides measures of mean thickness as well as distributions of thicknesses throughout the sample (Hildebrand & Ruegsegger, 1997) (Figure 12.4a). Inverting the analysis also enables direct assessment of the separation between trabeculae (Figure 12.4b). The development of (SR) micro-CT advanced in parallel with that of 3D morphometric trabecular measures and Borah and colleagues (Borah et al., 2001) presented an early review, focused on characterizing osteoporotic changes in trabecular bone. Many of the measures they reviewed, and the associated acronyms (Table 12.1) have become commonplace over the past two decades. Comparable measures for cortical bone (vascular) porosity (Cooper et al., 2003) (Table 12.1; Figure 12.4c) have also increasingly been applied. Moving to higher resolution, initial analyses of lacunar properties entailed lacunar density (number per unit volume) as well as volume (and volume fraction) were explored in mouse models using SR micro-CT (Schneider et al., 2007). Advancing beyond counts and volumes, shape, and orientation of lacunae have also been quantified from SR micro-CT scans (Carter, Thomas, Clement, & Cooper, 2013; Carter, Thomas, Clement, Peele, et al., 2013) (Figure 12.4d). SR nano-CT has enabled more detailed morphological analysis of lacunae and even canaliculi (Yu et al., 2020).

12.7 Anthropological Applications of (SR) Micro-CT for the Study of Bone

12.7.1 Taphonomy

Three-dimensional imaging and visualization can be used to identify physical changes to bone tissue microstructure caused by a variety of taphonomic factors (water penetration, bacterial and fungal infiltration, soil exposure, etc.). The ability to directly examine bone

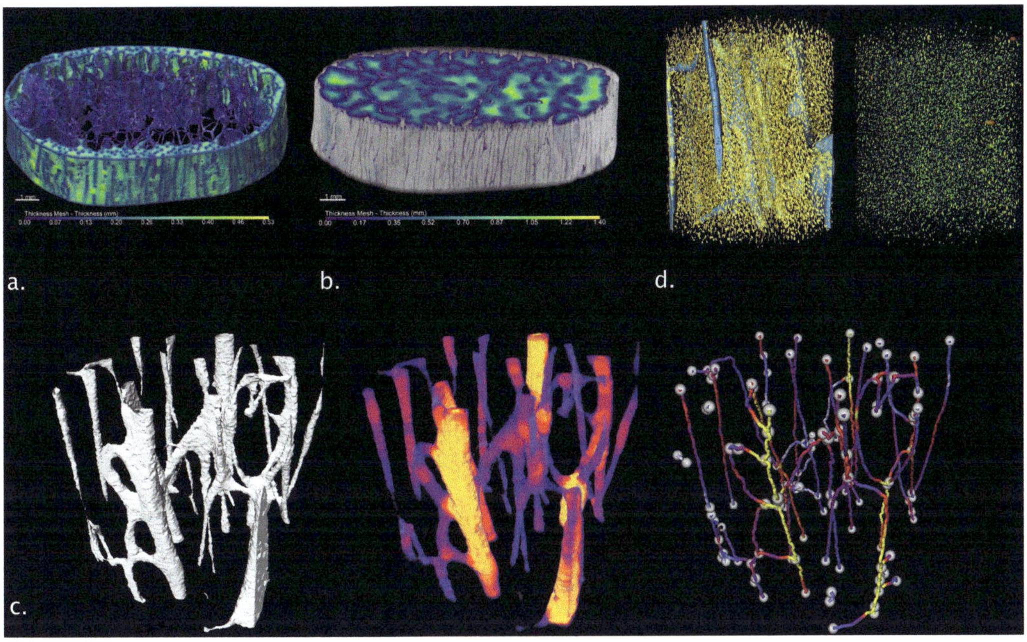

Figure 12.4 (a) 3D colorized representation of localized 3D trabecular and cortical thickness of a human rib; (b) corresponding visualization of trabecular and canal separation (image credit: Joshua T. Taylor & Janna M. Andronowski); (c) 3D rendering of osteocyte lacunae by SR micro-CT (left, with lacunae in gold and vascular canals in blue) with lacunar shape analysis reflected by a colorized map on the right (Carter et al. 2013); (d) 3D rendering of cortical bone porosity (left) with canal diameter mapped in 3D (center) and canal segment skeletonization and branching points (right) (Cooper et al., 2012).

porosity, given the high contrast of solid and porous states of bone tissue and vascular canals, represents a significant advancement for nondestructive diagenetic analyses. Assessment of 2D virtual sections via (SR) micro-CT or subsequent 3D visualizations of diagenetic bone microarchitecture offer novel avenues for investigating queries concerning bone's postmortem history. For example, such high-resolution investigations may enable solutions to challenging cases in forensic and archaeological research (e.g., burial environment), calling for a timely need for a better understanding of diagenetic changes at the microstructural level. Further novel benefits of employing (SR) micro-CT include the ability to nondestructively assess bone microstructural features impacted by diagenetic alterations, reduce specimen preparation time, as well as time consuming image segmentation methods (e.g., through application of noise-reducing and edge-preserving filters; (Duffett Carlson et al., 2022)). By creating 3D volume renderings of bone microarchitecture, researchers have successfully visualized and qualitatively assessed topographic patterns of bacterial bioerosion, including the invasion of cortical canals by microbiota (Duffett Carlson et al., 2022; Kral et al., 2021).

Some of the earliest micro-CT work evaluated bioerosion in bioarchaeological bone systematically and assessed the efficacy of the technique for describing bone diagenesis in subadult remains (Booth, 2016; Dal Sasso et al., 2014), and visualized microtaphonomy in archaeological samples using basic canal volume renders (Booth, 2016). Postmortem, bones are primarily influenced by a loss of structural collagen and further applications have explored the use of micro-CT as a prescreening tool for bone collagen content prior to

Table 12.1 Common 3D histomorphometric measures for trabecular bone and cortical porosity

	Trabecular Bone		
Parameter	Abbreviation	Units	Brief Description
Bone Volume Fraction	BV/TV	%	Volume of bone per total volume
Bone Surface/Tissue Volume	BS/TV	mm^{-1}	Surface of bone per total volume
Trabecular Thickness	Tb.Th	µm	Mean thickness of trabeculae via sphere fitting
Trabecular Separation	Tb.Sp	µm	Mean distance between trabeculae via sphere fitting
Connectivity Density	ConnD	mm^{-3}	Estimate of trabecular connections per unit volume
Trabecular Number	Tb.N	mm^{-3}	Estimate of trabeculae per unit volume
Structural Model Index	SMI	-	Estimate of plate vs. rod characteristics
Degree of Anisotropy	DA	-	Measure of preference of orientation
	Cortical Bone (Vascular Level) Porosity		
Parameter	Abbreviation	units	
Canal Volume Fraction (% Porosity)	Ca.V/TV	%	Volume of canals per total volume
Canal Surface/Tissue Volume	Ca.V/TV	mm^{-1}	Surface of canals per total volume
Canal Diameter	Ca.Dm	µm	Mean diameter of canals via sphere fitting
Canal Separation	Ca.Sp	µm	Mean distance between canals via sphere fitting
Canal Connectivity Density	Ca.ConnD	mm^{-3}	Number of canal intersections per total volume

radiocarbon dating through the assessment of cortical porosity (M. J. Collins et al., 2002; Tripp et al., 2010, 2018). Such studies highlight the importance of imaging techniques for sample preservation ahead of subsequent testing in addition to radiocarbon dating including DNA extraction and stable isotope analysis.

Recently micro-CT was employed to develop a new Virtual Histological Index (VHI) for diagnosing bioerosion in archaeological bone (Mandl et al., 2022). The authors compared thin ground sections from human femora rated with the Oxford Histological Index (OHI), with modifications following Jans and colleagues. (Hedges et al., 1995; Jans et al., 2002; Millard AM, 2001) to micro-CT-derived virtual sections from archaeological femoral bone specimens retrieved from Early Neolithic to Late Iron Age sites. Inter-evaluator result scores demonstrated that VHI evaluation of virtual sections was comparable with original OHI ratings of ground thin sections. Certain microscopic structures considered by the OHI, however, could not be differentiated using micro-CT (e.g., concentric lamellae, osteocyte lacunae), indicating a need for higher-resolution modalities (e.g., SR micro-CT). Though the VHI method remains qualitative and subjective, the congruent results demonstrate promise for employing a virtual histology approach for the scoring of diagenetic changes consistent with the OHI.

Current research employing synchrotron-based techniques for the study of bone diagenesis remains sparse, though there has been a recent rise in its use by anthropological scholars. Caruso and colleagues, for example, employed phase contrast SR micro-CT to understand how mineral and organic components influence one another as well as how diagenesis proceeds throughout bone's cortical envelopes (Booth et al., 2016; Caruso et al., 2020; Duffett

Carlson et al., 2022). There is disagreement among researchers, however, regarding the penetration of bone by osteolytic microbiota and how they gain access to the vascular network (e.g., endogenous versus exogenous initiation) (Bell et al., 1996; Booth et al., 2015, 2016; Caruso et al., 2020; Duffett Carlson et al., 2022). As exogenous initiation is anticipated to progress from the periosteum inward, 3D assessment and visualization of complete bone specimens may allow for the tracking of progress of microbial alterations through tissue envelopes to further explore these hypotheses.

Caruso et al. (Caruso et al., 2020) employed a multimodal approach (e.g., SR micro-CT, Fourier transform infrared spectroscopy, optical and scanning electron microscopy) to characterize the postmortem trajectory of diagenesis in a combined archaeological and modern human sample to address the complex interplay of factors influencing the diagenetic trajectory, including organic and mineral components. Their work is one of the few studies that presents a quantitative assessment of vascular canal volume, diameter, and connectivity retrieved from SR micro-CT data, and offers comparisons between total porosity and "diagenetic porosity". Stock and colleagues (Stock et al., 2022) further used both micro-CT and SR micro-CT to quantify microstructural parameters and detect bioerosion in archaeological metacarpal bones to infer both age-at-death and the assessment of 3D diagenetic alterations. This approach highlights the strength of combining imaging modalities for both the assignment of bones to age cohorts and clarifying the presence/absence and nature of diagenetic changes.

Such high-resolution investigations may enable solutions to challenging cases in forensic and archaeological research (e.g., burial environment), calling for a timely need for a better understanding of diagenetic changes at the microstructural level. Recent research in paleoradiology has gained traction, for example, demonstrating that (1) virtual 2D cross-sections obtained nondestructively via micro-CT are comparable to traditional hand-ground sections for the diagnosis and scoring of bone bioerosion (Mandl et al., 2022), and (2) volume render micro-CT data allows for the analysis of detailed 3D microstructural models and the potential tracking of bone diagenesis (Duffett Carlson et al., 2022).

12.7.2 Forensic and Bioarchaeological Applications

The study of both the bone remodeling process and forensically relevant queries has been facilitated by advances in high-resolution 3D imaging modalities. Techniques including (SR) micro-CT allow for macro- and microanalyses of bone using multi-planar and volumetric data at higher resolutions than previously possible. The former, for example, has been used for macro-scale modern forensic case inquiries related to traumatic fracture and identifying the timing of traumatic events. Examples include (1) evaluating cranial fractures in blunt force trauma and accident reconstruction (Baier et al., 2020; Brown et al., 2011), (2) identifying rib fractures in pediatric cases (Baier et al., 2019, 2021), and (3) fracture detection in the hyoid bone and thyroid cartilage in suspected cases of strangulation (Fais et al., 2016; Kettner et al., 2014). Micro-CT has further facilitated the evaluation of features associated with gunshot wound trauma on various skeletal elements initiated by projectiles of varying velocities (Caister et al., 2020; Jennings et al., 2018; Newton et al., 2020; Rickman & Shackel, 2019).

Accurately distinguishing ante-, peri-, and postmortem trauma using micro-CT has also proven useful for reconstructing biological profiles of individuals from bioarchaeological contexts. For example, during the osteological analysis of the remains of King Richard

III, a total of eleven cranial defects were found, nine of which were classified as perimortem and two as taphonomical alterations (Appleby et al., 2015). Additionally, in a reexamination of an arrowhead embedded in bone, micro-CT visualization revealed crushed and compacted trabecular bone with no signs of healing, contradicting an initial clinical CT diagnosis of sclerotic bone indicative of antemortem trauma (Flohr et al., 2015). In another investigation of projectile point trauma, Lieverse and colleagues (Lieverse et al., 2014) used several imaging techniques ranging from clinical CT to SR micro-CT to conduct a multilevel analysis of a mandibular injury (Figure 12.5a, b). In addition to diagnosing the injury as perimortem, the authors were also able to determine that missing central incisors were of congenital origin instead of due to traumatic injury.

Micro-CT analyses further assist in establishing the nature of the event(s) leading to injury (i.e., accidental vs. non-accidental traumatic fractures), the identification of the object(s) inflicting the injury, and the reconstruction of actions causing the injury. Distinguishing between non-accidental and accidental fractures is most challenging in infants and elderly individuals, for example, and high-resolution imaging can assist with the interpretation of complex trauma in such cases where bone fragility and fracture risk are high (K. A. Collins & Presnell, 2007; Dolinak, 2008; Murphy et al., 2013; Rosen et al., 2016; Rosenblatt et al., 1996). In another example from the archaeological record, Wheeler et al. (2013) reported a case of child abuse in ancient Egypt through analysis of a series of

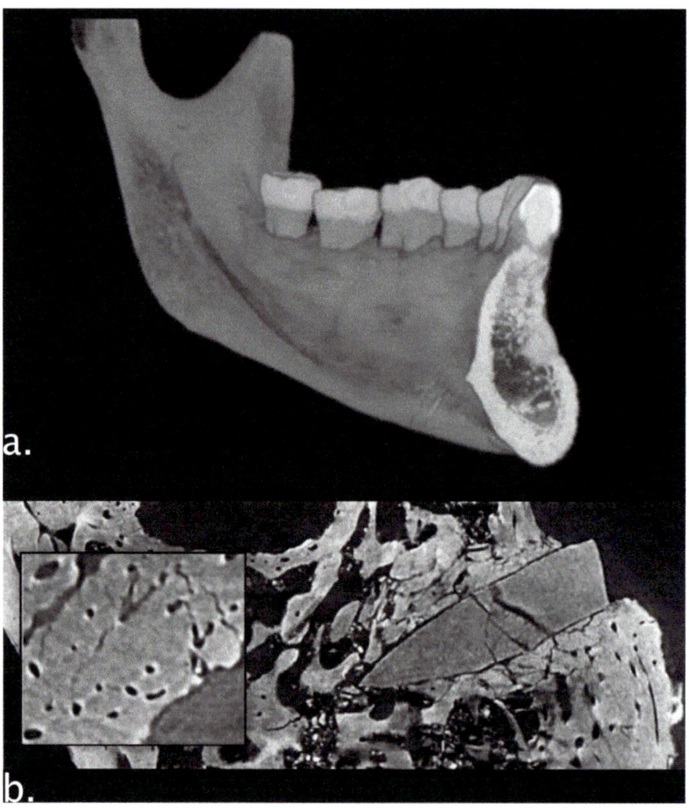

Figure 12.5 Projectile point embedded in mandible. (a) overview scan (HR-pQCT); (b) transverse section through the projectile point with an insert showing vascular canals (SR micro-CT, voxel size=10 µm, Lieverse et al., 2014).

ante- and perimortem fractures, including a humeral fracture, displaying new bone formation and a sclerotic margin that was not clearly visible on plain X-ray. Brinker and colleagues also identified distinctive fracture patterns associated with different penetrative injuries caused by different types of weapons (Brinker et al., 2018). Other bioarchaeological examples include distinguishing between similarly shaped weapons such as axes and adzes (Moreno-Ibáñez et al., 2021), and identifying a specific weapon type such as a dagger (Nicklisch et al., 2022) or sword (Zeppilli et al., 2023).

Tool mark analysis in bone has been another recent area of experimental inquiry in the forensic anthropological realm, given that the size and shape of tool mark characteristics may indicate the class of tool potentially employed to inflict injury. Thali et al. (Thali et al., 2003), for example, were the first to employ micro-CT to evaluate patterned injuries in bone produced by knives. Others have documented the morphology of sharp force injuries on bone and cartilage including the walls, floor, angles, shape of patterned injuries, and compared their findings to existing scanning electron microscopy data (Capuani et al., 2013; Stanley et al., 2018). The analysis of false starts (incomplete kerfs) produced by various hand saws has further allowed for accurate links between tool marks and saw class, and the prediction of saw blade thickness and tooth set (Alsop et al., 2021; Norman et al., 2018; Pelletti et al., 2017).

Recent research by the Andronowski Lab (see Chapter 10) has pioneered the application of SR micro-CT technology to forensic anthropological queries. Early research included evaluating differences in 3D bone microarchitecture to shed light on differential nuclear DNA yield rates among cortical and cancellous bone tissues and bone types (Andronowski, Mundorff, et al., 2017), and documenting the microstructural organization of osteon banding in a diverse human skeletal sample for species identification (Andronowski, Pratt, et al., 2017). Current work focuses on the impact of prolonged opioid use on 2D microstructural parameters (e.g., osteon population density, cortical area), 3D variables (e.g., intracortical porosity, osteocyte lacunar density), and the development of an animal model system for evaluating the longitudinal effects on bone turnover (Andronowski et al., 2021; Andronowski, Davis, & Cole, 2020; Andronowski et al., 2021, 2022; Andronowski & Depp, 2022). The primary objectives are to (1) provide foundational data related to longitudinal bone remodeling trajectories for insight into the clinical effects of prolonged opioid use in humans, (2) uncover potential mechanisms by which opioids impact bone quality, and (3) investigate potential impacts on forensic skeletal identification methods.

Preliminary research conducted via SR micro-CT technology using a known sample of opioid users (e.g., fentanyl, heroin) has yielded compelling results. In the anterior femur, for example, percent porosity and mean pore thickness of young drug users considerably exceeded young nonusers, and were more comparable to elderly nonusers (Figure 12.6); (Andronowski, Davis, & Cole, 2020; Andronowski & Depp, 2022). Osteocyte lacunar density and lacunar thickness of young drug users were further decreased compared to young nonusers, indicating further disruption to the osteocyte network (Andronowski, Davis, & Cole, 2020). Rib microarchitecture from these same individuals was further evaluated using micro-CT (Andronowski, Davis, & Cole, 2020). Percent porosity of two out of three young drug users exceeded the young nonuser and was more comparable to elderly nonusers. The ribs of both drug users and elderly nonusers further displayed elevated cortical pore density and pore connectivity density.

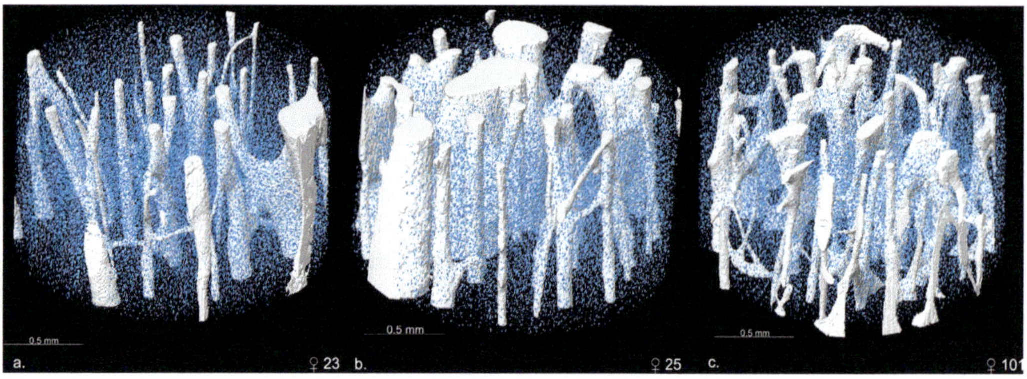

Figure 12.6 Three-dimensional SR micro-CT renders of midshaft femoral bone from a known opioid user (b) compared to younger nonuser (a) and older (c) controls. Both qualitatively and quantitatively, increased pore density and connectivity are evident in the former compared to the younger control. Scale bar = 0.5 mm (image credit: Joshua T. Taylor & Janna M. Andronowski).

These results suggest that chronic opioid use mimics age-associated bone loss, including increased cortical bone porosity, and pore thickness and separation. As such, current histological methods developed from healthy bone specimens may not be useful in forensic cases involving suspected substance abuse and result in the inaccurate reporting of age-at-death as part of the biological profile. Future planned work will leverage such 3D tools to examine bone microarchitecture from past populations and glean insights into bone microstructural variation without the addition of modern lifestyle variables, such as medication regimes and substance use.

12.7.3 Functional Adaptation

In recent decades (SR) micro-CT imaging has been employed to address bone adaptation queries in anthropologically relevant samples. Such imaging technologies possess capabilities to accurately assess whole bone or local volumes nondestructively from various skeletal sites for the analysis of human (or nonhuman) cortical and trabecular microarchitecture (Akbulut et al., 2021; Carter, Thomas, Clement, Peele, et al., 2013; Cooper et al., 2003; Ding et al., 1999; Fajardo & Müller, 2001; Kuhn et al., 2007; Le Garff et al., 2017).

Fajardo and Müller were among the first to employ micro-CT to investigate the relationship between locomotor behavior and trabecular structure in a nonhuman primate sample (Fajardo & Müller, 2001). Their preliminary work evaluated the potential of micro-CT in functional studies of primate trabecular architecture specifically and introduced relevant methodological issues pertinent to such comparative studies.

Investigating bone functional adaptation from the context of human behavior has long been a topic of interest within the anthropological community. Various researchers have focused on limb asymmetry in archaeological populations to infer behavior and study age-related change (Auerbach & Ruff, 2006; Pfeiffer, 1980; Ruff & Jones, 1981), while others have used modern athletes as subjects to better understand repetitive mechanical loading and thus variation in bone structural parameters (Macintosh & Stock, 2019; Payne et al., 2018; Shaw, 2011). Though certain researchers have studied upper limb bone asymmetry in an attempt to link macro- and micromorphology with handedness and favoring of a limb during habitual activities (Lazenby, 2002; Lazenby, Cooper, et al., 2008; Trinkaus et al., 1994), Perchalski and colleagues tracked asymmetry throughout ontogeny. The authors used micro-CT to analyze cortical and trabecular bone in developing paired humeri from a subset of individuals ranging in age from neonate to adult (Perchalski et al., 2018) from the Norris Farms #36 archaeological skeletal collection. Their work failed to identify significant directional asymmetry in trabecular bone in the humerus, though cortical bone displayed dominant directional asymmetry in the right humerus (Perchalski et al., 2018). This right-sided bias highlights the plasticity and capability of the humeral diaphysis to respond to mechanical stimuli. The lack of significant change in the trabecular bone and expression of increased cortical changes may be caused by the increased responsiveness of the periosteal surface of the humerus during pubertal growth (Perchalski et al., 2018).

Right-sided bias has further been identified in human metacarpals with a roughly equal sex distribution from a historically known 19th-century EuroCanadian site (Lazenby, Angus, et al., 2008). Lazenby and colleagues investigated inter- and intra-site epiphyseal variation in trabecular microarchitecture with foci on trabecular bone volume fraction, surface-to-volume ratio, trabecular number, connectivity, and structural model index (see Table 12.1), which all indicated a right-sided bias (Lazenby, Angus, et al., 2008). The SMI on the right side revealed plate-like trabeculae, which provide greater ultimate strength and resistance to compressive and shear strain (Lazenby, Angus, et al., 2008; Milovanovic et al., 2017). Further, the group determined that mechanical properties were most directly linked to bone mineral density, bone volume fraction, and SMI (Lazenby, Angus, et al., 2008). Further research has been conducted to study if this right-sided dominance is observable in other species. For example, evidence suggests that right-hand dominance emerged after the *Pan-Homo* split and is possibly linked to the use of tools (Stephens et al., 2016). The analysis revealed that *Pan* had a left-directional symmetry and *Homo* had a right-directional symmetry (Stephens et al., 2016).

Furthering the work by Perchalski, Milovanovic et al. (2017) studied the developmental trajectory of femora from individuals ranging from one month to twelve years from the late medieval graveyard of Stara Torina, Serbia. Using micro-CT, the authors analyzed biomechanically relevant subregions of the proximal femur (e.g., medial, intermediate, and lateral neck, intertrochanteric, and metaphyseal). Results revealed a decrease in bone volume fraction, trabecular number, and degree of anisotropy until six months of age, when the values began to increase (Milovanovic et al., 2017). The changes in trabecular morphology, especially the degree of anisotropy, are potentially linked to mechanical forces and initiating bipedalism. The SMI of the proximal femur indicated that rod-like trabeculae transitioned to a plate-like organization with age (Milovanovic et al., 2017), indicating increased compressive loads on the proximal femur. Previous studies further support that plate-like trabeculae may serve as an adaptative advantage to increased mechanical loads (Ding et al., 2002; Djuric et al., 2010; Milovanovic et al., 2012, 2017).

Micro-CT has also been used to investigate cultural and anthropologically relevant practices. For example, due to the mechanical adaptability of the skeleton and the weight carried by the lower limb, forced skeletal deformations may result in the redistribution of bones' microarchitecture. Foot binding, a prohibited historical practice in China, aimed to restrict the growth of the female foot structure and resulted in skeletal deformation (Al-Akl, 1932; Blake, 1994; Reznikov et al., 2017). Reznikov and colleagues examined a subset of bound-foot specimens from the 1800s that were donated to the Royal College of Surgeons of England. The authors found that the distal portion of bound calcanei had significantly lower trabecular thickness and no recognizable anisotropic pattern, but the degree of anisotropy was more consistent among long bones (Reznikov et al., 2017). Their study further demonstrates the ability of the calcaneus to adapt to unique mechanical constraints and indicates that anisotropic trabecular bone can be an acquired characteristic imposed by cultural practices.

To explore the early development of the talus, Figus et al. (2023) used micro-CT to examine trabecular and geometric morphometric parameters from an archaeological sample ranging from the Late Roman period to the 20th century. In the youngest age group (<6 months), the authors found elevated trabecular number and bone volume fraction when compared to the older, post-loading groups. Consequently, trabecular separation and thickness decreased (Figus et al., 2023). Most changes to the talus' cortical shell were experienced during the first year, likely initiated by the onset of bipedal locomotion (Figus et al., 2023).

12.7.4 Aging and Age Estimation

Aging of the skeleton is associated with thinning of the cortices, increases in cortical bone porosity (Figure 12.7) as well as thinning and loss of trabecular bone. Micro-CT has the capacity to directly assess these microstructural features, rather than inferring them from lower resolution or densitometric approaches, and thus has great potential for anthropological inquiry ranging from assisting age estimation to the study of aging in a more general sense.

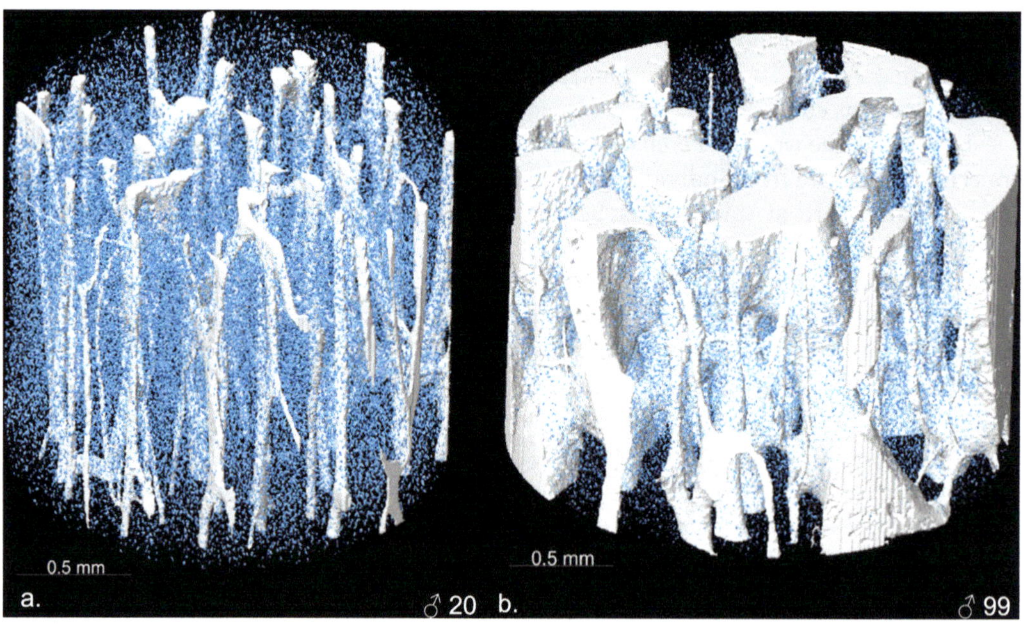

Figure 12.7 Three-dimensional SR micro-CT renders of human cortical bone from the Andronowski Skeletal Collection for Histological Research demonstrating age-associated changes of osteocyte lacunar and pore morphometry. Note the denser collection of osteocyte lacunae in the younger individual (a) compared to the older individual (b) The older individual further exhibits increased porosity from increasing pore diameter (but not by pore number), suggesting pore coalescence occurs with advancing age. Scale bars: (a) = 0.5 mm; (b) = 1 mm (image credit: Joshua T. Taylor & Janna M. Andronowski).

When discussing the topic of age-related changes in bone within an anthropological context, a topic that immediately comes to mind is estimation of age-at-death. With respect to microstructural aging, traditional light microscopy has been utilized as the basis for a variety of histomorphometric age estimation methods (see Chapter 8). Moving into the third dimension, Cooper (2005) employed micro-CT to explore cortical bone morphometric parameters as a basis for age-at-death estimation. The theoretical underpinning for this work was that the accumulation of secondary osteons via remodeling over time would be associated with an increase in the connectivity of the canal network in 3D (Cooper et al., 2003). It was revealed that age was responsible for 59% percent of the variation observed in canal connectivity density in combined males and females (samples derived from the Melbourne Femur Collection), and that canal number increased with age while degree of anisotropy, canal diameter, canal separation, and canal surface-to-volume ratio decreased. While this work was successful in developing statistically significant regression formulae, the overall precision of these formulae (standard errors of the estimate of 13.4, 10.7, and 14.2 years for the pooled sexes, females and males, respectively) did not improve upon values published for 2D histomorphometric techniques. While disappointing on one hand, the variation observed in the 3D cortical bone microstructural parameters is ultimately underpinned by variation in how we age as individuals and this is another area of inquiry of great interest to biological anthropologists.

A considerable amount of X-ray-based analyses of age-associated bone loss and osteoporosis in the bioarchaeological literature focus on BMD (Brickley et al., 2020; Stride et al., 2013; Van Spelde et al., 2021). Tarasiuk and colleagues (Tarasiuk et al., 2023) measured bone volume through proportional attenuation (CT value)-based segmentation, then calculated BMD via dividing weighted sample bone mass by bone volume (Tarasiuk et al., 2023). This may be, at least in part, due to the current definition of osteoporosis being based on BMD (WHO, 1994). Trabecular bone structure has been of particular interest since its gradual thinning, tissue loss, and structural connectivity have been associated with age-related imbalances in bone remodeling (Gosman et al., 2011). It further is readily visible without the need of high magnification (S. C. Agarwal et al., 2004), and when assessed with micro-CT, morphological changes can be both qualitatively and quantitatively assessed (Villa et al., 2019).

HR-pQCT has also been used to evaluate trabecular structures of fourth lumbar vertebral bodies in the Neolithic population of Çatalhöyük, Turkey and Imperial Roman individuals (S. Agarwal et al., 2011; Beauchesne & Agarwal, 2017). Trends of age-related bone loss similar to modern populations were observed, although the Imperial Romans displayed a less drastic decrease in bone volume and no sex differences. Teschner provided a multimodal analysis using both histomorphometry and micro-CT, offering higher resolution than HR-pQCT, to quantitatively assess a young Neolithic individual whose trabecular bone volume was significantly higher than individuals from the modern population (Teschner, 2007). Faccia and colleagues further measured cortical and trabecular parameters of Baikal foragers and discovered that their bone quality remained relatively stable throughout life (Faccia et al., 2014).

12.7.5 Paleopathology

Application of micro-CT imaging in qualitative and quantitative analysis has seen a substantial increase in paleopathology over the past 20 years. It is usually employed in conjunction with other methods including macroscopic examination, stable isotope analysis, histomorphometry, molecular, and chemical analysis, as well as other imaging modalities. The high resolution and less destructive nature of micro-CT make it a valuable tool to aid in the identification and diagnosis of internal microstructural skeletal changes as bone responds to disease.

Qualitative analysis (visual interpretation) represents a considerable part of micro-CT studies in bioarchaeology and paleopathology today, stemming from the emphasis on morphology and the descriptive nature of classic paleopathology. The diagnostic value of micro-CT has been confirmed through a series of comparisons between thin-section histology and micro-CT visualization of cranial and postcranial lesions including those linked to infection, trauma, metabolic disease, and neoplasm (Kuhn et al., 2007; Rühli et al., 2007). When features such as formative and resorptive reactions, trabecular orientation, and lesion morphology are the primary targets of the study, micro-CT may be preferred as its quality is comparable to histology and has the added benefits of minimal destructive sample preparation and the generation of 3D data. The latter means that micro-CT is increasingly used in place of histology where specimens are rare and/or precious. Examples of this include hominid fossils (Beaudet et al., 2021; Odes et al., 2018), samples derived from historical figures (Appleby et al., 2015), and cases of rare abnormalities/disorders (Coutinho-Nogueira et al., 2019; Čavka et al., 2022; Lieverse et al., 2012; Milella et al., 2015; Odes et al., 2018; van Duijvenbode et al., 2015; Zhao et al., 2023). Conversely, histological methods may be required in cases where fine microstructural details, such as lamellae and collagen orientation, need to be assessed.

Tuberculosis (TB) continues to generate paleopathological interest due to its contemporary clinical relevance (WHO, 2023), as well as its diverse skeletal manifestation and its elusive origin in light of new yet confounding molecular evidence. Complemented by micro-CT evidence suggestive of early signs of skeletal TB, Baker and colleagues (Baker et al., 2015, 2017) reported molecular analysis confirming the presence of the earliest known cases of modern human TB at a pre-domestication Neolithic site with no morphological evidence of animal TB. Nicklisch et al. evaluated rib lesions of TB using both micro-CT and histology (Nicklisch et al., 2012), and suggested an origin of soft tissue inflammation based on lesion morphology. Vekszler et al. examined multiple skeletal sites including vertebral bodies, femoral heads, and tibiae of known TB cases compared with non-TB controls (Vekszler et al., 2022, Figure 12.8a). The authors observed the most noticeable microstructural changes in tibiae, including decreased cortical and trabecular thickness, sclerosis, and pronounced cortical porosity and trabecularization. In another study of TB vertebral lesions in a young sub-adult, trabecular thickness showed a significant increase while trabecular separation and connectivity both showed a trend towards bone loss, and sclerosis was not noted (Coqueugniot et al., 2015, Figure 12.8b). The authors suggested the focal microarchitectural lesion was due to the pathogen reaching trabecular bone at an early stage of the disease through capillary endings.

Lamm and colleagues examined long bones with osteomyelitic lesions, where micro-CT revealed the internal structure of massive bone loss in the forms of cortical porosity, cortical thinning, and trabecularization, with peripheral compensatory osteoblastic activity (Lamm et al., 2015). Fraberger et al. evaluated internal alterations of cranial bone microstructure in syphilis and found extensive destructive lesions presenting as perforating defects, cortical porosity, and thinning or complete loss of outer and inner tables (Fraberger et al., 2021), which differed from the findings of Rühli et al. who observed cortical thickening with sclerosis (Ruhli et al., 2007) Importantly, although high-resolution 3D data were generated, Fraberger and colleagues still used a conventional, descriptive scoring-based system to

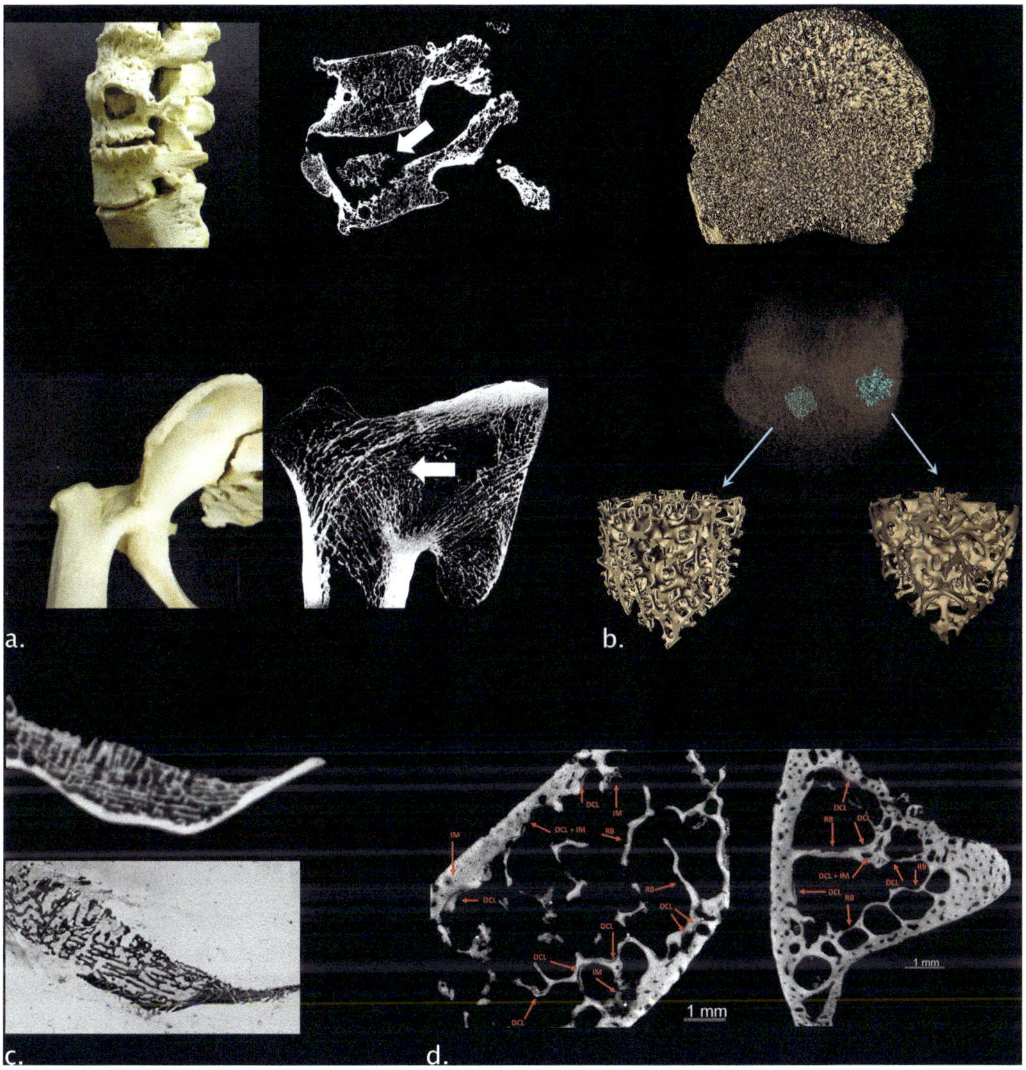

Figure 12.8 (a) Comparison of tuberculosis lesions between dry bone and micro-CT. top: A\a. sequester inside L3 vertebra, surrounded by a defective zone and a sclerotic zone; bottom: an ankylotic right hip joint showing trabecular defect (Vekszler et al., 2022). (b) 3D rendering of early signs of vertebral trabecular microstructural change (Coqueugniot et al., 2015). (c) micro-CT image (top) showing comparable microstructural details compared to histological thin sections (bottom) in *cribra orbitalia* lesions (Saint-Martin et al., 2015). (d) Features exhibiting defective mineralization seen in vitamin D deficient bone: defective areas adjacent to cement line (DCL), improper mineralization (IM), and resorptive bays (RB) (voxel size=6 μm, Welsh et al., 2020).

classify the presence and severity of the lesions (Fraberger et al., 2021) reflecting a blending of qualitative and quantitative applications of micro-CT.

Detailed evaluation of noninfectious inflammatory responses such as porosity, new bone formation, and erosive lesions can also benefit from micro-CT analysis. Dittmar and colleagues utilized micro-CT to evaluate the scooped-out, lytic lesion characteristic of gout, an inflammatory arthritis often associated with diets rich in alcohol and purine that could be linked to socioeconomic status (Dittmar et al., 2021). Cranial porotic lesions are a common occurrence in paleopathology that could indicate inflammatory response and have therefore been evaluated with CT-based imaging at various resolutions. Clinical CT, limited by its resolution, can only detect porosities with large diameters or having adjacent lesions extending into diploe (Anderson et al., 2021). Saint-Martin and colleagues compared the HR-pQCT scans (41 μm voxel size) and histological sections of cribra orbitalia specimens with similar macroscopic morphology (Saint-Martin et al., 2015). HR-pQCT produced visualization comparable to thin sections capable of identifying distinctive microstructural lesion patterns (Figure 12.5c). An even higher resolution study using micro-CT was conducted by Morgan to explore the differences between cranial and orbital porotic lesions and the underlying anatomical and physiological factors (Morgan, 2014). In an assessment of diffuse idiopathic skeletal hyperostosis (DISH) of the vertebral column, conventionally considered a noninflammatory degenerative condition, Holgate and colleagues described localized erosive cortical and formative trabecular and woven bone lesions, leading to the hypothesis of a potentially inflammatory origin, adding new information in the understanding of the pathophysiology of DISH (Holgate et al., 2023).

Micro-CT can be especially well suited for investigating conditions affecting bone remodeling dynamics. Welsh and colleagues compared the visibility of features representative of vitamin D deficiency between micro-CT, BSE-SEM, and light microscopy, and concluded that poor/incomplete mineralization and resorption activities are detectable at 6 μm voxel size (Welsh et al., 2020, Figure 12.58d). Trabecular thinning and perforation as a result of increased remodeling rate in vitamin D deficiency may also be observable on micro-CT (Brickley et al., 2020). Another disease that primarily affects remodeling dynamics is Paget's disease of bone (PDB). The distinctive macro- and micro-structural features of PDB mean that even with lower resolutions in plain radiography, clinical CT and micro-CT at relatively larger pixel sizes, diagnostic features such as cortical and trabecular thickening, diffuse sclerosis and uneven radiodensity can still be assessed (Wade et al., 2011). Using semiquantitative and quantitative methods at 40 μm pixel size, Nebot et al. were able to show regional differences in cortical and trabecular parameters in the femoral shaft, suggesting an uneven distribution of lesions that can potentially increase fracture risk (Nebot et al., 2019). Others have combined micro-CT and histomorphometric methods to show irregular osteon morphology, convoluted and highly vascularized trabeculae, as well as disorganized bone matrix, which supported the diagnosis of the first known case of PDB in the Middle East (Kesterke & Judd, 2019).

Paleopathologists face unique challenges due to the complex interplay between disease stage and mechanism, the mechanical strength requirement and anatomical position of the affected bone, as well as the age, health, and socioeconomic status of the affected individual, in addition to the absence of soft tissue and the associated clinical diagnostic tools. Therefore, micro-CT can be of great value in providing qualitative and quantitative, 3D high-resolution data to complement the conventional macroscopic and 2D microscopic analyses in the paleopathological diagnostic process.

12.8 Limitations and Technical Considerations

As described herein, micro-CT technology has proven invaluable for the characterization and visualization of bone microstructure. The outcomes achieved by X-ray micro-CT imaging systems, however, have intrinsic limitations determined by: (1) the interplay between resolution and field of view, (2) varied scan parameters across studies, (3) the imaging processing algorithms employed, and (4) access to the hardware/facilities necessary to acquire, process, and analyze the data. An important consideration of what resolution is needed to address one's research question and the size/nature of the specimens are critical.

12.8.1 Resolution vs. Field of View

A considerable limitation of micro-CT includes the limited FOV. There is generally an inverse relation between FOV and resolution, with smaller voxels (higher resolution) directly correlated with smaller FOVs. When pixel size decreases below 10 μm, for example, the FOV is typically limited to only a few millimeters in width and height. There are workarounds, however, that include sectional scans and subsequent computational tiling of images to generate larger FOVs with the trade-offs being increased scan time and data generation. The latter can also be a limitation in that micro-CT can consume tremendous computing power and storage. Further efforts to improve the resolution and FOV of micro-CT utilize large detector arrays and revised image acquisition and reconstruction techniques (e.g., spiral CT, distortion correction). The use of local CT (reconstructing a sub-portion of a target) and large FOV industrial micro-CT systems offer potential workarounds that can achieve high resolution while keeping larger (potentially precious anthropological) samples intact.

In a similar vein, although often described as "noninvasive", sample preparation for micro-CT is frequently destructive, at least to a degree. Apart from synchrotron facilities and industrial micro-CT systems, conventional scanners often have confined space within their gantries that may not be suitable for holding larger intact bones, given the trade-off between FOV and spatial resolution. Sampling may need to be opportunistic by prioritizing already fragmented bone if a specific element is not required. Otherwise, smaller regions need to be sectioned or cored from the larger sample (Andronowski, Davis, & Holyoke, 2020). Further, the intensity and radiation levels of the X-ray beams associated with synchrotron radiation can damage or alter specimens, particularly at high resolution. Loundagin and colleagues (Loundagin et al., 2020), for example, reported that SR micro-CT reduced the fatigue life of human femoral and tibial cortical bone samples scanned at 0.9 μm by 91% (Boone et al., 2012). With respect to bioarchaeological/paleoanthropological samples, SR micro-CT has been reported to result in discoloration (Richards et al., 2012), and ancient DNA damage (Immel et al., 2016).

12.8.2 Image Acquisition and Reconstruction Parameters

Due to the great variety in the shape, size, and structural integrity of anthropological specimens, as well as technological variations between scan configurations, there are no fixed protocols assigned when setting up imaging experiments. Various parameters can be chosen somewhat arbitrarily during the imaging setup including voltage, total exposure time, and beam filtering, which all influence results. Additionally, the reporting of scan parameters is not consistent across the published literature. Many researchers report details including the make/model of the machine, energy, and sometimes rotation step and filter selection; however, others neglect to report voxel/pixel size. This lack of reporting of scan parameters also makes comparison across studies less reliable and influences the design of future reproducible studies.

Besides operator-dependent image acquisition considerations, obtaining quantitative results from the 3D volume is crucial. Arguably the most critical operation in this analysis is the segmentation of the volume, which can be based on various thresholding methods, semiautomatic toolkits (Cole et al., 2021), and AI algorithms. Segmentation can be further complicated by, for example, phase contrast from SR micro-CT scans, where the underlying principles of density-based segmentation widely applied to micro-CT scans are not valid. Apparent increases in density at bone surfaces as well as high contrast (approaching that of bone) for soft-tissue remnants can greatly complicate conventional segmentation approaches. Phase retrieval approaches (discussed in Section 12.5.8) may ultimately prove useful in overcoming this complication.

Artifacts encountered during image processing include: (1) ring artifacts, for which there are many causes, but all are associated with pixel response (e.g., imperfect detector elements such as gain error (Sijbers & Postnov, 2004)), (2) streak artifacts, often caused by highly attenuating inclusions (e.g., metal fragment from sample preparations), and (3) movement artifacts caused by instability or movement of a sample during imaging experiments. Methods for overcoming these issues, again, vary and thus can introduce further differences between studies. Ultimately, the scientific validity of 3D digitization methods pertaining to their accuracy, precision, and reliability must be established, and indeed, methods utilizing micro-CT are no exception (White et al., 2018).

12.8.3 Image Processing

While imaging and related computing power are becoming increasingly accessible and affordable, the quality/resolution of imaging is also opening new possibilities for analysis. Image acquisition, processing, and analysis are all robust fields in their own rights and fertile collaborative ground for biological anthropologists seeking to capitalize upon cutting-edge technology to pursue their research questions. Advances in computing power have helped in this regard; however, but with ever-increasing resolution, larger detectors, and higher throughput systems, the concomitant computing demands continue to grow. The ability to visualize and resolve 3D internal structures of invaluable anthropological specimens nondestructively is a main advantage of micro-CT. Surface rendering software allows for the visual inspection of 3D volume-based data. Though such visual inspections offer qualitative information that have proven useful in numerous anthropological queries (e.g., diagenetic alterations, pathological lesions), quantitative results are often required. As described earlier in this chapter, quantitative 3D morphometric analyses of bone include data related to volume fractions, pore size parameters, and morphology (e.g., shape, sphericity), among numerous others (Table 12.1). In practice, however, quantitative 3D analyses are often not straightforward and can be prone to systematic errors introduced by noise, imaging artifacts, and operator-dependent decision making.

12.8.4 Cost/Access

Although micro-CT is being increasingly applied in anthropological inquiries, the availability and cost still limit accessibility. In many cases, specimens must be transported to and from imaging facilities to make use of regular laboratory units or take advantage of more advanced techniques such as SR micro-CT to achieve higher spatial resolution. Museum curators, for example, may be reluctant to agree to ship or relocate precious and fragile specimens due to the risk of damage. As such, logistic arrangements may be both challenging and cost prohibitive. Perhaps counterintuitively, SR micro-CT may have lower direct costs for researchers as time is generally awarded, free of charge, based upon scientific merit, via peer-review. That said, there are only about 70 synchrotrons in the world (currently operating or in development). Thus, synchrotron beam time is inherently limited and frequently involves travel for data collection, although remote/automated access for some applications is possible.

12.9 Future Directions

12.9.1 *In Vivo* Imaging of Animal Models

As already noted, HR-pQCT occupies a niche between micro-CT and clinical CT, enabling the *in vivo* detection of trabeculae and even cortical porosity in human subjects, although the resolution remains limited by radiation dose. A topic we have not explored to this point is what is possible in terms of imaging animal models *in vivo*? We have chosen to discuss this as a future direction for the field as the use of animal models – particularly imaged alive to track microstructural features – is an emerging area of inquiry for biological anthropologists. There are questions and mechanistic studies that simply cannot be pursued in human subjects or their remains and animal models provide a critical platform for such inquiry that we believe will become increasingly accessible in the future.

The development of *in vivo* micro-CT paralleled the overall development of micro-CT technology with an early application demonstrating the potential to track changes in the tibiae of rats, even down to the level of individual trabeculae using 20 μm voxels (Waarsing et al., 2004). Trabecular bone has since been the primary target of *in vivo* studies of animal models, with relatively few studies targeting cortical bone porosity in, for example, rats (Altman et al., 2015) and mice (Li et al., 2015) due to the smaller scale of cortical pores – necessitating higher resolution and thus radiation dose. Live animal imaging, like that of humans, is ultimately limited by radiation dose (Ford et al., 2003). One option for pushing the limits of *in vivo* imaging is the use of phase contrast SR micro-CT, which provides additional contrast at the same or even reduced radiation dose compared with conventional absorption-based micro-CT. This approach has been applied to visualize cortical porosity in rats (Harrison & Cooper, 2015; Pratt et al., 2015) as well as rabbits (Harrison et al., 2022), with the latter enabling the four-dimensional (4D) tracking of remodeling events as they create new osteons. As another example, the Andronowski Lab has future plans to pursue live rabbit 4D imaging to track the effects of opioid exposure and associated bone remodeling trajectories over time. Overall, the ability to longitudinally track remodeling in 4D holds tremendous potential for expanding our understanding of the spatiotemporal regulation of this process and its role in adaptation, aging, and disease.

12.9.2 Digital Collections

In addition to the complementary data micro-CT can provide, the scans are also a long-term digital data source. Indeed, collections of CT and micro-CT scans of paleopathological, forensic anthropological, and zoological specimens are already emerging as parts of digital databases that can be referenced in teaching, dissemination and research. The digital 3D scans of anatomical collections at the National Museum of Health and Medicine (NMHM) (Pearlstein et al., 2021) and Digital Morphology Library housed at the University of Texas at Austin (University of Texas, Austin, n.d.) are two of such examples. Additionally, a depository of over 700 SR micro-CT scans accompanying the Andronowski Skeletal Collection for Histological Research (ASCHR) (Andronowski & Taylor, 2022), a large and growing documented modern human skeletal collection, is expected to be available for online access soon (for details see Chapter 10 of this volume). Hassett also provided a list of digitized bioarchaeological data sources—although they may contain little to no micro-CT data as of now, they are nevertheless digitized human remains geared towards audiences ranging from the general public to researchers (Hassett, 2018). Digitization may be particularly suitable for rare and/or fragile specimens, or specimens that have to go through destructive sampling such as ancient DNA as both a permanent record and an easy-to-access source that benefits researchers and students around the world (Pearlstein et al., 2021; Sirak & Sedig, 2019). A critical issue associated with digital data of human remains, including micro-CT data, is ethics. The complexity here is the overlapping of ethical issues surrounding digital data such as use, ownership, and authorship, and the ethical issues surrounding human remains such as the origins of legacy anatomical collections, repatriation legislations and the rare mentioning of ethics in publications, among many others. The discussions surrounding the digitization of human remains have been going on for a long time, and will continue into the future (e.g., BABAO, 2019; Hassett, 2018; Schug et al., 2021). We are far from reaching a somewhat satisfactory consensus that can fit on the narrow balance point between the need in education and knowledge advancement, public engagement, and addressing the colonial roots of our field and the hurt that is still felt today, as well as diverse views different communities have towards the personhood of human remains and their images. However, we recognize the necessity to continue the conversation, and encourage bioarchaeologists to consider the ethical implications when conducting micro-CT imaging of human remains using the code of ethics pertaining to research, teaching, application, dissemination, and public engagement provided by anthropological associations around the world as frameworks (AAA, 2009; AABA, 2003; BABAO, 2019).

12.9.3 Artificial Intelligence

The emergence of artificial intelligence has the potential to transform data processing and segmentation techniques, with direct applications for medical imaging and biological anthropology. A subsection of AI, deep learning, learns through complex computer algorithms that are aligned in a network similar to neurons. The interlaying nodes ("neurons") work in coordination to detect and categorize structures based on their features without prior human input. Applying semantic segmentation and convolutional layers to deep learning has substantially improved image recognition and segmentation (Fan et al., 2020). The potential of AI in clinical fields has led to the development of U-Net (Ronneberger et al., 2015), U-Net++ (Zhou et al., 2018), and MA-Net (Fan et al., 2020), among other network architectures designed for medical imaging.

Of particular anthropological relevance, deep learning has been used to assess bone pathological conditions (He et al., 2020; Von Schacky et al., 2021), fracture detection (Chung et al., 2018; Dimililer, 2017; Lindsey et al., 2018; Rajpurkar et al., 2017), and skeletal age using radiographs (Larson et al., 2018; Lea et al., 2022; Lee et al., 2020). For example, deep learning outperformed junior radiologists in detecting bone tumors and performed comparably to experienced radiologists (He et al., 2020; Von Schacky et al., 2021). Deep learning has proven to perform equally well in fracture detection, correctly identifying fractures in 95% of cases (Chung et al., 2018; Dimililer, 2017). Deep learning algorithms have further been tested extensively for age and sex determination in forensic anthropology. Skeletal maturity of subadults, for example, can be assessed using radiographs of the hand. Age estimation via deep learning methods were found to be comparable to reviewers (Larson et al., 2018), with an error rate of 9.35 months (Lee et al., 2020). Deep learning is approximately 10% more accurate than observers when utilizing the pelvis in sex determination (Cao et al., 2021, 2022) and humerus (Venema et al., 2023). Similarly, artificial intelligence achieved a 90.3% accuracy in sex determination from knee radiographs (Oura et al., 2023) and a 95% accuracy utilizing photographs of human crania (Bewes et al., 2019). Artificial intelligence has demonstrated the potential to assist and improve image detection in the clinical and anthropological fields. Professionals can use such algorithms to reduce processing time, increase accuracy and precision for age and sex estimation, and eliminate interobserver error.

12.10 Conclusion

Biological anthropology is an inherently multidisciplinary field and one of its key strengths is its capacity and, indeed, desire to capitalize on technologies and theoretical approaches from many fields to extract as much information as possible from limited, and thus precious, specimens. In this vein, anthropologists have eagerly adopted imaging technologies, reaching back to the discovery of X-rays almost 130 years ago, microradiography 75 years ago, computed tomography 50 years ago, and micro-CT which was first developed 40 years ago – including the use of conventional and synchrotron X-ray sources. In more recent decades, micro-CT has been widely adopted in bone research, becoming the gold standard for 3D histomorphometric analysis. As the application of this technology continues to mature within biological anthropology we anticipate an increasing shift from complementary qualitative visualization, to increasingly quantitative analyses which improve inter-study comparisons within the field and beyond – particularly with respect to animal model research. The potential for the digital sharing of data is particularly exciting for its potential to accelerate discovery and facilitate more rigorous evaluation and re-evaluation into the future. The COVID-19 pandemic, for example, introduced a major shift to virtual collaboration and resulted in the increased availability and use of digital repositories and data dissemination. Continuing to build on this momentum has allowed for a great expansion of fruitful collaborations and accessible data sharing within biological anthropology and beyond.

Acknowledgments

Certain research presented in this manuscript was performed at the BMIT facility at the Canadian Light Source, which is supported by the Canada Foundation for Innovation, Natural Sciences and Engineering Research Council of Canada, the University of Saskatchewan, the Government of Saskatchewan, Western Economic Diversification Canada, the National Research Council Canada, and the Canadian Institutes of Health Research. The authors would like to thank the beamline scientists at the Canadian Light Source, particularly Adam Webb, Sergey Gasilov, Arash Panahifar, and Ning Zhu for their assistance and support. Andronowski also wishes to thank Beth Dalzell from The University of Toledo College of Medicine and Life Sciences, the Forensic Anthropology Unit at the NYC-Office of Chief Medical Examiner, and the Forensic Anthropology Center at The University Tennessee, Knoxville for access and permissions to bone specimens for bone imaging research. Cooper would like to acknowledge the pivotal role that the late Dr. John G. Clement played in inspiring and contributing to many of the manuscripts referenced in this chapter. He is grateful for John's mentorship and support, along with that of C. David L. Thomas.

References

AAA. (2009). *Code of ethics of the American Anthropological Association.* http://s3.amazonaws.com/rdcms-aaa/files/production/public/FileDownloads/pdfs/issues/policy-advocacy/upload/AAA-Ethics-Code-2009.pdf

AABA. (2003). *Code of ethics of the American Association of physical anthropologists.* https://bioanth.org/documents/3/ethics.pdf

Agarwal, S. C., Dumitriu, M., Tomlinson, G. A., & Grynpas, M. D. (2004). Medieval trabecular bone architecture: The influence of age, sex, and lifestyle. *American Journal of Physical Anthropology, 124*(1), 33–44. doi:10.1002/ajpa.10335

Agarwal, S., Glencross, B., & Beauchesne, P. (2011). *Bone growth, maintenance and loss in the Neolithic Community of Çatalhöyük.* Turkey: Preliminary Results.

Akbulut, N., Çetin, S., Bilecenoğlu, B., Altan, A., Ocak, M., Şen, E., Atakan, C., & Orhan, K. (2021). Evaluation of the detectability of early mandible fracture healing findings in terms of vitality aspect by using micro-CT technology in postmortem interval. *Legal Medicine, 52,* 101914. doi:10.1016/j.legalmed.2021.101914

Al-Akl, F. M. (1932). Bound feet in China. *The American Journal of Surgery, 18*(3), 545–550. doi:10.1016/S0002-9610(32)90976-3

Albu, I., & Georgia, R. (1984). Some aspects of the system of Haversian canals in the diaphysary compact bone in man. *Morphologie Et Embryologie, 30*(1), 17–20.

Albu, I., Georgia, R., Urlea, G., & Georoceanu, M. (1984). Comparative data on the haversian canal system in the human and horse femur. *Acta Anatomica, 120*(1–2), Article 1–2.

Alsop, K., Baier, W., Norman, D., Burnett, B., & Williams, M. A. (2021). Accurate prediction of saw blade thicknesses from false start measurements. *Forensic Science International, 318,* 110602. doi:10.1016/j.forsciint.2020.110602

Altman, A. R., Tseng, W.-J., de Bakker, C. M. J., Chandra, A., Lan, S., Huh, B. K., Luo, S., Leonard, M. B., Qin, L., & Liu, X. S. (2015). Quantification of skeletal growth, modeling, and remodeling by in vivo micro computed tomography. *Bone, 81,* 370–379. doi: 10.1016/j.bone.2015.07.037

Amprino, R. (1948). A contribution to the functional meaning of the substitution of primary by secondary bone tissue. *Acta Anat, 5,* 291–300.

Amstutz, H. C., & Sissons, H. A. (1969). The structure of the vertebral spongiosa. *The Journal of Bone and Joint Surgery. British Volume, 51*(3), 540–550.

Anderson, A. S., Sutherland, M. L., O'Donnell, L., Hill, E. C., Hunt, D. R., Blackwell, A. D., & Gurven, M. D. (2021). Do computed tomography findings agree with traditional osteological examination? The case of porous cranial lesions. *International Journal of Paleopathology, 33,* 209–219. doi: 10.1016/j.ijpp.2021.04.008

Andreasen, C. M., El-Masri, B. M., MacDonald, B., Laursen, K. S., Nielsen, M. H., Thomsen, J. S., Delaisse, J.-M., & Andersen, T. L. (2023). Local coordination between intracortical bone remodeling and vascular development in human juvenile bone. *Bone,* 116787. doi: 10.1016/j.bone.2023.116787

Andronowski, J. M., Cole, M. E., Davis, R. A., Schuller, A. J., Tubo, G. R., LaMarca, A. R., & Taylor, J. T. (2021). The longitudinal effects of prolonged opioid use on cortical bone remodeling in a rabbit model: part I – intraskeletal variability and regional differences detected via micro-computed tomography. *Proceedings of the American Academy of Forensic Sciences 2021 Annual Conference*, p. 62.

Andronowski, J. M., Cole, M. E., Davis, R. A., Schuller, A. J., Tubo, G. R., LaMarca, A. R., & Taylor, J. T. (2022). Longitudinal effects of prolonged opioid use on cortical bone remodeling in a rabbit model: part II—intraskeletal histological variability and regional differences. *Proceedings of the American Academy of Forensic Sciences 2022 Annual Conference*, p. 205.

Andronowski, J. M., Cole, M. E., Davis, R. A., Tubo, G. R., Taylor, J. T., & Cooper, D. M. L. (2023). A multimodal 3D imaging approach of pore networks in the human femur to assess age-associated vascular expansion and Lacuno-Canalicular reduction. *Anatomical Record (Hoboken, N.J.: 2007)*, *306*(3), 475–493. doi: 10.1002/ar.25089

Andronowski, J. M., Davis, R. A., & Cole, M. E. (2020). Investigating the impact of opioid abuse on intracortical porosity and bone cellular density: A synchrotron-radiation micro-computed tomography approach. *Journal of Visualized Experiments*, 160, e61081. doi:10.3791/61081

Andronowski, J. M., Davis, R. A., & Holyoke, C. W. (2020). A sectioning, coring, and image processing guide for high-throughput cortical bone sample procurement and analysis for synchrotron Micro-CT. *Journal of Visualized Experiments*, *160*, Article 160. doi: 10.3791/61081

Andronowski, J. M., & Depp, R. (2022). A multidisciplinary perspective on the role of marginalization in the identification of opioid users in medicolegal investigations. In Jennifer F. Byrnes and Iván Sandoval-Cervantes eds. *The Marginalized in Death" A Forensic Anthropology of Intersectional Identity in the Modern Era* (p. 281). Lexington Books.

Andronowski, J. M., Mundorff, A. Z., Pratt, I. V., Davoren, J. M., & Cooper, D. M. L. (2017). Evaluating differential nuclear DNA yield rates and osteocyte numbers among human bone tissue types: A synchrotron radiation micro-CT approach. *Forensic Science International: Genetics*, *28*, 211–218. doi: 10.1016/j.fsigen.2017.03.002

Andronowski, J. M., Pratt, I. V., & Cooper, D. M. L. (2017). Occurrence of osteon banding in adult human cortical bone. *American Journal of Physical Anthropology*, *164*(3), Article 3. doi: 10.1002/ajpa.23297

Andronowski, J. M., Schuller, A. J., Cole, M. E., LaMarca, A. R., Davis, R. A., & Tubo, G. R. (2021). Rabbits (Oryctolagus cuniculus) as a Model System for Longitudinal Experimental Opioid Treatments: Implications for Orthopedic and Biomedical Research. *Osteology*, *1*(4), Article 4. doi: 10.3390/osteology1040021

Andronowski, J. M., & Taylor, J. T. (2022). The Andronowski skeletal collection for histological research: A modern anatomical contribution. *Forensic Sciences*, *2*(1), 175–189. doi: 10.3390/forensicsci2010014

Appleby, J., Rutty, G. N., Hainsworth, S. V., Woosnam-Savage, R. C., Morgan, B., Brough, A., Earp, R. W., Robinson, C., King, T. E., Morris, M., & Buckley, R. (2015). Perimortem trauma in King Richard III: A skeletal analysis. *The Lancet*, *385*(9964), 253–259. doi: 10.1016/S0140-6736(14)60804-7

Arhatari, B. D., Cooper, D. M. L., Thomas, C. D. L., Clement, J. G., & Peele, A. G. (2011). Imaging the 3D structure of secondary osteons in human cortical bone using phase-retrieval tomography. *Physics in Medicine and Biology*, *56*(16), 5265–5274. doi: 10.1088/0031-9155/56/16/012

Arsenault, A. L. (1990). Vascular canals in bovine cortical bone studied by corrosion casting. *Calcified Tissue International*, *47*(5), 320–325. doi: 10.1007/BF02555916

Ashique, A. M., Hart, L. S., Thomas, C. D. L., Clement, J. G., Pivonka, P., Carter, Y., Mousseau, D. D., & Cooper, D. M. L. (2017). Lacunar-canalicular network in femoral cortical bone is reduced in aged women and is predominantly due to a loss of canalicular porosity. *Bone Reports*, *7*, 9–16. doi: 10.1016/j.bonr.2017.06.002

Auerbach, B. M., & Ruff, C. B. (2006). Limb bone bilateral asymmetry: Variability and commonality among modern humans. *Journal of Human Evolution*, *50*(2), Article 2. doi: 10.1016/j.jhevol.2005.09.004

BABAO. (2019). *British Association of Biological Anthropology and Osteoarchaeology Recommendations on the Ethical Issues Surrounding 2D and 3D Digital Imaging of Human Remains.* https://www.babao.org.uk/assets/Uploads/BABAO-Digital-imaging-code-2019.pdf

Bachman, C. H., & Ellis, E. H. (1965). Fluorescence of bone. *Nature, 206*(991), 1328–1331. doi: 10.1038/2061328a0

Baier, W., Donnelly, M. J., Payne, M., & Williams, M. A. (2020). A holistic multi-scale approach to using 3D scanning technology in accident reconstruction. *Journal of Forensic Sciences, 65*(5), 1774–1778. doi: 10.1111/1556-4029.14405

Baier, W., Mangham, C., Warnett, J. M., Payne, M., Painter, M., & Williams, M. A. (2019). Using histology to evaluate micro-CT findings of trauma in three post-mortem samples—First steps towards method validation. *Forensic Science International, 297*, 27–34. doi: 10.1016/j.forsciint.2019.01.027

Baier, W., Norman, D. G., & Williams, M. A. (2021). Micro-CT for the examination of paediatric rib injuries: A case series. *Forensic Science International, 325*, 110789. doi: 10.1016/j.forsciint.2021.110789

Baker, O., Coqueugniot, H., Dutour, O., Chamel, B., Coqueugniot, E., Stordeur, D., Le Mort, F., Khawam, R., Perrin, P., Pálfi, G., & Gourichon, L. (2017). Prehistory of human tuberculosis: Earliest evidence from the onset of animal husbandry in the Near East. *Paléorient, 43*(2), 35–51. https://www.jstor.org/stable/44988171?seq=4

Baker, O., Lee, O. Y.-C., Wu, H. H. T., Besra, G. S., Minnikin, D. E., Llewellyn, G., Williams, C. M., Maixner, F., O'Sullivan, N., Zink, A., Chamel, B., Khawam, R., Coqueugniot, E., Helmer, D., Le Mort, F., Perrin, P., Gourichon, L., Dutailly, B., Pálfi, G., … Dutour, O. (2015). Human tuberculosis predates domestication in ancient Syria. *Tuberculosis, 95*, S4–S12. doi: 10.1016/j.tube.2015.02.001

Bartsiokas, A. (2002). Hominid cranial bone structure: A histological study of Omo 1 specimens from Ethiopia using different microscopic techniques. *The Anatomical Record, 267*(1), 52–59. doi: 10.1002/ar.10083

Baud, C. A., & Morgenthaler, P. W. (1956). Recherches sur le Degrede Mineralisation de l'os Humain Fossile par la Methode Microradiographique. *Arch Suisses Anthropology General 21*, 79–86.

Baud, C.-A., Lagier, R., Boivin, G., & Boillat, M.-A. (1978). Value of the bone biopsy in the diagnosis of industrial fluorosis. *Virchows Archiv A Pathological Anatomy and Histology, 380*(4), 283–297. doi: 10.1007/BF00431313

Beauchesne, P., & Agarwal, S. C. (2017). A multi-method assessment of bone maintenance and loss in an Imperial R oman population: Implications for future studies of age-related bone loss in the past. *American Journal of Physical Anthropology, 164*(1), 41–61.

Beaudet, A., Atwood, R. C., Kockelmann, W., Fernandez, V., Connolley, T., Vo, N. T., Clarke, R., & Stratford, D. (2021). Preliminary paleohistological observations of the StW 573 ('Little Foot') skull. *eLife, 10*, e64804. doi: 10.7554/eLife.64804

Beck, J. D., Canfield, B. L., Haddock, S. M., Chen, T. J., Kothari, M., & Keaveny, T. M. (1997). Three-dimensional imaging of trabecular bone using the computer numerically controlled milling technique. *Bone, 21*(3), 281–287. doi: 10.1016/s8756-3282(97)00122-1

Bell, L. S., Skinner, M. F., & Jones, S. J. (1996). The speed of post mortem change to the human skeleton and its taphonomic significance. *Forensic Science International, 82*(2), 129–140. doi: 10.1016/0379-0738(96)01984-6

Bertelsen, P. K., Clement, J. G., & Thomas, C. D. (1995). A morphometric study of the cortex of the human femur from early childhood to advanced old age. *Forensic Science International, 74*(1–2), 63–77. doi: 10.1016/0379-0738(95)01738-5

Bewes, J., Low, A., Morphett, A., Pate, F. D., & Henneberg, M. (2019). Artificial intelligence for sex determination of skeletal remains: Application of a deep learning artificial neural network to human skulls. *Journal of Forensic and Legal Medicine, 62*, 40–43. doi: 10.1016/j.jflm.2019.01.004

Bigley, R. F., Singh, M., Hernandez, C. J., Kazakia, G. J., Martin, R. B., & Keaveny, T. M. (2008). Validity of serial milling-based imaging system for microdamage quantification. *Bone, 42*(1), 212–215. doi: 10.1016/j.bone.2007.09.043

Blake, C. F. (1994). Foot-binding in neo-confucian china and the appropriation of female labor. *Signs: Journal of Women in Culture and Society, 19*(3), 676–712. doi: 10.1086/494917

Boone, M. N., De Witte, Y., Dierick, M., Almeida, A., & Van Hoorebeke, L. (2012). Improved signal-to-noise ratio in laboratory-based phase contrast tomography. *Microscopy and Microanalysis, 18*(2), 399–405. doi: 10.1017/S1431927611012529

Booth, T. J. (2016). An investigation into the relationship between funerary treatment and bacterial bioerosion in European archaeological human bone. *Archaeometry, 58*(3), 484–499.

Booth, T. J., Chamberlain, A. T., & Pearson, M. P. (2015). Mummification in Bronze Age Britain. *Antiquity, 89*(347), 1155–1173. doi: 10.15184/aqy.2015.111

Booth, T. J., Redfern, R. C., & Gowland, R. L. (2016). Immaculate conceptions: Micro-CT analysis of diagenesis in Romano-British infant skeletons. *Journal of Archaeological Science, 74*, 124–134. doi: 10.1016/j.jas.2016.08.007

Borah, B., Gross, G. J., Dufresne, T. E., Smith, T. S., Cockman, M. D., Chmielewski, P. A., Lundy, M. W., Hartke, J. R., & Sod, E. W. (2001). Three-dimensional microimaging (MR mu I and mu CT), finite element modeling, and rapid prototyping provide unique insights into bone architecture in osteoporosis. *Anatomical Record, 265*(2), Article 2. doi: DOI 10.1002/ar.1060

Bousson, V., Bergot, C., Meunier, A., Barbot, F., Parlier-Cuau, C., Laval-Jeantet, A. M., & Laredo, J. D. (2000). CT of the middiaphyseal femur: Cortical bone mineral density and relation to porosity. *Radiology, 217*(1), Article 1. doi: 10.1148/radiology.217.1.r00se11179

Bousson, V., Peyrin, F., Bergot, C., Hausard, M., Sautet, A., & Laredo, J.-D. (2004). Cortical bone in the human femoral neck: Three-dimensional appearance and porosity using synchrotron radiation. *Journal of Bone and Mineral Research, 19*(5), Article 5. doi: 10.1359/jbmr.040124

Boutroy, S., Bouxsein, M. L., Munoz, F., & Delmas, P. D. (2005). In vivo assessment of trabecular bone microarchitecture by high-resolution peripheral quantitative computed tomography. *Journal of Clinical Endocrinology & Metabolism, 90*(12), Article 12. doi: 10.1210/jc.2005-1258

Bowen, D. K., Elliott, J. S., Stock, S. R., & Dover, S. D. (1986). *X-ray Microtomography with Synchrotron Radiation* (D. K. Bowen & L. V. Knight, Eds.; p. 94). doi: 10.1117/12.936625

Brickley, M. B., Ives, R., & Mays, S. (2020). *The bioarchaeology of metabolic bone disease*. Academic Press.

Brinker, U., Harten-Buga, H., Staude, A., Jantzen, D., & Orschiedt, J. (2018). Perimortem lesions on human bones from the bronze age battlefield in the Tollense valley: an interdisciplinary approach. In A. Dolfini, R. J. Crellin, C. Horn, & M. Uckelmann (Eds.), *Prehistoric Warfare and Violence* (pp. 39–60). Springer International Publishing. doi: 10.1007/978-3-319-78828-9_3

Brødholt, E. T., Gautvik, K. M., Günther, C.-C., Sjøvold, T., & Holck, P. (2022). Social stratification reflected in bone mineral density and stature: Spectral imaging and osteoarchaeological findings from medieval Norway. *PloS One, 17*(10), e0275448. doi: 10.1371/journal.pone.0275448

Bromage, T. G., Goldman, H. M., McFarlin, S. C., Perez Ochoa, A., & Boyde, A. (2009). Confocal scanning optical microscopy of a 3-million-year-old *Australopithecus afarensis* femur. *Scanning, 31*(1), 1–10. doi: 10.1002/sca.20139

Brooks, R. A., & Chiro, G. D. (1976). Beam hardening in X-ray reconstructive tomography. *Physics in Medicine and Biology, 21*(3), 390–398. doi: 10.1088/0031-9155/21/3/004

Brown, K. R., Silver, I. A., Musgrave, J. H., & Roberts, A. M. (2011). The use of μCT technology to identify skull fracture in a case involving blunt force trauma. *Forensic Science International, 206*(1–3), e8–e11. doi: 10.1016/j.forsciint.2010.06.013

Burghardt, A. J., Issever, A. S., Schwartz, A. V., Davis, K. A., Masharani, U., Majumdar, S., & Link, T. M. (2010). High-resolution peripheral quantitative computed tomographic imaging of cortical and trabecular bone microarchitecture in patients with type 2 diabetes mellitus. *Journal of Clinical Endocrinology & Metabolism, 95*(11), Article 11. doi: 10.1210/jc.2010-0226

Caister, A. J., Carr, D. J., Campbell, P. D., Brock, F., & Breeze, J. (2020). The ballistic performance of bone when impacted by fragments. *International Journal of Legal Medicine, 134*(4), 1387–1393. doi: 10.1007/s00414-020-02299-9

Cao, Y., Ma, Y., Vieira, D. N., Guo, Y., Wang, Y., Deng, K., Chen, Y., Zhang, J., Qin, Z., Chen, F., Huang, P., & Zhang, J. (2021). A potential method for sex estimation of human skeletons using deep learning and three-dimensional surface scanning. *International Journal of Legal Medicine, 135*(6), 2409–2421. doi: 10.1007/s00414-021-02675-z

Cao, Y., Ma, Y., Yang, X., Xiong, J., Wang, Y., Zhang, J., Qin, Z., Chen, Y., Vieira, D. N., Chen, F., Zhang, J., & Huang, P. (2022). Use of deep learning in forensic sex estimation of virtual pelvic models from the Han population. *Forensic Sciences Research, 7*(3), 540–549. doi: 10.1080/20961790.2021.2024369

Capasso, L., D'Anastasio, R., Guarnieri, S., Viciano, J., & Mariggiò, M. (2017). Bone natural autofluorescence and confocal laser scanning microscopy: Preliminary results of a novel useful tool to distinguish between forensic and ancient human skeletal remains. *Forensic Science International, 272,* 87–96. doi: 10.1016/j.forsciint.2017.01.017

Capuani, C., Rouquette, J., Payré, B., Moscovici, J., Delisle, M. B., Telmon, N., & Guilbeau-Frugier, C. (2013). Deciphering the elusive nature of sharp bone trauma using epifluorescence macroscopy: A comparison study multiplexing classical imaging approaches. *International Journal of Legal Medicine, 127*(1), 169–176. doi: 10.1007/s00414-012-0678-6

Carter, Y., Thomas, C. D., Clement, J. G., Peele, A. G., Hannah, K., & Cooper, D. M. (2013). Variation in osteocyte lacunar morphology and density in the human femur—A synchrotron radiation micro-CT study. *Bone, 52*(1), Article 1. doi: 10.1016/j.bone.2012.09.010

Carter, Y., Thomas, C. D. L., Clement, J. G., & Cooper, D. M. L. (2013). Femoral osteocyte lacunar density, volume and morphology in women across the lifespan. *Journal of Structural Biology, 183*(3), 519–526. doi: 10.1016/j.jsb.2013.07.004

Caruso, V., Marinoni, N., Diella, V., Berna, F., Cantaluppi, M., Mancini, L., Trombino, L., Cattaneo, C., Pastero, L., & Pavese, A. (2020). Bone diagenesis in archaeological and contemporary human remains: An investigation of bone 3D microstructure and minero-chemical assessment. *Archaeological and Anthropological Sciences, 12*(8), 162. doi: 10.1007/s12520-020-01090-6

Čavka, M., Erjavec, I., Seiwerth, S., Carić, M., Janković, I., Krznar, S., Rimpf, A., Brkić, H., Pavičin, I. S., Vodanović, M., & Novak, M. (2022). A case of a large pedunculated-type osteochondroma from late medieval Ilok, eastern Croatia: Bioarchaeological, paleoradiological and histological study. *Journal of Archaeological Science: Reports, 45,* 103574. doi: 10.1016/j.jasrep.2022.103574

Chung, S. W., Han, S. S., Lee, J. W., Oh, K.-S., Kim, N. R., Yoon, J. P., Kim, J. Y., Moon, S. H., Kwon, J., Lee, H.-J., Noh, Y.-M., & Kim, Y. (2018). Automated detection and classification of the proximal humerus fracture by using deep learning algorithm. *Acta Orthopaedica, 89*(4), 468–473. doi: 10.1080/17453674.2018.1453714

Cohen, J., & Harris, W. H. (1958). The three-dimensional anatomy of haversian systems. *The Journal of Bone and Joint Surgery. American Volume, 40*(2), 419–434.

Cole, M. E., Stout, S. D., & Agnew, A. M. (2021). Pore Extractor 2D: An ImageJ plugin for identification, classification, and regional characterization of cortical pores on histological bone images. *American Journal of Physical Anthropology, 174*(S71), 20–21.

Collins, K. A., & Presnell, S. E. (2007). Elder neglect and the pathophysiology of aging. *The American Journal of Forensic Medicine and Pathology, 28*(2), Article 2. doi: 10.1097/PAF.0b013e31805c93eb

Collins, M. J., Nielsen-Marsh, C. M., Hiller, J., Smith, C. I., Roberts, J. P., Prigodich, R. V., Wess, T. J., Csapo, J., Millard, A. R., & Turner-Walker, G. (2002). The survival of organic matter in bone: A review. *Archaeometry, 44*(3), 383–394. doi: 10.1111/1475-4754.t01-1-00071

Cooper, D. M. L. (2005). *3D micro-CT imaging of Human Cortical Bone Porosity: A Novel Method for Estimating Age at Death* [Dissertation, University of Calgary]. https://prism.ucalgary.ca/handle/1880/101397

Cooper, D. M. L., Erickson, B., Peele, A. G., Hannah, K., Thomas, C. D. L., & Clement, J. G. (2011). Visualization of 3D osteon morphology by synchrotron radiation micro-CT. *Journal of Anatomy, 219*(4), Article 4. doi: 10.1111/j.1469-7580.2011.01398.x

Cooper, D. M. L., Matyas, J. R., Katzenberg, M. A., & Hallgrimsson, B. (2004). Comparison of microcomputed tomographic and microradiographic measurements of cortical bone porosity. *Calcified Tissue International, 74*(5), Article 5. doi: 10.1007/s00223-003-0071-z

Cooper, D. M. L., Thomas, C. D. L., & Clement, J. G. (2012). Technological developments in the analysis of cortical bone histology: The third dimension and its potential in anthropology. In C. M. Crowder & S. D. Stout (Eds.), *Bone Histology: An Anthropological Perspective* (pp. 361–375). CRC Press.

Cooper, D. M. L., Turinsky, A. L., Sensen, C. W., & Hallgrímsson, B. (2003). Quantitative 3D analysis of the canal network in cortical bone by micro-computed tomography. *Anatomical Record. Part B, New Anatomist, 274*(1), Article 1. doi: 10.1002/ar.b.10024

Coqueugniot, H., Dutailly, B., Desbarats, P., Boulestin, B., Pap, I., Szikossy, I., Baker, O., Montaudon, M., Panuel, M., Karlinger, K., Kovács, B., Kristóf, L. A., Pálfi, G., & Dutour, O. (2015). Three-dimensional imaging of past skeletal TB: From lesion to process. *Tuberculosis, 95*, S73–S79. doi: 10.1016/j.tube.2015.02.004

Coutinho-Nogueira, D., Dutour, O., Coqueugniot, H., & Tillier, A.-M. (2019). Qafzeh 9 mandible (ca 90–100 kyrs BP, Israel) revisited: μ-CT and 3D reveal new pathological conditions. *International Journal of Paleopathology, 26*, 104–110. doi: 10.1016/j.ijpp.2019.06.002

Dal Sasso, G., Maritan, L., Usai, D., Angelini, I., & Artioli, G. (2014). Bone diagenesis at the micro-scale: Bone alteration patterns during multiple burial phases at Al Khiday (Khartoum, Sudan) between the Early Holocene and the II century AD. *Palaeogeography, Palaeoclimatology, Palaeoecology, 416*, 30–42. doi: 10.1016/j.palaeo.2014.06.034

Dallas, S. L., & Moore, D. S. (2020). Using confocal imaging approaches to understand the structure and function of osteocytes and the lacunocanalicular network. *Bone, 138*, 115463. doi: 10.1016/j.bone.2020.115463

Dalstra, M., Cattaneo, P. M., & Beckmann, F. (2006). Synchrotron radiation-based microtomography of alveolar support tissues. *Orthodontics & Craniofacial Research, 9*(4), 199–205. doi: 10.1111/j.1601-6343.2006.00376.x

Dalstra, M., Karaj, E., Beckmann, F., Andersen, T., & Cattaneo, P. M. (2004). Osteonal mineralization patterns in cortical bone studied by synchrotron-radiation-based computed microtomography and scanning acoustic microscopy. *Developments in X-Ray Tomography IV, 5535*, 143–151. doi: 10.1117/12.559583

de Froidmont, S., Grabherr, S., Vaucher, P., De Cesare, M., Egger, C., Papageorgopoulou, C., Roth, V., Morand, G., Mangin, P., & Uldin, T. (2013). Virtual anthropology: A comparison between the performance of conventional X-ray and MDCT in investigating the trabecular structure of long bones. *Forensic Science International, 225*(1–3), 53–59. doi: 10.1016/j.forsciint.2012.10.029

DeHoff, R. T. (1983). Quantitative serial sectioning analysis: Preview. *Journal of Microscopy, 131*(3), 259–263. doi: 10.1111/j.1365-2818.1983.tb04254.x

Dempster, D. W., Compston, J. E., Drezner, M. K., Glorieux, F. H., Kanis, J. A., Malluche, H., Meunier, P. J., Ott, S. M., Recker, R. R., & Parfitt, A. M. (2013). Standardized nomenclature, symbols, and units for bone histomorphometry: A 2012 update of the report of the ASBMR Histomorphometry Nomenclature Committee. *Journal of Bone and Mineral Research: The Official Journal of the American Society for Bone and Mineral Research, 28*(1), 2–17. doi: 10.1002/jbmr.1805

Dempster, W. T., & Enlow, D. H. (1959). Patterns of vascular channels in the cortex of the human mandible. *The Anatomical Record, 135*, 189–205. doi: 10.1002/ar.1091350305

Dimililer, K. (2017). IBFDS: Intelligent bone fracture detection system. *Procedia Computer Science, 120*, 260–267. doi: 10.1016/j.procs.2017.11.237

Ding, M., Odgaard, A., Danielsen, C. C., & Hvid, I. (2002). Mutual associations among microstructural, physical and mechanical properties of human cancellous bone. *The Journal of Bone and Joint Surgery. British Volume, 84-B*(6), 900–907. doi: 10.1302/0301-620X.84B6.0840900

Ding, M., Odgaard, A., & Hvid, I. (1999). Accuracy of cancellous bone volume fraction measured by micro-CT scanning. *Journal of Biomechanics, 32*(3), 323–326. doi: 10.1016/S0021-9290(98)00176-6

Dittmar, J. M., Mitchell, P. D., Jones, P. M., Mulder, B., Inskip, S. A., Cessford, C., & Robb, J. E. (2021). Gout and 'Podagra' in medieval Cambridge, England. *International Journal of Paleopathology, 33*, 170–181. doi: 10.1016/j.ijpp.2021.04.007

Djuric, M., Djonic, D., Milovanovic, P., Nikolic, S., Marshall, R., Marinkovic, J., & Hahn, M. (2010). Region-specific sex-dependent pattern of age-related changes of proximal femoral cancellous bone and its implications on differential bone fragility. *Calcified Tissue International*, *86*(3), 192–201. doi: 10.1007/s00223-009-9325-8

Dolinak, D. (2008). Review of the significance of various low force fractures in the elderly. *The American Journal of Forensic Medicine and Pathology*, *29*(2), Article 2. doi: 10.1097/PAF.0b013e31817440ac

Drusini, A. (1987). Refinements of two methods for the histomorphometric determination of age in human bone. *Zeitschrift Für Morphologie Und Anthropologie*, *77*(2), Article 2. JSTOR. https://www.jstor.org/stable/25757226

Duffett Carlson, K. S., Mandl, K., McCall, A., Brönnimann, D., Teschler-Nicola, M., Weiss-Krejci, E., & Metscher, B. (2022). 3D visualization of bioerosion in archaeological bone. *Journal of Archaeological Science*, *145*, 105646. doi: 10.1016/j.jas.2022.105646

Elliott, A. D. (2020). Confocal Microscopy: Principles and Modern Practices. *Current Protocols in Cytometry*, *92*(1), e68. doi: 10.1002/cpcy.68

Elliott, J. C., Bowen, D. K., Dover, S. D., & Davies, S. T. (1987). X-ray microtomography of biological tissues using laboratory and synchrotron sources. *Biological Trace Element Research*, *13*(1), 219–227. doi: 10.1007/BF02796634

Elliott, J. C., & Dover, S. D. (1982). X-ray microtomography. *Journal of Microscopy*, *126*(2), 211–213. doi: 10.1111/j.1365-2818.1982.tb00376.x

Elliott, J. C., & Dover, S. D. (1984). Three-dimensional distribution of mineral in bone at a resolution of 15 micron determined by x-ray microtomography. *Metabolic Bone Disease & Related Research*, *5*(5), 219–221. doi: 10.1016/0221-8747(84)90062-6

Faccia, K., Buie, H., Weber, A., Bazaliiskii, V. I., Goriunova, O. I., Boyd, S., Hallgrímsson, B., & Katzenberg, M. A. (2014). Bone quality in prehistoric, cis-baikal forager femora: A micro-CT analysis of cortical canal microstructure: Micro-CT Analysis of Forager Femora Bone Quality. *American Journal of Physical Anthropology*, *154*(4), 486–497. doi: 10.1002/ajpa.22533

Fais, P., Giraudo, C., Viero, A., Miotto, D., Bortolotti, F., Tagliaro, F., Montisci, M., & Cecchetto, G. (2016). Micro computed tomography features of laryngeal fractures in a case of fatal manual strangulation. *Legal Medicine*, *18*, 85–89. doi: 10.1016/j.legalmed.2016.01.001

Fajardo, R. J., & Müller, R. (2001). Three-dimensional analysis of nonhuman primate trabecular architecture using micro-computed tomography: Primate Trabecular Bone and the 3-D µCT Method. *American Journal of Physical Anthropology*, *115*(4), 327–336. doi: 10.1002/ajpa.1089

Fan, T., Wang, G., Li, Y., & Wang, H. (2020). MA-Net: A multi-scale attention network for liver and tumor segmentation. *IEEE Access*, *8*, 179656–179665. doi: 10.1109/ACCESS.2020.3025372

Feik, S. A., Thomas, C. D., Bruns, R., & Clement, J. G. (2000). Regional variations in cortical modeling in the femoral mid-shaft: Sex and age differences. *American Journal of Physical Anthropology*, *112*(2), Article 2. doi: 10.1002/(SICI)1096-8644(2000)112:2<191::AID-AJPA6>3.0.CO;2-3

Feik, S. A., Thomas, C. D., & Clement, J. G. (1997). Age-related changes in cortical porosity of the midshaft of the human femur. *Journal of Anatomy*, *191*(Pt 3), 407–416. doi: 10.1046/j.1469-7580.1997.19130407.x

Feldkamp, L. A., Goldstein, S. A., Parfitt, A. M., Jesion, G., & Kleerekoper, M. (1989). The direct examination of 3-dimensional bone architecture invitro by computed-tomography. *Journal of Bone and Mineral Research*, *4*(1), Article 1. doi: 10.1002/jbmr.5650040103

Feng, D., Marshburn, D., Jen, D., Weinberg, R. J., Taylor, R. M., & Burette, A. (2007). Stepping into the third dimension. *The Journal of Neuroscience: The Official Journal of the Society for Neuroscience*, *27*(47), 12757–12760. doi: 10.1523/JNEUROSCI.2846-07.2007

Figus, C., Stephens, N. B., Sorrentino, R., Bortolini, E., Arrighi, S., Higgins, O. A., Lugli, F., Marciani, G., Oxilia, G., Romandini, M., Silvestrini, S., Baruffaldi, F., Belcastro, M. G., Bernardini, F., Festa, A., Hajdu, T., Mateovics-László, O., Pap, I., Szeniczey, T., … Benazzi, S. (2023). Morphologies in-between: The impact of the first steps on the human talus. *The Anatomical Record*, *306*(1), 124–142. doi: 10.1002/ar.25010

Flannery, B. P., & Roberge, W. G. (1987). Observational strategies for three-dimensional synchrotron microtomography. *Journal of Applied Physics*, *62*(12), 4668–4674. doi: 10.1063/1.339016

Flohr, S., Brinker, U., Schramm, A., Kierdorf, U., Staude, A., Piek, J., Jantzen, D., Hauenstein, K., & Orschiedt, J. (2015). Flint arrowhead embedded in a human humerus from the Bronze Age site in the Tollense valley, Germany – A high-resolution micro-CT study to distinguish antemortem from perimortem projectile trauma to bone. *International Journal of Paleopathology*, *9*, 76–81. doi: 10.1016/j.ijpp.2015.02.004

Ford, N. L., Thornton, M. M., & Holdsworth, D. W. (2003). Fundamental image quality limits for microcomputed tomography in small animals. *Medical Physics*, *30*(11), 2869–2877. doi: 10.1118/1.1617353

Fraberger, S., Dockner, M., Winter, E., Pretterklieber, M., Weber, G. W., Teschler-Nicola, M., & Pietschmann, P. (2021). Micro-CT evaluation of historical human skulls presenting signs of syphilitic infection. *Wiener Klinische Wochenschrift*, *133*(11–12), 602–609. doi: 10.1007/s00508-021-01832-z

Gauthier, R., Follet, H., Olivier, C., Mitton, D., & Peyrin, F. (2019). 3D analysis of the osteonal and interstitial tissue in human radii cortical bone. *Bone*, *127*, 526–536. doi: 10.1016/j.bone.2019.07.028

Gauthier, R., Langer, M., Follet, H., Olivier, C., Gouttenoire, P.-J., Helfen, L., Rongiéras, F., Mitton, D., & Peyrin, F. (2018). 3D micro structural analysis of human cortical bone in paired femoral diaphysis, femoral neck and radial diaphysis. *Journal of Structural Biology*, *204*(2), Article 2. doi: 10.1016/j.jsb.2018.08.006

Genant, H. K., Engelke, K., Fuerst, T., Glüer, C. C., Grampp, S., Harris, S. T., Jergas, M., Lang, T., Lu, Y., Majumdar, S., Mathur, A., & Takada, M. (1996). Noninvasive assessment of bone mineral and structure: State of the art. *Journal of Bone and Mineral Research: The Official Journal of the American Society for Bone and Mineral Research*, *11*(6), 707–730. doi: 10.1002/jbmr.5650110602

Georgia, R., & Albu, I. (1988). The Haversian canal network in the femoral compact bone in some vertebrates. *Morphologie Et Embryologie*, *34*(3), 155–159.

Georgia, R., Albu, I., Sicoe, M., & Georoceanu, M. (1982). Comparative aspects of the density and diameter of Haversian canals in the diaphyseal compact bone of man and dog. *Morphologie Et Embryologie*, *28*(1), 11–14.

Gosman, J. H., Stout, S. D., & Larsen, C. S. (2011). Skeletal biology over the life span: A view from the surfaces. *American Journal of Physical Anthropology*, *146*(S53), 86–98. doi: 10.1002/ajpa.21612

Hannah, K. M., Thomas, C. D. L., Clement, J. G., De Carlo, F., & Peele, A. G. (2010). Bimodal distribution of osteocyte lacunar size in the human femoral cortex as revealed by micro-CT. *Bone*, *47*(5), Article 5. doi: 10.1016/j.bone.2010.07.025

Harrison, K. D., & Cooper, D. M. (2015). Modalities for visualization of cortical bone remodeling: The past, present, and future. *Front Endocrinol (Lausanne)*, *6*, 122. doi: 10.3389/fendo.2015.00122

Harrison, K. D., Sales, E., Hiebert, B. D., Panahifar, A., Zhu, N., Arnason, T., Swekla, K. J., Pivonka, P., Chapman, L. D., & Cooper, D. M. (2022). Direct assessment of rabbit cortical bone basic multicellular unit longitudinal erosion rate: A 4D synchrotron-based approach. *Journal of Bone and Mineral Research: The Official Journal of the American Society for Bone and Mineral Research*, *37*(11), 2244–2258. doi: 10.1002/jbmr.4700

Hassett, B. R. (2018). Which bone to pick: Creation, curation, and dissemination of online 3D digital bioarchaeological data. *Archaeologies*, *14*(2), 231–249. doi: 10.1007/s11759-018-9344-z

Havers, C. (1691). An account of a book. Osteologia Nova, or some observations of the bones, &c. Communicated to the Royal Society, in several discourses, read at their meetings, by the learned and accurate Author, Clopton Havers, M. D. and R. S. Soc. Lond. 1691, Octavo, for Sam. Smith. *Philosophical Transactions of the Royal Society of London*, *17*(194), Article 194. doi: 10.1098/rstl.1686.0104

He, Y., Pan, I., Bao, B., Halsey, K., Chang, M., Liu, H., Peng, S., Sebro, R. A., Guan, J., Yi, T., Delworth, A. T., Eweje, F., States, L. J., Zhang, P. J., Zhang, Z., Wu, J., Peng, X., & Bai, H. X. (2020). Deep learning-based classification of primary bone tumors on radiographs: A preliminary study. *eBioMedicine*, *62*, 103121. doi: 10.1016/j.ebiom.2020.103121

Hedges, R. E. M., Millard, A. R., & Pike, A. W. G. (1995). Measurements and relationships of diagenetic alteration of bone from three archaeological sites. *Journal of Archaeological Science*, *22*(2), 201–209. doi: 10.1006/jasc.1995.0022

Hert, J., Fiala, P., & Petrtýl, M. (1994). Osteon orientation of the diaphysis of the long bones in man. *Bone*, *15*(3), Article 3. doi: 10.1016/8756-3282(94)90288-7

Hildebrand, T., & Ruegsegger, P. (1997). Quantification of bone microarchitecture with the structure model index. *Comput Methods Biomech Biomed Engin*, *1*(1), Article 1. doi: 10.1080/01495739708936692

Holgate, R., L'Abbé, E., & Steyn, M. (2023). Diffuse Idiopathic Skeletal Hyperostosis (DISH): New evidence from micro-XCT scanning. *International Journal of Paleopathology*, *40*, 48–55. doi: 10.1016/j.ijpp.2022.12.003

Hounsfield, G. N. (1973). Computerized transverse axial scanning (tomography): Part I. Description of system. *The British Journal of Radiology*, *46*, 1016–1022.

Hsieh, J., & Flohr, T. (2021). Computed tomography recent history and future perspectives. *Journal of Medical Imaging (Bellingham, Wash.)*, *8*(5), 052109. doi: 10.1117/1.JMI.8.5.052109

Immel, A., Le Cabec, A., Bonazzi, M., Herbig, A., Temming, H., Schuenemann, V. J., Bos, K. I., Langbein, F., Harvati, K., Bridault, A., Pion, G., Julien, M.-A., Krotova, O., Conard, N. J., Münzel, S. C., Drucker, D. G., Viola, B., Hublin, J.-J., Tafforeau, P., & Krause, J. (2016). Effect of X-ray irradiation on ancient DNA in sub-fossil bones—Guidelines for safe X-ray imaging. *Scientific Reports*, *6*, 32969. doi: 10.1038/srep32969

Jans, M. M. E., Kars, H., Nielsen-Marsh, C. M., Smith, C. I., Nord, A. G., Arthur, P., & Earl, N. (2002). In situ preservation of archaeological bone: A histological study within a multidisciplinary approach. *Archaeometry*, *44*(3), 343–352. doi: 10.1111/1475-4754.t01-1-00067

Jennings, R. M., Malbon, C., Brock, F., Harrisson, S., & Carr, D. J. (2018). A preliminary study into injuries due to non-perforating ballistic impacts into soft body armour over the spine. *Injury*, *49*(7), 1251–1257. doi: 10.1016/j.injury.2018.05.015

Jowsey, J. (1960). Age changes in human bone. *Clinical Orthopaedics and Related Research*, *17*, 210. https://journals.lww.com/clinorthop/Citation/1960/00170/Age_Changes_in_Human_Bone.22.aspx

Kazakia, G. J., Lee, J. J., Singh, M., Bigley, R. F., Martin, R. B., & Keaveny, T. M. (2007). Automated high-resolution three-dimensional fluorescence imaging of large biological specimens. *Journal of Microscopy*, *225*(Pt 2), 109–117. doi: 10.1111/j.1365-2818.2007.01721.x

Kesterke, M. J., & Judd, M. A. (2019). A microscopic evaluation of Paget's disease of bone from a Byzantine monastic crypt in Jordan. *International Journal of Paleopathology*, *24*, 293–298. doi: 10.1016/j.ijpp.2018.08.005

Kettner, M., Potente, S., Schulz, B., Knauff, P., Schmidt, P. H., & Ramsthaler, F. (2014). Analysis of laryngeal fractures in decomposed bodies using microfocus computed tomography (mfCT). *Forensic Science, Medicine, and Pathology*, *10*(4), 607–612. doi: 10.1007/s12024-014-9584-1

Kneissel, M., Boyde, A., Hahn, M., Teschler-Nicola, M., Kalchhauser, G., & Plenk, H. (1994). Age- and sex-dependent cancellous bone changes in a 4000y BP population. *Bone*, *15*(5), 539–545. doi: 10.1016/8756-3282(94)90278-x

Kragstrup, J., & Melsen, F. (1983). Three-dimensional morphology of trabecular bone osteons reconstructed from serial sections. *Metabolic Bone Disease & Related Research*, *5*(3), 127–130. doi: 10.1016/0221-8747(83)90013-9

Kral, A. G., Ziegler, A., Tütken, T., & Geisler, T. (2021). Experimental aqueous alteration of cortical bone microarchitecture analyzed by quantitative micro-computed tomography. *Frontiers in Earth Science*, *9*, 609496. doi: 10.3389/feart.2021.609496

Kuhn, G., Schultz, M., Muller, R., & Ruhli, F. J. (2007). Diagnostic value of micro-CT in comparison with histology in the qualitative assessment of historical human postcranial bone pathologies. *Homo-Journal of Comparative Human Biology*, *58*(2), Article 2. doi: 10.1016/j.jchb.2006.11.002

Lagier, R., Baud, C. A., Arnaud, G., Arnaud, S., & Menk, R. (1982). Lesions characteristic of infection or malignant tumor in Paleo-Eskimo skulls: An anatomical and radiological study of two specimens. *Virchows Archiv A Pathological Anatomy and Histology*, *395*(3), 237–243. doi: 10.1007/BF00429350

Lamm, C., Dockner, M., Pospischek, B., Winter, E., Patzak, B., Pretterklieber, M., Weber, G. W., & Pietschmann, P. (2015). Micro-CT analyses of historical bone samples presenting with osteomyelitis. *Skeletal Radiology*, *44*(10), 1507–1514. doi: 10.1007/s00256-015-2203-8

Langer, M., Pacureanu, A., Suhonen, H., Grimal, Q., Cloetens, P., & Peyrin, F. (2012). X-ray phase nanotomography resolves the 3D human bone ultrastructure. *PloS One*, *7*(8), e35691. doi: 10.1371/journal.pone.0035691

Langer, M., & Peyrin, F. (2016). 3D X-ray ultra-microscopy of bone tissue. *Osteoporosis International*, *27*(2), 441–455. doi: 10.1007/s00198-015-3257-0

Larson, D. B., Chen, M. C., Lungren, M. P., Halabi, S. S., Stence, N. V., & Langlotz, C. P. (2018). Performance of a deep-learning neural network model in assessing skeletal maturity on pediatric hand radiographs. *Radiology*, *287*(1), 313–322. doi: 10.1148/radiol.2017170236

Lassen, N. E., Andersen, T. L., Pløen, G. G., Søe, K., Hauge, E. M., Harving, S., Eschen, G. E. T., & Delaisse, J.-M. (2017). Coupling of bone resorption and formation in real time: New knowledge gained from human haversian BMUs. *Journal of Bone and Mineral Research: The Official Journal of the American Society for Bone and Mineral Research*, *32*(7), 1395–1405. doi: 10.1002/jbmr.3091

Lazenby, R. A. (2002). Population variation in second metacarpal sexual size dimorphism. *American Journal of Physical Anthropology*, *118*(4), 378–384. doi: 10.1002/ajpa.10110

Lazenby, R. A., Angus, S., Cooper, D. M. L., & Hallgrímsson, B. (2008). A three-dimensional microcomputed tomographic study of site-specific variation in trabecular microarchitecture in the human second metacarpal. *Journal of Anatomy*, *213*(6), 698–705. doi: 10.1111/j.1469-7580.2008.00991.x

Lazenby, R. A., Cooper, D. M. L., Angus, S., & Hallgrímsson, B. (2008). Articular constraint, handedness, and directional asymmetry in the human second metacarpal. *Journal of Human Evolution*, *54*(6), 875–885. doi: 10.1016/j.jhevol.2007.12.001

Le Garff, E., Mesli, V., Delannoy, Y., Colard, T., De Jonckheere, J., Demondion, X., & Hédouin, V. (2017). The precision of micro-tomography in bone taphonomic experiments and the importance of registration. *Forensic Science International*, *273*, 161–167. doi: 10.1016/j.forsciint.2017.02.005

Lea, W. W., Hong, S.-J., Nam, H.-K., Kang, W.-Y., Yang, Z.-P., & Noh, E.-J. (2022). External validation of deep learning-based bone-age software: A preliminary study with real world data. *Scientific Reports*, *12*(1), 1232. doi: 10.1038/s41598-022-05282-z

Lee, J. H., Kim, Y. J., & Kim, K. G. (2020). Bone age estimation using deep learning and hand X-ray images. *Biomedical Engineering Letters*, *10*(3), 323–331. doi: 10.1007/s13534-020-00151-y

Leeuwenhoek, A. (1677). Microscopical observations of the structure of Teeth and other Bones: Made and communicated, in a letter by Mr. Anthony Leeuwenhoeck. *Philosophical Transactions of the Royal Society of London*, *12*(140), 1002–1003. doi: 10.1098/rstl.1677.0052

Lengeler, B. (2001). Coherence in X-ray physics. *Die Naturwissenschaften*, *88*(6), 249–260. doi: 10.1007/s001140100221

Li, Z., Kuhn, G., von Salis-Soglio, M., Cooke, S. J., Schirmer, M., Müller, R., & Ruffoni, D. (2015). In vivo monitoring of bone architecture and remodeling after implant insertion: The different responses of cortical and trabecular bone. *Bone*, *81*, 468–477. doi: 10.1016/j.bone.2015.08.017

Lieverse, A. R., Cooper, D. M. L., & Goriunova, O. I. (2012). A rare naviculocuneiform I coalition from Bronze Age Siberia. *International Journal of Paleopathology*, *2*(1), 25–30. doi: 10.1016/j.ijpp.2012.03.002

Lieverse, A. R., Pratt, I. V., Schulting, R. J., Cooper, D. M. L., Bazaliiskii, V. I., & Weber, A. W. (2014). Point taken: An unusual case of incisor agenesis and mandibular trauma in Early Bronze Age Siberia. *International Journal of Paleopathology*, *6*, 53–59. doi: 10.1016/j.ijpp.2014.04.004

Lindsey, R., Daluiski, A., Chopra, S., Lachapelle, A., Mozer, M., Sicular, S., Hanel, D., Gardner, M., Gupta, A., Hotchkiss, R., & Potter, H. (2018). Deep neural network improves fracture detection by clinicians. *Proceedings of the National Academy of Sciences, 115*(45), 11591–11596. doi: 10.1073/pnas.1806905115

Liu, Z. Q., Austin, T., Thomas, C. D., & Clement, J. G. (1996). Bone feature analysis using image processing techniques. *Computers in Biology and Medicine, 26*(1), 65–76. doi: 10.1016/0010-4825(95)00044-5

Liu, Z. Q., Liew, H. L., Clement, J. G., & Thomas, C. D. (1999). Bone image segmentation. *IEEE Transactions on Bio-Medical Engineering, 46*(5), 565–573. doi: 10.1109/10.759057

Locke, M. (2004). Structure of long bones in mammals. *Journal of Morphology, 262*(2), 546–565. doi: 10.1002/jmor.10282

Loundagin, L. L., Haider, I. T., Cooper, D. M. L., & Edwards, W. B. (2020). Association between intracortical microarchitecture and the compressive fatigue life of human bone: A pilot study. *Bone Reports, 12*, 100254. doi: 10.1016/j.bonr.2020.100254

Macintosh, A. A., & Stock, J. T. (2019). Intensive terrestrial or marine locomotor strategies are associated with inter- and intra-limb bone functional adaptation in living female athletes. *American Journal of Physical Anthropology, 168*(3), 566–581. doi: 10.1002/ajpa.23773

Mader, K. S., Donahue, L. R., Müller, R., & Stampanoni, M. (2015). High-throughput phenotyping and genetic linkage of cortical bone microstructure in the mouse. *BMC Genomics, 16*(1), 493. doi: 10.1186/s12864-015-1617-y

Maggiano, C., Dupras, T., Schultz, M., & Biggerstaff, J. (2006). Spectral and photobleaching analysis using confocal laser scanning microscopy: A comparison of modern and archaeological bone fluorescence. *Molecular and Cellular Probes, 20*(3), Article 3. doi: 10.1016/j.mcp.2005.11.009

Maggiano, C., Dupras, T., Schultz, M., & Biggerstaff, J. (2009). Confocal laser scanning microscopy: A flexible tool for simultaneous polarization and three-dimensional fluorescence imaging of archaeological compact bone. *Journal of Archaeological Science, 36*(10), Article 10. doi: 10.1016/j.jas.2009.06.021

Maggiano, I. S., Maggiano, C. M., Clement, J. G., Thomas, C. D. L., Carter, Y., & Cooper, D. M. L. (2016). Three-dimensional reconstruction of Haversian systems in human cortical bone using synchrotron radiation-based micro-CT: Morphology and quantification of branching and transverse connections across age. *Journal of Anatomy, 228*(5), Article 5. doi: 10.1111/joa.12430

Maggiano, I. S., Maggiano, C. M., & Cooper, D. M. (2021). Osteon circularity and longitudinal morphology: Quantitative and qualitative three-dimensional perspectives on human Haversian systems. *Micron (Oxford, England: 1993), 140*, 102955. doi: 10.1016/j.micron.2020.102955

Mandl, K., Carlson, K. S. D., Brönnimann, D., McCall, A., Grassberger, M., Teschler-Nicola, M., Weiss-Krejci, E., & Metscher, B. (2022). Substantiating microCT for diagnosing bioerosion in archaeological bone using a new Virtual Histological Index (VHI). *Archaeological and Anthropological Sciences, 14*(6), 104. doi: 10.1007/s12520-022-01563-w

Manske, S. L., Davison, E. M., Burt, L. A., Raymond, D. A., & Boyd, S. K. (2017). The estimation of second-generation HR-pQCT from first-generation HR-pQCT using in vivo cross-calibration. *Journal of Bone and Mineral Research: The Official Journal of the American Society for Bone and Mineral Research, 32*(7), 1514–1524. doi: 10.1002/jbmr.3128

Marchiafava, V., Bonucci, E., & Ascenzi, A. (1974). Fungal osteoclasia: A model of dead bone resorption. *Calcified Tissue Research, 14*(1), 195–210. doi: 10.1007/BF02060295

Marotti, G., & Zallone, A. Z. (1980). Changes in the vascular network during the formation of Haversian systems. *Acta Anatomica, 106*(1), 84–100. doi: 10.1159/000145171

Matsumoto, T., Yoshino, M., Asano, T., Uesugi, K., Todoh, M., & Tanaka, M. (2006). Monochromatic synchrotron radiation muCT reveals disuse-mediated canal network rarefaction in cortical bone of growing rat tibiae. *Journal of Applied Physiology (Bethesda, Md.: 1985), 100*(1), 274–280. doi: 10.1152/japplphysiol.00495.2005

Matsumoto, T., Yoshino, M., Uesugi, K., & Tanaka, M. (2007). Biphasic change and disuse-mediated regression of canal network structure in cortical bone of growing rats. *Bone, 41*(2), 239–246. doi: 10.1016/j.bone.2007.04.192

McBroom, R. J., Hayes, W. C., Edwards, W. T., Goldberg, R. P., & White, A. A. (1985). Prediction of vertebral body compressive fracture using quantitative computed tomography. *The Journal of Bone and Joint Surgery. American Volume*, *67*(8), 1206–1214.

Milella, M., Zollikofer, C. P. E., & Ponce de León, M. S. (2015). Virtual reconstruction and geometric morphometrics as tools for paleopathology: A new approach to study rare developmental disorders of the skeleton: New tools for paleopathology. *The Anatomical Record*, *298*(2), 335–345. doi: 10.1002/ar.23020

Millard AM. (2001). The deterioration of bone. In A.M. Pollard, & D.R. Brothwell (Eds.), *Handbook of archaeological science* (pp. 633–643). Wiley.

Milovanovic, P., Djonic, D., Hahn, M., Amling, M., Busse, B., & Djuric, M. (2017). Region-dependent patterns of trabecular bone growth in the human proximal femur: A study of 3D bone microarchitecture from early postnatal to late childhood period: MILOVANOVIC ET AL. *American Journal of Physical Anthropology*, *164*(2), 281–291. doi: 10.1002/ajpa.23268

Milovanovic, P., Djonic, D., Marshall, R. P., Hahn, M., Nikolic, S., Zivkovic, V., Amling, M., & Djuric, M. (2012). Micro-structural basis for particular vulnerability of the superolateral neck trabecular bone in the postmenopausal women with hip fractures. *Bone*, *50*(1), 63–68. doi: 10.1016/j.bone.2011.09.044

Miszkiewicz, J. J., Valentin, F., Vrahnas, C., Sims, N. A., Vongsvivut, J., Tobin, M. J., & Clark, G. (2021). Bone loss markers in the earliest Pacific Islanders. *Scientific Reports*, *11*(1), 3981. doi: 10.1038/s41598-021-83264-3

Mohsin, S., Taylor, D., & Lee, T. C. (2002). Three-dimensional reconstruction of Haversian systems in ovine compact bone. *European Journal of Morphology*, *40*(5), 309–315. doi: 10.1076/ejom.40.5.309.28901

Moreno-Ibáñez, M. Á., Saladié, P., Morales, J. I., Cebrià, A., & Fullola, J. M. (2021). Was it an axe or an adze? A cranial trauma case study from the Late Neolithic – Chalcolithic site of Cova Foradada (Calafell, Spain). *International Journal of Paleopathology*, *32*, 23–30. doi: 10.1016/j.ijpp.2020.11.002

Morgan, J. A. (2014). The Methodological and Diagnostic Applications of Micro-CT to Palaeopathology: A Quantitative Study of Porotic Hyperostosis [Ph.D., The University of Western Ontario (Canada)]. In *ProQuest Dissertations and Theses* (2701131312). ProQuest Dissertations & Theses Global. http://cyber.usask.ca/login?url=https://www.proquest.com/dissertations-theses/methodological-diagnostic-applications-micro-ct/docview/2701131312/se-2?accountid=14739

Murphy, K., Waa, S., Jaffer, H., Sauter, A., & Chan, A. (2013). A literature review of findings in physical elder abuse. *Canadian Association of Radiologists Journal*, *64*(1), Article 1. doi: 10.1016/j.carj.2012.12.001

Nebot, E., Heimel, P., Tangl, S., Dockner, M., Patsch, J., Weber, G. W., Pretterklieber, M., Teschler-Nicola, M., & Pietschmann, P. (2019). Paget's disease of long bones: Microstructural analyses of historical bone samples. *Calcified Tissue International*, *105*(1), 15–25. doi: 10.1007/s00223-019-00539-8

Newton, J., Savage, A., Coupar, N., & Fraser, J. (2020). Preliminary investigation into the use of Micro-CT scanning on impact damage to fabric, tissue and bone caused by both round and flat nosed bullets. *Science & Justice*, *60*(2), 151–159. doi: 10.1016/j.scijus.2019.09.006

Nicklisch, N., Maixner, F., Ganslmeier, R., Friederich, S., Dresely, V., Meller, H., Zink, A., & Alt, K. W. (2012). Rib lesions in skeletons from early neolithic sites in Central Germany: On the trail of tuberculosis at the onset of agriculture. *American Journal of Physical Anthropology*, *149*(3), 391–404. doi: 10.1002/ajpa.22137

Nicklisch, N., Ramsthaler, F., Bunnefeld, J.-H., Schulz, G., Friedrich, R., Alt, K. W., & Meller, H. (2022). Bioarchaeological investigations of the princely grave at Helmsdorf attesting to the violent death of an Early Bronze Age leader. *Scientific Reports*, *12*(1), 16139. doi: 10.1038/s41598-022-20720-8

Nishiyama, K. K., Macdonald, H. M., Buie, H. R., Hanley, D. A., & Boyd, S. K. (2009). Postmenopausal women with osteopenia have higher cortical porosity and thinner cortices at the distal radius and tibia than women with normal aBMD: An *In Vivo* HR-pQCT study. *Journal of Bone and Mineral Research*, 091019190442039–30. doi: 10.1359/jbmr.091020

Norman, D. G., Baier, W., Watson, D. G., Burnett, B., Painter, M., & Williams, M. A. (2018). Micro-CT for saw mark analysis on human bone. *Forensic Science International, 293*, 91–100. doi: 10.1016/j.forsciint.2018.10.027

Odes, E. J., Delezene, L. K., Randolph-Quinney, P. S., Smilg, J. S., Augustine, T. N., Jakata, K., & Berger, L. R. (2018). A case of benign osteogenic tumour in Homo naledi: Evidence for peripheral osteoma in the U.W. 101-1142 mandible. *International Journal of Paleopathology, 21*, 47–55. doi: 10.1016/j.ijpp.2017.05.003

Odgaard, A. (1997). Three-dimensional methods for quantification of cancellous bone architecture. *Bone, 20*(4), 315–328. doi: 10.1016/s8756-3282(97)00007-0

Odgaard, A., Andersen, K., Melsen, F., & Gundersen, H. J. (1990). A direct method for fast three-dimensional serial reconstruction. *Journal of Microscopy, 159*(Pt 3), 335–342. doi: 10.1111/j.1365-2818.1990.tb03038.x

Odgaard, A., Andersen, K., Ullerup, R., Frich, L. H., & Melsen, F. (1994). Three-dimensional reconstruction of entire vertebral bodies. *Bone, 15*(3), 335–342. doi: 10.1016/8756-3282(94) 90297-6

Oura, P., Junno, J.-A., Hunt, D., Lehenkari, P., Tuukkanen, J., & Maijanen, H. (2023). Deep learning in sex estimation from knee radiographs – A proof-of-concept study utilizing the Terry Anatomical Collection. *Legal Medicine, 61*, 102211. doi: 10.1016/j.legalmed.2023.102211

Pacureanu, A., Langer, M., Boller, E., Tafforeau, P., & Peyrin, F. (2012). Nanoscale imaging of the bone cell network with synchrotron X-ray tomography: Optimization of acquisition setup. *Medical Physics, 39*(4), 2229–2238. doi: 10.1118/1.3697525

Paganin, D., Mayo, S. C., Gureyev, T. E., Miller, P. R., & Wilkins, S. W. (2002). Simultaneous phase and amplitude extraction from a single defocused image of a homogeneous object. *Journal of Microscopy, 206*(Pt 1), 33–40. doi: 10.1046/j.1365-2818.2002.01010.x

Papageorgopoulou, C., Kuhn, G., Ziegler, U., & Rühli, F. J. (2010). Diagnostic morphometric applicability of confocal laser scanning microscopy in Osteoarchaeology. *International Journal of Osteoarchaeology, 20*(6), Article 6. doi: 10.1002/oa.1078

Parfitt, A. M., Drezner, M. K., Glorieux, F. H., Kanis, J. A., Malluche, H., Meunier, P. J., Ott, S. M., & Recker, R. R. (1987). Bone histomorphometry: Standardization of nomenclature, symbols, and units. Report of the ASBMR Histomorphometry Nomenclature Committee. *J Bone Miner Res, 2*(6), Article 6. doi: 10.1002/jbmr.5650020617

Payne, S., Kumar Bc, R., Pomeroy, E., Macintosh, A., & Stock, J. (2018). Thrifty phenotype versus cold adaptation: Trade-offs in upper limb proportions of Himalayan populations of Nepal. *Royal Society Open Science, 5*(6), 172174. doi: 10.1098/rsos.172174

Pazzaglia, U. E., & Congiu, T. (2013). The cast imaging of the osteon lacunar-canalicular system and the implications with functional models of intracanalicular flow. *Journal of Anatomy, 222*(2), 193–202. doi: 10.1111/joa.12004

Pearlstein, K. E., Simmons-Ehrhardt, T., Spatola, B. F., Means, B. K., & Mani, M. R. (2021). Modernizing medical museums through the 3D digitization of pathological specimens. *Biomedical Visualisation: Volume 10*, 181–204.

Pelletti, G., Viel, G., Fais, P., Viero, A., Visentin, S., Miotto, D., Montisci, M., Cecchetto, G., & Giraudo, C. (2017). Micro-computed tomography of false starts produced on bone by different hand-saws. *Legal Medicine, 26*, 1–5. doi: 10.1016/j.legalmed.2017.01.009

Perchalski, B., Placke, A., Sukhdeo, S. M., Shaw, C. N., Gosman, J. H., Raichlen, D. A., & Ryan, T. M. (2018). Asymmetry in the cortical and trabecular bone of the human humerus during development. *The Anatomical Record, 301*(6), 1012–1025. doi: 10.1002/ar.23705

Perzigian, A. J. (1973). The antiquity of age-associated bone demineralization in man. *Journal of the American Geriatrics Society, 21*(3), 100–105. doi: 10.1111/j.1532-5415.1973.tb00853.x

Peter, Z., Bousson, V., Bergot, C., & Peyrin, F. (2007). Segmentation of low contrast features in bone micro-CT images by a constrained region growing approach based on watershed. *2007 4th*

IEEE International Symposium on Biomedical Imaging: From Nano to Macro, 968–971. doi: 10.1109/ISBI.2007.357015

Petrtyl, M., Hert, J., & Fiala, P. (1996). Spatial organization of the haversian bone in man. *Journal of Biomechanics*, *29*(2), Article 2. doi: 10.1016/0021-9290(94)00035-2

Peyrin, F., Salome, M., Cloetens, P., Laval-Jeantet, A. M., Ritman, E., & Rüegsegger, P. (1998). Micro-CT examinations of trabecular bone samples at different resolutions: 14, 7 and 2 micron level. *Technology and Health Care: Official Journal of the European Society for Engineering and Medicine*, *6*(5–6), 391–401.

Peyrin, F., Salomé, M., Nuzzo, S., Cloetens, P., Laval-Jeantet, A. M., & Baruchel, J. (2000). Perspectives in three-dimensional analysis of bone samples using synchrotron radiation microtomography. *Cellular and Molecular Biology (Noisy-Le-Grand, France)*, *46*(6), 1089–1102.

Pfeiffer, S. (1980). Bone-remodeling age estimates compared with estimates by other techniques. *Current Anthropology*, *21*(6), Article 6. doi: 10.1086/202571

Piepenbrink, H. (1986). Two examples of biogenous dead bone decomposition and their consequences for taphonomic interpretation. *Journal of Archaeological Science*, *13*(5), 417–430. doi: 10.1016/0305-4403(86)90012-9

Pilolli, G. P., Lucchese, A., Maiorano, E., & Favia, G. (2008). New approach for static bone histomorphometry: Confocal laser scanning microscopy of maxillo-facial normal bone. *Ultrastructural Pathology*, *32*(5), 189–192. doi: 10.1080/01913120802397836

Pratt, I. V., Belev, G., Zhu, N., Chapman, L. D., & Cooper, D. M. L. (2015). In vivo imaging of rat cortical bone porosity by synchrotron phase contrast micro computed tomography. *Physics in Medicine and Biology*, *60*(1), Article 1. doi: 10.1088/0031-9155/60/1/211

Prentice, A. I. (1967). Autofluorescence of bone tissues. *Journal of Clinical Pathology*, *20*(5), 717–719. doi: 10.1136/jcp.20.5.717

Rai, M. F., Wu, C.-L., Capellini, T. D., Guilak, F., Dicks, A. R., Muthuirulan, P., Grandi, F., Bhutani, N., & Westendorf, J. J. (2021). Single cell omics for musculoskeletal research. *Current Osteoporosis Reports*, *19*(2), 131–140. doi: 10.1007/s11914-021-00662-2

Rajpurkar, P., Irvin, J., Bagul, A., Ding, D., Duan, T., Mehta, H., Yang, B., Zhu, K., Laird, D., Ball, R. L., Langlotz, C., Shpanskaya, K., Lungren, M. P., & Ng, A. Y. (2017). *MURA: Large Dataset for Abnormality Detection in Musculoskeletal Radiographs*. doi: 10.48550/ARXIV.1712.06957

Reznikov, N., Phillips, C., Cooke, M., Garbout, A., Ahmed, F., & Stevens, M. M. (2017). Functional adaptation of the calcaneus in historical foot binding: golden lotus: Anisotropy of trabecular bone reveals its loading manner. *Journal of Bone and Mineral Research*, *32*(9), 1915–1925. doi: 10.1002/jbmr.3185

Richards, G. D., Jabbour, R. S., Horton, C. F., Ibarra, C. L., & MacDowell, A. A. (2012). Color changes in modern and fossil teeth induced by synchrotron microtomography. *American Journal of Physical Anthropology*, *149*(2), 172–180. doi: 10.1002/ajpa.22103

Rickman, J. M., & Shackel, J. (2019). Crack propagation through sandwich bones due to low-velocity projectile impact. *International Journal of Legal Medicine*, *133*(5), 1443–1459. doi: 10.1007/s00414-019-02086-1

Robling, A. G., & Stout, S. D. (1999). Morphology of the drifting osteon. *Cells, Tissues, Organs*, *164*(4), Article 4. doi: 10.1159/000016659

Ronneberger, O., Fischer, P., & Brox, T. (2015). U-Net: Convolutional networks for biomedical image segmentation. In Nassir Navab, Joachim Hornegger, William M. Wells, Alejandro F. Frangi (eds.), *Medical Image Computing and Computer-Assisted Intervention—MICCAI 2015* (Vol. 3, pp. 234–241). Springer.

Röntgen, W. (1895). Ueber eine neue Art von Strahlen: Vorläufige Mitteilung. *Aus Den Sitzungsberichten Der Würzburger Physik.-Medic. Gesellschaft Würzburg*, 137–147.

Rosen, T., Bloemen, E. M., LoFaso, V. M., Clark, S., Flomenbaum, N., & Lachs, M. S. (2016). Emergency department presentations for injuries in older adults independently known to be victims of elder abuse. *The Journal of Emergency Medicine*, *50*(3), Article 3. doi: 10.1016/j.jemermed.2015.10.037

Rosenblatt, D. E., Cho, K. H., & Durance, P. W. (1996). Reporting mistreatment of older adults: The role of physicians. *Journal of the American Geriatrics Society*, *44*(1), Article 1. doi: 10.1111/j.1532-5415.1996.tb05640.x

Ruff, C. B., & Jones, H. H. (1981). Bilateral asymmetry in cortical bone of the humerus and tibia—Sex and age factors. *Human Biology*, *53*(1), 69–86. JSTOR. http://www.jstor.org/stable/41464595

Rühli, F. J., Kuhn, G., Evison, R., Müller, R., & Schultz, M. (2007). Diagnostic value of micro-CT in comparison with histology in the qualitative assessment of historical human skull bone pathologies. *American Journal of Physical Anthropology*, *133*(4), 1099–1111. doi: 10.1002/ajpa.20611

Ruth, E. B. (1947). Gross demonstration of the vascular channels in bone. *The Anatomical Record*, *98*(1), 59–65. doi: 10.1002/ar.1090980108

Saint-Martin, P., Dedouit, F., Rérolle, C., Guilbeau-Frugier, C., Dabernat, H., Rougé, D., Telmon, N., & Crubézy, E. (2015). Diagnostic value of high-resolution peripheral quantitative computed tomography (HR-pQCT) in the qualitative assessment of cribra orbitalia – A preliminary study. *HOMO*, *66*(1), 38–43. doi: 10.1016/j.jchb.2014.09.005

Saito, S. (1984). [Osteon patterns in the diaphysis of the human femur]. *Anatomischer Anzeiger*, *155*(1-5), 297–301.

Scheffer, M. K., & Jansen, H. W. (1987). Visualization of vascular channels in the mandibular cortex. *Journal De Biologie Buccale*, *15*(4), 235–238.

Schnapper, A., Reumann, K., & Meyer, W. (2002). The architecture of growing compact bone in the dog: Visualization by 3D-reconstruction of histological sections. *Annals of Anatomy-Anatomischer Anzeiger*, *184*(3), Article 3. doi: 10.1016/S0940-9602(02)80110-8

Schneider, P., Stauber, M., Voide, R., Stampanoni, M., Donahue, L. R., & Müller, R. (2007). Ultrastructural properties in cortical bone vary greatly in two inbred strains of mice as assessed by synchrotron light based micro- and nano-CT. *Journal of Bone and Mineral Research*, *22*(10), 1557–1570. doi: 10.1359/jbmr.070703

Schug, G. R., Killgrove, K., Atkin, A., & Barona, K. (2021). 3D dead: Ethical considerations in digital human osteology. *Bioarchaeol. Int*, *4*, 217–230.

Schultz, M. (2001). Paleohistopathology of bone: A new approach to the study of ancient diseases. *American Journal of Physical Anthropology*, *Suppl 33*, 106–147. doi: 10.1002/ajpa.10024.abs

Schulz, R. A., Stein, J. A., & Pelc, N. J. (2021). How CT happened: The early development of medical computed tomography. *Journal of Medical Imaging (Bellingham, Wash.)*, *8*(5), 052110. doi: 10.1117/1.JMI.8.5.052110

Sergi, S., Ascenzi, A., & Bonucci, E. (1972). Torus palatinus in the Neandertal Circeo I skull. A histologic, microradiographic and electron microscopic investigation. *American Journal of Physical Anthropology*, *36*(2), 189–197. doi: 10.1002/ajpa.1330360207

Shah, F. A., Ruscsák, K., & Palmquist, A. (2019). 50 years of scanning electron microscopy of bone-a comprehensive overview of the important discoveries made and insights gained into bone material properties in health, disease, and taphonomy. *Bone Research*, *7*, 15. doi: 10.1038/s41413-019-0053-z

Shaw, C. N. (2011). Is 'hand preference' coded in the hominin skeleton? An in-vivo study of bilateral morphological variation. *Journal of Human Evolution*, *61*(4), 480–487. doi: 10.1016/j.jhevol.2011.06.004

Sijbers, J., & Postnov, A. (2004). Reduction of ring artefacts in high resolution micro-CT reconstructions. *Physics in Medicine and Biology*, *49*(14), N247–N253. doi: 10.1088/0031-9155/49/14/N06

Sirak, K. A., & Sedig, J. W. (2019). Balancing analytical goals and anthropological stewardship in the midst of the paleogenomics revolution. *World Archaeology*, *51*(4), 560–573. doi: 10.1080/00438243.2019.1617190

Slyfield, C. R., Niemeyer, K. E., Tkachenko, E. V., Tomlinson, R. E., Steyer, G. G., Patthanacharoenphon, C. G., Kazakia, G. J., Wilson, D. L., & Hernandez, C. J. (2009). Three-dimensional surface texture visualization of bone tissue through epifluorescence-based serial block face imaging. *Journal of Microscopy*, *236*(1), 52–59. doi: 10.1111/j.1365-2818.2009.03204.x

Slyfield, C. R., Tkachenko, E. V., Wilson, D. L., & Hernandez, C. J. (2012). Three-dimensional dynamic bone histomorphometry. *Journal of Bone and Mineral Research: The Official Journal of the American Society for Bone and Mineral Research*, *27*(2), Article 2. doi: 10.1002/jbmr.553

Smith, A. C., Watamaniuk, L., & Rogers, T. L. (2022). Use of laser-scanning confocal microscopy in the detection of diagenesis in bone. *Journal of Forensic Sciences*, *67*(1), 92–101. doi: 10.1111/1556-4029.14902

Soltan, N., Kawalilak, C. E., Cooper, D. M., Kontulainen, S. A., & Johnston, J. D. (2019). Cortical porosity assessment in the distal radius: A comparison of HR-pQCT measures with Synchrotron-Radiation micro-CT-based measures. *Bone*, *120*, 439–445. doi: 10.1016/j.bone.2018.12.008

Stanley, S. A., Hainsworth, S. V., & Rutty, G. N. (2018). How taphonomic alteration affects the detection and imaging of striations in stab wounds. *International Journal of Legal Medicine*, *132*(2), 463–475. doi: 10.1007/s00414-017-1715-2

Stein, M. S., Feik, S. A., Thomas, C. D. L., Clement, J. G., & Wark, J. D. (1999). An automated analysis of intracortical porosity in human femoral bone across age. *Journal of Bone and Mineral Research*, *14*(4), Article 4. doi: 10.1359/jbmr.1999.14.4.624

Stephens, N. B., Kivell, T. L., Gross, T., Pahr, D. H., Lazenby, R. A., Hublin, J., Hershkovitz, I., & Skinner, M. M. (2016). Trabecular architecture in the thumb of *Pan* and *Homo*: Implications for investigating hand use, loading, and hand preference in the fossil record. *American Journal of Physical Anthropology*, *161*(4), 603–619. doi: 10.1002/ajpa.23061

Stock, S. R., Mays, S., Turner-Walker, G., & Soriano, C. (2022). Microcomputed tomography (laboratory and synchrotron) of intact archeological human second metacarpal bones and age at death. *International Journal of Osteoarchaeology*, *32*(1), 120–131. doi: 10.1002/oa.3049

Stout, S. D. (1978). Histological structure and its preservation in ancient bone. *Current Anthropology*, *19*(3), Article 3. doi: 10.1086/202141

Stout, S. D., Brunsden, B. S., Hildebolt, C. F., Commean, P. K., Smith, K. E., & Tappen, N. C. (1999). Computer-assisted 3D reconstruction of serial sections of cortical bone to determine the 3D structure of osteons. *Calcified Tissue International*, *65*(4), Article 4. doi: 10.1007/s002239900699

Stout, S. D., & Simmons, D. J. (1979). Use of histology in ancient bone research. *Yearbook of Physical Anthropology 22*, 228–249.

Stride, P., Patel, N., & Kingston, D. (2013). The history of osteoporosis: Why do Egyptian mummies have porotic bones? *The Journal of the Royal College of Physicians of Edinburgh*, *43*(3), 254–261.

Sugawara, Y., Kamioka, H., Honjo, T., Tezuka, K., & Takano-Yamamoto, T. (2005). Three-dimensional reconstruction of chick calvarial osteocytes and their cell processes using confocal microscopy. *Bone*, *36*(5), 877–883. doi: 10.1016/j.bone.2004.10.008

Tappen, N. C. (1977). Three-dimensional studies on resorption spaces and developing osteons. *The American Journal of Anatomy*, *149*(3), Article 3. doi: 10.1002/aja.1001490302

Tarasiuk, J., Mnich, B., Wroński, S., Lisowska-Gaczorek, A., & Szostek, K. (2023). Application of the microtomography technique in density studies of prehistoric and historical human skeletal materials. *International Journal of Osteoarchaeology*, oa.3231. doi: 10.1002/oa.3231

Teschner, M. (2007). Histomorphometric analysis of bone tissue from a neolithic skeleton. *Journal of Archaeological Science*, *34*(7), 1130–1134. doi: 10.1016/j.jas.2006.10.005

Thali, M. J., Taubenreuther, U., Karolczak, M., Braun, M., Brueschweiler, W., Kalender, W. A., & Dirnhofer, R. (2003). Forensic microradiology: Micro-computed tomography (Micro-CT) and analysis of patterned injuries inside of bone. *Journal of Forensic Sciences*, *48*(6), 1336–1342.

Thomas, C. D. L., Feik, S. A., & Clement, J. G. (2005). Regional variation of intracortical porosity in the midshaft of the human femur: Age and sex differences. *Journal of Anatomy*, *206*(2), Article 2. doi: 10.1111/j.1469-7580.2005.00384.x

Thomas, C. D. L., Feik, S. A., & Clement, J. G. (2006). Increase in pore area, and not pore density, is the main determinant in the development of porosity in human cortical bone. *Journal of Anatomy*, *209*(2), Article 2. doi: 10.1111/j.1469-7580.2006.00589.x

Thomas, C. D. L., Stein, M. S., Feik, S. A., Wark, J. D., & Clement, J. G. (2000). Determination of age at death using combined morphology and histology of the femur. *Journal of Anatomy*, *196*(Pt 3), Article Pt 3. doi: 10.1046/j.1469-7580.2000.19630463.x

Tkachenko, E. V., Slyfield, C. R., Tomlinson, R. E., Daggett, J. R., Wilson, D. L., & Hernandez, C. J. (2009). Voxel size and measures of individual resorption cavities in three-dimensional images of cancellous bone. *Bone*, *45*(3), 487–492. doi: 10.1016/j.bone.2009.05.019

Trinkaus, E., Churchill, S. E., & Ruff, C. B. (1994). Postcranial robusticity in Homo. II: Humeral bilateral asymmetry and bone plasticity. *American Journal of Physical Anthropology*, *93*(1), Article 1. doi: 10.1002/ajpa.1330930102

Tripp, J. A., Squire, M. E., Hamilton, J., & Hedges, R. E. M. (2010). A nondestructive prescreening method for bone collagen content using micro-computed tomography. *Radiocarbon*, *52*(2), 612–619. doi: 10.1017/S0033822200045641

Tripp, J. A., Squire, M. E., Hedges, R. E. M., & Stevens, R. E. (2018). Use of micro-computed tomography imaging and porosity measurements as indicators of collagen preservation in archaeological bone. *Palaeogeography, Palaeoclimatology, Palaeoecology*, *511*, 462–471. doi: 10.1016/j.palaeo.2018.09.012

University of Texas, Austin. (n.d.). *Digital Morphology Library. A National Science Foundation digital library at the University of Texas at Austin* [dataset]. http://www.digimorph.org/index.phtml

van Duijvenbode, A., Herschensohn, O. J., & Morgan, M. E. (2015). A severe case of congenital aural atresia in pre-Columbian Venezuela. *International Journal of Paleopathology*, *9*, 15–19. doi: 10.1016/j.ijpp.2014.11.002

Van Spelde, A.-M., Schroeder, H., Kjellström, A., & Lidén, K. (2021). Approaches to osteoporosis in paleopathology: How did methodology shape bone loss research? *International Journal of Paleopathology*, *33*, 245–257. doi: 10.1016/j.ijpp.2021.05.001

Vasciaveo, F., & Bartoli, E. (1961). Vascular channels and resorption cavities in the long bone cortex the bovine bone. *Acta Anatomica*, *47*, 1–33.

Vekszler, G., Granner, M., Nebot Valenzuela, E., Winter, E., Dockner, M., Weber, G. W., Pretterklieber, M., Teschler-Nicola, M., & Pietschmann, P. (2022). Microarchitecture of historic bone samples with tuberculosis. *Wiener Klinische Wochenschrift*, *134*(11–12), 449–457. doi: 10.1007/s00508-022-02017-y

Venema, J., Peula, D., Irurita, J., & Mesejo, P. (2023). Employing deep learning for sex estimation of adult individuals using 2D images of the humerus. *Neural Computing and Applications*, *35*(8), 5987–5998. doi: 10.1007/s00521-022-07981-0

Villa, C., Buckberry, J., & Lynnerup, N. (2019). Evaluating osteological ageing from digital data. *Journal of Anatomy*, *235*(2), 386–395. doi: 10.1111/joa.12544

Von Schacky, C. E., Wilhelm, N. J., Schäfer, V. S., Leonhardt, Y., Gassert, F. G., Foreman, S. C., Gassert, F. T., Jung, M., Jungmann, P. M., Russe, M. F., Mogler, C., Knebel, C., Von Eisenhart-Rothe, R., Makowski, M. R., Woertler, K., Burgkart, R., & Gersing, A. S. (2021). Multitask deep learning for segmentation and classification of primary bone tumors on radiographs. *Radiology*, *301*(2), 398–406. doi: 10.1148/radiol.2021204531

Waarsing, J. H., Day, J. S., van der Linden, J. C., Ederveen, A. G., Spanjers, C., De Clerck, N., Sasov, A., Verhaar, J. A. N., & Weinans, H. (2004). Detecting and tracking local changes in the tibiae of individual rats: A novel method to analyse longitudinal in vivo micro-CT data. *Bone*, *34*(1), Article 1. doi: 10.1016/j.bone.2003.08.012

Wade, A. D., Holdsworth, D. W., & Garvin, G. J. (2011). CT and micro-CT analysis of a case of Paget's disease (osteitis deformans) in the Grant skeletal collection: CT and Micro-CT Analysis of Paget's Disease of Bone. *International Journal of Osteoarchaeology*, *21*(2), 127–135. doi: 10.1002/oa.1111

Weber, G. W. (2015). Virtual anthropology. *American Journal of Physical Anthropology*, *156 Suppl 59*, 22–42. doi: 10.1002/ajpa.22658

Welsh, H., Nelson, A. J., van der Merwe, A. E., de Boer, H. H., & Brickley, M. B. (2020). An investigation of micro-CT analysis of bone as a new diagnostic method for paleopathological cases of osteomalacia. *International Journal of Paleopathology*, *31*, 23–33. doi: 10.1016/j.ijpp.2020.08.004

Wheeler, S. M., Williams, L., Beauchesne, P., & Dupras, T. L. (2013). Shattered lives and broken childhoods: Evidence of physical child abuse in ancient Egypt. *International Journal of Paleopathology*, *3*(2), 71–82. doi: 10.1016/j.ijpp.2013.03.009

White, S., Hirst, C., & Smith, S. E. (2018). The suitability of 3D data: 3D digitisation of human remains. *Archaeologies*, *14*(2), 250–271. doi: 10.1007/s11759-018-9347-9

WHO. (1994). *Assessment of fracture risk and its application to screening for postmenopausal osteoporosis. WHO Study Group on Assessment of Fracture Risk* (Issues 843–849). World Health Organization.

WHO. (2023). *World Health Organization tuberculosis fact sheet*. https://www.who.int/news-room/fact-sheets/detail/tuberculosis

Wilkins, S. W., Gureyev, T. E., Gao, D., Pogany, A., & Stevenson, A. W. (1996). Phase-contrast imaging using polychromatic hard X-rays. *Nature*, *384*(6607), 335–338. doi: 10.1038/384335a0

Wolff, J. (1892). *Das Gesetz der Transformation der Knochen*. Verlag von August Hirschwald.

Yoshino, M., Imaizumi, K., Miyasaka, S., & Seta, S. (1994). Histological estimation of age at death using microradiographs of humeral compact bone. *Forensic Science International*, *64*(2–3), 191–198. doi: 10.1016/0379-0738(94)90231-3

Yu, B., Pacureanu, A., Olivier, C., Cloetens, P., & Peyrin, F. (2020). Assessment of the human bone lacuno-canalicular network at the nanoscale and impact of spatial resolution. *Scientific Reports*, *10*(1), 4567. doi: 10.1038/s41598-020-61269-8

Zeppilli, C., Micarelli, I., Bernardini, S., Profico, A., Di Giannantonio, S., Giostra, C., Paine, R. R., Manzi, G., & Tafuri, M. A. (2023). Stabbed to death: An osteobiography example of violence among Longobards (Povegliano Veronese, Italy, 6th-8th c CE). *Archaeological and Anthropological Sciences*, *15*(5), 54. doi: 10.1007/s12520-023-01742-3

Zhang, Y., Wang, J., Yu, C., Xia, K., Yang, B., Zhang, Y., Ying, L., Wang, C., Huang, X., Chen, Q., Shen, L., Li, F., & Liang, C. (2022). Advances in single-cell sequencing and its application to musculoskeletal system research. *Cell Proliferation*, *55*(1), e13161. doi: 10.1111/cpr.13161

Zhao, D., He, L., & Xing, F. (2023). Skeletal evidence for two external auditory canal disorder cases from medieval China. *International Journal of Osteoarchaeology*, *33*(1), 170–177. doi: 10.1002/oa.3187

Zhou, Z., Siddiquee, M. M. R., Tajbakhsh, N., & Liang, J. (2018). UNet++: A Nested U-Net Architecture for Medical Image Segmentation. *Deep Learning in Medical Image Analysis and Multimodal Learning for Clinical Decision Support: 4th International Workshop, DLMIA 2018, and 8th International Workshop, ML-CDS 2018, Held in Conjunction with MICCAI 2018, Granada, Spain, S…*, *11045*, 3–11. doi: 10.1007/978-3-030-00889-5_1

Visualization and Interpretation of Cortical Porosity

13

MARY E. COLE

13.1 Introduction

Early microscopists recognized that the seemingly solid cortical walls of bone are actually perforated with a complex network of pores (Havers, 1691; Leeuwenhoeck, 1674; Todd & Bowman, 1845; Tomes & Morgan, 1853). Cortical pores are created by bone modeling, where new bone forms around blood vessels lying on the bone surface, and by bone remodeling, where vascular structures are extended as bone cells tunnel into existing bone tissue (Chen et al., 2021; Eriksen, 2010). These are sometimes referred to as "vascular pores" to distinguish them from a "cellular pore" network that also exists within cortical bone (Harrison et al., 2020). An estimated 2% of porosity of the bone matrix is contributed by the osteocyte lacunar-canalicular network (LCN) (Buenzli & Sims, 2015). Osteocytes are mechanosensory bone cells that initiate modeling or remodeling of local bone tissue in response to changing mechanical strain or microscopic damage. As a differentiating osteocyte becomes embedded in forming bone, its cell body is enclosed in an ovoid space called a lacuna. Its numerous dendritic processes extend through narrow channels called canaliculi, connecting it with the dendrites of other osteocytes (Varga et al., 2015). Cortical pores form complex and dense 3D network structures through the branching, coalescence, and transverse interconnection of discrete remodeling events (Maggiano et al., 2016). The osteocyte LCN is also densely interconnected, which is vital for osteocyte mechanosensing and cell-cell communication (Himeno-Ando et al., 2012). Cortical pore networks and the osteocyte LCN are interdependent, morphometrically and functionally. Both pore networks are formed and reshaped by osteocyte-initiated remodeling. In turn, the vascular structures extended through cortical pores supply trapped osteocytes with nutrients, and transport cell signaling factors and bone cell progenitors for subsequent remodeling (Chen et al., 2021; Eriksen, 2010).

Cortical pores and the osteocyte LCN are not uniformly structured throughout the skeleton. Instead, their morphometry and distribution reflect the local mechanical environment of the bone tissue, which controls the remodeling activity that creates these vascular and cellular spaces. Aspects of pore density, distribution, size, and shape are further altered by age and pathology, which exert physiological influences on the rate and extent of remodeling. Characterizing pore morphometry has broad application in biological anthropology. Variable pore morphometry within the skeleton is a record of mechanical loading history, aiding in interpreting the functional behavior of a skeletal element or tissue region. Morphometric variation in porosity between individuals can reveal the physiological influences of age, sex, or pathology. Moreover, cortical porosity is a key marker of bone quality and an important predictor of bone fragility. Cortical pores are stress concentrators where

 DOI: 10.4324/9781003385608-13

microcracks can initiate and propagate into fracture (Ebacher et al., 2007). Osteocyte lacunae also concentrate stress (Kola et al., 2020) and have been observed to initiate microcracks (Reilly & Currey, 2000). However, bone tissue tolerance of such defects appears dependent on lacunar morphometry and distribution (Buccino et al., 2023). The osteocyte LCN contributes to bone fragility through age-associated degradation of its density, size, and connectivity (Andronowski et al., 2023; Ashique et al., 2017). This deterioration impairs cell-cell communication and decreases osteocyte sensitivity to mechanical signals for bone tissue maintenance and repair (Niroobakhsh et al., 2023).

The purpose of this chapter is to provide an overview of pore morphometry as it is shaped by, and in turn alters, the mechanical demands of its local tissue environment. This chapter will largely focus on cortical bone, which encloses the vascular network required to nourish its less permeable tissue. The first section of this chapter will discuss how mechanical stimuli regulate the rate of bone remodeling, which determines how its porous products are distributed and shaped. The second section will describe cellular processes that form the cortical pore network. The third section will summarize microscopic and radiographic imaging techniques for visualizing pore morphometry in bone tissue. The fourth section will discuss specific aspects of pore morphometry and interpretations of their possible mechanical origins.

13.2 How Mechanical Stimuli Shape Remodeling Activity

13.2.1 Mechanostat Model for Modeling and Remodeling

The underlying theoretical paradigm of bone biology is bone functional adaptation, which we explore in greater detail in Chapter 1 of this volume. Broadly, bone functional adaptation is the concept that bone dynamically adjusts its shape and structure to withstand changes in its mechanical loading environment (Ruff et al., 2006). The response of bone tissue to a mechanical stimulus can be quantified using strain, its fractional change in length during deformation (Pivonka et al., 2018). The strain stimulus for bone adaptation is proportional to strain rate, which is the product of strain magnitude, quantified in microstrains (με), and strain frequency, quantified in in cycles/second (Hertz or Hz). Bone adaptation is stimulated both by high-magnitude strains, even if they occur at low frequency, and by high-frequency strains, even if they are low magnitude (Turner, 1998). While bone experiences relatively few daily high-magnitude (2,000–3,000 με) low-frequency (1–3 Hz) strains, it is continually loaded by low-magnitude (<5 με), high-frequency (10–50 Hz) strains associated with muscular maintenance of body posture (Ozcivici et al., 2010).

Bone functional adaptation is a feedback-based model, where bone is added or removed to maintain its tissue deformation within an "optimum customary strain level" (Lanyon et al., 1982). This optimal range is likely specific to the skeletal location, genetically regulated and physiologically modified over the lifespan (Skerry, 2006). Harold Frost's mechanostat model (Frost, 1983, 1987a, 1987b, 2003a) demonstrates how bone modeling and remodeling might be activated to maintain bone tissue strain within a consistent range, or "setpoint".

(1) When mechanical stimuli are reduced below the customary range, bone deformation is also reduced. If the strain that quantifies this deformation falls below the adapted window's lower threshold, the minimum effective strain for remodeling

(MESr = 50–100 με), the mechanostat enters the disuse window. Bone is removed until the weakened tissue, subjected to the reduced mechanical stimuli, again deforms within the adapted window's strain range. This bone removal is accomplished either through **resorptive modeling**, which preferentially resorbs bone from the endosteal surface, or **disuse-mode remodeling**, which resorbs more bone than it subsequently replaces through bone formation. When activated within the cortex, disuse-mode remodeling increases cortical porosity. It is preferentially located within the endocortical region, closest to the low-strain endosteal surface (Frost, 2003a; Hughes et al., 2020; Hughes & Petit, 2010).

(2) When mechanical stimuli are consistent with the customary range, bone deformation is also consistent. The strains that quantify this deformation fall within the adapted window. **Conservation-mode remodeling** maintains bone tissue quality by turning it over, resorbing and forming bone in roughly equal amounts. There is no significant change in net bone mass, except the space occupied by new Haversian canals (Frost, 2003a; Hughes et al., 2020; Hughes & Petit, 2010).

(3) When mechanical stimuli are increased above a customary range, bone deformation is also increased. If the strain that quantifies this deformation falls above the adapted window's upper threshold, the minimum effective strain for modeling (MESm = 1,000–1,500 με), the mechanostat enters the mild overload window. Bone is added until the strengthened tissue, subjected to the increased mechanical stimuli, again deforms within the adapted window's strain range. This bone formation is accomplished through **formative modeling**, which preferentially adds bone to the periosteal surface. Conservation-mode remodeling is suppressed. Bone tissue also dissipates the increased strain through the formation of microcracks within the cortex (Ritchie, 2011). **Targeted remodeling** is activated within a 100–300 μm radius of a microcrack to resorb the damage and form new bone (Kennedy et al., 2012). Targeted remodeling is estimated to comprise 10–30% of all remodeling activity (Burr & Martin, 1993; Li et al., 2001), although it is theoretically possible that all remodeling is "steering" toward microdamage (Martin, 2002).

(4) Further strain increases enter the pathological overload window, above the microdamage threshold (MESp = ~3,000 μE), where microscopic tissue damage accumulates faster than it can be remodeled, until bone failure (~25,000 με in young adults).

13.2.2 Importance of Remodeling in Human Bone Tissue

An important consideration for understanding human bone tissue microstructure is our reliance on remodeling. Smaller and shorter-lived vertebrate species, such as mice and rats, rarely engage in remodeling, except under environmental stress and in old age. Their bone tissue is dominated by primary modeled bone and its primary vascular canals (Piemontese et al., 2017; Ross & Sumner, 2017). Without frequent remodeling to repair damage, bone strength is more reliant on stiffness, the resistance of its tissue to elastic (nonpermanent) deformation (Ritchie, 2011). In larger, longer-lived species that do frequently remodel, it is advantageous to incur limited plastic (permanent) deformation in the form of microdamage, which can be later repaired. Microdamage allows the bone tissue to release energy, increasing its toughness, the total energy it can absorb before failure (Martin, 2003). Discrete microcracks (40–100 μm long, 1–2 μm wide, running 300–500 μm longitudinally) are repaired through targeted remodeling (Burr & Martin, 1983; Gauthier et al., 2018; Taylor & Clive Lee, 1998). Matrix rebonding may repair the more mesh-like diffuse microdamage (Seref-Ferlengez et al., 2015). Microcracking is itself a "toughening mechanism" to resist bone failure. The secondary osteons created by remodeling contain further toughening mechanisms to inhibit the propagation of these microcracks into fracture. The decline in stiffness associated with microdamage is most strongly associated with growing crack length and alignment with its propagating force, as opposed to the number of cracks (Akkus et al., 2003). The highly mineralized cement line that bounds a secondary osteon can trap and deflect a propagating microcrack, forcing it to follow the orientation of the Haversian system (Koester et al., 2008; Norman & Wang, 1997). Secondary osteons can also debond at the cement line and pull out from the matrix in a telescoping fashion to "bridge" a developing crack (Hiller et al., 2003). Cortical pores and osteocyte lacunae also interact with microcracks, as will be discussed in greater detail.

Both microdamage and cortical porosity are products of bone's toughening mechanism and its subsequent remodeling. The typical in vivo level of microdamage is likely not mechanically detrimental to bone unless it is allowed to accumulate without repair (Burr, 2011). Microdamage more frequently initiates in aged tissue because it is brittle, due to higher mineralization, and because it contains more cortical porosity, which concentrates stress for microcrack initiation. Age-associated declines in bone cell functionality and survival may also impair their ability to detect and remodel microdamage (Norman & Wang, 1997; Schaffler et al., 1995; Turnbull et al., 2014; Wasserman et al., 2005). Aged individuals also form fewer, longer linear microcracks, as opposed to the more numerous and small diffuse cracks of younger individuals (Allen & Burr, 2008). In humans, microdamage has been shown to accumulate with age in the femoral diaphyses, head, and neck, the tibial diaphyses, and the vertebral body (Koszyca et al., 1989; Mori et al., 1997; Norman & Wang, 1997; Schaffler et al., 1995; Vogel et al., 1993). An early principle of toxicology, often quoted from the 15th-century physician Paracelsus, is that "The right dose differentiates a poison and a remedy". The same concept may be applied to the bone tissue voids formed by microcracks and cortical pores, which are essential in moderation but fragile in excess.

13.2.3 Intraskeletal and Age-Associated Variation in Remodeling Rate

Remodeling can be activated under low strain (disuse-mode remodeling), moderate strain (conservation-mode remodeling), and high strain (targeted remodeling). This complicates predictions of intraskeletal variation in remodeling rate based on the local mechanical loading environment. Remodeling rate has been observed to increase toward the higher strain environment of the distal tapered limb bones of cursorial quadrupeds, potentially due to targeted remodeling (Lieberman et al., 2003; Skedros et al., 2003). Yet remodeling rate is also elevated in the relatively lower-strain environment of the rib, compared to weight-bearing limb bones such as the femur (Cho & Stout, 2011; Fahy et al., 2017; Mulhern, 2000; Mulhern & Van Gerven, 1997; Skedros et al., 2003; Stout, 1984; Vajda et al., 1999; Wilson et al., 1998). This could result from permissiveness to disuse-mode or conservation-mode remodeling in the relative absence of weight-bearing by the rib (Frost, 2003a; Skedros et al., 2013). Alternatively, this could result from more frequent targeted remodeling of microdamage as the rib is continually cyclically loaded by respiration (Cho & Stout, 2011; Skedros et al., 2013). It could also be an artifact of a potentially earlier effective age of adult compacta (Mulhern, 2000; Mulhern & Van Gerven, 1997). Skeletal elements with similar mechanical loading environments do appear more consistent in remodeling rate, without significant variation between the third through eighth human ribs (Crowder & Rosella, 2007) or between the weight-bearing femur and tibia (Drapeau & Streeter, 2006).

Mechanical strain also varies regionally within the cross-section of a skeletal element. Bending exerts approximately 70% of longitudinal forces in bone (Biewener & Bertram, 1993). Within the plane of bending, one side of the cortex is compressed, and the other side is tensed. Compressive and tensile stresses both decrease toward the neutral axis where they meet (Robling, 2014). In bone, regions under compression experience higher strain (x1.9) than regions under tension (Lanyon & Baggott, 1976). Within the cross-sections of many human and quadruped bones, regions that experience high-strain compression have a significantly elevated remodeling rate compared to regions that experience lower-strain tension (Lanyon & Bourn, 1979; Mason et al., 1995; Mayya et al., 2013; Riggs et al., 1993; Skedros et al., 1994, 1997, 2009; Zedda et al., 2015). However, this patterning is not universal. In some species and skeletal elements, cross-sectional regions under tension develop remodeling rates that exceed regions under compression. This has been attributed to combination with torsion or shear (Gocha & Agnew, 2016; Mason et al., 1995; Portigliatti Barbos et al., 1984; Skedros et al., 2001; Skedros, Sybrowsky, Anderson, et al., 2011) or to a complex loading environment that shifts the neutral axis or alters the magnitudes of strain direction as the gait changes (Skedros et al., 2009; Skedros, Kiser, & Mendenhall, 2011; Skedros, Kiser, Keenan, et al., 2011). Mechanical strain also varies across bone envelopes, decreasing toward the endosteum where strains are lowest due to proximity to the neutral axis (Martin, 1993).

Remodeling rate is further modified by physiological changes across the lifespan. High remodeling rates during childhood growth slow toward adulthood, reach a minimum around age 35, and then start to increase again (Martin, 1993). However, the high remodeling rates of youth produce a positive bone balance, while the high remodeling rates of older age result in net bone loss (Seeman & Delmas, 2006). As early as this third decade of life, bone resorption begins to outpace bone formation due to reduced physical activity, sarcopenia (muscle wasting), declining sex steroids, and senescence of bone cell sensitivity and function (Demontiero et al., 2012; Infante & Rodríguez, 2018). This age-associated bone loss is exacerbated in females. Estrogen promotes bone formation at the endosteum (Ferretti et al., 1998), while testosterone stimulates bone formation at the periosteum (Turner et al., 1990). Males exceed females in long bone cross-sectional area across the lifespan (Feik et al., 2000; Schlecht et al., 2015). The peri-menopausal decline of estrogen in aging females results in a 33% increase of resorption activation at the endosteum (Han et al., 1997). In males, sustained testosterone into older age promotes periosteal bone apposition, capable of sustaining or even increasing cortical area (Burr & Martin, 1983; Martin & Atkinson, 1977; Mosekilde & Mosekilde, 1990; Ruff & Hayes, 1988). While males and females experience similar declines in the material strength of bone, this periosteal apposition preserves structural strength in males, compared to females (Martin & Atkinson, 1977). Bone is stronger when distributed further from the neutral axis. From an engineering perspective, only ~30% of the bone resorbed on the endosteal surface must be added to the periosteal surface to maintain resistance to bending or torsion (Martin, 1993). After age 50, clinically low bone mass (osteopenia) can be detected at the hip or lumbar spine in 33.5% of males, compared to 51.5% of females in the United States. Low bone mass with risk of fracture (osteoporosis) is clinically diagnosed in 4.4% of males and 19.6% of females (Sarafrazi et al., 2021). This helps to explain why females experience about 1.6 times as many osteoporotic fractures as males (Johnell & Kanis, 2006).

13.3 Cellular Processes Forming the Cortical Pore Network

Bone modeling refers to modification of the bone surface, either by osteoblasts forming new bone (formative modeling) or osteoclasts removing existing bone without replacement (resorptive modeling). These cellular activities, although both initiated by osteocytes, are uncoupled and independent (Frost, 2003b). When formative modeling encloses a blood vessel lying on the bone surface, it forms a cortical pore called a **primary canal**, with the surrounding bone termed a primary osteon (Khoury et al., 2015).

Bone remodeling refers to the replacement of existing bone tissue by osteoclastic bone resorption followed by osteoblastic bone formation. The activity of these bone cells is coordinated at the same tissue location in a Basic Multicellular Unit (BMU). The phases of bone remodeling include (1) **activation** by osteocytes, (2) bone **resorption** by osteoclasts (2–3 weeks), (3) **reversal** to bone formation (4–5 weeks), (4) osteoid **formation** by osteoblasts (3–4 months), (5) osteoid **mineralization** by osteoblasts and differentiating osteocytes (1–4 week lag), and (6) **quiescence**, or resting. The remodeling cycle is shorter in cortical bone (~120 days) than cancellous bone (~200 days) (Agerbaek et al., 1991; Eriksen et al., 1984; Kenkre & Bassett, 2018; Sims & Martin, 2020). For a more detailed illustration of the cell signaling that regulates this cellular activity, see Chapter 1 of this same volume, or our previous work (Cole et al., 2023).

Remodeling produces distinct tissue microstructures in cortical and cancellous bone. In cancellous bone remodeling, the BMU sits on the exposed trabecular surface, covered by a canopy of retracted bone lining cells. This canopy is penetrated by marrow capillaries that transmit cell signaling factors and bone cell progenitors (Kristensen et al., 2014). Within the cancellous BMU, osteoclasts resorb and osteoblasts subsequently fill a 60–70 μm deep trench. Osteoblastic deposition of layers of lamellae within this trench produces a semicircular hemi-osteon, without a vascular pore (Trichilo & Pivonka, 2018).

In cortical bone remodeling, the BMU tunnels into the cortex, with osteoclasts resorbing existing bone in a "cutting cone", and osteoblasts following behind to form new bone in a "closing cone". The cortical BMU is covered by a canopy of cells of mesenchymal origin draped over its closing cone. This canopy is penetrated by a branch of cortical vasculature extending into the resorbed canal, transmitting bone cell signaling factors and progenitors (Eriksen, 2010). Although often simplified to a single blood vessel in textbook diagrams, this vasculature is actually complex, intertwining with neuronal structures, and forming vascular loops 35–40 μm behind the tip of the cutting cone. The vascular structures are more numerous when the pore is actively remodeling, especially when modifying a pre-existing or larger pore (Andreasen et al., 2023). While the eroded canal remains unfilled, it appears in cross-section as a large void, approximately 150–350 μm in diameter, with an irregular, scalloped border (Agerbaek et al., 1991; Britz et al., 2009; Brockstedt et al., 1996; Skedros, Clark, Sorenson, et al., 2011). Histologically, this is termed a **resorption space**, **resorptive bay**, or **Howship's lacuna** (Stout et al., 2019). The dimensions of the resorbed space and the resulting osteon are determined by the timing of the reversal to bone formation. This is not a clear transition, but a mixed "reversal-resorption" phase. "Primary" osteoclasts are located in the cutting cone's tip, where they elongate the canal. These are followed by "secondary" osteoclasts, scattered on the eroded surface among the osteoblast-lineage reversal cells that clean the demineralized collagen. Secondary osteoclasts widen the canal through radial resorption, which comprises on average 83% of the BMU's overall resorption. A corresponding cell arrangement exists in cancellous bone, where primary osteoclasts extend the edge of the trench on the trabecula, while secondary osteoclasts deepen the trench (Lassen et al., 2017). A complex interplay of cell signaling factors controls the timing of this reversal, "coupling" bone resorption to bone formation. Factors released from the resorbed matrix slow osteoclast function and trigger osteoblast differentiation and migration. Adjacent osteoclasts and osteoblast-lineage cells can slow or speed reversal with secreted or membrane-expressed factors that promote or inhibit differentiation of the opposing cell type (Sims & Martin, 2020). Reversal cells and osteoblasts must also pack to a critical cell density to start forming osteoid, halting bone resorption (Abdelgawad et al., 2016; Andersen et al., 2013; Lassen et al., 2017). At the maximum extent of bone resorption, reversal cells deposit a highly mineralized cement line, marking the reversal to bone formation. Osteoblasts deposit unmineralized osteoid from this periphery in concentric lamellae, then secrete factors that stimulate its mineralization. The finished product is a **secondary osteon**. The vasculature of the BMU is enclosed by these concentric lamellae in a cortical pore most commonly called a **Haversian canal**. The secondary osteon's cement line signifies the bone resorption that precedes bone formation, distinguishing this product of remodeling from the unbounded primary osteons created by formative modeling (Eriksen, 2010).

Mean Haversian canal size has been reported as ranging from approximately 20 µm to 200 µm in diameter, depending on tissue location (Agerbaek et al., 1991; Bala et al., 2016; Brockstedt et al., 1996; Gauthier et al., 2018; Jowsey, 1966; Milovanovic et al., 2017; Zimmermann et al., 2011). Larger Haversian canals are associated with the convergence of adjacent osteons (Bell et al., 2000) or the erosion of preexisting canals through repeated remodeling (Andreasen et al., 2020) (Figure 13.1). This is exacerbated by the lower strains closer to the endosteum, where the erosion and coalescence of adjacent canals can produce huge "trabecularized" pores (Zebaze & Seeman, 2015) (Figure 13.2).

13.4 Techniques for Imaging Cortical Pore Systems

13.4.1 Ex Vivo Histological Imaging

The first descriptions of cortical pores were microscopic observations of histological sections (Leeuwenhoeck, 1674). In a two-dimensional cross-section, these cylindrical canals appear as roughly circular voids. Descriptions of cortical pore size and shape in terms of area, diameter, and perimeter are derived from this two-dimensional approach. The complex three-dimensional structures of pore systems, with frequent branching and convergence, can pose a challenge for two-dimensional histological interpretation (Figure 13.3). Sections through a canal system that is tilted relative to the long axis of the bone will be skewed in cross-section, producing artificially elongated, ovoid pores. Separate cortical pores in cross-section do not necessarily represent discrete remodeling events, but may result from branching of the same pore system. The convergence of existing, adjacent pore systems increases pore size and shape beyond the true extent of resorption during a single remodeling event. The plane of section also influences cortical pore morphometry. Sections through transverse interconnections appear as long, thin streaks (Figure 13.4). Sections at the level of canal branching or bifurcation can create a pinched "dumbbell"-shaped osteon (Maggiano et al., 2016; Stout et al., 1999) (Figure 13.5). Even before the advent of X-ray-based three-dimensional imaging, this branching complexity was observed by

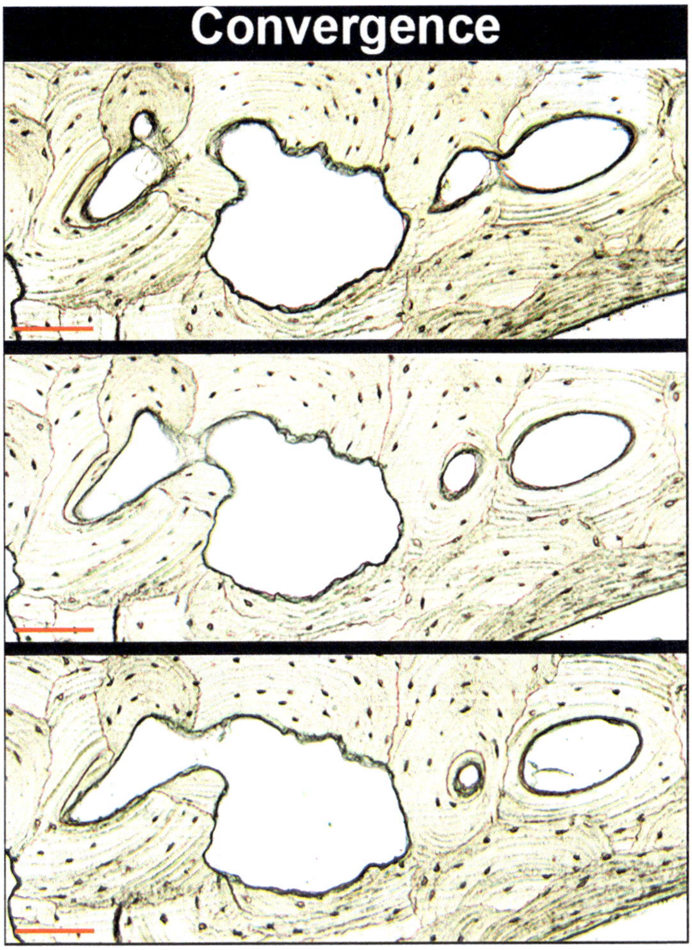

Figure 13.1 Serial histological cryosections of a human midshaft fourth rib, each 30 μm thick, show convergence of a resorption space with an existing adjacent Haversian canal. Scale bars are 100 μm.

following canals through serial histological sections (Cohen & Harris, 1958; Hert et al., 1994; Tappen, 1977).

Despite these caveats, histological analyses of porosity have several unique advantages. Even with its two-dimensional limitations, histological quantification of percent porosity, mean pore diameter, and mean pore separation is strongly correlated with measurements from three-dimensional micro-computed tomography (micro-CT) (Particelli et al., 2012; Wachter et al., 2001). Histology is typically lower-cost and more broadly accessible for labs that do not have access to their own desktop micro-CT system. Additionally, the transverse field of view is not limited in histology, as it is in X-ray-based imaging such as micro-CT. Overlapping microscopic images at high magnification (typically 100x) can be acquired across large bone cross-sections, providing a complete picture of porosity across variable strain regions. Another advantage of histology is its concurrent visualization of other cortical tissue microstructures, such as osteons and osteocyte lacunae. Diffuse and linear microcracks can be observed with contrast from histological staining (Dominguez & Agnew, 2019). While these microstructures can also be visualized three-dimensionally with

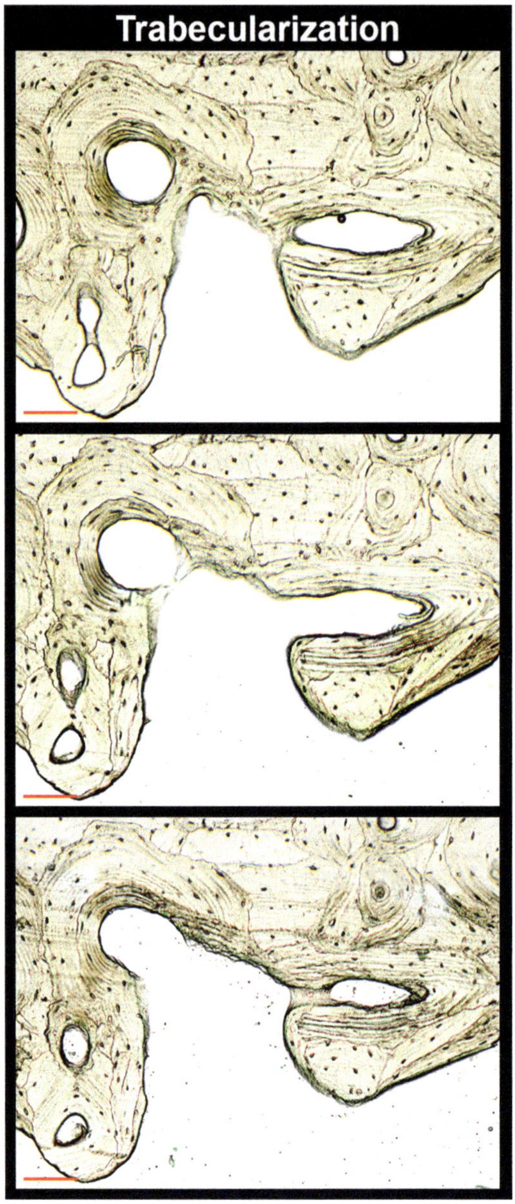

Figure 13.2 Serial histological cryosections of a human midshaft fourth rib, each 30 μm thick, show trabecularization of cortical pores adjacent to the endosteum. Scale bars are 100 μm.

synchrotron radiation micro-CT, the high resolution limits the field of view to a small bone core (Andronowski & Cole, 2021). Collagen fiber orientation, an indicator of local strain history, is also visible histologically under linear or circular polarization (Bromage et al., 2003; Skedros et al., 2009; Skedros, Kiser, Keenan, et al., 2011). The cement lines of secondary osteons are particularly important for contextualizing cortical pores as either the result of new remodeling events, or the remodeling of existing canals (Andreasen et al., 2020). In combination, these tissue microstructures interpret remodeling activity, demonstrate the

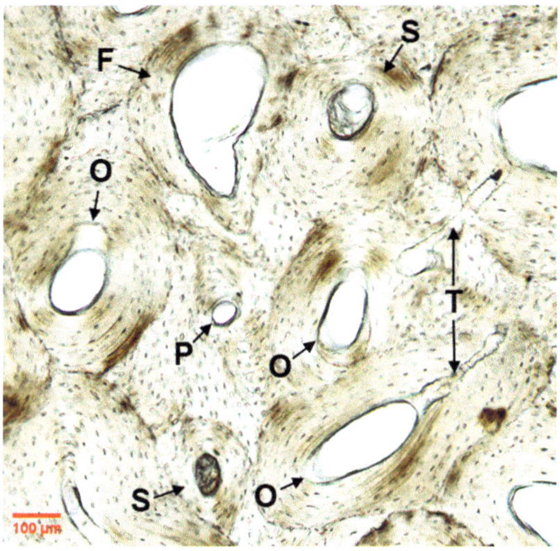

Figure 13.3 A histological image from a human midshaft femur illustrates some peculiarities of cortical pores in cross-section. Primary osteons (P) contain a cortical pore, but have no cement line and form fewer concentric lamellae, which are more integrated into the flow of the surrounding interstitial lamellae. Secondary osteons (S) are larger, have more numerous concentric lamellae, and are bounded by a distinct cement line representing the extent of resorption before reversal. Cortical pores of secondary osteons may contain clear space or be filled with the debris of soft tissue or embedding medium. While a secondary osteon is still forming (F), the incomplete infilling of concentric lamellae creates a transiently larger cortical pore. If the Haversian canal's longitudinal orientation is slightly oblique (O) to the transverse plane of section, it appears elongated in cross-section. It may also display a faintly shadowed lip, representing a section into the oblique tilt whose thickness is partially bone and partially cortical pore space. Sectioning through a transverse interconnection (T) creates a long, thin streak. This may appear as a discrete canal, or, if only the very edge is sectioned, a trailing overlay from a connected cortical pore. Scale bar is 100 µm.

interaction of localized tissue damage and repair, and provide a broader picture of bone quality and fragility.

In modern anthropological research, histological analyses of porosity remain limited by the time-consuming task of manually identifying, tracing, and measuring cortical pores on digital microscopic images. Within a single cross-section, cortical pores number in the hundreds for human ribs and in the thousands for human femora and tibiae. We have recently distributed an open-source software toolkit to expedite cortical pore analysis on histological images. Pore Extractor 2D, developed for the free image processing software ImageJ (NIH), provides computer-assisted, semiautomated cortical pore identification and fully automated morphometric analysis. Divisions of cortical pore morphometry by anatomical region and cortical or trabecularized pore "type" are also fully automated (Cole et al., 2022). There has also been recent interest in computer-assisted identification of cortical pores using AI-driven deep learning (Hoshi-Numahata et al., 2023).

Three-dimensional visualization into cut cross-sectional surfaces is possible with confocal laser scanning microscopy (CLSM). This is a form of fluorescence microscopy that uses a scanning laser to precisely excite the fluorophore stain, blocking out-of-focus light with a pinhole or spinning disc. Fluorescently stained bone spaces, including cortical pores, microcracks, and osteocyte lacunar-canalicular networks, can be precisely resolved in X, Y,

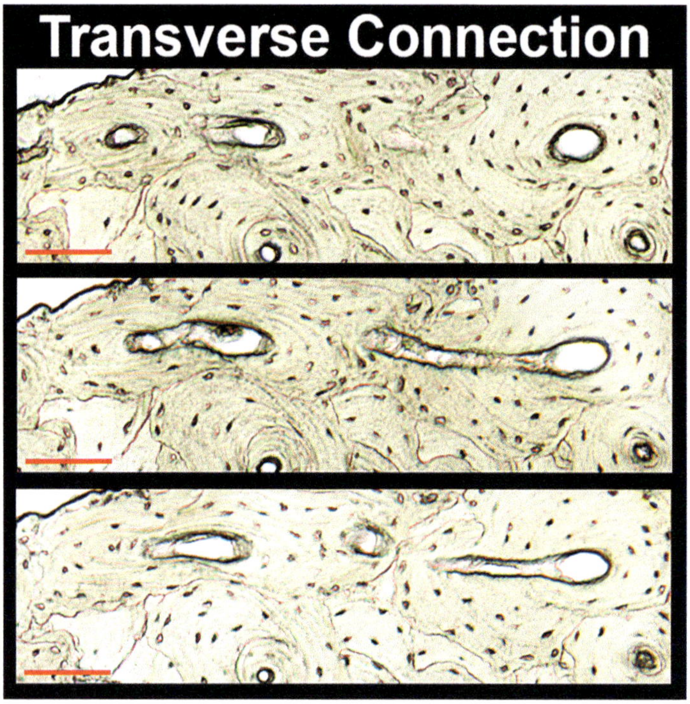

Figure 13.4 Serial histological cryosections of a human midshaft fourth rib, each 30 µm thick, show a transverse connection between adjacent pores, represented by thin streak. Scale bars are 100 µm.

and Z dimensions, reaching a potential spatial resolution of 200 nanometers (Goggin et al., 2016). Three-dimensional visualization of these structures is possible to a depth of approximately 100–150 µm, limited by the high mineralization of bone tissue (Dallas & Moore, 2020). Penetration may be improved with multi-photon confocal microscopy through its use of near-infrared photons (Cole & Stout, 2017). Like two-dimensional histology, CLSM can be applied across large tissue regions through merging many overlapping images (Ciani et al., 2009). CLSM has been used to characterize the interaction of cortical pores with osteocyte lacunar-canalicular systems (Andronowski et al., 2023; Ashique et al., 2017) and microcracks (Ebacher et al., 2007, 2012) in human bone.

13.4.2 Ex Vivo X-Ray-Based Imaging

Three-dimensional visualization of cortical pores is possible with X-ray-based imaging. Micro-computed tomography (micro-CT) has become the gold standard for three-dimensional visualization of bone tissue morphology since its first application to bone in the 1980s. Multiple X-ray projections are taken at numerous angles, accomplished by rotation of the sample between the X-ray source and detector (ex vivo micro-CT), or through rotation of the X-ray source and detector around a live small animal (in vivo micro-CT). A computer algorithm reconstructs these X-ray projections into two-dimensional trans-axial slices, which are stacked into a three-dimensional volume. X-rays are variably attenuated by thickness and density of the sample material as they pass through to the detector. Bone mineral, being relatively denser, attenuates X-ray energy more than the soft tissue, marrow,

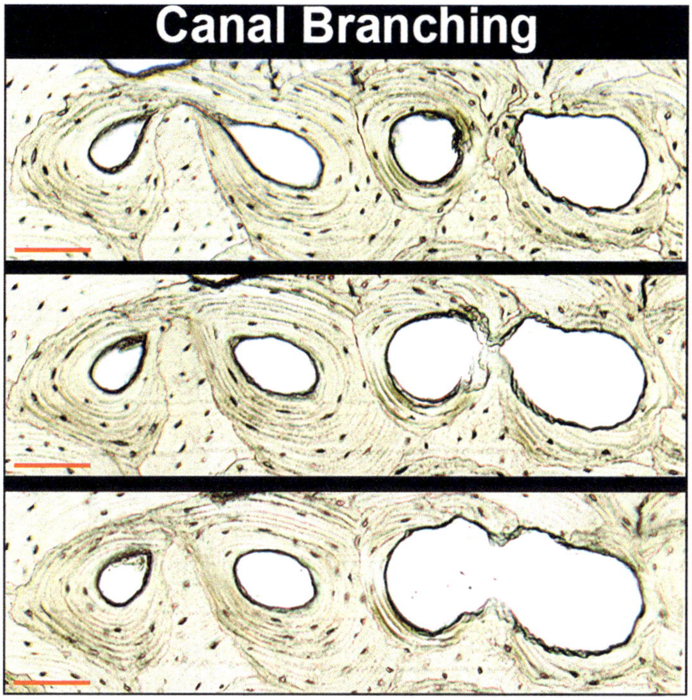

Figure 13.5 Serial histological cryosections of a human midshaft fourth rib, each 30 μm thick, show cortical pore branching, appearing as pinched "dumbbell"-shaped osteons in cross-section. Scale bars are 100 μm.

or air that fills spaces within and around bone tissue. This variable attenuation is represented by the grayscale brightness value of the pixels on the image slice, with bone pixels appearing brighter than non-bone pixels. In three-dimensional space, the pixels become volume elements called voxels, with the X-Y plane representing the image slice, and the Z-dimension representing the longitudinal slice thickness. Scanning the sample with a micro-CT hydroxyapatite phantom, containing rods of known mineral density, calibrates the sample volume for calculation of volumetric Bone Mineral Density (vBMD). Setting a threshold for the desired pixel brightness range, either globally or in the local neighborhood of each pixel, produces a binary volume representing either bone or space. This facilitates three-dimensional morphometric analysis of bone tissue structures, such as cortical geometry or trabecular architecture, or spaces within bone tissue, such as cortical pore systems (Boerckel et al., 2014; Kim et al., 2021) (Figure 13.6). While segmentation of bone and space can be accomplished with pixel brightness thresholding, segmentation of cortical bone and trabecular bone is more challenging. The cortical-trabecular interface is not a distinct sharp edge, but a gradual transition from enlarged "trabecularized" cortical pores to trabecular struts (Zebaze & Seeman, 2015). This ambiguity is exacerbated at the higher imaging resolutions required for accurate analysis of pore morphometry, where cortical pores appear as discrete structures. Deep learning approaches have successfully trained fully convolutional neural networks for accurate segmentation of this cortical-trabecular boundary (Jepsen et al., 2023).

A key parameter of micro-CT image quality is voxel size. This corresponds to the nominal spatial resolution, or the minimum size of objects that can be separated from

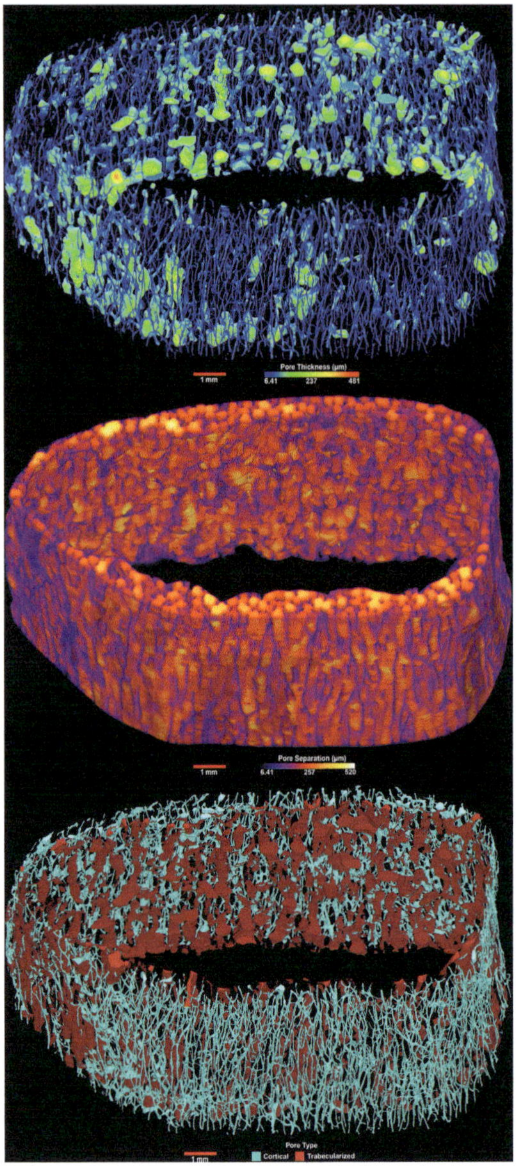

Figure 13.6 Micro-computed tomography of a human midshaft fourth rib at 6.41 μm voxel size facilitates visualization of cortical pore morphometry, including local pore thickness (top), local pore separation (middle), and differentiation between cortical and trabecularized pore types (bottom). Scale bars are 1 mm.

neighboring objects. A smaller voxel size is "higher resolution" and can distinguish smaller microstructures in bone tissue (Kim et al., 2021). At lower resolutions, with larger voxel sizes, visible structures will appear artificially thick, increasing cortical thickness, trabecular thickness, and cortical pore canal diameter. Smaller cortical pore systems and thinner trabecular elements will also be lost, decreasing canal number and trabecular number. This loss occurs through the partial volume effect, where structures smaller than the voxel size are averaged with the attenuation of the surrounding material, blurring cortical pores with cortical tissue, and trabecular struts with the marrow space (Engelke et al., 2013).

In human bone, the recommendation for accurate cortical pore morphometry is a voxel size of at least 5–10 μm, although < 5 μm is likely required to resolve all of the smallest cortical pores (Cooper et al., 2007). Resolving cortical porosity to the desired resolution limits the width of the cross-section that can be visualized. Voxel size is calculated as the field of view divided by the matrix size of the detector array. Field of view, in turn, depends on the distance between the X-ray source and the sample, and between the sample and the detector (Kim et al., 2021). Field of view is therefore limited by how close the sample can be physically positioned to the X-ray source. At the desired voxel size, larger samples such as human femora may be wider than the associated field of view. These out-of-field regions are not imaged and are also not accounted for by the computer reconstruction. This can cause brightness streaks or shading on the imaged regions within the field of view, an artifact known as truncated data or incomplete projection. Some micro-CT scanners have the capability for "region of interest" scans, where the whole sample is scanned at lower resolution to correct for this out-of-field attenuation, and a sub-region is scanned at the desired higher resolution (du Plessis et al., 2017). Adjacent regions-of-interest may be captured to cover the entire cross-section at this higher resolution. We have applied this technique successfully to reach < 10 μm voxel size for cortical pore morphometry in plastic-embedded cross-sections, and in wider skeletal regions such as the femoral neck.

Larger samples and smaller microstructures may be visualized through recent advances in computed tomography imaging. Nano-computed tomography (nano-CT) uses a nano-focused X-ray beam to improve image resolution within the 5–15 μm voxel size range, sharpening edges and decreasing sample blurring. A mobile detector also increases the field of view with multiple regions-of-interest, capable of scanning human radius cross-sections at 5 μm and human proximal femora at 27 μm (Jepsen et al., 2023; Khoury et al., 2015). Smaller microstructures can be visualized with synchrotron radiation micro-CT (SR-μCT). In synchrotron facilities, acceleration of electrons close to the speed of light releases highly intense and brilliant synchrotron light, capable of sub-micron resolution. The roughly threefold higher flux of X-ray photons over time also facilitates rapid scanning (Andronowski & Cole, 2021; Obata et al., 2020). The primary limitation of SR-μCT in bone tissue imaging is the small field of view required for this high resolution. At a 1.49 μm voxel size, appropriate for visualizing osteocyte lacunae, the field of view is approximately 2 mm wide, requiring extraction of small bone core for imaging (Andronowski et al., 2020). SR-μCT has been applied to quantify osteocyte lacunar morphometry in association with cortical pore morphometry in human bone (Andronowski et al., 2023; Carter, Thomas, Clement, & Cooper, 2013; Carter, Thomas, Clement, Peele, et al., 2013). SR-μCT is also used to characterize microcrack initiation and progression in interaction with cortical pore systems (Koester et al., 2008, 2011; Peña Fernández et al., 2020, 2021; Zimmermann et al., 2011). Phase contrast enhancement of SR-μCT can further visualize osteon cement lines, interpreting the remodeling activity that creates cortical pore systems (Cooper et al., 2011; Maggiano et al., 2017, 2016).

13.4.3 In Vivo X-Ray-Based Imaging

In vivo micro-CT imaging allows longitudinal tracking of bone microstructure through repeated imaging. While the radiation dose of micro-CT is prohibitive in living humans, it has been broadly applied in small animal models, whose smaller skeletal elements allow quicker scanning. This can be considered four-dimensional imaging, with the fourth dimension being time. Mouse and rat models are typically used to observe changes in bone microstructure with respect to aging, pharmacological treatment, and mechanical loading (Oliviero et al., 2019). However, these murine models are not ideal for characterizing cortical porosity, as they retain primary canals throughout the lifespan and rarely remodel (Ross & Sumner, 2017). Additionally, in vivo micro-CT typically focuses on trabecular architecture, which allows a lower resolution with a lower radiation dose. Mice have smaller trabeculae than humans, averaging 40 μm in thickness, so the recommended voxel size is no more than 15–20 μm, and ideally closer to 6–10 μm (Kim et al., 2021). Ex vivo imaging shows that vascular canals in mice are best resolved at 1 μm, substantially reduced at 4 μm, and largely disappear at 8 μm (Palacio-Mancheno et al., 2014). While cortical pore systems are theoretically resolvable in vivo, the exponentially higher radiation dose has been shown to exert adverse effects on bone formation. Even the typical trabecular bone image resolutions of 9 to 18 μm can result in reduced bone volume fraction in mice. An alternative approach is SR-μCT phase contrast imaging, which employs X-ray refraction or scattering, rather than absorption. This improves image contrast at lower radiation doses. SR-μCT phase contrast imaging has recently been applied in rabbits, the smallest laboratory animal model with remodeling similar to humans, without overt effects on cortical bone parameters. Cortical pore detection was comparable to those resolved at 10 μm voxel size with ex vivo micro-CT, appropriate for ~100 μm diameter remodeling spaces in rabbits (Harrison et al., 2022). The rabbit model has been used to visualize three-dimensional BMU morphometry and spatial progression over time (Cooper et al., 2023; Harrison et al., 2022; Loundagin et al., 2023).

The highest in vivo resolution achievable in living humans is through High Resolution Peripheral Quantitative Computed Tomography (HR-pQCT), applied at distal peripheral skeletal sites. The newest generation of scanners can reach a voxel size of 61 μm (Soltan et al., 2019), which far exceeds the recommended 5–10 μm for accurately resolving cortical pore morphometry. A comparison of HR-pQCT scanning at 82 μm and ex vivo SR-μCT scanning at 9 μm found that HR-pQCT consistently underestimated cortical porosity by an average of 6.7%. However, HR-pQCT was highly accurate at detecting larger pores greater than 140 μm in diameter (Jorgenson et al., 2015). The fatigue life of bone in respect to porosity is primarily driven by canals exceeding 120 μm in diameter, so HR-pQCT is likely capable of quantifying pores with the greatest mechanical importance (Loundagin et al., 2021). Indeed, HR-pQCT-derived cortical porosity of the distal tibia (Bjørnerem et al., 2013; Sundh et al., 2015) and distal radius (Bala et al., 2014, 2015) is an independent predictor of bone fracture. HR-pQCT also shows potential for tracking remodeling in human bone, where resorption spaces average 250 μm in diameter (Harrison & Cooper, 2015).

13.5 Sources of Morphometric Variation in Cortical Porosity

13.5.1 Pore Morphometry Provides Context for the Mechanical Stimulus of Remodeling

Intraskeletal patterns in remodeling rate are difficult to predict because remodeling can be activated by low strains (disuse-mode remodeling), moderate strains (conservation-mode remodeling), or high strains (targeted remodeling). Patterns in the microstructural products of remodeling, including porosity, present the same challenge. Histological studies have found no significant differences in percent porosity between the human midshaft rib and the more dynamically loaded or weight-bearing distal radius, midshaft femur, or midshaft tibia (Cole et al., 2022; Hunter & Agnew, 2016). Yet percent porosity is an aggregate metric, calculated as summed cortical pore areas or volumes divided by the cortical area. Percent porosity can be increased by a higher density of pore spaces, resulting from a higher remodeling rate, or by a larger pore size, resulting from delays or arrests in the remodeling cycle (Andreasen et al., 2018). The specific type of remodeling, and its originating mechanical stimulus, is more closely reflected in pore morphometry than percent porosity. For example, previous histological study found no significant difference in percent porosity between the human midshaft rib, femur, and tibia. However, the rib showed a significant elevation in total pore density, reflecting its higher remodeling rate. The rib also preferentially formed "trabecularized" porosity and pore density, while cortical porosity and pore density were more prevalent in the femur and tibia. This reflects the greater permissiveness of the relatively unloaded rib to disuse-mode remodeling, which resorbs more bone than is subsequently filled adjacent to the endosteum (Cole et al., 2022).

Cross-sectional patterns in porosity tend to be more consistently aligned with strain magnitude. Early studies recognized that porosity increases toward the endosteum, where strains are lowest due to proximity to the neutral axis (Atkinson, 1965; Bousson et al., 2001; Jowsey, 1960; Martin, 1984; Martin et al., 1980; Thomas et al., 2005; Zebaze et al., 2010). This suggests greater permissiveness to disuse-mode remodeling, with incomplete infilling of resorption spaces at the endosteum (Hughes et al., 2020). Another simplified strain gradient can be observed in the human femoral neck. Due to bipedal locomotion, the superior cortex of the femoral neck experiences minimal compression, which increases toward a maximum in the inferior cortex (Lotz et al., 1995; Lovejoy, 1988; Ohman et al., 1997). In older age, these low-strain superior regions develop "focal defects" of abnormally large pores (Bell, Loveridge, Power, Garrahan, Meggitt, et al., 1999; Bell, Loveridge, Power, Garrahan, Stanton, et al., 1999; Cole et al., 2020; Poole et al., 2012) (Figure 13.7). Micro-CT of the femoral neck in the author's laboratory (Cole, 2019; Cole et al., 2020) further suggests that pore morphometry reflects the type of remodeling activity. The higher porosity of the superior femoral neck is created by the dense interconnection of larger, more obliquely oriented pore systems. This pore morphometry reflects the erosion and convergence typical of disuse-mode remodeling, as would be expected for these lower-strain superior regions. The lower porosity of the inferior femoral neck is associated with smaller, poorly connected, more longitudinally oriented pore systems. These limited and isolated pore systems suggest the precision of targeted remodeling, as would be expected in these higher strain inferior regions (Cole, 2019; Cole et al., 2020).

Strain mode can also shape pore morphometry within a cross-section. Tension exerts lower strains than compression in compact bone (Lanyon & Baggott, 1976). Lower-strain tensed regions could be predicted to favor disuse-mode remodeling, where resorption spaces are left unfilled, resulting in a net increase in porosity. Higher-strain compressed regions could be predicted to favor targeted remodeling, as the linear microcracks that trigger targeted remodeling tend to initiate under compression in bending (Boyce et al., 1998; Diab & Vashishth, 2005). Tension in bending preferentially develops diffuse microdamage, which is repaired not by bone remodeling but by some other mechanism, potentially matrix rebonding (Seref-Ferlengez et al., 2015). The higher porosity expected under lower-strain tension has been observed in the calcanei of Rocky Mountain Mule Deer (Skedros et al., 1994). The human midthoracic rib is another potential example of this pattern. In a simplified model of rib bending, the rib is bent inward, along its curvature, as the thoracic cage expands laterally during inspiration (Moore et al., 2013). At the midshaft, this inward

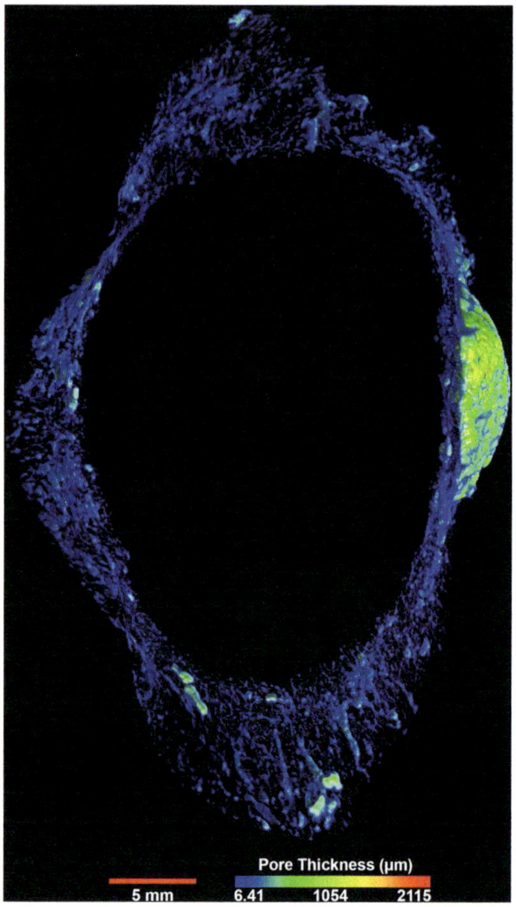

Figure 13.7 Micro-computed tomography of the cortical pore network of a human femoral neck at 6.41 μm voxel size shows the development of a focal defect in the superior-anterior cortex of a 70-year-old female. Scale bars are 1 mm.

bending would place the cutaneous (skin-side) cortex into lower-strain tension, predicting higher porosity, and the pleural (lung-side) cortex into higher-strain compression, predicting lower porosity. Histological studies confirm this significant elevation of porosity in the cutaneous cortex, compared to the pleural cortex (Agnew et al., 2013; Agnew & Stout, 2012; Cole et al., 2022; Dominguez & Agnew, 2016). However, this is not the only type of bending potentially experienced by human midthoracic ribs. The described lateral expansion of the thoracic cage is focused in the lower ribs (primarily 7th–10th). The upper ribs (primarily 2nd–6th) are bent in the opposite manner, against their curvature, to facilitate anterior movement of the sternum and anterior-posterior expansion of the thoracic cage. This outward bending would reverse strain mode at the midshaft, placing the cutaneous cortex into high-strain compression and the pleural cortex into low-strain tension. Moreover, during forced inspiration, the thoracic cage expands in both transverse and anteroposterior dimensions (Moore et al., 2013). Looking only at percent porosity, it is unclear whether the cutaneous cortex is reflecting the disuse-mode remodeling of low-strain tension, the targeted remodeling of high-strain compression, or some combination. Again, pore morphometry provides necessary context for the mechanical stimulus of remodeling and its resulting porosity.

Our micro-CT study of the midshaft fourth rib suggests that the cutaneous cortex derives its high porosity from more convergent and more obliquely oriented pore systems. This resembles the lower-strain regions of the femoral neck, which are permissive to disuse-mode remodeling. However, the cutaneous cortex lacks some aspects of pore morphometry seen in the lower-strain femoral neck, such as larger pore size and increased pore density. This may reflect the more complex and varied bending environment of the midthoracic human rib (Cole, 2019; Cole et al., 2020).

13.6 Conclusion

Cortical porosity is a key driver of age associated bone loss, and an important component of bone quality and bone fragility. Moreover, cortical pore morphometry and distribution reflect remodeling activity and its associated mechanical stimuli. This is relevant for anthropological interpretation of the dynamic loading and functional use of skeletal elements, especially when investigating variation in physical activity and physiological changes associated with age, sex, nutrition, pathology, and frailty. With X-ray-based 3D and 4D imaging of cortical porosity, these questions can also be explored nondestructively and over time. Future methodological research should further explore deep learning approaches for automated and consistent cortical pore segmentation and morphological classification.

References

Abdelgawad, M. E., Delaisse, J.-M., Hinge, M., Jensen, P. R., Alnaimi, R. W., Rolighed, L., Engelholm, L. H., Marcussen, N., & Andersen, T. L. (2016). Early reversal cells in adult human bone remodeling: Osteoblastic nature, catabolic functions and interactions with osteoclasts. *Histochemistry and Cell Biology, 145*(6), 603–615. doi: 10.1007/s00418-016-1414-y

Agerbaek, M. O., Eriksen, E. F., Kragstrup, J., Mosekilde, L., & Melsen, F. (1991). A reconstruction of the remodelling cycle in normal human cortical iliac bone. *Bone and Mineral, 12*(2), 101–112. doi: 10.1016/0169-6009(91)90039-3

Agnew, A. M., & Stout, S. D. (2012). Brief communication: Reevaluating osteoporosis in human ribs: The role of intracortical porosity. *American Journal of Physical Anthropology, 148*(3), 462–466. doi: 10.1002/ajpa.22048

Agnew, A. M., Moorhouse, K., Kang, Y.-S., Donnelly, B. R., Pfefferle, K., Manning, A. X., Litsky, A. S., Herriott, R., Abdel-Rasoul, M., & Bolte, J. H. (2013). The response of pediatric ribs to quasi-static loading: Mechanical properties and microstructure. *Annals of Biomedical Engineering, 41*(12), 2501–2514. doi: 10.1007/s10439-013-0875-6

Akkus, O., Polyakova-Akkus, A., Adar, F., & Schaffler, M. B. (2003). Aging of microstructural compartments in human compact bone. *Journal of Bone and Mineral Research, 18*(6), 1012–1019. doi: 10.1359/jbmr.2003.18.6.1012

Allen, M. R., & Burr, D. B. (2008). Skeletal microdamage: Less about biomechanics and more about remodeling. *Clinical Reviews in Bone and Mineral Metabolism, 6*, 24–30. doi: 10.1007/s12018-008-9015-5

Andersen, T. L., Abdelgawad, M. E., Kristensen, H. B., Hauge, E. M., Rolighed, L., Bollerslev, J., Kjærsgaard-Andersen, P., & Delaisse, J.-M. (2013). Understanding coupling between bone resorption and formation: Are reversal cells the missing link? *The American Journal of Pathology, 183*(1), 235–246. doi: 10.1016/j.ajpath.2013.03.006

Andreasen, C. M., Delaisse, J.-M., van der Eerden, B. C., van Leeuwen, J. P., Ding, M., & Andersen, T. L. (2018). understanding age-induced cortical porosity in women: The accumulation and coalescence of eroded cavities upon existing intracortical canals is the main contributor. *Journal of Bone and Mineral Research, 33*(4), 606–620. doi: 10.1002/jbmr.3354

Andreasen, C. M., Bakalova, L. P., Brüel, A., Hauge, E. M., Kiil, B. J., Delaisse, J.-M., Kersh, M. E., Thomsen, J. S., & Andersen, T. L. (2020). The generation of enlarged eroded pores upon existing intracortical canals is a major contributor to endocortical trabecularization. *Bone, 130*, 115127. doi: 10.1016/j.bone.2019.115127

Andreasen, C. M., El-Masri, B. M., MacDonald, B., Laursen, K. S., Nielsen, M. H., Thomsen, J. S., Delaisse, J.-M., & Andersen, T. L. (2023). Local coordination between intracortical bone remodeling and vascular development in human juvenile bone. *Bone, 173*, 116787. doi: 10.1016/j.bone.2023.116787

Andronowski, J. M., & Cole, M. E. (2021). Current and emerging histomorphometric and imaging techniques for assessing age-at-death and cortical bone quality. *WIREs Forensic Science, 3*(2), e1399. doi: 10.1002/wfs2.1399

Andronowski, J. M., Davis, R. A., & Holyoke, C. W. (2020). A sectioning, coring, and image processing guide for high-throughput cortical bone sample procurement and analysis for synchrotron micro-CT. *Journal of Visualized Experiments: JoVE, 160*. doi: 10.3791/61081

Andronowski, J. M., Cole, M. E., Davis, R. A., Tubo, G. R., Taylor, J. T., & Cooper, D. M. L. (2023). A multimodal 3D imaging approach of pore networks in the human femur to assess age-associated vascular expansion and Lacuno-Canalicular reduction. *Anatomical Record (Hoboken, N.J.: 2007), 306*(3), 475–493. doi: 10.1002/ar.25089

Ashique, A. M., Hart, L. S., Thomas, C. D. L., Clement, J. G., Pivonka, P., Carter, Y., Mousseau, D. D., & Cooper, D. M. L. (2017). Lacunar-canalicular network in femoral cortical bone is reduced in aged women and is predominantly due to a loss of canalicular porosity. *Bone Reports, 7*, 9–16. doi: 10.1016/j.bonr.2017.06.002

Atkinson, P. J. (1965). Changes in resorption spaces in femoral cortical bone with age. *The Journal of Pathology and Bacteriology, 89*(1), 173–178. doi: 10.1002/path.1700890118

Bala, Y., Zebaze, R., Ghasem-Zadeh, A., Atkinson, E. J., Iuliano, S., Peterson, J. M., Amin, S., Bjørnerem, Å., Melton, L. J., Johansson, H., Kanis, J. A., Khosla, S., & Seeman, E. (2014). Cortical porosity identifies women with osteopenia at increased risk for forearm fractures. *Journal of Bone and Mineral Research, 29*(6), 1356–1362. doi: 10.1002/jbmr.2167

Bala, Y., Bui, Q. M., Wang, X.-F., Iuliano, S., Wang, Q., Ghasem-Zadeh, A., Rozental, T. D., Bouxsein, M. L., Zebaze, R. M. D., & Seeman, E. (2015). Trabecular and cortical microstructure and fragility of the distal radius in women. *Journal of Bone and Mineral Research, 30*(4), 621–629. doi: 10.1002/jbmr.2388

Bala, Y., Lefèvre, E., Roux, J.-P., Baron, C., Lasaygues, P., Pithioux, M., Kaftandjian, V., & Follet, H. (2016). Pore network microarchitecture influences human cortical bone elasticity during growth and aging. *Journal of the Mechanical Behavior of Biomedical Materials, 63*, 164–173. doi: 10.1016/j.jmbbm.2016.05.018

Bell, K. L., Loveridge, N., Power, J., Garrahan, N., Meggitt, B. F., & Reeve, J. (1999). Regional differences in cortical porosity in the fractured femoral neck. *Bone, 24*(1), 57–64. doi: 10.1016/s8756-3282(98)00143-4

Bell, K. L., Loveridge, N., Power, J., Garrahan, N., Stanton, M., Lunt, M., Meggitt, B. F., & Reeve, J. (1999). Structure of the femoral neck in hip fracture: Cortical bone loss in the inferoanterior to superoposterior axis. *Journal of Bone and Mineral Research: The Official Journal of the American Society for Bone and Mineral Research, 14*(1), 111–119. doi: 10.1359/jbmr.1999.14.1.111

Bell, K. L., Loveridge, N., Jordan, G. R., Power, J., Constant, C. R., & Reeve, J. (2000). A novel mechanism for induction of increased cortical porosity in cases of intracapsular hip fracture. *Bone, 27*(2), 297–304. doi: 10.1016/S8756-3282(00)00318-5

Biewener, A. A., & Bertram, J. E. (1993). Mechanical loading and bone growth in vivo. In B. K. Hall (Ed.), *Bone, Volume VII*. CRC Press, LLC (Boca Raton, FL) (pp. 1–36).

Bjørnerem, Å., Bui, Q. M., Ghasem-Zadeh, A., Hopper, J. L., Zebaze, R., & Seeman, E. (2013). Fracture risk and height: An association partly accounted for by cortical porosity of relatively thinner cortices. *Journal of Bone and Mineral Research, 28*(9), 2017–2026. doi: 10.1002/jbmr.1934

Boerckel, J. D., Mason, D. E., McDermott, A. M., & Alsberg, E. (2014). Microcomputed tomography: Approaches and applications in bioengineering. *Stem Cell Research & Therapy*, *5*(6), 144. doi: 10.1186/scrt534

Bousson, V., Meunier, A., Bergot, C., Vicaut, E., Rocha, M. A., Morais, M. H., Laval-Jeantet, A. M., & Laredo, J. D. (2001). Distribution of intracortical porosity in human midfemoral cortex by age and gender. *Journal of Bone and Mineral Research*, *16*(7), 1308–1317. doi: 10.1359/jbmr.2001.16.7.1308

Boyce, T. M., Fyhrie, D. P., Glotkowski, M. C., Radin, E. L., & Schaffler, M. B. (1998). Damage type and strain mode associations in human compact bone bending fatigue. *Journal of Orthopaedic Research: Official Publication of the Orthopaedic Research Society*, *16*(3), 322–329. doi: 10.1002/jor.1100160308

Britz, H. M., Thomas, C. D., Clement, J. G., & Cooper, D. M. (2009). The relation of femoral osteon geometry to age, sex, height and weight. *Bone*, *45*(1), 77–83. doi: 10.1016/j.bone.2009.03.654

Brockstedt, H., Bollerslev, J., Melsen, F., & Mosekilde, L. (1996). Cortical bone remodeling in autosomal dominant osteopetrosis: A study of two different phenotypes. *Bone*, *18*(1), 67–72. doi: 10.1016/8756-3282(95)00424-6

Bromage, T. G., Goldman, H. M., McFarlin, S. C., Warshaw, J., Boyde, A., & Riggs, C. M. (2003). Circularly polarized light standards for investigations of collagen fiber orientation in bone. *The Anatomical Record Part B: The New Anatomist*, *274B*(1), 157–168. doi: 10.1002/ar.b.10031

Buccino, F., Zagra, L., Longo, E., D'Amico, L., Banfi, G., Berto, F., Tromba, G., & Vergani, L. M. (2023). Osteoporosis and Covid-19: Detected similarities in bone lacunar-level alterations via combined AI and advanced synchrotron testing. *Materials & Design*, *231*, 112087. doi: 10.1016/j.matdes.2023.112087

Buenzli, P. R., & Sims, N. A. (2015). Quantifying the osteocyte network in the human skeleton. *Bone*, *75*, 144–150. doi: 10.1016/j.bone.2015.02.016

Burr, D. B. (2011). Why bones bend but don't break. *Journal of Musculoskeletal & Neuronal Interactions*, *11*(4), 270–285.

Burr, D. B., & Martin, R. B. (1983). The effects of composition, structure and age on the torsional properties of the human radius. *Journal of Biomechanics*, *16*(8), 603–608. doi: 10.1016/0021-9290(83)90110-0

Carter, Y., Thomas, C. D., Clement, J. G., Peele, A. G., Hannah, K., & Cooper, D. M. (2013). Variation in osteocyte lacunar morphology and density in the human femur—A synchrotron radiation micro-CT study. *Bone*, *52*(1), 126–132. doi: 10.1016/j.bone.2012.09.010

Carter, Y., Thomas, C. D. L., Clement, J. G., & Cooper, D. M. L. (2013). Femoral osteocyte lacunar density, volume and morphology in women across the lifespan. *Journal of Structural Biology*, *183*(3), 519–526. doi: 10.1016/j.jsb.2013.07.004

Chen, M., Li, Y., Huang, X., Gu, Y., Li, S., Yin, P., Zhang, L., & Tang, P. (2021). Skeleton-vasculature chain reaction: A novel insight into the mystery of homeostasis. *Bone Research*, *9*(1), 1–20. doi: 10.1038/s41413-021-00138-0

Cho, H., & Stout, S. D. (2011). Age-associated bone loss and intraskeletal variability in the Imperial Romans. *Journal of Anthropological Sciences*, *89*, 109–125. doi: 10.4436/jass.89007

Ciani, C., Doty, S. B., & Fritton, S. P. (2009). An effective histological staining process to visualize bone interstitial fluid space using confocal microscopy. *Bone*, *44*(5), 1015–1017. doi: 10.1016/j.bone.2009.01.376

Cohen, J., & Harris, W. H. (1958). The three-dimensional anatomy of haversian systems. *The Journal of Bone and Joint Surgery. American Volume*, *40-A*(2), 419–434.

Cole, M., & Stout, S. D. (2017). *Optimization of a Protocol for Visualizing Vascular and Cellular Pore Networks in Human Bone Using Multiphoton Confocal Microscopy.*

Cole, M. E. (2019). *Optimizing Bone Loss Across the Lifespan: The Three-Dimensional Structure of Porosity in the Human Femoral Neck and Rib As a Metric of Bone Fragility* [The Ohio State University]. http://rave.ohiolink.edu/etdc/view?acc_num=osu1559033566505566

Cole, M. E., Stout, S. D., & Agnew, A. M. (2020). Three- dimensional cortical pore networks are morphologically optimized for localized mechanical strain in the human femoral neck and rib. *American Journal of Physical Anthropology, 171*, 55.

Cole, M. E., Stout, S. D., Dominguez, V. M., & Agnew, A. M. (2022). Pore Extractor 2D: An ImageJ toolkit for quantifying cortical pore morphometry on histological bone images, with application to intraskeletal and regional patterning. *American Journal of Biological Anthropology, 179*(3), 365–385. doi: 10.1002/ajpa.24618

Cole, M. E., Gosman, J. H., and, & Stout, S. D. (2023). Current concepts in bone biology. In *A Companion to Biological Anthropology* (pp. 527–543). John Wiley & Sons, Ltd. doi: 10.1002/9781119828075.ch31

Cooper, D. M. L., Turinsky, A., Sensen, C., & Hallgrimsson, B. (2007). Effect of Voxel Size on 3D micro-CT Analysis of Cortical Bone Porosity. *Calcified Tissue International, 80*(3), 211–219. doi: 10.1007/s00223-005-0274-6

Cooper, D. M. L., Erickson, B., Peele, A. G., Hannah, K., Thomas, C. D. L., & Clement, J. G. (2011). Visualization of 3D osteon morphology by synchrotron radiation micro-CT. *Journal of Anatomy, 219*(4), 481–489. doi: 10.1111/j.1469-7580.2011.01398.x

Cooper, D. M. L., Harrison, K. D., Hiebert, B. D., King, G. A., Panahifar, A., Zhu, N., Swekla, K. J., Pivonka, P., Chapman, L. D., & Arnason, T. (2023). Daily administration of parathyroid hormone slows the progression of basic multicellular units in the cortical bone of the rabbit distal tibia. *Bone, 176*, 116864. doi: 10.1016/j.bone.2023.116864

Crowder, C., & Rosella, L. (2007). Assessment of Intra- and Intercostal Variation in Rib Histomorphometry: Its Impact on Evidentiary Examination. *Journal of Forensic Sciences, 52*(2), 271–276. doi: 10.1111/j.1556-4029.2007.00388.x

Dallas, S. L., & Moore, D. S. (2020). Using confocal imaging approaches to understand the structure and function of osteocytes and the lacunocanalicular network. *Bone, 138*, 115463. doi: 10.1016/j.bone.2020.115463

Demontiero, O., Vidal, C., & Duque, G. (2012). Aging and bone loss: New insights for the clinician. *Therapeutic Advances in Musculoskeletal Disease, 4*(2), 61–76. doi: 10.1177/1759720X11430858

Diab, T., & Vashishth, D. (2005). Effects of damage morphology on cortical bone fragility. *Bone, 37*(1), 96–102. doi: 10.1016/j.bone.2005.03.014

Dominguez, V. M., & Agnew, A. M. (2016). Examination of factors potentially influencing osteon size in the human Rib. *Anatomical Record, 299*(3), 313–324. doi: 10.1002/ar.23305

Dominguez, V. M., & Agnew, A. M. (2019). Microdamage as a bone quality component: practical guidelines for the two-dimensional analysis of linear microcracks in human cortical bone. *Journal of Bone and Mineral Research Plus, 3*(6), e10203. doi: 10.1002/jbm4.10203

Drapeau, M. S. M., & Streeter, M. A. (2006). Modeling and remodeling responses to normal loading in the human lower limb. *American Journal of Physical Anthropology, 129*(3), 403–409. doi: 10.1002/ajpa.20336

du Plessis, A., Broeckhoven, C., Guelpa, A., & le Roux, S. G. (2017). Laboratory x-ray micro-computed tomography: A user guideline for biological samples. *GigaScience, 6*(6), 1–11. doi: 10.1093/gigascience/gix027

Ebacher, V., Tang, C., McKay, H., Oxland, T. R., Guy, P., & Wang, R. (2007). Strain redistribution and cracking behavior of human bone during bending. *Bone, 40*(5), 1265–1275. doi: 10.1016/j.bone.2006.12.065

Ebacher, V., Guy, P., Oxland, T. R., & Wang, R. (2012). Sub-lamellar microcracking and roles of canaliculi in human cortical bone. *Acta Biomaterialia, 8*(3), 1093–1100. doi: 10.1016/j.actbio.2011.11.013

Engelke, K., Libanati, C., Fuerst, T., Zysset, P., & Genant, H. K. (2013). Advanced CT based In Vivo methods for the assessment of bone density, structure, and strength. *Current Osteoporosis Reports, 11*(3), 246–255. doi: 10.1007/s11914-013-0147-2

Eriksen, E. F. (2010). Cellular mechanisms of bone remodeling. *Reviews in Endocrine & Metabolic Disorders, 11*(4), 219–227. doi: 10.1007/s11154-010-9153-1

Eriksen, E. F., Melsen, F., & Mosekilde, L. (1984). Reconstruction of the resorptive site in iliac trabecular bone: A kinetic model for bone resorption in 20 normal individuals. *Metabolic Bone Disease and Related Research*, *5*(5), 235–242. doi: 10.1016/0221-8747(84)90065-1

Fahy, G. E., Deter, C., Pitfield, R., Miszkiewicz, J. J., & Mahoney, P. (2017). Bone deep: Variation in stable isotope ratios and histomorphometric measurements of bone remodelling within adult humans. *Journal of Archaeological Science*, *87*, 10–16. doi: 10.1016/j.jas.2017.09.009

Feik, S. A., Thomas, C. D., Bruns, R., & Clement, J. G. (2000). Regional variations in cortical modeling in the femoral mid-shaft: Sex and age differences. *American Journal of Physical Anthropology*, *112*(2), 191–205. doi: 10.1002/(SICI)1096-8644(2000)112:2<191::AID-AJPA6>3.0.CO;2-3

Ferretti, J. L., Capozza, R. F., Cointry, G. R., García, S. L., Plotkin, H., Alvarez Filgueira, M. L., & Zanchetta, J. R. (1998). Gender-related differences in the relationship between densitometric values of whole-body bone mineral content and lean body mass in humans between 2 and 87 years of Age. *Bone*, *22*(6), 683–690. doi: 10.1016/S8756-3282(98)00046-5

Frost, H. M. (1983). A determinant of bone architecture—the minimum effective strain. *Clinical Orthopaedics and Related Research*, *175*, 286–292.

Frost, H. M. (1987a). The mechanostat—a proposed pathogenic mechanism of osteoporoses and the bone mass effects of mechanical and nonmechanical agents. *Bone and Mineral*, *2*(2), 73–85.

Frost, H. M. (1987b). Bone mass and the mechanostat—a proposal. *Anatomical Record*, *219*(1), 1–9. doi: 10.1002/ar.1092190104

Frost, H. M. (2003a). Bone's mechanostat: A 2003 update. *Anatomical Record Part A-Discoveries in Molecular Cellular and Evolutionary Biology*, *275a*(2), 1081–1101. doi: 10.1002/ar.a.10119

Frost, H. M. (2003b). Bone's mechanostat: A 2003 update. *Anatomical Record Part A-Discoveries in Molecular Cellular and Evolutionary Biology*, *275a*(2), 1081–1101. doi: 10.1002/ar.a.10119

Gauthier, R., Langer, M., Follet, H., Olivier, C., Gouttenoire, P.-J., Helfen, L., Rongiéras, F., Mitton, D., & Peyrin, F. (2018). 3D micro structural analysis of human cortical bone in paired femoral diaphysis, femoral neck and radial diaphysis. *Journal of Structural Biology*, *204*(2), 182–190. doi: 10.1016/j.jsb.2018.08.006

Gocha, T. P., & Agnew, A. M. (2016). Spatial variation in osteon population density at the human femoral midshaft: Histomorphometric adaptations to habitual load environment. *Journal of Anatomy*, *228*(5), 733–745. doi: 10.1111/joa.12433

Goggin, P. M., Zygalakis, K. C., Oreffo, R. O., & Schneider, P. (2016). High-resolution 3D imaging of osteocytes and computational modelling in mechanobiology: Insights on bone development, ageing, health and disease. *European Cells & Materials*, *31*, 264–295.

Han, Z.-H., Palnitkar, S., Rao, D. S., Nelson, D., & Parfitt, A. M. (1997). Effects of Ethnicity and Age or Menopause on the Remodeling and Turnover of Iliac Bone: Implications for Mechanisms of Bone Loss. *Journal of Bone and Mineral Research*, *12*(4), 498–508. doi: 10.1359/jbmr.1997.12.4.498

Harrison, K. D., & Cooper, D. M. (2015). modalities for visualization of cortical bone remodeling: The past, present, and future. *Front Endocrinol (Lausanne)*, *6*, 122. doi: 10.3389/fendo.2015.00122

Harrison, K. D., Sales, E., Hiebert, B. D., Panahifar, A., Zhu, N., Arnason, T., Swekla, K. J., Pivonka, P., Chapman, L. D., & Cooper, D. M. (2022). Direct assessment of rabbit cortical bone basic multicellular unit longitudinal erosion rate: A 4D synchrotron-based approach. *Journal of Bone and Mineral Research*, *37*(11), 2244–2258. doi: 10.1002/jbmr.4700

Harrison et al. (2020). Cortical bone porosity in rabbit models of osteoporosis. *Journal of Bone and Mineral Research*, *35* (11), 2211-2228. doi: 10.1002/jbmr.4124

Havers, C. (1691). An account of a book. Osteologia Nova, or some observations of the bones, &c. Communicated to the Royal Society, in several discourses, read at their meetings, by the learned and accurate Author, Clopton Havers, M. D. and R. S. Soc. Lond. 1691, Octavo, for Sam. Smith. *Philosophical Transactions of the Royal Society of London*, *17*(194), 544–554. doi: 10.1098/rstl.1686.0104

Hert, J., Fiala, P., & Petrtýl, M. (1994). Osteon orientation of the diaphysis of the long bones in man. *Bone*, *15*(3), 269–277. doi: 10.1016/8756-3282(94)90288-7

Hiller, L. P., Stover, S. M., Gibson, V. A., Gibeling, J. C., Prater, C. S., Hazelwood, S. J., Yeh, O. C., & Martin, R. B. (2003). Osteon pullout in the equine third metacarpal bone: Effects of ex vivo fatigue. *Journal of Orthopaedic Research: Official Publication of the Orthopaedic Research Society, 21*(3), 481–488. doi: 10.1016/S0736-0266(02)00232-2

Himeno-Ando, A., Izumi, Y., Yamaguchi, A., & Iimura, T. (2012). Structural Differences in the Osteocyte Network Between the Calvaria and Long Bone Revealed by Three-Dimensional Fluorescence Morphometry, Possibly Reflecting Distinct Mechano-Adaptations and Sensitivities. *Biochemical and Biophysical Research Communications, 417*(2), 765–770. doi: 10.1016/j.bbrc.2011.12.031

Hoshi-Numahata, M., Takakura, A., Nakanishi-Kimura, A., Watanabe, H., Takada, K., Nishiura, M., Sato, Y., Takao-Kawabata, R., & Iimura, T. (2023). Evaluation of cortical bone remodeling in canines treated with daily and weekly administrations of teriparatide by establishing AI-driven morphometric analyses and GIS-based spatial mapping. *Bone Reports, 19*, 101720. doi: 10.1016/j.bonr.2023.101720

Hughes, J. M., & Petit, M. A. (2010). Biological underpinnings of Frost's mechanostat thresholds: The important role of osteocytes. *Journal of Musculoskeletal & Neuronal Interactions, 10*(2), 128–135.

Hughes, J. M., Castellani, C. M., Popp, K. L., Guerriere, K. I., Matheny, R. W., Nindl, B. C., & Bouxsein, M. L. (2020). The central role of osteocytes in the four adaptive pathways of Bone's Mechanostat. *Exercise and Sport Sciences Reviews, 48*(3), 140–148. doi: 10.1249/JES.0000000000000225

Hunter, R. L., & Agnew, A. M. (2016). Intraskeletal variation in human cortical osteocyte lacunar density: Implications for bone quality assessment. *Bone Reports, 5*, 252–261. doi: 10.1016/j.bonr.2016.09.002

Infante, A., & Rodríguez, C. I. (2018). Osteogenesis and aging: Lessons from mesenchymal stem cells. *Stem Cell Research & Therapy, 9*(1), 244. doi: 10.1186/s13287-018-0995-x

Jepsen, K. J., Bigelow, E. M., Casden, M. A., Goulet, R. W., Kennedy, K., Hertz, S., Kadur, C., Nolan, B. T., Richards-McCullough, K., Merillat, S., Karvonen-Gutierrez, C. A., Clines, G., & Bredbenner, T. L. (2023). Associations among hip structure, bone mineral density, and strength vary with external bone size in white women. *JBMR Plus, 7*(3), e10715. doi: 10.1002/jbm4.10715

Johnell, O., & Kanis, J. A. (2006). An estimate of the worldwide prevalence and disability associated with osteoporotic fractures. *Osteoporosis International, 17*(12), 1726–1733. doi: 10.1007/s00198-006-0172-4

Jorgenson, B. L., Buie, H. R., McErlain, D. D., Sandino, C., & Boyd, S. K. (2015). A comparison of methods for in vivo assessment of cortical porosity in the human appendicular skeleton. *Bone, 73*, 167–175. doi: 10.1016/j.bone.2014.11.023

Jowsey, J. (1960). Age changes in human bone. *Clinical Orthopaedics and Related Research, 17*, 210.

Jowsey, J. (1966). Studies of Haversian systems in man and some animals. *Journal of Anatomy, 100*(Pt 4), 857–864.

Kenkre, J. S., & Bassett, J. (2018). The bone remodelling cycle. *Annals of Clinical Biochemistry, 55*(3), 308–327. doi: 10.1177/0004563218759371

Kennedy, O. D., Herman, B. C., Laudier, D. M., Majeska, R. J., Sun, H. B., & Schaffler, M. B. (2012). Activation of resorption in fatigue-loaded bone involves both apoptosis and active pro-osteoclastogenic signaling by distinct osteocyte populations. *Bone, 50*(5), 1115–1122. doi: 10.1016/j.bone.2012.01.025

Khoury, B. M., Bigelow, E. M. R., Smith, L. M., Schlecht, S. H., Scheller, E. L., Andarawis-Puri, N., & Jepsen, K. J. (2015). The use of nano-computed tomography to enhance musculoskeletal research. *Connective Tissue Research, 56*(2), 106–119. doi: 10.3109/03008207.2015.1005211

Kim, Y., Brodt, M. D., Tang, S. Y., & Silva, M. J. (2021). MicroCT for scanning and analysis of mouse bones. *Methods in Molecular Biology (Clifton, N.J.), 2230*, 169–198. doi: 10.1007/978-1-0716-1028-2_11

Koester, K. J., Ager, J. W., & Ritchie, R. O. (2008). The true toughness of human cortical bone measured with realistically short cracks. *Nature Materials, 7*(8), 672–677. doi: 10.1038/nmat2221

Koester, K. J., Barth, H. D., & Ritchie, R. O. (2011). Effect of aging on the transverse toughness of human cortical bone: Evaluation by R-curves. *Journal of the Mechanical Behavior of Biomedical Materials, 4*(7), 1504–1513. doi: 10.1016/j.jmbbm.2011.05.020

Kola, S. K., Begonia, M. T., Tiede-Lewis, L. M., Laughrey, L. E., Dallas, S. L., Johnson, M. L., & Ganesh, T. (2020). Osteocyte lacunar strain determination using multiscale finite element analysis. *Bone Reports, 12*, 100277. doi: 10.1016/j.bonr.2020.100277

Koszyca, B., Fazzalari, N. L., & Vernon-Roberts, B. (1989). Trabecular microfractures. Nature and distribution in the proximal femur. *Clinical Orthopaedics and Related Research, 244*, 208–216.

Kristensen, H. B., Andersen, T. L., Marcussen, N., Rolighed, L., & Delaisse, J.-M. (2014). Osteoblast recruitment routes in human cancellous bone remodeling. *The American Journal of Pathology, 184*(3), 778–789. doi: 10.1016/j.ajpath.2013.11.022

Lanyon, L. E., & Baggott, D. G. (1976). Mechanical function as an influence on the structure and form of bone. *The Journal of Bone and Joint Surgery. British Volume, 58-B*(4), 436–443.

Lanyon, L. E., & Bourn, S. (1979). The influence of mechanical function on the development and remodeling of the tibia. An experimental study in sheep. *The Journal of Bone and Joint Surgery. American Volume, 61*(2), 263–273.

Lanyon, L. E., Goodship, A. E., Pye, C. J., & MacFie, J. H. (1982). Mechanically adaptive bone remodelling. *Journal of Biomechanics, 15*(3), 141–154. doi: 10.1016/0021-9290(82)90246-9

Lassen, N. E., Andersen, T. L., Pløen, G. G., Søe, K., Hauge, E. M., Harving, S., Eschen, G. E. T., & Delaisse, J.-M. (2017). Coupling of bone resorption and formation in real time: New knowledge gained from human haversian BMUs. *Journal of Bone and Mineral Research, 32*(7), 1395–1405. doi: 10.1002/jbmr.3091

Leeuwenhoeck, M. (1674). Microscopical observations from Leeuwenhoeck, concerning blood, milk, bones, the brain, spitle, and cuticula, &c. Communicated by the said observer to the Publisher in a letter, dated June 1. 1674. *Philosophical Transactions of the Royal Society of London, 9*(106), 121–131. doi: 10.1098/rstl.1674.0030

Li, J., Mashiba, T., & Burr, D. B. (2001). Bisphosphonate treatment suppresses not only stochastic remodeling but also the targeted repair of microdamage. *Calcified Tissue International*, 69(5), 281–286. https://doi.org/10.1007/s002230010036

Lieberman, D. E., Pearson, O. M., Polk, J. D., Demes, B., & Crompton, A. W. (2003). Optimization of bone growth and remodeling in response to loading in tapered mammalian limbs. *Journal of Experimental Biology, 206*(18), 3125–3138. doi: 10.1242/jeb.00514

Lotz, J. C., Cheal, E. J., & Hayes, W. C. (1995). Stress distributions within the proximal femur during gait and falls: Implications for osteoporotic fracture. *Osteoporosis International, 5*(4), 252–261. doi: 10.1007/BF01774015

Loundagin, L. L., Pohl, A. J., & Edwards, W. B. (2021). Stressed volume estimated by finite element analysis predicts the fatigue life of human cortical bone: The role of vascular canals as stress concentrators. *Bone, 143*, 115647. doi: 10.1016/j.bone.2020.115647

Loundagin, L. L., Harrison, K. D., Wei, X., & Cooper, D. M. L. (2023). Understanding basic multicellular unit activity in cortical bone through 3D morphological analysis: New methods to define zones of the remodeling space. *Bone, 179*, 116960. doi: 10.1016/j.bone.2023.116960

Lovejoy, C. O. (1988). Evolution of human walking. *Scientific American, 259*(5), 118–125. doi: 10.1038/scientificamerican1188-118

Maggiano, I., Maggiano, C., Clement, J., Thomas, D., & Cooper, D. (2017). Longitudinal variation of osteon circularity in three-dimensional reconstructions of Haversian networks. *American Journal of Physical Anthropology, 162*, 273–273.

Maggiano, I. S., Maggiano, C. M., Clement, J. G., Thomas, C. D. L., Carter, Y., & Cooper, D. M. L. (2016). Three-dimensional reconstruction of Haversian systems in human cortical bone using synchrotron radiation-based micro-CT: Morphology and quantification of branching and transverse connections across age. *Journal of Anatomy, 228*(5), 719–732. doi: 10.1111/joa.12430

Martin, B. (1993). Aging and strength of bone as a structural material. *Calcified Tissue International, 53 Suppl 1*, S34–S39; discussion S39–S40. doi: 10.1007/bf01673400

Martin, R. B. (1984). Porosity and specific surface of bone. *Critical Reviews in Biomedical Engineering, 10*(3), 179–222.

Martin, R. B. (2002). Is all cortical bone remodeling initiated by microdamage? *Bone, 30*(1), 8–13. doi: 10.1016/S8756-3282(01)00620-2

Martin, R. B. (2003). Fatigue microdamage as an essential element of bone mechanics and biology. *Calcified Tissue International, 73*(2), 101–107. doi: 10.1007/s00223-002-1059-9

Martin, R. B., & Atkinson, P. J. (1977). Age and sex-related changes in the structure and strength of the human femoral shaft. *Journal of Biomechanics, 10*(4), 223–231. doi: 10.1016/0021-9290(77)90045-8

Martin, R. B., Pickett, J. C., & Zinaich, S. (1980). Studies of skeletal remodeling in aging men. *Clinical Orthopaedics and Related Research, 149*, 268–282.

Mason, M. W., Skedros, J. G., & Bloebaum, R. D. (1995). Evidence of strain-mode-related cortical adaptation in the diaphysis of the horse radius. *Bone, 17*(3), 229–237. doi: 10.1016/8756-3282(95)00213-w

Mayya, A., Banerjee, A., & Rajesh, R. (2013). Mammalian cortical bone in tension is non-Haversian. *Scientific Reports, 3*, 2533. doi: 10.1038/srep02533

Milovanovic, P., Vukovic, Z., Antonijevic, D., Djonic, D., Zivkovic, V., Nikolic, S., & Djuric, M. (2017). Porotic paradox: Distribution of cortical bone pore sizes at nano- and micro-levels in healthy vs. fragile human bone. *Journal of Materials Science. Materials in Medicine, 28*(5), 71. doi: 10.1007/s10856-017-5878-7

Moore, K. L., Agur, A. M. R., & Dalley, A. F. (2013). *Clinically Oriented Anatomy* (7th edition). LWW.

Mori, S., Harruff, R., Ambrosius, W., & Burr, D. B. (1997). Trabecular bone volume and microdamage accumulation in the femoral heads of women with and without femoral neck fractures. *Bone, 21*(6), 521–526. doi: 10.1016/s8756-3282(97)00200-7

Mosekilde, L., & Mosekilde, L. (1990). Sex differences in age-related changes in vertebral body size, density and biomechanical competence in normal individuals. *Bone, 11*(2), 67–73. doi: 10.1016/8756-3282(90)90052-z

Mulhern, D. M. (2000). Rib remodeling dynamics in a skeletal population from Kulubnarti, Nubia. *American Journal of Physical Anthropology, 111*(4), 519–530. doi: 10.1002/(SICI)1096-8644(200004)111:4<519::AID-AJPA7>3.0.CO;2-7

Mulhern, D. M., & Van Gerven, D. P. (1997). Patterns of femoral bone remodeling dynamics in a Medieval Nubian population. *American Journal of Physical Anthropology, 104*(1), 133–146. doi: 10.1002/(SICI)1096-8644(199709)104:1<133::AID-AJPA9>3.0.CO;2-S

Niroobakhsh, M., Laughrey, L. E., Dallas, S. L., Johnson, M. L., & Ganesh, T. (2023). Computational modeling based on confocal imaging predicts changes in osteocyte and dendrite shear stress due to canalicular loss with aging. *Biomechanics and Modeling in Mechanobiology*. doi: 10.1007/s10237-023-01763-w

Norman, T. L., & Wang, Z. (1997). Microdamage of human cortical bone: Incidence and morphology in long bones. *Bone, 20*(4), 375–379. doi: 10.1016/s8756-3282(97)00004-5

Obata, Y., Bale, H. A., Barnard, H. S., Parkinson, D. Y., Alliston, T., & Acevedo, C. (2020). Quantitative and qualitative bone imaging: A review of synchrotron radiation microtomography analysis in bone research. *Journal of the Mechanical Behavior of Biomedical Materials, 110*, 103887. doi: 10.1016/j.jmbbm.2020.103887

Ohman, J. C., Krochta, T. J., Lovejoy, C. O., Mensforth, R. P., & Latimer, B. (1997). Cortical bone distribution in the femoral neck of hominoids: Implications for the locomotion of Australopithecus afarensis. *American Journal of Physical Anthropology, 104*(1), 117–131. doi: 10.1002/(SICI)1096 8644(199709)104:1<117::AID-AJPA8>3.0.CO;2-O

Oliviero, S., Giorgi, M., Laud, P. J., & Dall'Ara, E. (2019). Effect of repeated in vivo microCT imaging on the properties of the mouse tibia. *PLOS ONE, 14*(11), e0225127. doi: 10.1371/journal.pone.0225127

Ozcivici, E., Luu, Y. K., Adler, B., Qin, Y.-X., Rubin, J., Judex, S., & Rubin, C. T. (2010). Mechanical signals as anabolic agents in bone. *Nature Reviews. Rheumatology, 6*(1), 50–59. doi: 10.1038/nrrheum.2009.239

Palacio-Mancheno, P. E., Larriera, A. I., Doty, S. B., Cardoso, L., & Fritton, S. P. (2014). 3D assessment of cortical bone porosity and tissue mineral density using high-resolution micro-CT: Effects of resolution and threshold method. *Journal of Bone and Mineral Research: The Official Journal of the American Society for Bone and Mineral Research, 29*(1). doi: 10.1002/jbmr.2012

Particelli, F., Mecozzi, L., Beraudi, A., Montesi, M., Baruffaldi, F., & Viceconti, M. (2012). A comparison between micro-CT and histology for the evaluation of cortical bone: Effect of polymethylmethacrylate embedding on structural parameters. *Journal of Microscopy, 245*(3), 302–310. doi: 10.1111/j.1365-2818.2011.03573.x

Peña Fernández, M., Kao, A. P., Witte, F., Arora, H., & Tozzi, G. (2020). Low-cycle full-field residual strains in cortical bone and their influence on tissue fracture evaluated via in situ stepwise and continuous X-ray computed tomography. *Journal of Biomechanics, 113*, 110105. doi: 10.1016/j.jbiomech.2020.110105

Peña Fernández, M., Kao, A. P., Bonithon, R., Howells, D., Bodey, A. J., Wanelik, K., Witte, F., Johnston, R., Arora, H., & Tozzi, G. (2021). Time-resolved in situ synchrotron-microCT: 4D deformation of bone and bone analogues using digital volume correlation. *Acta Biomaterialia, 131*, 424–439. doi: 10.1016/j.actbio.2021.06.014

Piemontese, M., Almeida, M., Robling, A. G., Kim, H.-N., Xiong, J., Thostenson, J. D., Weinstein, R. S., Manolagas, S. C., O'Brien, C. A., & Jilka, R. L. (2017). Old age causes de novo intracortical bone remodeling and porosity in mice. *JCI Insight, 2*(17). doi: 10.1172/jci.insight.93771

Pivonka, P., Park, A., & Forwood, M. R. (2018). Functional adaptation of bone: The mechanostat and beyond. In P. Pivonka (Ed.), *Multiscale Mechanobiology of Bone Remodeling and Adaptation* (pp. 1–60). Springer International Publishing. doi: 10.1007/978-3-319-58845-2_1

Poole, K. E. S., Treece, G. M., Mayhew, P. M., Vaculík, J., Dungl, P., Horák, M., Štěpán, J. J., & Gee, A. H. (2012). Cortical thickness mapping to identify focal osteoporosis in patients with hip fracture. *PloS One, 7*(6), e38466. doi: 10.1371/journal.pone.0038466

Portigliatti Barbos, M., Bianco, P., Ascenzi, A., & Boyde, A. (1984). Collagen orientation in compact bone: II. Distribution of lamellae in the whole of the human femoral shaft with reference to its mechanical properties. *Metabolic Bone Disease and Related Research, 5*(6), 309–315. doi: 10.1016/0221-8747(84)90018-3

Reilly, G. C., & Currey, J. D. (2000). The effects of damage and microcracking on the impact strength of bone. *Journal of Biomechanics, 33*(3), 337–343. doi: 10.1016/s0021-9290(99)00167-0

Riggs, C. M., Lanyon, L. E., & Boyde, A. (1993). Functional associations between collagen fibre orientation and locomotor strain direction in cortical bone of the equine radius. *Anatomy and Embryology, 187*(3), 231–238. doi: 10.1007/BF00195760

Ritchie, R. O. (2011). The conflicts between strength and toughness. *Nature Materials, 10*(11), 817–822. doi: 10.1038/nmat3115

Robling, A. (2014). Mechanical adaptation. In D. Burr & M. Allen (Eds.), *Basic and Applied Bone Biology* (pp. 175–203). Academic Press.

Ross, R. D., & Sumner, D. R. (2017). Bone matrix maturation in a rat model of intra-cortical bone remodeling. *Calcified Tissue International, 101*(2), 193–203. doi: 10.1007/s00223-017-0270-7

Ruff, C., Holt, B., & Trinkaus, E. (2006). Who's afraid of the big bad Wolff?: "Wolff's law" and bone functional adaptation. *American Journal of Physical Anthropology, 129*(4), 484–498. doi: 10.1002/ajpa.20371

Ruff, C. B., & Hayes, W. C. (1988). Sex differences in age-related remodeling of the femur and tibia. *Journal of Orthopaedic Research, 6*(6), 886–896. doi: 10.1002/jor.1100060613

Sarafrazi, N., Wambogo, E. A., & Shepherd, J. A. (2021). *Osteoporosis or Low Bone Mass in Older Adults: United States, 2017-2018*. National Center for Health Statistics. doi: 10.15620/cdc:103477

Schaffler, M. B., Choi, K., & Milgrom, C. (1995). Aging and matrix microdamage accumulation in human compact bone. *Bone, 17*(6), 521–525. doi: 10.1016/8756-3282(95)00370-3

Schlecht, S. H., Bigelow, E. M. R., & Jepsen, K. J. (2015). How does bone strength compare across sex, site, and ethnicity? *Clinical Orthopaedics and Related Research, 473*(8), 2540–2547. doi: 10.1007/s11999-015-4229-6

Seeman, E., & Delmas, P. D. (2006). Bone quality—The material and structural basis of bone strength and fragility. *The New England Journal of Medicine, 354*(21), 2250–2261. doi: 10.1056/NEJMra053077

Seref-Ferlengez, Z., Kennedy, O. D., & Schaffler, M. B. (2015). Bone microdamage, remodeling and bone fragility: How much damage is too much damage? *BoneKEy Reports, 4*. doi: 10.1038/bonekey.2015.11

Sims, N. A., & Martin, T. J. (2020). Osteoclasts provide coupling signals to osteoblast lineage cells through multiple mechanisms. *Annual Review of Physiology, 82*(1), 507–529. doi: 10.1146/annurev-physiol-021119-034425

Skedros, J. G., Mason, M. W., & Bloebaum, R. D. (1994). Differences in osteonal micromorphology between tensile and compressive cortices of a bending skeletal system: Indications of potential strain-specific differences in bone microstructure. *The Anatomical Record, 239*(4), 405–413. doi: 10.1002/ar.1092390407

Skedros, J. G., Su, S. C., & Bloebaum, R. D. (1997). Biomechanical implications of mineral content and microstructural variations in cortical bone of horse, elk, and sheep calcanei. *The Anatomical Record, 249*(3), 297–316. doi: 10.1002/(SICI)1097-0185(199711)249:3<297::AID-AR1>3.0.CO;2-S

Skedros, J. G., Mason, M. W., & Bloebaum, R. D. (2001). Modeling and remodeling in a developing artiodactyl calcaneus: A model for evaluating Frost's Mechanostat hypothesis and its corollaries. *The Anatomical Record, 263*(2), 167–185. doi: 10.1002/ar.1094

Skedros, J. G., Sybrowsky, C. L., Parry, T. R., & Bloebaum, R. D. (2003). Regional differences in cortical bone organization and microdamage prevalence in Rocky Mountain mule deer. *The Anatomical Record Part A: Discoveries in Molecular, Cellular, and Evolutionary Biology, 274*(1), 837–850. doi: 10.1002/ar.a.10102

Skedros, J. G., Mendenhall, S. D., Kiser, C. J., & Winet, H. (2009). Interpreting cortical bone adaptation and load history by quantifying osteon morphotypes in circularly polarized light images. *Bone, 44*(3), 392–403. doi: 10.1016/j.bone.2008.10.053

Skedros, J. G., Clark, G. C., Sorenson, S. M., Taylor, K. W., & Qiu, S. (2011). Analysis of the effect of osteon diameter on the potential relationship of osteocyte lacuna density and osteon wall thickness. *Anatomical Record (Hoboken, N.J.: 2007), 294*(9), 1472–1485. doi: 10.1002/ar.21452

Skedros, J. G., Kiser, C. J., Keenan, K. E., & Thomas, S. C. (2011). Analysis of osteon morphotype scoring schemes for interpreting load history: Evaluation in the chimpanzee femur. *Journal of Anatomy, 218*(5), 480–499. doi: 10.1111/j.1469-7580.2011.01348.x

Skedros, J. G., Kiser, C. J., & Mendenhall, S. D. (2011). A weighted osteon morphotype score outperforms regional osteon percent prevalence calculations for interpreting cortical bone adaptation. *American Journal of Physical Anthropology, 144*(1), 41–50. doi: 10.1002/ajpa.21365

Skedros, J. G., Sybrowsky, C. L., Anderson, W. E., & Chow, F. (2011). Relationships between in vivo microdamage and the remarkable regional material and strain heterogeneity of cortical bone of adult deer, elk, sheep and horse calcanei. *Journal of Anatomy, 219*(6), 722–733. doi: 10.1111/j.1469-7580.2011.01428.x

Skedros, J. G., Knight, A. N., Clark, G. C., Crowder, C. M., Dominguez, V. M., Qiu, S., Mulhern, D. M., Donahue, S. W., Busse, B., Hulsey, B. I., Zedda, M., & Sorenson, S. M. (2013). Scaling of Haversian canal surface area to secondary osteon bone volume in ribs and limb bones. *American Journal of Physical Anthropology, 151*(2), 230–244. doi: 10.1002/ajpa.22270

Skerry, T. M. (2006). One mechanostat or many? Modifications of the site-specific response of bone to mechanical loading by nature and nurture. *Journal of Musculoskeletal & Neuronal Interactions, 6*(2), 122–127.

Soltan, N., Kawalilak, C. E., Cooper, D. M., Kontulainen, S. A., & Johnston, J. D. (2019). Cortical porosity assessment in the distal radius: A comparison of HR-pQCT measures with Synchrotron-Radiation micro-CT-based measures. *Bone, 120*, 439–445. doi: 10.1016/j.bone.2018.12.008

Stout, S. D. (1984). Intraskeletal variation in osteon remodeling. *American Journal of Physical Anthropology, 63*, 223.

Stout, S. D., Brunsden, B. S., Hildebolt, C. F., Commean, P. K., Smith, K. E., & Tappen, N. C. (1999). Computer-assisted 3D reconstruction of serial sections of cortical bone to determine the 3D structure of osteons. *Calcified Tissue International*, *65*(4), 280–284. doi: 10.1007/s002239900699

Stout, S. D., Cole, M. E., & Agnew, A. M. (2019). Histomorphology: Deciphering the metabolic record. In J. E. Buikstra (Ed.), *Ortner's Identification of Pathological Conditions in Human Skeletal Remains (Third Edition)* (pp. 91–167). Academic Press. doi: 10.1016/B978-0-12-809738-0.00006-5

Sundh, D., Mellström, D., Nilsson, M., Karlsson, M., Ohlsson, C., & Lorentzon, M. (2015). Increased cortical porosity in older men with fracture. *Journal of Bone and Mineral Research*, *30*(9), 1692–1700. doi: 10.1002/jbmr.2509

Tappen, N. C. (1977). Three-dimensional studies on resorption spaces and developing osteons. *The American Journal of Anatomy*, *149*(3), 301–317. doi: 10.1002/aja.1001490302

Taylor, D., & Clive Lee, T. (1998). Measuring the shape and size of microcracks in bone. *Journal of Biomechanics*, *31*(12), 1177–1180. doi: 10.1016/S0021-9290(98)00133-X

Thomas, C. D. L., Feik, S. A., & Clement, J. G. (2005). Regional variation of intracortical porosity in the midshaft of the human femur: Age and sex differences. *Journal of Anatomy*, *206*(2), 115–125. doi: 10.1111/j.1469-7580.2005.00384.x

Todd, R. B., & Bowman, W. (1845). *The Physiological Anatomy and Physiology of Man*. Blanchard and Lea.

Tomes, J., & Morgan, C. (1853). Observations on the Structure and Development of Bone. *Philosophical Transactions of the Royal Society B: Biological Sciences*, *143*, 109–139.

Trichilo, S., & Pivonka, P. (2018). Application of disease system analysis to osteoporosis: From temporal to Spatio-temporal assessment of disease progression and intervention. In P. Pivonka (Ed.), *Multiscale Mechanobiology of Bone Remodeling and Adaptation* (pp. 61–121). Springer International Publishing. doi: 10.1007/978-3-319-58845-2_2

Turnbull, T. L., Baumann, A. P., & Roeder, R. K. (2014). Fatigue microcracks that initiate fracture are located near elevated intracortical porosity but not elevated mineralization. *Journal of Biomechanics*, *47*(12), 3135–3142. doi: 10.1016/j.jbiomech.2014.06.022

Turner, C. H. (1998). Three rules for bone adaptation to mechanical stimuli. *Bone*, *23*(5), 399–407. doi: 10.1016/S8756-3282(98)00118-5

Turner, R. T., Wakley, G. K., & Hannon, K. S. (1990). Differential effects of androgens on cortical bone histomorphometry in gonadectomized male and female rats. *Journal of Orthopaedic Research*, *8*(4), 612–617. doi: 10.1002/jor.1100080418

Vajda, E. G., Kneissel, M., Muggenburg, B., & Miller, S. C. (1999). Increased intracortical bone remodeling during lactation in beagle dogs. *Biology of Reproduction*, *61*(6), 1439–1444. doi: 10.1095/biolreprod61.6.1439

Varga, P., Hesse, B., Langer, M., Schrof, S., Männicke, N., Suhonen, H., Pacureanu, A., Pahr, D., Peyrin, F., & Raum, K. (2015). Synchrotron X-ray phase nano-tomography-based analysis of the lacunar–canalicular network morphology and its relation to the strains experienced by osteocytes in situ as predicted by case-specific finite element analysis. *Biomechanics and Modeling in Mechanobiology*, *14*(2), 267–282. doi: 10.1007/s10237-014-0601-9

Vogel, M., Hahn, M., & Delling, G. (1993). Relation between 2- and 3-dimensional architecture of trabecular bone in the human spine. *Bone*, *14*(3), 199–203. doi: 10.1016/8756-3282(93)90141-v

Wachter, N. J., Augat, P., Krischak, G. D., Mentzel, M., Kinzl, L., & Claes, L. (2001). Prediction of cortical bone porosity in vitro by microcomputed tomography. *Calcified Tissue International*, *68*(1), 38–42. doi: 10.1007/s002230001182

Wasserman, N., Yerramshetty, J., & Akkus, O. (2005). Microcracks colocalize within highly mineralized regions of cortical bone tissue. *European Journal of Morphology*, *42*(1–2), 43–51. doi: 10.1080/09243860500095471

Wilson, A. K., Bhattacharyya, M. H., Miller, S., Mani, A., & Sacco-Gibson, N. (1998). Ovariectomy-induced changes in aged beagles: Histomorphometry of rib cortical bone. *Calcified Tissue International*, *62*(3), 237–243. doi: 10.1007/s002239900423

Zebaze, R., & Seeman, E. (2015). Cortical bone: A challenging geography. *Journal of Bone and Mineral Research*, *30*(1), 24–29. doi: 10.1002/jbmr.2419

Zebaze, R. M. D., Ghasem-Zadeh, A., Bohte, A., Iuliano-Burns, S., Mirams, M., Price, R. I., Mackie, E. J., & Seeman, E. (2010). Intracortical remodelling and porosity in the distal radius and post-mortem femurs of women: A cross-sectional study. *Lancet*, *375*(9727), 1729–1736. doi: Doi 10.1016/S0140-6736(10)60320-0

Zedda, M., Lepore, G., Biggio, G. P., Gadau, S., Mura, E., & Farina, V. (2015). Morphology, morphometry and spatial distribution of secondary osteons in equine femur. *Anatomia, Histologia, Embryologia*, *44*(5), 328–332. doi: 10.1111/ahe.12141

Zimmermann, E. A., Schaible, E., Bale, H., Barth, H. D., Tang, S. Y., Reichert, P., Busse, B., Alliston, T., Ager, J. W., & Ritchie, R. O. (2011). Age-related changes in the plasticity and toughness of human cortical bone at multiple length scales. *Proceedings of the National Academy of Sciences of the United States of America*, *108*(35), 14416–14421. doi: 10.1073/pnas.1107966108

Index

Note: Page numbers in *italics* and **bold** refer to figures and tables, respectively.

D

I

J

K

L

M

N

O

P

Q

R

For Product Safety Concerns and Information please contact our EU representative GPSR@taylorandfrancis.com Taylor & Francis Verlag GmbH, Kaufingerstraße 24, 80331 München, Germany

* 9 7 8 1 0 3 2 4 7 3 2 7 7 *

T - #0062 - 280825 - C498 - 254/178/23 [25] - CB - 9781032473277 - Gloss Lamination